Cell Biology Monographs

Volume 9

Springer-Verlag Wien GmbH

Igor B. Raikov

The Protozoan Nucleus

Morphology and Evolution

Translated from the Russian
by Nicholas Bobrov
and Marina Verkhovtseva

Springer-Verlag Wien GmbH

Dr. Igor Borissovich Raikov
Institute of Cytology
of the Academy of Sciences of the USSR
Leningrad, USSR

Nicholas Bobrov and Marina Verkhovtseva
Moscow, USSR

Revised from the 1978 Russian edition
Original title:
Райков, И. Б.: Ядро простейших. Морфология и эволюция.
Издательство „Наука", Ленинград, 1978. ©Издательство „Наука", 1978 г.

Originally published by Springer-Verlag Wien New York in 1982
Softcover reprint of the hardcover 1st edition 1982

With 1 Portrait and 116 Figures

Library of Congress Cataloging in Publication Data. Raĭkov, Igor Borissovich. The protozoan nucleus, morphology and evolution. (Cell biology monographs; v. 9.) Translation of: Iadro prosteĭshikh, morfologija i ėvolutsija. Includes bibliographical references and indexes. 1. Protozoa – Morphology. 2. Protozoa – Evolution. 3. Cell nuclei. I. Title. II. Series. QL369.R33513. 593. 1'08732. 81-21380

ISSN 0172-4665
ISBN 978-3-7091-4138-0 ISBN 978-3-7091-4136-6 (eBook)
DOI 10.1007/978-3-7091-4136-6

Karl Bělăr (1895 – 1929), a pioneer of protozoan karyology

Dedicated to the memory of my teacher,
Professor Valentin A. Dogiel (1882—1955),
on the occasion of the centenary of his birthday

Preface

This book continues, somewhat, the author's monograph *Karyology of Protozoa*, which was published in Russian in 1967. Investigations of the nuclear apparatus of protozoans developed very rapidly in the 1960s and 1970s, when virtually the whole morphology of protozoan nuclei was transferred to the ultrastructural level. Our knowledge of protozoan mitosis also reached a newer, qualitatively higher level with electron microscopy playing the main role in this respect, also. A peculiar mechanism of mitosis in mesokaryotic protozoans (Dinoflagellida) was revealed, and extremely important discoveries were made in the field of ciliate macronuclei.

Hence, this book is not the second edition of *Karyology of Protozoa*, although it has the same plan, in part. Because an enormous amount of new material became available and many points had to be critically reviewed, the monograph was virtually rewritten for the 1978 Russian edition. When the text of this book was being prepared for the English translation, it was supplemented and worked over in accordance with the data obtained during 1977–1981. New illustrations were added, and some of the old ones were removed.

The author hopes this book will make it possible to use protozoan nuclei on a wider scale and give a better understanding of the karyology of lower eukaryotes within the framework of cell biology.

Leningrad, August 1982 I. B. Raikov

Acknowledgements

The author is very grateful to his teacher, Prof. G. Poljansky, for sedulously supervising the investigations in this book and for the advice he gave as a result of reading the manuscript. The author is also grateful to the staff of the Laboratory of Cytology of Unicellular Organisms of the Institute of Cytology of the USSR Academy of Sciences for its assistance in producing this book and especially to N. N. Bobyleva, N. V. Vinnikova, M. N. Golikova, V. G. Kovaleva, and M. A. Frenkel for their help in preparing the manuscript for the press.

The author takes this opportunity to thank his colleagues who kindly supplied prints from the original negatives of their published and sometimes unpublished light and electron micrographs as well as some original drawings. These colleagues are: Masamichi Aikawa (Figs. 9 *b*, 10 *b*, 16 *e*), Dieter Ammermann (Fig. 109 *a*–*g*), Jean André (Fig. 99 *d*), Christian F. Bardele (Fig. 102 *a*), Jacques Baudoin (Fig. 6 *c*), H. W. Beams (Fig. 2 *h*), Jacques Bohatier (Figs. 92 *c* and *d*, 97 *b* and *c*), Blair Bowers (Fig. 16 *a*), Guy Brugerolle (Fig. 33 *d*), Nina Carasso and Pierre Favard (Fig. 105 *c*), Renée Charret (Fig. 94 *a*, *b*, and *d*), Edward W. Daniels (Figs. 2 *d*–*f*, 4 *c*, 12 *a*–*c*), Isabelle Desportes (Fig. 25 *b*), Pierre Didier (Figs. 91 *a*, 97 *b* and *c*), John D. Dodge (Figs. 38 *a*, 39 *a*, 44 *b*), Jean-François Dubremetz (Fig. 16 *c* and *d*), Marina A. Frenkel (Fig. 116), Claude R. Gicquaud (Fig. 13 *a* and *b*), Thomas K. Golder (Fig. 2 *c*), Margaret N. Golikova (Fig. 114 *c* and *d*), Karl G. Grell (Figs. 63 *a*–*c*, 64 *a*–*e*), H.-A. Guénin and José Fahrni (Figs. 96 *a* and *b*, 106), Manfred Hauser (Figs. 10 *d*, 102 *c* and *d*), Håkan Hedin (Figs. 4 *e*, 11 *b*), André Hollande (Figs. 10 *a*, 33 *b* and *c*, 34 *a*–*d*), Maria A. Jareño (Fig. 109 *h*), Robert A. Jenkins (Fig. 105 *d*), Louis Joyon (Fig. 6 *a*), A. Jurand (Fig. 105 *b*), Bela P. Karadzhan (Figs. 30 *a*–*e*, 92 *a* and *b*), John R. Kennedy (Fig. 95 *b*), John A. Kloetzel (Fig. 110 *a* and *b*), Valentine G. Kovaleva (Figs. 10 *e*, 82 *c*, 91 *b*, 93 *b*), Donna F. Kubai (Figs. 34 *e*–*g*, 44 *a*), J. G. Lafontaine (Fig. 31 *d*), Gordon F. Leedale (Fig. 9 *d*), Claude Loubès (Fig. 25 *a*), Denis Lynn (Fig. 93 *a*), Jean-Pierre Mignot (Figs. 17 *a*–*d*, 56 *a*–*c*), Lyndell L. Millecchia and Maria A. Rudzinska (Figs. 95 *d*, 102 *b*), I. Minassian (Figs. 4 *d*, 13 *c* and *d*), Gilles Moyne (Fig. 11 *d*), Kuruganti G. Murti (Fig. 111 *a* and *b*), Jytte R. Nilsson (Figs. 94 *c*, 95 *a*), Frank O. Perkins (Fig. 56 *d*), Eliane Porchet (Fig. 11 *c*), David M. Prescott (Figs. 99 *a*–*c* and *e*, 111 *c*), Chandra K. Pyne (Figs. 97 *a*, 111 *d*), Hans Ris (Fig. 41 *a*–*d*), L. Evans Roth (Fig. 4 *a* and *b*), August Ruthmann (Fig. 100 *a*), Erich Scholtyseck (Fig. 9 *f*), Frederick L. Schuster (Fig. 9 *c*), Dieter Schwab (Figs. 11 *a*, 25 *d*), Serge O. Skarlato (Fig. 114 *a* and *b*), A. J. Solari (Fig. 25 *c*), Marie-Odile Soyer (Figs. 38 *b*–*d*, 39 *b* and *c*, 47 *a*–*d*, 52 *a*–*c*), John B. Tucker (Fig. 10 *c*), Nathalie V.

Vinnikova (Figs. 56 *e*, 105 *a*, 114 *d*), and Emile Vivier (Figs. 2 *a*, *b*, and *g*, 6 *b*, 9 *a*, 95 *c*).

The photograph of Karl Bělař (frontispiece) is a gift from Dr. Vassil Golemanski, of the Zoological Institute of the Bulgarian Academy of Sciences, Sofia.

The author also wishes to thank many other colleagues who kindly provided micrographs which, unfortunately, were not included due to limited space. The author especially wishes to thank G. Benjamin Bouck, Donald J. Cummings, Colette Febvre-Chevalier, Joseph G. Call, Gérard de Haller, Francelyne Marano, Regine Milder, Guy Morat, Michel Nouzarède, Stefan Radzikowski, Ian Stevenson, Keith Vickerman, and Glenn K. Walker.

The following publishers, societies, and journals have kindly granted permission to reproduce illustrations: Academic Press Inc. (Figs. 2 *c* and *h*, 7, 9 *d*, 10 *b*, 11 *d*, 16 *d*, 22, 45, 94 *a*, *b*, and *d*), Académie des Sciences, Paris (6 *a* and *c*, 26, 33 *b* and *c*, 48, 65, 67, 70, 95 *c*), Acta Protozoologica (Fig. 75 *c* and *d*, 76 *a* and *b*, 82 *b*, 83), American Academy of Arts and Sciences (Fig. 32), American Philosophical Society (Fig. 62), American Society of Tropical Medicine and Hygiene (Fig. 9 *b*), Année Biologique (Figs. 19, 20, 24, 40, 43), Annual Reviews Inc. (Fig. 100 *d*), Archives du Muséum de l'Histoire Naturelle, Paris (Fig. 71), Botanical Review (Fig. 39 *a*), Cahíers de Biologie Marine (Fig. 77, 78), Carlsberg Laboratory, Copenhagen (Figs. 94 *c*, 95 *a*), Company of Biologists Ltd. (Figs. 4 *d*, 10 *c*, 13 *c* and *d*, 31 *d*, 34 *e*–*g*, 111 *a* and *b*), Éditions du CNRS, Paris (Figs. 2 *b*, 5, 6 *b*, 9 *a*, 10 *a* and *f*, 11 *c*, 17 *a*–*c*, 25 *a* and *b*, 27, 30 *a*–*e*, 31 *a*–*c*, 33 *d*, 34 *a*–*d*, 56 *a*, *b*, and *e*, 57, 60, 79 *a*–*c*, 80 *a* and *b*, 82 *a*, 92 *c* and *d*, 109 *h*), Histochemical Society, Inc. (Fig. 99 *b* and *c*), Macmillan Journals Ltd. (Fig. 49), Marine Biological Laboratory, Woods Hole (Fig. 115), Masson & Cie, Éditeurs (Figs. 1 *i* and *l*, 46, 66, 68, 69, 99 *d*), Pergamon Press Ltd. (Fig. 75 *b* and *e*, 89, 90), Plenum Press (Fig. 112), Revue Suisse de Zoologie (Fig. 96 *b*), Rockefeller University Press (Figs. 4 *a* and *b*, 16 *a*, *b*, and *e*, 41 *a*–*d*, 42, 44 *a* and *b*, 110 *a* and *b*), Societé Française de Microscopie Électronique (Figs. 2 *a* and *g*, 3 *b*, 13 *a* and *b*, 16 *c*, 39 *b* and *c*, 47 *a* and *b*, 52 *a* and *b*, 53, 105 *c*), Society for General Microbiology (Fig. 105 *b*), Society of Protozoologists (Figs. 9 *c*, 33 *a*, 56 *d*, 93 *a*, 95 *b*, 99 *e*, 105 *d*), Springer-Verlag Berlin-Heidelberg (Figs. 2 *d*–*f*, 4 *c*, 9 *f*, 10 *d* and *e*, 11 *a*, 12 *a*–*c*, 14, 15, 25 *c*, 36, 37, 38 *b*, *c*, and *d*, 47 *c* and *d*, 50, 51, 52 *c*, 54, 82 *c*, 95 *d*, 102 *a*–*d*, 103, 107, 108, 109 *a* and *c*–*g*, 111 *c*), Université de Clermont-Ferrand (Figs. 18, 23), Uppsala University Press (Figs. 4 *e*, 11 *b*), VEB Gustav Fischer Verlag, Jena (Figs. 1 *a*–*h*, *k*, *m*, and *n*, 8, 21, 28, 38 *a*, 55, 63 *a*–*c*, 64 *a*–*e*, 72, 73 *a*–*d*, 81 *b*, 86, 88, 98, 100 *a* and *b*, 101, 104, 109 *b*, 113), Wissenschaftliche Verlagsgesellschaft m.b.H., Stuttgart (Figs. 97 *a*, 111 *d*), and the Wistar Institute of Anatomy and Biology (Figs. 58, 61).

In addition, the author is very thankful to N. N. Bobrov and M. I. Verkhovtseva for translating this book and to Dr. E. Ungersbäck and the staff of Springer-Verlag, Vienna, for their cooperation.

Contents

Introduction

KARL BĚLAŘ, an eminent protozoologist and cytologist, wrote the book *Der Formwechsel der Protistenkerne*, which was published in 1926. In it, he summed up the results of early research on the nuclear apparatus of protozoans and criticized the theories of the "dualism of the nuclear substance" and "chromidia", which were widely accepted at the time and which upheld the existence of anucleate (chromidial) stages in the life cycle of many protozoans. The publication of BĚLAŘ's book became a turning point in the development of the karyology of protozoans, since he was the first to consider and critically assess an enormous amount of very diverse literary material from a strictly scientific standpoint of general karyology and cytogenetics. BĚLAŘ drew the extremely important conclusion that the general laws of the replication and segregation of chromosomes can be fully applied to protozoans. Although the nondividing nuclei and the types of mitosis of protozoans seem to be very diverse, their chromosomes are characterized by a constant number, individuality, and regular segregation in mitosis.

The study of protozoan nuclei has greatly advanced since then. During the era of light microscopy, a thorough investigation was made of mitosis, meiosis, and the cycles of spiralization of chromosomes in some flagellates. The discoveries of that period included the alternation of the haploid and diploid generations in the life cycle of foraminiferans and the phenomenon of polyploidy in protozoans. The data obtained were included in textbooks and manuals of protozoology (DOFLEIN and REICHENOW 1953, GRELL 1956a, DOGIEL 1965, etc.) and in RAIKOV's book (1967a), which specifically dealt with the karyology of protozoans.

The methods of studying protozoan nuclei have substantially changed in the past 15 years. Prior to 1965–1967, the nuclear structure was investigated mainly by photomicroscopy, although some data were obtained by methods involving electron microscopy, autoradiography, and biochemistry. In the late 1960s and the 1970s, however, these methods underwent the greatest development. At present, all types of protozoan nuclei have an ultrastructural characterization. Instead of the old classification of the types of mitosis, a new one based on electron microscopy is now available. The fine structural cytochemistry of protozoan nuclei has also been markedly developed. The autoradiographic recording of the synthesis of certain substances in protozoan nuclei has been largely transferred to the ultrastructural level. Moreover, the biochemical studies of protozoan nuclei have advanced very rapidly. Some forms of Protozoa, especially the ciliate *Tetrahymena pyriformis*, are now frequently used by biochemists as a model of the eukaryotic cell in general. This field is now very large. Unfortunately, space does not allow this author to thoroughly discuss some biochemical aspects of protozoan karyology, such as the nucleotide composition of the DNA of protozoans, the composition and properties of the histones of protozoan nuclei, and the arrangement of unique and repetitive DNA sequences in their genomes. These aspects are considered, for instance, in the multivolume treatise *Biochemistry and Physiology of Protozoa* (Second edition, Academic Press, 1979). We will deal mainly with the morphological and morphofunctional aspects of protozoan karyology. Molecular biochemical data will be given only when they help to explain the structural organization of the nucleus (e.g. the ciliate macronucleus).

No special monographs on protozoan karyology have been published recently. However, some structural and functional aspects of protozoan nuclei have been discussed in reviews of general protozoology (GRELL 1973, SLEIGH 1973). In this book, various aspects of protozoan karyology are considered to different extents. For instance, the behaviour of nuclei during sexual processes (copulation, conjugation, and autogamy) is hardly discussed. As for ciliates, the author wrote a special review (RAIKOV 1972d) concerning those processes. Among the phenomena concerned, only meiosis is considered in this book. Ample space is given to nuclear dualism and the macronuclear structure of ciliates. In this respect, the author presents literary as well as his own data.

Protozoans are cells as well as organisms. Their evolution (as regards the origin of some taxa from others) is the subject of protozoology as a constituent of zoology. Besides the purely "zoological" evolution, however, protozoans as cells definitely evolved with respect to their cellular structures and organoids, and this is the subject of evolutionary cytology. For instance, it has now been firmly established that not all protozoans are on the same level of evolution as regards their cell nucleus. Most protozoans are typical *eukaryotic* organisms with a formed nucleus that is covered with a two-membrane envelope and contains typical chromosomes rich in DNA and histones. This nucleus divides by mitosis, forming an achromatic apparatus of microtubules. But in at least one group of protozoans (Dinoflagellida) chromosomes have a structure similar to that of the nucleoids of *prokaryotic* organisms (bacteria, cyanophyceans, actinomycetes): they do not have the histone component and lack nucleosomal subunits. But the Dinoflagellida do have a formed cell nucleus with a typical envelope; it divides by mitosis, although a very unusual one (membranes participate in the segregation of chromatids). Because of this unusual aspect, this level of the organization of the protozoan cell was called *mesokaryotic* (DODGE 1965). Thus, protozoans are either eukaryotic or mesokaryotic. Prokaryotic protozoans have not been discovered yet. These two groups of protozoans should be considered separately, beginning with the eukaryotic forms.

For the reader's convenience, the classification of protozoans used in this book is given below. It is the system proposed by the Committee on Systematics and Evolution of the Society of Protozoologists (see LEVINE *et al.* 1980). The system is presented here with some minor changes and in a simplified version (e.g. the enumeration of orders is omitted when it is not essential). The classes of myxomycetes (Acrasea, Eumycetozoea, and Plasmodiophorea), which cannot be regarded as true protozoans, are not discussed in this book. Widely used synonyms and trivial names of some taxa are given in parentheses.

Subkingdom *Protozoa*

Phylum I. *Sarcomastigophora*
Subphylum *Mastigophora* (Flagellata, flagellates)
Class 1. *Phytomastigophorea* ("plant" flagellates)
Orders: *Chrysomonadida, Prymnesiida, Heterochlorida, Cryptomonadida, Dinoflagellida, Euglenida, Chloromonadida, Volvocida (Phytomonadida), Prasinomonadida, Silicoflagellida*
Class 2. *Zoomastigophorea* ("animal" flagellates)
Orders: *Choanoflagellida* (collar flagellates), *Proteromonadida, Bicosoecida, Retortamonadida, Rhizomastigida, Kinetoplastida, Diplomonadida, Oxymonadida, Trichomonadida, Hypermastigida*
Subphylum *Opalinata*
Subphylum *Sarcodina*
Superclass *Rhizopoda*
Class 1. *Lobosea*
Subclass *Gymnamoebia* (naked amoebae)
Orders: *Schizopyrenida, Amoebida, Pelobiontida*
Subclass *Testacealobosia* (Testacea, testate amoebae)
Orders: *Arcellinida, Trichosida*
Class 2. *Acarpomyxea*
Orders: *Leptomyxida, Stereomyxida*
Class 3. *Acrasea*
Class 4. *Eumycetozoea*
Class 5. *Plasmodiophorea*
Class 6. *Filosea*
Orders: *Aconchulida, Gromiida*
Class 7. *Granuloreticulosea*
Orders: *Athalamida, Monothalamida, Foraminiferida*
Class 8. *Xenophyophorea*
Superclass *Actinopoda* (classes 1–3 are radiolarians)
Class 1. *Acantharea*
Class 2. *Polycystinea*
Orders: *Spumellarida, Nassellarida*
Class 3. *Phaeodarea*
Class 4. *Heliozoea* (heliozoans)
Orders: *Desmothoracida, Actinophryida, Taxopodida (Sticholonchida), Centrohelida*
Phylum II. *Labyrinthomorpha* (labyrinthulids)
Phylum III. *Apicomplexa*
Class 1. *Perkinsea*
Class 2. *Sporozoea* (sporozoans)
Subclass *Gregarinia* (gregarines)
Orders: *Archigregarinida, Eugregarinida, Neogregarinida*
Subclass *Coccidia*
Orders: *Agamococcidiida, Protococcidiida, Eucoccidiida* [with suborders *Adeleina, Eimeriina* (coccidians *s. str.*, toxoplasmids and sarcosporidians), *Haemosporina* (malaria parasites)]
Subclass *Piroplasmia* (piroplasmids)
Phylum IV. *Microspora* (microsporidians)
Class 1. *Rudimicrosporea*
Order: *Metchnikovellida*

Class 2. *Microsporea*
Orders: *Minisporida*, *Microsporida*
Phylum V. *Ascetospora*
Class 1. *Stellatosporea* (Haplosporea, haplosporidians)
Orders: *Occlusosporida*, *Balanosporida*
Class 2. *Paramyxea*
Phylum VI. *Myxozoa*
Class 1. *Myxosporea* (myxosporidians)
Orders: *Bivalvulida*, *Multivalvulida*
Class 2. *Actinosporea* (actinomyxidians)
Phylum VII. *Ciliophora* (ciliates, infusorians)
Class 1. *Kinetofragminophorea*
Subclass *Gymnostomatia*
Orders: *Primociliatida*, *Karyorelictida*, *Prostomatida*, *Pleurostomatida*
Subclass *Vestibuliferia*
Orders: *Trichostomatida*, *Colpodida*, *Bursariida*, *Platyophryida*, *Entodiniomorphida*
Subclass *Hypostomatia*
Orders: *Synhymeniida*, *Nassulida*, *Cyrtophorida*, *Chonotrichida*, *Rhynchodida*, *Apostomatida*
Subclass *Suctoria*
Class 2. *Oligohymenophorea*
Subclass *Hymenostomatia*
Orders: *Hymenostomatida*, *Scuticociliatida*, *Astomatida*
Subclass *Peritrichia*
Class 3. *Polyhymenophorea*
Subclass *Spirotrichia*
Orders: *Heterotrichida*, *Odontostomatida*, *Oligotrichida*, *Hypotrichida*

1. Morphology of Eukaryotic Protozoan Nuclei

All Protozoa have at least one nucleus. Descriptions of anucleate (chromidial) protozoans found in the old literature are now of purely historical interest. Protozoan nuclei consist of the same structural components as the nuclei of metazoan cells: the nuclear envelope, chromatin, nucleoli, karyolymph, and various intranuclear inclusions.

The nuclei of mesokaryotic protozoans (dinoflagellates) are not considered in this chapter because they will be dealt with in Chapter 4. Highly polyploid nuclei of radiolarians and the macronuclei of ciliates are also discussed in other chapters (Chaps. 6–8), but they will be dealt with in this chapter in terms of the structure of their nuclear envelopes.

Number, Shape, and Size of Nuclei

Most Protozoa are uninucleate. Almost all Mastigophora (except some forms, mainly parasitic ones, of the orders Diplomonadida, Trichomonadida, and Oxymonadida) and many Sarcodina have one nucleus. Moreover, nearly all Sporozoea have uninucleate stages in their life cycles. There are, however, also multinucleate forms with several identical nuclei in almost every group of protozoans. These forms are usually larger than their uninucleate relatives. In such cases, *nuclear polymerization* occurs.

In many protozoans with a complex life cycle, uninucleate stages alternate with multinucleate ones. In Coccidia, for instance, sporozoites and merozoites are uninucleate, while schizonts and microgametocytes are multinucleate. In such cases, multinuclearity is a temporary phenomenon related to the fact that nuclear divisions surpass cytoplasmic ones. The multinuclearity of some stages of the life cycle is more stable in the Foraminiferida: the specimens of the haploid sexual generation (the gamonts) are often uninucleate, while the specimens of the diploid asexual generation (the agamonts) are multinucleate.

In multinucleate protozoans, all the nuclei are usually morphologically alike and capable of undergoing mitosis. This applies, for instance, to the multinucleate amoebae *Chaos* and *Pelomyxa*, the heliozoan *Actinosphaerium*, and the flagellates of the subphylum Opalinata. Such protozoans are said to be *homokaryotic*. But the nuclei can greatly differ from each other forming two categories—*generative* and *somatic* nuclei. The former are usually small and divide by mitosis, while the latter are large and either do not divide at all or divide by direct pinching in two or by budding, which externally resembles amitosis. Both types of nuclei, however, have a common origin. In this case, *nuclear differentiation* or *nuclear dualism* occurs and the protozoans involved are said to be *heterokaryotic*. The most widely known examples of heterokaryotic protozoans are ciliates in which the generative (small) nucleus is called the *micronucleus*, while the somatic (large) nucleus is called the *macronucleus*. Nuclear dualism in ciliates and other protozoans will be discussed in detail in Chapters 7 and 8.

The shape of protozoan nuclei is very diverse. Spherical and ellipsoidal nuclei are found most frequently (Fig. 1), while spindle-shaped, cap-shaped, elongated, lobulated, and other nuclei are seen more rarely. The nucleus of *Amoeba proteus* is disc-shaped (Fig. 4*c*) while that of the radiolarian *Thalassophysa* has sacculiform pockets on the surface (Fig. 66). The macronuclei of ciliates have the most diverse shapes. They can be sausage-shaped, moniliform, branched, etc. (Fig. 90).

The dimensions of protozoan nuclei vary greatly. Nuclei with a diameter of about 1 μm and even less occur in minute intracellular parasites: the flagellates *Leishmania*, malaria parasites, etc. The diameter of the nucleus of the smallest free-living protozoans (e.g. the flagellates *Pedinomonas*, *Heteromastix*, and *Sphaleromantis*) is also within the range of 1–2 μm. The micronuclei of ciliates are also small; they range from 1 μm in *Blepharisma wardsi* to 7–8 μm (rarely to 20 μm) in *Paramecium bursaria*. The largest spherical protozoan nuclei have a diameter of 400 μm (in the radiolarians *Planktonetta atlantica* and *Thalassophysa sanguinolenta*) and even 1 mm (in *Thalassophysa melacapsa*; Hollande and Enjumet 1953). Thus, the diameter of the largest protozoan nuclei is 1000 times greater than that of the smallest ones, while the volume of the largest ones is 1000^3, i.e. a billion times greater than that of the smallest ones! The macronuclei of some ciliates (e.g. *Bursaria*) are also very large, but they usually have a complex shape; therefore, it is difficult to compare their volume with that of the other nuclei. It should be noted that usually the largest protozoan nuclei either are polyploid, i.e. consist of several identical chromosome sets (radiolarian nuclei, macronuclei of some ciliates) or

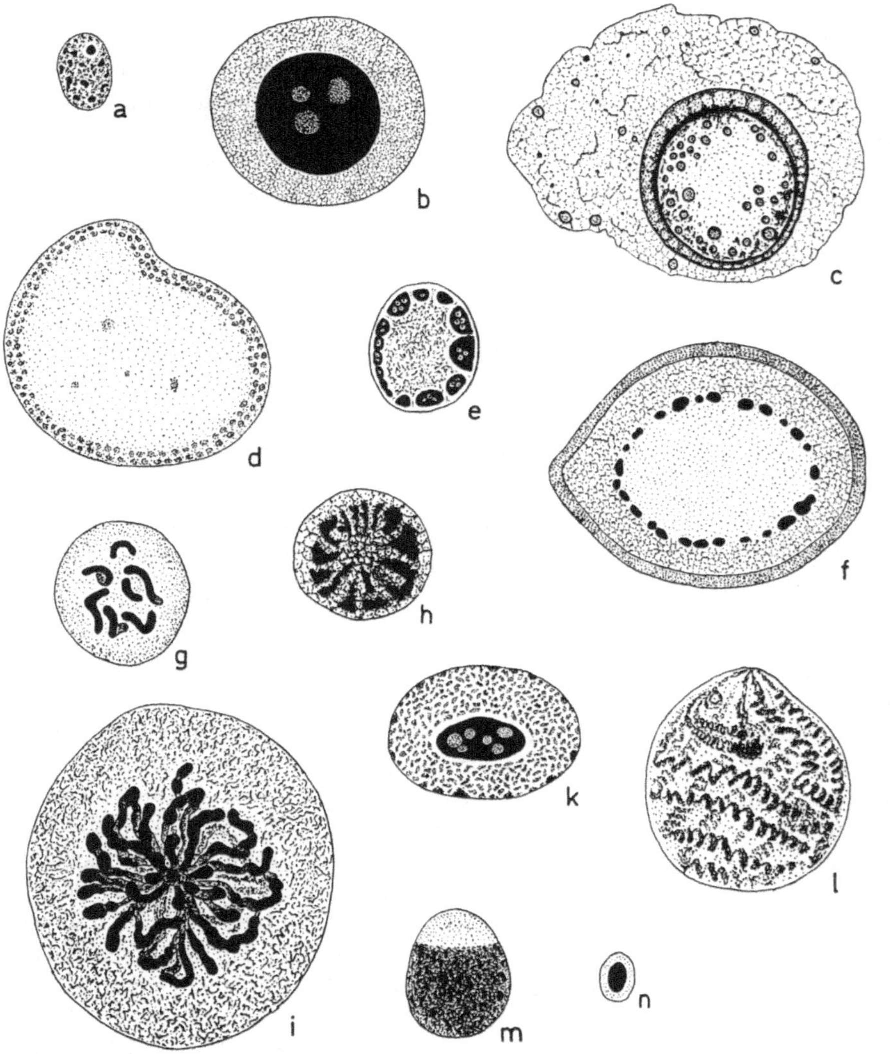

Fig. 1. Main types of nonpolyploid nuclei in protozoans. *a Trichomonas muris*, 2500 × ; *b Amoeba sphaeronucleolus*, 1500 × ; *c Aggregata eberthi*, a macrogamete, 1000 × ; *d Amoeba crystalligera*, 1500 × ; *e Amoeba terricola*, 1500 × ; *f Endamoeba blattae*, 1500 × ; *g Euglypha* sp. (sausage-shaped nucleoli), 1000 × ; *h Actinosphaerium eichhorni* (rosette-like arrangement of nucleoli), 1450 × ; *i Duboscquella aspida* (a branched central nucleolus), 1200 × ; *k Euglena* sp., 1200 × ; *l Holomastigotoides psammotermitidis*, 1500 × ; *m* micronucleus of *Paramecium caudatum*, 1500 × ; *n* micronucleus of *Paramecium aurelia*, 1500 ×. *a–h*, *k*, *m*, and *n* after BĚLAŘ (1926a); *i* after CACHON (1964); *l* after GRASSÉ and HOLLANDE (1963).

have many isolated gene copies that are selectively multiplied as a result of a process that resembles gene amplification (macronuclei of other ciliates). Such nuclei will be considered in detail in Chapters 6 and 8.

Structural Components of Nuclei

Nuclear Envelope

Typical Structure. All protozoan nuclei are surrounded by a nuclear envelope. Basically, this envelope has the same structure as that of the cell nuclei of other eukaryotes (FRANKE and SCHEER 1974), i.e. it consists of two parallel *unit membranes* (outer and inner ones), each of which is about 7.5 nm thick and separated from the other by *perinuclear space*. The nuclear envelope is perforated with *pores* at whose periphery the outer nuclear membrane is continuous with the inner one (Fig. 2*a*, *f*, *g*). In sections that are tangential to the nuclear surface, the pores usually have the form of dense rings or *annuli* (Fig. 2*b*, *c*), while the *central granules* can be seen in the pore openings. The annulus is formed by a ring of dense material along the pore edges. The nuclear pore with the accompanying structures (annulus, central granule, etc.) is called the *pore complex*. The inner surface of the nuclear envelope is often covered with a thin layer of amorphous or indistinctly fibrillar material called *fibrous lamina*.

This classic structure of the nuclear envelope has been observed in many eukaryotic Protozoa belonging to all protozoan phyla (see the column Nuclear Envelope in Table 1). Among the Sarcodina, however, the nuclear envelope is of a typical structure only in forms with rather small nuclei. As a rule, the nuclear envelope of sarcodinians with large nuclei (i.e. foraminiferans, radiolarians and large uninucleate amoebae) has various complications. Among the Sporozoea, the nuclei of trophozoites of only some gregarines have a "classic" envelope (Fig. 2*a*); this envelope is more complex among other Gregarinia and occasionally in the gamonts of the same species (Table 1). As for the Ciliophora, their generative nuclei (micronuclei) as well as their somatic nuclei (macronuclei) are invariably surrounded by a classic two-membrane envelope in which more or less frequent pores are regularly seen (Tables 1 and 6).

Pore Complexes. The number of pore complexes per unit surface of the protozoan nucleus varies greatly. In general, it is proportional to the size of the nucleus: as a rule, there are many more pores per square μm on the surface of large nuclei than in the same area on the surface of small nuclei. Very sparse pores are characteristic of the nuclei of some flagellates, i.e. the chrysomonadid *Pseudodendromonas* and the cryptomonadid *Cyathomonas* (MIGNOT 1965, 1974a), the nuclei of the merozoites of the gregarine *Selenidium* (SCHRÉVEL 1971b), the nuclei of the trophozoites of avian malaria parasites (AIKAWA 1966), etc. Pores are rather sparse in the envelope of the micronucleus of the ciliate *Nassula ornata*, numbering 12 per square μm, while they number more than 80 per square μm in the envelope of the macronucleus (RAIKOV 1966). There are, however, 30–40 pores (RAIKOV and DRAGESCO 1969) and even 60 pores per square μm (RAIKOV and KOVALEVA 1978) in the micronuclei of the karyorelictids belonging to the genus *Tracheloraphis*. A moderate pore frequency (13–15 per square μm) has been observed in rather small nuclei (6 μm in diameter) of the rhizopod *Gromia* (HEDLEY and WAKEFIELD 1969).

The nucleus of the trophozoite of the protococcidian *Coelotropha* has about 20 pores per square μm (VIVIER and HENNERÉ 1965; Fig. 2*b*).

Considerably greater frequency of the nuclear pores (30–35 per square μm) has been observed in the dysenteric amoeba *Entamoeba histolytica* (MARTINEZ-PALOMO *et al.* 1976). On the same area of the nucleus, there are about 40 pores in the multinucleate free-living amoebae of the genus *Chaos* and in *Pelomyxa palustris* (DANIELS and ROTH 1964, DANIELS *et al.* 1966, 1969), 50–60 pores in the flagellate *Hypotrichomonas acosta* (MATTERN *et al.* 1969), about 80 pores in the hypermastigid *Trichonympha* (GRIMSTONE 1959), and about 100 pores in the foraminiferan *Globigerina* (FEBVRE-CHEVALIER 1971). The pore frequency is not constant for a given species: in *Euglena* it depends on the cultivation temperature (LOTT *et al.* 1977), while in *Acanthamoeba* it sharply decreases during encystment (WILLAERT *et al.* 1978a).

In ciliates, the pores are very numerous on the envelopes of the macronuclei belonging to the diploid and especially to the polyploid type. As regards the former, one square μm of the surface of the nucleus has about 40 pores in *Tracheloraphis caudatus* (RAIKOV and DRAGESCO 1969), about 60 pores in *Loxodes magnus* (RAIKOV 1975a), and about 85 pores in *Tracheloraphis phoenicopterus* (RAIKOV and KOVALEVA 1978). As regards the latter, the figure comes to about 35 in *Chilodonella cucullulus* (RADZIKOWSKI 1973a), about 50 in *Neobursaridium gigas* (NILSSON 1969), about 75 in *Woodruffia metabolica* (GOLDER 1976; Fig. 2*c*), 60–80 in *Paramecium primaurelia* (STEVENSON and LLOYD 1971b), about 100 in *Didinium nasutum* (RIEDER 1971, KARADZHAN and RAIKOV 1977a), and up to 140 in *Nassula* sp. (TUCKER 1967). It has been shown that in *Tetrahymena pyriformis*, pore frequency depends on the physiological state of the ciliates: there are 70–145 pores in the stationary phase of culture growth and 180–190 pores in the logarithmic phase per square μm of the surface of the macronucleus (WUNDERLICH and FRANKE 1968, WUNDERLICH 1969a). The diameter of the pore openings is less in the logarithmic phase than in the stationary phase, being 47 and 56–65 nm, respectively. When the pores are very numerous, they are arranged on the nuclear surface in a regular order (usually hexagonally).

A thorough study of the structure of the pore complexes in some protozoans has shown that it is not always the same. As a rule, the *annulus* is revealed on the tangential sections of the nucleus not as a homogeneous ring, but as a crown of

(*Continued on p. 38*)

Fig. 2. Ultrastructure of the nuclear envelope and pore complexes in protozoans. *a* part of the trophozoite nucleus of the gregarine *Selenidium* (*1* nucleolus; *2* pores of the nuclear envelope), 30 000 ×; *b* pore complexes in a tangential section through the nuclear envelope of the coccidian *Coelotropha*, 70 000 ×; *c* same in the macronuclear envelope of *Woodruffia*, 56 000 ×; *d, e* tangential sections of pore complexes of the amoeba *Chaos carolinensis* (*d* 47 500 ×; *e* 160 000 ×); *f* same, a cross section (the arrows indicate the subunits of the annulus), 72 000 ×; *g* cross section of a pore complex of the gregarine *Selenidium* (the arrows indicate the annulus, the double arrow shows the central element), 100 000 ×; *h* honeycomb layer on the outer surface of the nuclear envelope in *Endamoeba blattae* (the arrows indicate pore complexes), 25 000 ×. *a* after VIVIER, SCHRÉVEL (1964); *b* after VIVIER and HENNERÉ (1965); *c* after GOLDER (1976); *d–f* after DANIELS and others (1969); *g* after VIVIER (1967); *h* after BEAMS and others (1959).

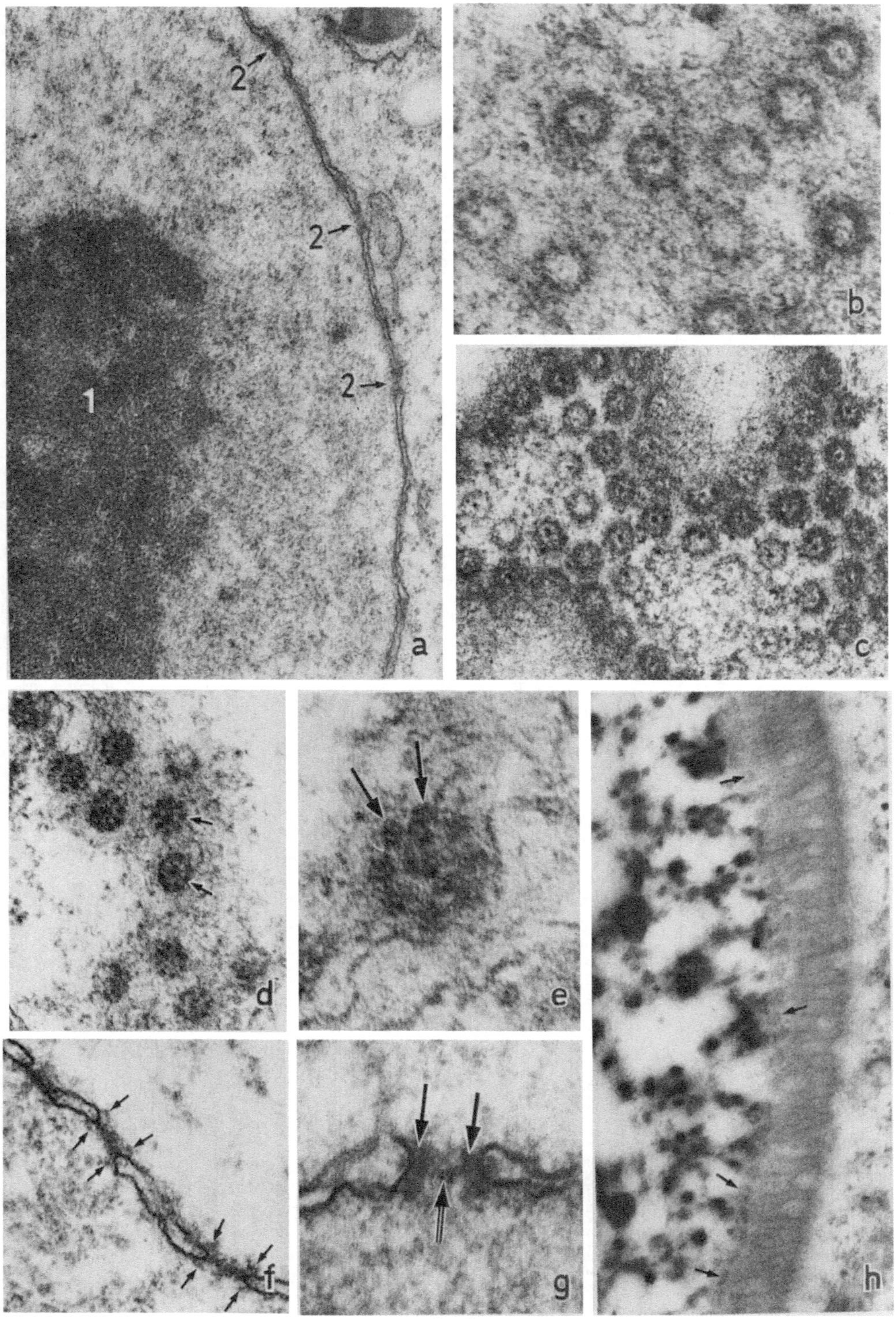

Fig. 2

Table 1. *Fine structural data on the nuclei of eukaryotic Protozoa (except radiolarian nuclei and ciliate macronuclei)*

Species	Number of nuclei	Type of nucleus	Nucleoli	Chromatin	Nuclear envelope	Other components	References
1	2	3	4	5	6	7	8
Phylum Sarcomastigophora Subphylum Mastigophora Class Phytomastigophorea Order Chrysomonadida							
Chromulina psammobia	1	ves.	1, cent.	ret.	cl.	–	Rouiller and Fauré-Fremiet 1958
Ochromonas (3 species)	1	ves.	1, cent.	disp.	cl. + chl., GA, mast.	–	Gibbs 1962, Hibberd 1970, Bouck 1971, Slankis and Gibbs 1972, Hill and Outka 1974
Hydrurus foetidus	1	ves.	1, cent.	disp.	cl. + chl., ER	–	Joyon 1963
Dinobryon divergens	1	ves.	1, cent.	disp.	cl. + chl., ER	–	Joyon 1963, Wujek 1969
Anthophysa vegetans	1	ves.	1, cent.	disp.	cl. + chl., ER	–	Belcher and Swale 1972
Chrysamoeba radians	1	ves.	1, cent.	disp.	cl. + ER, mast.	–	Hibberd 1971
Sphaleromantis tetragona	1	ves.	1, cent.	disp.	cl. + chl.	–	Manton and Harris 1966
Paraphysomonas vestita	1	ves.	1, cent.	disp.	?	–	Manton and Leedale 1961
Uroglena, *Uroglenopsis*	1	ves.	1, cent.	disp.	cl. + chl., ER	–	Wujek 1976
Pseudodendromonas vlkii	1	ves.	1, cent.	ret.	cl.	–	Mignot 1974a
Cyathobodo (3 species)	1	ves.	1, cent.	per. m.	cl.	–	Hibberd 1976
Spumella (= *Monas*) sp.	1	ves.	1, cent.	disp.	cl. + ER, mast.	–	Hill and Outka 1974, Mignot 1977
S. elongata	1	ves.	1, cent.	disp.	cl. + mast.	–	Belcher and Swale 1976

Order Prymnesiida							
Prymnesium parvum	1	ves.	1, cent.	disp.	cl. + chl., ER	–	Manton 1964, 1966
Chrysochromulina chiton	1	ves.	1, cent.	ret.	cl.	–	Manton 1967a, 1967b
Coccolithus pelagicus	1	ves.	1, exc.	ret.	cl.	–	Manton and Leedale 1969
Order Cryptomonadida							
Cryptomonas sp.	1	ves.	1, cent.	ret.	cl. + chl.	–	Gibbs 1962, Joyon 1963
Rhodomonas lens	1	ves.	1, cent.	ret.	cl. + chl.	–	Gibbs 1962
Chilomonas paramecium	1	ves.	1, cent.	ret.	cl. + chl.	–	Joyon 1963
Cyathomonas truncata	1	chr.	1, cent.	chr.	cl. + ER, incl.	–	Mignot 1965
Order Euglenida							
Euglena (3 species)	1	chr.	1, cent.	chr.	cl. + ER	PG or U	Leedale *et al.* 1965, O'Donnell 1965, Mignot 1966, Pyne 1967, Leedale 1968, Rogers *et al.* 1972, Moyne *et al.* 1975, Kluyeva *et al.* 1978
Astasia longa	1	chr.	1, cent.	chr.	cl.	PG	Sommer and Blum 1965, Chaly *et al.* 1977, 1979
Entosiphon sulcatum	1	chr.	1, cent.	chr.	cl.	–	Mignot 1966
Distigma proteus	1	chr.	1, cent.	chr.	cl.	sph.	Mignot 1966, 1975
Trachelomonas volvocina	1	chr.	1, cent.	chr.	cl.	–	Dodge 1975
Isonema nigricans	1	chr.	1, cent.	chr.	cl.	–	Schuster *et al.* 1968
Khawkinea quartana	1	chr.	1, cent.	chr.	cl.	–	Schuster and Hershenow 1974
Order Chloromonadida							
Gonyostomum semen	1	ves.	sev. exc.	ret.	cl. + out. l., GA	PG	Mignot 1967, Heywood 1976
Vacuolaria virescens	1	ves.	sev. cent.	ret.	cl. + out. l., GA	–	Mignot 1967
Colponema loxodes	1	ves.	1, cent.	disp.	cl. + ER, GA	–	Mignot and Brugerolle 1975b
Chattonella subsala	1	ves.	1–2, cent.	ret.	cl. + GA	–	Mignot 1976

Table 1 (continued)

Species	Number of nuclei	Type of nucleus	Nucleoli	Chromatin	Nuclear envelope	Other compo-nents	References
1	2	3	4	5	6	7	8
Order Volvocida							
Chlamydomonas (2 species)	1	ves.	1, cent.	disp. or per. m.	cl. + ER	–	Sager and Palade 1957, Ringo 1967, Johnson and Porter 1968, Triemer and Brown 1974
Dunaliella (2 species)	1	ves.	1, cent.	per. m.	cl. + ER	–	Eyden 1975, Marano 1976
Polytoma (2 species)	1	ves.	1, cent.	per. m.	cl. + ER	(sph.)	Siu *et al.* 1976, Gaffal and Schneider 1978
Polytomella agilis	1	ves.	1, cent.	disp.	cl. + ER, GA	–	Moore *et al.* 1970, Brown *et al.* 1976
Carteria acidicola	1	ves.	1, cent.	disp.	cl.	–	Joyon and Fott 1964
Haematococcus pluvialis	1	ves.	1, cent.	disp.	cl. + ER	–	Joyon 1963, 1965
Stephanosphaera pluvialis	1	ves.	1, cent.	disp.	cl. + ER	–	Joyon 1963, 1965
Volvox aureus,							Soyer 1973
somatic cells	1	ves.	2, subm. and exc.	disp.	cl. + ER	–	
macrogametes	1	l. ves.	2, cent. and exc.	disp.	cl. + ER	–	
Order Prasinomonadida							
Prasinocladus marinus	1	ves.	1, cent.	disp.	cl. + ext.	–	Parke and Manton 1965
Platymonas (3 species)	1	ves.	1, cent.	disp.	cl. + ER	–	Manton and Parke 1965, Parke and Manton 1967
Pyramimonas (4 species)	1	ves.	1, cent.	ret.	cl. + ER	–	Manton *et al.* 1963, Norris and Pearson 1975
Heteromastix rotunda	1	ves.	1, cent.	disp.	cl. + ER	–	Manton *et al.* 1965
Scourfieldia (2 species)	1	ves.	1, cent.	disp.	cl.	–	Manton 1975

Order Choanoflagellida							
Salpingoeca pelagica	1	ves.	1, cent.	disp.	cl.	—	Laval 1971
Codonosiga (2 species)	1	ves.	1, cent.	disp.	cl. + ER	—	Leadbeater and Morton 1974, Hibberd 1975
Choanoeca perplexa	1	ves.	1, cent.	disp.	cl. + ER	—	Leadbeater 1977
Order Proteromonadida							
Proteromonas lacertae-viridis	1	ves.	1, subm.	per. m.	cl. + ER	—	Joyon 1963, Brugerolle and Joyon 1975
Karotomorpha bufonis	1	ves.	?	disp.	cl.	—	Brugerolle and Joyon 1975
Order Bicosoecida							
Bicosoeca (4 species)	1	ves.	1, cent.	ret.	cl. + ER, mast.	—	Mignot 1974b, Belcher 1975, Moestrup and Thomsen 1976
Order Retortamonadida							
Chilomastix sp.	1	ves.	2, subm.	per. m.	cl.	—	Brugerolle 1973
Retortamonas sp.	1	ves.	1, cent.	per. m.	cl.	—	Brugerolle 1977
Order Rhizomastigida							
Cercomonas (2 species)	1	ves.	1, cent.	disp. or per. m.	cl. + GA	—	Mignot and Brugerolle 1975a, Schuster and Pollak 1978
Order Kinetoplastida							
Crithidia fasciculata	1	ves.	1, cent.	per. m.	cl.	--	Kusel *et al.* 1967, Muse and Roberts 1973
Herpetomonas megaseliae	1	ves.	1, cent.	per. m.	cl.	--	Janovy *et al.* 1974
Leishmania (7 species)	1	ves.	1 (rarely 2), cent.	per. m.	cl. + ER	(cr.)	Rudzinska *et al.* 1964, Jadin and Creemers 1966, Avakyan and Manukyan 1967, Manukyan 1967, Creemers and Jadin 1967, Jadin 1971, Avakyan 1976, Croft 1979
Trypanosoma (10 species)	1	ves.	1, cent.	per. m.	cl. + ER	—	Mühlpfordt and Bayer 1961, Wery and Groodt-Lasseel 1966, Sanabria 1966, Milder and Deane 1967, Baker and Bird 1968, Preston 1969, Vickerman 1969, 1970, 1973, Jadin 1971, De Souza and Meyer 1974, Heywood *et al.* 1974, Watson and Lee 1975, Scholtyseck 1979

Table 1 (continued)

Species	Number of nuclei	Type of nucleus	Nucleoli	Chromatin	Nuclear envelope	Other compo-nents	References
1	2	3	4	5	6	7	8
Endotrypanum schaudinni	1	ves.	1, cent.	ret.	cl. + ER	–	Jadin 1971
Trypanoplasma borreli	1	ves.	1, cent.	per. m.	cl.	–	Brugerolle *et al.* 1979
Bodo (4 species)	1	ves.	1, cent.	per. m.	cl.	(sph.)	Brooker 1971, Brooker and Odgen 1972, Burzell 1975, Eyden 1977, Brugerolle *et al.* 1979
Rhynchomonas metabolita	1	ves.	1, cent.	per. m.	cl.	–	Burzell 1973
Ichthyobodo (Costia) necatrix	1	ves.	1, cent.	per. m.	cl.	–	Joyon and Lom 1966, 1969
Cryptobia (4 species)	1	ves.	1, cent.	per. m.	cl. + ER	–	Pyne 1959, Vickerman 1977, Brugerolle *et al.* 1979
Spiromonas angusta	1	ves.	1, cent.	per. m.	cl.	–	MacDonald and Darbyshire 1977
S. perforans	1	ves.	1, exc.	ret.	cl. + ER	–	Brugerolle and Mignot 1979
Order Diplomonadida							
Enteromonas sp.	1-2	ves.	–	disp.	cl.	–	Brugerolle 1975a
Trepomonas (2 species)	2	ves.	1, subm.	disp.	cl.	–	Brugerolle *et al.* 1973
Hexamita muris	2	ves.	1, exc.	disp.	cl.	–	Brugerolle 1974
Octomitus (2 species)	2	ves.	1, subm.	disp.	cl.	–	Brugerolle *et al.* 1974
Lamblia (Giardia) muris	2	ves.	?	kar.	cl. + ER	–	Solovjev and Chentsov 1966, Friend 1966
L. intestinalis	2	ves.	?	disp.	cl.	–	Avakyan 1976
Order Oxymonadida							
Monocercomonoides melolonthae	1	ves.	1, exc.	per. m.	cl.	–	Brugerolle and Joyon 1973
Pyrsonympha vertens	1	ves.	2–3, exc.	ret.	cl.	–	Grassé 1956, Bloodgood *et al.* 1974
Microrhopalodina inflata	sev.	chr.	1, exc.	chr.	cl.	–	Lavette 1973, 1975

Order Trichomonadida							
Trimitus termitis	1	ves.	?	disp.	?	–	Lavette 1975
Hexamastix termitis	1	ves.	?	disp.	?	–	Lavette 1975
Histomonas meleagridis	1	ves.	1, exc.	disp.	cl.	–	Schuster 1968
Monocercomonas sp.	1	ves.	?	disp.	cl.	–	Mattern *et al.* 1972
Hypotichomonas acosta	1	ves.	–	disp.	cl.	–	Mattern *et al.* 1969
Trichomitus (2 species)	1	ves.	1, exc.	disp.	cl.	–	Brugerolle 1971, Honigberg *et al.* 1972
Tritrichomonas muris	1	ves.	1, cent.	disp.	cl.	–	Anderson 1955, Anderson and Beams 1959, Osada 1962
The same	1	ves.	–	per. m.	cl.	–	Selukaite and Semenov 1976
Trichomonas gallinae	1	ves.	1, exc.	disp.	cl.	–	Mattern *et al.* 1967
T. lacertae	1	ves.	?	per. m.	cl.	–	Joyon 1963
T. vaginalis	1	ves.	1 – sev., exc.	per. m.	cl.	–	Inoki *et al.* 1959, Nielsen *et al.* 1966, Smith and Stewart 1966, Avakyan 1976
Pentatrichomonas hominis	1	ves.	1, exc.	ret.	cl.	–	Honigberg *et al.* 1968
Dientamoeba fragilis	2	ves.	1, exc.	disp.	cl.	–	Camp *et al.* 1974
Mixotricha paradoxa	1	chr.	?	chr.	cl.	–	Cleveland and Grimstone 1964
Stephanonympha chagasi	sev.	ves.	1, cent.	disp.	cl.	–	Lavette 1971
Order Hypermastigida							
Lophomonas blattarum	1	ves.	sev., exc.	disp.	cl. + inn. l., n. tub.	–	Beams *et al.* 1961, Beams and Sekhon 1969
Projoenia sawayai	1	ves.	1, cent.	disp.	cl. + out. l.	–	Lavette 1970
Joenia (2 species)	1	chr.	1, exc.	chr.	cl. + out. l.	–	Hollande and Valentin 1969, Hollande 1979
Trichonympha (3 species)	1	chr.	1, exc.	chr.	cl.	–	Grimstone 1959
Urinympha talea	1	l. ves.	1, exc.	disp.	?	–	Hollande and Carruette-Valentin 1971
Staurojoenina caulleryi	1	l. ves.	1, exc.	disp.	?	–	ibid.
Barbulanympha ufalula	1	chr.	?	chr.	?	–	ibid.
Spirotrichonympha psammotermitidis	1	chr.	?	chr.	cl.	–	ibid.
Holomastigotoides hemigymnum	1	chr.	sev., per.	chr.	cl.	–	ibid.

Table 1 (continued)

Species	Number of nuclei	Type of nucleus	Nucleoli	Chromatin	Nuclear envelope	Other compo-nents	References
1	2	3	4	5	6	7	8
Subphylum Opalinata							
Opalina ranarum	many	ves.	sev., per.	disp.	cl.		Noirot-Timothée 1959
Cepedea dimidiata	many	ves.	sev., per.	disp.	cl.		Noirot-Timothée 1967
Subphylum Sarcodina							
Superclass Rhizopoda							
Class Lobosea							
Order Schizopyrenida							
Naegleria gruberi	1	ves.	1, cent.	disp.	cl. + ER	–	Vickerman 1962, Schuster 1963, 1975a, 1975b, Fulton 1970, Carosi *et al.* 1977
N. fowleri	1	ves.	1, cent.	disp. or ret.	cl. + ER, incl.	–	Visvesvara and Callaway 1974, Schuster 1975b, Rondanelli *et al.* 1976, Carosi *et al.* 1977
N. jadini	1	ves.	1, cent.	disp. or ret.	cl. + ER	–	Schuster 1975b, Carosi *et al.* 1976, 1977
Schizopyrenus sp.	1	ves.	1, cent.	disp.	cl. + ER	–	Vickerman 1962
Vahlkampfia damariscottae	1	ves.	1, cent.	disp.	cl.	–	Page 1980
Order Amoebida							
Hartmanella (4 species)	1	ves.	1, cent.	disp. or ret.	cl. + ER, GA	–	Deutsch and Swann 1959, Vickerman 1962, Avakyan 1976, Page 1980
Amoeba proteus (and 3 related species)	1	ov.	many per.	disp.	cl. + inn. l.	hel.	Harris and James 1952, Greider *et al.* 1956, 1958, Pappas 1956a, 1956b, 1959, Cohen 1957, Mercer 1959, Roth *et al.* 1960, Daniels and Breyer 1966, Wise and Goldstein 1972, Daniels 1973, Flickinger 1974

Chaos (2 species)	many	ov.	many per.	disp.	cl.	hel.	Roth and Daniels 1962, Daniels and Roth 1964, Daniels *et al.* 1969, Daniels 1973
Polychaos fasciculatum	1	l. ves.	1, per., bran.	disp.	cl.	—	Page and Baldock 1980
Entamoeba (4 species)	1	ves.	many per.	disp. or kar. + per. m.	cl. + (inn. l.)	no or: sph., cr., d. b., ves. and/or res. sp.	Osada 1959, Deutsch and Zaman 1959, Miller *et al.* 1961, Avakyan and Sarkisyan 1965, Sarkisyan 1965, Siddiqui and Rudzinska 1965, Rondanelli *et al.* 1967a, 1967b, Avakyan 1976, Gicquaud 1979, Injeyan *et al.* 1979, Injeyan and Hübner 1979
Malamoeba scolyti	1 or sev.	ves.	1, cent.	disp.	cl.	—	Purrini 1980
Trichamoeba villosa	1	ves.	1. cent.	disp.	cl.	—	Bhowmick 1967
Thecamoeba sphaeronucleolus	1	l. ves.	1, cent.	disp.	cl. + inn. l.	—	Houssay and Prenant 1970
Th. proteoides	1	ov.	many per.	disp.	cl. + inn. l.	—	Page 1978
Endamoeba blattae	1	ov.	many per.	disp.	cl. + out. l.	—	Beams *et al.* 1959
Vannella (2 species)	1	ves.	1–4, subm.	ret.	cl.	—	Page 1979c
Sappinia diploidea	2	ves.	1, cent.	disp.	cl.	—	Goodfellow *et al.* 1974
Flabellula (3 species)	1–3	ves.	1, cent.	disp.	cl.	—	Page 1980
Paramoeba eilhardi	1	ves.	1, cent.	disp.	cl.	—	Grell and Benwitz 1970
P. perniciosa	1	ves.	1, exc.	ret.	cl.	—	Perkins and Castagna 1971
Vexillifera (2 species)	1	ves.	1, cent.	disp.	cl.	—	Page 1979a, 1979b
Pseudoparamoeba pagei	1	ves.	1, cent.	ret.	cl.	—	Page 1979c
Janickina pigmentifera	1	l. ves.	1, cent.	disp.	cl.	—	Hollande 1980
Mayorella pussardi	1	l. ves.	1, cent.	ret.	cl. + inn. l., out. l.	—	Hollande *et al.* 1981
Acanthamoeba (3 species)	1	ves.	1, cent.	disp.	cl.	—	Vickerman 1962, Bowers and Korn 1968, 1969, Carosi *et al.* 1977, Willaert *et al.* 1978a, 1978b
A. palestinensis	1	ves.	1, cent.	disp.	cl.	d. b.	Lasman 1977
Comandonia operculata	1	ves.	1, cent.	disp.	cl. + ER	—	Pernin and Pussard 1979

Table 1 (continued)

Species	Number of nuclei	Type of nucleus	Nucleoli	Chromatin	Nuclear envelope	Other components	References
1	2	3	4	5	6	7	8
Order Pelobiontida							
Pelomyxa palustris	many	ov.	many per. or cent.	disp.	cl. + out. l., (inn. l.)	d. b., sph., cr.	Daniels *et al.* 1966, Daniels and Breyer 1967, Andresen *et al.* 1968, Daniels 1973
Order Arcellinida							
Chlamydophrys major	1	ves.	1, cent.	ret.	cl.	–	Peschkoff and Akinshina 1963
Arcella polypora	5-7	ves.	1, cent. and sev. per.	ret.	cl. + inn. l.	–	Charret and Vivier 1964, Charret 1967
Gocevia fonbrunei	1	ves.	1, cent.	disp.	cl.	–	Pussard *et al.* 1977
Hyalosphenia papilio	1	ves.	1, cent.	disp.	cl. + out. l.	–	Joyon and Charret 1962
Order Trichosida							
Trichosphaerium (2 species)	many	ves.	1, cent.	ret.	cl.	d. b.	Schuster 1976, Angell 1976
Class Acarpomyxea							
Order Leptomyxida							
Rhizamoeba saxonica	1	ves.	1, cent.	disp.	cl.	–	Page 1980
Order Stereomyxida							
Stereomyxa ramosa	1	ves.	1, cent.	disp. + CC	cl.	–	Benwitz and Grell 1971a
S. angulosa	1	ves.	1 – sev.,	disp. + CC	cl.	cyt. incl.	Benwitz and Grell 1971b
Corallomyxa chattoni	many	ves.	1, exc.	disp.	cl.	–	Grell and Benwitz 1978
Class Filosea							
Order Aconchulida							
Nuclearia simplex	1	ves.	1, cent.	disp.	cl. + out. l.	–	Mignot and Savoie 1979
Nucleosphaerium tuckeri	1	ves.	1, cent.	disp.	cl.	–	Cann and Page 1979

Order Gromiida							
Gromia oviformis	many	ves.	1, exc.	disp.	cl. + ER	–	HEDLEY and WAKEFIELD 1969
Trinema lineare	1	ves.	1, exc.	ret.	cl. + ER	–	HEDLEY and OGDEN 1974
Euglypha (3 species)	1	ves.	1, cent.	ret.	cl. + ER	–	HEDLEY and OGDEN 1973, OGDEN 1979
Amphitrema flavum	1	l. ves.	1, cent. + sev. per.	disp.	cl. + ER	d. b.	BONNET *et al.* 1981
Lesquereusia spiralis	1	l. ves.	many per.	disp.	cl. + inn. l.	–	HÆRRISON *et al.* 1976
Class Granuloreticulosea							
Order Athalamida							
Arachnula impatiens	many	ves.	1, exc.	ret.	cl.	–	OLD and DARBYSHIRE 1980
Order Foraminiferida							
Myxotheca arenilega							SCHWAB 1969
agamont	many	ves.	1, subm.	per. m.	cl. + GA	–	
gamont	1	ov.	many per.	disp.	cl. + out. l.	ves.	
gamete	1	s. ves.	–	per. m.	cl. + GA	–	
Allogromia laticollaris							
agamont	many	ov.	many per.	disp.	cl. + out. l.	fib.	LENGSFELD 1969, SCHWAB 1970
agamete	1–3	ves.	1, cent. or sev. per.	ret. or disp.	cl.	fib.	SCHWAB 1974
gamete	1	chr.	–	chr.	cl.	–	SCHWAB 1976
Ovammina opaca							DAHLGREN 1964, 1967a
agamont	8–16	ov.	many per.	disp.	?	–	
gamont	1	ov.	many per.	disp.	cl. + inn. l., out. l.	–	
Hippocrepinella alba	sev.	ov.	many per. and cent.	disp.	cl. + inn. l., out. l.	–	DAHLGREN 1967b
Nemogullmia longevariabilis (agamont)	1–sev.	ov.	many per. and cent.	disp.	cl. + out. l.	–	NYHOLM and NYHOLM 1975
Iridia lucida							
gamont	1	ov.	many per. and cent.	disp.	cl. + inn. l.	–	CÉSANA 1975
gamete	1	ves.	–	per. m.	cl. + GA, incl.	–	CÉSANA 1972

Table 1 (continued)

Species	Number of nuclei	Type of nucleus	Nucleoli	Chromatin	Nuclear envelope	Other compo-nents	References
1	2	3	4	5	6	7	8
Boderia albicollaris (agamont)	sev.	ov.	many per. and cent.	disp.	cl. + inn. l., out. l.	—	Schwab 1977
Kibisidytes sp. (agamont)	many	ves.	1, cent.	ret.	cl.	—	Schwab and Schwab-Stey 1979
Shepheardella taeniformis	1	ov.	many per.	disp.	cl. + out. l.	—	Hedley *et al.* 1967
Pontomyxa flava	1	ov.	many per.	disp.	inn. l.	—	Cachon and Cachon-Enjumet 1963
Sorites marginalis						—	Müller-Merz and Lee 1976
generative nuclei	many	ves.	1, exc.	per. m.	?	—	
somatic nuclei	many	ov.	many per.	disp.	?	—	
Globobulimina turgida (gamont)	1	ov.	many subm.	disp.	cl. + inn. l., ER	—	Dahlgren 1967b
Globigerina bulloides (gamont)	1	l. ves.	1, subm.	disp.	cl. + inn. l., out. l.	—	Febvre-Chevalier 1971
Hastigerina pelagica (gamete)	1	s. ves.	—	per. m.	cl.	—	Spindler *et al.* 1978
Superclass Actinopoda							
Class Heliozoea							
Order Actinophryida							
Actinophrys sol (trophont)	1	ov.	many per.	disp. and per. m.	cl. + incl.	—	Mignot 1979, Patterson 1979
Ibid., gamete	1	ves.	sev., per.	ret.	cl.	—	Mignot 1980b
Actinosphaerium nucleofilum	many	ov.	many per.	disp.	cl.	—	Tilney and Porter 1965

Order Centrohelida							
Heterophrys marina	1	ves.	1, exc.	disp.	cl.	–	Bardele 1975
Raphidiophrys elegans	1	ves.	1, cent.	disp.	cl.	–	Hovasse 1965
Hedraiophrys hovassei	1	l. ves.	sev., per.	disp.	cl. + inn. l.	d. b.	Febvre-Chevalier 1973
Gymnosphaera albida	many	ves.	sev., per.	disp.	cl.	–	Febvre-Chevalier 1975
Order Taxopodida							
Sticholonche zanclea	1	l. ves.	1, cent.	disp.	cl. + inn. l., out. l.	–	Hollande *et al.* 1967, Cachon and Cachon 1978
Phylum Labyrinthomorpha							
Order Labyrinthulida							
Labyrinthula sp.	many	ves.	1, cent.	ret.	cl. + ER	–	Perkins and Amon 1969
Labyrinthula coenocystis	many	ves.	1, cent.	ret.	cl.	–	Stey 1969
Phylum Apicomplexa							
Class Sporozoea							
Subclass Gregarinia*							
Order Eugregarinida							
Selenidium (3 species)							
trophozoites	1	l. ves.	1, cent.	disp.	cl. + inn. l.	–	Vivier and Schrével 1964, MacGregor and Thomasson 1965, Schrével 1971a
merozoites	1	s. ves.	–	per. m.	cl. + ER	–	Schrével 1971b
Lecudina pellucida	1	l. ves.	1, cent.	disp.	cl. + inn. l.	–	Vivier 1967
Monocystis sp.	1	l. ves.	1, cent.	disp.	cl.	–	Vinckier 1969
Rhynchocystis pilosa	1	l. ves.	1, exc.	disp.	cl.	–	Warner 1968
Zeylanocystis burti	1	l. ves.	1, cent.	disp.	cl.	–	Sathananthan 1977
Lankesteria (3 species)	1	l. ves.	1 or sev., cent.	disp.	cl.	–	Vavra 1969, Walsh and Callaway 1969, Sanders and Poinar 1973
Lankesteria culicis (sporozoite)	1	s. ves.	1, cent.	per. m.	cl.	–	Sheffield *et al.* 1971
Diplauxis (2 species)							Porchet-Henneré and Fischer 1973, Prensier 1970
gamont	1	l. ves.	1, cent.	disp.	?	–	
sporozoite	1	s. ves.	–	per. m.	cl.	–	

* If not otherwise indicated, the data relate to the trophozoite.

Table 1 (continued)

Species	Number of nuclei	Type of nucleus	Nucleoli	Chromatin	Nuclear envelope	Other components	References
1	2	3	4	5	6	7	8
Gregarina (3 species)	1	l. ves.	1, cent.	disp.	cl. + inn. l.	–	DEVAUCHELLE 1968
Gregarina blaberae							TRONCHIN and SCHRÉVEL 1977
trophozoite	1	l. ves.	1, exc.	disp.	cl. + ER, GA	–	
sporozoite	1	s. ves.	1, exc.	per. m.	cl.	–	
Gonospora beloneides	1	l. ves.	1, exc.	disp.	cl. + inn. l., out. l.	–	CORBEL *et al.* 1979
Stylocephalus (3 species)							GRASSÉ and THÉODORIDÈS 1957, 1959
trophozoite	1	l. ves.	1, cent.	disp.	cl. + inn. l.	d. b.	DESPORTES 1966, 1967, 1969, 1970
gametes	1	s. ves.	1, exc.	per. m.	cl.	–	
Grebnickiella gracilis							MOLON-NOBLOT 1977, 1978
macrogamete	1	s. ves.	1, exc.	kar.	cl.	–	
microgamete	1	sp.	–	sol.	cl.	–	
Enterocystis fungoides	1	l. ves.	1, cent., bran.	disp.	cl. + inn. l., bran. ext.	–	DESPORTES 1974
Callyntrochlamys phronimae	1	l. ves.	sev.	disp.	cl. + inn. l., ext.	–	DESPORTES and THÉODORIDÈS 1969
Didymophyes chaudefouri	1	l. ves.	1, cent.	disp.	cl. + bran. ext.	–	BAUDOIN and ORMIÈRES 1973
Didymophyes gigantea							HILDEBRAND 1976, 1978
young trophozoite	1	ves.	1, exc.	disp.	cl.	–	
adult trophozoite	many	ves.	1, exc.	disp.	cl. + ER, GA	–	
Order Neogregarinida							
Gigaductus anchi							ORMIÈRES 1971
trophozoite	1	ves.	1, exc.	disp.	cl.	–	
merozoite	1	s. ves.	1, exc.	per. m.	cl.	–	

Farinocystis tribolii							Canning and Sinden 1975, Žižka 1978a, 1978b, 1978c
schizont	many	ves.	1, exc.	ret.	cl.	–	
merozoite	1	s. ves.	1, exc.	ret.	cl. + incl.	–	
gametocyte	1–sev.	ves.	1, cent.	ret.	cl.	–	
Subclass Coccidia							
Order Protococcidiida							
Coelotropha durchoni							Vivier 1963a, Vivier and Henneré 1965, Porchet-Henneré 1967a, 1967b, 1970
young trophozoite	1	ves.	1, cent.	per. m.	cl.	PG	
mature trophozoite	1	l. ves.	1, cent.	disp.	cl.	PG	
microgamete	1	sp.	–	spon.	cl.	–	
Eucoccidium dinophili							Bardele 1966
trophozoite	1	ves.	1, cent.	ret. or disp.	cl.	–	
macrogamete	1	l. ves.	1, cent.	disp.	cl.	–	
microgamete	1	sp.	–	sol.	cl.	–	
sporozoite	1	s. ves.	–	per. m.	cl.	–	
Order Eucoccidiida							
Suborder Adeleina							
Klossia helicina							Volkmann 1967, Scholtyseck 1979, Moltmann 1981
trophozoite	1	ves.	1, cent.	per. m.	cl. + ER	–	
macrogamont	1	l. ves.	1, cent.	disp.	cl.	–	
sporozoite	1	s. ves.	1, exc.	per. m.	cl. + ER	–	
Adelina tribolii (macrogamont)	1	ves.	1, exc.	disp.	cl.	–	Žižka 1969
Haemogregarina sp. (gametocyte)	1	ves.	1, cent.	per. m.	?	–	Desser and Weller 1973
Karyolysus sp. (merozoite)	1	s. ves.	1, exc.	per. m.	cl.	–	Beyer 1977
Suborder Eimeriina							
Aggregata eberthi							Porchet-Henneré and Richard 1969, 1970a, 1970b, 1971a, 1971b, Heller 1969, 1970
trophozoite	1	l. ves.	1, cent.	disp.	cl. + ER	PG, fib.	
merozoite	1	s. ves.	–	per. m.	cl. + GA	–	
macrogamete	1	l. ves.	1, exc.	disp.	cl. + ER	–	
microgamete	1	sp.	–	sol.	cl.	–	
sporozoite	1	s. ves.	–	per. m.	cl.	–	

Table 1 (continued)

Species	Number of nuclei	Type of nucleus	Nucleoli	Chromatin	Nuclear envelope	Other components	References
1	2	3	4	5	6	7	8
Myriosporides amphiglenae (trophozoite)	1	ves.	2, subm.	disp.	cl.	–	Henneré 1967
Dehornia sthenelais (sporozoite)	1	s. ves.	–	per. m.	cl.	–	Porchet-Henneré 1972a
Grasseella microcosmi							Porchet-Henneré and Ormières 1973, Ormières and Porchet-Henneré 1974
trophozoite	1	ves.	1, cent.	disp.	cl.	–	
merozoite	1	s. ves.	1, exc.	per. m.	cl.	–	
macrogamont	1	l. ves.	1, exc.	disp.	cl.	–	
Eimeria – parasites of rodents (*E. intestinalis*, *E. magna*, *E. stiedae*, *E. perforans*, *E. nieschulzi*, *E. pragensis*, *E. callospermophili*, *E. ferrisi*, *E. falciformis*, *E. papillata*)							Mosevich and Cheissin 1961, Scholtyseck 1962, 1965a, 1965b, Scholtyseck and Spiecker 1964, Cheissin 1964, 1965, Cheissin and Snigirevskaya 1965, Scholtyseck and Piekarski 1965, Scholtyseck *et al.* 1966, 1972, Hammond *et al.* 1967, Colley 1967a, 1967b, 1968, Snigirevskaya 1969a, 1969b, 1969c, Sénaud and Černa 1969, 1970, Roberts *et al.* 1970a, 1970b, Danforth and Hammond 1972, Youssef and Hammond 1972, Mehlhorn *et al.* 1973, Speer *et al.* 1973, Chobotar, Scholtyseck *et al.* 1975, Chobotar, Sénaud *et al.* 1975, Speer and Danforth 1976, Beyer *et al.* 1978, Scholtyseck 1979, Chobotar *et al.* 1980, Pittilo *et al.* 1980
trophozoite	1	ves.	1, cent. or. exc.	ret.	cl. + ER, GA	–	
merozoite	1	s. ves.	1. exc.	per. m.	cl. + ER	–	
macrogamete	1	l. ves.	1, cent.	disp.	cl. + ER	–	
microgamete	1	sp.	–	sol.	cl.	–	
sporozoite	1	s. ves.	–	per. m.	cl.	–	

Eimeria—parasites of ungulates (*E. alabamensis*, *E. ninakohlyakimovae*, *E. ellipsoidalis*, *E. auburnensis*, *E. bovis*)							Sheffield and Hammond 1966, Scholtyseck *et al.* 1966, 1972, Hammond *et al.* 1969, Roberts and Hammond 1970, Kelley and Hammond 1972, 1973, Sampson and Hammond 1972, Dubremetz and Elsner 1979, Scholtyseck 1979
trophozoite	1	ves.	1, exc.	ret.	cl. + GA	–	
merozoite	1	s. ves.	1, exc.	per. m.	cl. + GA	–	
macrogamete	1	l. ves.	1, cent.	disp.	cl.	–	
microgamete	1	sp.	–	sol.	cl.	–	
sporozoite	1	s. ves.	no or 1, exc.	per. m.	cl.	–	
Eimeria—parasites of birds (*E. tenella*, *E. maxima*, *E. necatrix*, *E. praecox*, *E. acervulina*, *E. kotlani*, *E. labbeana*)							Scholtyseck 1963, Seliverstova 1969, McLaren 1969, Ryley 1969, Scholtyseck *et al.* 1969, 1972, Sénaud and Černa 1969, Strout and Scholtyseck 1970, Lee and Millard 1971, Mehlhorn 1972a, 1972b, Dubremetz 1973, 1975, Fernando 1973, 1974, Shibalova 1974, Shibalova *et al.* 1974a, 1974b, Fernando and Stockdale 1974, Hoppe 1974, 1976, Pacheco *et al.* 1975, Varghese 1975, Antukhaev 1977, Kravtsov 1977, Varghese 1977, Beyer *et al.* 1978, Scholtyseck 1979, Sénaud *et al.* 1980
trophozoite	1	ves.	1, cent. or exc.	ret.	cl. + ER	–	
merozoite	1	s. ves.	1, exc.	per. m.	cl. + ER, GA	–	
macrogamete	1	l. ves.	1, cent.	disp.	cl. + ER	–	
microgamete	1	sp.	–	sol.	cl.	–	
sporozoite	1	s. ves.	–	per. m.	cl. + GA	–	
Tyzzeria parvula							Shibalova and Morozova 1979
trophozoite	1	ves.	1, exc.	disp.	cl.	–	
merozoite	1	s. ves.	1, per.	per. m.	cl.	–	
Globidium gilruthi (merozoite)	1	s. ves.	1, exc.	per. m.	cl.	–	Porchet-Henneré 1977, Scholtyseck 1979

Table 1 (continued)

Species	Number of nuclei	Type of nucleus	Nucleoli	Chromatin	Nuclear envelope	Other components	References
1	2	3	4	5	6	7	8
Isospora (4 species)							Pelster 1973, Mehlhorn and Markus 1976, Milde 1979
trophozoite	1	ves.	1, exc.	ret.	cl. + ER	–	
merozoite	1	s. ves.	1, exc.	per. m.	cl. + ER, GA	–	
microgamont	many	s. ves.	–	per. m.	cl.	–	
microgamete	1	sp.	–	sol.	cl.	–	
macrogamete	1	l. ves.	1, cent. or exc.	disp. or ret.	cl.	–	
sporozoite	1	s. ves.	1, subm.	per. m.	cl. + ER	–	
Toxoplasma gondii (stages in intermediate hosts):							Gustafson *et al.* 1954, Ludvik 1956, 1958a, 1963, Gavin *et al.* 1962, Garnham *et al.* 1962b, Akinshina and Bykovsky 1964, Scholtyseck and Piekarski 1965, Jadin and Creemers 1967, 1968, Sénaud 1967, Zypen and Piekarski 1967a, 1967b, Sheffield and Melton 1968, Vivier 1970, Porchet-Henneré and Vivier 1971, Vivier and Petitprez 1972a, Avakyan 1976, Beyer *et al.* 1978, Scholtyseck 1979
endodyocytes (endozoites)	1	s. ves.	1–2, exc. or subm.	per. m.	cl. + ER, GA	–	
Toxoplasma gondii (stages in definitive host)							Sheffield 1970, Pelster and Piekarski 1971, Ferguson *et al.* 1974, 1975, Christie *et al.* 1978, Scholtyseck 1979
trophozoite	1	ves.	1, cent.	ret.	cl.	–	
merozoite	1	s. ves.	1, exc.	per. m.	cl.	–	

macrogamete	1	l. ves.	1, exc.	disp.	cl. + GA	–	
microgamete	1	sp.	–	sol.	cl.	–	
sporozoite	1	s. ves.	–	per. m.	cl. + GA	–	
Besnoitia jellisoni							Sheffield 1966, 1968, Sénaud 1969a, 1969b, Scholtyseck *et al.* 1973, Sénaud and Mehlhorn 1978, Fedoseyenko 1979
endodyocyte	1	s. ves.	1, exc. or subm.	per. m. or ret.	cl. + ER, GA	–	
cystozoite	1	s. ves.	–	per. m.	cl.	–	
Frenkelia sp.							Ludvik 1963, Kepka and Scholtyseck 1970, Scholtyseck *et al.* 1970, 1973
metrocyte	1	ves.	1, cent.	ret.	cl. + ER, GA	–	
merozoite	1	s. ves.	1, exc.	per. m.	cl. + ER, GA	–	
Sarcocystis (2 species), stages in definitive host							Vetterling *et al.* 1973, Scholtyseck 1979, Dubey *et al.* 1980
trophozoite	1	ves.	1, cent.	ret.	cl. + ER	–	
merozoite	1	s. ves.	1, exc.	per. m.	cl.	–	
microgamont	sev.	ves.	1, exc.	ret.	cl. + ER	–	
microgamete	1	sp.	–	sol.	cl.	–	
macrogamont	1	ves.	1, cent.	ret.	cl.	–	
macrogamete	1	l. ves.	fragm.	disp.	cl.	–	
Sarcocystis (7 species), stages in intermediate hosts							Ludvik 1956, 1958b, 1960, 1963, Simpson 1966, Sénaud 1967, Mehlhorn and Scholtyseck 1973, Scholtyseck *et al.* 1973, Mehlhorn *et al.* 1975, Rzepczyk and Scholtyseck 1976, Pacheco and Fayer 1977, Sénaud and Černa 1978, Fedoseyenko and Levit 1979
trophozoite	1	ves.	1, cent.	ret.	cl.	–	
schizont	sev.	ves.	sev., exc.	per. m.	cl.	–	
merozoite	1	s. ves.	1, exc.	per. m.	cl.	–	
metrocyte	1	ves.	1, exc.	ret.	cl. + ER	–	
cystozoite	1	s. ves.	(1, subm.)	per. m.	cl. + ER	–	
Cryptosporidium wrairi							Vetterling *et al.* 1971
trophozoite	1	ves.	1, cent.	disp.	cl. + ER, GA	–	
merozoite	1	s. ves.	1, cent.	per. m.	cl. + ER	–	
Lankesterella (2 species)							Lainson *et al.* 1961, Garnham *et al.* 1962a, Stehbens 1966
trophozoite	1	ves.	1, exc.	ret.	cl.	–	
sporozoite	1	s. ves.	1, exc.	per. m.	cl.	–	

Table 1 (continued)

Species	Number of nuclei	Type of nucleus	Nucleoli	Chromatin	Nuclear envelope	Other components	References
1	2	3	4	5	6	7	8
Suborder Haemosporina							
Plasmodium – avian and reptilian parasites (*P. gallinaceum*, *P. fallax*, *P. lophurae*, *P. cathemerium*, *P. elongatum*, *P. pinotti*, *P. relictum*, *P. floridense*, *P. mexicanum*, *P. gabaldoni*)							GARNHAM *et al.* 1960, 1962c, RUDZINSKA and TRAGER 1961, AIKAWA 1966, 1971, HEPLER *et al.* 1966, AIKAWA *et al.* 1967, 1968, TERZAKIS *et al.* 1967, AIKAWA and JORDAN 1968, AIKAWA and BEAUDOIN 1969, AIKAWA, HUFF, and SPRINZ 1969, RUDZINSKA 1969, LANGRETH and TRAGER 1973, STERLING and AIKAWA 1973, MOORE and SINDEN 1974, GARNHAM 1977, MEHLHORN *et al.* 1980
trophozoite	1	ves.	1, subm. or exc.	ret. or per. m.	cl. + ER	–	
merozoite	1	s. ves.	(1, subm.)	per. m.	cl. + ER	–	
macrogametocyte	1	ves.	1, exc.	disp.	cl.	–	
microgametocyte	1	ves.	no or 1, exc.	disp. or ret.	cl.	–	
ookinete and oocyst	1	ves.	1, exc.	disp.	cl. + ER	–	
sporozoite	1	s. ves.	–	per. m.	cl.	–	
Plasmodium – mammalian parasites (*P. berghei*, *P. vinckei*, *P. chabaudi*, *P. knowlesi*, *P. coatneyi*, *P. cynomolgi*, *P. falciparum*)							RUDZINSKA and TRAGER 1959, 1968, GARNHAM *et al.* 1961, 1962c, 1967, VANDERBERG *et al.* 1967, VICKERMAN and COX 1967, JADIN *et al.* 1968, SCALZI and BAHR 1968, AIKAWA, COOK *et al.* 1969, AIKAWA, HUFF, and

trophozoite	1	ves.	–	disp. or ret.	cl. + ER	–	Sprinz 1969, Killby and Silverman 1969, Ladda 1969, Rudzinska 1969, Aikawa 1971, Terzakis 1971, Canning and Sinden 1973a, Langreth *et al.* 1978
merozoite	1	s. ves.	–	ret. or per. m.	cl. + GA	–	
macrogametocyte	1	ves.	–	disp.	cl. + ER	–	
microgametocyte	1	ves.	–	disp.	cl.	–	
microgamete	1	sp.	–	sol.	cl.	–	
ookinete	1	ves.	–	disp.	cl.	–	
sporozoite	1	sp.	–	sol.	cl. + ER	–	
Haemoproteus (3 species)							Bradbury and Trager 1968a, 1968b, Bradbury and Roberts 1970, Bradbury and Gallucci 1971, 1972, Desser 1972, Klei 1972, Sterling 1972, Sterling and De Guisti 1972, 1974, Sterling and Aikawa 1973, Gallucci 1974a, 1974b
trophozoite	1	ves.	1, exc.	ret.	cl. + ER	–	
merozoite	1	s. ves.	–	per. m.	cl.	–	
macrogametocyte	1	ves.	1, exc.	disp.	cl. + ER	–	
macrogamete	1	ves.	fragm.	disp.	cl. + ER	–	
microgametocyte	1	ves.	–	disp.	cl.	–	
microgamete	1	sp.	–	sol.	cl.	–	
zygote and ookinete	1	ves.	–	disp.	cl. + ER	–	
sporozoite	1	s. ves.	–	per. m.	cl. + ER	–	
Leucocytozoon (2 species)							Garnham *et al.* 1967, Desser *et al.* 1970, Aikawa *et al.* 1970, Sterling and Aikawa 1973, Wong and Desser 1976, 1978, 1981
trophozoite	1	ves.	1, exc.	ret.	cl.	–	
merozoite	1	s. ves.	–	per. m.	cl.	–	
macrogametocyte	1	ves.	1, exc.	disp.	cl.	PG	
microgametocyte	1	ves.	–	disp.	cl.	PG	
microgamete	1	sp.	–	sol.	cl.	–	
oocyst	1	ves.	–	disp.	cl.	–	
sporozoite	1	s. ves.	–	per. m.	cl.	–	
Subclass Piroplasmia							
Babesia (6 species) (trophozoite, merozoite, gametocyte)	1	ves.	no or 1, exc.	disp. or ret.	cl. + ER	–	Bayer and Dennig 1961, Rudzinska and Trager 1962, Simpson *et al.* 1963, Buttner 1968, Friedhoff and Scholtyseck 1968, 1969, 1977, Molyneux and Bafort 1970, Friedhoff *et al.* 1972, Rudzinska 1976, Rudzinska *et al.* 1979, Scholtyseck 1979

Table 1 (continued)

Species	Number of nuclei	Type of nucleus	Nucleoli	Chromatin	Nuclear envelope	Other compo-nents	References
1	2	3	4	5	6	7	8
Theileria (5 species)							BUTTNER 1966, 1967a, 1967b, MEHLHORN and SCHEIN 1977, SCHEIN *et al.* 1978, MEHLHORN *et al.* 1979
trophozoite, merozoite	1	ves.	—	disp.	cl.	—	
zygote, ookinete	1	ves.	1, exc.	disp.	cl.	—	
Anthemosoma garnhami	1	ves.	—	disp.	cl.	—	VIVIER and PETITPREZ 1969, 1972b, VIVIER *et al.* 1974
Phylum Microspora* (trophozoites, sporonts, sporoblasts)							
Class Rudimicrosporea							
Order Metchnikovellida							
Metchnikovella (2 species)	1–many	ves.	—	disp.	cl.	—	VIVIER 1965, HILDEBRAND and VIVIER 1971, VIVIER and SCHRÉVEL 1973, HILDEBRAND 1974
Amphiamblys laubieri	1–many	ves.	1, exc.	ret.	cl.	—	DESPORTES and THÉODORIDÈS 1979
Class Microsporea							
Order Minisporida							
Hessea squamosa	[2]	ves.	1, exc.	disp.	cl.	—	ORMIÈRES and SPRAGUE 1973
Order Microsporida							
Nosema (5 species)	[2]	ves.	no or 1, exc.	disp.	cl.	—	SPRAGUE and VERNICK 1969, WEINDER 1970, WALKER and HINSCH 1972, CANNING and SINDEN 1973b, TAKIZAWA *et al.* 1973

* In microsporidians, the number of nuclei is given for the vegetative stages (trophozoites, meronts, schizonts); [2] indicates a close union of two nuclei of the trophozoite into the so-called diplokaryon, many [2] means many diplokaryons in the plasmodium.

Spraguea lophii	1 or [2]	ves.	–	disp.	cl.	–	Loubès *et al.* 1979
Nosemoides (2 species)	1 many	ves.	–	disp.	cl.	–	Vinckier *et al.* 1970, 1971, Vinckier 1975, Loubès and Akbarieh 1977
Encephalitozoon cuniculi	1	ves.	–	disp.	cl.	–	Sprague and Vernick 1971, Pakes *et al.* 1975
Perezia lankesteriae	[2]	ves.	1, cent.	disp.	cl.	–	Ormières *et al.* 1977
Ichthyosporidium sp.	[2]	ves.	–	disp.	cl.	–	Sprague and Vernick 1974
Glugea (2 species)	many	ves.	–	ret.	cl.	–	Sprague and Vernick 1968, Loubès *et al.* 1976a
Toxoglugea variabilis	[2]	ves.	–	disp.	cl. + ER	–	Larsson 1980a
Buxtehudea scaniae	many	ves.	–	disp.	cl. + ER	–	Larsson 1980b
Telomyxa glugeiformis	1	ves.	1, cent.	disp.	cl.	–	Codreanu and Vávra 1970
Thelohania (7 species)	[2]	ves.	–	disp.	cl.	–	Kudo and Daniels 1963, Liu and Davies 1972, Maurand and Vey 1973, Vivarès 1975, 1980, Maurand and Loubès 1978
Gurleya chironomi	1	ves.	–	disp.	cl.	–	Loubès and Maurand 1975
Amblyospora sp.	[2]	ves.	–	ret.	cl.	–	Andreadis and Hall 1979
Ormieresia carcini	[2]	ves.	?	disp.	cl.	–	Vivarès *et al.* 1977
Orthothelohania octospora	[2]	ves.	–	ret.	cl.	–	Codreanu *et al.* 1974
Stempellia mutabilis	1-2	ves.	–	disp.	cl.	–	Desportes 1976
S. simulii	[2]	ves.	–	disp.	cl.	–	Maurand and Loubès 1978
Pleistophora hyphessobriconis	1-many	ves.	–	disp.	cl.	–	Lom and Corliss 1967
P. simulii	many [2]	ves.	–	disp.	cl.	–	Maurand and Loubès 1978
Tuzetia debaisieuxi	[2]	ves.	(1, exc.)	disp. or per. m.	cl.	–	Vávra *et al.* 1966, Weiser and Žižka 1975, Loubès and Maurand 1976, Maurand and Loubès 1978
Octosporea muscaedomesticae	1-[2]	ves.	–	disp.	cl.	–	Ormières *et al.* 1976
Baculea daphniae	1-many	ves.	–	disp.	cl.	–	Loubès and Akbarieh 1978
Phylum Ascetospora							
Class Stellatosporea							
Order Occlusosporida							
Marteilia refringens	1	ves.	1-3, exc.	disp.	cl.	–	Perkins 1976
Paramarteilia orchestiae	1	ves.	1, cent.	disp.	cl.	–	Ginsburger-Vogel and Desportes 1979

Table 1 (continued)

Species	Number of nuclei	Type of nucleus	Nucleoli	Chromatin	Nuclear envelope	Other components	References
1	2	3	4	5	6	7	8
Order Balanosporida							
Haplosporidium ascidiarum	1	ves.	1, subm.	disp.	cl.	–	ORMIÈRES and PUYTORAC 1968
Minchinia (4 species)	1, [2] or sev.	ves.	1, exc.	per. m. or disp.	cl. + ER	res. sp.	PERKINS 1968, 1969, 1975, ROSENFELD *et al.* 1969, MAURAND and SPRAGUE 1979
Urosporidium crescens	1	ves.	1, subm.	disp.	cl.	res. sp.	PERKINS 1971, 1975
Phylum Myxozoa							
Class Myxosporea*							
Order Bivalvulida							
Sphaeromyxa (2 species)							GRASSÉ 1960, LOM 1969, GRASSÉ and LAVETTE 1978
plasmodium	many	ves.	1, exc.	disp.	cl.	cr.	
generative cells, 3 types	1	ves.	1, exc.	ret. or per. m.	cl.	–	
Zschokkella nova	many	ves.	1, exc.	disp.	cl.	–	LOM and PUYTORAC 1965
Myxidium lieberkühni	many	ves.	1, exc.	disp.	cl.	–	LOM and PUYTORAC 1965, SCHULMAN 1966
M. gasterostei	many	ves.	1, exc.	disp.	cl.	–	USPENSKAJA 1969
Chloromyxum cristatum	many	ves.	1, exc.	disp.	cl.	–	LOM and PUYTORAC 1965
Myxosoma funduli							CURRENT *et al.* 1979
plasmodium	many	ves.	–	disp.	cl.	–	
generative cells	1	ves.	1, subm.	disp.	cl.	–	
pansporoblast	sev.	ves.	1, subm.	disp.	cl.	–	

* When the stage of the life cycle is not given, data relate to the nuclei of the plasmodium, those of the generative cells and those of the sporoblasts.

Myxobolus sp.							Lom and Puytorac 1965, Desser and Patterson 1978
plasmodium	many	ves.	1, exc.	disp.	cl.	–	
generative cells	1	ves.	1, exc.	ret.	cl.	–	
Henneguya (3 species)	many	ves.	1, exc.	ret.	cl. + GA	–	Schubert 1968, Current and Janovy 1977, Current 1979
H. psorospermica	many	ves.	1, exc.	disp.	cl.	–	Lom and Puytorac 1965a, 1965b
Class Actinosporea							
Aurantiactinomyxon eiseniellae (sporont)	many	ves.	1, exc.	disp.	cl.	–	Ormières 1970
Phylum Ciliophora (micronuclei only)							
Class Kinetofragminophorea							
Subclass Gymnostomatia							
Order Karyorelictida							
Trachelocerca geopetiti	2-4	sp.	–	spon.	cl.	kch.	Raikov and Kovaleva 1981
Tracheloraphis (4 species)	2-many	sp.	–	spon. or sol.	cl.	–	Raikov and Dragesco 1969, Raikov 1972a, 1974, Raikov and Kovaleva 1978, 1980, Kovaleva *et al.* 1979
Trachelonema sulcata	many	sp.	–	sol.	cl.	–	Kovaleva and Raikov 1970, 1973
Loxodes magnus	many	sp.	–	spon.	cl.	sph.	Puytorac and Njiné 1970, Raikov 1975a
Remanella multinucleata	many	sp.	–	spon.	cl.	–	Raikov 1979
Kentrophoros (2 species)	many or sev.	sp.	–	sol. or spon.	cl. + (inn. l.)	–	Raikov 1972b, 1972c, 1973a
Avelia martinicense	1	sp.	–	sol.	cl.	–	Nouzarède 1975, 1976
Geleia orbis	1	sp.	–	sol.	cl.	–	Nouzarède 1976, Raikov 1981
Order Prostomatida							
Didesmis quadrata	1	sp.	–	sev. CC	cl.	res. sp.	Grain 1966
Metacystis elongata	1	sp.	–	spon.	cl. + Ma, ER	–	Detcheva *et al.* 1979
Coleps hirtus	1	s. ves.	–	ret.	cl.	–	Rodrigues de Santa Rosa 1976
Didinium nasutum	2	s. ves.	–	ret.	cl.	MTOC	Karadzhan and Raikov 1977a
Lagynophrya fusidens	1	sp.	–	spon.	cl.	–	Grain 1970
Dileptus anser	many	s. ves.	–	kar.	cl.	–	Vinnikova 1974a, 1976, Bohatier 1977a

Table 1 (continued)

Species	Number of nuclei	Type of nucleus	Nucleoli	Chromatin	Nuclear envelope	Other components	References
1	2	3	4	5	6	7	8
Spathidium sp.	1	chr.	–	chr.	cl.	–	Ivashchenko 1980
Bryophyllum tegularum	1	chr.	–	chr.	cl.	–	Bohatier *et al.* 1978
Homalozoon vermiculare	sev.	s. ves.	–	kar.	cl.	res. sp.	Saxena *et al.* 1978
Subclass Vestibuliferia							
Order Trichostomatida							
Isotricha sp.	1	s. ves.	–	kar.	cl.	res. sp.	Noirot-Timothée 1960, Grain 1966
Balantidium caviae	1	s. ves.	–	kar.	cl.	–	Paulin and Krascheninnikow 1973
Balantidium coli	1	chr.	–	chr.	cl.	–	Krascheninnikow and Scholtyseck 1975, Avakyan 1976
Order Colpodida							
Colpoda maupasi	1	sp.	–	spon.	cl. + ER	–	Rudzinska *et al.* 1966, Lynn 1976a
C. steini	1	chr.	–	chr.	cl.	–	Lynn 1976b, Frenkel 1980
C. simulans	1	chr.	–	chr.	cl.	–	Lynn 1976d
Tillina magna	sev.	chr.	–	chr.	cl. + out. l.	–	Lynn 1976c
Order Platyophryida							
Platyophrya spumacola	1	s. ves.	–	ret.	cl. + Ma	–	Dragesco *et al.* 1977
Cyrtolophosis mucicola	1	s. ves.	–	spon.	cl. + Ma	–	Detcheva 1976
Woodruffia metabolica	1	s. ves.	–	ret.	cl. + Ma	–	Golder 1976
Aristerostoma marinum	1	sp.	–	spon.	cl. + Ma	–	Detcheva and Puytorac 1979
Order Entodiniomorphida							
Ophryoscolecidae gen. sp.	1	chr.	–	chr.	cl. + inn. l.	–	Noirot-Timothée 1960
Subclass Hypostomatia							
Order Nassulida							
Nassula (2 species)	sev.	s. ves.	–	ret.	cl. + ER	–	Raikov 1966, Tucker 1967

Order Cyrtophorida							
Chilodonella cucullulus	1	sp.	–	spon.	cl.	–	Radzikowski 1973a
Ch. uncinata	1	chr.	–	chr.	cl.	–	Pyne 1978
Order Chonotrichida							
Cavichona elegans	2	sp.	–	spon.	cl.	–	Karadzhan 1976
Order Rhynchodida							
Ancistrocoma pelseneeri	1	sp.	–	sol.	cl. + ER	res. sp.	Khan 1969
Subclass Suctoria							
Podophrya parameciorum	1	chr.	–	chr.	cl.	–	Jurand and Bomford 1965
Acineta tuberosa	sev.	s. ves.	–	ret.	cl.	res. sp.	Bardele 1968
Paracineta limbata	1	s. ves.	–	kar.	cl.	MTOC	Hauser 1968, 1970, 1972
Tokophrya infusionum	1	s. ves.	–	ret.	cl.	res. sp.	Millecchia and Rudzinska 1971
Phalacrocleptes verruciformis	1	s. ves.	–	ret.	cl.	res. sp.	Lom and Kozloff 1967
Class Oligohymenophorea							
Order Hymenostomatida							
Tetrahymena (*pyriformis* group)	1	sp.	–	spon.	cl.	–	Roth and Minick 1961, Elliott 1963, Flickinger 1965, Gorovsky 1970a
Paramecium (*aurelia* group)	2	s. ves.	–	kar.	cl. + ER	res. sp.	Jurand *et al.* 1962, Jurand and Selman 1969, 1970, Stevenson and Lloyd 1971a
P. multimicronucleatum	4	s. ves.	–	kar.	cl.	–	Inaba and Kudo 1972
P. bursaria	1	chr.	–	chr.	cl.	res. sp.	Schwartz 1976
Neobursaridium gigas	sev.	s. ves.	–	ret.	cl.	–	Nilsson 1969
Ophryoglena bacterocaryon	1	sp.	–	spon.	cl.	–	Roque *et al.* 1965
Ichthyophthirius multifiliis	1	chr.	–	chr.	cl.	res. sp., sph.	Peshkov and Tikhomirova 1970, Hauser 1972
Order Scuticociliatida							
Thigmophrya macomae	1	sp.	–	spon.	cl.	–	Khan 1970
Uronema sp.	1	sp.	–	spon.	cl. + ER	–	Kaneshiro and Holz 1976
Class Polyhymenophorea							
Order Heterotrichida							
Blepharisma intermedium	6–30	sp.	–	spon.	?	–	Seshachar 1964
B. undulans	11–13	sp.	–	spon.	cl.	–	Kennedy 1965

Table 1 (continued)

Species	Number of nuclei	Type of nucleus	Nucleoli	Chromatin	Nuclear envelope	Other compo-nents	References
1	2	3	4	5	6	7	8
B. wardsi	sev.	sp.	–	spon.	cl.	–	INABA and SOTOKAWA 1968
Blepharisma (4 species)	sev.	s. ves.	–	kar.	cl.	–	JENKINS 1967
Spirostomum ambiguum	many	sp.	–	spon.	cl.	–	SESHACHAR 1965, ZAZULINA and POLYAKOV 1976
Condylostoma magnum	sev.	s. ves.	–	kar.	cl.	–	BOHATIER 1978a
Stentor coeruleus	many	s. ves.	–	kar.	cl.	–	SKARLATO 1977
Order Oligotrichida							
Favella ehrenbergii	2	sp.	–	spon.	?	–	HEDIN 1975
Parafavella gigantea	2	sp.	–	spon.	?	–	HEDIN 1975
Order Hypotrichida							
Urostyla (2 species)	sev.	sp.	–	spon.	cl.	–	INABA and SUGANUMA 1966, SUGANUMA 1969
Keronopsis rubra	many	sp.	–	spon.	cl.	–	RUTHMANN 1973
Gastrostyla steini	4	sp.	–	spon.	cl.	–	WALKER 1976
Euplotes (2 species)	1	sp.	–	spon.	cl.	–	ROTH 1957, RINGERTZ *et al.* 1967b
E. eurystomus	1	chr.	–	chr.	cl.	–	RUFFOLO 1978

Abbreviations:

In all columns: sev. = several; brackets indicate facultative occurrence of the structure being considered.

Type of nuclei: ves. = typical vesicular; s. ves. = small vesicular; l. ves. = large vesicular; ov. = ovular; chr. = chromosomal; sp. = spermal.

Nucleoli (figures refer to their number): cent. = central; exc. = eccentric; per. = peripheral; subm. = submembranous; bran. = branching; fragm. = fragmented.

Chromatin: disp. = dispersed; per. m. = condensed into peripheral masses; ret. = condensed into reticulum of threads; chr. = condensed as individual chromosomes; CC = condensed into chromocentres; kar. = condensed into a single central karyosome; sol. = solid mass of condensed chromatin fills the entire nucleus; spon. = same but chromatin mass is spongy.

Nuclear envelope: cl. = classic (two-membraned with pores); +ER = same plus continuities of the outer membrane with endoplasmic reticulum; +Ma = common outer membrane encircling the micronucleus and the macronucleus; +chl. = extensions of the outer membrane encircle chloroplasts; +GA = nuclear envelope serving as forming surface for Golgi bodies; +inn. l. = additional inner layer(s) (honeycomb or other); +out. l. = additional outer layer(s); +n. tub. = radial extensions of the outer membrane into the cytoplasm (nuclear tubules); +ext. = simple or branching (bran.) extensions of both nuclear membranes; +mast. = mastigonemes are assembled in perinuclear space; +incl. = various inclusions in perinuclear space.

Other components (in the karyolymph): sph. = microfibrillar sphere; d. b. = various dense bodies, cr. = crystalloid(s); PG = perichromatin granules; U = U-shaped particles; hel. = RNP helices; cyt. incl. = inclusions or deep invaginations of cytoplasm inside the nucleus; ves. = membrane-bound vesicles; fib. = bundles of various fibrils; res. sp. = "residual spindle" of microtubules; MTOC = microtubule-organizing centre; kch. = persisting kinetochores.

discrete subunits. In the hypermastigid *Trichonympha*, these subunits are globular and number 10–12 (GRIMSTONE 1959). In the multinucleate amoeba *Chaos carolinensis* (DANIELS *et al.* 1969), the subunits are also globular, but they number only 8 (Fig. 2*d*, *e*). They form two crowns on the cytoplasmic and nuclear sides of the pore; they are not in the opening of the pore, but beyond its edges (Fig. 2*f*). Thus, the inner diameter of the annulus (about 65 nm) in *Chaos* is more or less equal to that of the pore proper, while the outer diameter (about 115 nm) is much larger than the pore opening (Fig. 3*a*). Pore complexes of a similar structure are found in *Amoeba proteus* (LEESON and BHATNAGAR 1975a) and *Polychaos fasciculatum* (PAGE and BALDOCK 1980). This type of structure of the pore complex corresponds to the scheme proposed by AFZELIUS (1955) and WATSON (1959) for metazoan nuclei.

In the gregarines *Selenidium* (VIVIER 1967, SCHRÉVEL 1971a) and *Lecudina* (VIVIER 1967), as well as the coccidians *Coelotropha* (VIVIER 1963a, VIVIER and HENNERÉ 1965) and *Sarcocystis* (SÉNAUD 1967), the annulus also consists of eight subunits (Fig. 2*b*), but it is arranged differently. In this case, the material of the annulus is a cylinder or a truncated cone inserted into the pore opening (Figs. 2*g*, 3*b*). Moreover, the subunits of the annulus of these sporozoans are not globules, but sections of tubules 12–15 nm thick and 80–90 nm high and are "cemented" into the cylinder wall (Fig. 3*b*). Thus, the inner diameter of the annulus (50–60 nm) of sporozoans is much less than the pore diameter (about 100 nm). As for the outer diameter of the annulus (100–120 nm), it may be either equal to (if the annulus is cylindrical) or somewhat greater than the pore diameter (if the annulus is conical, see Fig. 3*b*). This construction of the pore complex corresponds to the schema proposed by WISCHNITZER (1958) for the nuclei of the amphibian oocytes. It apparently includes the annuli of the macronuclear envelope of *Tetrahymena*, whose diameter of the pore opening (47–65 nm)[1] is close to the outer diameter of the annulus (60–70 nm), while the inner diameter of the annulus is only 20 nm (WUNDERLICH and FRANKE 1968).

The nuclear pores of some protozoans lack clearly defined annuli. Then, the edge of the pore is seen on the tangential sections as a thin circular line, and such pores seem to be "open" when they are viewed in profile. Pores without annuli have been observed in some autotrophic flagellates (*Chrysochromulina*, MANTON 1967a, *Euglena*, LEEDALE *et al.* 1965, *Platymonas*, MANTON and PARKE 1965) and on the envelope of the micronuclei of some primitive ciliates: *Tracheloraphis caudatus* (RAIKOV and DRAGESCO 1969), *Trachelonema sulcata* (KOVALEVA and RAIKOV 1970, 1973), and *Kentrophoros fistulosum* (RAIKOV 1972b). Most protozoan nuclei, however, have annuli, including the micronuclei of the higher ciliates (ROQUE *et al.* 1965, RAIKOV 1966, TUCKER 1967, JURAND and SELMAN 1970, VINNIKOVA 1974a, 1976, GOLDER 1976) and the macronuclei of almost all ciliates (Fig. 2*c*). The presence of annuli around the pores in the macronuclear envelope of *Tetrahymena*, however, depends on the phase of cultural growth: in the logarithmic phase, all pores have annuli, while in the stationary phase, only some have them (WUNDERLICH and FRANKE 1968).

---

[1] The freeze-etching method has shown that the diameter of the pore opening is even 78 nm (SPETH and WUNDERLICH 1970).

As for the diameter of the nuclear pores, figures ranging from 30 to 120 nm are found in the literature. Unfortunately, many authors do not say what their figures stand for, i.e. whether for the outer diameter of the annulus, its inner diameter, or the width of the pore itself. As we have seen, the relationships between these three measurements can vary. The outer diameter of protozoan annuli is usually 80–120 nm, while their inner diameter is 35–65 nm. The diameter of the opening itself in the nuclear envelope varies to the greatest extent, i.e. from 40 to 100 nm.

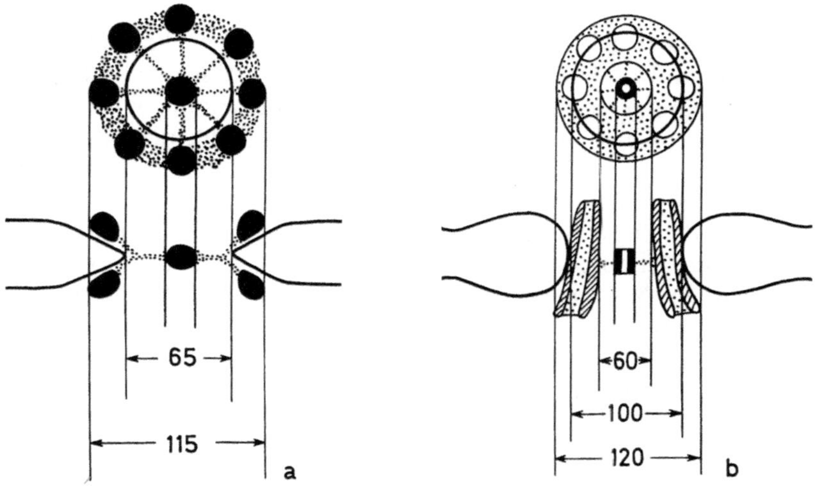

Fig. 3. Diagrams of the structure of pore complexes (the size is given in nm). *a* in amoebae, drawn on the basis of the data of DANIELS and others (1969); *b* in sporozoans, after VIVIER (1967), modified.

The central element of the pore complex usually has the form of a granule 10–20 nm in diameter (Fig. 2*c*). Some pore complexes may occasionally lack central granules, e.g. in the macronucleus of *Tetrahymena* (WUNDERLICH 1969b). The central granules appear also in pores without annuli, e.g. in *Euglena* (LEEDALE *et al.* 1965), the foraminiferan *Globigerina* (FEBVRE-CHEVALIER 1971), and in the micronucleus and macronucleus of the lower ciliate *Tracheloraphis caudatus* (RAIKOV and DRAGESCO 1969). In amoebae (*Amoeba proteus* and *Chaos carolinensis*), DANIELS and his colleagues (1969) observed fine radial fibrils that link the central granule with the globular subunits of the annulus (Fig. 2*d*, *e*).

The functional significance of the central granules has not been fully ascertained. Some authors regard them as transit particles that either come out of the nucleus (RNP?) or enter it. They base their theory on the fact that these granules are not found in all pores and that their number increases under the conditions of intense metabolism (WUNDERLICH 1969b, GOMEZ *et al.* 1975).

A more intricate structure of the central elements of the pore complexes was observed in Gregarinia and Coccidia (VIVIER 1967). In these sporozoans, the central element seems to be a short tubule suspended on fibrils or lamellae in the pore opening (Figs. 2*g*, 3*b*). Such a structure of the central element makes it more likely to be a permanent component of the pore complex, which possibly ensures the selective passage of certain macromolecules through the pore (FRANKE and SCHEER 1974). For instance, particles with a size of 2-5 nm are known to enter the nucleus of amoebae through the axial part of the pore, not near its edge (FELDHERR 1970). RNP helices (Fig. 4*d*) behave in exactly the same way when they leave the nucleus (LEESON and BHATNAGAR 1975a, MINASSIAN and BELL 1976a).

Additional Laminae of the Nuclear Envelope. As Table 1 shows, the structure of the nuclear envelope of most protozoans is quite typical. It is among protozoans, however, that cases of extreme structural complexity are found, cases that have no analogues among metazoans. The structure of the classic nuclear envelope is usually preserved, but extra laminae are added to it from inside and/or outside.

The fibrous lamina on the inner side of the nuclear envelope, while usually thin and hardly noticeable, develops greatly and distinctly structuralizes in many cases. In *Amoeba proteus*, for instance, the intertwining bundles of microfibrillar material form the walls of regular hexagonal alveoli that are up to 500 nm high, 150-180 nm wide and resemble a honeycomb; they are therefore called the *honeycomb layer*. The alveoli are perpendicular to the nuclear envelope (Fig. 4*a-c*), and each of them has a nuclear pore at the bottom (HARRIS and JAMES 1952, GREIDER *et al.* 1956, 1958, PAPPAS 1956b, COHEN 1957, ROTH *et al.* 1960, LEESON and BHATNAGAR 1975a). The enhancement of the mechanical strength of the nuclear envelope is usually regarded as a probable function of the honeycomb layer of the envelope, although it may also participate in the transport of particles and macromolecules from the nucleus into the cytoplasm and back (FELDHERR 1962, 1966, 1968a, etc.). In any case, the honeycomb layer does not prevent the RNP helices, which come out of the nucleus into the cytoplasm, from passing to the nuclear pores and through them (WISE *et al.* 1972, LEESON and BHATNAGAR 1975a, MINASSIAN and BELL 1976a). It may be that the fibrils that form the alveolar walls are contractile (LEESON and BHATNAGAR 1975b, OPAS *et al.* 1978).

The honeycomb layer on the inner side of the nuclear envelope has also been described in other large free-living uninucleate Amoebida (HOUSSAY and PRENANT 1970, FLICKINGER 1974) and the parasitic *Entamoeba invadens* (SIDDIQUI and RUDZINSKA 1965). It is also found in the multinucleate *Pelomyxa palustris* (Fig. 12*b*), but only in some nuclei (DANIELS *et al.* 1966, DANIELS and BREYER 1967). It does not exist in the multinucleate amoebae of the genus *Chaos* (*Ch. carolinensis*, *Ch. illinoisensis*: Table 1). Among other Sarcodina a similar

---

Fig. 4. Honeycomb layer on the inner surface of the nuclear envelope (electron micrographs). *a Amoeba proteus*, cross section, 18000×; *b* same, tangential section, 68000×; *c* section through the disc-shaped nucleus of *A. proteus* (*1* honeycomb layer of the envelope, *2* nucleoli), 4600×; *d* extrusion of an RNP helix (*1*) through an alveolus of the honeycomb layer (*2*) and a nuclear pore (*3*) of *A. proteus*, 50000×; *e* honeycomb layer (*1*) and spongy layer (*2*) on the inner surface, and exonuclear vacuome (*3*) on the outer surface of the nuclear envelope in the foraminiferan *Hippocrepinella alba*, 15500×. *a* and *b* after ROTH and others (1960); *c* after DANIELS and BREYER (1966); *d* after MINASSIAN and BELL (1976a); *e* after DAHLGREN (1967b).

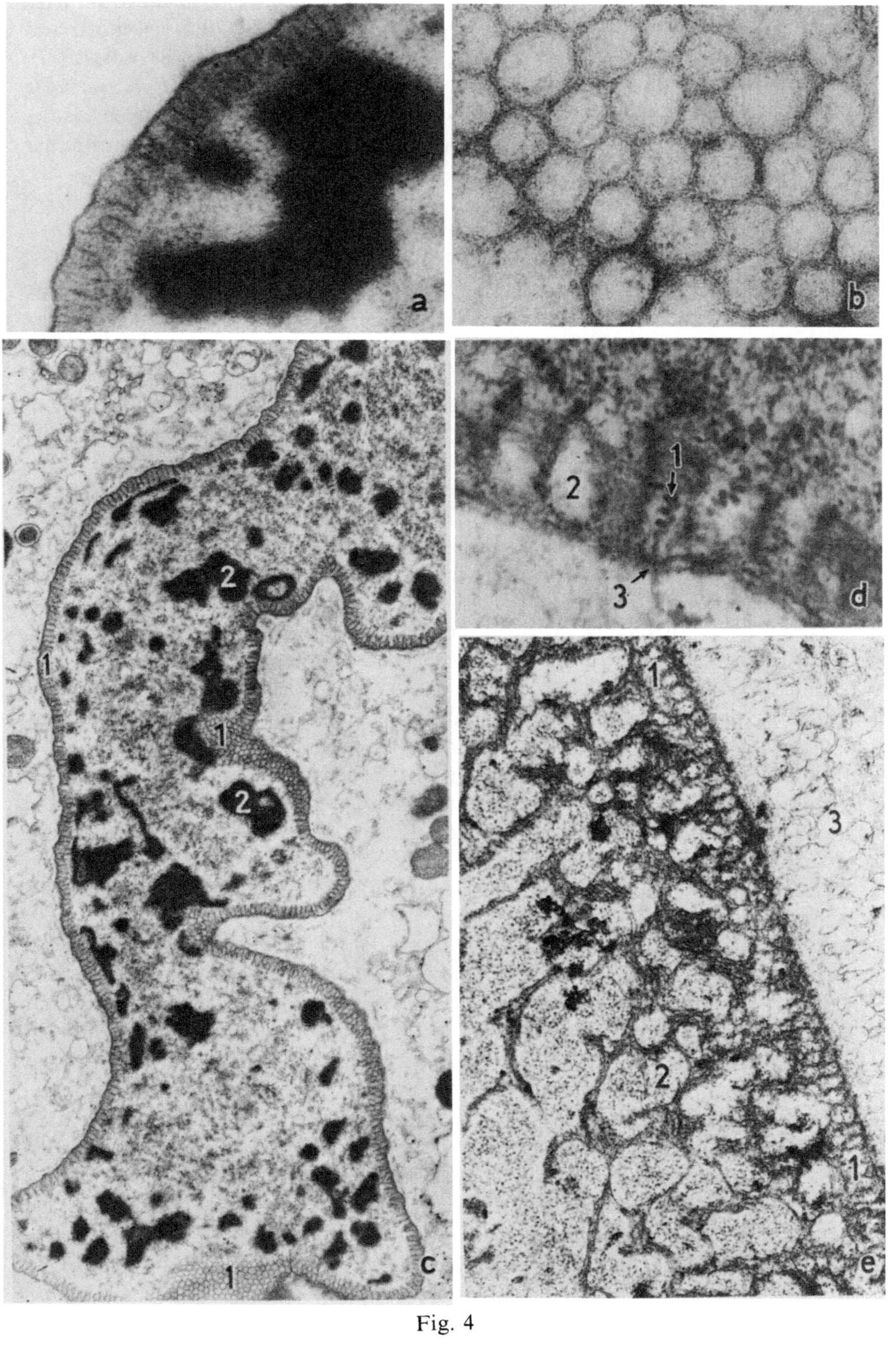

Fig. 4

honeycomb layer is found in the heliozoan *Hedraiophrys* (FEBVRE-CHEVALIER 1973) and the foraminiferans *Iridia* (CÉSANA 1975), *Gloiogullmia* (NYHOLM and NYHOLM 1975), *Hippocrepinella* (DAHLGREN 1967b), and *Boderia* (SCHWAB 1977), although the alveoli of the latter have a far less regular shape. In *Hippocrepinella* and *Boderia*, the honeycomb layer passes, toward the centre of the nucleus, into a thick (up to 5 μm) *spongy layer* consisting of irregular alveoli with microfibrillar walls (Fig. 4*e*). Such a powerful development of the fibrillar armature of the nuclear envelope of *Hippocrepinella* and *Boderia* is probably due to the very large dimensions (up to 100–110 μm) of the nuclei of these foraminiferans (Fig. 11*b*).

The foraminiferan *Pontomyxa* has no honeycomb layer, but its spongy layer is highly developed (CACHON and CACHON-ENJUMET 1963). In this case, the nuclear surface is covered with papillae 2–3 μm high that are capable of periodically expanding and contracting, i.e. pulsating. The nuclear envelope covers the papillae from the outside while they are full of alveoli of the spongy layer inside. The ability of the papillae to "pulsate" points at the contractility of the microfilaments that form the alveolar walls. Moreover, the spongy layer of *Pontomyxa* is separated from the contents of the nucleus by a layer of fibrils parallel to the nuclear surface. Papillae exist also on the nuclear surface of the foraminiferan *Ovammina* (DAHLGREN 1967a), and they are likewise filled with microfibrillar material that does not, however, form alveoli. A rather thick fibrous lamina that does not form alveoli has also been discovered in other foraminiferans: *Globobulimina* (DAHLGREN 1967b), *Globigerina* (FEBVRE-CHEVALIER 1971), etc.

The structure of the nuclear envelope of the aberrant marine heliozoan of the order Taxopodida, *Sticholonche zanclea* (HOLLANDE *et al.* 1967, CACHON and CACHON 1978), turned out to be very intricate. Outside, it has cup-shaped "menisci" filled with electron-dense material and in which the bases of axopodia are movably fixed like joints (Fig. 5). The inner side of the classic two-membrane nuclear envelope has a fibrous lamina that is differentiated into five strata (1–5) whose overall thickness is about 0.3 μm. These strata undoubtedly increase the mechanical rigidity of the nucleus, whose surface is the base of the motile axopodia in *Sticholonche*. In any case, the fibrous lamina is not stratified in the region of the nucleus without axopodia: there, this lamina is homogeneous and is only about 0.1 μm thick.

A rather thick structureless fibrous lamina is found on the inner side of the nuclear envelope of the amoeba *Thecamoeba proteoides* (PAGE 1978), the testacean *Arcella* (CHARRET 1967), the flagellate *Lophomonas* (BEAMS and SEKHON 1969), and the gregarines *Selenidium* (SCHRÉVEL 1971a), *Gregarina* (DEVAUCHELLE 1968), *Lecudina* (VIVIER 1967), *Callyntrochlamys* and *Thalicola* (DESPORTES and THÉODORIDÈS 1969), *Enterocystis* (DESPORTES 1974), and *Gonospora* (CORBEL *et al.* 1979). In other gregarines, however, it is more structured and assumes the form of a honeycomb layer, although not a very regular one, such as in *Gregarina melanopli* (BEAMS *et al.* 1957), *Grebnickiella gracilis* (MOLON-NOBLOT and DESPORTES 1980), and the species of the genus *Stylocephalus* (GRASSÉ and THÉODORIDÈS 1957, DESPORTES 1970).

In the micronuclei and macronuclei of Ciliophora, the fibrous lamina is usually not developed and the nuclear envelope has no complications. A single exception is the micronucleus of Ophryoscolecidae (NOIROT-TIMOTHÉE 1960) and both types of nuclei of *Kentrophoros fistulosum* (RAIKOV 1972b), which have inner structureless fibrous laminae.

Laminae on the external side of the nuclear envelope are also found rather frequently, especially in large nuclei. A layer of microfilaments may be contiguous to the outer membrane, such as in the chloromonadids *Vacuolaria* and *Gonyostomum* (MIGNOT 1967), the hypermastigids Joeniidae (HOLLANDE and VALENTIN 1969, LAVETTE 1970), the foraminiferan *Globigerina* (FEBVRE-CHEVALIER 1971), and the gregarine *Gonospora* (CORBEL *et al.* 1979); in the last two forms, fibrous laminae are found on both sides of the nuclear envelope. These laminae are sometimes structuralized in the form of a honeycomb layer. For instance, it has

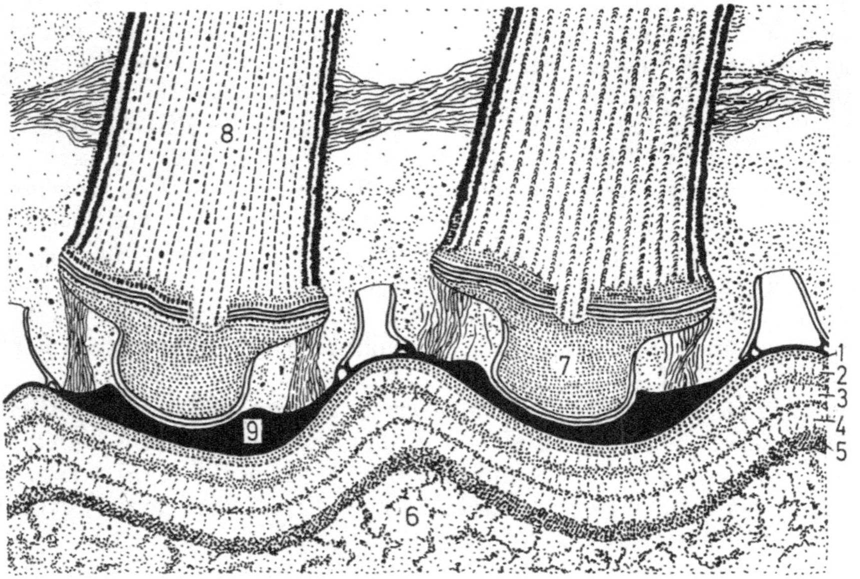

Fig. 5. Diagram of the attachment of axopodia to the nuclear envelope of the heliozoan *Sticholonche* (after HOLLANDE *et al.* 1967). *1 5* additional layers on the inner side of the nuclear envelope; *6* nucleus; *7* base of an axopodium; *8* axopodium; *9* electron-dense material on the outer side of the nuclear envelope; the envelope proper is not shown; it is located between layers *1* and *9*.

long been known that the nuclear envelope of the amoeba *Endamoeba blattae* (Fig. 1*f*) is very thick (1.5 μm) and mechanically strong (SASSUCHIN 1929, 1930, EPSTEIN 1941). Electron microscopy (BEAMS *et al.* 1959) has shown that this is due to the presence of a thick (0.5 μm) external honeycomb layer consisting of high and narrow hexagonal alveoli with a pore at the bottom of each alveolus (Fig. 2*h*).

The external side of the nuclear envelope occasionally has several additional layers of different nature. In the testacean *Hyalosphenia* (JOYON and CHARRET 1962), the first layer consists of intersecting bundles of filaments that limit shallow alveoli that have nuclear pores at the bottom. The second layer is granular, and the third consists of minute vesicles with a homogeneous content (Fig. 6*a*). In the amoeba *Pelomyxa palustris*, the external additional laminae as well as the internal structure of nuclei greatly vary from one specimen to another; this is apparently due to some unknown peculiarities of the life cycle of *Pelomyxa* (Daniels *et al.* 1966, Daniels and Breyer 1967). In most specimens, the lamina nearest the nuclear envelope is homogeneous. It is followed by a layer of minute vesicles with a transparent content, and then comes the layer of symbiotic bacteria (Fig. 12*a*). The internal honeycomb layer of these nuclei may be either present (Fig. 12*b*) or absent (Fig. 12*a*). In some specimens, however, the nuclear envelope of all nuclei is covered with a thick layer of concentrically arranged lamellae, outside of which is a layer of vesicles with dense inclusions. Such nuclei have neither an internal honeycomb layer nor associated symbiotic bacteria (Fig. 12*c*).

In many Foraminiferida, the nuclei are surrounded by a vesicular zone of cytoplasm that consists of closely packed vacuoles with agranular membranes. This "exonuclear vacuome" has been observed in *Ovammina* (Dahlgren 1967a), *Hippocrepinella* (Dahlgren 1967b; Fig. 4*e*), *Shepheardella* (Hedley *et al.* 1967), *Myxotheca*, *Allogromia* (Schwab 1969, 1970), and *Nemogullmia* (Nyholm and Nyholm 1975). In *Allogromia*, the vesicular zone (Fig. 11*a*) can turn into a perinuclear lacuna (apparently, by the fusion of vesicles). Then, the nucleus is covered with only a thin layer of the perinuclear cytoplasm and seems to be suspended by thin cytoplasmic filaments in the cavity of such a lacuna (Lengsfeld 1969, Schwab 1970, 1974). Similar perinuclear lacunae are characteristic of the heliozoans *Hedraiophrys* and *Raphidiophrys* where they even communicate with the external environment through a common lacunary system of the cytoplasm (Febvre-Chevalier 1973, 1975).

In some Heliozoea, such as *Actinophrys sol*, the external side of the nuclear envelope serves as an organizing centre for microtubules, which constitute the axonemes of axopodia. The microtubules grow perpendicular to the surface of the external membrane of the nucleus (Hovasse 1965, Ockleford and Tucker 1973, Mignot 1979, Patterson 1979).

Extensions of Nuclear Membranes. It is widely held that the nuclear envelope is a part of the general system of the endoplasmic reticulum. Evidence of this is the presence of ribosomes and polyribosomes on the cytoplasmic side of the outer nuclear membrane as well as the frequent continuities between it and the membranes of the rough endoplasmic reticulum (Franke and Scheer 1974). Such continuities are also common among protozoans (Figs. 6*b*, 9*a*, *c*), especially among the Mastigophora and Sporozoea (Table 1). In many Chrysomonadida and Cryptomonadida, the outgrowths of the outer membrane of the nuclear

Fig. 6. Additional structures of the nuclear envelope (electron micrographs). *a* three perinuclear layers in *Hyalosphenia* (tangential section through the nuclear surface, *1* pores of the nuclear envelope, *2* bundles of microfilaments, *3* layer of granules, *4* layer of vesicles), 27 500 × ; *b* formation of a dictyosome (*1*) on the nuclear envelope surface in *Toxoplasma* (the arrowheads indicate transition vesicles, *2* transition between the outer membrane of the envelope and a channel of the endoplasmic reticulum), 42 000 × ; *c* anastomosing extensions on the nuclear surface of the gregarine *Didymophyes chaudefouri* (*1* nucleus, *2* cytoplasm, *3* extension of the nucleus), 8000 × . *a* after Joyon and Charret (1962); *b* after Vivier and Petitprez (1972a); *c* after Baudoin and Ormières (1973).

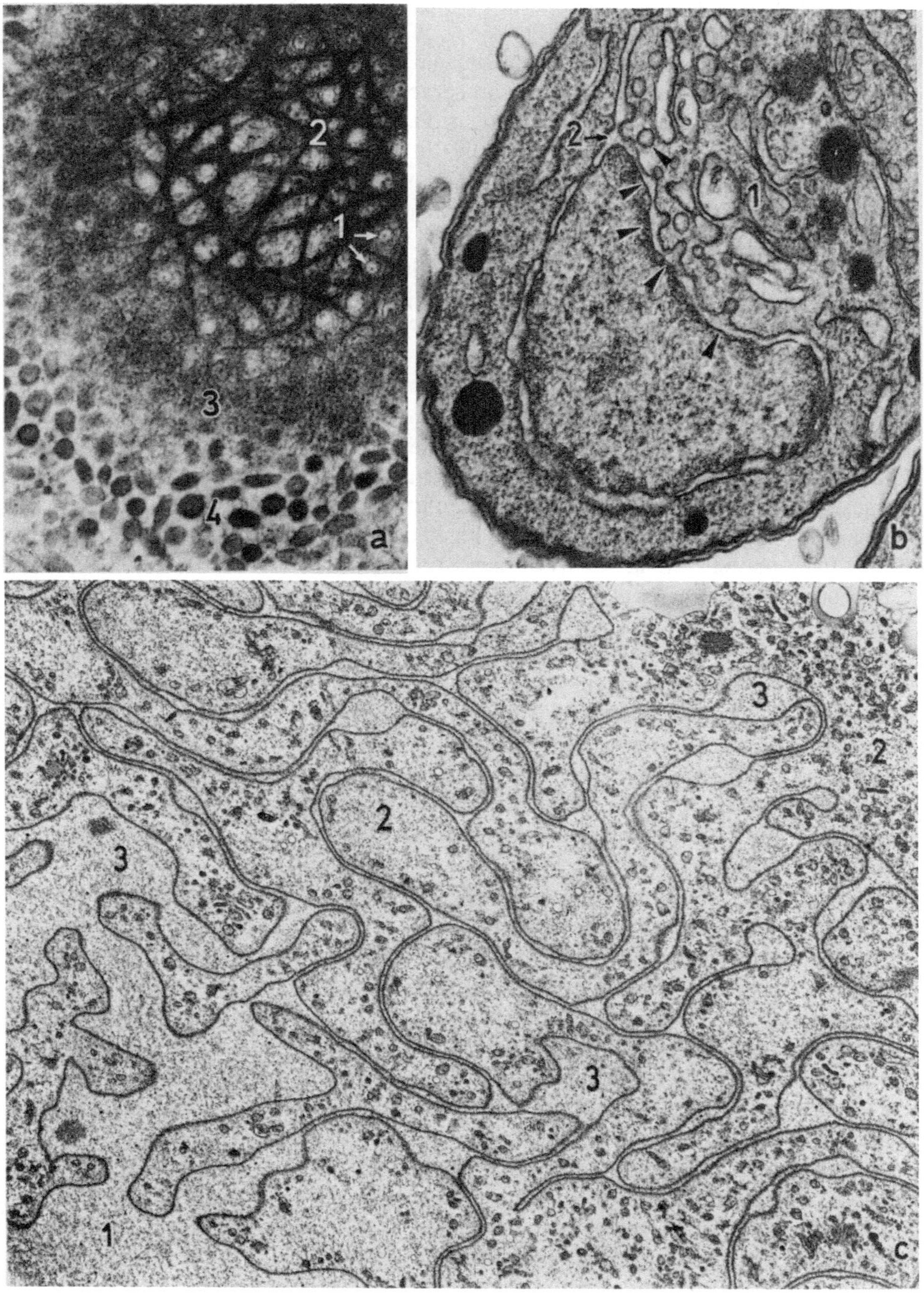

Fig. 6

envelope embrace the chloroplasts, covering them with two additional membranes over their own envelope, such as in *Ochromonas* (GIBBS 1962, 1979) and others (Table 1; reviewed by GIBBS 1981).

Like the elements of the rough endoplasmic reticulum, the outer membrane of the nuclear envelope of protozoans can serve as the forming surface of the Golgi apparatus (dictyosomes). Minute "transitional" vesicles bud from it. When these vesicles fuse with one another, they form the proximal cisterna of the dictyosome (Fig. 6*b*). As a result, the dictyosome (which is often the only one in the cell) "sits" right on the surface of the nucleus, such as in the rhizomastigid *Cercomonas* (MIGNOT and BRUGEROLLE 1975a), the gametes of the foraminiferan *Myxotheca* (SCHWAB 1969), the trophozoites and merozoites of some coccidians (SAMPSON and HAMMOND 1972), the endozoites of toxoplasms (VIVIER and PETITPREZ 1972a), and others (Table 1). Numerous dictyosomes of the chloromonadids *Gonyostomum* and *Vacuolaria* are formed similarly and fuse with each other to compose "supranuclear caps" (MIGNOT 1967).

Some inclusions, especially the proteinaceous ones, are assembled both in the perinuclear space and in the cisternae of the endoplasmic reticulum. For instance, in the chrysomonad *Olisthodiscus*, which has hairy flagella, the flagellar hairs (mastigonemes) are formed in the saccular expansions of perinuclear space. Apparently the mastigonemes (whose basis consists of microtubules) are assembled from protein molecules synthesized on the ribosomes of the outer nuclear membrane. The bundles of mastigonemes are then transported through the system of canals of the endoplasmic reticulum to the cell surface (LEEDALE *et al.* 1970, etc.). This process is also observed among many other Chrysomonadida (Table 1). Inclusions in the form of granules of an unknown nature have been observed in expansions of the perinuclear space in the cysts of the schizopyrenid *Naegleria fowleri* (SCHUSTER 1975*b*) and in the gametes of the foraminiferan *Iridia* (CÉSANA 1972). Granules are also formed in the perinuclear space of the macronucleus of the ciliate *Mesnilella*. These granules then get into minute vesicles that separate from the outer membrane of the nucleus and enter the cytoplasm (PUYTORAC 1963b).

Numerous radial extensions of only the outer nuclear membrane, in the form of elongated sacs, also develop on the surface of the nucleus of the growing macrogamont of the coccidian *Eimeria acervulina*. In the mature macrogamete, these sacs become detached from the nuclear envelope (SÉNAUD *et al.* 1980).

A special case of the growth of the outer membrane of the nuclear envelope is observed in the hypermastigid *Lophomonas blattarum* (BEAMS *et al.* 1961, BEAMS and SEKHON 1969). It forms numerous radial tubular outgrowths (nuclear tubules). The ends of these outgrowths are closed and expanded in the form of ampullae filled with electron dense material. The tubules have many ribosomes on the external side of their membrane closer to the surface of the nucleus (Fig. 7). The cavities of the nuclear tubules communicate with the perinuclear space. Certain substances that accumulate in the ampullae are probably synthesized in nuclear tubules with the participation of ribosomes.

The inner membrane of the nuclear envelope differs biochemically from the outer one (FRANKE and SCHEER 1974). Pore complexes are formed only between membranes of different kinds (outer and inner ones) and not between two membranes of the same kind. For instance, there are no pore complexes on the outgrowths formed by only the external membrane (e.g. on the nuclear tubules of *Lophomonas* and on the cisternae that surround the chloroplasts of chrysomonads). The same occurs in the ciliates *Platyophrya* (Dragesco *et al.* 1977), *Woodruffia* (Golder 1976), *Cyrtolophosis* (Detcheva 1976, Didier *et al.* 1980) and *Aristerostoma* (Detcheva and Puytorac 1979), members of the order Platyophryida, where the macronucleus and the micronucleus have a common outer membrane but separate inner ones (Fig. 91*a*). In this case, the pore complexes are

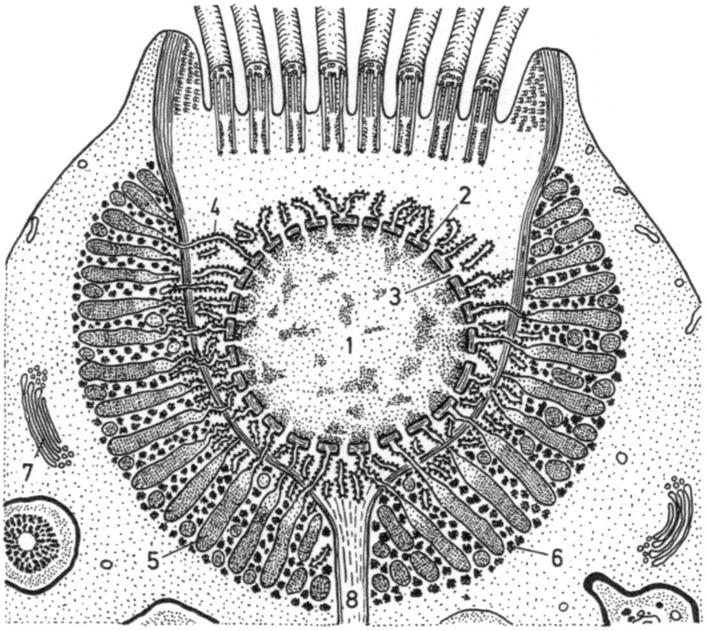

Fig. 7. Diagram of the ultrastructure of the perinuclear region in *Lophomonas* (after Beams and Sekhon 1969). *1* nucleus; *2* nuclear envelope; *3* pores; *4* proximal parts of nuclear tubules; *5* ampullae at the ends of nuclear tubules; *6* glycogen; *7* Golgi apparatus; *8* axostyle.

formed only on the outer surfaces of both nuclei (i.e. where the outer membrane is contiguous to the inner one), but they do not exist at the interface of the two nuclei where two inner membranes come into contact with one another.

In some protozoans, the protrusions of the nuclear envelope involve both of its membranes at once. They are especially characteristic of the Gregarinia. The first such case was observed by Dogiel (1910a) in *Callyntrochlamys phronimae*: as the trophozoite grows, numerous tubular protrusions rise from the surface of its nucleus. These protrusions form a crown of canaliculi around the nucleus (Fig. 8); the surface of the nucleus increases by hundreds of times. The nucleus of *Callyntrochlamys* (as well as *Thalicola*) was recently re-examined with the electron microscope (Desportes and Théodoridès 1969). It turned out that the tubular protrusions with a diameter of about 1 μm are covered with a normal two-membrane nuclear envelope. Thus, they are not outgrowths of perinuclear space (as in *Lophomonas*), but are processes of the nucleus itself. In *Enterocystis*, the nuclear processes are branched, and they fill the whole perinuclear cytoplasm as they grow (DESPORTES 1974). Nucleolar material and multilamellar systems are found in them. The walls of the processes have no pores, however. In the gregarine *Didymophyes*, such outgrowths of the nuclear envelope are lamellar; they have many pores in their walls and anastomose with one another (BAUDOIN and

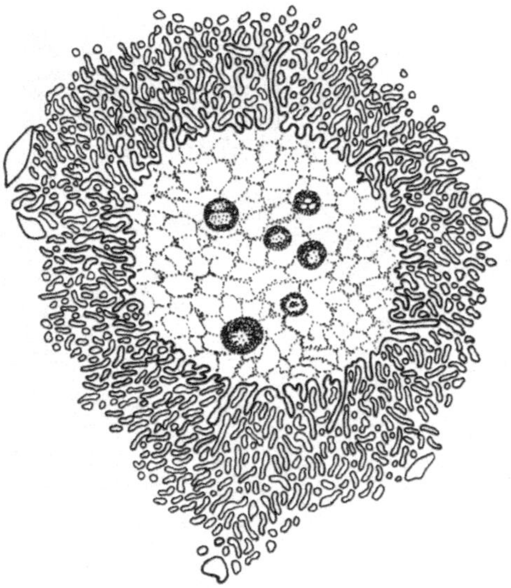

Fig. 8. Nucleus of the gregarine *Callyntrochlamys* with tubular extensions of the nuclear envelope (after DOGIEL 1910a).

ORMIÈRES 1973). Consequently a large canalicular zone develops around the nucleus (Fig. 6*c*). Functionally, all these extensions apparently intensify nuclear-cytoplasmic exchanges as a result of a sharp increase in the nuclear surface. However, the absence of pores in the walls of the nuclear processes of *Enterocystis* suggests that the processes have some other functions, too.

In the green flagellate *Prasinocladus marinus*, an outgrowth of the nuclear envelope, which is formed by both nuclear membranes, penetrates deep into the pyrenoid (PARKE and MANTON 1965). Such an increase in the surface of contact between the nucleus and the pyrenoid probably indicates active exchange between them.

Invaginations of the cytoplasm into the nucleus are also observed among protozoans. Such regions of the cytoplasm are limited by a normal nuclear envelope. They have been observed in *Stereomyxa* (BENWITZ and GRELL 1971b), *Labyrinthula* (PERKINS and AMON 1969), the foraminiferan *Ovammina* (DAHLGREN 1967a), and other species.

Chromatin Structures

Chromatin is now defined as a complex of DNA with histones and non-histone proteins. In eukaryotes, chromatin is organized into fibrils about 10 nm thick that look like chains of globular subunits or nucleosomes. Every nucleosome is a complex of four types of histones with a DNA molecule winding around it (review by Ris and Korenberg 1979). When nucleosomal fibrils combine with non-histone proteins, they form thicker filaments (25–30 nm). The nucleosomal organization of chromatin has also been revealed in protozoans, including the flagellates *Olisthodiscus*, *Trypanosoma*, and *Euglena* (Rizzo, Burghardt 1980, Belnat and Riou 1979, Haapala and Soyer 1975, Bré *et al.* 1980), the macronuclei and micronuclei of ciliates (Lipps and Morris 1977, Gorovsky *et al.* 1978), etc.

At the same time, the old definition of chromatin (Bělař 1926a) as a substance from which chromosomes are formed in mitosis is still mainly valid. In other words, chromatin is chromosome material that is more or less decondensed in the interphase nucleus. It should be noted, however, that Bělař's definition does not apply to extrachromosomal chromatin, which contains DNA that has been synthesized as a result of gene amplification (e.g. nucleolar chromatin of the oocytes of many animals).

Under the light microscope, chromatin in the nondividing protozoan nuclei is usually seen as a loose granuloreticulate structure with occasionally more compact Feulgen-positive *chromocentres*. The latter are often contiguous to the nuclear envelope from inside. Loose reticulate structures correspond to decondensed, and compact ones correspond to condensed chromosomes or their regions.

In protozoans, the degree of condensation of the chromosomes varies greatly during the interphase. In some forms, we see nuclei with finely dispersed chromatin that produce a faint diffuse Feulgen reaction (Fig. 1*b*–*f*). In large nuclei of this type, chromatin is not seen at all, while the concentration of DNA decreases so much that the Feulgen reaction becomes negative, such as in the nuclei of the gamonts of some Gregarinia and the macrogametes of the Coccidia (Cheissin 1958, Scholtyseck 1963, Vivier and Henneré 1964, Gaibova 1972). The fluorescence method, however, which is more sensitive, reveals DNA even in such nuclei (Cheissin *et al.* 1968).

It is difficult to distinguish decondensed chromatin from other components of the karyolymph in electron micrographs. Fibrils 10–30 nm thick and assumed to be chromatinic can usually be seen in the karyolymph (Figs. 2*a*, 12*a*). More rarely, the contents of the nucleus appear granular (especially in trichomonadids). In most protozoans, decondensed chromatin is more or less uniformly distributed throughout the nucleus (see Table 1).

In *Amoeba proteus* and *Chaos illinoisensis*, the fibrils of dispersed chromatin form a definite layer in the nucleus during centrifugation (Daniels and Breyer 1966). If the DNA of amoebae is preliminarily labelled with ^{3}H-thymidine, the entire label will be in this layer. The chromatin fibrils of amoebae thus identified form loose strands 80–90 nm thick (Wise and Goldstein 1972). In *A. proteus*, individual chromatin fibrils, which also incorporate ^{3}H-thymidine, have a diameter of about 10 nm (Dickinson 1970).

It is easier to study the ultrastructure of compact chromatin. Often, it has a peripheral position in the nucleus and is arranged in a continuous layer or in the form of individual chromocentres close to the inner membrane of the nuclear envelope (Fig. 9*a*, *b*). Peripheral chromatin always has a finely fibrillar structure. Such a state of chromatin is very characteristic of the nuclei of Kinetoplastida, the gametes of Foraminiferida, and the dispersive stages of the life cycle of Sporozoea, i.e. their sporozoites, merozoites, endodyocytes, etc. (Table 1). The peripheral chromocentres decondense, as a rule, when sporozoites or merozoites are transformed into the actively growing stages of the given parasites (trophozoites, gametocytes). This is apparently connected with the activation of the transcription processes in chromatin. The relation of the peripheral condensations of chromatin to chromosomes is not quite clear. They may be parts of chromosomes, but more likely they are whole chromosomes condensed on the inner nuclear membrane.

In other protozoans, chromosomes condense in the form of filaments that permeate the entire karyolymph. For instance, filamentous chromosomes can be seen with the light microscope in the central zone of the nuclei of some Foraminiferida, e.g. in the gamont of *Myxotheca* (GRELL 1956a) and in the agamonts of *Allogromia* (McENERY and LEE 1976). In smaller nuclei, filamentous chromosomes are usually seen as a Feulgen-positive reticulum. At the ultrastructural level, blocks and strands of compact chromatin with a diameter of usually about 0.1 μm, are seen in the karyolymph of these nuclei; they are apparently sections of filamentous chromosomes (Fig. 9*c*). Some strands are attached to the nucleolus, while others are attached to the nuclear envelope; they have a finely fibrillar ultrastructure. The chromatin strands can decondense under certain conditions, e.g. during the excystment of the schizopyrenid amoeba *Naegleria fowleri* (SCHUSTER 1975b). Ultrastructural investigations have shown that nuclei with reticulate chromatin are characteristic of several phytoflagellates, especially the Cryptomonadida, some Sarcodina, and the trophozoites of many Coccidia and Haemosporina (Table 1).

Chromosomes are condensed even more in the nondividing nuclei of many Euglenida (LEEDALE 1958a, 1967, 1968). They are seen with the electron microscope as very dense rods of compact chromatin (Fig. 9*d*) whose ultrastructure is similar to that of mitotic chromosomes (LEEDALE *et al.* 1965, MIGNOT 1966, ROGERS *et al.* 1972, SCHUSTER and HERSHENOW 1974, MOYNE *et al.* 1975, etc.). The

---

Fig. 9. Nuclei with condensed chromatin: *a–c* vesicular, *d–e* "chromosomal", *f* formation of a "spermal" nucleus. *a* nucleus of *Toxoplasma* (*1* nucleolus, *2* condensed chromatin, *3* transition between the outer membrane of the nuclear envelope and the endoplasmic reticulum), 30 000 ×; *b* nucleus of the merozoite of *Plasmodium gallinaceum*, 56 000 ×; *c* nucleus of the cyst of *Naegleria jadini* (same designations as in *a*), 22 000 ×; *d* nucleus of *Euglena gracilis* (*1* nucleolus, *2* chromosomes), 10 000 ×; *e* chromosomes in the nondividing nucleus of *Trichonympha turkestanica*, Feulgen reaction, 900 ×; *f* separation of a nucleus of the microgametocyte of *Eimeria falciformis* into the future spermal nucleus of a microgamete (*1*) and the achromatic part to be resorbed (*2*), 41 000 ×. *a–d* and *f* electron micrographs; *e* photomicrograph. *a* after VIVIER and PETITPREZ (1972a); *b* after AIKAWA and BEAUDOIN (1969); *c* after SCHUSTER (1975b); *d* after LEEDALE (1968); *e* after RAIKOV (1967a); *f* after SCHOLTYSECK and others (1972).

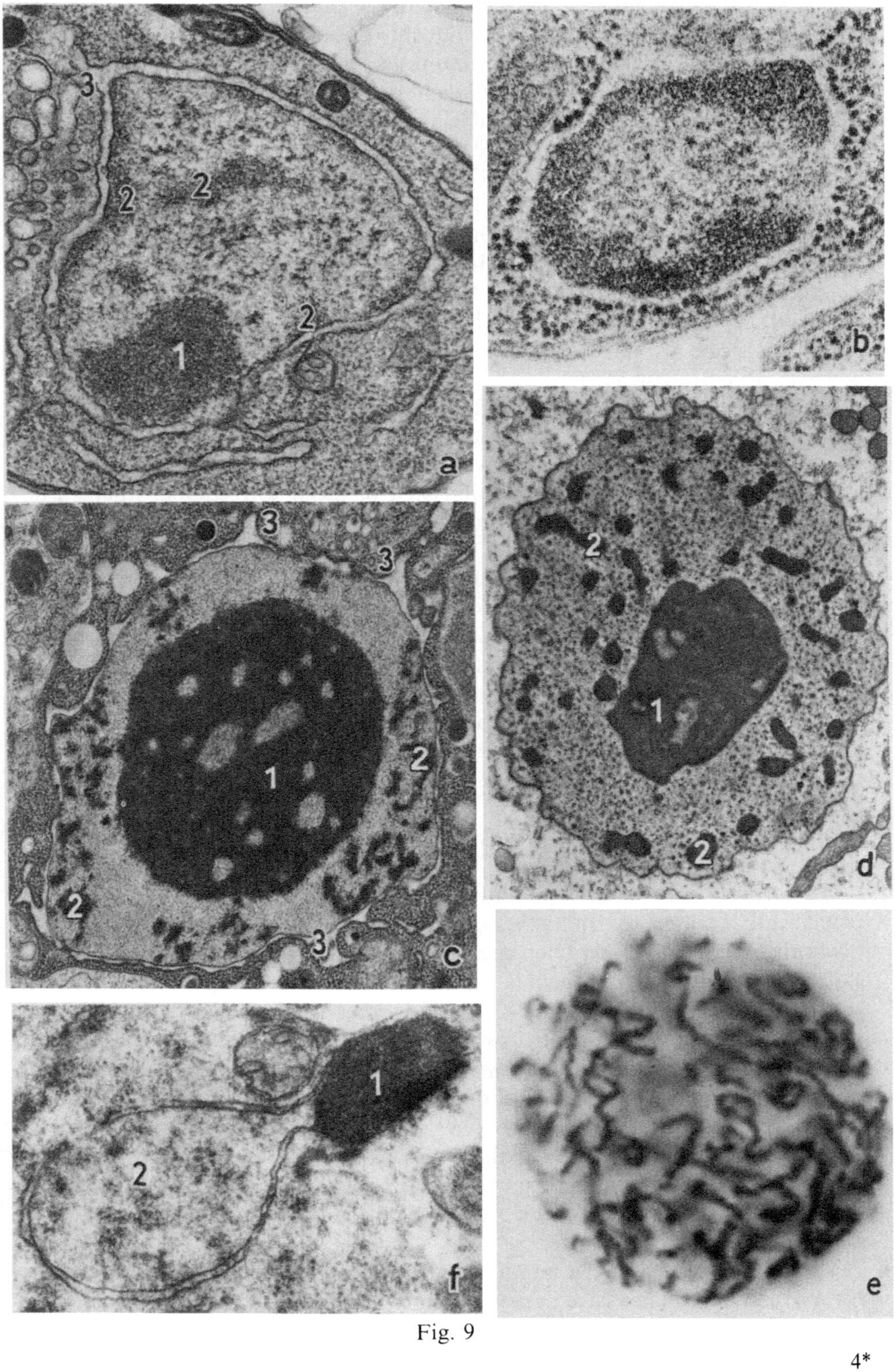

Fig. 9

telomeres of these chromosomes contact the nuclear envelope (KLUYEVA *et al.* 1978). The degree of condensation of the chromosomes of euglenids can change, however, depending on the culture conditions; in the light, the chromosomes of *Euglena gracilis* become very loose, while in the dark, they become compact (PYNE 1967). The amount of DNA in the nuclei of euglenids also decreases when they are transferred into darkness; in some clones, it decreases to almost one-third (OVCHINNIKOVA and SUKHANOVA 1976, OVCHINNIKOVA *et al.* 1976). Under the conditions of carbon starvation, the DNA in the nuclei of streptomycin-bleached euglenids decreases in two stages: the first stage coincides with the cessation of divisions and the blocking of the cells in the G_1 period; the second stage sets in considerably later and consists in a sudden loss of about 30% of DNA from the nuclei (BLUM and BUETOW 1963). At the same time, the electron density of the peripheral zone of all chromosomes substantially decreases (LEEDALE 1975, LEEDALE and BUETOW 1976).

These facts show that two DNA fractions with different stabilities exist in the chromosomes of euglenids. While studying these chromosomes spread on the water-air interface, HAAPALA and SOYER (1975) discovered that two types of DNP fibrils—axial and chromomere—really exist in them. Chromomere fibrils can either form loops or condense into chromomeres. Moreover, the DNP fibrils of euglenids are studded with glomeruli of the nucleosome type having a diameter of 10–12 nm that are characteristic of eukaryotes (HAAPALA and SOYER 1975, BRÉ *et al.* 1980). The chromosomes of euglenids also function according to eukaryotic principles: it has been discovered biochemically that besides compact chromatin, the nucleus has decondensed chromatin that contains only 14% of all DNA but is associated with more than 80% of the activity of RNA polymerase. In other words, decondensed chromatin is active in transcription (LYNCH *et al.* 1975, MAGNAVAL *et al.* 1979). Using ultracytochemical tests for DNA, MOYNE and his colleagues (1975) discovered chromatin microfibrils that come out of the chromosomes of *Euglena* and enter the karyolymph. They are probably loops of chromomere fibrils, and they correspond to the fraction of decondensed chromatin in the nucleus of *Euglena*.

The chromosomes of many (although not all) Hypermastigida (Table 1) are also permanently condensed. Under the light microscope, they are seen as thick coiled filaments (Fig. 9*e*). The chromosomes are permanently attached by their kinetochores to the inner side of the nuclear envelope (CLEVELAND *et al.* 1934, GRASSÉ and HOLLANDE 1951, 1963, etc.). The nondividing nuclei of hypermastigids still have not been studied sufficiently with the electron microscope. In *Trichonympha*, *Barbulanympha*, and *Spirotrichonympha*, the chromosomes in the nondividing nucleus remain clearly defined, although they are loose in comparison with the mitotic chromosomes (GRIMSTONE 1959, HOLLANDE and CARRUETTE-VALENTIN 1971). In the interphase, the kinetochores have the form of plates of finely fibrillar material that are inserted between the chromosome body proper and the inner membrane of the nuclear envelope (HOLLANDE and CARRUETTE-VALENTIN 1971, KUBAI 1973).

The chromosomes in the nondividing nucleus of *Holomastigotoides* are especially compact. They have been thoroughly studied with the light microscope (CLEVELAND 1949a, GRASSÉ and HOLLANDE 1963); a clearly defined spiral structure is seen in them (Fig. 1*l*). Electron micrographs show that the chromosomes consist of two relationally coiled chromatids and are surrounded by a common chromosomal matrix of moderate density, clearly differing from the karyolymph (Fig. 10*a*). The nature of the matrix is still unknown (HOLLANDE and CARRUETTE-VALENTIN 1971).

Finally, protozoans have nuclei in which all the chromatin is compact; it is so densely packed in a small nuclear volume that the boundaries of individual chromosomes cannot be seen. These nuclei are always very small and highly Feulgen-positive (Fig. 1*m*, *n*). Electron microscopy has shown that compact chromatin that fills the entire nucleus is characteristic of the microgametes of almost all Sporozoea (Fig. 10*b*) and the micronuclei of many ciliates (Table 1). As for micronuclei, the density of the packing of their chromatin greatly varies. They include nuclei with a rather loose network of chromatin strands (Fig. 10*c*), nuclei containing condensed chromosomes that are partially fused (Figs. 10*d*, 91*a*, 93*a*), nuclei in which chromatin has the form of a spongy mass and individual chromosomes are indistinguishable (Figs. 10*f*, 76*a*, 80*a*), and nuclei in which chromatin has the form of a single compact mass even on electron micrographs (Figs. 10*e*, 79*a*). There are also micronuclei whose chromatin is packed into one solid body (the *karyosome*) that does not occupy the entire nucleus; the periphery of such a micronucleus is free of chromatin. These variants apparently reflect the different density of chromatin packing in a small volume of the micronucleus (for references, see Table 1).

A looser chromatinic karyosome that occupies the centre of the nucleus is characteristic of many parasitic amoebae—especially the genus *Entamoeba* (EPSTEIN 1941)—and of *Hydramoeba hydroxena* (WERMEL 1925). It is very large in *Endamoeba blattae* (EPSTEIN 1941). This karyosome is Feulgen-positive, as has been shown in *Hydramoeba* (BEERS 1964) and *Entamoeba* (GORDEYEVA and SOLOVJEV 1967). The Feulgen-positive karyosome is also typical of the heliozoan *Actinosphaerium* (BOGDANOWICZ 1930a), the flagellate *Lamblia* (SOLOVJEV 1963), etc. On electron micrographs, the karyosome of *Lamblia* is dense (SOLOVJEV and CHENTSOV 1966) while that of *Entamoeba* is looser (MILLER *et al.* 1961, SIDDIQUI and RUDZINSKA 1965, AVAKYAN 1976). It has been shown recently that the karyosome of *Entamoeba histolytica* is formed only before mitosis and is an accumulation of chromosomes around a microtubule-organizing centre (GICQUAUD 1979; Fig. 13*b*).

Nucleolar chromatin, which lies inside or on the surface of the nucleolus, is a special category of chromatin structures (see p. 59).

Nucleoli

Nucleoli are dense intranuclear bodies that mainly consist of *ribonucleoproteins* (RNP), which are the precursors of ribosomes. The composition of nucleolar RNP includes preribosomal RNA (with a sedimentation coefficient of 45 S) and the products of its processing, i.e. 28 S rRNA and 18 S rRNA, which are the components of the large and small subunits of ribosomes, respectively. Moreover, nucleoli contain, as a rule, *nucleolar chromatin*, which is a series of *ribosomal genes* on which nucleolar ribonucleic acids are synthesized (see the reviews by BUSCH and

SMETANA 1970, SMETANA and BUSCH 1974, etc.). Thus, nucleoli are the products of the transcription of ribosomal genes, the site of rRNA processing and the centre of the biosynthesis of ribosomes in the cell (although the final assembly of ribosomes occurs outside the nucleus).

Number and Arrangement of Nucleoli. The nuclei of most protozoans have nucleoli. Exceptions to this rule are the nuclei of the gametes in Foraminiferida (SCHWAB 1969, CÉSANA 1972), the nuclei of microgametes, sporozoites and occasionally merozoites of the Sporozoea (Figs. 9*b*, 10*b*), and the micronuclei of the Ciliophora (Fig. 10*c*, *e*), i.e. nuclei with densely packed and transcriptionally inactive chromatin (Table 1). The nuclei of the actively growing stages of some protozoans also have no nucleoli; such protozoans are always intracellular parasites. For instance, there are no nucleoli in the nuclei of the trophozoites and gametocytes of the mammalian malaria parasites, although the same stages of the avian malaria parasites have nucleoli (RUDZINSKA 1969, AIKAWA 1971), and there are none in the Piroplasmia and usually none in the Microsporida. Nucleoli are always formed in other intracellular parasites, however, such as *Leishmania*, haplosporidians, and coccidian trophozoites (AIKAWA and STERLING 1974). It could be that the parasite somehow uses the ribosomes of the host cell when it lacks its own nucleoli.

Table 1 shows that most protozoan nuclei have one central nucleolus (Figs. 1*b*, *k*, 9*c*, *d*). The nucleolus is sometimes positioned eccentrically in the nucleus (Fig. 1*c*), as in the Trichomonadida, Hypermastigida, coccidian merozoites, and the Myxosporea. Moreover, it can assume a submembranous position, adjoining the nuclear envelope from the inside (Fig. 9*a*). This variant is typical of the nuclei of the trophozoites of the avian malaria parasites, the endodyocytes of *Besnoitia*, some strains of *Toxoplasma*, and especially the Haplosporea (Table 1). A lobular peripheral nucleolus is characteristic of the amoeba *Polychaos fasciculatum* (PAGE and BALDOCK 1980).

Very few protozoans consistently have two nucleoli. Two nucleoli of different size and density lie next to one another in the centre of the nucleus of *Leishmania donovani*. They fuse when the flagellate is transformed into the leptomonad form (RUDZINSKA *et al.* 1964). Two submembranous nucleoli (which are on the opposite sides of the nucleus) have been observed in the flagellate *Chilomastix* (BRUGEROLLE 1973) and the trophozoites of the coccidian *Myriosporides* (HENNERÉ 1967).

When there are several nucleoli, they are usually arranged in the nucleus either peripherally (Fig. 10*a*) or submembranously, as in Opalinata (CHEN 1948, SUKHA-

---

Fig. 10. Nuclei with condensed chromatin (electron micrographs). *a* "chromosomal" nucleus of *Holomastigotoides* (*1* sections of chromatids, *2* matrix of chromosomes, *3* nucleoli), 6100×; *b* "spermal" nucleus (*1*) in a microgamete of the haemosporidian *Leucocytozoon* (*2* flagellar axoneme), 30500×; *c* vesicular micronucleus of the ciliate *Nassula* sp., 13700×; *d* "chromosomal" micronucleus of the ciliate *Ichthyophthirius* (*1* chromosomes, *2* sphere), 19000×; *e* micronucleus of *Trachelonema sulcata*, 17000×; *f* part of the *Trachelocerca geopetiti* micronucleus with persisting kinetochores (arrows), 22000×. *a* after HOLLANDE and CARRUETTE-VALENTIN (1971); *b* after AIKAWA and others (1970); *c* after TUCKER (1967); *d* after HAUSER (1972); *e* after KOVALEVA and RAIKOV (1978); *f* after RAIKOV and KOVALEVA (1981).

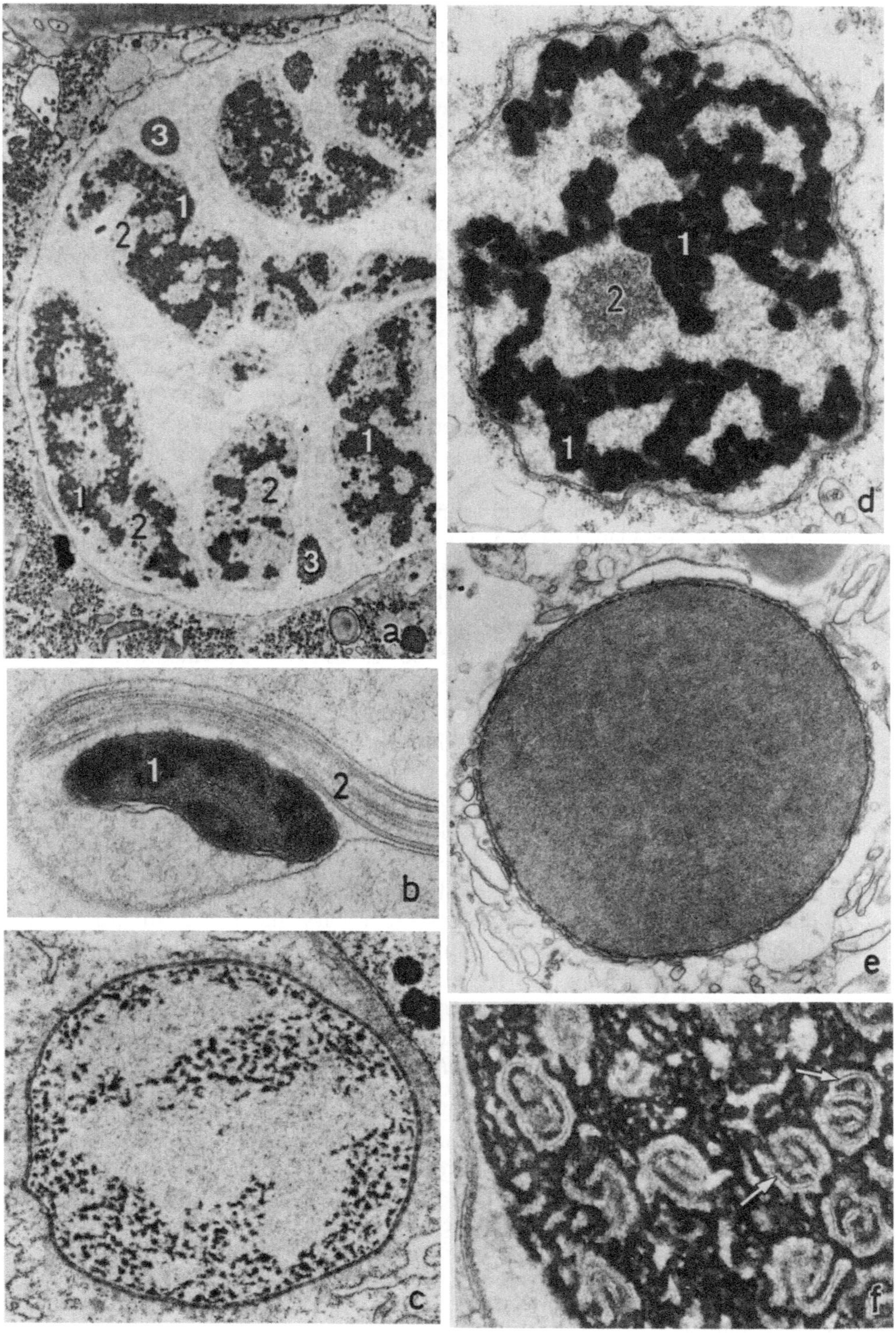

Fig. 10

NOVA 1959, 1961, NOIROT-TIMOTHÉE 1959, 1967), *Amoeba terricola* (Fig. 1*e*), the amoebae of the genus *Chaos* (KUDO 1947, MCCLELLAN 1959, ROTH and DANIELS 1962, DANIELS and ROTH 1964), and some Heliozoea (FEBVRE-CHEVALIER 1973, 1975, MIGNOT 1979, PATTERSON 1979). Peripheral nucleoli are especially numerous in the nuclei of the representatives of the genus *Amoeba* (Fig. 1*d*). In *A. proteus*, the nucleoli are contiguous to the honeycomb layer that underlies the nuclear envelope (Fig. 4*c*), while in *Endamoeba blattae*, they form a distinct layer at a certain distance from the nuclear envelope (Fig. 1*f*; SASSUCHIN 1929, 1930, EPSTEIN 1941). In *Entamoeba histolytica* (Fig. 13*a*), about ten macronucleoli and many micronucleoli are under the nuclear envelope (GICQUAUD 1979).

Nucleoli, however, multiply to the maximum in Foraminiferida. In the large nucleus of the gamont of *Myxotheca*, thousands of nucleoli form a thick peripheral layer (FØYN 1936, SCHWAB 1969). The nucleolar material is distributed similarly in the nuclei of *Allogromia* (Fig. 11*a*), *Ovammina*, and *Pontomyxa*; in *Hippocrepinella*, nucleoli are distributed throughout the nucleus without any zonality (Fig. 11*b*); in *Globobulimina*, they are all submembranous (for references, see Table 1).

Cytochemically, RNA and proteins are always found in high concentrations in nucleoli. As for intranucleolar DNA, it was discovered by Feulgen staining in only a few protozoans, e.g. in the foraminiferans *Ovammina* (DAHLGREN 1964) and *Allogromia* (MCENERY and LEE 1976). As can be expected, the labelled precursors of RNA are intensively incorporated into protozoan nucleoli (SCHUSTER and SVIHLA 1968, SCHRÉVEL and DHAINAUT 1968, GIBBS 1970, DICKINSON 1970).

Nucleoli are most frequently spherical, but occasionally they take on some other shape. The nucleoli are sausage-shaped in the testacean *Euglypha* (Fig. 1*g*), ribbon-shaped in the opalinids *Zelleriella* (CHEN 1972), and dumbbell-shaped in the euglenid *Distigma proteus* (MIGNOT 1966).

Ultrastructure of Nucleoli. Protozoan nucleoli have the same two main RNA-containing components as the nucleoli of other eukaryotes: fibrils that are 5–10 nm thick and granules whose size is 10–20 nm. It is accepted that the former contain 45 S preribosomal RNA, while the latter contain 18 S and 28 S rRNA in combination with proteins. As in other eukaryotes, the fibrils in protozoans usually occupy the centre of the nucleolus, while the granules occupy its periphery. Consequently the nucleolus has fibrillar and granular zones. However, both components are sometimes mixed, as in the nucleoli of *Eimeria* (YOUSSEF and HAMMOND 1972), or are poorly separated, as in the nucleolus of *Acanthamoeba* (BOWERS and KORN 1968, WILLAERT *et al.* 1978*a*) or *Thecamoeba sphaeronucleolus*

Fig. 11. Nucleoli of protozoans. *a* ovular nucleus in the agamont of the foraminiferan *Allogromia* (*1* nucleoli, *2* vesicular zone on the outside of the nuclear envelope), 4000 ×; *b* ovular nucleus of the foraminiferan *Hippocrepinella* (*1* nucleoli, *2* spongy layer on the inner surface of the nuclear envelope), gallocyanin staining, 1000 ×; *c* concentric structure of a nucleolus in a macrogametocyte of the coccidian *Grasseella* (*1* fibrillar zone, *2* granular zone), 35000 ×; *d* nucleolus of *Euglena gracilis* (*1* fibrillar zones, *2* granular zone, *3* chromosomes, *4* nuclear envelope), 33000 ×. *a, c, d* electron micrographs; *b* photomicrograph. *a* after SCHWAB (1970); *b* after DAHLGREN (1967b); *c* after PORCHET-HENNERÉ and ORMIÈRES (1973); *d* after MOYNE and others (1975).

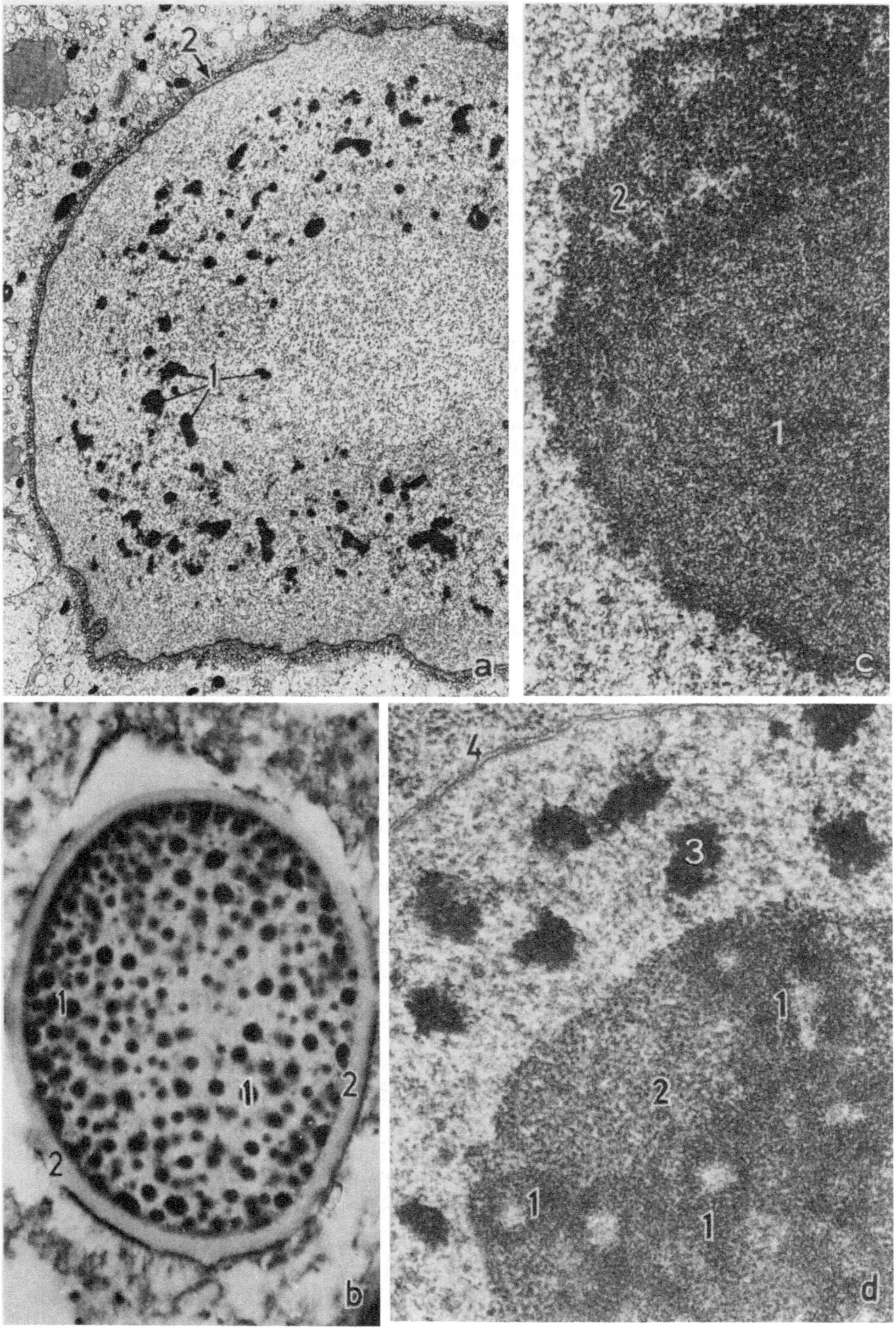

Fig. 11

(HOUSSAY and PRENANT 1970). In *Amoeba proteus*, the fibrillar and granular components of nucleoli are clearly segregated only after amoebae are treated with actinomycin D (FLICKINGER 1968). A typical concentric structure of nucleoli (a fibrillar zone in the centre and a granular zone at the periphery) has been observed in the opalinid *Cepedea* (NOIROT-TIMOTHÉE 1967), the schizopyrenid *Naegleria* (SCHUSTER 1975*a*; Fig. 9*c*), the foraminiferan *Globobulimina* (DAHLGREN 1967b), the gregarine *Selenidium* (SCHRÉVEL 1971*a*), the macrogametocyte of the coccidian *Grasseella* (Fig. 11*c*), and other protozoans. The granular cortex of the nucleolus often becomes permeated with lacunae (e.g. in *Selenidium*); individual granules and groups of granules may separate from it and get into the karyolymph (*Grasseella*; PORCHET-HENNERÉ and ORMIÈRES 1973). When many lacunae develop, a network of strands consisting of granular material is formed at the nucleolar surface. This network is called the nucleolonema. In the gregarine *Enterocystis* (DESPORTES 1974), it is easily fragmented; individual fragments of strands lie freely in the karyolymph and even enter the digitiform outgrowths of the nuclear envelope that have been described on p. 48.

Apparently, such patterns should be interpreted as an expression of the massive transport of nucleolar (preribosomal) granules into the cytoplasm. During the transformation from the amoeboid to the flagellate form in the schizopyrenid *Naegleria*, nucleolar fragments are extruded into the cytoplasm via vesicles that bud from the nucleus (SCHUSTER 1963). Before schizogony begins, the nucleolus of the trophozoite of the coccidian *Aggregata* fragments into irregular parts. During the first mitosis of schizogony, a large part of the nucleolar material gets into the cytoplasm (PORCHET-HENNERÉ and RICHARD 1971a). The nucleolus of the macrogamont of the protococcidian *Coelotropha* behaves similarly (VIVIER and HENNERÉ 1964, 1965). The passage of nucleolar material into the cytoplasm has been observed with the light microscope in the macrogametocytes of the coccidian *Eimeria maxima* (SCHOLTYSECK 1954, 1963) and in the precystic forms of the opalinid *Protoopalina intestinalis* (SUKHANOVA 1961). It has been shown autoradiographically that the larger part of the labelled RNA comes out from the nucleolus of the gregarine *Lecudina* into the cytoplasm in 24–48 hours (SCHRÉVEL and DHAINAUT 1968).

Sections of the nucleoli of *Euglena* and *Astasia* show several fibrillar zones (Fig. 11*d*). The fibrillar component is packed there in the form of a convoluted strand immersed in the granular material of the nucleolus (MOYNE *et al.* 1975, CHALY *et al.* 1977, 1979).

The structure of the nucleolus of the macrogametocyte of *Eucoccidium* is more complex (BARDELE 1966). There is a vacuole in the centre of the nucleolus. The granular component covers both the outer surface of the nucleolus and the surface of the vacuole, while the fibrillar component is between these two layers. In the foraminiferan *Ovammina* (DAHLGREN 1967a), the granular part of the nucleolus is cupped and permeated with lacunae (nucleolonema), while the fibrillar part has the form of a strand that lies in the cup.

More rarely, the arrangement of the nucleolar RNA-containing components is inverted. For instance, the centre of the nucleolus of the heliozoan *Gymnosphaera* consists of granules, while several fibrillar "caps" are on its surface (FEBVRE-CHEVALIER 1975). In another heliozoan, *Hedraiophrys*, this arrangement is more complicated: a number of fibrillar plaques covered with a granular monolayer adjoin the central, purely granular mass. These plaques also resemble a nucleolonema that is not granular but mainly fibrillar (FEBVRE-CHEVALIER 1973).

The structure of the nucleolar apparatus of the multinucleate pelobiontid amoeba *Pelomyxa palustris* (DANIELS *et al.* 1966, DANIELS and BREYER 1967) is very complicated and rather incomprehensible due to the absence of ultracytochemical data. In most specimens, the nuclei have only peripheral nucleoli – large and small (Fig. 12*a*). Clusters of spherical (50 nm) particles are found in some of the nucleoli or in the karyolymph next to them (Fig. 12*b*); the authors suppose these clusters are virus particles. Moreover, inclusions of finely granular or finely fibrillar material are formed in some nucleoli and are often found also in the karyolymph (Fig. 12*b*). They are much larger than the clusters of spherical particles, but their nature is unclear. Central aggregates of large and small nucleoli are found along with the peripheral ones in some specimens. In others, the entire nucleolar apparatus is collected in the centre of the nucleus, with spongy material in the middle. (The authors regard it as chromatin, but it is more likely a granular "nucleolonema" of a complex nucleolus.) At the same time, minute dense nucleoli and clusters of spheroids are collected at the periphery (Fig. 12*c*). The relationships between the intermediate stages in the life cycle of *Pelomyxa* are still unknown.

Nucleolar Chromatin. It is now known that nucleolar chromatin may be either a region of the chromosome that penetrates the nucleolus and carries a series of genes that code for rRNA (the nucleolar organizer proper) or an extrachromosomal accumulation of chromatin. In the latter case, this chromatin contains the "extra" DNA, i.e. free (separated from chromosomes) copies of the same ribosomal gene that have been preliminarily synthesized on the appropriate region of the chromosome as a result of *gene amplification.*

A link between nucleoli and definite chromosomes of a set has been observed (also in nondividing nuclei) under the light microscope in a number of eukaryotic protozoans, e.g. the opalinid *Zelleriella* (CHEN 1936b, 1948) and the hypermastigids *Spirotrichonympha* (CLEVELAND 1938c) and *Holomastigotoides* (CLEVELAND 1949a). ZHIVAGO (1948) observed the nucleolar organizer ("secretory apparatus") for the first time *in vivo* in the interphasic nucleus of *Amoeba polypodia.*

The nucleolar chromatin of the Euglenida has been studied mostly by electron microscopy. The nucleoli of euglenids have cavities filled with loose interlaced fibrils 6–10 nm thick (O'DONNELL 1965, PYNE 1967). These fibrils are similar to those of chromosomes and can be destroyed by deoxyribonuclease (O'DONNELL 1965). Two specific ultracytochemical tests for DNA (MOYNE *et al.* 1975, BERTAUX *et al.* 1978, CHALY *et al.* 1979) have been used to show that the DNA fibrils in the nucleolus of *Euglena gracilis* and *Astasia longa* are localized in the wall of a convoluted canal that penetrates the fibrillar zone of the nucleolus (Fig. 11*d*). Biochemical investigations showed that the nucleus of *E. gracilis* has a fraction of repetitive sequences of DNA that hybridize with rRNA (SCOTT 1973). A more detailed analysis demonstrated that this nucleus has about 800 genes for each of the two types of rRNA and about 760 genes for 4 S RNA (GRUOL and HASELKORN 1976). A similar DNA has been obtained from the isolated nucleoli of the colourless euglenid *Astasia* (MOROSOLI and LAFONTAINE 1976). The intranucleolar chromatin of euglenids thus contains ribosomal genes (rDNA). The link between the nucleolus of *Euglena* and a chromosome has often been noted (O'Donnell 1965, Pyne 1967, Moyne *et al.* 1975). At this site, the chromosome untwines into fine DNA-containing filaments that penetrate deep into the nucleolus (Moyne *et al.* 1975). Thus, the nucleolar chromatin of euglenids is more likely a chromosomal locus with a series of a few hundred ribosomal genes rather than amplified extra DNA.

In contrast, the ribosomal gene is probably amplified in *Chlamydomonas*. A DNA fraction, which hybridizes with rRNA and differs from the total DNA in its buoyant density (Sinclair 1972), accumulates in the nucleus as macrogametes grow. Such DNA is also synthesized in the form of extrachromosomal copies in the vegetative cell cycle (Howell 1972). Ribosomal genes in the extra DNA of *Chlamydomonas* are united into tandemic arrays (Marco and Rochaix 1980). It is very likely that they are localized in the nucleolus.

Nucleolar chromatin is undoubtedly present in the numerous peripheral nucleoli of *Amoeba proteus* that incorporate ^{3}H-thymidine, as has been shown by ultrastructural autoradiography (Dickinson 1970, Minassian and Bell 1976b). The transcription of the nucleolar genes of amoebae was recently demonstrated by the spreading technique and electron microscopy. Numerous rRNA molecules, which grow like brush bristles on the axis of nucleolar DNA, were revealed (Murti and Prescott 1978). It has been shown cytofluorimetrically that a considerable amount of DNA is synthesized in *A. proteus* after nuclear DNA is completely duplicated; it is then destroyed in the course of mitosis and must be produced anew in the next cell cycle (Makhlin *et al.* 1979a). Such overreplication of DNA does not occur in starving amoebae (Makhlin *et al.* 1979b). There are reasons to believe that this extra DNA is amplified nucleolar extrachromosomal DNA (Spear and Prescott 1980).

A body of compact chromatin has been discovered in each of the numerous nucleoli of the large nucleus of the foraminiferan *Ovammina* (Dahlgren 1964, 1967a). This body is Feulgen-positive; it is digested by deoxyribonuclease and has a fibrillar ultrastructure. It always closely contacts the fibrillar part of the nucleolus. The development of peripheral nucleoli on chromatin blocks that lie under the nuclear envelope has been described with the aid of the light microscope in the uninucleate agamonts of the foraminiferan *Allogromia* (McEnery and Lee 1976). Drawing a parallel with the peripheral nucleoli of the oocytes of fishes or amphibians, it can be supposed that intranucleolar chromatin contains amplified rDNA in both *Ovammina* and *Allogromia*. There is, however, still no direct evidence of this. Also, it is still unknown whether DNA is present in the numerous peripheral nucleoli of other foraminiferans (Table 1). It should merely be noted

---

Fig. 12. Ultrastructure of the *Pelomyxa palustris* nuclei (after Daniels *et al.* 1966). *a* nucleus with peripheral nucleoli (*1* large nucleoli, *2* small nucleoli, *3* perinuclear bacteria, *4* nuclear envelope, *5* perinuclear vesicles), 11 000 × ; *b* periphery of the nucleus (*1* inclusions of fibrillar material, *2* nucleolus, *3* karyolymph, *4* group of spheroids, *5* honeycomb layer of the nuclear envelope), 18 000 × ; *c* nucleus with a complex central nucleolus (*1*) and a lamellar layer on the outside of the nuclear envelope (*2*), 5500 × .

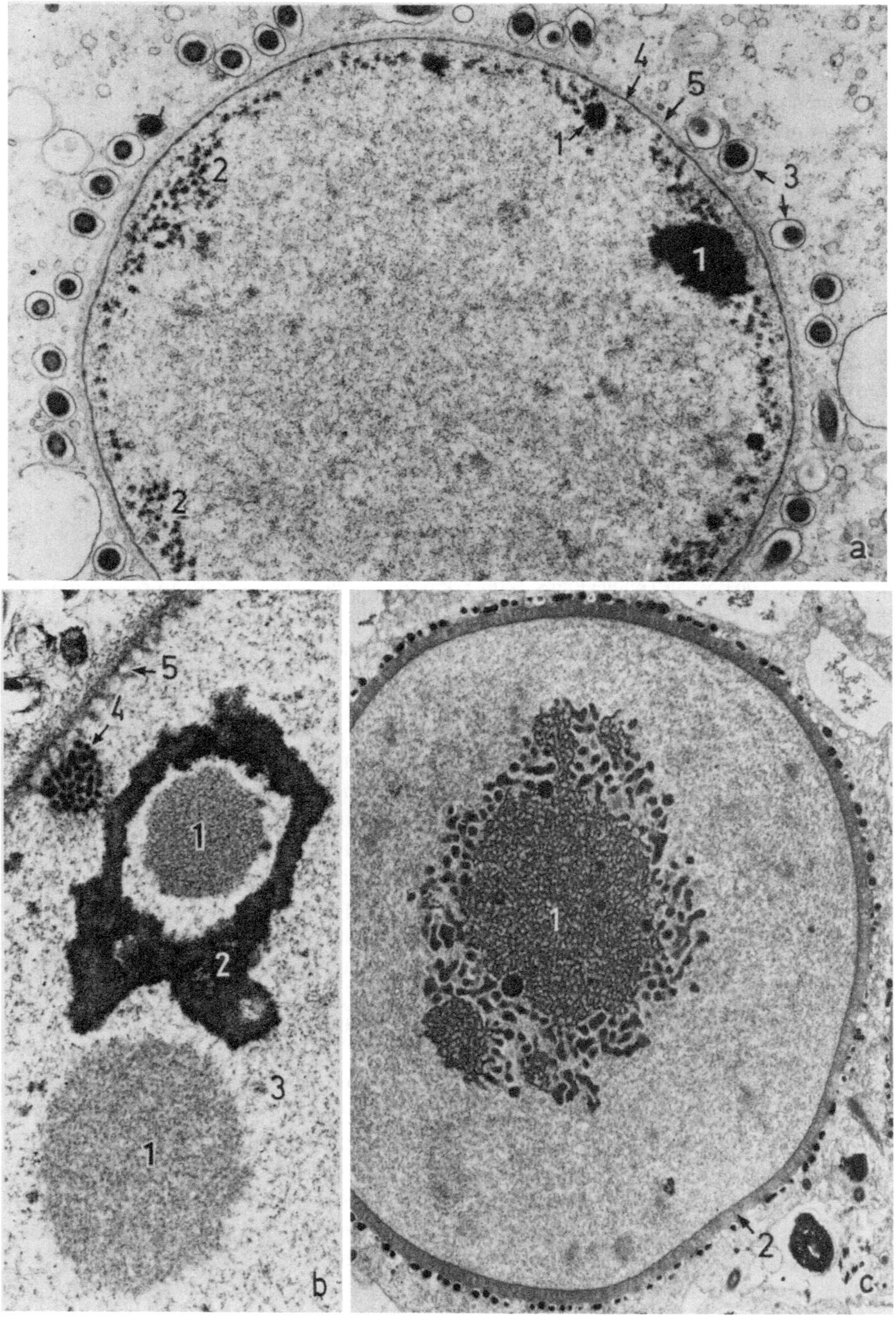

Fig. 12

that every nucleolus of *Hippocrepinella* has an outer layer of electron-dense, weakly Feulgen-positive material resembling chromatin (DAHLGREN 1967b). Conversely, each of the numerous nucleoli of the fresh-water rhizopod *Lesquereusia* has a central electron-dense body (HARRISON *et al.* 1976). Since the centre of every nucleolus is Feulgen-positive (HARRISON *et al.* 1975), it can be assumed that the body is a chromatinic one.

Structures that are morphologically similar to the nucleolar organizer (intranucleolar chromatin) have also been observed in other protozoans: *Volvox* (SOYER 1973), the aconchulid amoebae *Nuclearia* and *Nucleosphaerium* (MIGNOT and SAVOIE 1979, CANN and PAGE 1979), and the gregarines *Selenidium* (SCHRÉVEL 1971a) and *Gregarina* (DEVAUCHELLE 1968). They usually consist of a fibrillar axis which passes inside a canal that permeates the fibrillar zone of the nucleolus. In the coccidian *Eimeria magna*, however, the fibrils that can be destroyed by deoxyribonuclease are not spatially separated from other fibrillar and granular components of the nucleolus (YOUSSEF and HAMMOND 1972).

The nucleoli and nucleolar chromatin of the nuclei of radiolarians and the macronuclei of ciliates will be considered separately in Chapters 6 and 8.

Karyolymph and Inclusions in the Nucleus

The karyolymph (nuclear sap, karyoplasm) is a liquid or semiliquid medium in which the formed components of the nucleus are immersed. It is usually structureless in the living cells. On electron micrographs, it is structuralized in the form of fine filaments (less than 5–7 nm thick) that are difficult to distinguish from decondensed chromatin (Figs. 2*a*, 11*d*, 12*a*).

Many granules of various size are almost always suspended in the karyolymph. The only exception is the micronuclei of ciliates, which lack the granular component (Fig. 10*c*–*f*). Of course, some of the granules are nucleolar particles on their way to the cytoplasm. It is now known, however, that the karyolymph also has RNA-containing particles of non-nucleolar nature; these particles are perichromatin fibrils as well as perichromatin and interchromatin granules (reviewed by BOUTEILLE *et al.* 1974). It is assumed that these particles (especially perichromatin fibrils and the granules formed from them) contain the messenger RNA (mRNA) bound with proteins and correspond to its inactive (storage and/or transport) form.

Perichromatin granules, which are large (40–50 nm), have been found in the nuclei of Euglenida (PYNE 1967, CHALY *et al.* 1977, 1979, KLUYEVA *et al.* 1978), of the chloromonad *Gonyostomum* (HEYWOOD 1976), of the trophozoites of the coccidians *Coelotropha* (VIVIER and HENNERÉ 1965) and *Aggregata* (PORCHET-HENNERÉ and RICHARD 1970a), and in the gametocytes of the haemosporidian *Leucocytozoon* (STERLING and AIKAWA 1973). However, MOYNE and his colleagues (1975) have used an improved method of staining RNP in ultrathin frozen sections to show that the nucleus of *Euglena gracilis* contains not the typical perichromatin granules, but U-shaped particles (50 × 90 nm) formed by a bent fibril 20 nm thick. Apparently, these fibrils also contain mRNA. The karyolymph of *Euglena* also has thin (5–7 nm) fibrils that likewise contain RNA and form the fibrillar nuclear matrix. These fibrils resemble *perichromatin fibrils*, although they differ from the latter by their localization in places other than the periphery of condensed chromatin (KLUYEVA *et al.* 1978).

Another form of large nuclear RNP particles is characteristic of some Amoebida. These are the *nuclear helices* (Fig. 13*c*, *d*). They have been observed in the nucleus of *Amoeba proteus* (PAPPAS 1956a, 1959, MERCER 1959, ROTH *et al.* 1960, BRANDT and PAPPAS 1963, etc.) and in the nuclei of two related species (FLICKINGER 1974). They are found more rarely in *Chaos* (ROTH and DANIELS 1962, DANIELS and ROTH 1964). Each helix consists of a filament 12–14 nm thick that is twisted into coils with a diameter of 35–40 nm. The length of the helix averages 250–300 nm (it can reach 750 nm), and the filament makes 8–10 (up to 20) turns. The helices are usually arranged radially around a certain axis like lamp-brush bristles (Fig. 13*c*, *d*). They are present only in the interphase nuclei, appearing not earlier than one hour after the end of mitosis. At first, helices were usually regarded as the components of chromosomes, implying the presence of DNA in them (BRANDT and PAPPAS 1963, DICKINSON 1966, 1970). Since then, investigations involving ultrastructural autoradiography have clearly shown that helices do not incorporate DNA precursors, but actively incorporate ^{3}H-uridine, i.e. they contain RNA (STEVENS and PRESCOTT 1966, WOLSTENHOLME 1966, WISE and GOLDSTEIN 1972, WISE *et al.* 1972). Moreover, helices are destroyed by ribonuclease, but they are resistant to the action of deoxyribonuclease. After mitosis, their appearance is inhibited by actinomycin D (STEVENS and PRESCOTT 1965, 1966, WISE *et al.* 1972). It is also known that individual helices can separate from the common axis, migrate to the periphery of the nucleus, enter the alveoli of the honeycomb layer and go out into the cytoplasm through the central part of the nuclear pore. In this case, the helix temporarily uncoils (Fig. 4*d*). In the cytoplasm, helices soon disappear (STEVENS and PRESCOTT 1965, 1966, STEVENS 1967, WISE *et al.* 1972, LEESON and BHATNAGAR 1975a, MINASSIAN and BELL 1976a). These facts are in accord with STEVENS' and PRESCOTT's hypothesis that the nuclear helices of amoebae are an inactive (transfer) form of mRNA. They believe that the mRNA molecules are synthesized on the DNA axis (around which they are arranged) and are immediately covered with a protein sheath, twisting into helices. In the cytoplasm, the sheath is discarded (helices disappear), while the mRNA becomes attached to ribosomes and becomes the axis of a polyribosome. Thus, the helices of amoebae are probably related to perichromatin granules and informosomes (see the review by GEORGIEV 1974).

Other intranuclear inclusions of protozoans have been studied far less. In the euglenid *Distigma*, a dense proteinaceous *sphere* of a microfibrillar structure, which is digested by pepsin and pronase, is next to the dumbbell-shaped nucleolus. It contains neither DNA nor RNA (MIGNOT 1966, 1975). Dense spheres of an unknown nature are formed in the nuclei of the amoeba *Acanthamoeba palestinensis*, though only at the cyst stage (LASMAN 1977). Looser microfibrillar spheres (presumably proteinaceous ones) have been observed in the micronuclei of the ciliates *Loxodes* (RAIKOV 1975a) and *Ichthyophthirius* (HAUSER 1972; Fig. 10*d*). Spheres of a similar type in the nuclei of the pelobiontid amoeba *Pelomyxa palustris* (Fig. 12*b*) have already been mentioned. Protein spheres are frequently found also in the macronuclei of the diploid type in ciliates (Chapter 7). The physiological significance of the intranuclear spheres is not known. Similar protein spheres are found in the nuclei of the oogonia and oocytes of insects, fishes, amphibians, reptiles and other animals.

In some cases, intranuclear fibrillar bodies have the form not of spheres but of *bundles of parallel microfilaments*, such as in the coccidian *Aggregata* (PORCHET-HENNERÉ and RICHARD 1971a) or in the agamonts of the foraminiferan *Allogromia* (SCHWAB 1970). An interesting fact is that such bundles (although more compact) of very young agamonts of *Allogromia* reveal cross striation with a periodicity of 26-30 nm and resemble collagen fibrils (SCHWAB 1974).

Crystalloids are formed in the nuclei of some protozoans. They usually appear when the nucleus becomes infected with a virus, such as in the schizopyrenid *Naegleria* (SCHUSTER 1969). In this case, however, the material of the crystalloid (protein?) differs from the virus particles proper. Conversely, the crystalloid in the nucleus of the gregarine *Pachyspora* consists of the virus particles themselves, which have a diameter of 250 nm (VIVARÈS 1974). The intranuclear crystalloid in *Entamoeba histolytica* (Fig. 13*a*) may also be of a viral nature; it consists of parallel tubules with a diameter of 35 nm (GICQUAUD 1979). A crystalloid of helically arranged ribosome-like particles with a diameter of 15 17 nm is found in the nucleus of the flagellate *Leishmania hertigi* (CROFT 1979).

Along with other nuclear inclusions, mention should be made of the electron-dense globules (lipid?) in the nuclei of the amoeba *Pelomyxa* (DANIELS *et al.* 1966, DANIELS and BREYER 1967), a large osmiophilic body in the nucleus of *Entamoeba sp.* (INJEYAN *et al.* 1979), vacuolized bodies in the heliozoan *Hedraiophrys* (FEBVRE-CHEVALIER 1973), glomeruli of dense strands (RNP?) in the nuclei of gamonts of the gregarine *Stylocephalus* (DESPORTES 1970), clusters of closely packed tubular formations in the trichosid rhizopod *Trichosphaerium* (SCHUSTER 1976), various membranous vesicles, which are probably invaginations of the inner nuclear membrane in various forms, such as in the foraminiferan *Myxotheca* (SCHWAB 1969) and the amoeba *Entamoeba histolytica* (GICQUAUD 1979; Fig. 13*a*, *b*), and cytoplasmic islets in the nucleus of the stereomyxid rhizopod *Stereomyxa angulosa* (BENWITZ and GRELL 1971b).

Bacteria are frequently found in the nucleus, as in that of euglenids (LEEDALE 1969) and the hypermastigid *Joenia* (HOLLANDE 1979). The nuclei of ciliates of the genus *Paramecium* have symbiotic bacteria that are strictly specific with respect to both the species of paramecia and the type of nucleus: some bacteria species infect only micronuclei, while others infect only macronuclei (HAFKINE 1890, L. B. PREER 1969, OSSIPOV *et al.* 1976, 1980, etc.; reviewed by OSSIPOV 1981). Virus particles are found in the nucleus of the schizopyrenid amoeba *Naegleria gruberi*; they can pass into the cytoplasm through tubular outgrowths of the nuclear envelope (SCHUSTER 1969, SCHUSTER and DUNNEBACKE 1976).

Fig. 13. Ultrastructure of nuclei in amoebae. *a* interphase nucleus of *Entamoeba histolytica* (*1* macronucleolus, *2* micronucleoli, *3* crystalloid, *4* intranuclear vesicle, *5* karyolymph), 18000 ×; *b* preprophase nucleus of *E. histolytica* (*6* MTOC, *7* chromosomes, other designations are the same as in *a*), 18000 ×; *c* and *d* RNP helices (*1*) in *Amoeba proteus* (*c* 41000 ×; *d* 62500 ×). *a* and *b* after GICQUAUD (1979); *c* and *d* after MINASSIAN and BELL (1976a).

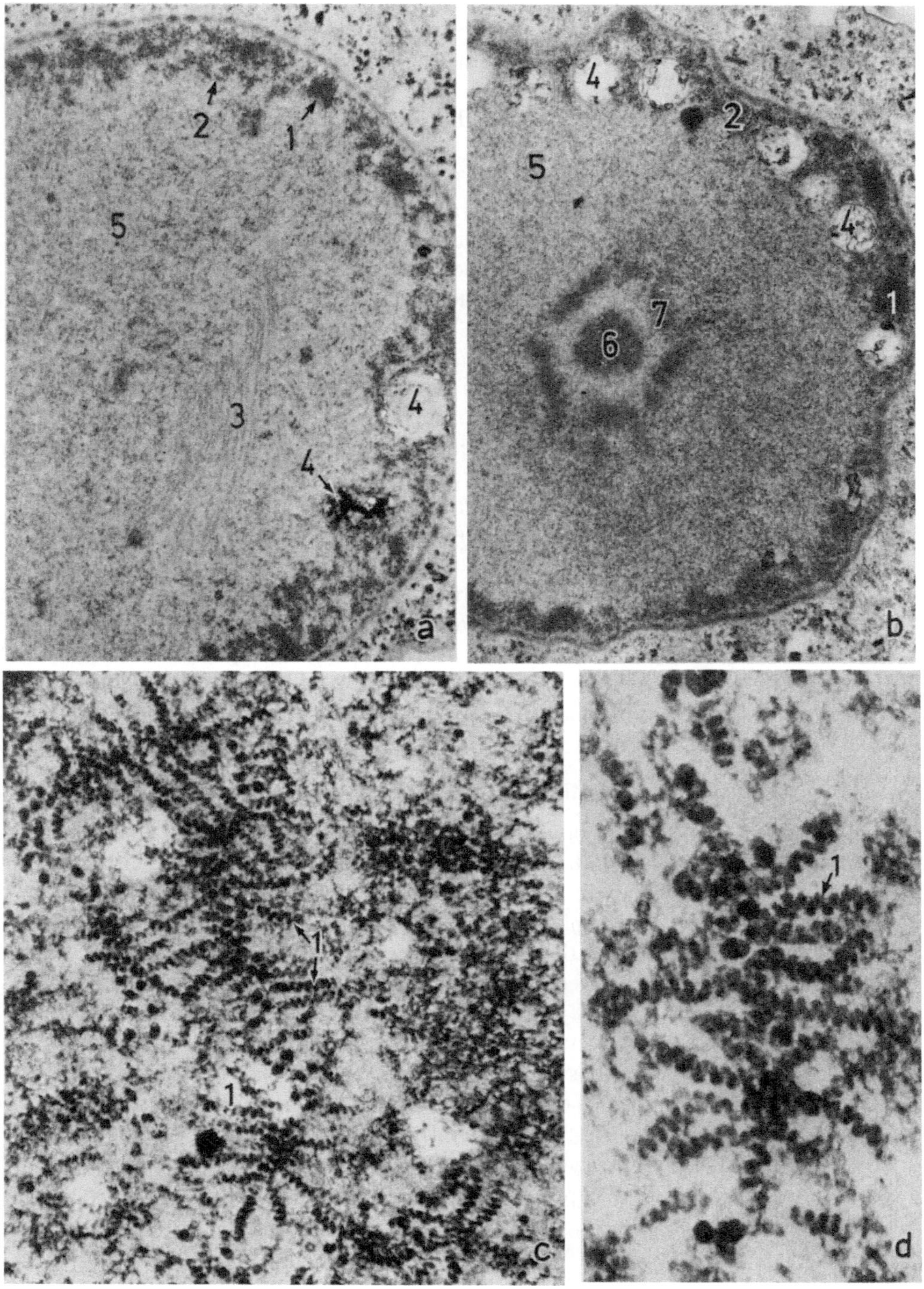

Fig. 13

Structures that are morphogenetically connected with the mitotic spindle hold a special place among intranuclear inclusions in protozoans. Sometimes, they are dense bodies that induce the assembly of microtubules in the prophase. PICKETT-HEAPS (1969, 1974) has called them the "microtubule-organizing centres" (MTOCs). They have been observed in the preprophase nuclei of the amoeba *Entamoeba histolytica* (Fig. 13*b*) and the interphase micronuclei of the ciliates *Paracineta* (HAUSER 1968, 1970, 1972) and *Didinium* (KARADZHAN and RAIKOV 1977a; Fig. 30*a*). In *Paracineta*, the MTOC has a paracrystalline structure. The formation of microtubules with the participation of MTOCs will be considered in Chapter 2. In other cases, the above-mentioned structures are the *residual spindle*, i.e. they consist of microtubules. A dense bundle of parallel microtubules (nuclear rod) is characteristic of the interphase nuclei of the Haplosporea (PERKINS 1975). A diffuse layer of microtubules under the nuclear envelope is found in the micronuclei of a number of ciliates: *Didesmis*, *Isotricha*, *Ancistrocoma*, *Acineta*, *Tokophrya*, *Phalacrocleptes*, *Paramecium aurelia*, *Ichthyophthirius*, etc. (Table 1). In *Paramecium bursaria*, the residual spindle has been observed even with the light microscope (SCHWARTZ 1965). It is evident also on electron micrographs; however, it consists not of microtubules, but microfilaments with a diameter of about 6 nm that compose lamellae that are intercalated between individual chromosomes (SCHWARTZ 1976). Nevertheless, these filaments and lamellae consist of tubulins (HAUSER and BEINBRECH 1973) and are transformed into microtubules in mitosis.

Moreover, an instance has been described when chromosomal *kinetochores* are preserved in the interphase nucleus, namely, in the micronucleus of the lower ciliate *Trachelocerca geopetiti* (RAIKOV and KOVALEVA 1981). In the interphase, these kinetochores have the form of lamellae that contain RNP and are situated in "clefts" that permeate the mass of condensed chromatin of the micronucleus (Fig. 10*f*).

Classification of Protozoan Nuclei

It is extremely difficult to build up a natural classification of protozoan nuclei. In almost all protozoans with a complex life cycle (Foraminiferida, Gregarinia, Coccidia, Haemosporina, etc.), the same nucleus passes consecutively through several stages that differ sharply from one another morphologically and probably biochemically. For instance, the volume of the nucleus increases tens and even hundreds of times, the parietal condensations of chromatin decondense and nucleoli are formed when a trophozoite or a macrogamete develops from a sporozoite or a merozoite. Nevertheless, diverse protozoan nuclei can be nominally divided into several groups if account is taken of the lability of the types of nuclei and the possibility of their mutual transitions, that is, if these types are not regarded as distinct categories. This was done by GRELL (1956a) as well as DOGIEL (1965) on the basis of knowledge acquired by light microscopy. This classification can be kept for eukaryotic protozoans even now, although some corrections should be made.

Vesicular Nuclei

Vesicular nuclei of diverse size and shape (they are not always spherical!) are characteristic of most protozoans. Their common feature is the presence of more or less large zones filled with karyolymph. The state of the chromatin varies from being decondensed to being compact. They usually have nucleoli.

At least three subtypes of vesicular nuclei, which are not sharply delimited from one another, can be distinguished. In Table 1, these variants are specified as "small", "typical", and "large" vesicular nuclei.

Small vesicular nuclei usually have small dimensions (1–3 μm) and are characterized by the prevalence of compact chromatin (in the form of peripheral condensations, a chromatin network or a central karyosome). The volume of the nucleus, taken up by the karyolymph, is not large. The nucleolus is either small (often eccentric or submembranous without a developed granular cortex) or absent altogether. Such nuclei undoubtedly either have a low level of transcription or do not synthesize RNA at all. Small vesicular nuclei are characteristic of the gametes of several foraminiferans (*Myxotheca*, *Iridia*) and gregarines (*Stylocephalus*, the macrogametes of *Grebnickiella*), the merozoites (Fig. 9*b*), endodyocytes (Fig. 9*a*), and sporozoites of the Gregarinia, Coccidia, and Haemosporina (Table 1), as well as the micronuclei of some Ciliophora (Fig. 10*c*). In sporozoans, chromatin in such nuclei is peripheral, as a rule, while in micronuclei, it either is gathered into a central karyosome (*Dileptus*, *Isotricha*, *Paracineta*, *Paramecium aurelia*, *P. multimicronucleatum*) or is reticulate, as is the case in *Coleps*, *Didinium*, *Nassula*, *Acineta*, *Tokophrya*, etc. (Table 1).

Typical vesicular nuclei are most commonly found in protozoans. Their dimensions vary from 1 to 2 μm to about 10 μm. Decondensed chromatin prevails in them, but they also often have compact chromatin in the form of a chromatin network or small peripheral condensations. A definite, relatively large nucleolus is usually central, but sometimes it is peripheral or even clearly submembranous. A few (two or three) nucleoli are occasionally found, but never many of them. The karyolymph is abundant, occupying a volume several times greater than that taken up by nucleoli and compact chromatin. Morphologically, typical vesicular nuclei are very similar to the nuclei of the nondifferentiated (embryonal, cambial) cells of metazoans and metaphytes. Among protozoans, this is apparently also the main nondifferentiated type of nucleus which is capable of becoming differentiated and transformed into all other types. In the state of the chromosomal and nucleolar apparatus, these nuclei correspond to those with active (but not intensive) transcription.

The nuclei of almost all eukaryotic Phytomastigophorea (except the euglenids) can be included in that group, as well as the nuclei of Kinetoplastida, Opalinata, and Schizopyrenida (Fig. 9*c*). The nuclei of Diplomonadida and Trichomonadida, with their granular karyolymph and minute nucleoli, are somewhat less typical. Along with other varieties of nuclei, typical vesicular nuclei are common among the Amoebida. Arcellinida, and Heliozoea, and are characteristic of the Labyrinthomorpha (Table 1). Among the Sporozoea, typical vesicular nuclei are characteristic of the growing, but not hypertrophically grown, stages of the life cycle: they are characteristic of young trophozoites of the Gregarinia (Fig. 2*a*), young and occasionally mature trophozoites of the Coccidia, and the trophozoites and gametocytes of avian Haemosporina. This applies also to the nuclei of the Haplosporea and Myxosporea (Table 1).

The nuclei of intracellular parasites with decondensed chromatin, which do not form nucleoli, are somewhat different. Such are the nuclei of the trophozoites and gametocytes of the mammalian malaria parasites as well as the nuclei of Piroplasmia and many Microsporida. Nevertheless, we relegate these nuclei to the rank of typical (not small) vesicular nuclei owing to the state of the chromatin in them. This instance again illustrates the arbitrary nature of dividing nuclei into types and subtypes. In general, the boundary between small and typical vesicular nuclei is arbitrary: for instance, the nuclei of the Kinetoplastida (*Leishmania*, *Trypanosoma*, etc.) can be related to the first group (since peripheral compact chromatin is usually developed in them), although they more readily belong to the second group by nucleolar development and the abundance of the karyolymph.

The nuclei of the parasitic amoebae of the genera *Hydramoeba* and *Entamoeba* as well as the nuclei of the diplomonadid *Lamblia*, which resemble "typical" vesicular nuclei but have a Feulgen-positive karyosome instead of the central nucleolus, hold an unclear position in our classification. Electron microscopy has shown the karyosome of *Entamoeba* to consist of a cluster of chromosomes around a MTOC (Fig. 13*b*), while the nucleoli are numerous and peripheral in this case (Fig. 13*a*). Thus, the nucleus of *Entamoeba* resembles a large vesicular nucleus in its structure (but not in its size).

Large vesicular nuclei are also vaguely delimited from the "typical" ones. They have an even greater amount of karyolymph, while chromatin is always completely dispersed. Earlier (Raikov 1967a), we included such nuclei in the group of ovular ones, but now we prefer to keep this term for the next category. The nucleolar apparatus is highly developed, but nucleoli are not very plentiful: usually, there is one large nucleolus (frequently vacuolated) with a prominent granular nucleolonema. More rarely, there are several nucleoli, generally peripheral. The nuclear envelope often has additional layers, or it forms digitiform outgrowths. The nuclei of this category are large but not gigantic (usually within the range of 10 to 40 μm).

"Large" vesicular nuclei are found especially at the high-growth uninucleate stages of protozoan life cycles: the macrogametes of *Volvox* (Soyer 1973), the gamonts of several multithalamous Foraminiferida (Myers 1935, 1936, Føyn 1937, Le Calvez 1938, 1953, Grell 1958a, 1958b, 1959, Voronova 1971, Febvre-Chevalier 1971), the gamonts of many Gregarinia, the adult trophozoites of some Coccidia, and the macrogametocytes and macrogametes of all coccidians (Table 1). In all these cases, a "large" vesicular nucleus develops directly from a "typical" one that is characteristic of the earlier stages of the development of the trophozoite, gamont, or macrogamete. There are, however, protozoans (without a complex life cycle) in which the nucleus of this type is always present. This applies to some Hypermastigida, several large amoebae (*Thecamoeba sphaeronucleolus*, etc.), and the heliozoan *Hedraiophrys* (Table 1). Apparently, "large" vesicular nuclei very actively synthesize RNA, especially ribosomal RNA. This is evident, for instance, from the high development of nucleoli, the abundance of the granules in them, and the active separation of granular clusters from the cortex of nucleoli in such nuclei.

"Ovular" Nuclei

We include large and very large (up to 100 μm in diameter) nuclei with many peripheral nucleoli, which resemble the nuclei of the oocytes of fishes or amphibians, in the category of *"ovular" nuclei.* Among protozoans, such nuclei are known to exist in two groups: several large amoebae (*Amoeba proteus*, *Thecamoeba proteoides*, etc.) and the gamonts (and occasionally the agamonts) of monothalamous foraminiferans: *Myxotheca*, *Allogromia*, *Ovammina*, etc. (Table 1). Ovular nuclei are apparently not polyploid (we do not include the polyploid nuclei of radiolarians and the macronuclei of ciliates in this category). The chromatin in ovular nuclei is usually dispersed; and there is a large amount of karyolymph. The chromosomes are generally invisible, but sometimes they can be seen in the central zone of the nucleus in a form resembling the lampbrush chromosomes of the oocytes of animals, e.g. in the gamont of *Myxotheca*. In general, the stratification of nuclei into concentric zones, i.e. the central zone and one or two peripheral zones, is characteristic of the ovular nuclei of protozoans (Figs. 1*d*,*f*, 11*a*) as well as the nuclei of the oocytes of metazoans. The nuclear envelope is almost always strengthened by the appearance of additional layer on the inside (*A. proteus*, foraminiferans) and sometimes on the outside.

As a rule, there are very many nucleoli (hundreds and even thousands) in ovular nuclei. In some cases, they are distributed throughout the nucleus (*Boderia*, *Hippocrepinella*; Fig. 11*b*). In others, they are situated either submembranously (*Globobulimina*) or in a rather thick peripheral zone of the nucleus, e.g. in *Myxotheca*, *Allogromia* (Fig. 11*a*) and *Amoeba proteus* (Fig. 4*c*). Sometimes they are found at the boundary between the central and peripheral zones of the nucleus (*Endamoeba blattae*, Fig. 1*f*). It has been shown that the nucleoli of *Ovammina* and *Allogromia* are formed on chromatin bodies that probably contain amplified extra DNA. Thus, it can be assumed that ovular nuclei are the site of the amplification of rDNA, which participates in the formation of numerous peripheral nucleoli, just as in the nuclei of the oocytes of fishes and amphibians. This assumption must be verified, however.

It should be noted that the category of ovular nuclei is also not clearly delimited from the category of "large" vesicular nuclei. An intermediate position is held by the nuclei of some amoebae with peripheral nucleoli that are not so numerous as in *A. proteus*, i.e. the nuclei of *Chaos carolinensis*, *Ch. illinoisensis*, *Pelomyxa palustris* (Fig. 12*a*), *Amoeba terricola* (Fig. 1*e*) and others, and also the nuclei of the heliozoans *Actinophrys* and *Actinosphaerium* (see Table 1). Nominally, they can all be regarded as ovular nuclei.

As for the central zone of the ovular nuclei of amoebae, it apparently contains the material of decondensed chromosomes. This zone usually produces a weak diffuse Feulgen reaction (Bogdanowicz 1930a, Liesche 1938, Kudo 1947, Heller and Kopac 1955, McClellan 1959).

In the life cycle of the Foraminiferida, ovular nuclei develop from the ordinary vesicular nuclei of agametes. The only nucleolus of the nucleus of *Allogromia* at first sharply hypertrophies and then fragments (Schwab 1974); probably, new peripheral nucleoli are also formed. When ovular nuclei enter mitosis, all the nonchromosomal material (karyolymph, nucleoli) that has accumulated in them is not included in the daughter nuclei but gets into the cytoplasm where it is resorbed. For instance, during the first division of the nucleus of the gamont of *Myxotheca*, all the chromosomes gather in the zone of a very small mitotic spindle at the periphery of the nucleus (Fig. 14*a*), then the envelope of the gigantic nucleus disintegrates (FØYN 1936). After mitosis, two minute vesicular nuclei are formed. These nuclei continue to divide (Fig. 14*b*, *c*). Similar phenomena have been observed in the gamonts of *Iridia* (LE CALVEZ 1938, CÉSANA 1971), the agamonts of

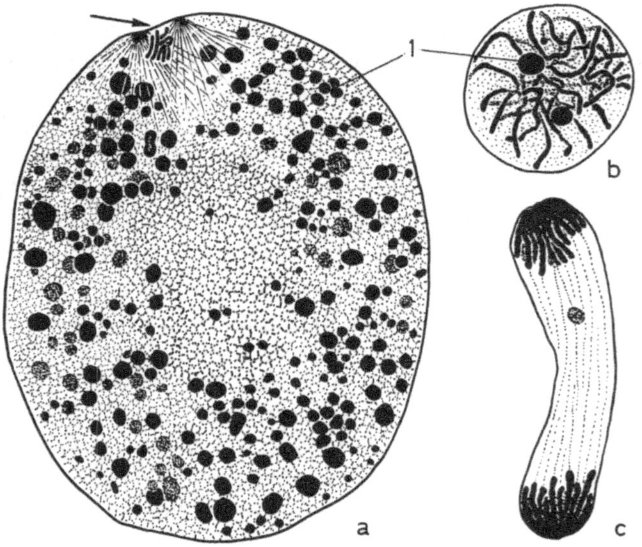

Fig. 14. Gamogony of the foraminiferan *Myxotheca* (after FØYN from GRELL 1956a). *a* first division of the nucleus in a gamont (the arrow indicates the mitotic spindle, *1* nucleoli), 1100 × ; *b* secondary small nucleus, 3000 × ; *c* mitosis of the secondary nucleus, 3000 × .

Allogromia (ARNOLD 1955, SCHWAB 1972, MCENERY and LEE 1976), and the gamonts of *Ovammina* (DAHLGREN 1964). The extrusion of not only nucleoli but also intranucleolar chromatin into the cytoplasm (when the ovular nucleus breaks down) has been shown in *Ovammina*, thus again confirming its extrachromosomal nature. Such a rejection of the nonchromosomal part of the nucleus is characteristic of not only ovular nuclei; it is also observed in the first mitotic division of some "large" vesicular nuclei, such as the nucleus of the gamont of the gregarine *Monocystis* (MULSOW 1911) or that of the microgametocyte of the coccidian *Coelotropha* (VIVIER and HENNERÉ 1964). The division of nuclei into chromosomal and nonchromosomal parts is observed also when the nuclei of the microgametes of coccidians are being formed (Fig. 9*f*).

"*Chromosomal*" *Nuclei*

This type of nuclei is characterized by the preservation of the condensed state of chromosomes in the nondividing nucleus. Usually, there is a nucleolus that is linked with a chromosome that decondenses when it enters the nucleolus. More rarely, the nucleolus does not exist. The karyolymph is abundant and is clearly delimited from condensed chromosomes. As a rule, the nuclear envelope does not undergo any complications.

As we have seen, the boundary between vesicular nuclei with reticular chromatin and chromosomal nuclei is arbitrary. While not discussing dinoflagellates for the time being (see Chapt. 4), it can be said that typical chromosomal nuclei (with nucleoli) are characteristic of two groups of eukaryotic protozoans: the Euglenida (Fig. 9*d*) and many Hypermastigida, e.g. *Trichonympha* (Fig. 9*e*), *Holomastigotoides* (Fig. 10*a*), etc. Chromosomal nuclei without a nucleolus are found among the micronuclei of ciliates, e.g. *Balantidium coli*, *Colpoda steini*, *Ichthyophthirius* (Fig. 10*d*) and some other species (Table 1). In this case, there is also no clear-cut boundary between "small" vesicular micronuclei with thin chromatin filaments (see, for instance, Fig. 10*c*) and the micronuclei of the "chromosomal" type.

"Spermal" Nuclei

"Spermal" nuclei are filled with compact chromatin, just as the nuclei of spermatozoa are. The chromatin mass is sometimes solid without any regions of karyolymph and sometimes "spongy", i.e. it is permeated with lacunae occupied by the karyolymph. Individual chromosomes, however, are indistinguishable. At the periphery of "spermal" nuclei, chromatin either adjoins the nuclear envelope or is separated from it by a narrow rim of karyolymph. Thus, there is a minimum amount of the karyolymph in these nuclei. There are no nucleoli in them, while the nuclear envelope is classic. As a rule, these nuclei are very small.

"Spermal" nuclei are characteristic of the microgametes of the Coccidia, Haemosporina (Fig. 10*b*), and some Gregarinia (Molon-Nobbot 1977, 1978). They are also found in some sporozoites, such as in the mammalian malaria parasites (Table 1). At these stages, the nuclei are usually elongated (according to the shape of a microgamete or a sporozoite), and chromatin is solid. Moreover, "spermal" nuclei include the micronuclei of many Ciliophora. It was believed earlier that all the micronuclei were spermal nuclei, but electron microscopy has disproved this. As we have seen, micronuclei include "small" vesicular nuclei and chromosomal nuclei, both of which have a substantial amount of the karyolymph. In true "spermal" micronuclei, chromatin is either solid (Fig. 10*e*) or, more frequently, spongy (Figs. 10*f*, 76*a*, 80*a*). Besides chromatin, they occasionally have inclusions in the form of spheres (Raikov 1975a), MTOCs, and, more rarely, persisting kinetochores (Fig. 10*f*). Although the chromatin of the micronucleus of *Tetrahymena* contains an aberrant set of histones (H_1 is absent and H_3 is modified, unlike in the macronucleus, see Allis *et al.* 1979, 1980), it has a typical nucleosomal structure (Gorovsky and Keevert 1975, Gorovsky *et al.* 1978).

In "spermal" nuclei, the level of transcription is apparently very low or even absent. No RNA has been discovered cytochemically in sporozoan gamete nuclei and ciliate micronuclei. However, the incorporation of ^{3}H-uridine into the micronuclei of some ciliates has been observed autoradiographically, although rather faintly (Pasternak 1967, Rao and Prescott 1967, Ammermann 1970, Murti and Prescott 1970, reviewed by O. N. Borchsenius 1975).

Apparently, ciliate micronuclei cannot be regarded as completely inert nuclei. They undoubtedly control certain functions, which are important in maintaining the life of the cell, not only during the sexual process (as is generally known) but also during the agamic period. This is borne out by the fact that the clones of *Paramecium bursaria*, which have spontaneously lost the micronucleus, are less viable and age more quickly than the normal ones (GOLIKOVA 1978). Moreover, experimental destruction of the micronucleus (by ultraviolet or laser microbeam irradiation) proved lethal for most species of *Paramecium* (FOKIN and BORCHSENIUS 1978, FOKIN and OSSIPOV 1981). Some viable amicronucleate clones were obtained with that technique only in *P. tetraurelia*, *P. bursaria*, and *P. putrinum*. Such clones of *P. tetraurelia*, however, showed reduced vitality (NG 1980), and fragments (pseudomicronuclei), which can partially replace the micronucleus functionally, always separated from the macronucleus in the amicronucleate specimens of *P. bursaria* and *P. putrinum* (SCHWARTZ 1958, FOKIN 1979). Pseudomicronuclei are also formed in the amicronucleate *Stylonychia* (AMMERMANN 1970). Amicronucleate cells of *Paramecium caudatum*, obtained by microsurgical elimination of the micronucleus, proved less viable than normal ones; their viability was fully restored upon renucleation with a normal micronucleus (FUJISHIMA and WATANABE 1981). For reviews of the problem, see OSSIPOV (1981) and FOKIN and OSSIPOV (1981).

An interesting fact is that the spermal nuclei of the microgametes of the Coccidia are formed from larger and looser nuclei of the multinucleate microgametocyte by the unequal division of every nucleus into two parts (Fig. 9*f*). All the chromatin is gathered in one part, which buds off the nucleus and becomes the nucleus of the microgamete. All the nonchromosomal material gets into the other part, which degenerates together with the residual body of the microgametocyte (SNIGIREVSKAYA 1969b, HAMMOND *et al.* 1969, MCLAREN 1969, PELSTER and PIEKARSKI 1971, MEHLHORN 1972b, SCHOLTYSECK *et al.* 1972, FERNANDO 1973, SÉNAUD *et al.* 1980).

Other Types of Nuclei

Some clearly defined types of protozoan nuclei have not been considered in this chapter, as they will be dealt with in other chapters. One such type is the highly polyploid radiolarian nuclei, which have many identical chromosome sets; they contain very much DNA and are capable of undergoing multiple division (Chapter 6). Another type is the ciliate macronuclei, which are also very rich in DNA and form numerous nucleoli (except the macronuclei of some primitive ciliates); they are often classified as the nuclei of a special massive type (Chapter 8). The third type is the very primitive nuclei of the Dinoflagellida, the dinokarya (Chapter 4).

2. Mitosis in Eukaryotic Protozoa

Mitosis is the principal type of division of protozoan nuclei, just as of metazoan nuclei. This assumption, first clearly substantiated by BĚLAŘ (1926a), still holds good. The main characteristic of mitosis is that the two copies of each replicated chromosome (its two chromatids) regularly migrate into the two daughter nuclei.

Consequently, daughter nuclei receive identical complete sets of chromosomes. Details of the mitotic mechanism may vary to a considerable extent, however, particularly in lower eukaryotes, including Protozoa.

In Metazoa, the nuclear envelope usually disintegrates and the nucleolus disappears in the course of mitosis; the spindle is generally well developed and has centrioles with astral rays at its poles. In Metaphyta, centrioles usually cannot be discerned and the spindle is blunt.

In Protozoa, the degree of development of the spindle, the centrioles or structures that are functionally similar to them, and the behaviour of nucleoli during mitosis vary greatly. The nuclear envelope is preserved very frequently during the whole mitosis. Nevertheless, whatever the variant is, the two chromatids of each replicated chromosome migrate into the daughter nuclei, i.e. the fundamental characteristic of mitosis is retained.

During the pre-Feulgen period, different forms of amitosis had often been described in protozoans (reviewed by Bělař 1926a). Later, nearly all these cases proved to be disguised mitosis. True amitosis (i.e. the division of a nucleus in the interphasic state resulting in the irregular distribution of chromosomes) is encountered in protozoans only as a facultative process concurrent with mitosis; amitosis is a more or less pathological way of nuclear division. In *Acanthamoeba*, for example, amitosis is induced by certain growth media (Band and Mohrlok 1973). In *Euglena*, amitosis is found only in cultures, and even then only in a small number of cells, but never in natural populations; often, the cytoplasm does not divide in amitosis, thus giving rise to binucleate cells (Leedale 1959a). Division of the polyploid macronuclei of ciliates is now no longer covered by the concept of "amitosis", since either whole chromosome sets (genomes) or free genes segregate during the division of the macronucleus (see Chapter 8).

Types of Mitosis

All the main types of mitosis found in protozoans can now be described in ultrastructural terms. This makes it necessary to revise the classification of protozoan mitoses that was based almost exclusively on the data of light microscopy only 15 years ago (Raikov 1967a).

Using data that pertained primarily to amoebic mitosis, Chatton (1910) proposed that a distinction be made between *promitosis* (mitosis in which the nuclear envelope is preserved and the nucleolus, to which the role of an intranuclear centrosome was ascribed, is pinched in two), *mesomitosis* (the nuclear envelope is preserved, "centrioles" are intranuclear, and the nucleolus dissolves), and *metamitosis* (the nuclear envelope and the nucleolus dissolve, the centrosomes are extranuclear). Alexeieff (1913) proposed a more detailed classification comprising 12 types of mitosis. Pussard (1973) tried to update the classification proposed by Chatton, at least for Sarcodina; he excluded the criterion of the mitotic centre position from the definition of the mitotic types and proceeded only from the behaviour of the nuclear envelope and the nucleolus.

Bělař (1926a) was quite justified in criticizing Chatton's and Alexeieff's classifications. Bělař indicated that their terms "chromatin", "centrosome", "centriole", and "nucleolo-centrosome" were too vague. Instead, he suggested that a distinction be made only between three types of mitosis: *eumitosis*, *paramitosis*, and *cryptomitosis* (a mixed group of mitoses that could not be precisely analyzed then). GRASSÉ (1952) distinguished between the two main types of mitosis: *orthomitosis* (corresponding to BĚLAŘ's eumitosis) and *pleuromitosis* (BĚLAŘ's paramitosis). According to GRASSÉ, orthomitosis involves an equatorial plate, and chromosomes are attached to the chromosomal fibres of the spindle. In pleuromitosis, there is no distinct equatorial plate, the chromosomal kinetochores are attached to the nuclear envelope, which is preserved during mitosis, and the central spindle is outside the nucleus.

The concept of pleuromitosis and, in particular, the structure of kinetochores and their relationships with the nuclear envelope and the spindle have been clarified through recent electron microscopic studies (HOLLANDE and VALENTIN 1968a, HOLLANDE 1972). As a result, the difference between orthomitosis and pleuromitosis diminished, and certain types of mitosis that fitted into neither of the two categories were found. HOLLANDE (1972) proposed the first classification of mitotic types based largely on ultrastructural data. He discriminated between five types of mitosis in protozoans: in one type, the nuclear envelope disintegrates and the spindle is of the metazoan type; in the other four types, the nuclear envelope is preserved (*cryptomitoses*). In classifying the "cryptomitoses", HOLLANDE considers where (inside or outside the nucleus) the spindle is located and where (in the nucleus or in the cytoplasm) the centres of organization of the spindle are found. Finally, HOLLANDE takes the shape of the intranuclear spindle into account: whether it looks like a monaxonic-symmetrical bipolar figure or two separate half-spindles often arranged at an angle to one another. According to these criteria, HOLLANDE distinguishes between cryptopleuromitosis (former pleuromitosis with the spindle located entirely outside the nucleus), cryptomitosis with intranuclear half-spindles connected to intranuclear centres, cryptomitosis with intranuclear half-spindles bound to extranuclear centres, and cryptoorthomitosis with a bipolar symmetrical intranuclear spindle. VIVIER and VICKERMAN (1974) also use this classification, though with certain specifications.

An analysis of recent electron microscopic descriptions of mitosis indicates that some of HOLLANDE's categories are actually heterogeneous. This is particularly true of his cryptoorthomitosis in which the mitotic centres are either intranuclear or extranuclear, that is, when they are not absent. Therefore, this book sets forth a somewhat more detailed classification of protozoan mitoses based on HOLLANDE's criteria but with certain corrections. To begin with, the term "cryptomitosis" is not suitable for any mitosis in which the nuclear envelope is preserved, since BĚLAŘ (1926a) has already used this term with a different meaning. Instead, we use the term *closed mitosis*, while mitosis with the dispersion of the nuclear envelope is referred to as *open mitosis* (terms coined by JENKINS 1967); mitosis in which the envelope dissolves only at the poles of the spindle, producing "polar gaps", is called *semiopen mitosis* (after TAYLOR 1976). We also propose that the term "pleuromitosis" cover not only mitosis with an extranuclear spindle lateral to the nucleus but also mitoses with a sharply asymmetrical intranuclear spindle (for instance, mitoses with half-spindles arranged at an angle to one another in the metaphase or anaphase). Accordingly, we distinguish six categories of mitosis in protozoans:

Open mitosis
1. Open orthomitosis (eumitosis)

Semiopen mitoses
2. Semiopen orthomitosis
3. Semiopen pleuromitosis

Closed mitoses with an intranuclear spindle
4. Intranuclear pleuromitosis
5. Intranuclear orthomitosis

Closed mitosis with an extranuclear spindle
6. Extranuclear pleuromitosis

TAYLOR (1976) uses similar criteria and discriminates between five types of mitosis in the flagellates. But he does not subdivide the groups of semiopen and closed intranuclear mitoses into orthomitotic and pleuromitotic types. Detailed reviews of the distribution of different types of mitosis among lower eukaryotes (protozoans, algae, fungi) were written by KUBAI (1975) and HEATH (1980).

Open Orthomitosis

This type of mitosis corresponds, in all the principal criteria, to the classic mitosis of metazoans and can therefore also be called eumitosis. In such mitosis, the nuclear envelope disintegrates, nucleoli disappear, while the spindle is bipolar, axially symmetrical, and consists of continuous and chromosomal fibres; the centres organizing the spindle are located in the cytoplasm (their nature, however, may differ). In the metaphase, the chromosomes form a distinct equatorial plate and are attached with their kinetochores to the chromosomal fibres of the spindle. Open orthomitosis has been observed in a number of Phytomastigophorea (Chrysomonadida, Prymnesiida, Cryptomonadida, Prasinomonadida), Sarcodina (amoebae, heliozoans), Labyrinthomorpha, and certain Gregarinia.

Gregarinia. According to light microscopic data, mitoses of the gamogony in the gregarines *Monocystis* (MULSOW 1911, BĚLAŘ 1926a) and *Stylocephalus* (GRELL 1940) are the closest to the metazoan type. In the prophase of *Stylocephalus*, a pair of "centrosomes" with astral rays (Fig. 15*a*) gradually migrate to the opposite poles of the nucleus (Fig. 15*b*, *c*). At the same time, the central spindle is formed between the "centrosomes" (Fig. 15*b*). The nuclear envelope dissolves in the late prophase. By the beginning of the metaphase, the spindle becomes axially symmetrical and the condensed chromosomes are arranged on the equatorial plate (Fig. 15*c*). In the anaphase, the daughter chromosomes move to the poles (Fig. 15*d*, *e*); their number is only 4 in *S. longicollis*, which is a haplont, and all of them have unequal arms (Fig. 15*f*). The spindle greatly elongates (Fig. 15*e*); in the late anaphase, the central granules (centrioles?) inside the polar "centrosomes" are doubled (Fig. 15*f*). In the telophase, the nuclear envelope and nucleoli are formed again (Fig. 15*g*, *h*); the doubled "centrosomes" slightly separate. The nucleus of *Monocystis* divides in a similar way.

DESPORTES (1970) studied mitosis in *Stylocephalus* by electron microscopy. She has shown that the cytoplasmic "centrosomes" consist, as a rule, of granular material without true centrioles in it. Centrioles are formed there only in the last

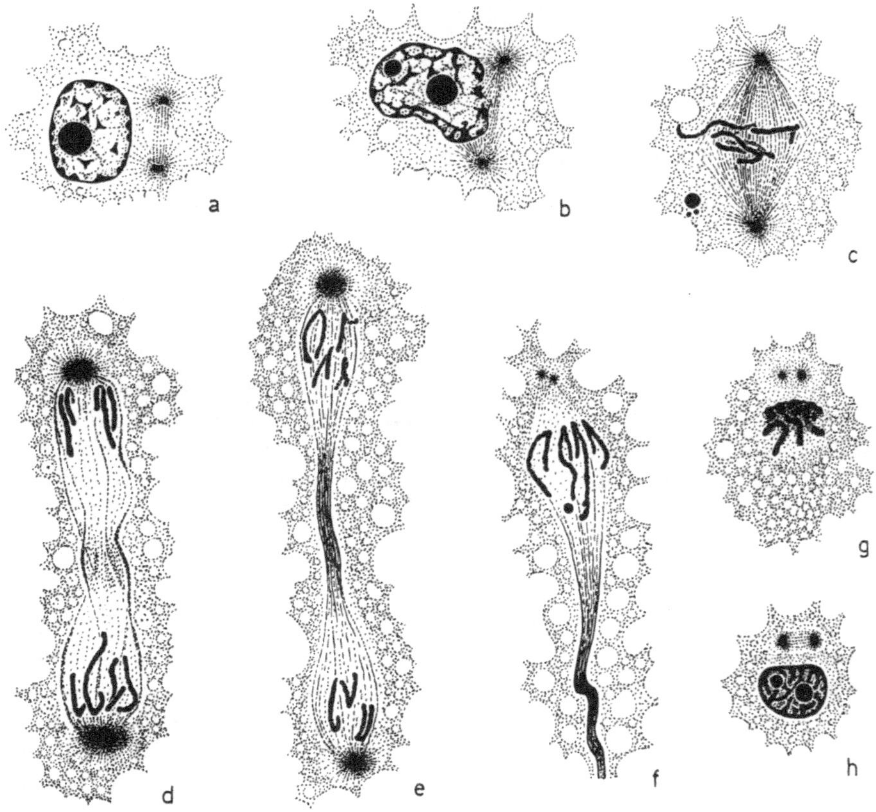

Fig. 15. Gamogonic mitosis in the gregarine *Stylocephalus*, 2700 × (after GRELL 1956a). *a* and *b* prophase; *c* metaphase; *d* and *e* anaphase; *f*, *g*, and *h* telophase.

mitoses of the male gamogony. The nuclear envelope disintegrates in the prometaphase, and microtubular cones emerging from the granular polar bodies join, forming a bipolar spindle. The fragments of the nuclear envelope, however, do not scatter throughout the cytoplasm; instead, they settle on the surface of the chromosomes, each of which is encased in a karyomere that splits into two daughter karyomeres in the anaphase. Only the chromosomal kinetochores of a typical trilaminar structure are devoid of the membranous envelope. The bundles of chromosomal microtubules are attached to the kinetochores. The karyomeres of separate chromosomes fuse in the telophase, re-forming the envelope of the daughter nucleus.

Sarcodina and Labyrinthomorpha. Mitosis with the disintegration of the nuclear envelope and with distinct centrosomes at the spindle poles was observed in the primitive heliozoan *Dimorpha mutans* by means of light microscopy (BĚLAŘ 1926a). In this case, the centrosome is always present in the cytoplasm; it has the central granule and astral rays. When the heliozoan is transformed into the flagellate form, the central granule of the centrosome becomes the basal body of the flagellum; hence, the central granule is a true centriole. The nucleus of

Cinetidomyxa chattoni, a form related to lower heliozoans, divides in a similar way (CACHON and CACHON-ENJUMET 1964a).

Open orthomitosis (with the centrioles at the poles) was examined by electron microscopy in two *Labyrinthula* species (PERKINS and AMON 1969, STEY 1969, PERKINS 1970); in a third species, the nuclear envelope disperses only partially (PORTER 1972). In all these species, the centrioles are formed *de novo* prior to the first division of sporulation; a pair of perpendicular centrioles (a diplosome) appears at each of the two opposite poles of the nucleus. In the prometaphase, the nuclear envelope falls into fragments, producing flattened cisternae, and the bundles of microtubules "grow" into the nucleus from the poles toward the equator. In the telophase, the cisternae fuse, re-forming the nuclear envelopes of the daughter nuclei; the centrioles remain outside the nucleus and are involved in the subsequent mitoses, after which they become the basal bodies (kinetosomes) of flagella in the zoospores. Apart from the protozoans, similar mitosis with the dispersion of the nuclear envelope and the centrioles at the poles was found in a number of myxomycetes, e.g. at the haploid amoeboid stage of the cycle in *Physarum* (reviewed by HEATH 1980).

Open orthomitosis with more or less distinct cytoplasmic centres was also investigated by light microscopy in small amoebae: *Hartmanella* (ARNDT 1924, BĚLAŘ 1926a, SINGH 1952, RAY and HAYES 1954, PAGE 1967a), *Acanthamoeba* (COMANDON and FONBRUNE 1937, SINGH 1952, PUSSARD 1966, PAGE 1967b, SAWYER and GRIFFIN 1971), *Rosculus* (HAWES 1963, PAGE 1974), and others. In *Acanthamoeba klitzkei*, for instance, the cell centre in the interphase looks like a centrosphere, i.e. a blurred zone of homogeneous cytoplasm without astral rays. In the prophase, the centrosphere divides in two and the nucleolus dissolves; in the metaphase, the two daughter centrospheres (with the astral rays beginning to show) are arranged at the spindle poles, the nuclear envelope disappears, and the chromosomes form an equatorial plate. Other amoebae of this group were described as having, in their "centrospheres" (also in the living cells), paired central granules (centrioles?) and occasionally astral rays or peculiar radial rods (COMANDON and FONBRUNE 1937, RAY and HAYES 1954, PUSSARD 1966, SAWYER and GRIFFIN 1971). The nature of most of these formations remains obscure, however, since mitosis in the amoebae of this group has not been studied by electron microscopy. The ultrastructure of the centrosphere was studied so far only during the interphase in *Acanthamoeba castellanii* (BOWERS and KORN 1968) and in two other *Acanthamoeba* species (PUSSARD and PONS 1978, WILLAERT *et al.* 1978b). This is a rod-like or disc-shaped electron-dense body, up to 1 μm in size, located in the cytoplasm near the nucleus and surrounded by dictyosomes; the body engenders cytoplasmic microtubules that are oriented more or less radially (Fig. 16*a*). Its ultrastructure is not typical of a centriole, and it is not connected with centrioles not found in *Acanthamoeba*. It might be designated by a more general term, i.e. the "microtubule organizing centre" (MTOC), as suggested by PICKETT-HEAPS (1969). In this case, the MTOC is located in the cytoplasm, although it can be intranuclear in other protozoans.

Data obtained by light microscopy show that open orthomitosis with cytoplasmic centres occurs also in certain testate amoebae, namely, in *Cochliopodium* (SAEDELEER 1930) and *Gocevia* (PUSSARD 1965).

Open, but acentric, orthomitosis occurs also in large fresh-water amoebae, *Amoeba proteus*, *Chaos carolinensis*, and *C. illinoisensis*. No centrosomes, centrioles, or MTOCs at the spindle poles were found in this case. Mitosis in these three species was studied by both light microscopy (HINCHEY 1937, LIESCHE 1938, KUDO 1947, McCLELLAN 1959) and electron microscopy (ROTH *et al.* 1960, ROTH and DANIELS 1962, DANIELS and ROTH 1964, ROTH 1967). Here, the spindle is blunt with parallel (not converging toward the poles) fibres. Small but very numerous (500–600 in *A. proteus*, according to LIESCHE) chromosomes form a distinct equatorial plate. Bundles of spindle microtubules are formed in the nucleus only after the first gaps appear in the nuclear envelope (possibly, the karyolymph and the cytoplasm must mix, i.e. the "mixoplasm" must be formed for microtubules to polymerize). The microtubules are sensitive to low temperature, which causes their depolymerization (ROTH 1967). GOODE and ROTH isolated the mitotic apparatus of *Ch. carolinensis* and showed that the spindle microtubules of this amoeba are stabilized by divalent cations and depolymerize in the presence of EDTA (GOODE 1965, GOODE and ROTH 1969).

In these three species of amoebae, the bundles of microtubules grow in the direction from the chromosomes to the poles, i.e. the chromosomal kinetochores, whose morphology was described with respect to *Ch. illinoisensis* (DANIELS and ROTH 1964), play the role of MTOCs in this case. However, the continuous bundles of microtubules that pass between the chromosomes are also formed in the spindle. Peripheral nucleoli disappear even prior to the fragmentation of the nuclear envelope. In the metaphase-anaphase, the fragments of the envelope are large and arranged at the periphery of the spindle. In the late anaphase, they begin to fuse again, forming the envelopes of the daughter nuclei. In the telophase, the continuous fibres of the spindle that connect the two daughter nuclei are resorbed in the cytoplasm (ROTH and DANIELS 1962).

In *Amoeba proteus*, the honeycomb layer, which underlies the nuclear envelope of the nondividing organisms, and the nuclear RNP helices disappear even before the nuclear envelope fragments. Both structures reappear only in the early interphase (ROTH *et al.* 1960, FELDHERR 1965, 1968b).

Phytomastigophorea. Open orthomitosis has been studied by electron microscopy in the prymnesiid *Prymnesium parvum* (MANTON 1964). In this case, the nuclear envelope falls into fragments in the late prophase, producing tubular cisternae arranged not only at the periphery of the spindle but also within it. At the same time, a spindle made of microtubules 20–24 nm in diameter is formed.

Fig. 16. Ultrastructure of a MTOC (*a*) and semiopen mitoses (*b*–*e*). *a* cytoplasmic MTOC (*1*) with adjacent microtubules (arrows) near the nucleus of *Acanthamoeba castellanii*, 48 000 ×; *b* metaphase spindle pole of *Chlamydomonas* (*1* microtubules, *2* polar gap in the nuclear envelope), 40 000 ×; *c* centrocone (*1*) and half-spindle (*2*) in the anaphase of semiopen pleuromitosis in *Eimeria necatrix* (*3* centrioles, *4* nuclear envelope, *5* kinetochore, *6* chromosome, *7* nucleolus), 34 000 ×; *d* centrocone in the telophase of mitosis in *E. necatrix* (same designations as in *c*), 49 000 ×; *e* early anaphase of schizogonic mitosis in *Plasmodium gallinaceum* (*1* "centriolar plaques", *2* kinetochores, *3* microtubules, *4* nuclear envelope), 71 000 ×. *a* after BOWERS and KORN (1968); *b* after JOHNSON and PORTER (1968); *c* after DUBREMETZ (1971); *d* after DUBREMETZ (1973); *e* after AIKAWA and BEAUDOIN (1968).

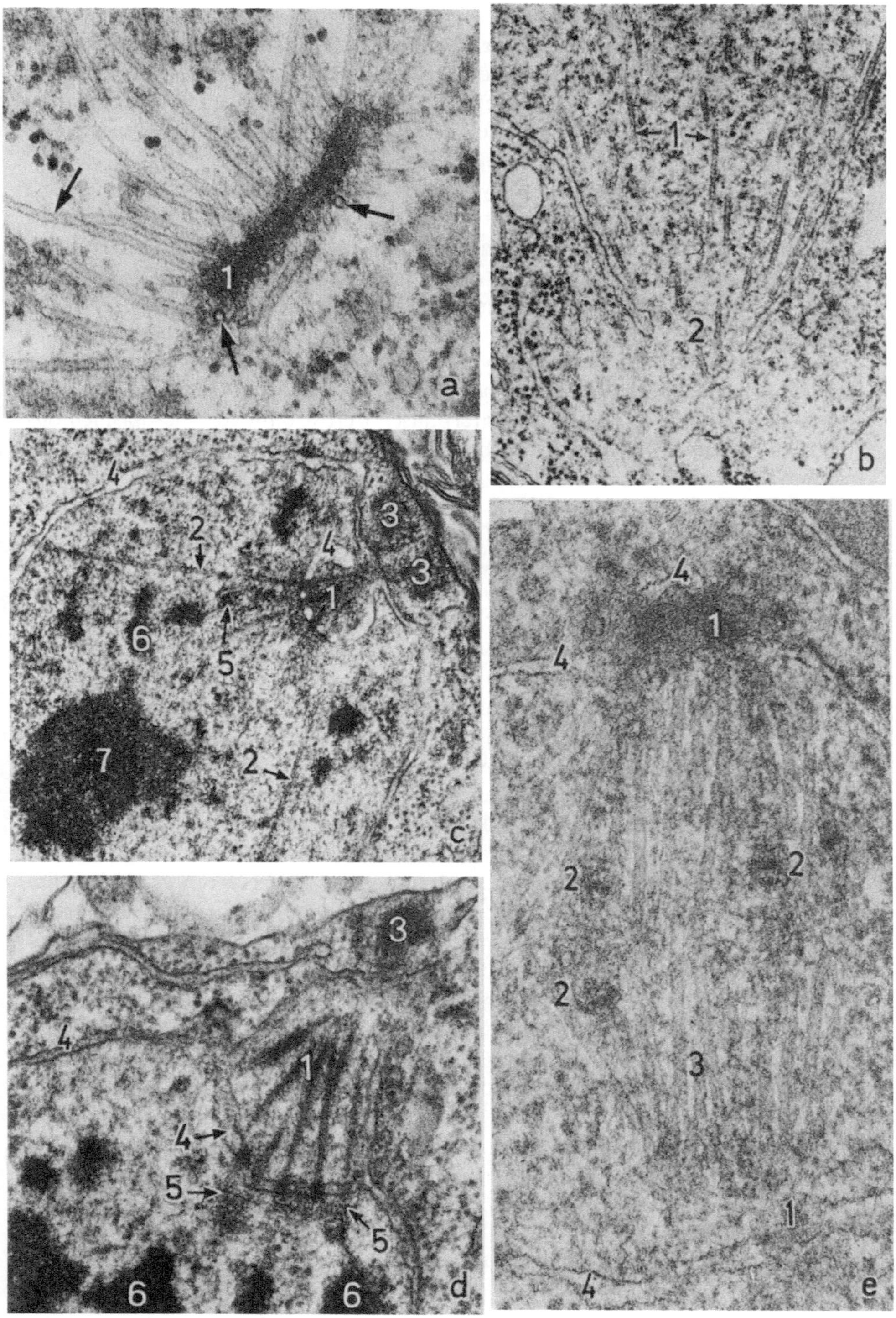

Fig. 16

Flagellar kinetosomes redouble and separate prior to nuclear division; however, pairs of daughter kinetosomes, though located not far from the spindle poles, are never found at the poles. The poles are occupied by zones of homogeneous cytoplasm without inclusions. Therefore, although homologues of the centrioles (the kinetosomes) are present in the cell, the spindle must be considered acentric. It resembles the mitotic spindle of higher plants in this respect.

The spindle of *Prymnesium* is symmetrical; the continuous bundles of the microtubules are clearly visible, but no distinct chromosomal fibres have been seen. The equatorial plate is very dense. The nucleolus disappears during mitosis. In the telophase, the envelopes of the daughter nuclei are formed *de novo*, while the central spindle remains in the cytoplasm and is resorbed there. The nucleolus is the last to be formed.

The ultrastructural relationships between the flagellar kinetosomes and the spindle of the chrysomonad *Ochromonas* are noteworthy (Slankis and Gibbs 1972, Bouck and Brown 1973). In the interphase, a cross-banded microfilamentous ribbon (a rhizoplast) extends from the bases of a pair of kinetosomes toward the nucleus. Cytoplasmic microtubules extend from the surface of the rhizoplast toward the back of the body. The kinetosomes replicate and separate before mitosis; each pair of daughter kinetosomes has its own rhizoplast. The nuclear envelope breaks down in the prophase, an exception being the zones adjacent to the chloroplasts. The spindle is formed not between the pairs of flagellar kinetosomes but between their derivatives, the rhizoplasts. Hence, the latter represent cytoplasmic MTOCs in *Ochromonas*. The equatorial plate is well defined, while the chromosomes are small, compact, and without clearly visible kinetochores. Three categories of microtubules are observed in the spindle (Tippit *et al.* 1980): polar (in which one end is attached to a MTOC, while the other end is free), free (in which both ends are free), and chromosomal (which are attached to a MTOC with one end and to a chromosome with the other). Continuous microtubules connecting one MTOC with the other were not found at all. Therefore, the central spindle of *Ochromonas* consists of two interdigitating half-spindles made of polar microtubules; each half-spindle starts from its MTOC. The central spindle also contains free microtubules in its middle part. In the course of the anaphase, the half-spindles apparently move apart due to the mutual sliding of the polar microtubules, which also grow in length, thus causing the elongation of the whole central spindle. Conversely, the chromosomal bundles of microtubules shorten and then disappear. Consequently, the daughter groups of chromosomes in the telophase are pulled almost up to the polar rhizoplasts. When the envelopes of the daughter nuclei are re-formed, the central spindle remains outside and is resorbed.

Acentric open orthomitosis (with a broad blunt spindle and numerous small chromosomes) that resembles mitosis in *Amoeba* and *Chaos* was observed in the cryptomonads *Chilomonas* (Hollande 1942) by light microscopy and *Chroomonas* (Oakley and Dodge 1973, 1976) and *Cryptomonas* (Oakley and Bisalputra 1977, Oakley and Heath 1978) by electron microscopy. In *Chroomonas*, flagellar kinetosomes that redouble prior to mitosis are surrounded by a zone of cytoplasmic microtubules in the prophase and are apparently involved in the early stages of spindle induction. Once the nuclear envelope is disrupted, the microtubules penetrate into the nucleus, growing from the poles (in contrast to mitosis in amoebae; see p. 78). In the metaphase, however, the kinetosomes are no longer connected with the spindle and no MTOCs can be seen. In *Chroomonas* as well as *Cryptomonas*, the equatorial plate is dense, since single chromosomes fuse in a uniform disc, while the microtubules of the central spindle pass through holes in it. The chromosomal bundles of microtubules are attached to the surface of the chromosomal disc at many points. In the anaphase, the daughter chromosome plates are also continuous. The nuclear envelope re-forms around these chromosomal masses in the telophase.

Just as in *Ochromonas* (see p. 80), three types of microtubules were found in *Cryptomonas*: polar, free, and chromosomal ones. It is quite likely that the anaphasic elongation of the central spindle is accomplished by the mechanochemical sliding of the polar microtubules along one another and along the free microtubules as well as by the growth of the polar microtubules in length (Oakley and Heath 1978).

Finally, open orthomitosis is typical of certain (but not all) lower green flagellates from the order Prasinomonadida: *Platymonas* (Stewart *et al.* 1974), *Pyramimonas* (Pearson and Norris 1975), and *Heteromastix* (Mattox and Stewart 1977). All three forms differ in the ultrastructure of the spindle poles: *Pyramimonas* has a pair of centrioles (they are at the same time kinetosomes of the flagella) at each pole; the kinetosomes of *Heteromastix* lie in the vicinity of the poles, but not in the focus at which the microtubules meet (as in *Prymnesium*); in *Platymonas*, the spindle is formed in the prophase between the loose granular ends of two rhizoplasts that join the kinetosomes with their other ends. Therefore, the latter form has cytoplasmic MTOCs similar to those in *Ochromonas*. In *Platymonas*, however, the rhizoplasts break down in the metaphase so that only granular spheres remain at the spindle poles; new rhizoplasts are formed in the telophase.

In all three cases, the nuclear envelope completely disintegrates. The microtubules of the central spindle behave differently in the telophase: in *Pyramimonas* and *Heteromastix*, the spindle stretches out in length, forcing the daughter nuclei far apart; in *Platymonas*, a plate perpendicular to the telophase spindle (a phycoplast) is formed from the microtubules between the daughter nuclei.

Apart from Protozoa, open orthomitosis of a similar type is observed in multicellular green algae (*Ulothrix*, *Coleochaete*, etc.) in which the centrioles are usually located at the spindle poles or near them, while kinetochores, occasionally with a complex structure, are differentiated on the chromosomes (Leedale 1974, Kubai 1975, Heath 1980). In other green algae, such as *Chara*, the spindle is acentric (Pickett-Heaps 1967). This type of open orthomitosis is very close to mitosis in higher plants. Finally, in diatoms, the central spindle is formed between two dense plaques (resembling the MTOCs in *Acanthamoeba*) that are always present in the cytoplasm; once the nuclear envelope breaks down, the spindle enters the nuclear area, with the MTOCs at the poles and the central spindle in axial position, and chromosomal fibres appear at the periphery of the spindle (Manton *et al.* 1969, Tippit *et al.* 1975, Tippit and Pickett-Heaps 1977, Pickett-Heaps *et al.* 1978a, 1978b, 1980a, 1980b, etc.). Open orthomitosis is found also among lower fungi, such as Protostelidia (Heath 1980) and Plasmodiophorea (Garber and Aist 1979).

Semiopen Orthomitosis

Semiopen orthomitosis is characterized by the preservation of the nuclear envelope around the whole spindle except the polar zones, where large fenestrae or gaps appear in the envelope. Mitotic centres, if expressed, are in the cytoplasm; microtubules induced by them penetrate into the nucleus through the polar gaps in the nuclear envelope and form a bipolar symmetrical spindle, a large part of which is consequently intranuclear while the poles are extranuclear. Just as in open orthomitosis, the spindle consists of continuous and chromosomal fibres. As a rule, the nucleolus dissolves during mitosis. The extent of chromosome condensation and the expression of the equatorial plate may vary.

This type of mitosis is typical of three unrelated protozoan groups: the green flagellates of the orders Volvocida and Chloromonadida, certain heliozoans, and, in another variant, several gregarines. However, semiopen orthomitosis may also occur in other protozoan groups, though this can be definitely stated only when ultrastructural data are available.

Volvocida and Chloromonadida. Semiopen orthomitosis was studied by electron microscopy in the volvocids *Chlamydomonas reinhardi* (JOHNSON and PORTER 1968, COSS 1974), *Ch. moewusii* (TRIEMER and BROWN 1974) and *Dunaliella bioculata* (MARANO 1976). It appears to be similar in all three species. Prior to mitosis, the two flagellar kinetosomes of the cell start to migrate in opposite directions; at this stage, they are in close contact with the surface of the nucleus (MARANO 1976). Two disc-shaped plaques of electron-dense material (extranuclear MTOCs) differentiate at the outer surface of the nuclear envelope adjacent to the kinetosomes. The kinetosomes soon replicate and become paired again. They serve as the basal bodies of flagella and are no longer involved in mitosis.

In the prophase, all the pores of the nuclear envelope expand and look "open"; the chromosomes condense gradually and the nucleolar material disperses throughout the karyolymph. The extranuclear MTOCs migrate along the outer surface of the nuclear envelope after the diverging pairs of kinetosomes until they occupy opposite sides of the nucleus. Then (or perhaps somewhat earlier) the nuclear envelope disintegrates in the regions adjacent to the MTOCs, and polar gaps 0.2–0.5 μm in diameter appear in the envelope. The bundles of microtubules protruding from the MTOCs penetrate through these gaps into the nucleus. The MTOCs swell and are converted into zones of homogeneous cytoplasm that plug the polar gaps in the nuclear envelope (Fig. 16*b*). A Golgi body is usually adjacent to each such zone.

In the metaphase, the spindle is bipolar, occasionally crescent shaped, and consists mainly of chromosomal microtubules; the equatorial plate is very compact. A pair of kinetosomes still remains near each pole, although at the side (COSS 1974, TRIEMER and BROWN 1974, MARANO 1976). Kinetochores are clearly visible on the chromosomes in the anaphase (TRIEMER and BROWN 1974); kinetosomes "lag" more and more behind the poles of the elongating division figure until they entirely lose contact with it in the telophase. Only one bundle of continuous microtubules can be discerned in the telophasic spindle; the chromosomes fuse, forming a compact mass. The polar gaps close gradually. At the end of the telophase, both daughter nuclei separate from the middle part of the spindle, which is resorbed in the cytoplasm; nucleoli are formed in them *de novo* in the early interphase. Therefore, mitosis in Volvocida may be described as "semiopen" with very obscure cytoplasmic MTOCs, which later look like amorphous spheres. The kinetosomes are not directly involved in the formation of the spindle, although they may induce the formation of extranuclear MTOCs in the early prophase.

Mitosis in the chloromonads *Gonyostomum* and *Vacuolaria* was investigated by both light microscopy (HOVASSE 1945, MIGNOT 1967, HEYWOOD 1974, HEYWOOD and GODWARD 1974) and electron microscopy (HEYWOOD and GODWARD 1972, HEYWOOD 1978). In *Vacuolaria*, the microtubules of the spindle are formed in the cytoplasm near the kinetosomes of the flagella and penetrate into the nucleus through the polar gaps that appear in the nuclear envelope; otherwise, the nuclear envelope is preserved until the telophase. Later, the spindle becomes acentric (the kinetosomes lose contact with it). In the prophase, the chromosomes are large, distinctly bichromatidal, and the nucleolus disperses. In the metaphase, a clearly visible equatorial plate is formed, and kinetochores resembling amorphous discs are seen on the chromosomes. In the telophase, the nuclear envelope becomes reorganized: new envelopes around both daughter nuclei are formed inside the old envelope surrounding the whole spindle by the fusion of the membrane-bound vesicles that pinched off into the nucleus from the old envelope. Only then does the old envelope disintegrate, releasing both daughter nuclei (HEYWOOD 1978). Such an "intranuclear" or "indirect" mode of the telophasic reorganization of the nuclear envelope was also found in *Actinophrys* (p. 84) as well as in the closed orthomitosis of the micronuclei of ciliates. It differs essentially from the "direct" way typical of mitosis in *Chlamydomonas* in which the envelope of the parent nucleus simply constricts in two in the telophase.

Besides Protozoa, the above variety of semiopen orthomitosis was found by electron microscopy in a number of multicellular green algae (*Ulva*, *Microspora*, *Spirogyra*, *Tetraspora*, etc.), brown algae (*Fucus*, *Dictyota*, *Zonaria*, etc.), and red algae (*Membranoptera*, *Polysiphonia*, etc.) (see the reviews by LEEDALE 1970, 1974, KUBAI 1975, HEATH 1980). Such mitosis occurs also in lower fungi, e.g. chytrids (*Harpochytrium*, *Phlyctochytrium*) and myxomycetes (*Sorosphaera*, *Polysphondilium*, etc.) (see the reviews by KUBAI 1975, FULLER 1976, HEATH 1975, 1980, 1981). However, the structure of the mitotic centres in the above organisms may vary greatly from the distinct pairs of centrioles at the spindle poles in *Fucus* or *Harpochytrium* through the disc-shaped extranuclear MTOCs in *Polysphondilium* (ROOS 1975) to the centriole-like bodies that appear for a short time near the spindle pole in *Ulva*. Apparently, the presence or absence of the centrioles (kinetosomes) at the poles is not fundamentally important in characterizing this group of mitoses; it is likely that even if the centrioles are present at the poles, the spindle is organized not by them but by their derivatives of the MTOC type. Therefore, some authors believe that the association of the kinetosomes (centrioles) with the poles of the division figure ensures the regular segregation of the kinetosomes (centrioles) between the two daughter cells by the mitotic apparatus and not spindle induction (PICKETT-HEAPS 1969, 1971).

Heliozoea. Progamic mitosis of the heliozoan *Actinophrys sol*, which occurs when the specimen (in a cyst) divides into two gamonts, was described in the classic work of Bělař (1922). Recently, it was studied again with the electron microscope by Mignot (1979) and, although in less detail, by Patterson (1979). This mitosis is semiopen orthomitosis (Bělař believed it to be closed): the nuclear envelope is preserved, but it becomes perforated with many small holes at the poles during the metaphase and anaphase. Large centrospheres (polar caps) consisting of finely vesicular cytoplasm are adjacent to the poles (Fig. 17*a*). These are diffuse cytoplasmic MTOCs; the bundles of microtubules are formed in them and penetrate into the nucleus through the perforations in the nuclear envelope (Fig. 17*b*) to produce an axial bipolar spindle whose middle part is located in the nucleus, while the poles are in the cytoplasm. The spindle is acentric and the ends of the microtubules do not converge at any formed body but disappear within the centrosphere. Certain microtubules are attached to the chromosomes arranged in a typical equatorial plate; the material of the peripheral nucleoli disperses and is visible at both sides of the chromosome plate (Fig. 17*a*). In the anaphase, the daughter chromosomes migrate to the poles, the perforations in the nuclear envelope widen, the number of the microtubules that penetrate into the nucleus increases, and many microtubules attach themselves to the nuclear envelope on the outside (Fig. 17*c*). Just as in *Vacuolaria*, the nuclear envelope of *Actinophrys* is reorganized during the telophase in the "indirect" (intranuclear) way: a large number of membrane-bound vesicles and cisternae are pinched off the inner membrane of the nuclear envelope into the nucleus (Fig. 17*c*, *d*) and fuse at the surface of the chromatin mass, producing a new envelope (Fig. 17*d*). The old nuclear envelope then falls into fragments. All the microtubules and a large part of the membrane-bound vesicles remain outside the envelopes of the two daughter nuclei.

Gregarinia. Semiopen orthomitosis is found in gregarines together with open orthomitosis. Its ultrastructure was studied during gamogony in *Diplauxis hatti* (Prensier 1973) and *Lecudina tuzetae* (Besse and Schrével 1973).

In *Diplauxis*, a mass of granular material, which soon splits in two, appears in a groove of the outer surface of the nucleus in the prophase. These are extranuclear MTOCs. No centrioles can be found near them. Bundles of microtubules begin to grow from each MTOC toward the nuclear envelope; such extranuclear primordia of the spindle (or two half-spindles, to be more exact) are referred to as *centrocones* (Fig. 18*a*). At the same time, the centrocones migrate to the opposite sides of the nucleus; the microtubules press in the nuclear envelope deeply (Fig. 18*b*), then break through it and join with one another to form the central spindle (Fig. 18*c*). Other bundles of microtubules interact with the chromosomal kinetochores and,

---

Fig. 17. Ultrastructure of progamic mitosis in *Actinophrys*. *a* metaphase, 5200 × ; *b* pole of the metaphase nucleus (microtubules passing through perforations in the nuclear envelope at arrows), 13 000 × ; *c* pole of the anaphase nucleus, perforations in the nuclear envelope widen (arrows), 9500 × ; *d* one of the daughter nuclei in telophase (formation of a new nuclear envelope), 10 400 × ; *1* old nuclear envelope, *2* polar caps, *3* chromosomes, *4* nucleolar material, *5* intranuclear membranous cisternae, *6* new envelope of the daughter nucleus, *7* its chromatin. *a–c* after Mignot (1979); *d* unpublished micrograph, courtesy of J.-P. Mignot.

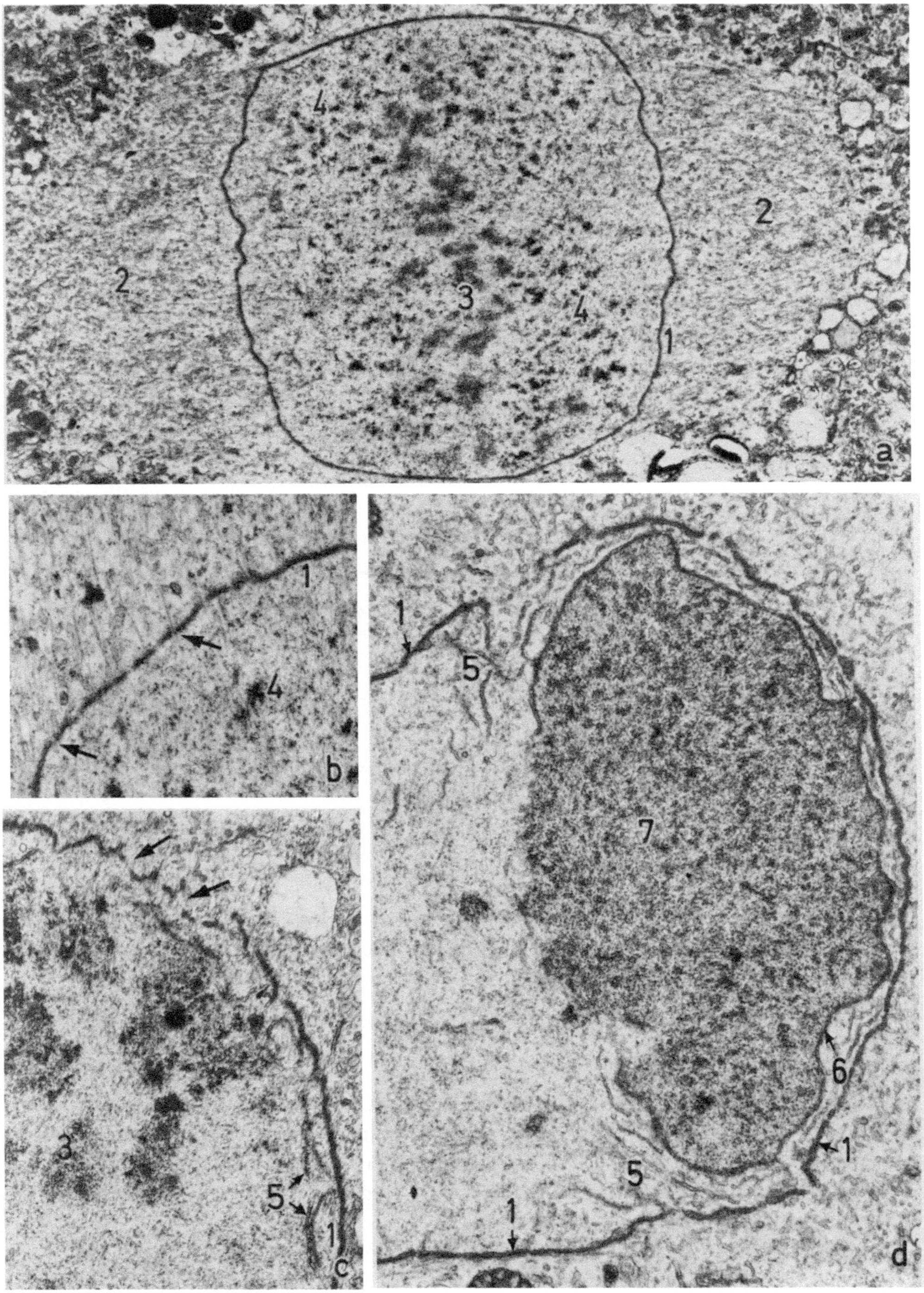

Fig. 17

upon shortening, pull the kinetochores to the poles (Fig. 18*c*). The chromosomal arms are poorly coiled; consequently, no distinct equatorial plate is formed (only the centromere regions of the chromosomes lie on the plane of the metaphase and, later, anaphase plates). Mitosis ends with the stretching and pinching in two of the whole nucleus as the fibres of the central spindle elongate (Fig. 18*d*).

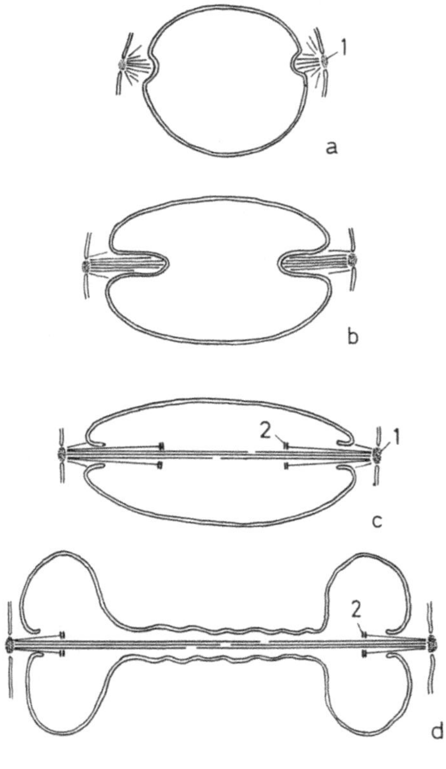

Fig. 18. Diagram of mitosis in the gregarine *Diplauxis* (after Prensier from Vivier and Vickerman 1974). *a* prophase; *b* prometaphase; *c* anaphase; *d* telophase; *1* cytoplasmic MTOC, *2* kinetochores (the chromosomes are not shown).

The course of mitosis is similar in *Lecudina*, in which centrocones are more conspicuous and the polar gaps in the nuclear envelope are wider (Fig. 19). The gaps in the nuclear envelope appear in such mitosis only after the centrocones migrate to the opposite sides of the nucleus; therefore, the nascent spindle has an axially symmetrical shape from the very beginning and encompasses both continuous and chromosomal fibres.

Semiopen orthomitosis of gregarines differs from that of volvocids, chloromonadids, and *Actinophrys* mainly in the peculiar structure of the extranuclear MTOCs (centrocones) and in the poorer condensation of the metaphasic chromosomes, as a result of which the equatorial plate is absent.

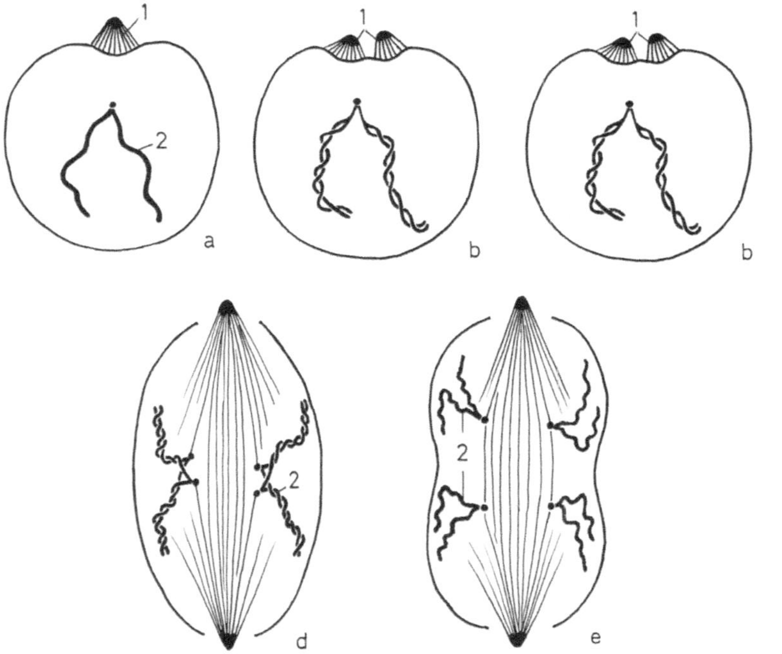

Fig. 19. Diagram of semiopen orthomitosis in the gregarine *Lecudina* (after BESSE and SCHRÉVEL from HOLLANDE 1972). *a* and *b* prophase; *c* prometaphase; *d* metaphase; *e* anaphase; *1* centrocones, *2* chromosomes (one in *a–c*, two in *d–e*).

Semiopen Pleuromitosis

This type of mitosis differs from semiopen orthomitosis of gregarines in that, instead of a bipolar spindle, two independent half-spindles are present. Consequently, the division figure is bilaterally symmetrical (Fig. 20). HOLLANDE (1972) designates this type of mitosis as "cryptomitosis with half-spindles depending on cytoplasmic centres". The nuclear envelope does not dissolve in semiopen pleuromitosis, with the exception of two small regions adjacent to the mitotic centres. These centres look like cytoplasmic centrocones with the MTOCs at the apices (Fig. 20*a*, *b*), near which pairs of kinetosomes or centrioles usually lie. Unlike in semiopen orthomitosis, fenestrae in the nuclear envelope appear long before the centrocones occupy the diametrically opposite positions on the nucleus; accordingly, a nascent spindle consists of two half-spindles whose poles are near each other at the beginning (Fig. 20*b*) and at an angle later (Fig. 20*c*). The half-spindles are composed of radial microtubules, some of which are attached to the chromosomal kinetochores (Fig. 20*b*). As the centrocones and the half-spindles separate, the centromere regions of the daughter chromosomes separate, too (Fig. 20*c*). Apparently this stage corresponds already to the anaphase; there is no metaphase stage and no equatorial plate (even the centromere regions of the chromosomes do not lie on the same plane at any of the stages). All these peculiarities of mitosis stem directly from the *premature* formation of the half-spindles and the premature beginning of the anaphasic separation of chromosomes. The chromosome arms are usually long, and they completely separate only at the very end of mitosis (Fig. 20*d*, *e*). At these stages, which correspond to the late anaphase, the centrocones reach the opposite poles of the nucleus in certain cases (in relatively small nuclei), and the two half-spindles join, producing one bipolar central spindle (Fig. 20*e*); the latter elongates, thus finally separating the nuclei. In larger nuclei, however, the half-spindles usually remain at the side of the nucleus until the very end of mitosis, and no central spindle is formed (Fig. 22*e*).

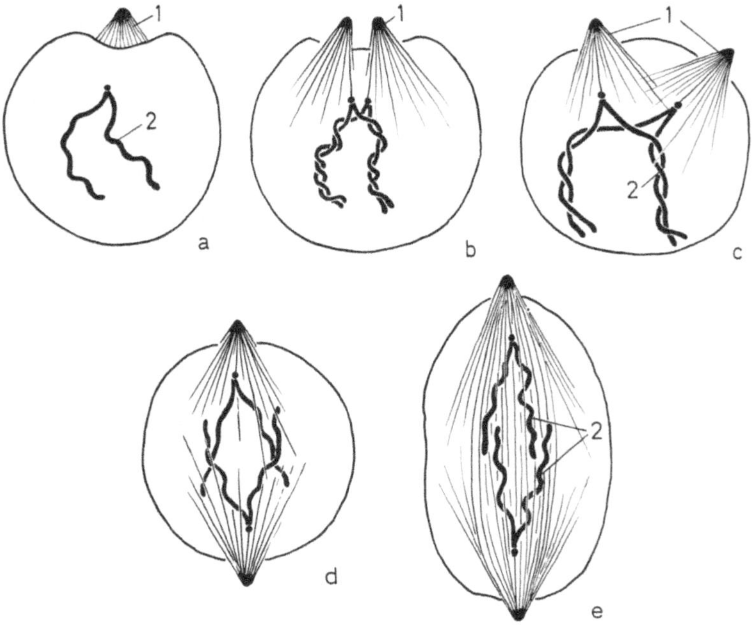

Fig. 20. Diagram of semiopen pleuromitosis (after HOLLANDE 1972). *a* and *b* prophase; *c–e* anaphase; *1* centrocones, *2* chromosomes (only one of the set is shown).

Semiopen pleuromitosis occurs during the gamogony of at least one gregarine: *Grebnickiella gracilis* (MOLON-NOBLOT and DESPORTES 1980). As we have seen, most gregarines display semiopen orthomitosis, but in *G. gracilis*, the microtubules of the conspicuous cytoplasmic centrocones penetrate into the nucleus through the forming polar gaps even before the daughter centrocones migrate to the opposite poles of the nucleus, thus imparting bilateral symmetry to the mitotic figure. This peculiarity may be due to the accelerated rate of the gamogonic mitoses of *G. gracilis*, and it also demonstrates the high plasticity of the mitotic patterns in Gregarinia. Unlike most gregarines, centrioles are found at the summits of the centrocones of *G. gracilis* (MOLON-NOBLOT 1976), which is apparently related to the flagellated type of its male gametes (MOLON-NOBLOT 1977, 1978). The trilaminar kinetochores, which were submembranous during the prophase, contact the spindle microtubules after the breakdown of the nuclear envelope in the centrocone regions. The nuclear envelopes of the daughter nuclei are re-formed in the telophase by the deposition of two-membrane cisternae on the fusing chromosomes, i.e. they are formed inside the old nuclear envelope which is still intact (except the polar gap regions). This new envelope encloses the kinetochores and separates them from the spindle microtubules, which are excluded (MOLON-NOBLOT and DESPORTES 1980).

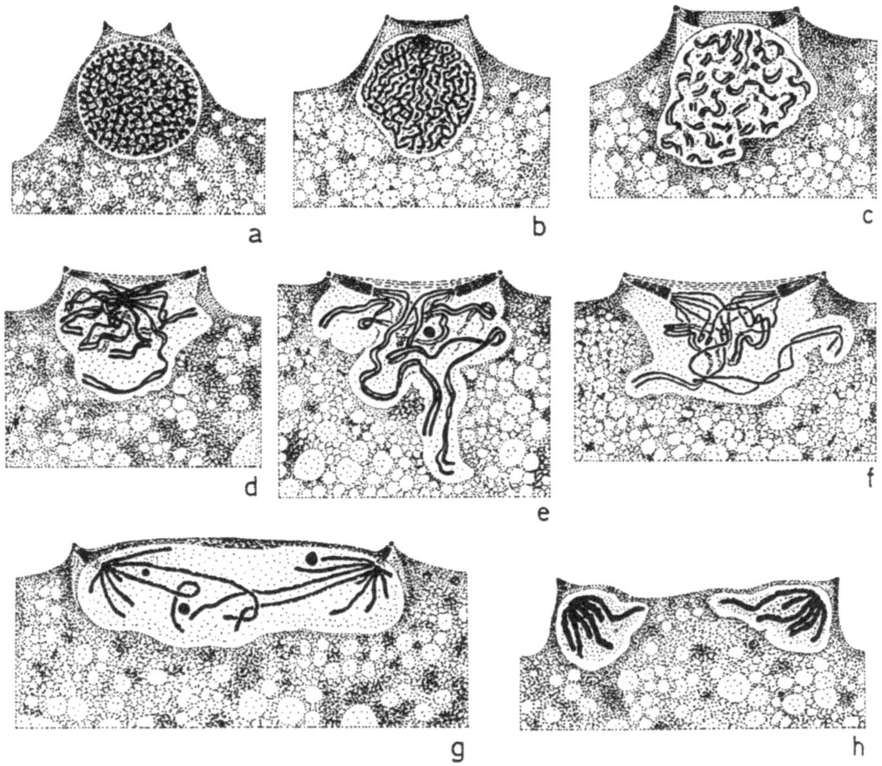

Fig. 21. Sporogonic mitoses of the coccidian *Aggregata*, 2300× (after BĚLAŘ 1926b). *a* interphase; *b* and *c* prophase (individualization of chromatids); *d*–*g* anaphase (gradual separation of chromatids); *h* telophase.

Semiopen pleuromitosis is most typical of Coccidia, including the toxoplasmids and sarcosporidians; in a modified form, it is also typical of Haemosporina and Piroplasmia (see the reviews by PORCHET-HENNERÉ 1972b, VIVIER and VICKERMAN 1974, BEYER *et al.* 1978). Mitosis in these organisms can hardly be analyzed with the light microscope because their centrocones and half-spindles are very small and chromosomal condensation is poor. That is why BĚLAŘ classified most mitoses of this type as cryptomitoses. Even the most accurate descriptions of mitoses in coccidians, e.g. in *Aggregata* (Fig. 21), were inadequate for establishing precisely where the mitotic centres are located, how the spindle is formed, and whether the spindle is intranuclear or extranuclear. Nevertheless, even these descriptions have clearly shown that chromosomal condensation was atypical, the equatorial plate was absent, the chromosome arms separated very late, and the two half-spindles as well as the whole mitotic apparatus occupied an extremely eccentric (lateral) position with respect to the nucleus (BĚLAŘ 1926b). BĚLAŘ (1926a) classified such mitoses as paramitoses.

Coccidia. A clear insight into the mechanism of mitosis in coccidians was gained only in the 1970s owing to electron microscopy. DUBREMETZ (1971) was the first to give a detailed description of the ultrastructure of the centrocones in *Eimeria necatrix*. The complete cycle of mitosis in the schizogony of *E. necatrix*, described later by DUBREMETZ (1973), comprises the following stages (Fig. 22). In the early prophase, the extranuclear MTOC is an electron-dense crescent lying in a deep furrow of the nuclear envelope; both ends of the crescent are in contact with the centriole pairs. The kinetochores of all chromosomes are located near the MTOC on the inner surface of the nuclear envelope (Fig. 22a). Somewhat later (in the "proanaphase", the MTOC splits into two centrocones; the nuclear envelope at the bottom of the furrow dissolves and the half-spindles of the microtubules penetrate into the nucleus; some of them connect the centrocones with one another (the central spindle), others link up with the kinetochores (Fig. 22*b*). The chromatids immediately begin to separate (Fig. 22*c*); the central spindle rapidly disappears and only the half-spindles remain (Fig. 22*d*). The centrocones then lie in deep invaginations of the nuclear envelope (Fig. 22*d*); the apex of each centrocone is connected by strands of amorphous material to a pair of cytoplasmic centrioles. Microtubules starting from the centrocones penetrate into the nucleus through multiple small gaps in the nuclear envelope at the bottom of such an invagination. The centrioles are atypical in schizogony and look like short cylinders with nine single microtubules (singlets).

At this stage, the chromosomes (at least their kinetochores) are almost contiguous to the centrocones (Figs. 16*c*, 22*d*). At the end of mitosis, the centrocones "evaginate" from the nucleus; the intranuclear microtubules disappear, the nuclear envelope between the nucleus and the centrocones is restored, and the kinetochores of all chromosomes gather on the inner surface of the envelope (Fig. 16*d*). Microtubules remain only in the extranuclear centrocone (Fig. 22*e*). There is no central spindle at the late stages of schizogonial mitosis in *E. necatrix*; the nucleus is divided in two with a septum which, at the beginning, consists of only two inner membranes of the nuclear envelope; the nucleolus also divides in two during mitosis (Fig. 22*e*). The two centrocones separate probably due to the differential growth of the nuclear envelope between them.

Semiopen pleuromitosis of a similar type has been observed during schizogony also in other Eimeriid coccidians: *Eimeria falciformis* (MEHLHORN *et al.* 1973), *E. callospermophili* and *E. magna* (HAMMOND *et al.* 1973), *E. ninakohlyakimovae* (KELLEY and HAMMOND 1972, 1973), *E. tenella* (PACHECO *et al.* 1975, HOPPE 1976), *E. bovis* (DUBREMETZ and ELSNER 1979), *Globidium gilruthi* (PORCHET-HENNERÉ 1977). Centrocones were also found in the mitoses of *Toxoplasma* (SHEFFIELD and MELTON 1968), *Sarcocystis* (PACHECO and FAYER 1977), and the Adeleid coccidian *Hepatozoon* (VIVIER *et al.* 1972).

The schizogonial mitosis in *Aggregata eberthi* differs somewhat from other coccidial mitoses (PORCHET-HENNERÉ and RICHARD 1971a). In this case, an

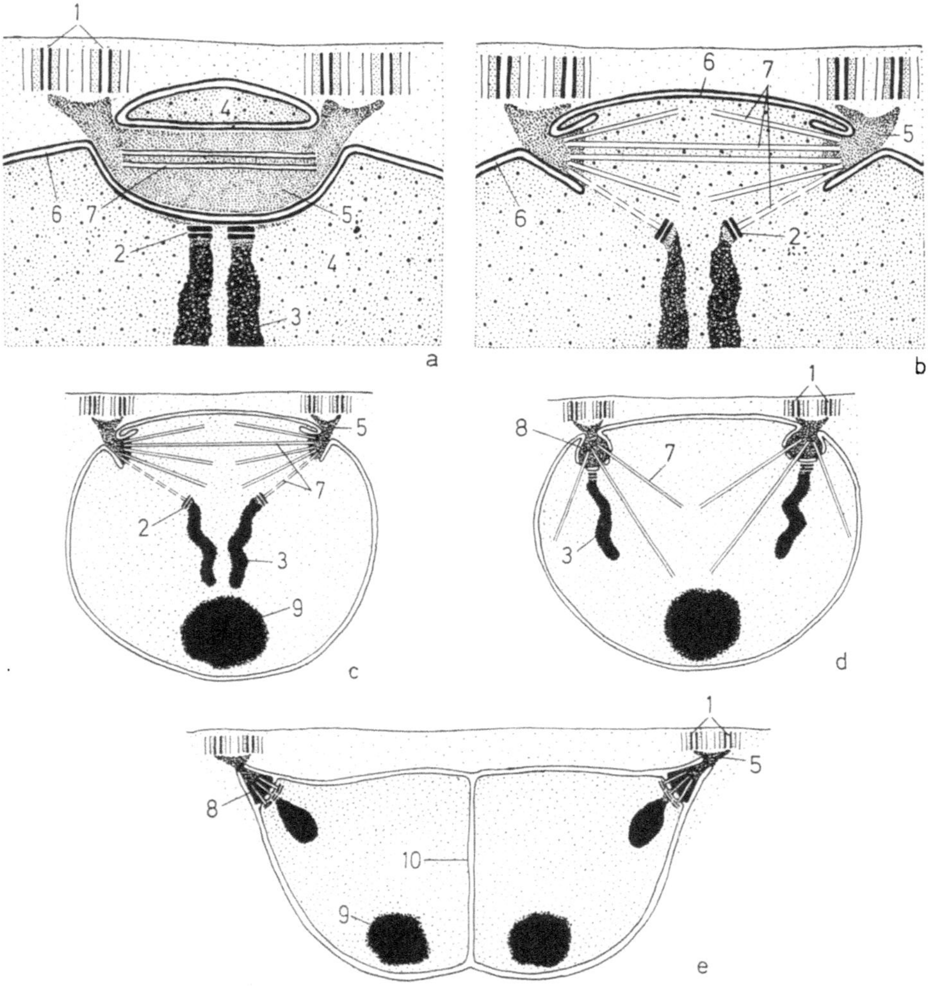

Fig. 22. Scheme of the ultrastructure of schizogonic mitosis in *Eimeria necatrix* (after DUBREMETZ 1973). *a* prophase; *b* "proanaphase"; *c* and *d* anaphase; *e* telophase. Part of the nucleus is shown in *a* and *b*, and the whole nucleus is shown in *c*–*e*. *1* pairs of centrioles, *2* kinetochores, *3* chromosomes, *4* nucleus, *5* MTOC, *6* nuclear envelope, *7* spindle microtubules, *8* centrocones, *9* nucleolus, *10* septum separating daughter nuclei.

extranuclear central spindle, or "centrodesmose", appears between the bases of two centriole pairs (i.e. between the apices of two centrocones). Apparently, the growth of the "centrodesmose" in length is what drives the daughter centrocones apart (cf. Fig. 21*b*–*f*). In other coccidians, such as *Eimeria ninakohlyakimovae* (KELLEY and HAMMOND 1972, 1973) and *E. bovis* (DUBREMETZ and ELSNER 1979), the central spindle, if there is one, is formed inside the nucleus.

Haemosporidians and Piroplasmids. In malaria parasites (suborder Haemosporina), the MTOCs are always incorporated in the expanded pores of the nuclear envelope and look like finely granular discs or plugs (Figs. 16*e*, 23), being often referred to as "spindle pole bodies" or "kinetic centres". They are not isolated from the cytoplasm and in *Leucocytozoon* even protrude from a pore into the cytoplasm (WONG and DESSER 1978). Therefore, the spindle pole bodies may be regarded as reduced centrocones of the coccidial type; this is consistent with the generally accepted origin of haemosporodians from coccidians. Haemosporidians have no centrioles near their MTOCs, at least in schizogonial and sporogonial mitoses.

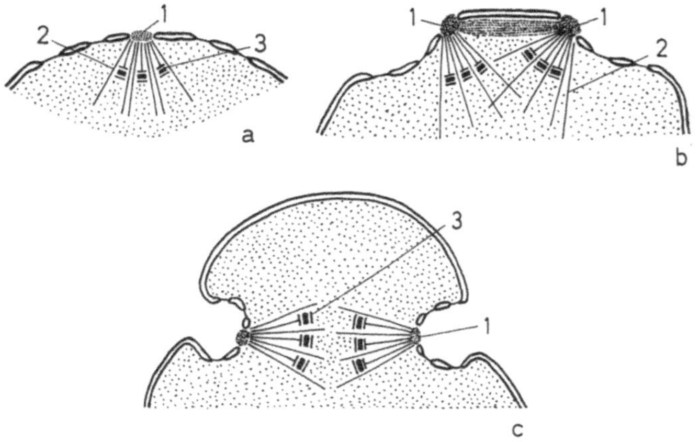

Fig. 23. Scheme of mitosis in *Plasmodium berghei* (after FOUCHER and SCHRÉVEL from VIVIER and VICKERMAN 1974). *a* single half-spindle; *b* duplication of half-spindles; *c* anaphase (the nucleus does not divide); *1* MTOC, *2* microtubules, *3* kinetochores (the chromosomes are not shown).

Mitosis in *Plasmodium* begins with the division of a single spindle pole body (Fig. 23*a*) in two; the cone of the intranuclear microtubules (the half-spindle) connected to it divides, too (Fig. 23*b*). A strand of fibrogranular material, which passes along the inner surface of the nuclear envelope, is preserved for some time between the daughter MTOCs. The daughter half-spindles consist of radial (with their free ends in the karyolymph) and chromosomal microtubules; the latter are attached to the kinetochores of the daughter chromosomes. The spindle is usually very small in comparison with the nucleus, and its location in the nucleus is eccentric; often, it is on a lateral bulge of the nuclear surface, and then the half-spindles may lie opposite one another (Fig. 23*c*). A few interpolar microtubules are occasionally observed together with the chromosomal microtubules at this stage (Fig. 16*e*). The kinetochores of the daughter chromosomes look like discs with three dense layers separated by less dense interlayers (Fig. 16*e*); only one microtubule seems to be attached to each kinetochore. The chromosomes of *Plasmodium* hardly condense in mitosis and cannot be discerned on electron micrographs. The chromosomal microtubules shorten at the end of mitosis and the kinetochores are pulled to the daughter MTOCs; the separation of the chromosome arms is greatly delayed (just as in coccidians). The two half-spindles then completely separate from one another (Fig. 23*a*) and each of them is ready to undergo another mitosis. Gamogonial and sporogonial mitoses in haemosporidians do not involve the immediate division of the nucleus: the latter temporarily becomes polyploid and lobed with numerous small spindles and half-spindles at the periphery. Later, such a nucleus undergoes multiple division, providing nuclei for the future microgametes (BRADBURY and TRAGER 1968b) or sporozoites (GARNHAM *et al.* 1969, CANNING and SINDEN 1973a, VIVIER and VICKERMAN 1974).

Mitosis of the above type is found in the schizogony of *Plasmodium fallax*, *P. lophurae*, *P. cathemerium* (AIKAWA 1966, HEPLER *et al.* 1966), *P. elongatum* (AIKAWA *et al.* 1967), *P. gallinaceum* (AIKAWA and BEAUDOIN 1968, AIKAWA *et al.* 1968), *P. vinckei*, *P. chabaudi* (SCALZI and BAHR 1968), *P. berghei* (LADDA 1969), *P. mexicanum* (MOORE and SINDEN 1974), and *Leucocytozoon dubreuili* (WONG and DESSER 1978) as well as in the sporogony of *P. gallinaceum* (TERZAKIS *et al.* 1967, MEHLHORN *et al.* 1980), *P. cynomolgi* (TERZAKIS 1971), and *P. berghei* (HOWELLS and DAVIES 1971, CANNING and SINDEN 1973a, SCHRÉVEL *et al.* 1977). In early works (see the review by AIKAWA 1971), the trilaminar discs within the nucleus were regarded as whole chromosomes; however, they later proved to be the kinetochores of decondensed chromosomes (SCALZI and BAHR 1968, CANNING and SINDEN 1973a, SCHRÉVEL *et al.* 1977).

In Piroplasmia, mitosis has not been studied well enough. In *Theileria*, half-spindles with poles incorporated into the pores of the nuclear envelope were discovered in the nucleus of a schizont (BÜTTNER 1967b, SCHEIN *et al.* 1978). In *Anthemosoma*, numerous small spindles with the same poles as in *Theileria* are formed at the periphery of the lobed nucleus in a schizont. Then "buds" separate from the nucleus; these "buds" seem to contain individual genomes and are the nuclei of the future merozoites (VIVIER *et al.* 1974).

Closed Intranuclear Pleuromitosis

Mitoses of this type are usually very small and, therefore, nothing definite was known about most of them prior to the era of electron microscopy. Closed intranuclear pleuromitosis resembles semiopen pleuromitosis by the general topography of the mitotic apparatus, the presence of two independent half-spindles (which gradually diverge from each other during the anaphase) inside the nucleus, the low extent of chromosome condensation, and the absence of the equatorial plate (Fig. 24). Just as in semiopen pleuromitosis, the spindle has bilateral symmetry. These two types of mitosis differ mainly in the position of the MTOCs, which are located on the inner surface of the nuclear envelope in closed intranuclear pleuromitosis. Accordingly, the spindle is entirely intranuclear. Centrioles (kinetosomes) occasionally occur in the cytoplasm opposite the poles. Just as HOLLANDE (1972), we believe that these variations are not important in characterizing the type of mitosis and are determined only by the presence or absence of the flagellated stages in the life cycle of an organism. In closed intranuclear pleuromitosis, the spindle is organized by special MTOCs that underlie the nuclear envelope, not by centrioles.

According to HOLLANDE's classification, closed intranuclear pleuromitosis corresponds to "cryptomitosis with half-spindles depending on intranuclear centres". This type of mitosis is typical of microsporidians, kinetoplastids, oxymonadids, foraminiferans, radiolarians, and some green flagellates (Prasinomonadida).

Microspora and Haplosporea. The intranuclear MTOCs of microsporidians are plaques of granular material ("polar" or "centriolar" plaques) that underlie the intact nuclear envelope, which is usually cupped at these sites. Microsporidians have no centrioles in the cytoplasm opposite the MTOCs, but they usually have various other differentiations, such as an electron-dense disc in *Gurleya* (Fig. 25*a*), a group of vesicles in *Metchnikovella* (VIVIER 1965) and *Baculea* (LOUBÈS and

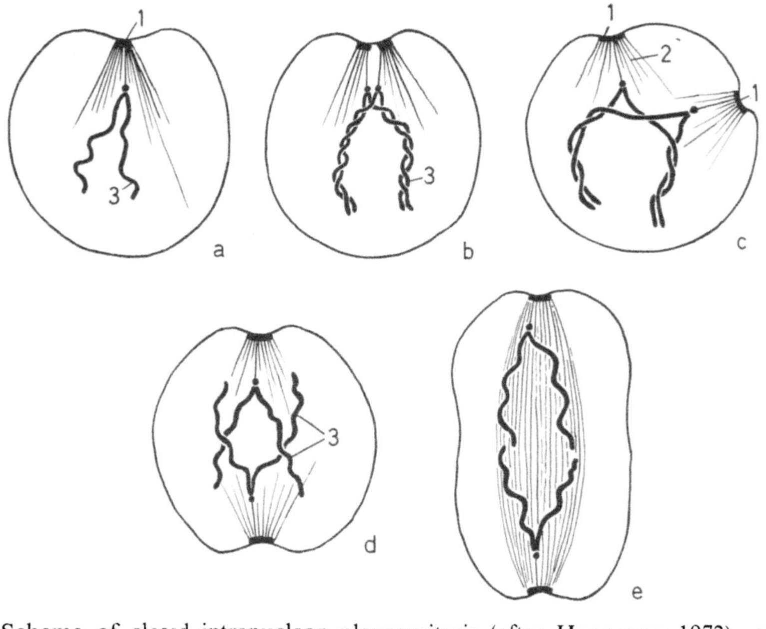

Fig. 24. Scheme of closed intranuclear pleuromitosis (after HOLLANDE 1972). *a* and *b* prophase; *c*-*e* anaphase; *1* MTOCs, *2* half-spindles, *3* chromosomes (only one is shown).

AKBARIEH 1978), a short cylinder without the typical centriolar structure in *Amphiamblys* (DESPORTES and THÉODORIDÈS 1979), two or three flattened cisternae in a dense matrix in *Stempellia*, *Glugea*, and *Ormieresia* (DESPORTES 1976, LOUBÈS *et al.* 1976a, VIVARÈS *et al.* 1977), and so forth. In these cases, the cytoplasmic differentiations are separated from the MTOC proper (the "centriolar plaque") by an intact nuclear envelope.

The course of mitosis in microsporidians fits well into the scheme shown in Fig. 24. At the beginning, the nucleus contains only one centriolar plaque, which is connected to the sole intranuclear half-spindle; some of the spindle microtubules are attached to the kinetochores of the chromosomes (Fig. 24*a*). As the centriolar plaque redoubles, the half-spindle divides in two and the centromere regions of the chromatids begin to separate (Fig. 24*b*, *c*). The separation of the centriolar plaques and the half-spindles connected to them appears to be caused by the intensive growth of the membranes of the nuclear envelope in the intermediate zone. At the end of mitosis, when the centriolar plaques reach the opposite poles of the nucleus, the central spindle is formed between them and begins to stretch out, thus completely separating the two chromosome sets (Fig. 24*d*, *e*), and then the nuclear envelope pinches in two.

This type of mitosis was first studied in the primitive microsporidian *Metchnikovella* (VIVIER 1965, VIVIER and SCHRÉVEL 1973, HILDEBRAND 1974) and later in *Nosema* (WEIDNER 1970, VINCKIER *et al.* 1971, CANNING and SINDEN 1973b), *Ichthyosporidium* (SPRAGUE and VERNICK 1974), etc. Very distinct patterns of mitosis were obtained for *Gurleya chironomi* (LOUBÈS and MAURAND 1975) and *Toxoglugea variabilis* (LARSSON 1980a), in which not only centriolar plaques and half-spindles but also the kinetochores of chromosomes are clearly discernible (Fig. 25*a*). In *Gurleya*, the kinetochores look like dense discs with several chromosomal microtubules attached to each. The chromosomes are poorly coiled; at the late stages of mitosis, the microtubules of the central spindle are well defined. Mitosis was studied in detail also in *Stempellia mutabilis*, a form with relatively large nuclei and chromosomes (DESPORTES 1976). In the course of mitosis, the centriolar plaques are distinct, the chromosomes condense to a considerable extent, and the half-spindles join early, producing a bipolar spindle (Fig. 25*b*).

In mitosis in haplosporidians, the intranuclear MTOCs (centriolar plaques, spindle pole bodies) seem to be deep in the nucleus at a certain distance from the envelope. The central spindle is well defined as a bundle of microtubules that connects the MTOCs ("nuclear rod"). The kinetochores of the chromosomes and chromosomal microtubules have not been found so far (PERKINS 1968, 1975).

Kinetoplastida. Mitosis in trypanosomids (and other kinetoplastids) has not been studied well enough. In some trypanosomes, however, it may definitely be regarded as a type of closed intranuclear pleuromitosis. In *Trypanosoma raiae*, for instance, two half-spindles made of microtubules that diverge in a fan-like way are visible inside the dividing nucleus with an intact nuclear envelope (VICKERMAN and PRESTON 1970). The half-spindles are arranged at an angle to one another (a feature of pleuromitosis) at the early stages of mitosis. The apices of the half-spindles are plaques of amorphous material on the inner surface of the nuclear envelope, i.e. intranuclear MTOCs or "centriolar plaques", which have also been found in *T. equiperdum* (PAULIN 1975). The chromosomes hardly condense in mitosis, but their trilaminar kinetochores resembling those in *Plasmodium* are clearly seen; the bundles of chromosomal microtubules of the half-spindles are attached to the kinetochores. At the late stages of mitosis, the MTOCs segregate to the opposite poles of the nucleus and a distinct central spindle is formed between them, elongating with the nucleus. Finally, the nucleus pinches in two in a dumbbell manner.

Mitosis in *Trypanosoma cruzi* is more or less similar to mitosis in *T. raiae* (DE SOUZA and MEYER 1974, SOLARI 1980a, 1980b). According to SOLARI, the chromosomal microtubules coming from both poles of the spindle (1–2 from each pole) are attached to each trilaminar kinetochore, which total ten at the equatorial plate stage (metaphase). Other microtubules (radial) bypass the kinetochores. In *T.*

cruzi, the stage of two independent half-spindles arranged at an angle to one another has not been found so far: at the earliest known stage, the half-spindles are already combined, forming a bipolar symmetrical spindle attached at both poles to the inner surface of the nuclear envelope (Fig. 25*c*). Apparently, this is due to the early migration of the daughter MTOCs to the opposite sides of the nucleus. Nevertheless, the spindle comprises only radial and chromosomal microtubules and no interpolar microtubules; thus, it consists of two diametrically opposite intranuclear half-spindles (Fig. 25*c*). In the anaphase, each kinetochore divides into two daughter kinetochores, which are attached to one bundle of chromosomal microtubules. At the end of the anaphase, however, peculiar interzonal microtubules appear; they connect the two daughter kinetochores of each pair and rapidly elongate. The nucleus becomes dumbbell-shaped and divides in two without any particular telophasic reorganization of the nuclear envelope (the "direct" way). The chromosomes remain decondensed and the nuclear envelope is continuous throughout mitosis.

JADIN (1971) believes that mitosis follows the above type in all trypanosomids (with the half-spindles and MTOCs on the inner surface of the nuclear envelope). Mitoses that are difficult to interpret, however, have also been reported in these flagellates. In *Trypanosoma rhodesiense*, for instance, only a single bundle of microtubules of the central spindle that penetrates through some dense bands (groups of kinetochores?) was found in the dividing nucleus; however, no chromosomal microtubules attached to the bands were seen. The microtubules butt against the nuclear envelope at the poles, but no visible MTOCs differentiate in these places (VICKERMAN and PRESTON 1970). More or less similar mitoses were reported in *T. brucei*, *T. vivax* (VICKERMAN 1974), *T. cyclops* (HEYWOOD and WEINMAN 1978), *Leishmania tropica* (BIANCHI *et al.* 1969), and *L. hertigi* (CROFT 1979). It is likely that the connection of the chromosomes with the half-spindles was not detected in many cases simply because only the late stages of mitosis (after a bundle of interzonal microtubules had been formed) were studied.

The central nucleolus, which is typical of trypanosomids, usually does not disappear in mitosis but elongates and divides in two in a dumbbell shape. It is frequently located in the middle of a bundle of the spindle microtubules, e.g. in *T. rhodesiense*. The nucleolus loosens up greatly or even fragments during division (Fig. 25*c*), and its material is distributed between the daughter nuclei in other species, such as *T. raiae* (VICKERMAN and PRESTON 1970), *T. cruzi* (DE SOUZA and MEYER 1974, SOLARI 1980a), *T. cyclops* (HEYWOOD and WEINMAN 1978), and *Leishmania hertigi* (CROFT 1979).

Fig. 25. Ultrastructure of closed intranuclear pleuromitosis. *a* half-spindle in the anaphase of mitosis in the microsporidian *Gurleya* (*1* centriolar plaque, *2* kinetochores, *3* microtubules, *4* nuclear envelope, *5* dense disc at its outer surface, *6* chromosomes), 24000 × ; *b* anaphase of mitosis in the microsporidian *Stempellia* (same designations as *a*), 20000 × ; *c* metaphase of mitosis in *Trypanosoma cruzi* (*1* MTOC, *2* nuclear envelope, *3* microtubules, *4* kinetochore, *5* material of the loosened nucleolus, *6* chromatin fibres), 30000 × ; *d* mitotic spindle pole in the foraminiferan *Allogromia* (*1* "centrosphere", *2* nuclear envelope, *3* nucleus, *4* chromosomes), 12000 × . *a* after LOUBÈS and MAURAND (1975); *b* after DESPORTES (1976); *c* after SOLARI (1980a); *d* after SCHWAB (1972).

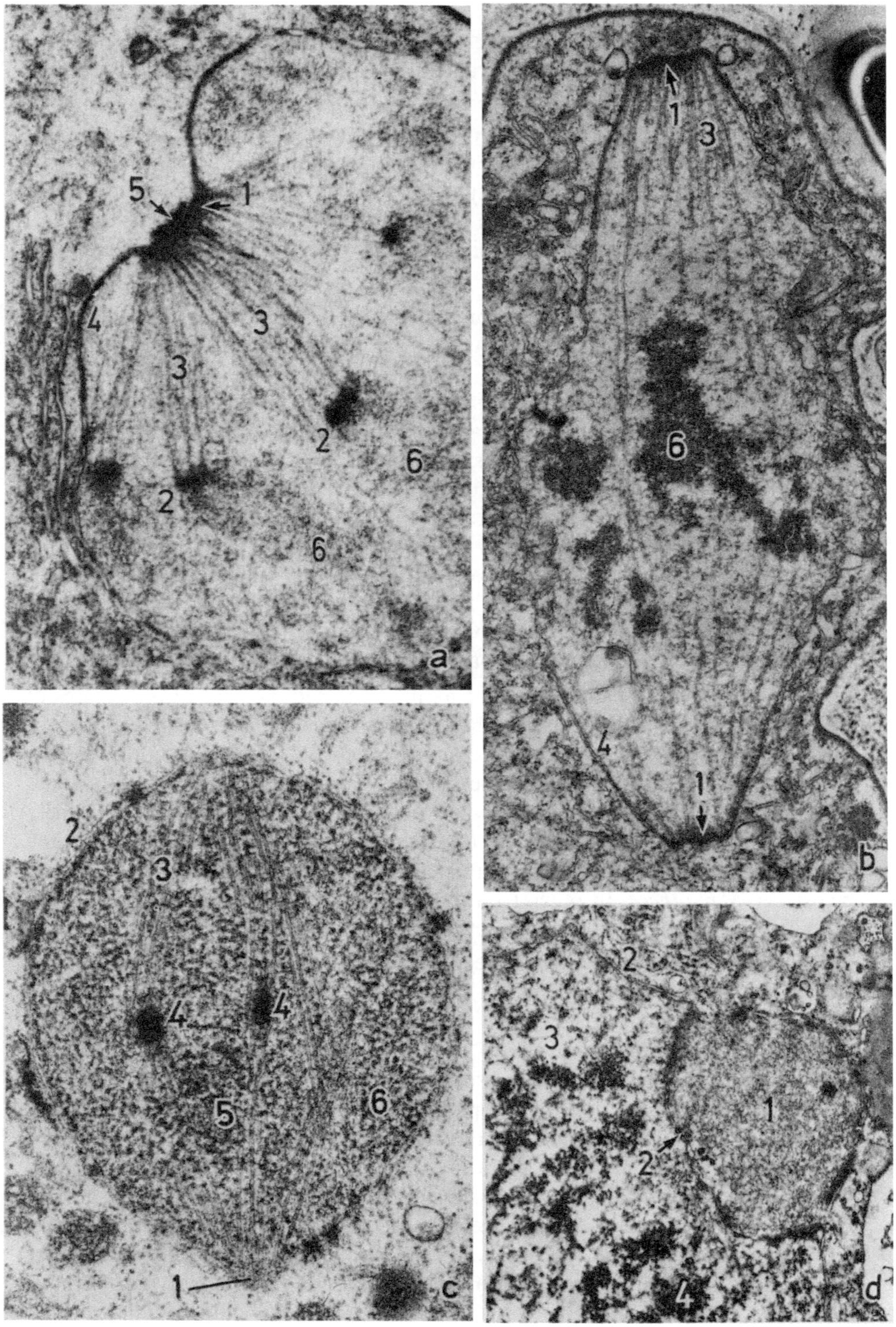

Fig. 25

Despite the fact that dividing trypanosomids always have kinetosomes (the basal bodies of the flagella), in none of the known cases are they located at the poles of the half-spindles or the central spindle. The intranuclear MTOCs seem to be entirely autonomous.

Oxymonadida. Mitosis with an intranuclear spindle and with poles on the nuclear envelope had been studied long ago in oxymonadids (*Oxymonas*, *Pyrsonympha*) by light microscopy (CLEVELAND *et al.* 1934, CLEVELAND 1935a, 1938a). The mitotic apparatus in *Oxymonas* proved to consist of two independent half-spindles at the beginning of mitosis (GRASSÉ 1952); these half-spindles were observed also in meiosis (Fig. 61).

Mitosis has been studied by electron microscopy only in *Pyrsonympha flagellata* so far (HOLLANDE and CARRUETTE-VALENTIN 1970a). In this case, the half-spindles are entirely intranuclear and are polarized toward the centriolar plaques (MTOCs), which are on the inner surface of the intact nuclear envelope. The chromosomes are tightly coiled. In early prophase, the nucleus contains only one centriolar plaque; the chromosomes are permanently bound to it through the chromosomal bundles of microtubules. The centriolar plaque then divides into two parts; two half-spindles are formed and gradually separate, and the sister chromatids gradually separate at the same time. There are also some fibres of the central spindle that pass from one pole to the other through the karyolymph. At the end of mitosis, the half-spindles reach opposite sides of the nucleus and the segregation of the chromatids is completed. Therefore, the mitosis of *Pyrsonympha* is an almost exact copy of microsporidian mitosis, although a much larger one.

Foraminiferida and Radiolaria. Just as in oxymonadids, the mitotic figures of foraminiferans may be quite large. Therefore, mitosis was investigated in many of them by light microscopy. In *Patellina corrugata*, for instance, the nuclear envelope does not dissolve, while the spindle centres are in the form of plaques under the nuclear envelope (LE CALVEZ 1938). It turned out, however, that these data refer to the meiotic division of the nucleus (Fig. 57). Half-spindles with the centres on the inner surface of the nuclear envelope were observed also during meiosis in *Myxotheca*. In other foraminiferans (with smaller mitoses), only the intranuclear position of the spindle was reported (GRELL 1957, 1958a, 1958b, etc.).

Mitosis was studied with the electron microscope in the gametogenesis of *Myxotheca* and *Allogromia* (SCHWAB 1969, 1976). Their nuclear envelopes do not disappear in mitosis and have, on their inner surface, two centriolar plaques (intranuclear MTOCs) from which the half-spindles of microtubules protude deep into the nucleus; the chromosomal microtubules are in contact with the condensed chromosomes. Centriole pairs lie in the cytoplasm near the poles; in *Myxotheca*, they later become the flagellar kinetosomes of gametes. However, centrioles are absent at other stages of the life cycle of *Myxotheca*, e.g. during meiosis in the agamonts (SCHWAB 1973). In this case, their place near the poles is occupied by homogeneous spheres covered with extensions of the outer membrane of the nuclear envelope. In other respects, the intranuclear spindle and MTOCs are the same as in mitosis. Thus, centrioles do not seem to be an indispensable component of the mitotic apparatus in foraminiferans. Apparently, they do not participate in spindle induction (this is the function of intranuclear MTOCs) but only "use" nuclear division for their own separation into the future gametes where their function is to form flagella (cf. p. 83).

Typical closed pleuromitosis with intranuclear half-spindles occurs during gamogony in *Iridia* (CÉSANA 1971, 1978). MTOCs that look like dense plaques on the inner side of the nuclear envelope gradually separate from one another during mitosis. Every MTOC is duplicated with a layer of dense material at the outer side of the nuclear envelope, which is cupped in these places; pairs of developing centrioles are opposite the MTOCs in the cytoplasm. Bundles of chromosomal microtubules protrude from the MTOCs into the nucleus; they become attached to the trilaminar kinetochores of the chromosomes and form typical intranuclear half-spindles. Other bundles of microtubules connect both MTOCs with one another as they pass parallel to the nuclear envelope; this is the intranuclear central spindle. At the later stages of mitosis, the half-spindles migrate to the opposite poles of the nucleus. The chromosomal microtubules shorten, while the central spindle elongates. The nucleus pinches in two without the reorganization of the nuclear envelope (the "direct" way). Later, centrioles become the flagellar kinetosomes of gametes.

Similar mitosis was reported in the diploid agamonts of *Allogromia* (SCHWAB 1972). But they have no centrioles in the cytoplasm. In these agamonts, a variety of cytoplasmic differentiations opposite the spindle poles is noteworthy. The differentiations referred to as nuclear pole bodies (NPB) may look like a vague amorphous "cap", a fibrogranular "centrosphere" separated from the remaining cytoplasm by a two-membrane cisterna (Fig. 25*d*), or they may be absent altogether. In another foraminiferan, *Kibisidytes*, NPB look like discs made of two layers of different density (SCHWAB and SCHWAB-STEY 1979). Conversely, the intranuclear MTOC in which the microtubules end is always the same.

In radiolarians, similar mitoses were reported during sporogenesis in the multinucleate "polycyttaric" forms of the order Spumellarida, e.g. *Collozoum* (HOLLANDE *et al.* 1969, RIS 1975, ANDERSON 1976a) and *Sphaerozoum* (ANDERSON 1976b) as well as in a number of Acantharea (FEBVRE 1977). A diagram of such mitoses is shown in Fig. 26. One can see that this is a typical closed pleuromitosis with MTOCs on the inner side of the nuclear envelope and with intranuclear half-spindles and a thick intranuclear central spindle. At the beginning, the central spindle is eccentric with respect to the nucleus and becomes diametral only at the end of mitosis. The chromosome arms are long and separate very late; consequently, there is no equatorial plate. The MTOCs appear as "centriolar plaques" on the inner side of the nuclear envelope. All the microtubules of the spindle terminate in this layer. Centrioles (the future flagellar kinetosomes of zoospores) are near the MTOCs in the cytoplasm. Bilaminar chromosomal kinetochores are clearly visible. A bundle of microtubules is attached to one dense layer of each kinetochore, while another such layer is connected with the chromosomal material (RIS 1975, ANDERSON 1976a). The limits of separate chromosomes in *Collozoum* cannot be discerned on electron micrographs because of their poor condensation; they are more condensed in *Sphaerozoum* (ANDERSON 1976b).

Prasinomonadida. Most of the green flagellates belonging to this order are characterized by open orthomitosis (see p. 81). However, a completely closed mitosis with intranuclear half-spindles that seem to be connected with MTOCs on the inner side of the nuclear envelope was discovered in the most primitive prasinomonad, *Pedinomonas*. Pairs of flagellar kinetosomes are in the cytoplasm opposite the MTOCs (Pickett-Heaps and Ott 1974).

Thus, closed intranuclear pleuromitosis is widespread among protozoans. Its mechanism is very uniform; only the time and extent of development of the central spindle in the nucleus can vary. It develops at the late stages of mitosis in

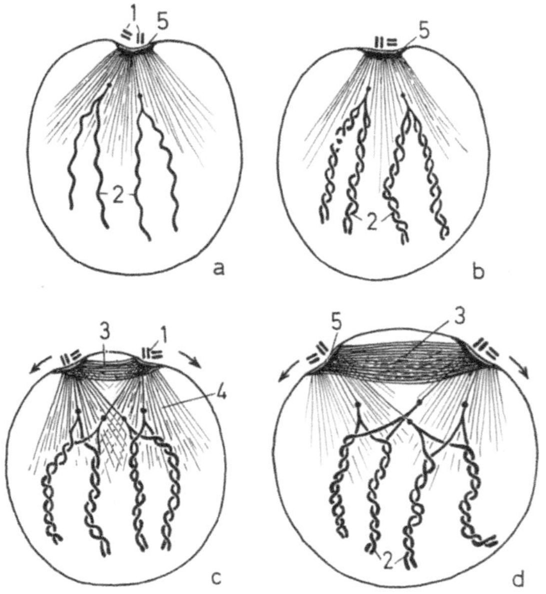

Fig. 26. Diagram of mitosis in sporogenesis of the radiolarian *Thalassophysa* (after Hollande *et al.* 1969). *a* early prophase; *b* individualization of two chromatids; *c* and *d* anaphase; *1* pairs of centrioles, *2* chromosomes (only two are shown), **3** central spindle, *4* half-spindles, *5* intranuclear MTOC.

microsporidians and trypanosomids; it appears comparatively early (and is therefore eccentric at the beginning) in oxymonadids, foraminiferans, and radiolarians. Variations of this type of mitosis can also stem from the presence or absence of centrioles and other cytoplasmic structures on the outer surface of the nucleus near the poles. This, however, bears no direct relationship to either the induction of the spindle or the mechanism of mitosis.

This type of mitosis is also widespread outside Protozoa, particularly in fungi: chytrids (*Catenaria*, *Blastocladiella*, *Allomyces*), oomycetes (*Saprolegnia*, *Thraustotheca*, *Aphanomyces*), zygomycetes (*Phycomyces*, *Pilobolus*, *Mucor*), ascomycetes (*Xylosphaera*, *Podospora*, *Neurospora*, *Ascobolus*, *Fusarium*), and others. It is also found in the plasmodia of myxomycetes (*Clastoderma*, *Didymium*) and in the xanthophycean alga *Vaucheria* (for reviews see Kubai 1975, Fuller 1976, Heath 1980). The occurrence of closed intranuclear pleuromitosis on such a wide scale indicates its primitiveness.

Closed Intranuclear Orthomitosis

In this type of mitosis, the spindle is, from the very beginning, bipolar, axially symmetrical, and wholly intranuclear. Chromosomes usually (but not always) form a distinct equatorial plate. Such mitosis is found in two variants.

In the first variant, the spindle is formed between two MTOCs (centriolar plaques) that lie on the inner side of the intact nuclear envelope (just as in closed intranuclear pleuromitosis) but are at the opposite poles of the nucleus from the beginning of spindle formation. The spindle is therefore orthomitotic from the very beginning. This variant resembles certain varieties of closed intranuclear pleuromitosis with the early migration of the daughter MTOCs to the poles of the nucleus (for instance, in *Trypanosoma cruzi*). The first variant is rather rare among protozoans, and is found in certain rhizopods and gregarines.

In the second variant, small bundles of microtubules are formed throughout the karyolymph; more rarely, they grow from an intranuclear MTOC not connected with the nuclear envelope. The spindle is acentric, and its microtubules either end free in the karyolymph or butt against the nuclear envelope from the inside. Unlike the first variant, contact between the spindle and the nuclear envelope is secondary in this case. This variant of closed orthomitoses is most typical of the micronuclei of ciliates, the nuclei of schizopyrenid amoebae, and, in a modified form (without a distinct equatorial plate), the nuclei of euglenids.

First Variant (Mitoses with Submembranous MTOCs)

Rhizopoda. This variant of mitosis was studied in *Euglypha* (order Gromiida) by SCHEWIAKOFF (1888) and BĚLAŘ (1926a) with the light microscope and by OGDEN (1979) with the electron microscope. In the prophase, a small "centrosphere" differentiates on the outer side of the nuclear envelope and soon divides into two parts. The daughter centrospheres migrate to the opposite poles of the nucleus. Dense spherical nuclear pole bodies (NPB) surrounded by a zone of radial cytoplasmic microtubules correspond to these centrospheres at the fine structural level. Some of the microtubules bend around the whole dividing nucleus on the outside and connect the two NPB (Fig. 27). Dense centriolar plaques appear on the inner side of the nuclear envelope opposite the NPB; the central spindle is formed between the plaques and chromosomal microtubules extend from them (Fig. 27). Therefore, *Euglypha* has two types of MTOC: submembranous intranuclear MTOCs, which organize the mitotic spindle, and cytoplasmic MTOCs (NPB), which organize the system of cytoplasmic microtubules. In the metaphase, the spindle is symmetrical with an equatorial plate made of condensed chromosomes that split lengthwise (Fig. 27). The nucleoli disperse in the prophase and are formed *de novo* in the telophase. The nuclear envelope remains intact during the whole course of mitosis and divides in a dumbbell-like manner in the telophase, producing the envelopes of the daughter nuclei (the "direct" way).

According to fragmentary ultrastructural data, closed intranuclear orthomitosis with an equatorial plate made of compact chromosomes and with submembranous MTOCs occurs also in the marine rhizopod *Trichosphaerium* of the order Trichosida (ANGELL 1976, SCHUSTER 1976).

Gregarinia. Open and semiopen mitoses predominate in gregarines. However, closed intranuclear orthomitosis with submembranous MTOCs was observed recently in the gamogony of *Didymophyes gigantea* (HILDEBRAND 1978). In this case, a narrow central spindle passes along the axis of the nucleus and is formed between two electron-dense conical masses located on the inner side of the nuclear envelope. Chromosomal microtubules emerge from them. Chromosomal conden-

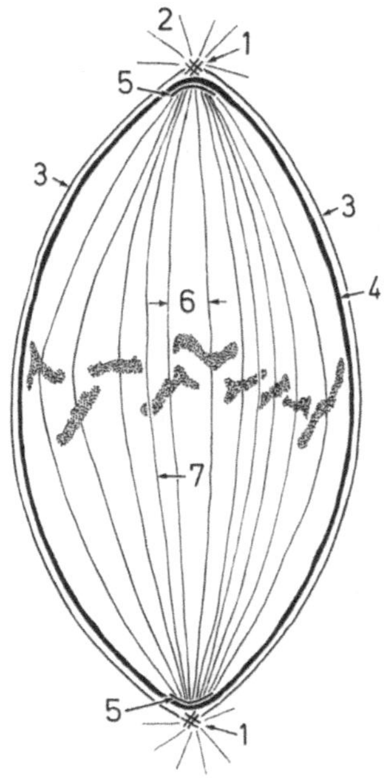

Fig. 27. Diagram of the ultrastructural organization of the *Euglypha* metaphase spindle (after OGDEN 1979; modified). *1* nuclear pole body, *2* radial microtubules in the cytoplasm, *3* cytoplasmic microtubules bending around the nucleus, *4* nuclear envelope, *5* centriolar plaques, *6* chromosomal microtubules, *7* continuous microtubule.

sation in mitosis is poor, and there is no equatorial plate; only the kinetochores of the chromosomes are visible.

Outside the Protozoa, closed intranuclear orthomitosis with submembranous MTOCs is rather rare. Mitosis of the green alga *Cladophora* is an example of this type of mitosis (see the review by HEATH 1980).

Second Variant (Mitoses without MTOCs or with Intranuclear MTOCs)

Micronuclei of Ciliophora. Mitoses of the micronuclei were studied in detail by both light and electron microscopies. A typical light microscopic pattern of mitosis is illustrated in Fig. 28. In the prophase, the chromosomes are arranged along the future spindle and are clearly doubled (Fig. 28*a*). In the metaphase, they form an equatorial plate; the spindle is axially symmetrical from the very beginning

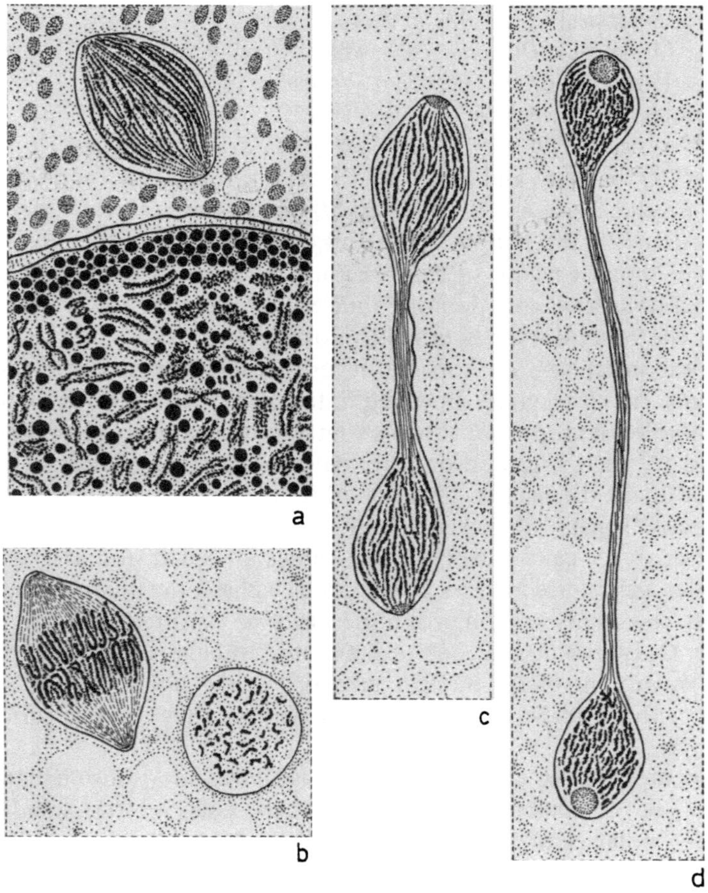

Fig. 28. Mitosis of the micronucleus in *Nassula*, 1850× (after RAIKOV 1962a). *a* prophase (part of the macronucleus at the bottom); *b* metaphase (side view and polar view); *c* anaphase; *d* telophase. Staining: *a* iron haematoxylin; *b*–*d* Feulgen with light green.

(Fig. 28*b*). In the anaphase and telophase, the spindle elongates and a long connecting strand appears between the daughter micronuclei (Fig. 28*c*–*d*). Proteinaceous bodies are visible in the daughter micronuclei near the poles of the division figure in the telophase (Fig. 28*d*). They were found also in other ciliates, such as *Paramecium caudatum* (CALKINS and CULL 1907) and *Stentor* (MULSOW 1913), and were often regarded as intranuclear centrioles or centrospheres; however, electron microscopy showed that, at least in *Nassula* (RAIKOV 1966, TUCKER 1967), these bodies correspond to the structureless zones of the karyolymph (Fig. 10*c*).

In the early works involving electron microscopy, it was reported only that the nuclear envelope did not disperse and that an intranuclear apparatus of microtubules appeared in the dividing micronucleus (ROTH and SHIGENAKA 1964, CARASSO and FAVARD 1965, GRAIN 1966, etc.). The complete cycle of micronuclear mitosis was examined at the ultrastructural level in the gymnostomes *Loxodes* (Raikov 1973, 1976a), *Didinium* (Karadzhan and Raikov 1977), and *Dileptus* (Vinnikova 1974a, 1976), the suctorian *Paracineta* (Hauser 1968, 1970, 1972), the hymenostomes *Tetrahymena* (Davidson and La Fountain 1975, 1977, La Fountain and Davidson 1979) and *Paramecium* (Jurand and Selman 1970, Stevenson and Lloyd 1971a, Inaba and Kudo 1972), the heterotrichs *Blepharisma* (Jenkins 1967, Inaba and Sotokawa 1968) and *Stentor* (Raikhel *et al.* 1981), and the hypotrichs *Urostyla* (Suganuma 1969) and *Gastrostyla* (Walker 1976). Observations of some separate mitotic stages concerning other species have also been made.

As an example, let us consider micronuclear mitosis in the lower ciliate *Loxodes magnus* of the order Karyorelictida (Raikov 1973b, 1976a).

During the interphase, the chromatin forms a spongy electron-dense mass. In the prophase, it disintegrates into separate compact chromosomes. Randomly oriented bundles of short microtubules simultaneously appear in the karyolymph. No centres of their organization can be detected either beneath the nuclear envelope or inside the nucleus. This is a "diffuse" type of spindle fibre formation. Proteinaceous "spheres" that existed in the interphase micronucleus dissolve in the preprophase; it is quite possible that tubulins needed for the assembly of microtubules were stored in them. In the prometaphase, the micronucleus becomes elliptic; all the bundles of microtubules become oriented along it and grow lengthwise. Only then do the ends of many microtubules come into contact with the inner membrane of the nuclear envelope at the poles. At the same time, kinetochores differentiate on the chromosomes, but they are not turned toward the poles yet.

In the metaphase, the chromosomes are assembled in a loose equatorial plate (Fig. 29*a*). The microtubules disassemble at the poles of the spindle and big amorphous "polar caps" are formed. Therefore, both chromosomal and continuous microtubules end free in the karyolymph. The intranuclear division spindle is blunt and surrounded by an intact nuclear envelope. All the chromosomes are now turned with their kinetochores to the poles. The chromosomes of *Loxodes* are small with short arms that are either equal or unequal in length. The centromere region occupies a considerable part of their length. Two disc-shaped kinetochores, 200–400 nm in diameter, differentiate at two opposite sides of each chromosome in the centromere region (Fig. 29*b*). A bundle consisting of 10–12 chromosomal microtubules is attached to each kinetochore. The interchromatidal cleft cannot be discerned in the metaphase.

At the beginning of the anaphase, the chromosomes duplicate and each daughter chromosome has only one kinetochore. In the late anaphase, the chromosomes migrate to the poles of the spindle, which slightly elongates. Chromosomal microtubules disappear. The ends of the microtubules do not reach

---

Fig. 29. Ultrastructure of micronuclear mitosis in *Loxodes magnus* (after Raikov 1976a). *a* metaphase, 12000×; *b* metaphase chromosome, 37000×; *c* telophase, 8000×; *1* chromosomes, *2* spindle "polar caps", *3* kinetochores, *4* chromosomal microtubules, *5* envelope of the mother micronucleus, *6* new envelope of the daughter micronucleus, *7* microtubules in the interzone.

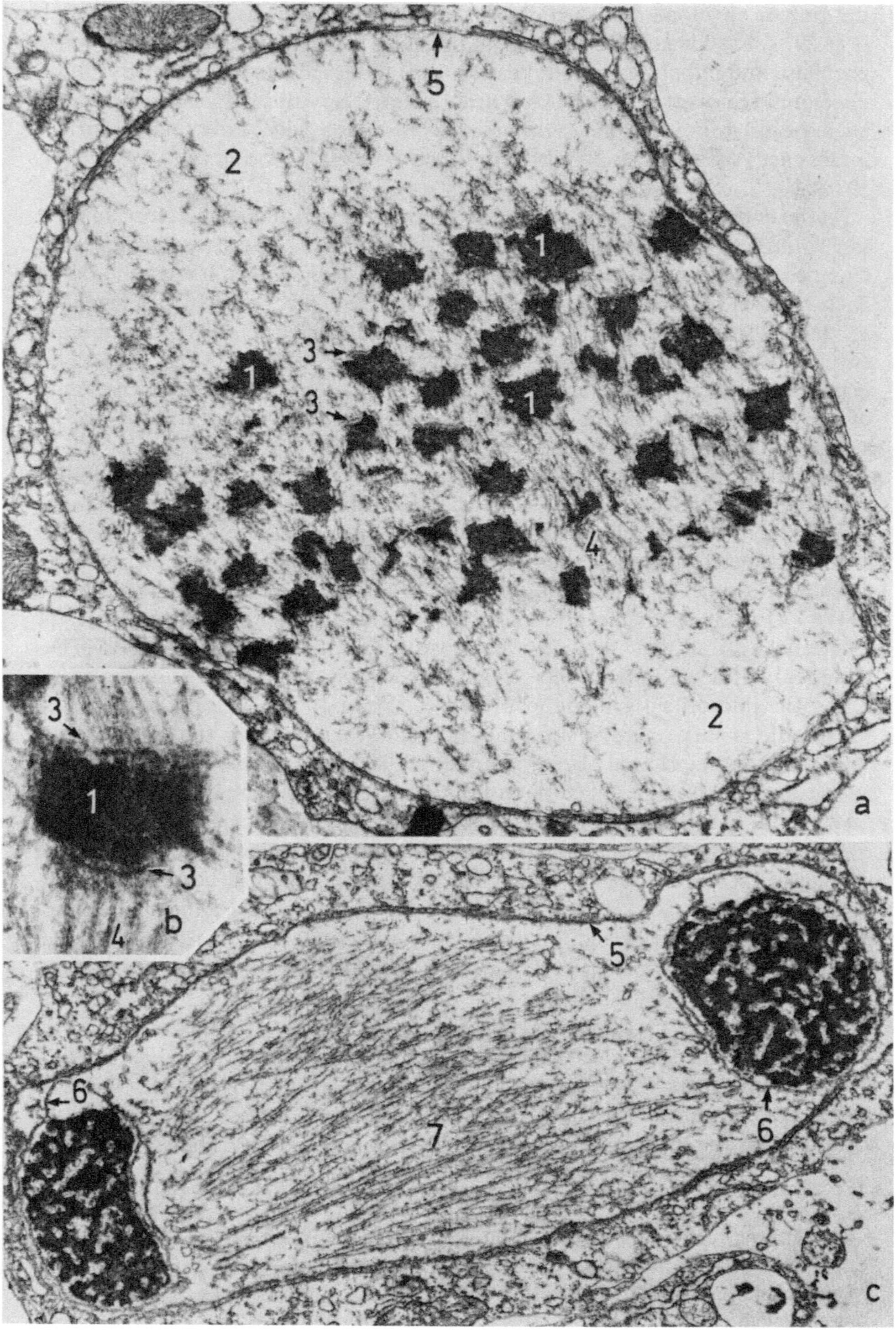

Fig. 29

the nuclear envelope. In the telophase, the chromosomes fuse into a spongy mass (Fig. 29*c*). The kinetochore material is still visible in the alveoli of this mass. In the anaphase and telophase, the number of microtubules increases considerably in the interzonal region of the spindle. Apparently, together with continuous microtubules, a special population of microtubules (which can be called interzonal) is formed in this case; they were also reported in *Tetrahymena* (DAVIDSON and LA FOUNTAIN 1975).

At the end of mitosis, the nuclear envelope of the micronucleus is reorganized in the "indirect" ("intranuclear") way (cf. pp. 83, 84). Numerous vesicles and flattened cisternae pinch off into the nucleus from the inner membrane of the nuclear envelope that surrounds the whole spindle. The cisternae then merge on the surface of the telophasic chromatin mass, producing a new two-membrane envelope that surrounds each daughter micronucleus separately (Fig. 29*c*). All the microtubules of the spindle remain outside these envelopes. The common envelope of the parent micronucleus then disintegrates, and the entire material of the spindle gets into the cytoplasm.

A different variant of intranuclear orthomitosis is typical of the micronucleus of *Didinium nasutum*, a ciliate of the order Prostomatida (KARADZHAN and RAIKOV 1977). The interphasic micronucleus contains a peculiar electron-dense body that is not connected with the nuclear envelope. In the early prophase, microtubules grow from this body in a fan-like manner; it thus appears to be an intranuclear MTOC (Fig. 30*a*). In the prophase, the microtubules scatter throughout the micronucleus and begin to align themselves parallel to one another; the MTOC disappears and the chromosomes begin to condense (Fig. 30*b*). In the prometaphase, the microtubules fill the whole micronucleus. In the metaphase (Fig. 30*c*), the spindle is surrounded by an intact micronuclear envelope; the chromosomes form a distinct equatorial plate and are connected with bundles of chromosomal microtubules coming from both poles. Other bundles of microtubules (continuous) pass between the chromosomes. The microtubules partially disassemble in the polar zones of the spindle, but no large amorphous polar caps appear. In the anaphase, the chromosomal bundles of microtubules can still be discerned, but, unlike in *Loxodes*, the kinetochores are poorly expressed in *Didinium* (Fig. 30*d*). The intranuclear spindle and the whole micronucleus become greatly elongated. Newly formed interzonal microtubules are added to the continuous microtubules in the middle part of the spindle.

In the telophase, the division figure looks like a very long connecting tube with two bulges at the ends, which are the future daughter micronuclei (Fig. 30*e*). All the microtubules of the spindle depolymerize and electron-dense bodies (MTOCs) typical of interphasic micronuclei reappear in the polar bulges. They seem to be formed from the material of the disintegrating microtubules. The chromosomes

---

Fig. 30. Ultrastructure of micronuclear mitosis in *Didinium nasutum* (after KARADZHAN and RAIKOV 1977). *a* outgrowth of microtubules from a MTOC in the early prophase, 40 000 × ; *b* late prophase, 16 000 × ; *c* metaphase, 25 000 × ; *1* chromosomes, *2* microtubules, *3* micronuclear envelope, *4* MTOC, attachment of the microtubules to the chromosomes is indicated by arrows.

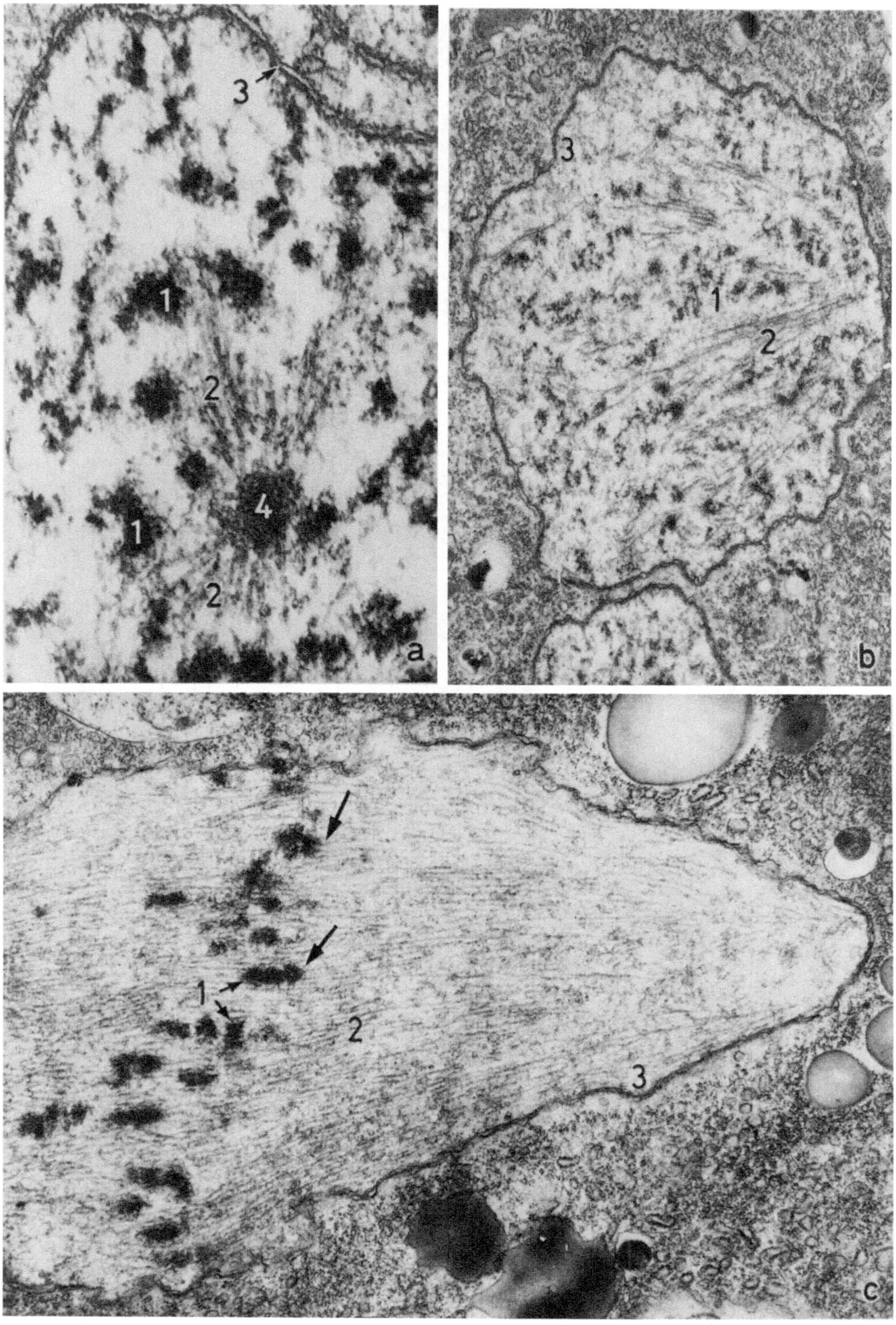

Fig. 30 *a*–*c*

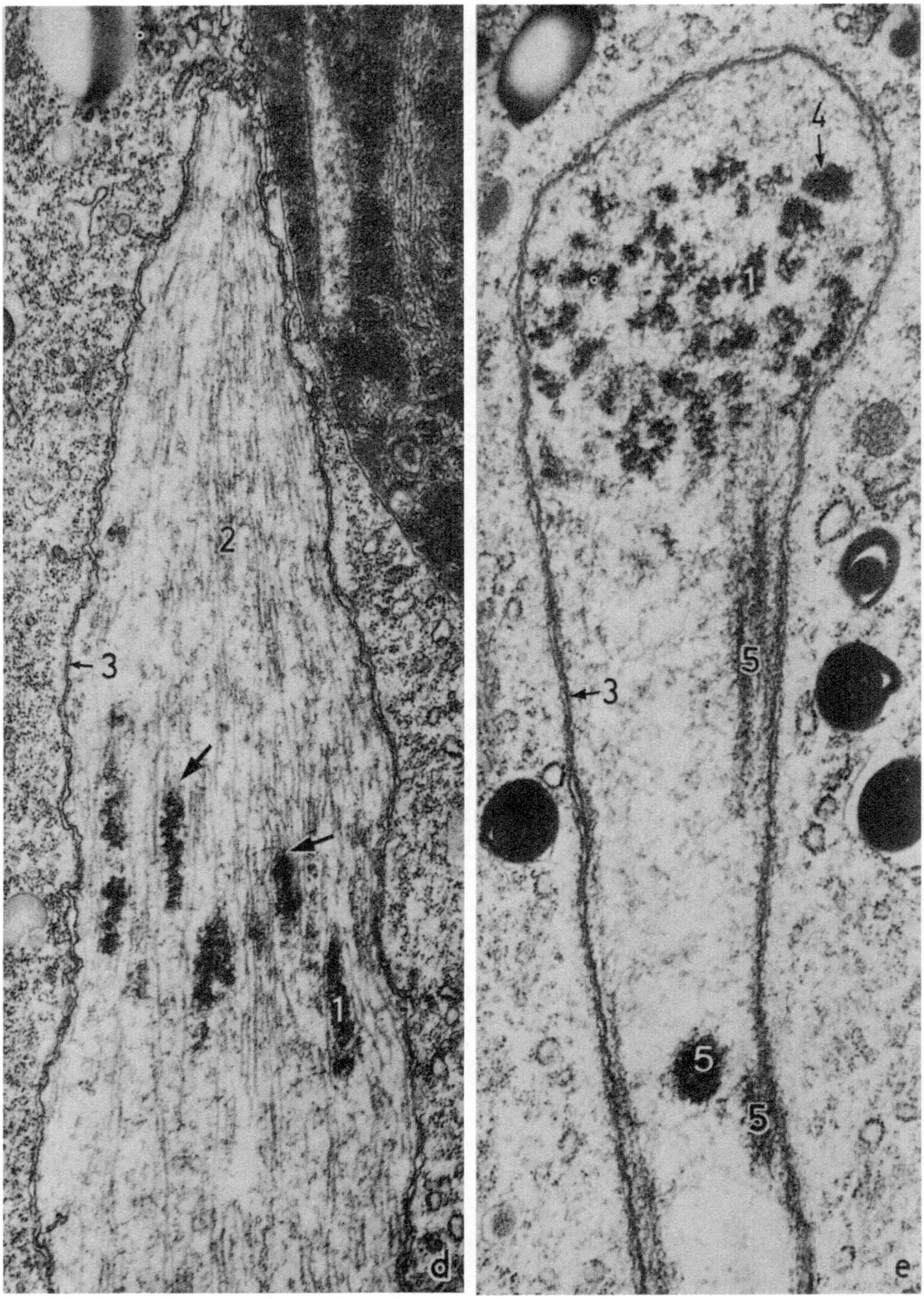

Fig. 30 *d-e*. Mitosis in *Didinium*. *d* part of the anaphase spindle, 36 000 × ; *e* telophase spindle pole, 27 000 × ; same designations as in *a-c*, *5* material of depolymerizing microtubules.

gradually decondense in the late telophase, the daughter micronuclei simply pinch off the middle of the spindle (the "direct" way of the telophasic reorganization of the nuclear envelope). The middle part of the telophasic spindle, together with its tubular envelope, is resorbed in the cytoplasm.

Other ciliates also vary in the way in which the microtubules are formed in the prophasic micronuclei. Their "diffuse" (just as in *Loxodes*) formation in the karyolymph predominates (*Dileptus*, *Stentor*, *Blepharisma*, *Urostyla*, *Gastrostyla*; references on p. 104). In other species, however, some microtubules are preserved, usually as short fragments, at the periphery of the micronucleus during the whole interphase and are referred to as a residual spindle (cf. p. 66). In the prophase, the fragments of microtubules grow, forming spindle fibres; thus, each residual fragment acts as a "crystallization centre" of tubulins or as an elementary MTOC. Such a mode of spindle formation is typical of *Tetrahymena* and *Paramecium*. In *P. bursaria*, however, "reserve" tubulins are stored in the nucleus not as microtubules, but as dense microfilaments 6 nm thick that are capable of undergoing paracrystalline aggregation (HAUSER and BEINBRECH 1973, SCHWARTZ 1976). A noteworthy fact is that in *Ichthyophthirius*, which has residual fragments of microtubules, a bundle of anomalously wide tubules (with an outer diameter of 30–35 nm) is formed first at the periphery of the micronucleus in the prophase. These "macrotubules", however, soon disappear and normal microtubules (20 nm) are formed throughout the nucleus (HAUSER 1972).

Some ciliates contain a compact MTOC in the micronucleus. Such a MTOC was first observed in the suctorian *Paracineta* by HAUSER (1968, 1970, 1972). In the interphasic micronucleus, it looks like a crystalloid, 0.25 μm long, consisting of fibrils 8 nm in diameter that are packed in a system with a periodicity of 13 nm. In the prophase, the crystalloid elongates, and then its ends are transformed into a bundle of wide tubules 35–40 nm in diameter. The latter grow intensively in length, stretching the nuclear envelope with their ends, while the crystalloid (MTOC) separates from the bundle of "macrotubules". Then, the "macrotubules" are gradually replaced by ordinary microtubules. Later, normal microtubules are also produced diffusely in the prophasic micronucleus, and the MTOC is conserved during the entire course of mitosis. In the metaphase or anaphase, the MTOC seems to duplicate, since both daughter micronuclei have it in the telophase.

The structure of the spindle in the metaphasic micronuclei may also vary among ciliates. In most species, continuous and chromosomal microtubules combine, producing a common spindle (just as in *Loxodes* and *Didinium*). This applies, for instance, to *Dileptus*, *Paracineta*, all the species of *Paramecium* examined, *Blepharisma*, *Stentor*, *Urostyla*, and *Gastrostyla* (references on p. 104), as well as to the hypostome *Nassula* (TUCKER 1967) and the peritrichs *Epistylis* and *Vorticella* (CARASSO and FAVARD 1965). In *Tetrahymena*, however, the spindle components are separated in space: continuous microtubules are located only at the periphery of the spindle (i.e. under the micronuclear envelope), while the bundles of chromosomal microtubules occupy a position closer to the centre; bundles of free (nonkinetochore) microtubules also occur among the latter (DAVIDSON and LA FOUNTAIN 1975, LA FOUNTAIN and DAVIDSON 1979). Crossbridges are observed between the microtubules of all three classes, suggesting that their mutual mechano-chemical sliding occurs.

The attachment of microtubules to the chromosomes was observed in *Dileptus*, *Nassula*, *Paracineta*, *Tetrahymena*, *Paramecium*, and *Blepharisma*, but the morphological differentiation of kinetochores is poor in most of these ciliates (just as in *Didinium*). The kinetochores are defined more clearly in *Paramecium primaurelia* (JURAND and SELMAN 1969, 1970) and *Tetrahymena* (DAVIDSON and LA FOUNTAIN 1975, LA FOUNTAIN and DAVIDSON 1979), though far less clearly than in *Loxodes*. Large lamellar kinetochores were detected only on the chromosomes of the karyorelictid *Trachelocerca geopetiti* (RAIKOV and KOVALEVA 1981).

The mitotic apparatus of the micronucleus of *Paramecium bursaria* has a peculiar structure (LEWIS 1974, 1975, LEWIS *et al.* 1976, SCHWARTZ 1976). According to LEWIS, bundles of chromosomal microtubules exist in this case only near the poles of the spindle; closer to the equator, they are replaced by strands of microfilaments that are in contact with the chromosomes and are likely to be contractile. According to SCHWARTZ, the microtubules are in direct contact with the chromosomes, and the spindle consists of both continuous and chromosomal bundles of microtubules. In the prophase, it develops from the reserve nontubular filaments preserved in the interphase mainly near one (primary) pole of the micronucleus. Neither author found typical kinetochores in *P. bursaria*, though they observed peculiar funnel-shaped microfilamentous structures covering the chromosomal ends turned toward the poles. SCHWARTZ termed these formations "kinosomes".

In the polar regions of the dividing micronucleus, direct contact between the ends of microtubules and the nuclear envelope is usually established in the prometaphase, i.e. when the microtubules rapidly elongate (*Loxodes*, *Didinium*, *Dileptus*, *Paracineta*, *Paramecium multimicronucleatum*, *Blepharisma undulans*). Such contact is usually lost in the metaphase and/or the anaphase because the microtubular ends are disassembled and amorphous polar caps are formed beneath the micronuclear envelope. The caps are small in *Didinium*, *Dileptus* (VINNIKOVA 1974a), *Paracineta* (HAUSER 1970), *Paramecium multimicronucleatum* (INABA and KUDO 1972), *Urostyla* (SUGANUMA 1969), and *Gastrostyla* (WALKER 1976) but are very large in *Loxodes* and *Paramecium primaurelia* (JURAND and SELMAN 1970). In the anaphase and/or telophase, the continuous microtubules grow again, sometimes very rapidly, and often butt their ends against the nuclear envelope, e.g. in *Didinium* (Fig. 30*d*) and *P. primaurelia* (STEVENSON and LLOYD 1971a). In *Blepharisma*, the microtubules remain in contact with the nuclear envelope from the prometaphase to the telophase (JENKINS 1967).

Most micronuclear mitoses are characterized by very intensive elongation of the spindle in the anaphase and telophase. This seems to be connected with the appearance of special interzonal microtubules other than the continuous ones (DAVIDSON and LA FOUNTAIN 1975). Apparently this is an adaptation typical of ciliates and is due to the fact that the micronucleus and its metaphase spindle are disproportionately small compared to the size of the entire cell; great elongation helps the daughter micronuclei to segregate during cytokinesis. The telophase spindles of the micronuclei remain comparatively short in those multinucleate ciliates, in which the daughter micronuclei segregate at random rather than in a regular way during cytokinesis, e.g. in *Loxodes* (RAIKOV 1959a). In *Tetrahymena*, the overall mass of microtubular material in the micronucleus increases 2.5 times during the anaphase, mainly at the expense of the interzonal microtubules (Davidson and La Fountain 1977). The interzonal microtubules apparently originate from the greatly elongated "free" microtubules that have been detected as early as in the metaphase. Nevertheless, the micronucleus of *Tetrahymena* is capable of stretching even if the number of microtubules in it is greatly reduced by colchicine inhibition; this fact suggests that other (e.g. membrane-related) mechanisms may be responsible for telophasic elongation (Jaeckel-Williams 1978). It has been observed in *Spirostomum* that the micronuclear spindle is intensively stretched in the telophase, being accompanied by its strong birefringence (Jenkins 1978).

As for the telophasic reorganization of the micronuclear envelope, its direct constriction (like in *Didinium*) has also been observed in *Nassula* sp. (Tucker 1967), *Paramecium primaurelia* (Stevenson and Lloyd 1971a), and *P. multimicronucleatum* (Inaba and Kudo 1972). The "indirect" (intranuclear) way, when the envelopes of the daughter micronuclei are formed inside the envelope of the mother micronucleus (as in *Loxodes*), is more widespread; it was investigated in *Dileptus*, *Nassula ornata*, two species of *Blepharisma*, *Stentor*, *Urostyla*, *Gastrostyla*, and especially *Euplotes* (Ruffolo 1978). A noteworthy fact is that both "direct" and "indirect" ways were found in different species within the genus *Nassula* (Raikov 1966, Tucker 1967).

The behaviour of chromosomes in micronuclear mitosis was debated in the literature. On the one hand, individual chromosomes that separate lengthwise into two chromatids were seen with the light microscope in a number of ciliates, such as *Paramecium caudatum* (Chen 1940a), *Tracheloraphis* (Raikov 1957a, 1958), and *Nassula* (Fig. 28); in other words, the behaviour of chromosomes was described as being quite typical. This was also confirmed by electron microscopy in *Loxodes*, *Didinium*, *Dileptus*, *Paracineta*, *Tetrahymena*, *Paramecium primaurelia*, and *Blepharisma* (references on p. 104). These forms have small monocentric chromosomes, attached to two bundles or one bundle of chromosomal microtubules, respectively, in the metaphase and the anaphase (Figs. 29*a*, *b*; 30*c*).

On the other hand, light microscopy has revealed that the micronuclei of certain ciliates do not display individual chromosomes in mitosis; instead, they show irregularly shaped chromatin masses whose number varies and that divide transversely, e.g. in *Colpidium*, *Chilodonella*, *Euplotes*, *Oxytricha*, *Stylonychia* (Devidé and Geitler 1947, Devidé 1951), *Vorticella* (Mügge 1957), etc. True chromosomes in the form of bivalents were detected in these organisms only in meiosis, and the number of chromatin masses was much less than that of the chromosomes; in *Colpidium campylum*, for instance, 2n = 42, while the number of the masses is only 8–10 (Devidé 1951). In the works just cited, these masses were interpreted as aggregates of several chromosomes. It is possible that in some cases the masses were simply fixation artifacts (the result of chromosome clumping). In the hypotrichids *Urostyla* (Suganuma 1969), *Gastrostyla* (Walker 1976), and *Stylonychia* (Chadha 1980), however, the micronuclear chromosomes are atypical even on the ultrastructural level. The amount of chromatin material is very large in this case as compared with the size of the spindle; the "chromosomes" are massive and poorly individualized during the metaphase and anaphase. Apparently they are aggregates of several chromosomes. A noteworthy fact is that in neither *Urostyla* nor *Gastrostyla* do the spindle fibres seem to be attached to the chromosomes; in other words, these forms seem to lack chromosomal microtubules. Therefore, while the micronuclei of most ciliates have typical chromosomes, it is still unknown whether some other ciliates (Hypotrichida, in particular) have them.

Schizopyrenida and Other Rhizopods. The mitosis of *Naegleria*, *Vahlkampfia*, *Tetramitus*, and other schizopyrenid amoebae, many of which have a flagellate phase in their life cycle, should also apparently be related to the group of closed intranuclear orthomitoses (a variant without membrane-bound MTOCs). It is in these protozoans that the allegedly very primitive "promitosis", with the nuclear envelope being preserved and the nucleolus being pinched in two, was described for the first time; in this case, the nucleolus was believed to play the role of an intranuclear "centrosome" (CHATTON 1910, ALEXEIEFF 1913). However, such a double function of the nucleolus was not confirmed later.

Mitosis of the "*Vahlkampfia* type" (BĚLAŘ 1926a) was studied in detail by light microscopy in *Naegleria gruberi* (RAFALKO 1947, SINGH 1952, PITTAM 1963, PAGE 1967a). Mitosis is found only in the amoeboid phase of the life cycle. Gradually condensing Feulgen-positive chromosomes are seen in the prophase. The nucleolus constricts in a dumbbell-like way; the fibres of the central spindle appear in the karyolymph. The chromosomes attached to the chromosomal fibres of the spindle later form a distinct equatorial plate that splits into two daughter ones. In the anaphase, the nucleolar Feulgen-negative material begins to accumulate in the interzonal region. The nuclear envelope simply pinches in two in the telophase.

"Promitosis" of the same type is observed in *Tetramitus* (HOLLANDE 1942, RAFALKO 1951) in which it occurs in the flagellated phase of the cycle, too. In this case, the basal bodies of the flagella are not involved in mitosis. Similar mitosis is typical of other small schizopyrenid amoebae, such as *Endolimax* (EPSTEIN 1941), *Vahlkampfia* (= *Schizopyrenus*) (SINGH 1952, PAGE 1967a), and *Sappinia* (PUSSARD 1973).

So far, mitosis in schizopyrenids has been studied with the electron microscope only in *Naegleria* (FULTON 1970, FULTON and DINGLE 1971, SCHUSTER 1975a). Despite all efforts, neither centrioles nor any formed nuclear pole bodies were found in the cytoplasm near the poles of the dividing nucleus (FULTON and DINGLE 1971). It is likely that *Naegleria* has no centrioles in the amoeboid phase of the cycle, and flagellar kinetosomes are formed *de novo* when the amoeboid form is transformed into the flagellated one (DINGLE and FULTON 1966, FULTON and DINGLE 1971). The nuclear envelope is preserved throughout mitosis and pinches in two in the telophase without any reorganization (SCHUSTER 1975a). The nucleolus loosens in the course of mitosis, and clumps of its material are distributed between the daughter nuclei; the nucleolus condenses once again in the telophase. The spindle is entirely intranuclear and acentric and consists of longitudinal bundles of microtubules that are either located at the periphery of the loosened nucleolus or partially pass through it. The microtubules either end free in the karyolymph near the poles of the nucleus or butt against the inner membrane of the nuclear envelope, which shows no differentiations like MTOCs or the centriolar plaques (FULTON 1970, FULTON and DINGLE 1971, SCHUSTER 1975a). Apparently, spindle microtubules develop diffusely in the karyolymph, just as in the micronuclei of many ciliates.

Chromosomes could not be detected at first in *Naegleria* mitosis by electron microscopy (FULTON 1970, SCHUSTER 1975a); this was inconsistent with the data of light microscopy, which indicated the presence of compact Feulgen-positive chromosomes and chromosomal spindle fibres. GRIFFIN (cited in GOODE 1975), however, used a new method of fixation and detected both small chromosomes and the microtubules attached to them. Therefore, the mitosis of *Naegleria* and other schizopyrenids should apparently be brought closer to that of the ciliate micronuclei in spite of the conspicuous difference with respect to the nucleolus. The latter pinches in two in schizopyrenids, while the micronuclei lack it.

Among other amoeboid protozoans, closed intranuclear orthomitosis in its acentric variant (with a broad, blunt-ended spindle enclosed in an intact nuclear envelope) has been reported in the multinucleate athalamid rhizopod *Arachnula impatiens* (OLD and DARBYSHIRE 1980).

Euglenida. For a long time, mitosis in euglenids was considered to be an independent type, i.e. closed "euglenoid" mitosis, because neither the chromosomal fibres nor any other connections between the spindle and the chromosomes could be found despite the presence of an axially symmetrical (orthomitotic), entirely intranuclear spindle without MTOCs beneath the nuclear envelope. Such connections were discovered only recently (GILLOTT and TRIEMER 1978), thus making it possible to reduce the mitosis of euglenids to closed intranuclear orthomitosis.

The light microscopic pattern of euglenoid mitosis is shown in Fig. 31 *a–c*. In the prophase, the chromosomes are filamentous and consist of two parallel chromatids. In the metaphase, there is no distinct equatorial plate; paired chromosomes are oriented mainly along the division figure; the nuclear envelope does not disperse; the spindle cannot be discerned either in the nucleus or in the cytoplasm; at the end of the prophase, the nucleolus elongates and begins to divide (Fig. 31 *a*). In the anaphase, the chromatids separate and gradually migrate to the poles (Fig. 31 *b*). Finally, the nucleolus and then the nucleus divide into two parts (Fig. 31 *c*).

Early descriptions of mitosis in euglenids are rather contradictory. Some authors observed the basal bodies of the flagella (kinetosomes) at the poles of the division figure (HALL 1923, 1937, BAKER 1926, RATCLIFFE 1927, BROWN 1930) while others did not (TSCHENZOFF 1916, HOLLANDE 1942, LEEDALE 1958, etc.). When other types of mitosis were discussed earlier, it was noted that the presence or absence of kinetosomes at the poles hardly bore any relation to the mitotic mechanism, particularly in closed intranuclear mitoses. Electron microscope studies have shown that the dividing nucleus of *Astasia* develops peculiar fingerlike extensions at the poles that are directed toward the flagellar kinetosomes. Bundles of microtubules were found in these extensions (BLUM *et al.* 1965, SOMMER and BLUM 1965). *Euglena* has no such extensions (LEEDALE 1958, 1968a), though the presence of kinetosomes near the poles of the dividing nucleus was confirmed, at least in *E. gracilis* (GILLOTT and TRIEMER 1978). The spindle cannot be revealed in euglenids by light microscopy, and it was long held that the spindle was absent.

The stage at which sister chromatids separate varies considerably. Sometimes, separation immediately follows mitosis, and then chromosomes look paired (bichromatidal) during the whole interphase (LEEDALE 1958, SAITO 1961a). Apparently, such an early separation of chromatids indicates that the chromosomes of Euglenida are oligotene or even polytene.

The anaphase movement of chromosomes to the poles of the dividing nucleus is atypical in Euglenida. The shape of the anaphase chromosomes varies from straight rods to bent threads turned to the pole not only with their bend, but also with their free ends. Therefore, the chromosomes of euglenids were often believed to lack localized centromere regions. Irradiation of euglenids does not induce the formation of anaphase bridges and chromosome fragments that greatly lag behind (LEEDALE 1958). It should be noted, however, that any acentric chromosome fragments in closed mitosis would inevitably be incorporated in one of the daughter nuclei. LEEDALE (1959b, 1968) studied mitosis in *Euglena gracilis* and *E. spirogyra in vivo* and showed that the anaphase takes more than an hour, while the movement of chromosomes is slow, asynchronous, discontinuous, and "staggering", as it were. Moreover, the movement is not inhibited by colchicine (LEEDALE 1958, 1968). All these findings seemed to indicate that the segregation of chromosomes involved a mechanism other than that of the microtubules attached to the chromosomes (even at many points).

There are only a few works in which mitosis of euglenids was examined with the electron microscope. Most of the authors found only an intranuclear central spindle consisting of microtubules (Fig. 31*d*). The latter form a few bundles situated near the surface of the elongating nucleolus. Such an intranuclear apparatus of continuous microtubules was reported in *Euglena gracilis* (LEEDALE 1968, 1970), *Astasia longa* (BLUM *et al.* 1965, SOMMER and BLUM 1965, CHALY *et al.* 1977), *Isonema nigricans* (SCHUSTER *et al.* 1968), and *Scytomonas pusilla* (MIGNOT 1966). Differences between the species are slight: the bundles of microtubules lie on the very surface of the nucleolus in *Scytomonas*, while they are at a certain distance away from it in *Euglena*, *Astasia*, and *Isonema* (Fig. 31*d*). In *Euglena* and *Isonema*, the microtubules at the poles reach the nuclear envelope; in *Astasia*, they even occasionally bulge it, forming finger-like extensions; in *Scytomonas*, the microtubules do not reach the envelope. In the prophase, microtubules are formed in the karyolymph, apparently in a diffuse way (LEEDALE 1968, CHALY *et al.* 1977); no MTOCs were found beneath the nuclear envelope or elsewhere.

However, small kinetochores that look like trilaminar discs were recently discovered on the metaphase chromosomes of *Euglena gracilis* (GILLOTT and TRIEMER 1978). Bundles of microtubules are attached to them; apparently, the spindle of *Euglena* consists not only of continuous, but also chromosomal microtubules. Consequently, the mitosis of Euglenida is similar to the acentric variant of closed intranuclear orthomitosis (such as in Schizopyrenida) and should not be regarded as an independent type.

Fig. 31. Mitosis in euglenids. *a*-*c* mitosis in *Euglena granulata* (*a* late prophase, *b* anaphase, *c* early telophase), 3200 × ; *d* electron micrograph of the dividing nucleus of *Astasia longa* in the late prophase, 23 500 × ; *1* chromosomes, *2* dividing nucleolus, *3* microtubules (many of them are close to the nucleolar surface), *4* nuclear envelope. *a*-*c* after HOLLANDE (1942); *d* after CHALY *et al.* (1977).

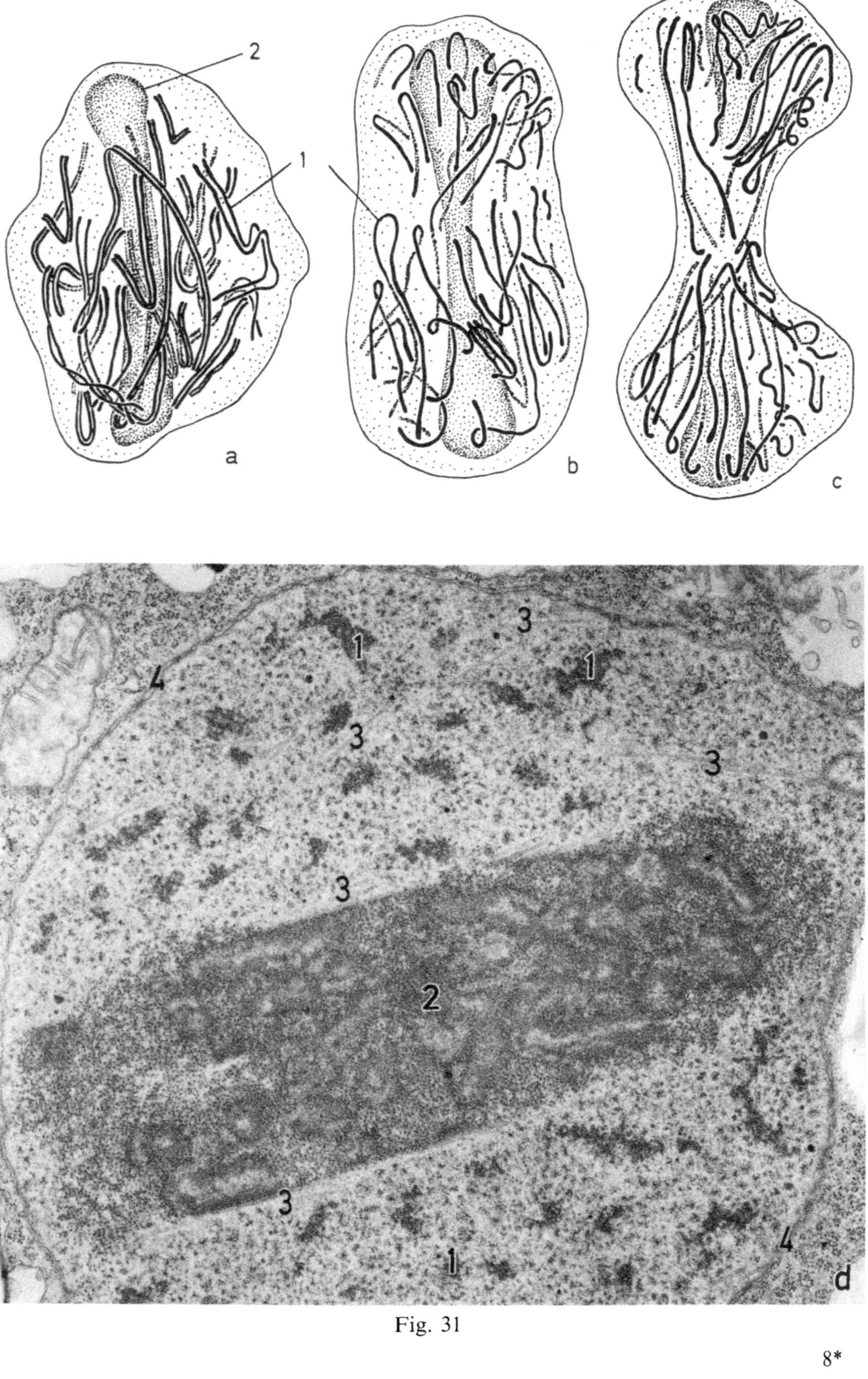

Fig. 31

The organization of the mitotic apparatus in euglenids is still obscure in many aspects. In particular, the spatial relationship between chromosomal and continuous microtubules is unknown; it is also quite possible that the chromosomes of euglenids are polycentric, i.e. each of them has several kinetochores (this would account for their strange behaviour in the anaphase). The resistance of euglenoid mitosis to colchicine might be attributed simply to its poor uptake by the cells (LEEDALE 1968).

In other respects, the data of electron microscopy corroborate the observations made with the light microscope. The nucleolus of euglenids always stretches into a rod during mitosis; then it pinches in two without changing its ultrastructure as compared with the interphase (CHALY *et al.* 1977). The nuclear envelope remains intact during the entire course of mitosis, though numerous folds are formed in it at the spindle poles (Fig. 31 *d*); in the telophase, the nucleus divides into two parts without the envelope being reorganized (CHALY *et al.* 1977, GILLOTT and TRIEMER 1978, etc.).

Other Protozoans. According to data obtained by light microscopy (CHEN 1936a, 1936b, 1948, 1972), closed intranuclear acentric orthomitosis is typical of the nuclei of the opalinids of the genus *Zelleriella*. The same type of mitosis was found in *Opalina* (DEVIDÉ 1951). Unfortunately, mitosis in Opalinata has still not been studied with the electron microscope.

Closed intranuclear orthomitosis with a free (not membrane-attached) MTOC was recently described by means of electron microscopy in *Entamoeba histolytica* (GICQUAND 1979). In this case, the MTOC appears in the early prophase as a dense body (about 1 μm in diameter) in the centre of the nucleus; microtubules grow radially from the MTOC and join the numerous chromosomes (Fig. 13 *b*). As a result of the prometaphasic reorganization of the spindle, the MTOC disappears, the microtubules reorient themselves as a single axial bundle whose ends do not reach the nuclear envelope, the chromosomes form a transient annular equatorial plate, and the peripheral nucleoli disperse. The nuclear envelope constricts in a "direct" manner in the telophase (cf. p. 83).

Closed Extranuclear Pleuromitosis

Closed extranuclear pleuromitosis, being a very peculiar and well-defined type of mitosis, is referred to simply as pleuromitosis by GRASSÉ (1952) or as cryptopleuromitosis by HOLLANDE (1972). In this case, the nuclear envelope remains intact while the spindle (the central one as well as the chromosomal fibres) is wholly extranuclear. The centres that organize the spindle are not flagellated kinetosomes, but their ribbon-shaped derivatives, the *atractophores*, occasionally with amorphous "centrospheres" (or, better, *atractospheres*) at the ends. The spindle is adjacent to the nucleus on the outside and is therefore always bilaterally symmetrical. The central spindle is either a narrow bundle of microtubules (centrodesmose, paradesmose) that connects the atractophores or a greatly developed apparatus made of continuous and/or polar microtubules; in the latter case, astral rays (radial cytoplasmic microtubules) are usually observed around the atractospheres. The chromosomal fibres of the spindle are completely separated from the central spindle and are arranged in two extranuclear half-spindles that contact the kinetochores of the chromosomes. In this type of mitosis, the kinetochores are built into the nuclear envelope. As a rule, there is no equatorial plate.

Pleuromitosis with an extranuclear spindle is most typical of two zoomastigophorean orders: Trichomonadida and Hypermastigida. We will see later that it is also typical of mesokaryotic Dinoflagellida (see Chapter 4).

Hypermastigida. Pleuromitosis was studied in detail in a number of hypermastigids with the light microscope (CLEVELAND *et al.* 1934, CLEVELAND 1935b, 1938b, 1949a, 1957, 1958a, 1958b, 1966a, 1966b, GRASSÉ and HOLLANDE 1951, 1963, GRASSÉ 1952). Owing to the large size of the mitotic apparatus, almost all the peculiarities of mitosis in hypermastigids were interpreted correctly even in the era of light microscopy. The schools of CLEVELAND and GRASSÉ differed, however, in the terminology and interpretation of certain details in mitosis. For instance, from the very beginning, CLEVELAND believed that chromosomal fibres were involved in the movement of chromosomes. Conversely, GRASSÉ long held that kinetochores slid "autonomously" along the inner side of the nuclear envelope. CLEVELAND's opinion turned out to be right when mitosis in hypermastigids was studied later in detail by electron microscopy (HOLLANDE and VALENTIN 1967a–c, 1968a–c, 1969, HOLLANDE and CARRUETTE-VALENTIN 1970b, 1971, 1972, KUBAI 1972, 1973).

Let us consider, for example, the light microscopic pattern of mitosis in *Barbulanympha* (CLEVELAND *et al.* 1934, CLEVELAND 1938, 1958a, RITTER *et al.* 1978, INOUÉ and RITTER 1978) while using the data of electron microscopy with respect to certain details of the mitotic apparatus (HOLLANDE and VALENTIN 1967a–c, 1968b, HOLLANDE and CARRUETTE-VALENTIN 1971, HOLLANDE 1972, RITTER *et al.* 1978).

Two ribbon-shaped strands up to 30 μm long are always seen in the cytoplasm at the anterior body end; each of them has a "terminal granule" at the front end and a sphere of homogeneous cytoplasm at the rear end (Fig. 32*a*). CLEVELAND referred to these strands as "centrioles", but the term must be discarded since electron microscopy has shown them to have a microfilamentous structure. HOLLANDE and VALENTIN called them *atractophores*, i.e. "carriers of the spindle". The atractophores are fibrillar derivatives of a "morphoplasmic" zone that, in turn, is in contact with a group of four "privileged" kinetosomes located in the foremost part of the body. These kinetosomes correspond to CLEVELAND's "terminal granules". The spheres (atractospheres) at the rear ends of the atractophores have a fine granular ultrastructure in the interphase.

In the prophase, the nucleus approaches the anterior end of the body and, consequently, the atractospheres. Astral rays, which are visible even *in vivo*, appear around the atractospheres. Electron microscopy has shown that they are radial bundles of microtubules induced by the atractosphere. The latter is therefore an extranuclear MTOC in *Barbulanympha*. The central spindle, which is also made of strong bundles of microtubules, then develops between the atractospheres (Fig. 32*b*). The chromosomes are not duplicated yet; nearly all of them are V-shaped (have two arms) and are attached with their kinetochores to the inner side of the nuclear envelope (Fig. 32*b'*). All the chromosomes and their kinetochores duplicate in the late prophase (Fig. 32*c*, *c'*). The central spindle, while being still "suspended" on the atractophores, is now next to the front pole of the nucleus

(Fig. 32*c*). The spindle then elongates and "presses" into the nucleus, setting itself in its furrow (Fig. 32*d*). The nuclear envelope does not dissolve, however, and the spindle remains entirely extranuclear (Fig. 33*a*). By this time, some of the astral rays passing from the atractospheres join the chromosomal kinetochores "through the nuclear envelope" and become the chromosomal fibres of the spindle (Fig. 32*c*, *c'*). A fibre coming from one pole is always attached to the kinetochore of one chromatid, while a fibre from the other pole joins the kinetochore of the other chromatid. Later, the centromere regions of the two chromatids of each chromosome segregate toward the respective poles of the spindle, although they remain bound to the nuclear envelope (Fig. 32*d*, *d'*). Then, the arms of the chromatids separate, too (Fig. 32*e*, *e'*), while the chromosomal fibres of the spindle shorten. This stage seems to be the anaphase. The metaphase stage is absent in *Barbulanympha* (and other hypermastigids) because the centromere regions of the chromosomes duplicate very early and chromatids immediately begin to segregate.

The ultrastructure of the kinetochores and their relationship with the nuclear envelope and the spindle fibres are noteworthy. In *Barbulanympha* (HOLLANDE and VALENTIN 1968b), the prophase kinetochore is an electron-dense disc about 2 μm in diameter located on the inner side of the intact nuclear envelope and bordering on the material of the respective chromosome (Fig. 33*b*). Later, the kinetochore splits into two daughter ones, which condense, become trilaminar (the outer layer being the thickest and the densest), and are built into the openings of the nuclear envelope (Fig. 33*c*). The outer layer of the kinetochore is now in contact with the cytoplasm and can attach a bundle of radial microtubules. We will see below how the kinetochore is built into the nuclear envelope by taking the example of *Trichonympha*.

Even under the traction of chromosomal fibres, the movement of the kinetochores built into the nuclear envelope implies that the membranes of the envelope dissolve in front of the kinetochore and re-form behind it (a sort of "fluidity" of the membranes).

The anaphase comprises two stages in *Barbulanympha* (INOUÉ and RITTER 1978). In the first stage, the chromosomal fibres of the spindle shorten considerably and all the kinetochores of each daughter group of chromosomes assemble on a small region of the nuclear envelope in the vicinity of the respective atractosphere; the length of the central spindle does not change (Fig. 32*e*). In the second stage, the nuclear envelope pinches in two and the daughter nuclei completely separate owing to the great elongation of the central spindle (Fig. 32*f*). In the telophase, the central spindle disappears (i.e. its microtubules depolymerize in the cytoplasm) and the daughter nuclei remain suspended each on its own atractophore

Fig. 32. Closed extranuclear pleuromitosis in *Barbulanympha* (after CLEVELAND *et al.* 1934). *a* anterior part of the body with interphase atractophores and atractospheres; *b* early prophase, development of the central spindle, chromatids have not yet individualized (*b'* one chromosome); *c* late prophase, individualization of chromatids (*c'* one chromosome, spindle fibres attached to the kinetochores of two chromatids); *d* anaphase (*d'* scheme showing how chromatids of one chromosome separate); *e* late anaphase (*e'* one pair of daughter chromosomes); *f* and *g* telophase. Iron haematoxylin staining, 1050 × (*b'*–*e'* 1600 ×).

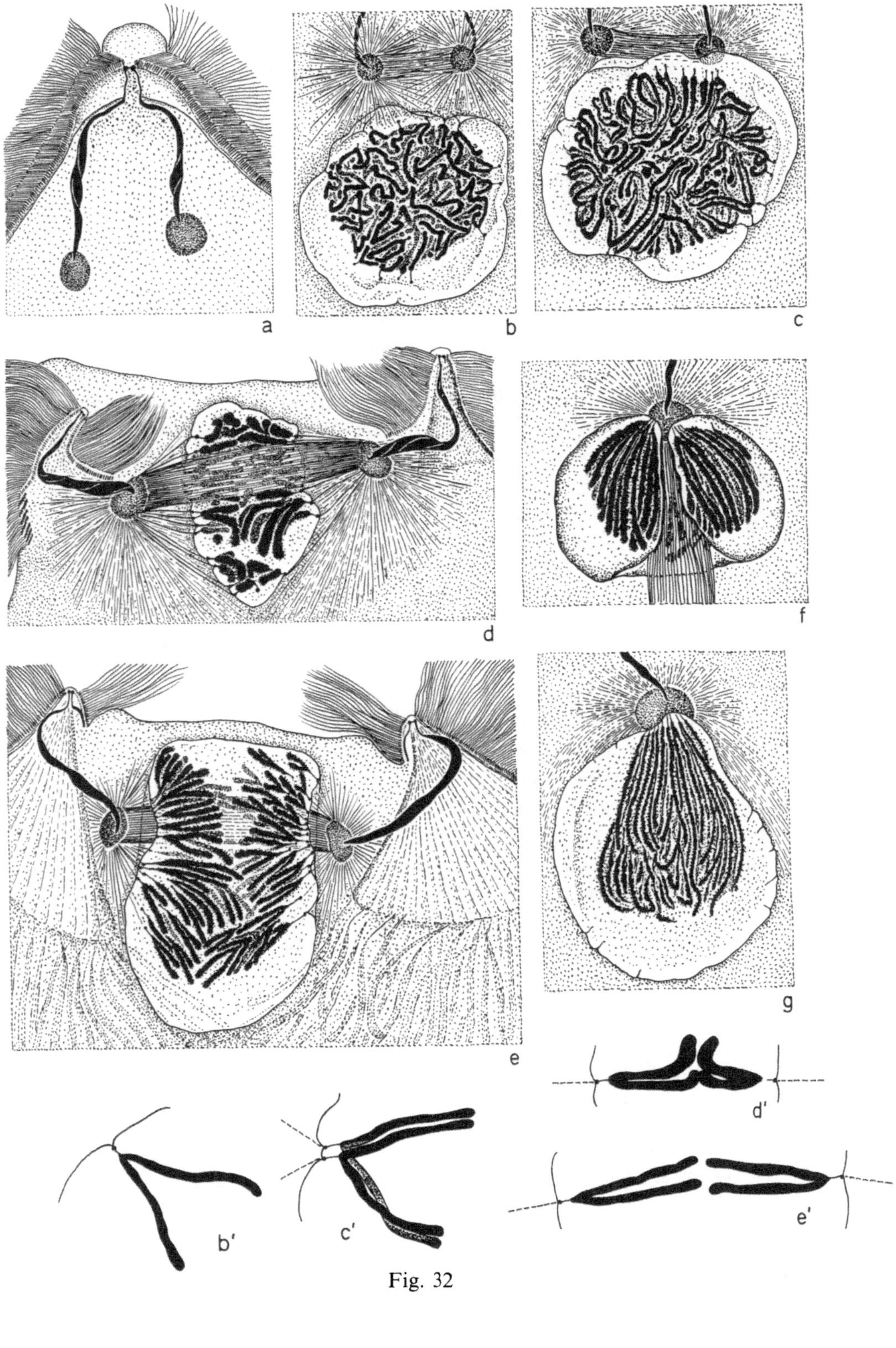

Fig. 32

(Fig. 32*g*). Finally, the nucleus separates from the atractophore and migrates deep into the body.

The atractophores double simultaneously with mitosis. "Privileged" kinetosomes duplicate as early as in the prophase (Fig. 32*d*); a second atractophore starts to grow from the anterior end of each nascent daughter specimen in the anaphase (Fig. 32*e*). The second atractophore becomes as long as the preexisting one and develops an atractosphere in the early interphase (Fig. 32*a*).

A similar pattern of mitosis is observed in *Rhynchonympha*, *Urinympha*, *Trichonympha*, *Deltotrichonympha*, *Koruga*, *Eucomonympha*, *Pseudotrichonympha*, *Spirotrichonympha* (CLEVELAND *et al.* 1934, CLEVELAND 1935b, 1958b, 1960, 1966a, 1966b, GRASSÉ and HOLLANDE 1951), *Lophomonas* (BĚLAŘ 1926a), *Joenia* (GRASSÉ 1952, HOLLANDE and VALENTIN 1969), and some other hypermastigids. Mitoses vary mainly in the behaviour of the atractophores (CLEVELAND's "centrioles") in the cell cycle (CLEVELAND 1957). In *Trichonympha*, for instance, the atractophores are generally short; they greatly differ in length in the interphase, becoming identical only in the prophase (CLEVELAND 1960). In *Pseudotrichonympha*, both atractophores and atractospheres are resorbed in the telophase and grow *de novo* from the "terminal granules" (kinetosomes) prior to every mitosis (GRASSÉ and HOLLANDE 1951). *Joenia* has no atractospheres, the central spindle being formed directly between the atractophores (HOLLANDE and VALENTIN 1969). The atractophores of *Lophomonas* are short and without atractospheres, being always connected with a primordium of the central spindle (HOLLANDE and CARRUETTE-VALENTIN 1972). The same is true of *Holomastigotoides* (Fig. 34*a*). But in *Staurojoenina*, the atractophores are very long (HOLLANDE and VALENTIN 1968c, HOLLANDE and CARRUETTE-VALENTIN 1971).

The fine structure of kinetochores in hypermastigids is of definite interest, especially their duplication in the cell cycle and the way in which they "grow" through the nuclear envelope. The duplication of kinetochores was examined in *Spirotrichonympha* (HOLLANDE and CARRUETTE-VALENTIN 1971). It occurs in the preprophase when the kinetochores are still on the inner side of the nuclear envelope. The kinetochore disc gradually condenses (Fig. 34*b*) and divides into two parts without losing contact with the inner nuclear membrane (Fig. 34*c*). Then, the daughter kinetochores separate from one another by a distance of about 1 μm and begin to raise the nuclear envelope above them into a cupule (Fig. 34*d*). It would be relevant to note that this primary separation of kinetochores occurs even before they "grow" into the nuclear envelope and attach chromosomal microtubules.

Fig. 33. Mitosis in *Barbulanympha* and *Trichomonas*. *a–c Barbulanympha*: *a* section through early anaphase, photomicrograph, 1000 ×; *b* prophase kinetochore, electron micrograph, 13 000 ×; *c* kinetochore built into the nuclear envelope, electron micrograph, 62 500 × (*1* atractosphere, *2* central spindle, *3* atractophore end, *4* chromosomes, *5* kinetochore, *6* nuclear envelope); *d* anaphase of mitosis in *Trichomonas vaginalis* (*1* kinetosomes, *2* atractophores, *3* central spindle, *4* chromosomal fibres, *5* kinetochore, *6* nucleus), electron micrograph, 24 000 ×. *a* after CLEVELAND (1958a); *b* and *c* after HOLLANDE and VALENTIN (1968b); *d* after BRUGEROLLE (1975b).

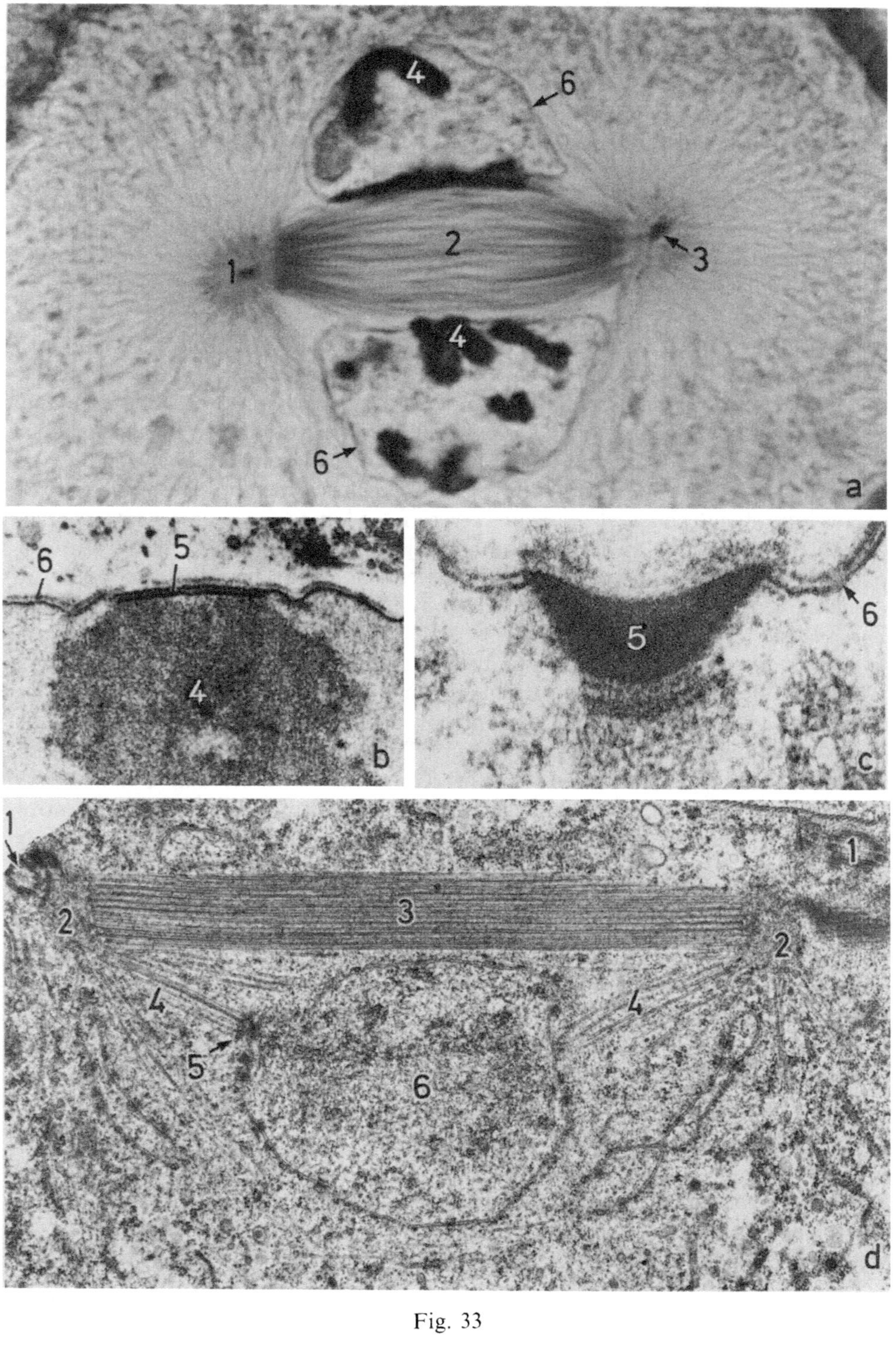

Fig. 33

Further stages in the behaviour of kinetochores are best studied in *Trichonympha* (KUBAI 1973). Its kinetochores also duplicate in the preprophase, but they are poorly condensed then. Using reconstructions from ultra-thin serial sections, KUBAI has shown that all the kinetochores are already paired in the prophase when they are condensed well enough but are still located on the inner side of the nuclear envelope. Since the kinetochores are still not attached to the chromosomal microtubules, KUBAI believes that the *primary separation of the kinetochores is ensured by a membrane mechanism* (namely, the growth of the membranes in the nuclear envelope or their "fluidity"). The classic mechanism of chromatid segregation, in which the chromosomal bundles of microtubules play the leading role, begins to operate later, i.e. as soon as the kinetochores "have grown" through the nuclear envelope (see also KUBAI 1975, RIS 1975).

After their primary separation, the kinetochores of *Trichonympha* are built into the nuclear envelope (KUBAI 1973). In the prophase, they consist of a dense disc adjacent to the chromosome and a fibrous layer underlying the nuclear envelope. Somewhat later, the kinetochore raises the nuclear envelope in a cupulate manner so that the dense disc comes to the level of the nuclear envelope surrounding the kinetochore (Fig. 34*e*). Finally, in the early anaphase, the cupulate portion of the envelope disintegrates, the fibrous layer of the kinetochore comes into contact with the cytoplasm, and the chromosomal microtubules attach themselves to it (Fig. 34*f*, *g*). The dense disc is now inserted like a plug into the opening in the nuclear envelope, a stage observed also by HOLLANDE and CARRUETTE-VALENTIN (1970b).

Trichomonadida. It is commonly held that trichomonadids are the direct ancestors of hypermastigids (HOLLANDE and VALENTIN 1968a). Mitosis can be examined in detail by light microscopy only in the largest trichomonadids, such as *Caduceia theobromae* (GRASSÉ 1939, 1952), *Trichomonas gigantea* (CLEVELAND 1949a, 1961a), *Mixotricha paradoxa* (CLEVELAND 1966c), and *Gigantomonas herculea* (CLEVELAND 1966d). In these forms, just as in hypermastigids, the central spindle appears in the cytoplasm between two atractophores (CLEVELAND's "centrioles"), which are laminar or rod-like appendages extending from the common base of a group of flagellar kinetosomes. The atractophores, however, do not have spheres at their ends; the central spindle is usually narrow and looks like a thread or a flat ribbon (in *Mixotricha*) that connects the atractophores (paradesmose, centrodesmose). Just as in hypermastigids, the chromosomes are attached with their kinetochores to the nuclear envelope. CLEVELAND has demonstrated in *Trichomonas* and especially *Mixotricha* that the chromosomal spindle fibres that extend from the atractophores are attached to the kinetochores.

Fig. 34. Ultrastructure of the spindle and kinetochores in Hypermastigida. *a* central spindle primordium in *Holomastigotoides*, 18 000 × ; *b*–*d* kinetochores of *Spirotrichonympha* (*b* in the preprophase, 30 000 × ; *c* kinetochore doubling, 33 000 × ; *d* separation of daughter kinetochores in the prophase, 28 000 ×); *e*–*g* kinetochores of *Trichonympha* (*e* incorporation of the kinetochore into the nuclear envelope; *f* and *g* attachment of microtubules to the kinetochores), 75 000 × ; *1* central spindle, *2* atractophores, *3* kinetochore, *4* chromosome, *5* nuclear envelope, *6* its pore, *7* chromosomal microtubules. *a*–*d* after HOLLANDE and CARRUETTE-VALENTIN (1971); *e*–*g* after KUBAI (1973).

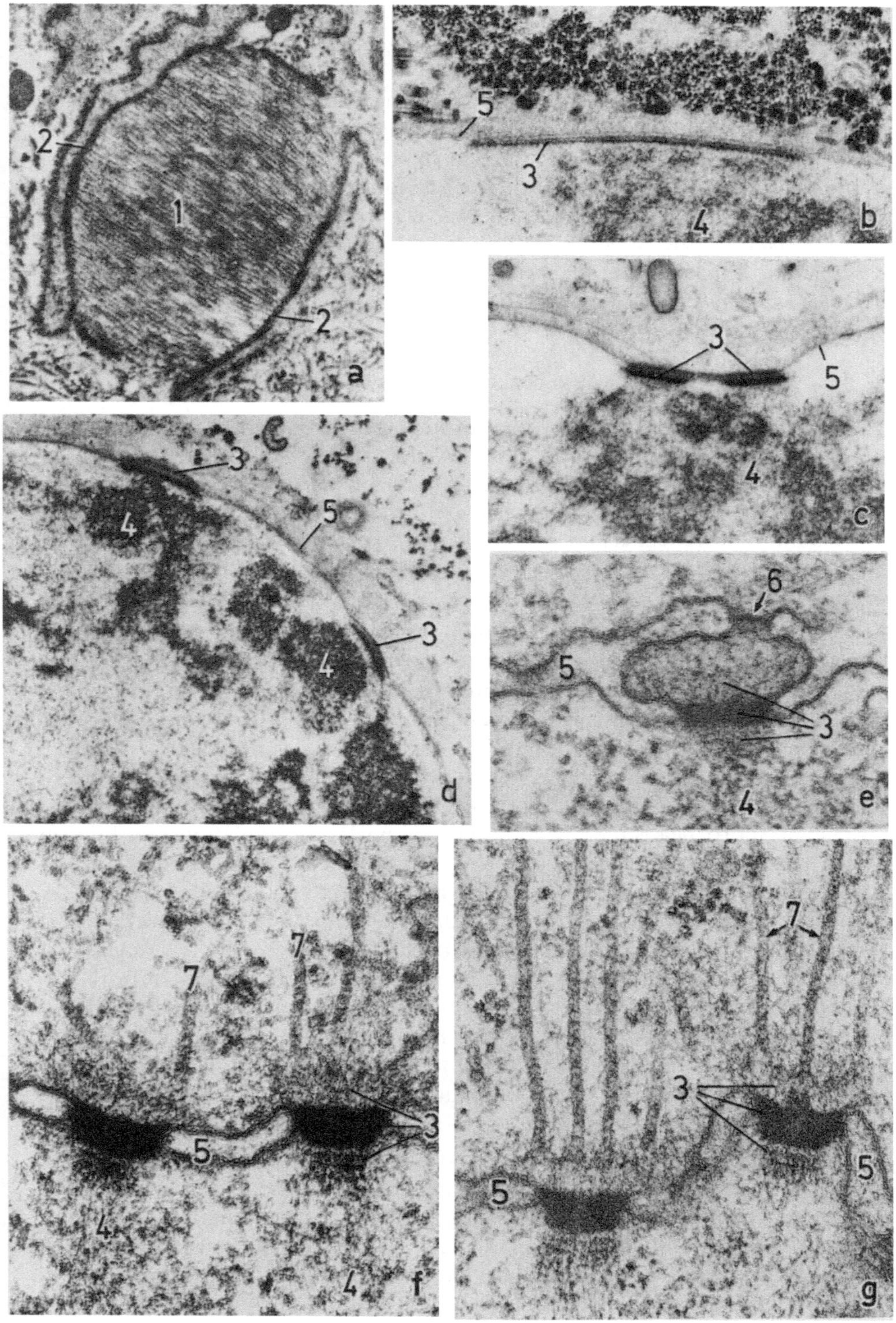

Fig. 34

Mitosis was studied with the electron microscope in *Trichomonas* by BRUGEROLLE (1975b). Its lamellar atractophores are attached to the bases of the kinetosomes (Fig. 33*d*); the central spindle has the form of a narrow bundle of microtubules that greatly elongates in the telophase. One can easily see the radial and chromosomal bundles of microtubules extending from the lower ends of the atractophores; the bundles are attached to small but typical kinetochores built into the nuclear envelope (Fig. 33*d*). The nuclear envelope is preserved throughout mitosis, while the nucleus pinches in two in the telophase. Therefore, mitosis in *Trichomonas* is exactly the same as in hypermastigids, although all the elements of the mitotic apparatus are much smaller. Mitosis of the same type occurs also in *Histomonas meleagridis* (SCHUSTER 1968).

Problem of the Evolution of Mitosis

The problem of evolutionary relationships between different types of mitosis has been discussed in a number of reviews (KUBAI 1975, GOODE 1975, RIS 1975, FULLER 1976, ALOV *et al.* 1977), especially in the review by HEATH (1980), who proposed a detailed hypothetical scheme of the origin and evolution of the mitotic apparatus in lower eukaryotes, including nine major and two side steps. Nevertheless, the problem of the evolution of mitosis, at least in protozoans, is far from being solved.

The "natural" scheme of the evolution of mitosis should, in principle, be consistent with the phylogeny of the respective protozoan taxa. However, this evolution undoubtedly involves repeated parallelisms that complicate the general pattern of the relationships between different types of mitosis. Moreover, protozoan phylogeny has been poorly worked out, too. For example, the exact origin of most zoomastigophorean taxa, even such important ones as Kinetoplastida and Trichomonadida together with Hypermastigida, is still unknown. The origin of Sporozoea is obscure, although the phylogenetic relationships between individual groups within this class are understood better now. The phylogenetic relationships between the phyla Microspora, Ascetospora, and Myxozoa are unknown. We still know hardly anything about the origin and phylogenetic relationships between individual groups of Sarcodina, especially the Actinopoda (radiolarians and heliozoans). The relationships of Amoebida with Labyrinthomorpha and Heliozoea as well as myxomycetes are obscure. We still do not know from what flagellates the Ciliophora have originated. Moreover, the most primitive taxon in the subkingdom Protozoa is still undecided; at different times, the candidates for it were Chrysomonadida, Dinoflagellida, small heterotrophic amoebae and Rhizomastigida.

Advances in electron microscopy in the last 10-15 years, however, have given a better insight into the behaviour of the nuclear envelope, the structure and position of the MTOCs, and other details of mitosis in many different protozoans, details that were only suspected earlier. As a result, a common type (semiopen pleuromitosis) was revealed despite the apparent variety of mitoses, e.g. in coccidians. Externally different variants of mitosis proved similar in most phytomastigophorean orders. Moreover, the mitotic mechanism was established in several protozoans (trypanosomes, malaria parasites, microsporidians, etc.) in which it could not be studied by light microscopy at all.

The comparative characteristics of the known major types of mitosis in Protozoa are given in Table 2. It can be seen from the table that the first criterion used in classifying the types of mitosis is the preservation or disintegration (complete or partial) of the nuclear envelope. The second criterion is the position of the spindle MTOCs in either the nucleus or the cytoplasm. The third criterion is the position (in either the nucleus or the cytoplasm) and the shape of the spindle (the spindle can be either "orthomitotic", i.e. with axial symmetry, or "pleuromitotic", i.e. with bilateral symmetry). The presence of a distinct equatorial plate is of secondary importance since it is correlated with the presence of the orthomitotic spindle. The data on the behaviour of the nucleolus are supplementary; this property can hardly be used as a criterion for classifiying mitosis because of its great variability.

At present, only a very hypothetical scheme of possible evolutionary relationships between the types of mitosis can be proposed (Fig. 35). They are denoted by the same numbers as in Table 2. The scheme takes into account the possible parallelisms in the evolution of mitosis; for instance, the series 4 → 5a → 2a → 1a → 1b comprises mitoses of flagellates and sarcodinids, while the series 4 → 3 → 2b → 1c consists of mitoses of sporozoans.

Closed intranuclear pleuromitosis (type 4) seems to occupy the central place in the evolution of protozoan mitoses. This type of mitosis is the most widespread: it occurs in the most primitive Phytomastigophorea (*Pedinomonas*) and in Zoomastigophorea (trypanosomids, oxymonadids), Sarcodina (foraminiferans, radiolarians), Microspora, and Ascetospora (haplosporidians). Besides Protozoa, this type of mitosis is very common in lower fungi, that are currently believed to have originated from some primitive flagellates (see the review by Fuller 1976).

The differential growth of the membranes of the nuclear envelope that causes the segregation of daughter MTOCs and, consequently, chromatids, plays a significant role in the mechanism of this mitosis. Apparently, this type of mitosis is very old and primitive since the membrane mechanism of the segregation of genetic material is a property of prokaryotes. It follows that the preservation of the nuclear envelope in mitosis is also a primitive feature. An intermediate link, i.e. a system of microtubules, was probably formed between the nuclear envelope and the chromosomes when multiple chromosomes appeared. The structure of the mesosome type, which attached the nucleoid of prokaryotes to a membrane, seemed to split into two elements, namely, a MTOC on the nuclear envelope and a kinetochore on the chromosome; they became connected with a microtubule, an evolutionary neoformation absent in prokaryotes. Indeed, in this type of mitosis, the intranuclear half-spindles are often very poorly developed and consist of a small number of microtubules; frequently, only one microtubule per chromatid is attached to the kinetochores. Most authors hold that closed intranuclear pleuromitosis is indeed a primitive type of mitosis (Pickett-Heaps 1969, 1972, 1974, Leedale 1970, Goode 1975, Kubai 1975, Ris 1975, Alov *et al.* 1977, Heath 1980).

A number of parallel evolutionary lines seems to stem from different variants of this type of mitosis, which are found in most protozoan phyla. Only two groups of these lines are shown in our scheme (Fig. 35). In one of them, the variant of closed

Table 2. *Comparative characteristics of the types of mitosis in Protozoa*

Types of mitosis	Nuclear envelope	Spindle organizing centres	Shape and position of the mitotic apparatus	Central spindle	Position of chromosomal fibres	Equatorial plate	Behaviour of the nucleolus
1. Open orthomitosis	disintegrates	various in C or absent	monaxonic in the mixoplasm	axial	together with CS	present	disperses
2. Semiopen orthomitosis	polar gaps	as above	monaxonic, mainly in N	axial	as above	present (sometimes indistinct)	as above
3. Semiopen pleuromitosis	as above	centrocones in C or MTOCs in NE	bilateral, mainly in N	absent or lateral in N, rarely in C	half-spindles in N	absent	disperses or divides
4. Closed intranuclear pleuromitosis	conserved	MTOCs beneath NE	bilateral in N	lateral or late axial in N	as above	absent	as above (or absent)
5. Closed intranuclear orthomitosis	as above	absent or MTOCs either beneath NE or in N	monaxonic in N	axial in N	together with CS	present (rarely absent)	absent, divides or disperses
6. Closed extranuclear pleuromitosis	as above	atractophores in C	bilateral in C	lateral in C	half-spindles in C	absent	disperses

N = nucleus; C = cytoplasm; NE = nuclear envelope; CS = central spindle; MTOC = microtubule-organizing centre.

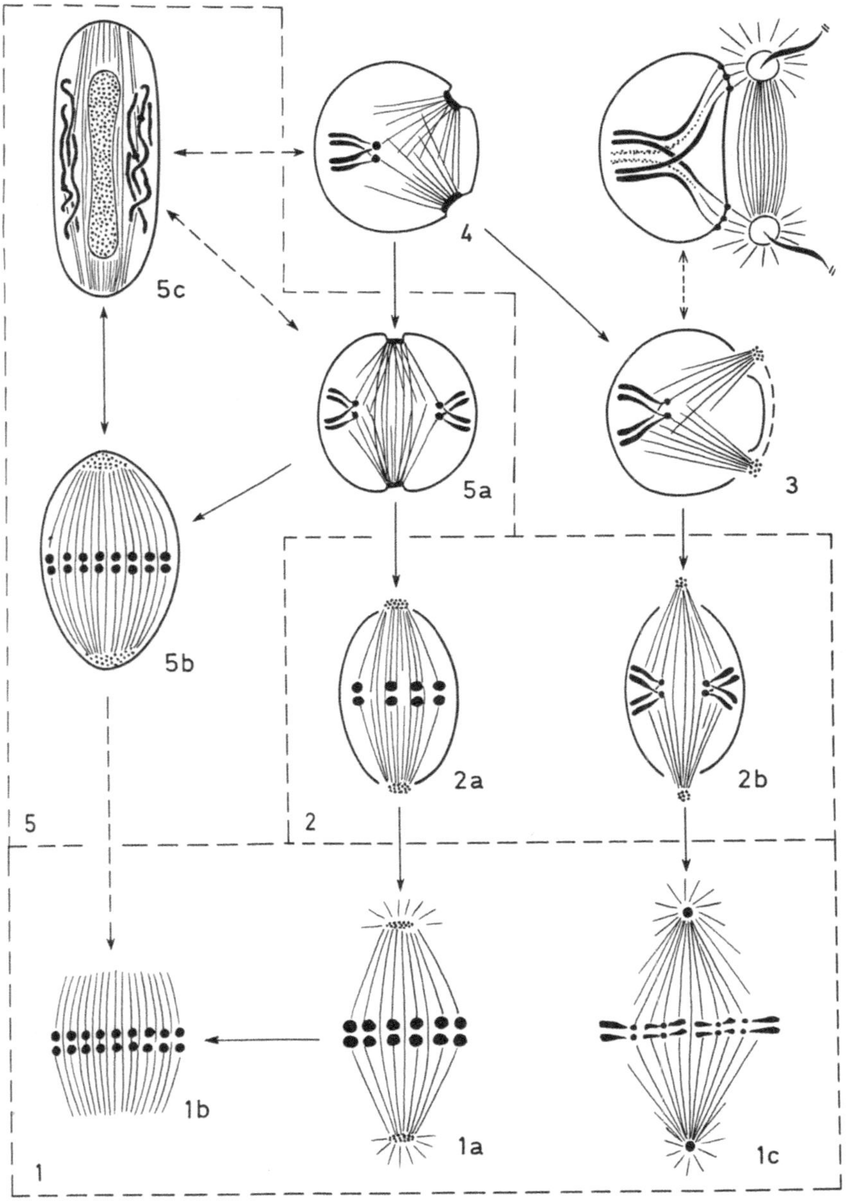

Fig. 35. Possible paths of evolution of the types of mitosis in protozoans. Original. Numbers in the scheme correspond to those of the types of mitosis in Table 2: *1*, *a*–*c* open orthomitoses (*a* and *c* with MTOCs, *b* acentric); *2*, *a* and *b* semiopen orthomitoses; *3* semiopen pleuromitosis; *4* closed intranuclear pleuromitosis; *5* closed intranuclear orthomitoses: *a* with submembranous MTOCs, *b* and *c* acentric (*b* micronuclei of ciliates, *c* euglenids and schizopyrenids); *6* closed extranuclear pleuromitosis.

intranuclear orthomitosis with submembranous intranuclear MTOCs, like that found in *Euglypha* (type 5a), holds an important place; in this case, the MTOCs reach the opposite poles of the nucleus before the half-spindles are formed. This form of mitosis is a bridge between pleuromitosis and orthomitosis; the nearly orthomitotic nuclear divisions of certain trypanosomes also approximate it.

Orthomitoses of flagellates, sarcodinids, and fungi could originate from such intermediate forms of mitosis. For example, the local dissolution of the nuclear envelope at the pole of the dividing nucleus might make the MTOC pass into the cytoplasm (PICKETT-HEAPS 1969, 1972), thus giving rise, in one line, to a group of semiopen orthomitoses in volvocids and some green algae, while in another line, to the *Actinophrys* type of mitosis in sarcodinids, and in a third line, to a group of similar orthomitoses in myxomycetes and phycomycetes, etc. (type 2a in our scheme). We believe that the sporadically occurring association of cytoplasmic MTOCs with flagellar kinetosomes is secondary (it was found also in closed intranuclear pleuromitosis; see pp. 98, 99, 100).

This group of evolutionary lines leads further to entirely open orthomitoses (type 1a in our scheme) characteristic of Prymnesiida, Chrysomonadida, and Prasinomonadida. In this case, MTOCs at the poles may vary ("morphoplasm" in the vicinity of kinetosomes, rhizoplasts, etc.) or be absent altogether (Cryptomonadida, type 1b in the scheme). In the parallel lines of sarcodinids, the amoebae of the Hartmannellidae family seem to have reached stage 1a (open mitosis with MTOCs of various types in the cytoplasm), whereas large amoebae reached stage 1b (acentric open orthomitosis).

The second group of evolutionary lines stemming from closed intranuclear pleuromitosis (type 4) is found in Sporozoea. In this case, the MTOCs are no longer wholly intranuclear; they are probably first "built" into the polar gaps in the nuclear envelope, and then they pass into the cytoplasm (centrocones). Mitosis, while retaining its bilateral symmetry, becomes "semiopen" (pleuromitosis of coccidians and some gregarines, type 3). It is likely that haemosporidians have retained a more primitive structure of their mitotic apparatus than the extant coccidians, a structure intermediate between types 4 and 3 (with the MTOCs in the very gaps of the nuclear envelope). Further processes in the second group of lines are the same as in the first: the spindle straightens out and becomes orthomitotic (semiopen orthomitosis of gregarines, type 2b); in certain lines, the nuclear envelope completely breaks down during mitosis (open orthomitosis of gregarines, type 1c).

Therefore, the following features of mitosis should be regarded as primitive ones in most protozoans: the preservation of the nuclear envelope, a bilaterally symmetric mitotic apparatus consisting of two intranuclear half-spindles, and the poor condensation and gradual separation of the chromosomes in mitosis. Conversely, the following features appear in the course of evolution: the nuclear envelope tends to disintegrate, the MTOCs are prone to pass from the nucleus into the cytoplasm, the half-spindles join producing a bipolar spindle, the chromosomes become more condensed, and the equatorial plate is formed. To put it differently, the evolution follows the route from closed intranuclear pleuromitosis to open orthomitosis in which no traces of the membrane mechanism of chromatid segregation can be found.

It is more difficult to explain the origin of the acentric variety of closed intranuclear orthomitosis, especially acentric mitosis of the micronuclei of ciliates (Fig. 35, type 5b) and mitosis in Schizopyrenida and Euglenida (type 5c). This variety can be deduced from closed intranuclear orthomitosis with submembranous MTOCs (Fig. 35, type 5a). To this end, it should be assumed that while the nuclear envelope is entirely preserved, the MTOCs on its inner surface have been reduced or have sunk deep into the nucleus and that the spindle began to develop in a diffuse manner in the karyolymph.

Pleuromitosis with an extranuclear spindle differs from other mitoses in the scheme (Fig. 35, type 6); this very peculiar and clear-cut type of mitosis is found in Trichomonadida, Hypermastigida, and, as we shall see in Chapter 4, Dinoflagellida. In this case, the chromosomes are permanently linked with the nuclear envelope and the primary separation of kinetochores involves a membrane mechanism. Since this type of mitosis is characteristic also of the mesokaryotic protozoans (Dinoflagellida), many authors regard pleuromitosis with an extranuclear spindle as an "unorthodox" and very primitive type of mitosis (Kubai 1973, 1975, Goode 1975, Ris 1975, Taylor 1976). It possibly developed independently of mitosis in most of the other protozoans. However, it slightly resembles some varieties of semiopen pleuromitosis in coccidians (e.g. mitosis of *Aggregata*, which also has extranuclear centrodesmose made of microtubules). But this similarity is most probably convergent since the half-spindles are almost entirely intranuclear in coccidians, whereas they are wholly extranuclear in trichomonadids and hypermastigids.

A review (based mainly on ultrastructural data) of the types of mitoses that are found in large protozoan taxa is given below. The types of mitoses are denoted by the same numbers and letters as in Fig. 35.

Sarcomastigophora
- Phytomastigophorea (except Dinoflagellida and Euglenida): 1a, 1b, 1c, 2a, 4
  - Dinoflagellida: 6
  - Euglenida: 5c
- Zoomastigophorea (except Trichomonadida and Hypermastigida): 4, 5a (?)
  - Trichomonadida, Hypermastigida: 6
- Opalinata: 5b (?)
- Rhizopoda, Lobosea: 1a, 1b, 5b, 5c
  - Filosea (Gromiida): 5a
  - Granuloreticulosea (Foraminiferida): 4
- Actinopoda, Acantharea and Polycystinea: 4
  - Heliozoea: 2a, 1c (?)

Labyrinthomorpha: 1c

Apicomplexa, Gregarinia: 1a, 1c, 2b, 3, 5a
- Coccidia and Piroplasmia: 3

Microspora: 4

Ascetospora: 4 (?)

Myxozoa: ?

Ciliophora (micronuclei): 5b

Mitosis of Protozoa and Hypotheses of the Chromosome Movement

There are a number of hypotheses that explain the anaphasic movement of chromosomes. In all of them, importance is attached to the microtubules of the mitotic apparatus (see the reviews by FORER 1969, LUYKX 1970, NICKLAS 1971, BAJER and MOLÈ-BAJER 1972).

According to the *hypothesis of the assembly-disassembly of the spindle microtubules* (INOUÉ and SATO 1967), chromosomal microtubules shorten in the anaphase as a result of their local disassembly at the poles and/or kinetochores, while continuous microtubules lengthen as the same tubulin dimers are added at their ends.

According to the *hypothesis of the mutual sliding of microtubules* (MCINTOSH *et al.* 1969, NICKLAS 1971), chromosomal microtubules in the anaphase slide along the microtubules of the central spindle toward the poles as a result of their mechano-chemical interaction, being connected with the formation of labile cross-bridges of the protein dynein, which displays ATPase activity, between parallel microtubules. The microtubules of the central spindle are also capable of sliding along each other in opposite directions because some of them belong to one pole, while the rest, to the other. Then the central spindle can be seen as two half-spindles that deeply interdigitate one another but slide apart during the anaphase, increasing the distance between the poles. NICKLAS (1971) believes the mechanism of assembly-disassembly and that of microtubule sliding to exist in various combinations.

According to the *two-component hypothesis* (FORER 1969, LUYKX 1970), the traction applied to the chromosomes is transmitted to them through the chromosomal microtubules only passively, while it is generated in the contractile proteins of the actomyosin type that are in the spindle matrix.

Taking these three hypotheses into account, it would be interesting to analyze the possible role of the microtubules in the chromosome movement during different types of protozoan mitosis. Certain views on this subject were expressed by KUBAI (1975).

In *closed intranuclear and semiopen pleuromitoses* (types 3 and 4 in Fig. 35), the chromosomal microtubules cannot slide along the fibres of the central spindle since the latter either is absent or develops after the chromosomes have segregated (for instance, in microsporidians), or it is spatially separated from the chromosomal half-spindles (e.g. in foraminiferans). The chromosomal microtubules may only shorten as a result of disassembly at the MTOC and/or at the kinetochore (RIS 1975). In some cases, the length of the chromosomal microtubules may undergo almost no change in these types of mitosis; then the chromosomes are separated exclusively by the membrane mechanism, which separates the daughter MTOCs (particularly in yeasts and other fungi; see KUBAI 1975). As for the later stages of mitosis, the elongation of the central spindle may be attributed to the assembly-disassembly of the microtubules and/or their mutual sliding (trypanosomids, radiolarians, foraminiferans, microsporidians, etc.).

In *semiopen orthomitoses* (type 2 in Fig. 35), continuous and chromosomal microtubules usually form a common spindle. Hence, in principle, the mechanism of mitosis in the anaphase can be explained in this case by any of the three hypotheses. The same is true of *open orthomitoses* (type 1). However, in *Ochromonas* (TIPPIT *et al.* 1980) and *Cryptomonas* (OAKLEY and HEATH 1978), the central spindle consists of two interdigitating half-spindles whose microtubules can most probably slide along each other. Furthermore, it is likely that the "polar" microtubules of the half-spindles also slide along the "free" microtubules that remain in the centre of the spindle, elongating it even more. In amoebae (*Amoeba, Chaos*), where the spindle is acentric and the polar ends of all microtubules are free, it is unlikely that the chromosomes move only due to the "disassembly" of the chromosomal fibres: the shortening of a fibre with an unattached end would not generate traction. In amoebae, the "combined" hypothesis of NICKLAS appears to be most plausible: presumably, continuous (free) microtubules elongate because of their assembly at the ends, while the chromosomal microtubules together with the chromosomes slide along the continuous ones toward the poles. The mere "sliding apart" of the continuous fibres in amoebae would not cause the central spindle to completely elongate, since its length at the end of the anaphase is much greater than its double length in the metaphase.

In *closed intranuclear orthomitoses* (type 5b in Fig. 35), the anaphasic movement of chromosomes can be explained, in most cases, by any of the three hypotheses. However, everything that has been said earlier about mitosis in large amoebae and cryptomonads holds true when the spindle poles have big "polar caps" (e.g. in the micronucleus of *Loxodes*). In this case, the most plausible explanation is probably that the chromosomal microtubules slide along the continuous ones as the continuous microtubules are assembled. Since the central spindle in micronuclear mitosis often stretches ten times its length and even more, this stretching cannot be attributed only to the sliding apart of the microtubules which already existed in metaphase. In the anaphase and the telophase, the microtubules of the central spindle are certainly assembled very intensively. This applies to not only the "continuous" (interpolar) ones but also to special "interzonal" microtubules that stretch from only one anaphasic group of chromosomes to another. In *Paramecium bursaria* (LEWIS *et al.* 1976), contractile proteins appear to be involved in pulling the chromosomes to the poles (the two-component hypothesis) because in the dividing micronucleus the chromosomes are attached to microfilaments and not to microtubules.

Finally, in *closed extranuclear pleuromitosis* (type 6 in Fig. 35), the chromosomal microtubules can shorten only as a result of their disassembly. They cannot slide because the central spindle is completely separated from the chromosomal fibres. For the central spindle to be greatly elongated (five times its length in *Barbulanympha*) in the telophase, continuous microtubules must be assembled at its ends; this cannot be explained merely by the "sliding apart" of two interdigitated half-spindles (INOUÉ and RITTER 1978).

Hence, an analysis of protozoan mitosis shows that none of the hypotheses of the anaphasic movement of chromosomes can explain all the cases. Apparently, there is no universal mechanism of mitosis common to all eukaryotes. There are several such mechanisms while the function of the microtubules of the mitotic apparatus substantially changed in the course of evolution (KUBAI 1975). It is likely that the changes in the length of the spindle fibres (the chromosomal ones, in particular) caused by their disassembly or assembly is an older (and less effective) mechanism than the mutual mechano-chemical sliding of the microtubules, which requires a more developed and complicated mitotic apparatus.

The elaboration of the *concept of the microtubule-organizing centres*, or MTOCs (PICKETT-HEAPS 1969, 1971), can be regarded as an important achievement in recent years, mainly with respect to lower eukaryotes (protozoans, fungi, algae). A comparative analysis of the mitotic spindles in many species has shown that centrioles are not always the organizing centres of the spindle. In lower organisms, MTOCs are frequently represented by more or less dense discs or globules of amorphous or finely granular material. They may be either in the nucleus or in the cytoplasm. The primary position of MTOC, however, appears to be nuclear (on the inner side of the nuclear envelope). In the most primitive types of mitosis (types 4 and 3 in Fig. 35), the morphology of MTOCs is comparatively uniform, although it may vary greatly in other types, particularly in open orthomitosis (type 1 in Fig. 35). In general, mitotic MTOCs should be regarded as a special case of the more general concept of the "*nematogenic matrix*" (morphoplasm, archoplasm), which implies any specialized cell structure that induces the formation of microtubules, both nuclear and cytoplasmic ones (FAURÉ-FREMIET 1970a). The MTOCs frequently associate with centrioles (usually flagellar kinetosomes); in closed mitoses, these two structures are separated by the nuclear envelope; in open mitoses, they may come into contact with each other either directly or via an intermediate link (such as rhizoplasts or atractophores). PICKETT-HEAPS (1969, 1971) believes that the association is secondary, i.e. it developed in the course of evolution, while the primary function of a centriole (kinetosome) in eukaryotes was to induce a flagellar axoneme and not the mitotic spindle. In the course of evolution, however, the cell began to use the mitotic spindle to correctly segregate the daughter kinetosomes, engendering a more or less stable association between the kinetosomes (centrioles) and the spindle poles. Only the transition from closed to open mitosis allowed close contact and, later, association between the mitotic MTOC and the flagellar MTOC (i.e. the kinetosome). According to PICKETT-HEAPS, the kinetosomes in the advanced eukaryotes are consequently arranged *within* the mitotic MTOC, which is a zone of homogeneous "morphoplasm" usually surrounding a pair of centrioles in classic metazoan mitosis. In many cases, kinetosomes (centrioles) have lost their primary function, i.e. the formation of a flagellum.

3. Chromosomes of Eukaryotic Protozoa

Protozoan nuclei contain chromosomes that can be seen at certain stages of the cell cycle. There is no doubt that the law of constancy of the chromosome number in the normal genome of the species can be applied to all protozoans. The chromosomes of protozoans also have individuality and display longitudinal differentiation.

Chromosome Number

There are many records of the number of chromosomes in some particular species of protozoans. A large part of this information is dubious, however, especially in old works. More reliable data are included in Table 3. It is usually known (even in agamic organisms) whether a given number is haploid (n) or diploid (2 n). Both the haploid and diploid numbers of chromosomes are indicated for protozoans that undergo the sexual process and, accordingly, reveal the alternation of nuclear phases. The number of chromosomes in protozoans varies greatly within not only large taxa, but also genera (*Cryptomonas*, *Euglena*, *Chlamydomonas*, *Volvox*, *Spirotrichosoma*, *Barbulanympha*, *Stylocephalus*, etc.). Sometimes, the chromosome numbers of related species are multiples of a certain

Table 3. *Chromosome numbers in eukaryotic Protozoa*

Species	Chromosome number[a] haploid	diploid	References
Phylum Sarcomastigophora			
Subphylum Mastigophora			
Order Cryptomonadida			
Chroomonas mesostigmatica	(~ 24)		M. Thakur (quoted from Godward 1966)
Ch. salina	(~ 42)		ibid.
Cryptomonas suberosa	(~ 34)		ibid.
C. biplastida	(~ 53)		ibid.
C. appendiculata	(~ 61)		ibid.
C. ovata var. palustris	(~ 86)		ibid.
Chilomonas paramecium	(~ 35)		ibid.
Order Euglenida			
Astasia klebsii	(~ 18)		Leedale 1958b, 1966, 1968
Colacium mucronatum	*~ 35*	—	ibid.
Phacus pyrum	*~ 30–40*	—	ibid.
Euglena viridis	(~ 42)		ibid.
E. gracilis	(~ 45)		ibid.
E. spirogyra	(~ 86)		ibid.
Eutreptia viridis	(~ 44)		ibid.
E. pertyi	(~ 90)		ibid.
Peranema trichophorum	(~ 177)		ibid.
Order Volvocida			
Chlamydomonas dysosmos	*16*	32	Schaechter and de Lamater 1955
Ch. zimbabwiensis	*18*	36	Heimke and Starr 1979
Ch. reinhardi	*18*	36	Schaechter and de Lamater 1955
Ch. eugametos	*36*	72	ibid.
Ch. moewusii	*36*	72	ibid.
Gonium pectorale	*17*	34	Cave and Pocock 1951, Shyam and Sarma 1975
Eudorina unicocca	*7*	14	Cave and Pocock 1951
E. elegans	*12*	24	Cave and Pocock 1951, Sarma and Shyam 1973
Pandorina morum[b]	*2–14*	4–28	Coleman and Zollner 1977

Table 3 (continued)

Species	Chromosome number[a] haploid	diploid	References
Volvox globator	*5*	10	CAVE and POCOCK 1951
V. tertius	*13*	26	ibid.
V. aureus	*14*	28	ibid.
V. powersii	*15*	30	ibid.
Order Kinetoplastida			
Trypanosoma equiperdum	*3*	–	ROSKIN and SCHISHLIAEIEWA 1928
T. lewisi	*3*	–	WOLCOTT 1952
T. cruzi[c]	*10*	–	SOLARI 1980a, 1980b
Order Oxymonadida			
Notila proteus	14	*28*	CLEVELAND 1950c
Oxymonas doroaxostylus	~ *25*	~ 50	CLEVELAND 1950a
Saccinobaculus ambloaxostylus	~ *30*	~ 60	CLEVELAND 1950b
Order Trichomonadida			
Trichomonas caviae	*4*	–	GRASSÉ and FAURE 1939
Tritrichomonas batrachorum	*6*	–	SAMUELS 1957
Dientamoeba fragilis	*4*	–	WENRICH 1944
Caduceia theobromae	(8)		GRASSÉ 1939
Order Hypermastigida			
Holomastigotoides tusilata	*2*	–	CLEVELAND 1949a
H. psammotermitidis	*2*	–	GRASSÉ and HOLLANDE 1963
Spirotrichonympha polygyra	*2*	–	CLEVELAND 1938c
Urinympha talea	8	*16*	CLEVELAND 1951b
Rhynchonympha tarda	10	*20*	CLEVELAND 1952
Leptospironympha wachula	*10*	20	CLEVELAND 1951a
Spirotrichosoma normum	*12*	–	CLEVELAND and DAY 1958
S. submagnum, S. promagnum	*24*	–	ibid.
S. paramagnum	*48*	–	ibid.
S. magnum	*60*	–	ibid.
Barbulanympha wenyoni	*12*	24	CLEVELAND 1954
B. estaboga	*16*	32	ibid.
B. laurabuda	*20*	40	ibid.
B. ufalula	*26*	52	ibid.
Trichonympha okolona (and other species)	*24*	48	CLEVELAND 1949b
Eucomonympha imla	~ *50*	~ 100	CLEVELAND 1950d
Subphylum Opalinata			
Zelleriella (5 species)	12	*24*	CHEN 1948, 1972
Opalina ranarum	~ *9*	~ *18*	KACZANOWSKI 1968, 1971
Subphylum Sarcodina			
Order Schizopyrenida			
Naegleria gruberi	(14–16)		RAFALKO 1947
Order Amoebida			
Rosculus ithacus	(4)		HAWES 1963

Species	Chromosome number[a] haploid	diploid	References
Entamoeba histolytica	(5-6)		Kofoid and Swezy 1925, Epstein 1941
Amoeba proteus	(> 500)		Liesche 1938
Order Foraminiferida			
Ovammina opaca	*3*	*6*	Dahlgren 1964
Discorbis vilardeboanus	*6*	*12*	Le Calvez 1950
Cibicides lobatulus	*6*	*12*	Voronova 1978
Rubratella intermedia	*8*	*16*	Grell 1958a
Rotaliella roscoffensis	*9*	*18*	Grell 1957
R. heterocaryotica	*18*	*36*	Grell 1954
Glabratella sulcata	*9*	*18*	Grell 1958b
Allogromia laticollaris	*10*	*20*	Arnold 1955
Patellina corrugata	*24*	*48*	Le Calvez 1950
Order Actinophryida			
Actinophrys sol	22	*44*	Bělař 1922
Phylum Labyrinthomorpha			
Labyrinthula sp.	9	*18*	Moens and Perkins 1969
Phylum Apicomplexa			
Subclass Gregarinia			
Mattesia trogodermae	*2*	4	Canning and Anwar 1969
Diplocystis schneideri	*3*	6	Jameson 1920
Gregarina blattarum	*3*	6	Sprague 1941
Actinocephalus parvus	*4*	8	Weschenfelder 1938
Stylocephalus longicollis	*4*	8	Grell 1940
S. mesomorphi	*4*	8	Uttangi and Desai 1963
S. conoides	*8*	16	Desai 1965a
S. elongatus	*9*	18	Desai 1965b
Zygosoma globosum	*6*	12	Noble 1938
Subclass Coccidia			
Coelotropha durchoni	*3*	6	Vivier and Henneré 1964
Klossia helicina	*4*	8	Naville 1927
Eimeria maxima	*5*	10	Scholtyseck 1963
E. tenella	*5*	10	Canning and Anwar 1968
Karyolysus lacertae	*5*	10	Reichenow 1921
Eucoccidium dinophili	*5*	10	Grell 1953a
Aggregata eberthi	*6*	12	Dobell 1925
Merocystis kathae	*6*	12	Patten 1935
Adelea cryptocerci	*8*	16	Yarwood 1937
Adelina deronis	*10*	20	Hauschka 1943
Plasmodium berghei[c]	*8*	16	Schrével *et al.* 1977
Phylum Ciliophora (micronuclei)[d]			
Order Karyorelictida			
Tracheloraphis phoenicopterus	17	*34*	Raikov 1958

Table 3 (continued)

Species	Chromosome number[a] haploid	diploid	References
T. caudatus	22	*44*	Dragesco and Raikov 1966
Trachelocerca coluber	~ 27	*~ 54*	Raikov 1963a
Trachelonema sulcata	.28	*56*	Kovaleva 1972
Order Nassulida			
Nassula ornata	~ 20	*~ 40*	Raikov 1967a
Order Hymenostomatida			
Tetrahymena thermophila	5	*10*	Ray 1956
Colpidium campylum	21	*42*	Devidé and Geitler 1947, Devidé 1951
Paramecium tetraurelia[e]	35–50	70–100	Dippell 1954
P. primaurelia[e]	43–63	86–126	Kościuszko 1965
P. triaurelia[e]	77–85	*~ 160*	Przyboś *et al.* 1979
P. jenningsi	~ 64	*~ 128*	Przyboś 1980
P. bursaria (race Fd)	~ 40	*~ 80*	Chen 1940a, 1940c
P. bursaria (2 strains)	~ 52	*~ 104*	Schwartz 1978
P. caudatum	~ 165	*~ 330*	Calkins and Cull 1907
Order Peritrichida			
Vorticella campanula	~ 75	*~ 150*	Mügge 1957
Order Heterotrichida			
Blepharisma japonicum	~ 59	*~ 118*	Miyake *et al.* 1979
Spirostomum ambiguum	18–24	*36–48*	Rao 1968a
Climacostomum virens	~ 60	*~ 120*	Peshkovskaya 1936
Order Hypotrichida			
Euplotes woodruffi	~ 16	*~ 32*	Rao 1964
E. patella	~ 24	*~ 48*	Katashima 1968
E. aediculatus	50–60	*100–120*	Ammermann 1971
Kahlia sp.	~ 25	*~ 50*	Rao 1966
Stylonychia mytilus	125–150	*250–300*	Ammermann 1965, 1971a

[a] The italicized figures stand for the chromosome number characteristic of the vegetative stages of the life cycle (i.e. the haploid number for haplonts, the diploid number for diplonts, and both numbers for the forms with intermediate meiosis). One figure is given in parentheses if it is not known whether it is haploid, diploid, or polyploid.

[b] The chromosome numbers are very dissimilar in different strains (which were collected from three continents), but are constant within a strain.

[c] The data were obtained by counting the kinetochores on electron micrographs.

[d] The figures were obtained only by counting the bivalents in the meiotic prophase. In vegetative mitoses, the chromosomes occasionally combine into aggregates in ciliates. Consequently, their number is greatly underestimated in some works.

[e] The figures indicate the limits of variability of the number of chromosomes in a group of strains of a given sibling species belonging to the *Paramecium aurelia* group.

figure; in such cases, it is very probable that speciation in a given genus has occurred through polyploidy. Such polyploid series are likely to exist, for instance, in the flagellate genera *Spirotrichosoma* and *Chlamydomonas*, in the foraminiferan genus *Rotaliella*, and in the gregarine genus *Stylocephalus* (Table 3). Nonmultiple relations of chromosome numbers, however, are found much more frequently within a genus (species of *Cryptomonas*, *Eudorina*, *Volvox*, *Barbulanympha*, *Euplotes*, etc.). It is quite possible that the nuclei of many Cryptomonadida and Euglenida and of *Amoeba proteus* are polyploid. This has not been proved, however, since no stages of the life cycle with a smaller chromosome number are known, and there is no clear individuality of chromosomes that would make it possible to determine the extent of polyploidy of the nuclei by the number of homologous chromosomes.

The chromosome number only slightly depends on the systematic position of the species. The numbers are usually small ($n < 10$) in trypanosomes, trichomonadids, foraminiferans, gregarines, and coccidians but are large ($n > 20$) in cryptomonadids, euglenids, and ciliates. Both categories have exceptions, however, e.g. the foraminiferan *Patellina corrugata* ($n = 24$) and the ciliate *Tetrahymena thermophila* ($n = 5$).

Morphology of Chromosomes

Individuality of Chromosomes

Efforts to reveal the individuality of most protozoan chromosomes fail, since they are too small and resemble one another. On favourable material, however, the chromosomes of the haploid genome differ with respect to the overall length, the site of the centromere region (i.e. the length of the arms), and the presence and site of the nucleolar organizer.

The individuality of the chromosomes in a protozoan was traced for the first time by DOBELL (1925) throughout the life cycle of the coccidian *Aggregata eberthi*. In spite of the different degree of condensation of the chromosomes at various stages of the cycle, all six chromosomes of the haploid set always differed in their overall length. Clearly defined distinctions between chromosomes with respect to their length as well as the site of the centromere region and the presence of nucleolar organizers have been described by CHEN (1936a, 1948, 1972) in five species of Opalinata of the genus *Zelleriella* (Fig. 36). The presence of two chromosomes of every individual species in the nucleus was so obvious that CHEN could prove the diploidy of *Zelleriella* even before meiosis was discovered in opalinids. The individuality of chromosomes is clearly expressed also in many Hypermastigida, e.g. in *Barbulanympha* during meiosis (Fig. 58). The two chromosomes of the haploid set in *Holomastigotoides tusilata* (CLEVELAND 1949a) differ with respect to the length and the number of nucleolar organizers (two on one chromosome and one on the other). Both chromosomes are one-armed (Fig. 37). In *H. psammotermitidis*, however, both chromosomes are two-armed; one has arms of equal length, while the other has arms of different lengths (GRASSÉ and HOLLANDE 1963).

It is often impossible to identify all the chromosomes in the haploid set, but one or two characteristic chromosomes can be recognized. This was used to determine whether a certain nucleus was haploid or diploid. In *Tritrichomonas batrachorum*, for instance, one chromosome out of six is linked with the nucleolus, while another chromosome is longer than the rest; apparently, this species is a haplont with n = 6 (SAMUELS 1957). Similar observations have been made with respect to the euglenids *Colacium* and *Phacus*, which can also be regarded as haplonts on this basis (LEEDALE 1958a, 1968), the gregarine *Stylocephalus* (UTTANGI and DESAI 1963), etc.

Longitudinal Differentiation

The longitudinal differentiation of chromosomes is expressed by the presence of the primary constriction (the centromere region or simply the centromere) as well as secondary constrictions, nucleolar organizers, satellites, and heterochromatin segments (especially at the chromosome ends).

Primary constrictions (centromeres) have been observed with the light microscope on the chromosomes of some protozoans as narrowings to which spindle fibres are attached (Fig. 36). The primary constriction divides the chromosome into two arms. If the arms are of equal length, the chromosomes are metacentric, if they are of different lengths, the chromosomes are submetacentric, if one arm is very short, the chromosomes are acrocentric, and if there is only one arm (i.e. the centromere is at the end of the chromosome), the chromosomes are telocentric. Submetacentric chromosomes are most commonly found in protozoans. For instance, nine of twelve chromosomes of the haploid set of the opalinid *Zelleriella louisianensis* (Fig. 36) are submetacentric. Telocentric chromosomes are characteristic of some hypermastigids: *Holomastigotoides tusilata* (Fig. 37), *Trichonympha*, etc. Other hypermastigids, however, have two-armed chromosomes, e.g. *Barbulanympha* (Fig. 32). Since the chromosomes of many protozoans become V-shaped in the anaphase (i.e. they move forward with the centromeres), localized centromeres are probably widely distributed in protozoans (Fig. 15).

Electron microscopy has shown that the surface of the centromere regions of chromosomes carries dense disc-shaped formations to which the spindle microtubules are attached. These formations are called *kinetochores*[1]. There are two kinetochores on the metaphasic chromosome, and they lie on the opposite sides of the centromere region. Naturally, there is only one kinetochore on the anaphasic chromosome. Kinetochores often have a characteristic laminated (trilaminar) structure (reviewed by FUGE 1974).

Among protozoans, kinetochores of this type exist in the Hypermastigida. They also have a trilaminar structure and are permanently linked with the nuclear envelope. We have already considered their ultrastructure as well as their incorporation into the nuclear envelope during mitosis (Chapter 2, Figs. 33, 34). Kinetochores in hypermastigids were described with the aid of even the light microscope in many of CLEVELAND's works as dense granules found where the

[1] In the era of the light microscope, the terms "centromere" and "kinetochore" were used as synonyms. The term "centromere", however, now often implies the whole centromere region of the chromosome where the kinetochore is found, while the term "kinetochore" implies the disc to which the microtubules are actually attached.

chromosomes are attached to the nuclear envelope (Fig. 32*d*, *e*); in *Barbulanympha*, they were observed even *in vivo* (CLEVELAND 1954).

Laminated disc-shaped kinetochores have also been found in *Trypanosoma raiae* (VICKERMAN and PRESTON 1970) and *T. cruzi* (DE SOUZA and MEYER 1974), the radiolarian *Collozoum* (HOLLANDE *et al.* 1969), the gregarine *Stylocephalus* (DESPORTES 1970), the coccidian *Eimeria necatrix* (DUBREMETZ 1973), and in some

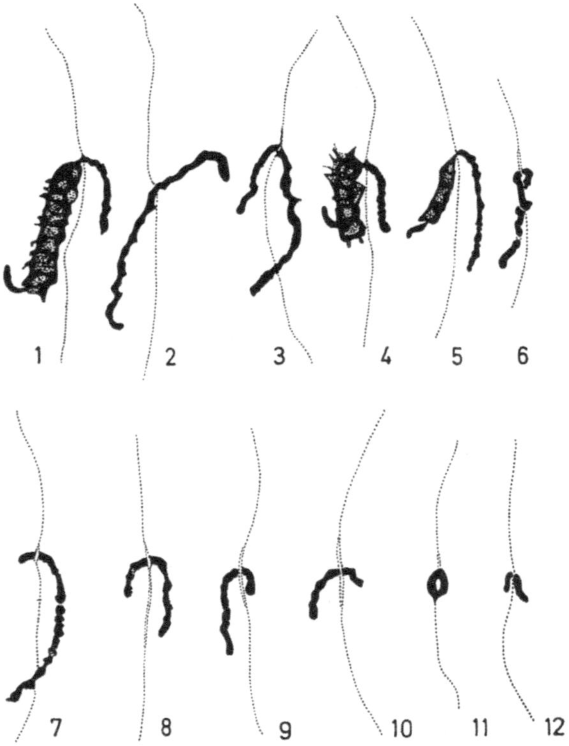

Fig. 36. Haploid chromosome set of *Zelleriella louisianensis*, 2600× (after CHEN from GRELL 1956a). *1–12* individual chromosomes (*1*, *4*, and *5* nucleolus-organizing chromosomes).

species of *Plasmodium* (Fig. 16*e*). In *Trypanosoma cruzi*, however, the kinetochores have a trilaminar structure only in the metaphase, while each of them is divided into two simple discs in the anaphase. Thus, metaphase "kinetochores" are, in essence, complexes of two kinetochores with an interlayer of the material of the chromosome proper (SOLARI 1980a, 1980b). This is probably also the same with *Plasmodium* (Fig. 16*e*) and *Collozoum*.

In the lower ciliates, *Loxodes magnus* (RAIKOV 1973, 1976a) and *Trachelocerca geopetiti* (RAIKOV and KOVALEVA 1981), there are two prominent lamellar kinetochores on each metaphase chromosome (Fig. 29*b*). In *T. geopetiti*, they are especially large and consist of three layers (a loose outer layer, a dense middle layer, and a transparent inner layer that is contiguous to the chromosome). In the

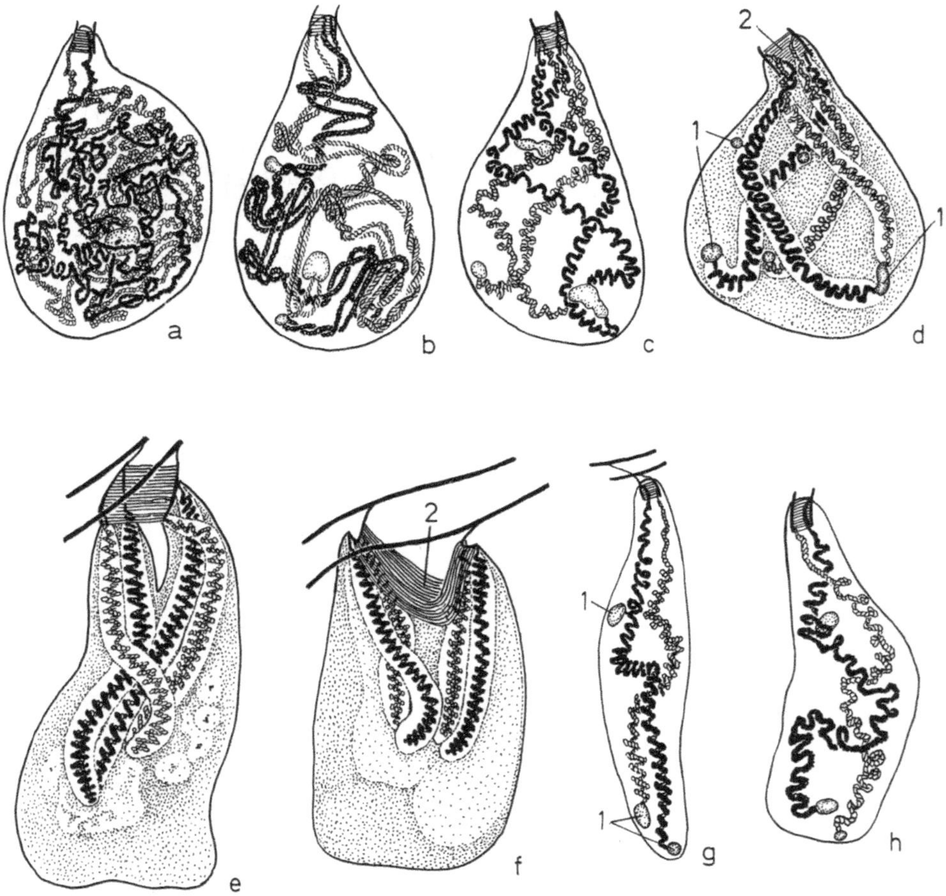

Fig. 37. Chromosome cycle in *Holomastigotoides tusilata*, 1200 × (after CLEVELAND from GRELL 1956a). *a* early prophase; *b* replication of chromosomes; *c* major coil formation; *d* "resting" nucleus (late prophase); *e* and *f* anaphase; *g* and *h* telophase; *1* nucleoli, *2* central spindle.

anaphase, one structurally similar kinetochore remains on every daughter chromosome.

Kinetochores in the form of dense discs without a definite laminar substructure have been observed in the microsporidian *Gurleya* (LOUBÈS and MAURAND 1975; Fig. 25*a*). Even simpler kinetochores in the form of globular electron dense bodies exist in *Labyrinthula* sp. (PORTER 1972). But most protozoans studied with the electron microscope seem to lack any definite disc-shaped kinetochores on their chromosomes; in this case, spindle microtubules end directly in the fibrillar mass of the chromosome that is only sometimes locally condensed. This applies to the chromosomes of Phytomastigophorea, Amoebida, many Foraminiferida, and the chromosomes of the micronuclei of most Ciliophora.

Secondary constrictions, which differ from the nucleolar organizers, have been found only on the chromosomes of *Zelleriella* (CHEN 1948, 1972). In *Z.*

louisianensis, for instance, they are seen on the longer arms of the sixth and seventh chromosomes (Fig. 36).

The presence of *heterochromatin segments* in the chromosomes of protozoans is still an open question. So far, they have been found only in the polyploid nuclei of the radiolarians *Thalassophysa* and *Collosphaera* (PÄTAU 1937a) and on the polytene chromosomes of the developing macronuclear anlagen of ciliates.

Nucleolar organizers have been studied most thoroughly in the opalinids of the genus *Zelleriella* (CHEN 1936b, 1948, 1972). In this case, the nucleoli are permanently linked with the chromosomes that form them; they do not disappear even during mitosis. The number and localization of the nucleolar organizers are species-specific, e.g. in *Z. louisianensis*, they exist on the first, fourth, and fifth chromosomes of the haploid set (Fig. 36). Since *Zelleriella* are diplonts, six ribbon-shaped nucleoli exist in the interphase nucleus. *Z. intermedia* also has three pairs of nucleolar chromosomes, but they are the fourth, sixth, and eighth pairs (in the order of their diminishing overall length). *Z. elliptica* has only two pairs of nucleolar chromosomes; accordingly, there are four nucleoli in the interphasic nucleus. However, secondary fusion of the nucleoli may also occur in the interphase nuclei.

The ultrastructure of the nucleolar organizers and intranucleolar chromatin in interphase nuclei has already been considered in Chapter 1.

Spiral Structure

Protozoans have a typical cycle of chromosome coiling (PÄTAU 1937b, LE CALVEZ 1938). In the foraminiferan *Patellina corrugata*, the spiral packing of the chromonema is revealed best of all in the first meiotic division of the agamont nucleus (LE CALVEZ 1938, 1950, GRELL 1959). Only the "large" spiral is clearly seen (Fig. 57*f*), although LE CALVEZ (1950) believes that a "small" spiral also exists. The spiral structure of chromosomes in *Euglena viridis* has been described by SAITO (1961a). A noteworthy fact is that not one but two relationally coiled chromonemata have been discovered in the anaphase chromosomes of both *Patellina* and *Euglena*. This seems to contradict the generally accepted concept of the single-strandedness of the anaphase chromosome. However, two strands (half-chromatids or subchromatids) have also sometimes been observed in the metazoan anaphase chromosomes.

The spiral structure of the chromosomes is extraordinarily clearly revealed in the hypermastigids of the genus *Holomastigotoides*. In this case, both "small" and "large" spirals are seen. The small spiral is preserved at all stages of the mitotic cycle (which includes no real interphase), while the large spiral exists only at some stages of chromosome condensation. Soon after the nuclear division in *H. tusilata* (CLEVELAND 1949a), both chromosomes become long and convoluted; they consist of a single chromonema that is packed into the small spiral, while the large spiral and the matrix are absent (Fig. 37*a*). The chromonema then doubles and the two daughter chromonemata are relationally coiled (Fig. 37*b*). Later, they separate by "uncoiling" and immediately begin to coil again, but independently of one another, forming the large spiral (Fig. 37*c*). Soon, the chromatids thus formed are surrounded by a matrix (Fig. 37*d*). This is the longest (resting) stage, which lasts most of the time between two divisions of the cell (compare the electron micrograph of a nondividing nucleus of *H. hemigymnum*, Fig. 10*a*, showing both the chromonemata and the matrix). But cytologically, this is the late prophase: the two chromatids or each chromosome are already completely separated. That is followed by mitosis proper, or by its anaphase, to be more exact, when the sister chromatids separate from one another and enter the daughter nuclei (Fig. 37*e,f*). Morphologically, mitosis in *Holomastigotoides* is a variant of pleuromitosis with an extranuclear spindle. In the telophase, the matrix disappears, while the large spiral begins to uncoil (Fig. 37*g,h*).

The chromonemata pass through a similar cycle of spiralization in *H. psammotermitidis* (GRASSÉ and HOLLANDE 1963), where there are also two orders of coiling—large and small spirals. In this case, however, the matrix is thinner; it adjoins the chromosome more closely and repeats the coils of the large spiral. Under the electron microscope, the matrix appears to be amorphous material of low density (Fig. 10*a*). The double spiralization of the chromonema (large and small spirals) has also been discovered in other Hypermastigida (CLEVELAND 1949a, 1954).

Chromosome Replication and Mitotic Cycles

The replication of the DNA fibril, which is the basis of the chromonema, occurs in protozoans, just as in metazoans, during a definite period of the interphase. The mitotic (or cell) cycle is known to include not only mitosis proper (the period M or D) but also the entire interphase, which is generally divided into three periods: the presynthetic period (G_1), the period of DNA synthesis (S), and the postsynthetic period (G_2). Theoretically, the amount of DNA in the nucleus in the periods G_2 and G_1 is in a ratio of 2 : 1. During mitosis, the DNA is equally distributed between the two nuclei. In protozoans, the duration of the whole mitotic cycle (being known as *generation time*) and of its individual periods greatly depends on culture conditions. However, the occurrence of the S period at the beginning, the middle, or the end of the interphase is apparently hereditary.

The mitotic cycles of amoebae and some flagellates as well as ciliate micronuclei are rather well known now. DNA synthesis in the division cycle of macronuclei will be considered in Chapter 8. Almost nothing is known about the mitotic cycles of sporozoans, in which every stage of the complex life cycle should certainly have its own type of DNA replication.

Sarcodina. Among sarcodinids, DNA synthesis in the mitotic cycle has been studied most thoroughly in *Amoeba proteus*. According to autoradiographic data (PRESCOTT and GOLDSTEIN 1967, ORD 1968, RON and PRESCOTT 1969), the period of DNA synthesis occurs in this case immediately after mitosis. In other words, the G_1 period is absent. In the experiments carried out by PRESCOTT and his associates, the generation time was about 36 hours, the S period lasted 3–6 hours, and the G_2 period, about 30 hours. In the experiments carried out by ORD, the generation time was longer (about 48 hours). The S period lasted up to 12 hours and included two phases. At its beginning (0–6 hours of generation time), the greater part of DNA was synthesized, then synthesis diminished; but in about ten hours, it briefly reached a peak, during which about 15% of all DNA was synthesized. It has been shown recently by ultrastructural autoradiography that the late peak of ^{3}H-thymidine incorporation occurred in nucleoli and apparently corresponded to the replication or amplification of rDNA (MINASSIAN and BELL 1976b). *A. proteus* should have much rDNA, since its nucleus has many peripheral nucleoli (see p. 56). Indeed, it has been discovered cytofluorometrically that there is a regular "over-replication" of DNA in every cell cycle of *A. proteus*: its amount in the G_2 period is not twice but 2.2–2.9 times that in the G_1 period (MAKHLIN *et al.* 1979a). The over-replication of DNA is absent in starving amoebae, which are unable to divide. When re-fed, they divide only after completing the synthesis of "extra" DNA, which is most probably rDNA (MAKHLIN *et al.* 1979b). "Extra" DNA is not transferred to the nuclei of the daughter cells, but is degraded or somehow lost during mitosis. According to SPEAR and PRESCOTT (1980), the DNA synthesized upon re-feeding of the amoebae is a satellite DNA with high buoyant density.

It is still unknown what factors start and stop DNA synthesis in *A. proteus*. PRESCOTT and his associates reported that both of these occurrences depended on cytoplasmic "signals". In their experiments, the transplantation of a replicating nucleus into the "postsynthetic" cytoplasm blocked DNA synthesis in the transplanted nucleus, while the transplantation of a postsynthetic nucleus into the cytoplasm of a specimen in the S period caused the renewal of DNA synthesis in the completely replicated nucleus (PRESCOTT and GOLDSTEIN 1967, GOLDSTEIN and PRESCOTT 1967). However, neither the cessation nor the renewal of DNA synthesis was discovered in the transplanted nucleus in similar experiments carried out by Ord (1969, 1971).

Similar mitotic cycles without the G_1 period and with a long G_2 period have also been found in the multinucleate amoebae *Chaos illinoisensis* (McCLELLAN 1964) and *Ch. carolinensis* (HODGE and McCLELLAN 1966), where mitoses and S periods are synchronous in all the nuclei of a specimen.

It seems likely that mitotic cycles with complete reduction of the G_1 period are characteristic of all Amoebida. In the small soil amoebae *Acanthamoeba castellanii*, for instance, DNA synthesis begins already in the telophase and lasts only 20 minutes (the generation time is 18 hours). The remaining part of the interphase is occupied by the G_2 period (BAND and MOHRLOK 1973). However, in the amoebae of the order Schizopyrenida, e.g. in *Naegleria gruberi*, all three periods of the interphase are clearly expressed: G_1 and S last about three hours each, G_2 lasts 1.5 hours, while D takes 0.5 hour (CORFF and YUYAMA 1976).

Mastigophora. All three periods of the interphase are well expressed in the cultural forms of trypanosomids. According to STEINERT and STEINERT (1962), in *Trypanosoma mega*, the G_1 period lasted 8.5 hours, S lasted about 7 hours, and G_2, about 2 hours, while mitosis took 1.4 hour. In *Crithidia luciliae*, the periods G_1, S, and G_2 lasted 2, 3.3, and 1.2 hours, respectively (STEINERT and VAN ASSEL 1967). At optimum cultivation temperature (25 °C) in *Crithidia fasciculata*, the S period was relatively short (1.3 hour), while the periods G_1 and G_2 were 3.3 and 1.2 hours, respectively (COSGROVE and SKEEN 1970). In the case of lower (15 °C) and higher (32 °C) cultivation temperatures, all the periods became longer, especially S (6 and 3.6 hours, respectively). Hence, the rate of DNA replication clearly depends on temperature.

In the euglenid *Astasia longa*, DNA synthesis was studied in cultures that were synchronized by a temperature cycle (BLUM and PADILLA 1962); the generation time set by thermoperiodism was 24 hours. Under these conditions, the G_1 period was long (the whole "cold" part of the thermocycle and the beginning of the "warm" part), the S period preceded mitosis, while the G_2 period was absent. It should be noted, however, that the procedure of the thermal synchronization of cell division could have affected the duration of the periods of the mitotic cycle. In a 12-hour cell cycle in *Euglena gracilis*, the G_1 period lasted four hours, S lasted six hours, and G_2, about one hour (LEDOIGT and CALVAYRAC 1979).

In Volvocida, which reproduce themselves by *palintomy* (rapid repeated cleavage-type divisions in a cyst), e.g. in *Eudorina* and *Pandorina*, the period of DNA synthesis precedes every cytokinesis. However, since the intervals between the successive cell divisions are reduced to a minimum (15–20 minutes), the S period takes up virtually the whole interphase in palintomy (LEE and KEMP 1975, COLEMAN 1979).

In the nuclei of the multinucleate *Opalina ranarum*, the S period lasts about seven hours in the summer and the periods G_1 and G_2 also exist. As the rate of mitoses slows down (by autumn), the duration of the periods G_1 and G_2 greatly increases, while in the winter, the nuclei do not divide at all and do not synthesize DNA. In *O. ranarum*, both nuclear mitoses and the S periods are asynchronous in different nuclei of one specimen (NILOVA and SUKHANOVA 1964, 1969). In *Opalina*, every nucleus should apparently have its own autonomous mechanism for starting and ending DNA synthesis, unlike the multinucleate amoebae in which DNA replication starts simultaneously in all nuclei, probably by a cytoplasmic "signal".

Micronuclei of Ciliophora. The S period can take different positions in the interphase of micronuclei (see Table 8, p. 298). It is rarely in the middle of the interphase, or at least it is not right at its beginning or end. In such cases, the mitotic cycle of the micronucleus includes all four periods: G_1, S, G_2, and D (mitosis). In *Chilodonella cucullulus*, for instance, G_1 is equal to 1–2 hours; S, to one hour; G_2, to 8–9 hours, and D, to about 30 minutes (RADZIKOWSKI 1969, 1976). According to autoradiographic data (RAO and PRESCOTT 1967), the corresponding periods in *Paramecium caudatum* are 3 hours, 3.5 hours, one hour and 0.5 hour (or 37%, 45%, 12%, and 6% of the generation time). According to less reliable photometric data, the S period in *P. tetraurelia* is also near the middle of the interphase, but it takes up only about 10% of the generation time of the micronucleus, 50% of which corresponds to G_1 and about 40% to $G_2 + D$ (WOODARD *et al.* 1961). In *Euplotes crassus*, G_1 in the micronucleus lasts about 2 hours; S, about 3 hours; and $G_2 + D$, about 13 hours (SALVANO 1974a).

More frequently, the S period is at the beginning of the interphase, i.e. DNA synthesis begins immediately after the mitosis of the micronucleus (just as in amoebae). The period G_1 is absent then, while G_2 takes up the greater part of the interphase. Such mitotic cycles of micronuclei have been revealed in *Loxodes* (RAIKOV and MORAT 1977, BOBYLEVA *et al.* 1980), *Nassulopsis* (RAIKOV 1964), *Chilodonella* (SESHACHAR 1950a, PYNE *et al.* 1974), *Tetrahymena* (MCDONALD 1962, 1973, FLICKINGER 1965, PRESCOTT and STONE 1966, WOODARD *et al.* 1972), *Glaucoma* (PIERI 1966), *Colpidium* (MORAT 1970), *Paramecium bursaria* (JUST 1973), and *Euplotes* (Prescott *et al.* 1962). In these cases, the S period usually lasts less than one hour (it is longer only in *P. bursaria*), while the remaining part of the interphase is taken up by the G_2 period.

Finally, there is a group of ciliates in whose micronuclei the DNA replicates at the end of the interphase, just before mitosis. Then the G_1 period is very long, while the G_2 period is either short or absent. This applies to *Paramecium trichium* (PIERI *et al.* 1968), *P. putrinum* (KUDRJAVTSEV 1966), and many hypotrichids: *Urostyla* (JERKA-DZIADOSZ and FRANKEL 1970), *Keronopsis* (RUTHMANN 1972), *Stylonychia* (AMMERMANN 1970, CHADHA *et al.* 1978), *Onychodromus* (SAXENA and SAXENA 1978), *Laurentia* (TORRES *et al.* 1979), *Oxytricha* and *Diophrys* (DINI *et al.* 1975), and *Aspidisca* (DINI and BRACCHI 1976). According to BLANC (1965), that applies also to *Paramecium caudatum*. This, however, contradicts the above-mentioned results obtained by RAO and PRESCOTT, who found a clearly expressed, though brief, G_2 period in *P. caudatum*. The micronuclear mitotic cycles could possibly be different in various races (syngens?) of *P. caudatum*.

Finally, in the ciliates that reproduce themselves by palintomy inside cysts, e.g. in *Colpoda steini* (FRENKEL 1978), the S period in the micronucleus takes up the whole interphase between two rapidly succeeding mitoses (just as in Volvocida). Conversely, the micronucleus of *Colpoda* enters a long G_1 period after palintomy is over.

The relationships between the timing of DNA synthesis in the macronucleus and micronucleus of ciliates will be considered in Chapter 8 (see also Table 8).

Polyploidy in Protozoa

In protozoans, there are two types of polyploidy, i.e. multiplication of chromosome sets within a nucleus (POLJANSKY and RAIKOV 1960). In the polyploidy of the first type, or *generative* polyploidy, the extent of the multiplication of the chromosome sets is relatively small, and the nuclei continue to divide by mitosis. Having once appeared, such nuclear polyploidy becomes hereditary, i.e. it is steadily preserved at all the stages of the life cycle. This type of polyploidy originates most often spontaneously in individual specimens when the mechanism of mitosis or the nuclear behaviour in the sexual process is disturbed. If such specimens are viable, they initiate polyploid clones and strains within a species. Thus, generative polyploidy undoubtedly plays an important role in intraspecies differentiation in protozoans. It occurs in many taxa and can involve both the nuclei of homokaryotic protozoans and the generative nuclei of heterokaryotic forms, such as the ciliate micronuclei. This type of polyploidy could also have served as the basis for the separation of polyploid strains into independent species within a genus (Table 3). In this respect, it is quite similar to the generative polyploidy of the angiosperm plants, in which it plays a very important role in speciation.

The polyploidy of the second type, or *hyperploidy*, usually involves very high degrees of chromosome multiplicity. It is always connected with the disappearance of mitosis and the appearance of a special type of nuclear division: *genome segregation* (*depolyploidization*). Such nuclei may undergo division with the simple distribution of the available chromosome sets; they may divide not only in two, but also into many sister nuclei simultaneously. Hence, they must contain several relatively autonomous genomes. Therefore, the polyploid nuclei of the second type are also called *polyenergid* (HARTMANN 1909, 1952) or, to be more exact, *polygenomic* (MOSHKOVSKI 1957, 1963). The polyploidy of this type characterizes not all but merely some stages of the life cycle of a species, and it develops every time in ontogenesis due to a series of *endomitoses* (autonomous duplications of chromosomes without nuclear division). It is inherent not in the intraspecies categories (clones, races, etc.) or individual species but in large protozoan taxa. Polygenomic nuclei are characteristic especially of Radiolaria in which the sole nucleus temporarily becomes polygenomic (and consequently, depolyploidization, i.e. a return to the nonpolyploid stages, also occurs in the life cycle). Such a form of polyploidy can be called a *cyclic* one. The macronuclei (somatic nuclei) of some ciliates are also polygenomic. But since ciliates possess nuclear dualism, no return from the polyploid state of the macronucleus to the nonpolyploid one is observed in them. In this case, the situation is similar to *somatic polyploidy* in metazoans when highly polyploid nuclei are found in the cells of some specialized tissues in the generally nonpolyploid organism (reviewed by GEITLER 1953). In ciliates, however, the polygenomic macronucleus and the nonpolyploid micronucleus (or the micronucleus with only "generative" polyploidy) are in the common cytoplasm.

This section will be, in the main, only on generative polyploidy, since the polygenomic nuclei of radiolarians and the macronuclei of ciliates are specially dealt with in Chapters 6 and 8.

Polyploidy in Mastigophora

The substantial variability of the DNA content of the nuclei of euglenids indicates that polyploidy possibly occurs in these flagellates. In *Euglena gracilis*, for instance, some individual cells in a clone may have from two to five times more DNA than the normal cells in the G_2 period (OVCHINNIKOVA and SUKHANOVA 1976). Conversely, a shortage of some elements in the culture medium roughly halves not only the DNA content of the nucleus but also the chromosome number (SUKHANOVA *et al.* 1979). Judging from the great differences in chromosome numbers, most strains of the volvocid *Pandorina morum* display generative polyploidy. Only the strains with the same chromosome number are capable of mating, i.e. the individual biological categories within the species *P. morum* are already isolated reproductively (COLEMAN and ZOLLNER 1977).

A clear case of generative polyploidy has been observed in *Tritrichomonas batrachorum* (SAMUELS 1959). Diploid and triploid forms with 12 and 18 chromosomes, respectively, spontaneously occur in that normally haploid species ($n = 6$). However, generative polyploidy has been studied most thoroughly in three species of hypermastigids of the genus *Holomastigotoides* (CLEVELAND 1949a, GRASSÉ and HOLLANDE 1963). All the species of this genus are normally haplonts with $n = 2$. In *H. tusilata* and *diversa*, however, races with four chromosomes (diplonts) are known to exist, while in *H. psammotermitidis* there are also races with eight chromosomes (tetraploids). In this case, polyploidy apparently originates spontaneously when the chromosome duplication and nuclear division are uncoupled, i.e. endomitosis occurs. In all three species, both haplonts and diplonts are found in two variants, i.e. with either one or two chromonemata in the anaphase chromosomes. The appearance of a double chromonema in the anaphase can be interpreted as *polyteny*, which is the rudiment of the next step of nuclear polyploidization. It is still unknown how constant the polyploid races of *Holomastigotoides* are.

As many as 60 chromosomes (6–8 n) are observed in some individual nuclei in the multinucleate diplont *Opalina ranarum* (2 n = 18). Sometimes, all the nuclei of some specimens are polyploid (Kaczanowski 1968).

A case of *cyclic polyploidy* (which occurs at only some stages of the life cycle) is known in the oxymonadid *Pyrsonympha*, which is an intestinal parasite of termites (Hollande and Carruette-Valentin 1970a). The trophozoites of this flagellate undergo several rapid cell divisions before the host moults. They give rise to minute dispersive forms that infect new hosts. Earlier, these forms were believed to belong to another genus (*Dinenympha*). It has been discovered that the nucleus of the trophozoite has 16–32 times more DNA than that of the minute form, i.e. the former is rather highly polyploid. Apparently, when the trophozoite develops, endomitoses occur in its nucleus, and when the trophozoite rapidly divides, its nucleus is depolyploidized (whole chromosome sets are segregated without new replication of the chromosomes). This interpretation is substantiated by the fact that *synaptonemal complexes*, i.e. structures that normally unite the homologous chromosomes during the prophase of meiosis, are observed in the polyploid nucleus before the rapid divisions. Apparently, chromosome pairing in *Pyrsonympha* precedes not true meiosis (since the sexual process is unknown in them) but *somatic reduction*, i.e. the separation of homologous chromosomes and their segregation between the daughter nuclei. Unfortunately, the chromosomes in *Pyrsonympha* are small and numerous, making it difficult to observe their behaviour during endomitosis and depolyploidization and even to determine whether the nuclei of the dispersive stages are haploid.

Polyploidy in Sarcodina

If account is not taken of the polyploid nuclei of radiolarians (see Chap. 6), very little is known about polyploidy in sarcodinids. It has often been suspected that the nucleus of *Amoeba proteus*, which contains over 500 chromosomes, is polyploid (Liesche 1938). There is, however, still no evidence of this.

The large (60 μm) primary nucleus of the gamont in the rhizopod *Gromia oviformis* (order Gromiida) has been known to undergo fragmentation; as a result, many secondary nuclei were formed through several intermediate stages (Arnold 1972). According to Arnold, such fragmentation is preceded by the polyploidization of the primary nucleus, i.e. this is a case of cyclic polyploidy. There is, however, still no conclusive evidence of this. The great diversity of the structures observed both in the primary nuclei (pleistokarya) and in their fragmentation products suggests that some intranuclear parasites were taken for the stages of the nuclear cycle of *Gromia* in Arnold's work, especially since some fragmentation products possess, according to his data, their own amoeboid cytoplasm (amoebokarya). Unfortunately, no electron microscopic observations of these fragments were included. It is quite likely that the "fragmentation" of the primary nucleus is merely the ejection of its nonchromosomal material (cf. Fig. 14) together with the intranuclear parasites into the cytoplasm, while the initial mitoses of the minute secondary nucleus were unnoticed. Anyhow, electron microscopy of the same species (HEDLEY and WAKEFIELD 1969) has revealed only minute (6 μm) nuclei, which correspond to the numerous secondary nuclei in ARNOLD's work.

The situation is just as unclear with respect to another example of a fragmenting primary nucleus, that of the sessile heliozoan *Wagnerella borealis* (ZUELZER 1909). The fragments of this nucleus, i.e. the secondary nuclei, become the nuclei of the dispersive stages of the life cycle. Hence, HARTMANN (1952) believes that the primary nucleus of *Wagnerella* is polyploid (polyenergid). However, ZUELZER's observations should be verified by modern methods.

Cyclic polyploidy is more distinctly involved in the life cycle of the aberrant heliozoan *Sticholonche zanclea* (order Taxopodida). Its large primary nucleus also undergoes fragmentation during sporogenesis (CACHON and CACHON 1969b). No chromosomes are seen in the primary nucleus, which is virtually Feulgen-negative. A bundle of chromosomes condenses in it, however, before sporogenesis; this bundle is attached to a "centrosome" (MTOC?) located on the nuclear envelope. Both the centrosome and the chromosomes (possibly attached to a half-spindle) duplicate many times in rapid sequence, so that hundreds of chromosome bundles (genomes?) originate from the single bundle. Thus, the primary nucleus of *Sticholonche* very probably becomes polyploid (polygenomic), though for a short time only. Then it fragments so that each chromosome bundle gets into a small secondary nucleus (CACHON and CACHON 1969b). This is obviously a case of genome segregation.

Polyploidy of Ciliate Micronuclei

The generative polyploidy of the micronucleus was observed for the first time by CHEN (1940a, 1940c, 1950d) in *Paramecium bursaria*. The micronuclei of several races of this species had different sizes, stained differently by the Feulgen method and had different chromosome numbers in the meiotic prophase. CHEN believed that the smallest chromosome number he found, being about 80 in the race Fd, was diploid. This, however, has never been proved. In other races, the chromosome number came to a few hundred. SCHWARTZ (1978) also discovered a two- or three-fold difference in the number of chromosomes in the micronuclei of various strains of *P. bursaria*. The polyploidy of the micronuclei of this species usually originates as a result of various anomalies in the behaviour of pronuclei in conjugation. e.g. by the fusion of three and even four pronuclei (instead of two) following the unilateral passage of a migratory pronucleus into the partner and/or the differentiation of "excess" pronuclei (CHEN 1940a, 1940b, 1940e).

The races of *P. bursaria* with heteroploid micronuclei can conjugate with one another. However, they form not haploid but polyploid pronuclei. In normal conjugation, the ploidy degrees of the micronuclei of the two partners are equalized as a result of the mutual exchange of migratory pronuclei. For instance, when a diploid race is crossed with an octoploid one ($2n \times 8n$), the former produces haploid pronuclei (n), while the latter produces tetraploid ($4n$) pronuclei. Then, both synkarya and their progeny, i.e. the micronuclei, will be pentaploid ($5n$). However, the heteroploidy of the micronuclei can also increase if anomalies occur again in conjugation. The feasibility of these ways of changing the degree of polyploidy of the micronucleus was experimentally confirmed by OVCHINNIKOVA (1970) by photometrically evaluating the micronuclei of the two conjugants and their progeny in *P. bursaria*. According to CHEN, the ploidy degree of the micronucleus can be halved when a polyploid race is crossed with an amicronucleate race. In both partners, the nuclear apparatus is restored then from a single pronucleus (hemikaryon), which is stationary in the initially micronucleate partner and migratory in the initially amicronucleate one (CHEN 1940b, 1940e, 1940f).

GOLIKOVA (1972) has also discovered substantial polymorphism of micronuclei in the natural populations of *P. bursaria*; during cloning, however, the type of the micronucleus remained constant. According to cytophotometric data, the differences in the DNA content between the micronuclei of various clones can go up to four-fold (OVCHINNIKOVA 1970) and roughly form four classes with a ratio of 1:2:3:4 (GOLIKOVA 1974). If the first class of micronuclei is diploid, the other classes will be tetraploid, hexaploid, and octoploid. Just as in CHEN's works, however, the diploidy of the micronuclei of the first class has not been proved.

The polymorphism of micronuclei has been observed also in other species of *Paramecium* (reviewed by OSSIPOV and BORCHSENIUS 1973). In *P. caudatum*, for instance, the DNA contents of the micronuclei of individual clones form the polyploid series 1:2:4:6:8 (CHEISSIN *et al.* 1964, O. BORCHSENIUS and OSSIPOV 1971). Various morphological types of micronuclei, which are stable in clones, correspond to these classes (O. BORCHSENIUS *et al.* 1968). Thus, the degree of generative polyploidy of the micronuclei of some strains of *P. caudatum* is at least 16n. Wide-scale heteroploidy of the micronuclei of *P. putrinum* has been described by JANKOWSKI (1972). In this case, the chromosome numbers vary in different clones from about 15 to approximately 180, i.e. they differ 12-fold.

JANKOWSKI (1972) expressed the view that all the massive "wild type" micronuclei that prevailed in the natural clones of *Paramecium putrinum*, *P. bursaria*, *P. caudatum*, and some other species were polyploid and that this polyploidy was a species character. In the course of evolution, "massive" polyploid micronuclei originated from vesicular nonpolyploid micronuclei found in *P. aurelia*, *P. woodruffi*, *P. jenningsi*, and *P. calkinsi*, which display no substantial polymorphism and contain much less DNA. Indeed, CHEISSIN and OVCHINNIKOVA (1964) have shown that the DNA contents of the micronuclei of *P. woodruffi* and *P. calkinsi* were 8 and 5 units, respectively, while the DNA content of the "wild type" micronuclei was 140 units in *P. caudatum*, 130 units in *P. putrinum*, and 120 units in *P. bursaria*. However, the minimum DNA content found in a clone of *P. caudatum* was only 19 units. Similar data are presented by DUPY-BLANC (1969b), who showed that the DNA contents of the micronuclei of *P. jenningsi*, *P. caudatum*, and *P. bursaria* ("wild" clones) were in a ratio of 1:8:16.

Micronuclear polyploidy is rare outside the genus *Paramecium*. A tetraploid clone ($4n = 20$) has been found in *Tetrahymena pyriformis* (RAY 1958). A polyploid series of species has been observed in the astomatid ciliates of the genus *Metaradiophrya* (MORAT 1965). The populations of the heterotrichid *Plagiotoma lumbrici*, which are parasitic in two species of earthworms, have micronuclei with DNA contents that are in a ratio of 3:2, indicating triploidy (MORAT 1979).

Conversely, the micronucleus of *Oxytricha* sp. is clearly not polyploid: the size of the true haploid genome of this nucleus (according to renaturation data) is only one-half its overall DNA content. Hence, the micronucleus is diploid (LAUTH *et al.* 1976).

The polyploidy of the micronucleus apparently either does not influence or only slightly influences the phenotype of the ciliate, e.g. the size of the body and of the macronucleus, etc. (CHEN 1940a, RAY 1958, OVCHINNIKOVA 1970, OSSIPOV and BORCHSENIUS 1973). This is undoubtedly caused by the ciliate's nuclear dualism. SONNEBORN (1947) has shown that it was not the genes of the micronucleus, but those of the macronucleus that determined all the phenotypic characters of ciliates. The former are in an inactive state and are not phenotypically expressed. Naturally, the differences in the ploidy of the micronuclei of various clones and races are also not phenotypically expressed. However, recent data on transcription in ciliate micronuclei (see the review by O. BORCHSENIUS 1975) suggest that, in some cases, micronuclear polyploidy may be of certain significance in the vegetative period of clonal life and, consequently, in natural selection. This would explain the origin of the polyploid species of paramecia or the polyploid series of *Metaradiophrya*. Further research, however, should be carried out in this field.

4. Nuclear Apparatus of Mesokaryotic Protozoa

At present, only one order of protozoans is known to have a mesokaryotic nucleus, namely, the Dinoflagellida. Almost all dinoflagellates are uninucleate or have a uninucleate stage in their life cycle. A few cases of apparent nuclear dualism among them will be considered in Chapter 7. For reviews on dinoflagellate karyology, see DODGE (1966, 1971a), KUBAI (1975), and LOEBLICH (1976).

Structure of Nondividing Nuclei

Types of Nuclei and the Organization of Chromatin in Them

Most dinoflagellates, particularly the free-living forms (with the exception of the family Noctilucidae), have very peculiar nuclei that are often referred to as *dinokarya*. These nuclei are rather small, and they always contain compact chromosomes that do not undergo any noticeable decondensation in the nondividing nucleus. Therefore, dinokarya are often said to belong to the group of "chromosomal" nuclei that also includes the nuclei of euglenids, many hypermastigids, etc. (see p. 70). However, as we shall see, the chromosomes of Dinoflagellida differ very much from those of eukaryotic protozoans, and therefore dinokarya must be regarded as an independent type of protozoan nucleus.

The chromosomes of the dinokarya stain brightly with the Feulgen reaction or acetocarmine and look like thick threads or rods (Fig. 38*a*). Under the electron microscope, the chromosomes are seen as dense bodies with an ordered fibrillar structure. Dinokarya usually contain one nucleolus or several nucleoli and are surrounded by a typical nuclear envelope (Fig. 39*a*). Dinokarya have been studied in detail by electron microscopy, particularly in *Amphidinium* (GRASSÉ and

DRAGESCO 1957, GRELL and WOHLFARTH-BOTTERMANN 1957, GIESBRECHT 1961, 1962, HALLER 1963, HALLER *et al.* 1964, GRELL and SCHWALBACH 1965, DODGE and CRAWFORD 1968, MATTHYS-ROCHON 1979), *Gymnodinium* (GRASSÉ and DRAGESCO 1957, DODGE and CRAWFORD 1969, SOUSA e SILVA 1977), *Ceratium* (DODGE and CRAWFORD 1970), *Crypthecodinium* (= *Gyrodinium*) (KUBAI and RIS 1969), *Woloszynskia* (LEADBEATER and DODGE 1966, 1967), *Peridinium* (HOLLANDE *et al.* 1963, ZINGMARK 1970), *Prorocentrum* (DODGE and BIBBY 1973, SOYER and HAAPALA 1973, 1974a, 1974b), *Endodinium* (HOLLANDE and CARRÉ 1974), etc.

However, other types of nuclei are also found in Dinoflagellida. In some large planktonic forms of the family Noctilucidae, such as *Leptospathium* and *Leptophyllus*, the nuclei are rather bulky and contain several nucleoli and condensed chromosomes submerged in the abundant karyolymph (CACHON and CACHON-ENJUMET 1964b). But in very large nuclei of other Noctilucidae, numerous polymorphic nucleoli develop and chromosomes completely decondense; the nucleus as a whole becomes a "large vesicular" or "ovular" one (Fig. 38b). Such is the nuclear structure of the trophozoite of *Noctiluca* studied by SOYER (1970a, 1972) and ZINGMARK (1970). ZINGMARK coined a new term, *noctikarya*, for such nuclei of Dinoflagellida to distinguish them from dinokarya. The nuclei of some other Noctilucidae, such as *Kofoidinium* and *Leptodiscus*, have a similar structure (CACHON and CACHON 1967, 1969a).

The nucleus of the free-living dinoflagellate *Oxyrrhis marina* holds an intermediate place between "dinokarya" and "noctikarya" as regards chromosomal condensation. In this case, the chromosomes of the nondividing nucleus are filamentous and are seen under the light microscope as chains of chromomeres (HALL 1925a). Only their condensed regions can be seen on electron micrographs (GRELL and SCHWALBACH 1965, DODGE and CRAWFORD 1971, CACHON *et al.* 1979). A nucleus with greatly decondensed chromosomes has also been found by electron microscopy at certain stages of development in *Peridinium* sp. (MIGNOT 1970).

Vesicular nuclei with greatly decondensed chromosomes, being rare among free-living Dinoflagellida, are typical of the trophozoites of parasitic forms characterized by hypertrophic growth (CHATTON 1920a, CACHON 1964, SOYER 1971). In *Duboscquella aspida*, for instance, the dinokaryon of a zoospore only 1–2 μm in diameter grows to a large vesicular nucleus (up to 40 μm) without visible chromosomes in a trophozoite, and a branching nucleolus develops in the nucleus (Fig. 1*i*). During sporogenesis, the nucleus undergoes several consecutive divisions, and the chromosomes condense progressively in the nuclei of every successive generation of sporocytes. Similar vesicular nuclei occur also in the trophozoites of other parasitic Dinoflagellida (CACHON 1964, CACHON and CACHON 1977).

SOYER (1971) used the electron microscope to study the nuclei of *Blastodinium* (a parasite of planktonic crustaceans), whose trophozoite undergoes multiple division after a period of hypertrophic growth and gives rise to a colony of cells. The latter consists of a feeding cell (trophocyte) and several generations of sporocytes, the last of which produce flagellated zoospores. The large nucleus of the trophocyte is cupped, permeated with cytoplasmic channels (plasmodendrites), and filled with a granular matrix in which chromosomal elements are hardly visible (Fig. 38*c*). The Feulgen reaction in this nucleus is negative, apparently due to the great "dilution" of chromatin; the nucleus has well-developed nucleoli that contain RNA. Chromosomes gradually begin to condense in the parent cell of a series of sporocytes. The process goes on in sporocytes, whose nuclei become smaller and smaller until they eventually look like typical dinokarya (Fig. 38*d*). This is also observed during the sporogenesis of free-living *Noctiluca*, in which the noctikaryon of the trophozoite is connected by a series of gradual transitions with typical dinokarya of late sporocytes (SOYER 1970a, 1972). In those parasitic dinoflagellates, however, where the trophozoite rapidly becomes multinucleate and plasmodial (such as in *Solenodinium* and *Syndinium*), the nuclei of the plasmodium always remain small and dinokarya-like (CHATTON 1920a). Electron microscopy shows that they have condensed chromosomes whose fibrillar structure is, however, not so orderly as in free-living dinoflagellates (RIS and KUBAI 1974, HOLLANDE 1974a). The same is typical of *Haplozoon*, a colonial parasitic dinoflagellate (DOGIEL 1910b, SIEBERT and WEST 1974).

Therefore, the condensed state of chromosomes predominates in Dinoflagellida. Chromosomes decondense only in nuclei that function very actively and do not divide for a long time. Such nuclei are found in particularly large cells (*Noctiluca*, uninucleate trophocytes of parasitic Dinoflagellida). Another peculiarity of mesokaryotic nuclei is that their chromosomes hardly change morphologically with the phase of a mitotic cycle. The extent of their condensation changes only gradually in the course of several cell generations.

Nuclear Envelope

The nuclear envelope of all dinokarya and most of the other dinoflagellid nuclei has a classic structure (see Chapt. 1). It consists of two membranes with a narrow perinuclear space between them and has numerous pore complexes (HALLER *et al.* 1964, GRELL and SCHWALBACH 1965, DODGE and CRAWFORD 1968, 1971, DODGE 1971a, DODGE and BIBBY 1973, etc.). The freeze-etching technique (WECKE and GIESBRECHT 1970, SOYER and ESCAIG 1980) has shown that they were arranged randomly in *Glenodinium* and hexagonally in *Prorocentrum*. The outer membrane of the nuclear envelope has continuities with the endoplasmic reticulum in *Amphidinium* (DODGE and CRAWFORD 1968). *Ceratium* has an additional cisterna of the endoplasmic reticulum that is parallel to the outer surface of the nucleus (DODGE and CRAWFORD 1970). Several such flattened cisternae stratify, forming a special "perinuclear capsule" in *Peridinium* sp. (MIGNOT 1970). *Oxyrrhis* and the Warnowiidae (*Warnowia*, *Erythropsis*) have similar perinuclear cisternae, though the cisternae have pores like those on the nuclear envelope. These cisternae are assembled in stacks, i.e. they form cytoplasmic annulate lamellae (GREUET 1972).

Fig. 38. Dinoflagellate nuclei (photomicrographs). *a Prorocentrum micans* cell with chromosomes visible in the nondividing dinokaryon, acetocarmine staining, 2000×; *b* noctikaryon of *Noctiluca*, semithin section (*1* nucleoli, *2* archoplasmic sphere), azure II-methylene blue staining, 2800×; *c* section through the nucleus of a *Blastodinium* trophocyte (*1* nucleus, *2* cytoplasm, *3* archoplasmic sphere, the arrows indicate plasmodendrites), gallocyanin staining, 700×; *d Blastodinium* sporocytes, azure II-methylene blue staining, 950×. *a* after DODGE (1963a); *b* after SOYER (1972); *c* and *d* after SOYER (1971).

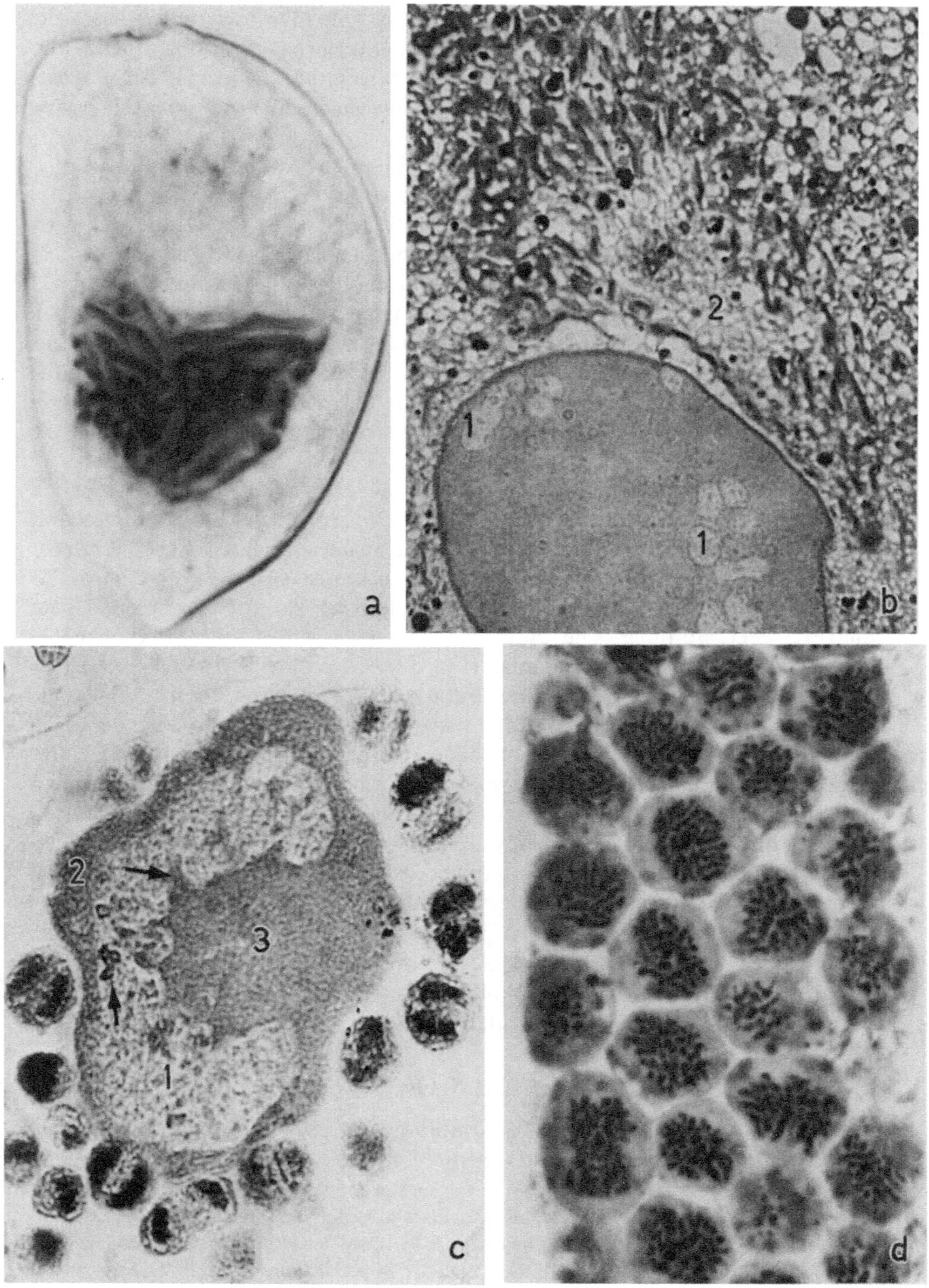

Fig. 38

A nuclear envelope of the classic structure is also typical of parasitic Dinoflagellida. The large noctikaryon of the trophozoite of *Blastodinium* has many pores in the envelope through which RNP granules can pass from the nucleus into the cytoplasm (SOYER 1971). Pores are more rare in the envelope of small nuclei with condensed chromosomes, typical of *Solenodinium* and *Syndinium* (HOLLANDE 1974a, RIS and KUBAI 1974).

The envelope of the noctikaryon in the trophozoite of *Noctiluca* is noteworthy. AFZELIUS (1963) was the first to discover that it consists of two membranes and has an underlying fibrous lamina but completely lacks pores. Nevertheless, many (8000–11 000) "annulated vesicles" or "porous ampullae" with a diameter less than 1 μm are adjacent to the inner side of the envelope; these vesicles are limited by two concentric membranes with numerous typical pores (Fig. 39*b*, *c*). According to AFZELIUS, the annulated vesicles are formed within the nucleus without the direct participation of its outer envelope. Substances released from the nucleus pass through the pores into the vesicles. The wall of each annulated vesicle then fuses with the nuclear envelope, and its contents are discharged into the cytoplasm through the opening thus formed.

The porous ampullae (annulated vesicles) of *Noctiluca* were also studied by ZINGMARK (1970) and especially by SOYER (1969a, 1969b, 1972). SOYER has shown that the ampullae are not formed *de novo* inside the nucleus, but are permanent invaginations of the two membranes of the nuclear envelope, being comparable with the intranuclear annulate lamellae. There are, on the average, 94 pore complexes per ampulla. The ampullae are permanently connected with the cytoplasm via a narrow outlet channel (Fig. 39*c*). Moreover, SOYER has shown that the ampullae gradually disappear in the course of sporogenesis: their walls become incorporated into the nuclear envelope while their pores come to the surface of the nucleus. Therefore, the porous ampullae not only greatly increase the active surface of the nucleus but also represent a stock of the membrane material consumed in numerous nuclear divisions during sporogenesis. RNP granules are released from the nucleus into the cytoplasm of *Noctiluca* through the annulated vesicles, particularly at the beginning of sporogenesis when nucleoli disperse in the nucleus (SOYER 1972).

It would be relevant to note that, apart from *Noctiluca*, porous ampullae on the inner side of the nuclear envelope were found in *Gymnodinium fuscum*, a comparatively large free-living flagellate (DODGE and CRAWFORD 1969). Such structures are so far unknown outside the Dinoflagellida.

Nucleoli and Nuclear Matrix

Typical dinokarya of free-living Dinoflagellida usually have one nucleolus (Fig. 39*a*) or, less frequently, two or three nucleoli of relatively simple shape

Fig. 39. Ultrastructure of dinoflagellate nuclei. *a* section through the dinokaryon of *Heterocapsa triquetra* (*1* chromosomes, *2* nucleolus, *3* nucleolar organizer, *4* nuclear envelope), 19 000 ×; *b* periphery of the nucleus of a *Noctiluca* trophozoite (*1* porous ampullae of the nuclear envelope, *2* nucleus, *3* cytoplasm, *4* vacuolar zone around the nucleus), 12 150 ×; *c* the same at higher magnification (same designations as in *b*), 41 500 ×. *a* after DODGE (1971a); *b* and *c* after SOYER (1969a).

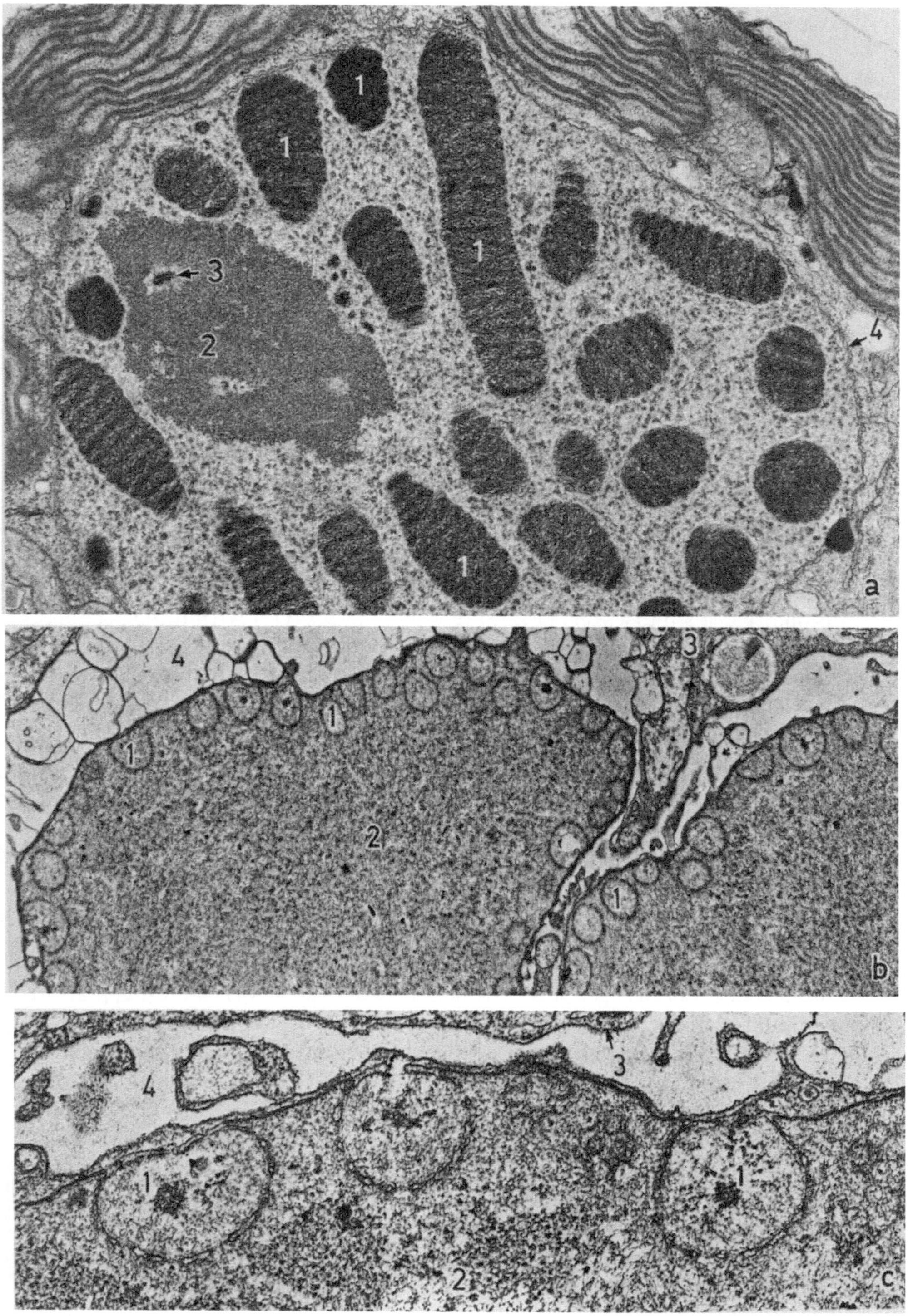

Fig. 39

(HOLLANDE *et al.* 1963, HALLER *et al.* 1964, DODGE and CRAWFORD 1968, 1970, 1971, KUBAI and RIS 1969, DODGE 1971a, DODGE and BIBBY 1973, etc.). The nucleolus of *Amphidinium* is submembranous, being adjacent to the inside of the nuclear envelope (MATTHYS-ROCHON 1979). In parasitic dinoflagellates, however, nucleoli often have an intricate shape, as in the *Duboscquella aspida* trophozoites (Fig. 1*i*), or become multiple, as in the *Blastodinium* trophocytes (SOYER 1971). Multiple polymorphous nucleoli are also typical of the nuclei of the trophozoites of large planktonic dinoflagellates, such as *Noctiluca* (SOYER 1969a, 1970a, 1972, Fig. 38*b*), *Leptospathium* and *Leptophyllus* (CACHON and CACHON-ENJUMET 1964b). Both in *Noctiluca* and in parasitic forms, the nucleoli fragment prior to sporogenesis and their granular material passes into the karyolymph. The RNP granules migrate from the nucleus through nuclear pores either directly to the cytoplasm (in *Blastodinium*) or first into "annulated vesicles" and then to the cytoplasm (in *Noctiluca*). The nuclei of the late sporocytes of these flagellates lack nucleoli (SOYER 1971, 1972).

The ultrastructure of nucleoli in Dinoflagellida hardly differs from that in eukaryotic protozoans (see Chapt. 1). A central fibrillar zone and a peripheral granular zone are usually discernible in nucleoli. In *Crypthecodinium*, the granular zone contains crystalloids of an unknown nature. A clear channel can often be observed within the fibrillar core of the nucleolus; it contains the nucleolar organizer, i.e. a region of one of the chromosomes (Fig. 39*a*). This region occasionally coils into a skein in a more or less large cavity within the nucleolus. Such a structure is found in the nucleoli of typical dinokarya, e.g. in *Peridinium* (HOLLANDE *et al.* 1963), *Crypthecodinium* (KUBAI and RIS 1969, RAE 1970) and *Ceratium* (DODGE and CRAWFORD 1970), and in the nucleoli of noctikarya, such as in *Noctiluca* (SOYER 1972) and *Blastodinium* (SOYER 1971). Light microscopy has shown that in parasitic dinoflagellates, the nucleolar-organizing regions may be located either on one chromosome of the set, e.g. in *Merodinium* (PÄTAU 1937b), *Ichthyodinium*, and *Amoebophrya acanthometrae* (CACHON 1964), or on several chromosomes, e.g., in Duboscquellidae (CACHON 1964). In the latter case, several nucleolar organizers wind into a single skein and produce a common nucleolus.

A detailed study of the ultrastructure of serially sectioned *Prorocentrum* nucleoli has shown that in the region of the nucleolar organizer, most of the DNA fibrils constituting the chromosome produce long lateral loops that pass deep into the fibrillar zone of the nucleolus (SOYER and HAAPALA 1974a); rRNA appears to be synthesized on these decondensed loops.

Just as in eukaryotes, the treatment of dinoflagellates with actinomycin D makes nucleolar components segregate (BABILLOT 1970a). ^{3}H-uridine is actively incorporated into the nucleolus of *Amphidinium* (BABILLOT 1970b). Biochemical studies have revealed the eukaryotic-type processing of the preribosomal RNA in the nucleolus of *Crypthecodinium*: the molecule of 38 S RNA dissociates into 16 S and 25 S rRNA (RAE 1970). These forms of rRNA are the same as in plants, while 18 S and 28 S rRNAs are found in animals.

The interchromosomal nuclear matrix (i.e. the karyolymph) of dinoflagellate nuclei is rich in granular material and, just as nucleoli, synthesizes RNA (BABILLOT 1970b). RNP particles with a diameter of 16, 32, and 50 nm have been found in *Blastodinium* (SOYER 1971). It could be that some of these particles contain messenger RNA rather than ribosomal RNA, since the largest of them resemble perichromatin granules. Furthermore, dense accumulations of a protein nature (Fig. 52*b*), being dissolved by pronase (SOYER and HAAPALA 1974b), are arranged in the nuclear matrix at the periphery of the *Prorocentrum* and *Crypthecodinium* chromosomes. These accumulations cover plexuses of extrachromosomal DNA loops (see p. 178).

The karyolymph of *Gymnodinium* and *Glenodinium* contains bacteria (SOUSA e SILVA 1978), while that of *Blastodinium* has groups of viral particles (SOYER 1978).

Mitosis

For a long time, the nuclear division of free-living Dinoflagellida was regarded as a special type of mitosis with an "autonomous" separation of chromosomes, related to the mitosis of Euglenida (DODGE 1963a, 1964b, 1966, RAIKOV 1967a). Mitosis of parasitic Dinoflagellida was regarded as a variety of pleuromitosis (paramitosis) with a more or less distinct extranuclear spindle, the preservation of the nuclear envelope, and the fixation of the chromosomal kinetochores to the envelope. Therefore, mitoses of free-living and parasitic dinoflagellates were divided into different categories (e.g. in RAIKOV 1967a).

The situation has radically changed in the past 15 years. Using electron microscopy, DODGE and his associates (1971a) in England, KUBAI (1975) and RIS (1975) in the U.S.A., SOYER (1971, 1972), HOLLANDE (1972, 1974a), and J. CACHON and M. CACHON (1974, 1977, 1979) in France gained a better insight into the mitosis of both free-living and parasitic dinoflagellates. As they have shown, the mitotic mechanism in free-living dinoflagellates basically differs from that in euglenids: the microtubular mitotic spindle is completely extranuclear in the former, while it is intranuclear in the latter. At the same time, the mitosis of free-living dinoflagellates proved to be similar, in principle, to that of parasitic forms and to differ from it only in details. Therefore, the term *dinomitosis*, introduced many years ago by CHATTON (1920a), can now be used again to designate mitosis of all dinoflagellates. As we shall see later, dinomitosis in its major characteristics is identical with closed extranuclear pleuromitosis of eukaryotic protozoans (see Chapter 2 and Table 2).

Let us first consider the dinomitosis of parasitic dinoflagellates, which is relatively simple.

Dinomitosis of Parasitic Forms

This variety of dinomitosis was, in general, correctly described and interpreted even in the era of light microscopy. Complete data are available for *Merodinium belari*, a parasite of the radiolarian *Collozoum* (BĚLAŘ 1926a, HOLLANDE and ENJUMET 1953, HOLLANDE 1972). Two granules (centrioles) connected by means of a centrodesmose, i.e. a threadlike central spindle, can be discerned in the perinuclear cytoplasm during mitosis (Fig. 40). The nuclear envelope is preserved in mitosis, and the kinetochores of all chromosomes are permanently attached to the envelope. In the prophase, the kinetochores are duplicated and short chromosomal fibres extending from the opposite ends of the central spindle are fixed to the sister kinetochores from the cytoplasmic side (Fig. 40*a*). The next stage may be called the "proanaphase", since the metaphase is not expressed in this case: the central spindle elongates and the centromere regions of chromatids begin to separate. However, the chromosome arms are condensed very poorly and, consequently, their separation takes a long time (Fig. 40*b*). It is only in the late anaphase that the chromatids condense and separate and the central spindle elongates, while the chromosomal fibres remain short (Fig. 40*c*). In the telophase,

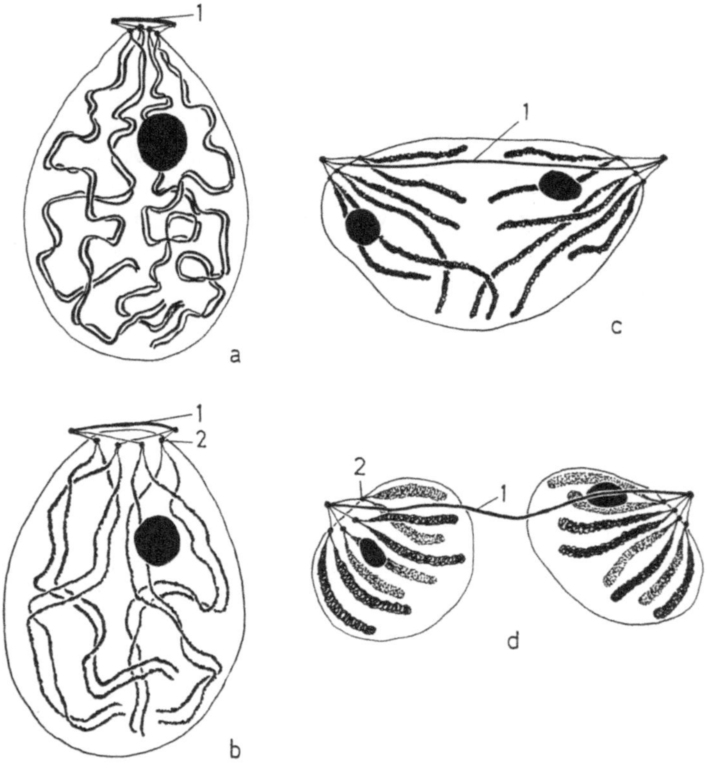

Fig. 40. Diagram of mitosis in *Merodinium belari* (after HOLLANDE 1972). *a* prophase; *b* proanaphase; *c* anaphase; *d* telophase; *1* centrodesmose, *2* kinetochores.

the nucleus pinches in two (Fig. 40*d*) and the chromosomes reach the maximum degree of condensation.

Even this description shows that dinomitosis in *Merodinium* essentially coincides with the varieties of closed pleuromitosis involving an extranuclear central spindle represented by a narrow centrodesmose, e.g. with the mitosis of *Trichomonas* (Fig. 33*d*). Similar dinomitosis with a centrodesmose occurs in *Atlanticellodinium* (CACHON and CACHON-ENJUMET 1965a).

If this type of dinomitosis is assumed to be the initial one, two morphological series could start from it. In one of them, no centrodesmoses can be observed, at least with the light microscope. In *Amoebophrya acanthometrae*, *Duboscquella melo* (CACHON 1964), and *Neresheimeria catenata* (CACHON and CACHON-ENJUMET

1964c), for instance, "centrioles" appear to be closely adjacent to the nuclear envelope on the outside, while all chromosomes are suspended at the same site of the envelope from the inside; the dividing nucleus has a peculiar "bicorn" shape. The nuclear apex in *Neresheimeria* was studied by electron microscopy (CACHON and CACHON 1966). The "centriole" turned out to be a pair of centrioles (kinetosomes of the future flagella) surrounded by a common dense "archoplasm". Two types of cytoplasmic microtubules, namely long radial ones (which, during mitosis, apparently form the central spindle that is invisible under the light microscope) and very short chromosomal microtubules, emanate from the "archoplasm" (extranuclear MTOC). The nucleus is suspended on the chromosomal microtubules, although no kinetochores were detected. The extranuclear spindle of *Amoebophrya* has a more or less similar ultrastructure (CACHON and CACHON 1970).

Finally, neither "centrioles" nor a centrodesmose can be discerned by light microscopy in *Syndinium*. This is the "acentric dinomitosis" or "syndinial mitosis" (CHATTON 1920a, 1921). Five V-shaped chromosomes whose centromere regions converge on one pole condense in the nucleus of *S. turbo* during the transition from the interphase to multiple mitoses. During mitosis, all chromosomes "split" lengthwise, the single pole divides in two, and sister chromatids begin to "deviate" from one another owing to the gradual separation of the poles. Eventually, the poles take up the opposite positions on the nucleus, while sister chromatids remain connected only by their telomeres. Then, this connection also breaks and the nuclear envelope constricts, producing two nuclei. The last stages of syndinial mitosis can be mistaken for the transverse division of the chromosomes.

The forces that caused the migration of the poles and the segregation of the chromatids in that case were unknown for a long time. Only recent electron microscopic studies have cleared this up. It has been shown that mitosis in *Syndinium*, *Solenodinium* (RIS and KUBAI 1973, 1974, HOLLANDE 1974a), and *Haplozoon* (SIEBERT and WEST 1974) involves centrioles, a central spindle, and chromosomal microtubules. Actually, all these structures are very small and are buried in a deep furrow on the surface of the dividing nucleus or even in a "tunnel" that penetrates the nucleus (consequently, they are invisible under the light microscope). In other respects, however, the mechanism of syndinial mitosis resembles that of the dinomitosis of the *Merodinium* type (Fig. 40).

RIS and KUBAI as well as HOLLANDE have shown that there was a pit on the surface of the nondividing nucleus in both *Syndinium* and *Solenodinium* (Fig. 41*a*). Two centrioles lie in the pit, while the nuclear envelope covering it incorporates the kinetochores of all chromosomes. The kinetochores are small and consist of two dense bands 25 nm thick and a less dense interlayer about 15 nm thick; they stopper the gaps in the nuclear envelope whose diameter (70 nm) is similar to that of the nuclear pores (Fig. 41*b*). Every kinetochore is permanently bound to the bases of both centrioles through a bundle of several chromosomal microtubules. In the early prophase, the kinetochores duplicate (each chromatid gets its own kinetochore) and, from then on, each of the two sister kinetochores remains connected only with one of the two centrioles (Fig. 41*c*). At the same time, the centrioles begin to separate and a bundle of microtubules, or the central spindle, which is very short at the beginning, appears between the bases of the centrioles.

The pit on the nucleus extends into a furrow that cuts deeper and deeper into the nucleus and eventually becomes a cylindrical channel or tunnel penetrating the nucleus and filled with cytoplasm. The tunnel encloses the central spindle (centrodesmose), which consists of only 20–25 microtubules. The segregating centrioles finally replicate to form a pair of centrioles at either pole. The same events are shown diagrammatically in Fig. 42.

Since the bundles of chromosomal microtubules had connected the duplicated sister kinetochores with different centrioles, the mere separation of the centrioles is sufficient to drive the kinetochores apart and cause the chromatids to segregate. Chromosomal microtubules are short (about 0.25 μm in *Syndinium*), completely extranuclear, and always at an angle to the continuous microtubules (Fig. 41 *d*). Their length does not change during mitosis. Conversely, the central spindle rapidly elongates (apparently because its microtubules are assembled at the ends), driving apart the two pairs of centrioles and the kinetochores of two chromatid sets bound to them through the chromosomal fibres of constant length (Fig. 42). Therefore, in syndinial mitosis, continuous microtubules play an active role, while chromosomal microtubules play only a passive one. In this case, the chromosomes do not actually approach the poles (Ris and Kubai 1974). The nuclear envelope of the furrow (or tunnel) should apparently grow in the zone between the sister kinetochores. The arms of the chromosomes separate later than their centromere regions. The nuclear envelope then constricts (Fig. 42 *e*).

Electron microscopy has shown that chromosomes replicate again as early as in the anaphase, and from then on to another mitosis, they consist of two relationally coiled, condensed chromatids (Hollande 1974a). The "splitting" of chromosomes in mitosis, described by Chatton (1920a, 1921), is nothing but the separation of these chromatids after the replication of kinetochores.

In *Haplozoon*, the sole tunnel that penetrates the nucleus is rather wide and encloses various cytoplasmic organelles besides the continuous microtubules. It is formed as an invagination at a pole of the nucleus and grows through the nucleus. No centrioles or MTOCs were found at the poles of the extranuclear spindle located in the tunnel. The chromosomes are attached to the nuclear envelope, which lines the tunnel, by means of dense kinetochores. These kinetochores, however, lack the typical trilaminar structure; bundles of chromosomal microtubules are attached to them from the side of the tunnel lumen (Siebert and West 1974). Therefore, mitosis in *Syndinium* and *Haplozoon* may be regarded as dinomitosis with a cytoplasmic channel immersed in the nucleus and with the central spindle (centrodesmose) located in the channel (*single-channelled dinomitosis*).

Fig. 41. Ultrastructure of single-channelled dinomitosis in *Syndinium* sp. (after Ris and Kubai 1974). *a* part of the nucleus (*1*) with a centriolar pit (*2*) in the interphase, 33 600 × ; *b* centriolar pit, the nuclear envelope (*3*) incorporates the kinetochores (*4*) of three chromosomes, 48 000 × ; *c* duplication of the kinetochores (*4*) in the early prophase, 48 000 × ; *d* longitudinal section through the centriolar tunnel (*8*) in the anaphase, 27 000 × ; *5* centrioles, *6* microtubules, *7* chromosomes, *9* central spindle, *10* bundles of chromosomal microtubules.

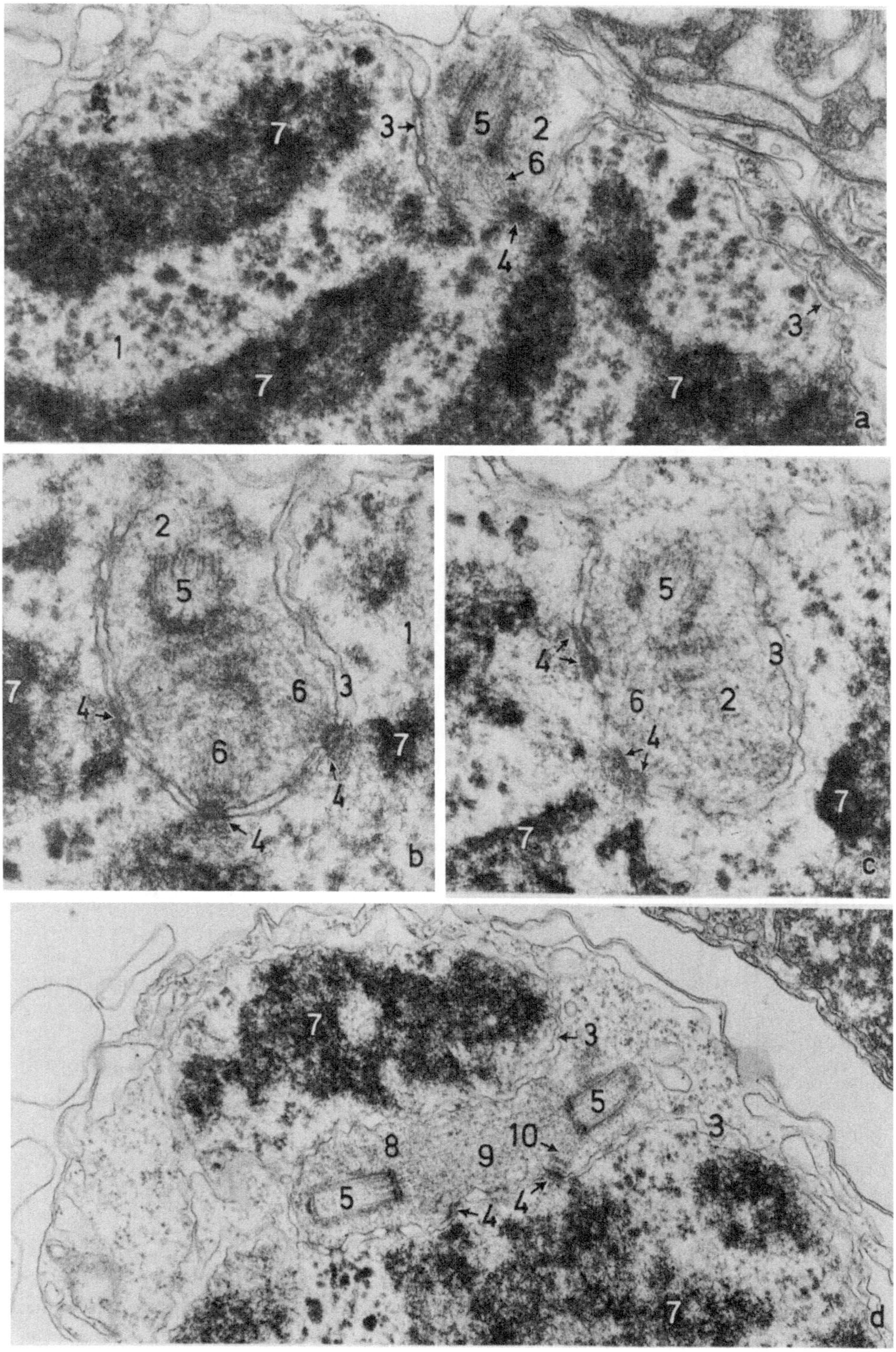

Fig. 41

The second morphological series, starting from the dinomitoses of the *Merodinium* type, is characterized by the high development of the central spindle that is often formed between two bulky atractospheres and greatly resembles the mitotic apparatus of *Barbulanympha*. Such dinomitoses are observed, for instance, during the first sporogony divisions in *Diplomorpha* (CACHON 1953), *Duboscquella*, *Hollandella* (CACHON 1964), *Oodinium* (CACHON and CACHON 1977), etc. In

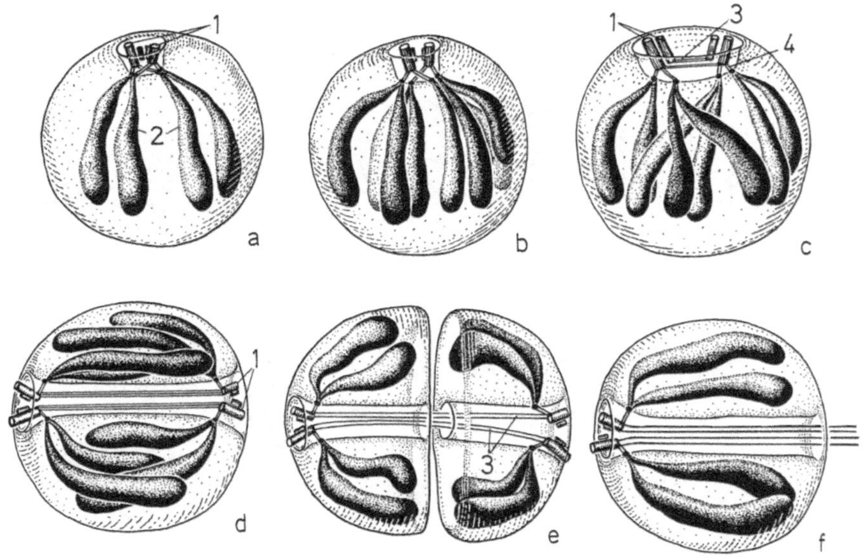

Fig. 42. Diagram of single-channelled dinomitosis in *Syndinium* sp. (after RIS and KUBAI 1974). *a* nondividing nucleus (two chromosomes are shown); *b* individualization of chromatids; *c* replication of the centrioles and the beginning of their separation (early anaphase); *d* late anaphase, transformation of the furrow on the nucleus into a closed tunnel; *e* pinching of the nucleus in two; *f* daughter nucleus with a tunnel (telophase); *1* centrioles, *2* chromosomes, *3* central spindle, *4* chromosomal fibres of the spindle.

Diplomorpha, the central extranuclear spindle that connects two atractospheres is located in a nuclear furrow whose walls incorporate the kinetochores of long despiralized chromosomes (Fig. 43).

The central spindle is especially prominent during the first sporogony mitoses of *Blastodinium* (CHATTON 1920a, SOYER 1971). In this case, a large "archoplasmic sphere" (atractosphere) is always located in a depression of the nondividing nucleus (Fig. 38*c*). In terms of the ultrastructure, this is a zone of uniformly vesiculate cytoplasm without any centrioles (GRELL and SCHWALBACH 1965, SOYER 1970b). The atractosphere splits in two in the prophase of mitosis. The central spindle develops between the two daughter atractospheres, while bichromatidal chromosomes condense in the nucleus. As the daughter spheres move apart, the central spindle, made up of numerous bundles of microtubules, presses into the nucleus, becoming localized in a nuclear furrow. Just as in *Syndinium*, the latter can close and become a tunnel that penetrates the nucleus. The chromosomes are now arranged radially around this furrow or tunnel, which is called the *main plasmodendrite*. In the anaphase, the main plasmodendrite subdivides into several narrower tunnels, or *secondary plasmodendrites*. Each of them encloses a bundle of microtubules. At the end of the anaphase, daughter chromosomes assemble around the daughter atractospheres, which sink again into the polar depressions on the nucleus. The nuclear envelope pinches in two and the secondary plasmodendrites are preserved in the daughter nuclei for a long time.

Electron microscopy has revealed that chromosomes are in close contact with the inner membrane of the nuclear envelope, which lines both the main and

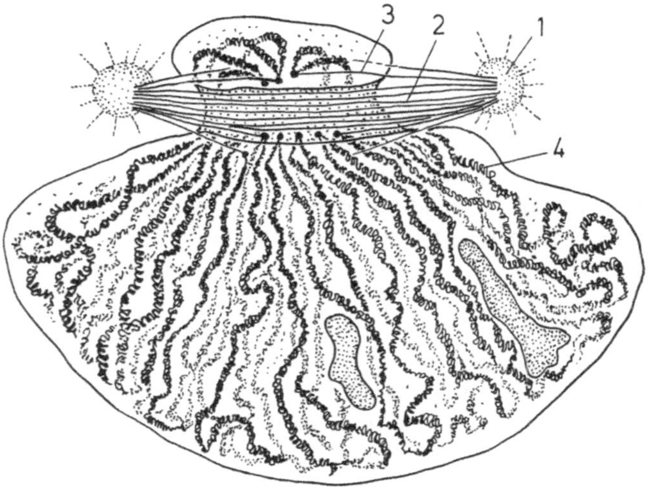

Fig. 43. Diagram of mitosis in *Diplomorpha paradoxa* (after CACHON from HOLLANDE 1972). *1* atractospheres, *2* central spindle, *3* chromosomal fibres, *4* chromosomes.

secondary plasmodendrites; however, no special kinetochores or chromosomal microtubules have been found yet in *Blastodinium* (SOYER 1969c, 1971). SOYER presumes that chromatids segregate in this case by a purely membranous mechanism (owing to the intensive growth of the nuclear membranes between the sites at which two chromatids of one chromosome are attached).

A noteworthy fact is that all the above forms (*Diplomorpha*, *Duboscquella*, *Hollandella*, *Blastodinium*) have a highly developed central spindle only during the first sporogony divisions. During the last divisions, the nuclei become much smaller and the central spindle can no longer be detected, while mitosis approaches the syndinial type. These transformations of mitosis were recently studied with the electron microscope in *Oodinium* (CACHON and CACHON 1974, 1977). During the initial mitoses of sporogony, the nuclei are large with loose chromosomes and permeated with numerous plasmodendrites containing bundles of microtubules of the central spindle. During the last mitoses, the chromosomes greatly condense and acquire an ordered fibrillar structure typical of dinokarya; the number of plasmodendrites decreases and, just as in *Syndinium*, only one tunnel containing a bundle of cytoplasmic microtubules eventually remains in every dividing nucleus. Although centrioles cannot be discerned during the initial mitoses of sporogony (microtubules of the central spindle end in vast zones of homogeneous "archoplasm"), a single bundle of microtubules (centrodesmose) is formed during the last mitoses between two typical pairs of centrioles, which are the future flagellar kinetosomes. What is most important, however, is that chromatids are attached to the walls of plasmodendrites via typical kinetochores built into pores of the nuclear envelope during both early and late mitoses. Only one chromosomal microtubule attaches itself to every kinetochore from the side of the plasmodendrite lumen. However, it pulls the kinetochore after it, since the nuclear envelope bulges in this place in a peculiar manner.

The mitotic apparatus of *Apodinium* has a similar structure (Cachon and Cachon 1979). It has been shown by enzymatic treatments that the kinetochores built into the pores of the nuclear envelope are made up of two parts: an inner one consisting of DNA and an outer one consisting of protein.

Observations of mitosis in *Oodinium* and *Apodinium* have thus shown that mitosis with several plasmodendrites (*multichannelled dinomitosis*) is nothing but a complicated variety of syndinial mitosis (single-channelled dinomitosis) with one plasmodendrite. The mitotic mechanism appears to be the same in both cases.

Dinomitosis of Free-Living Forms

As we have seen, the chromosomes of most Dinoflagellida remain condensed during the interphase. They split into two parallel daughter chromosomes in the prophase. The process was studied many times by light microscopy (Hall 1925a, 1925b, Skoczylas 1958, Dodge 1963a, 1964b, Chunosoff and Hirshfield 1967, Tomas 1974, etc.) and by electron microscopy. The nucleus then elongates, while the nucleolus, if any, pinches in two (Hall 1925b). The chromosomes are oriented more or less parallel to the axis of the division figure. At this moment, the daughter chromosomes must separate, since the segregating chromosomes are no longer double. The mechanism of their disjunction and separation, however, could not be analysed by light microscopy. Dodge (1963a, 1963b, 1964a, 1964b) described this process as the "slow, staggering, asynchronous, autonomous" anaphase movement of chromosomes with no visible spindle, localized centromeres, or mitotic centres. Nevertheless, localized centromeres were found in *Ceratium* (Skoczylas 1958). A noteworthy fact is that mitosis of Dinoflagellida is not inhibited by colchicine (Dodge 1963a).

Leadbeater and Dodge (1967) were the first to examine mitosis in the free-living dinoflagellate *Woloszynskia micra* by electron microscopy. Kubai and Ris (1969) then made an extensive study of mitosis in *Crypthecodinium* (= *Gyrodinium*) *cohnii* by using reconstructions of the dividing nuclei based on serial thin sections.

Both groups of authors have shown that the nuclear envelope is preserved throughout mitosis. Soon after the chromosomes are duplicated, the nuclear envelope develops deep digitiform invaginations at one of the poles of the dividing nucleus. The invaginations penetrate through the nucleus and become cytoplasmic tunnels (or plasmodendrites). They number 3-8 in *Woloszynskia* and 8-15 in

Crypthecodinium per nucleus. The tunnels are lined with an ordinary nuclear envelope. A bundle of 8–20 (up to 40) microtubules passes through every tunnel. The microtubules end free in the cytoplasm outside the nucleus. The flagellar kinetosomes of *Woloszynskia* and *Crypthecodinium* are not involved in the formation of microtubules, which seem to be a homologue of the extranuclear central spindle. Then the nucleus and all its tunnels stretch out, the nucleolus pinches in two, and the chromosomes segregate in the form of two anaphasic groups. The nuclear envelope pinches in two in the middle and the nucleus divides, producing two daughter nuclei that remain connected for some time with bundles of microtubules that pass from the tunnels of one nucleus directly into the corresponding tunnels of the other nucleus. Finally, the microtubules and, later, the tunnels in the daughter nuclei disappear.

LEADBEATER and DODGE have found that many chromosomes contact the nuclear envelope that lines the tunnels, but they refrained from interpreting this fact. Using the technique of three-dimensional reconstruction, KUBAI and RIS have shown that all the V-shaped chromosomes of *Crypthecodinium* are attached with their apices (bending points) to the nuclear envelope of the tunnels (Fig. 44*a*). However, they failed to detect any kinetochores at the points of attachment or any chromosomal microtubules in the tunnels. Consequently, taking the resistance of dinomitosis to colchicine into account, KUBAI and RIS (1969) assumed that the *membranes* of the nuclear envelope play the leading role in separating the daughter chromosomes in free-living dinoflagellates (membranous mechanism of mitosis). Indeed, the selective growth of the membranes of the tunnel in a zone between the sites at which two sister chromatids are attached could result in their segregation. These authors believe that the bundles of continuous microtubules in the tunnels only perform the "directing" function in nuclear elongation and keep the tunnels from collapsing. However, we will see later that this interpretation is not quite adequate.

The formation of cytoplasmic tunnels in the dividing nuclei and the connection of chromosomes with the envelopes of these tunnels were observed, though in less detail, in other free-living dinoflagellates, namely, *Amphidinium* (DODGE and CRAWFORD 1968), *Heterocapsa* (DODGE 1971a), *Glenodinium* (DODGE 1971b), and *Prorocentrum* (DODGE and BIBBY 1973). The attachment of chromosomes to the nuclear envelope was found also in the nondividing nuclei of *Amphidinium* (BOULIGAND *et al.* 1968a). However, only OAKLEY and DODGE (1974, 1976) and, independently, MATTHYS-ROCHON and her co-workers (MATTHYS-ROCHON and DURRAND-ROUSSY 1974, MATTHYS-ROCHON 1979) have successfully shown that kinetochores, unnoticed by KUBAI and RIS, were at the sites of contact between the chromosomes and the nuclear envelope of the tunnels during mitosis in *Amphidinium*; only one chromosomal microtubule was attached to every kinetochore from the side of the tunnel lumen (Fig. 44*b*).

Each kinetochore of *Amphidinium* looks like a small condensation on the outer surface of the nuclear envelope (i.e. in the lumen of the tunnel). The only microtubule passing to it ends in this condensation. The narrowed centromere region of a chromosome is attached to the same site of the nuclear envelope from the inside of the nucleus. In this place, the nuclear envelope has no opening. Apparently, the microtubule really draws the kinetochore after it, since the nuclear envelope forms a peculiar bulge around the kinetochore (Fig. 44*b*). Traction is transmitted to the chromosome through a region of the apparently intact nuclear envelope. The length of chromosomal microtubules hardly changes during the anaphase, while continuous microtubules greatly elongate (OAKLEY and DODGE 1976).

According to MATTHYS-ROCHON (1979), when the nucleus of *Amphidinium* elongates and constricts, the chromosomes are attached not only by their centromere regions to the membranes of the tunnels, but also by their telomeres to the free regions on the envelopes of the future daughter nuclei (outside the tunnels). Such a "double anchoring" of chromosomes seems to make their migration into the daughter nuclei more reliable. Close contact between the nuclear envelope and both the telomeres and the lateral surfaces of many chromosomes is established during mitosis also in *Prorocentrum*; the contact zones lose their pore complexes, which become relocated to the free zones of the nuclear envelope (SOYER and ESCAIG 1980). "Archoplasmic spheres", i.e. zones of finely vesicular cytoplasm surrounded by Golgi bodies, are located at the poles of the division figure in *Amphidinium* (MATTHYS-ROCHON 1979). They seem to be diffuse extranuclear MTOCs, the analogues of "polar caps" and atractospheres. It is from them that bundles of microtubules start to grow at the beginning of mitosis. The bundles first deeply invaginate the nuclear envelope, forming wells, then they penetrate through the nucleus, forming cytoplasmic tunnels. All microtubules protruding from the tunnels into the cytoplasm end in these spheres at the later stages of mitosis.

Therefore, KUBAI and RIS (1969) were apparently mistaken in denying the existence of kinetochores in free-living Dinoflagellida. Since we now know what they look like, we can find them even on the micrographs of *Crypthecodinium* published by KUBAI and RIS themselves (Fig. 44*a*). Apparently, spindle microtubules play the leading role in the segregation of sister chromatids. Consequently, we should reconsider the basic concept proposed by KUBAI and RIS about the exclusively membranous mechanism of mitosis in free-living dinoflagellates. This does not imply, however, that membranes cannot perform a significant auxiliary function in chromosome segregation; for instance, synthesis of new membranes and certain membrane "fluidity" are undoubtedly necessary for allowing microtubules to drive sister kinetochores apart. It could be that, at the later stages of mitosis, the daughter chromosomes are forced apart also by their membrane-attached telomeres.

The kinetochores and the spindle of *Amphidinium* have almost the same morphology as those of the parasitic dinoflagellate *Oodinium* (with the only

---

Fig. 44. Ultrastructure of multichannelled dinomitosis. *a* late stage of mitosis in *Crypthecodinium* in a longitudinal section through one of the tunnels that penetrate the daughter nucleus (the pole of the mitotic figure is at the top; the solid arrows indicate the attachment of the chromosomes to the tunnel envelope, the open arrows indicate kinetochores), 44 500 × ; *b* kinetochores (arrows) in the wall of a cytoplasmic tunnel in *Amphidinium*, 35 000 × ; *1* cytoplasmic tunnel, *2* nucleus, *3* chromosomes, *4* nuclear envelope, *5* continuous microtubules. *a* after KUBAI and RIS (1969); *b* after OAKLEY and DODGE (1974).

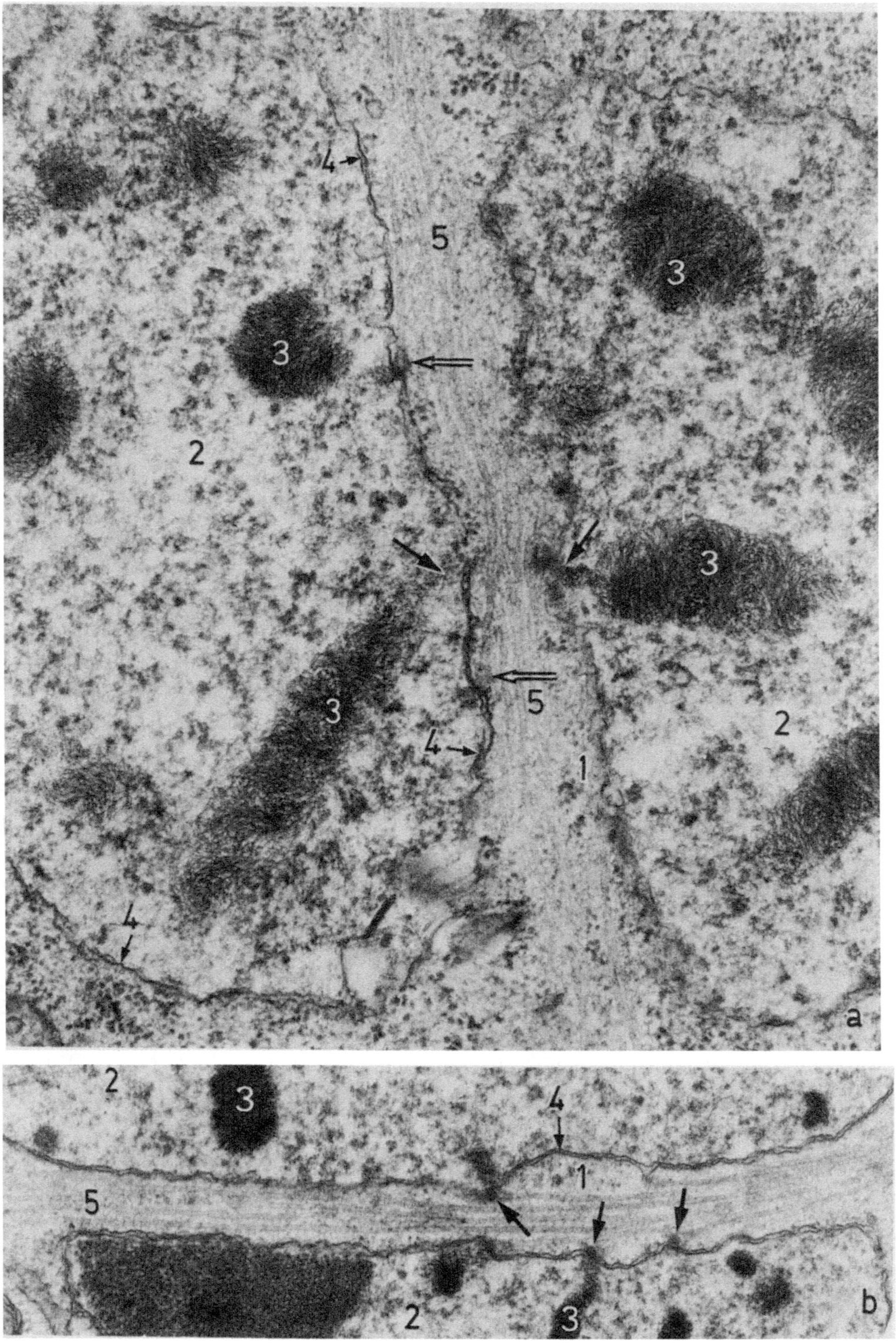

Fig. 44

exception being that the kinetochore of *Amphidinium* is divided into two parts with an intact nuclear envelope, whereas the kinetochore of *Oodinium* is built into its pore, see p. 164). This also shows that the mechanism of dinomitosis is fundamentally the same in all dinoflagellates. Just as the initial sporogony mitoses of *Oodinium*, the mitosis of many free-living Dinoflagellida is multi-channelled dinomitosis without clearly defined mitotic centres (Fig. 45).

It is still maintained, however, that certain dinoflagellates may have a purely membranous mechanism of mitosis. In *Prorocentrum micans* (FILFILAN and SIGEE 1978) and *Oxyrrhis marina* (CACHON *et al.* 1979), neither chromosomal nor

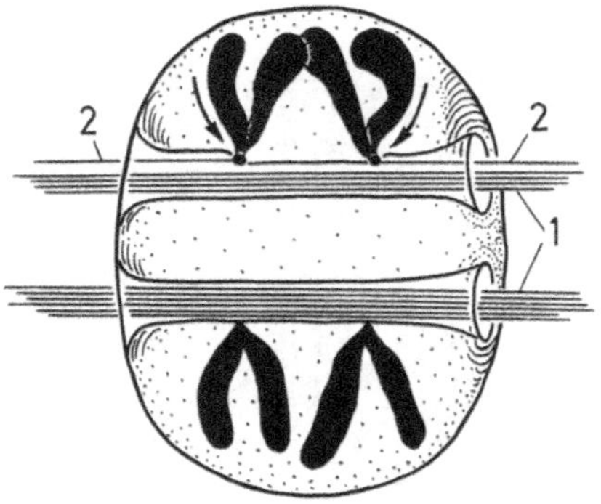

Fig. 45. Scheme of multichannelled dinomitosis (after KUBAI 1975). Only two tunnels in the anaphase nucleus and two pairs of the daughter chromosomes are shown; *1* continuous microtubules, *2* chromosomal microtubules (the kinetochores are indicated by arrows).

continuous microtubules were found in the tunnels that penetrate the dividing nucleus. Hence, the authors believe that chromatids, which are attached to the tunnel membrane, separate only due to the growth of the membranes of the nuclear envelope. In both cases, however, we cannot say for sure that there are no microtubules in the tunnels: they could have depolymerized either during or before fixation.

Another aberrant form of mitosis was found in *Gonyaulax*. In this case, light microscopy has revealed a homogeneous "central body" whose halves migrate to the poles of the nucleus (DODGE 1964b) and "lead apart", as it were, the daughter chromosomes attached to them (CHUNOSOFF and HIRSHFIELD 1967, TOMAS 1974). DODGE (1964b) believed the central body to be extranuclear, but in two electron microscopic works it was identified with a nucleolus (ZINGMARK 1970, TOMAS 1974). Apparently, DODGE's "central body" and the nucleolus are not identical since SCHMITTER (1971) found, by means of electron microscopy, a submembranous nucleolus in the nucleus of *Gonyaulax* as well as a zone of homogeneous cytoplasm rich in dictyosomes opposite the nucleolus; this zone resembles the atractosphere of *Blastodinium* and, as we will see later, *Noctiluca*. Moreover, SCHMITTER described the cytoplasmic tunnels in the dividing nucleus of *Gonyaulax*. Hence, mitosis in *Gonyaulax* is likely to be eventually reduced to multichannelled dinomitosis with large archoplasmic spheres.

Mitosis in Noctilucidae

The largest forms among free-living dinoflagellates, namely, *Noctiluca* and its relatives, have mitosis with a highly developed extranuclear central spindle. Therefore, it resembles mitosis in both large parasitic dinoflagellates (*Blastodinium*) and hypermastigids (*Barbulanympha*).

In mitosis that accompanies binary fission in *Noctiluca* (CALKINS 1899), a thick central spindle is formed between two very large atractospheres. CALKINS detected paired granules (which he regarded as centrioles) within the atractospheres ("centrospheres"). However, no centrioles were found there by electron microscopy: the atractospheres ("archoplasmic spheres") of *Noctiluca* are formed by a dense cytoplasm with a large number of Golgi bodies, while flagellar kinetosomes are not involved in mitosis (SOYER 1970b).

Unfortunately, mitosis during the binary fission of *Noctiluca* has still not been studied with the electron microscope. Only mitoses of sporogony have been examined (SOYER 1969b, 1972). At these stages, the nuclei become small, atractospheres are poorly defined, no large central spindle is formed, and numerous cytoplasmic tunnels containing bundles of microtubules are visible in the dividing nuclei. In other words, mitosis follows the well-known type of multichannelled dinomitosis. No kinetochores or chromosomal microtubules have been detected yet in *Noctiluca*.

In other Noctilucidae, such as *Kofoidinium* (CACHON and CACHON 1967) and especially *Leptodiscus* (CACHON and CACHON 1969a), mitosis occurs during binary fission in the same way as in *Noctiluca*, though the atractospheres are less developed in this case. Light microscopy has shown that *Leptodiscus* possesses a thick central spindle that passes through the nucleus in a wide cytoplasmic tunnel. The cross-sections show that the tunnel is actually a furrow that cuts deeply into the nucleus. All the chromosomes are very poorly condensed and are attached to the nuclear envelope of the tunnel along the perimeter of an equatorial ring (Fig. 46*a*). Their kinetochores are presumably located here (mitosis still has not been studied with the electron microscope). At the end of the anaphase, the attachment sites of all daughter chromosomes shift to the two ends of the tunnel, which narrows down and bends (Fig. 46*b*). A C-shaped nucleolus appears at the equator of the nucleus and divides when the nucleus is "cut" in two by a ring-like invagination of the nuclear envelope. Therefore, mitosis of *Leptodiscus* is single-channelled dinomitosis of almost the same type as in *Syndinium*, only it is much larger.

The above examples show that mitoses in all Dinoflagellida are of the same type. Mitosis with an extranuclear centrodesmose that is not immersed in the nucleus (*Merodinium*) easily becomes single-channelled dinomitosis if the centrodesmose is pressed into a furrow or a tunnel in the nucleus (*Syndinium*, *Haplozoon*). If the central spindle is more developed, we get dinomitoses of the *Leptodiscus* or *Noctiluca* type, which, in principle, are also single-channelled. Mitosis in *Blastodinium* shows how single-channelled dinomitosis becomes a multichannelled one due to the splitting of a single central spindle immersed in the nucleus. The example of *Oodinium* illustrates that the difference between multichannelled and single-channelled dinomitoses is purely quantitative (since all the intermediate forms between them exist in the sporogony cycle). Moreover, the example of *Oodinium* indicates that when dinomitosis is being described, too much importance should not be attached to the presence or absence of centrioles at the spindle poles. This association is obviously secondary.

There is no doubt that dinomitosis is phylogenetically related to *closed extranuclear pleuromitosis* of eukaryotic protozoans, sharing the following fundamental characteristics with the latter: the preservation of the nuclear envelope, the incorporation of kinetochores into it, the existence of a completely extranuclear mitotic apparatus at the side of the nucleus, and the spatial separation of the central spindle and the chromosomal extranuclear half-spindles.

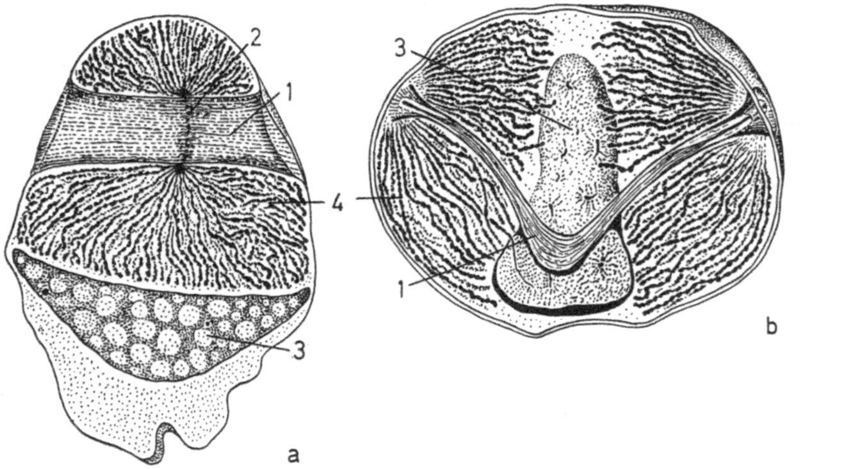

Fig. 46. Mitosis in *Leptodiscus*, 1000 × (after CACHON 1964). *a* metaphase; *b* late anaphase; *1* cytoplasmic tunnel, *2* ring of kinetochores (?), *3* nucleolus, *4* chromosomes.

Chromosomes

Chromosome Number

The number of chromosomes in Dinoflagellida varies within a very wide range: from 4 to 270 (Table 4). A noteworthy fact is that all the free-living forms have high chromosome numbers (not less than 20). Since the individuality of chromosomes is poorly expressed, it is difficult to say whether these high numbers result from the polyploidy of the nuclei. The ratios between the chromosome numbers are aliquant, as a rule, in different species of the same genus. The nuclei of some free-living dinoflagellates (*Cochlodinium*, *Goniodoma*, *Gymnodinium*, *Crypthecodinium*, *Glenodinium*, *Gonyaulax*, *Prorocentrum*, etc.) are capable of undergoing unequal amitotic division or budding that results in small nuclei with a reduced chromosome number. Cells with such nuclei are also much smaller than the normal ones, but they are quite viable and can give rise to ordinary cells after several generations. This indirectly indicates polyploidy and/or aneuploidy of the dinoflagellate nuclei (SOUSA e SILVA 1967, 1969, 1971, 1977). However, genetic data on the inheritance of certain mutations during the sexual process and meiosis

Table 4. *Chromosome numbers in Dinoflagellida*

Species	Chromosome number	References
	Free-living forms	
Exuviaella baltica	~ 20	Dodge 1963c, 1966
E. pusilla	~ 24	ibid.
Prorocentrum triestinum	~ 24	ibid.
P. mariaelebourae	~ 32	ibid.
P. micans	~ 68	ibid.
Amphidinium klebsi	~ 32	ibid.
A. carteri	~ 32	Dodge and Crawford 1968
Gymnodinium vitiligo	44	Dodge 1963c, 1966
Peridinium trochoideum	44	ibid.
Oxyrrhis marina	~ 55	ibid.
Massartia rotundata	~ 68	ibid.
Crypthecodinium cohnii	~ 100	Kubai and Ris 1969
Gonyaulax tamarensis	~ 144	Dodge 1963c, 1966
Ceratium hirundinella	~ 270	Entz 1921
	Parasitic forms	
Syndinium sp.	4	Ris and Kubai 1974
S. borgerti	4	Hollande 1974a
S. globiforme	5	ibid.
S. turbo	5	Chatton 1920a, 1921
S. brandti	5	Chatton 1952
S. vernale	6	ibid.
S. oikopleurae	7	Hollande 1974a
Solenodinium fallax	5	ibid.
Merodinium belari	5	Hollande and Enjumet 1953, Hollande 1974a
Merodinium sp.	6	Pätau 1937b
Duboscquella aspida	5	Cachon 1964
D. melo	8	ibid.
Amoebophrya grassei	6	ibid.
A. acanthometrae	6	ibid.
Hollandella mycetoides	8	ibid.
Myxodinium pipiens	~ 40	Cachon *et al.* 1969
Blastodinium sp.	> 60	Soyer 1971
Haplozoon armatum	> 100	Chatton 1920a

in *Crypthecodinium cohnii* show that its nucleus with about 100 chromosomes behaves like a haploid one (Roberts *et al.* 1974, Himes and Beam 1975, Loeblich 1976, Tuttle and Loeblich 1977).

Conversely, chromosome numbers are small (4–8) in most of the parasitic Dinoflagellida (Table 4). The high incidence of odd numbers (5, 7) and the individuality of chromosomes expressed in certain species (*Merodinium* sp. and *Amoebophrya acanthometrae*) indicate that all these numbers are true haploid ones. However, some parasitic dinoflagellates (*Myxodinium*, *Blastodinium*, *Haplozoon*) also have high chromosome numbers. Apparently, these species originated independently of the *Syndinium*-like forms from some free-living Dinoflagellida (*Blastodinium*, for instance, has not lost pigmented chloroplasts yet; see SOYER 1970b).

Morphology of Chromosomes

As we have seen, the nondividing nuclei of most dinoflagellates contain condensed chromosomes that usually look like dense rods under the light microscope. They decondense to the maximum in the prophase rather than in the interphase. The chromosomes show no longitudinal differentiation except in the centromere regions. This was confirmed cytochemically by HAAPALA and SOYER (1974b) who used three modern techniques to detect the heterochromatin segments in the *Prorocentrum* chromosomes, i.e. the fluorescent technique for Q bands and Giemsa staining for G and C bands; the chromosomal regions did not differentially fluoresce or became stained.

DODGE (1963b) revealed a helical thread in the *Prorocentrum* chromosomes by means of certain treatments (dehydration, etc.) that somewhat elongated the chromosomes. He interpreted the thread as a chromonema that is packed in a tight helix under normal conditions but is somewhat uncoiled when the chromosomes are stretched out. HAAPALA and SOYER (1974a) also detected spiral structures that appeared when the chromosomes were treated with NaOH. The spiral structure of the chromosomes was clearly seen in *Merodinium* (PÄTAU 1937b) and *Ceratium* (SKOCZYLAS 1958).

Ultrastructure of Chromosomes

The ultrastructure of the chromosomes of most Dinoflagellida (including all the free-living forms) proved to be very different from that of the eukaryotic chromosomes. The arrangement of unit chromosomal DNA fibrils is highly ordered in this case. The chromosomes show transverse periodicity in the form of dense discs or bands (Figs. 39*a*, 47*a*, 52*b*). A large number of very thin fibrils can be seen passing arc-wise from one disc to another. In every disc, they are arranged on a plane perpendicular to the chromosome axis (Figs. 47*a*, 52*b*).

Such a fine structure of the dinoflagellate chromosomes was described with minor variations in many technically advanced studies (GRASSÉ and DRAGESCO 1957, GRELL and WOHLFARTH-BOTTERMANN 1957, GIESBRECHT 1961, 1962, RIS 1962, HALLER *et al.* 1964, GRASSÉ *et al.* 1965b, GRELL and SCHWALBACH 1965, BOULIGAND *et al.* 1968a, 1968b, KUBAI and RIS 1969, SOYER 1970a, 1971, 1972, SOYER and HAAPALA 1973, 1974a, 1974b, HAMKALO and RATTNER 1977, OAKLEY

Fig. 47. Ultrastructure of *Prorocentrum micans* chromosomes. *a* attachment of the end of a chromosome (*1*) to the nuclear envelope (*2*) via extrachromosomal DNA loops (*3*), 33 000 × ; *b* part of a spread chromosome, 26 500 × ; *c* accumulations of extrachromosomal DNA loops (*4*) connected with one or two of the neighbouring chromosomes (arrows), 34 000 × ; *d* the same, a detail, 87 000 × . *a* after SOYER and HAAPALA (1974a); *b* after SOYER and HAAPALA (1973); *c* and *d* after SOYER and HAAPALA (1974b).

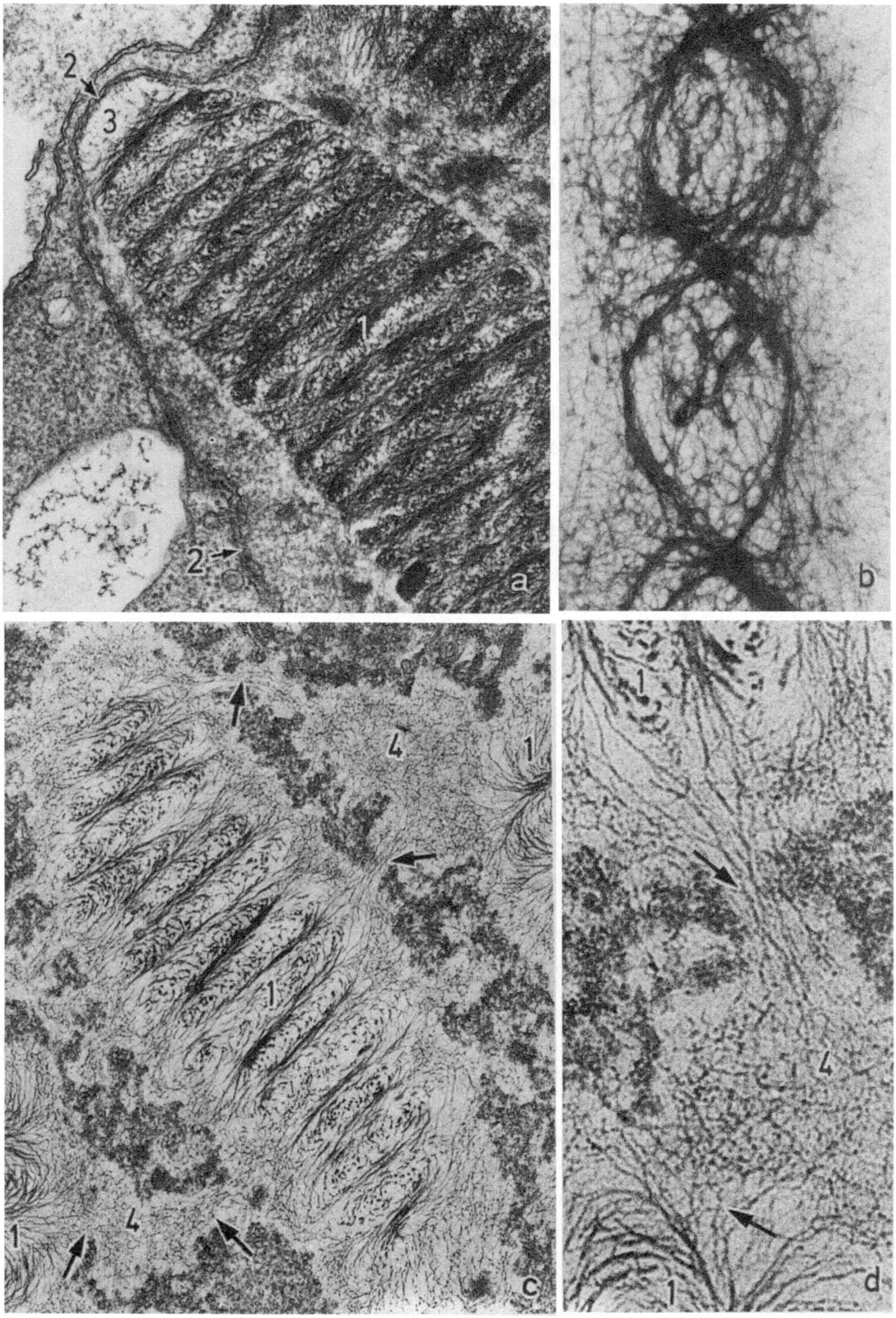

Fig. 47

and DODGE 1979, etc.). It is very difficult, however, to interpret the three-dimensional structure of these chromosomes.

The simplest interpretation is that the transverse interbands of a chromosome are coils of a "chromonema" packed in a tight helix like a spiral spring. In turn, the "chromonema" is a *polytenic* structure, namely, a bundle that consists of dozens of unit chromosomal DNA fibrils, or *genofibrils*, intertwined like strands in a cable (GRASSÉ and DRAGESCO 1957). The hypothesis of helical polytene chromonema was consistent with the data obtained by DODGE (1963b) by means of light microscopy on the helical structure of the dinoflagellate chromosomes and was universally accepted until 1965 (GIESBRECHT 1961, RIS 1962, DODGE 1963b, GRELL and SCHWALBACH 1965). Using the technique of the stereoscopic high-voltage electron microscopy of thick sections, RIS (1969) corroborated the helical packing of unit chromosomal fibrils into the "chromonema".

Another hypothesis concerning the structure of chromosomes in dinoflagellates was proposed by GIESBRECHT (1965), who held that the chromosome was uninemic rather than polytenic. The only strand has two orders of a zigzag or "serpentine" packing—a minor one and a major one. To explain the integrity of the chromosome, GIESBRECHT postulated the existence of two hypothetical "axial structures": one for the minor "serpentine" and the other for the major "serpentine".

At the same time, GRASSÉ and his associates (1965a) put forward a third hypothesis. He discarded his own concept of the helical packing of a genofibrillar "cable" and suggested that the transverse bands on the chromosome were constrictions in the unified bundle of genofibrils that constituted the chromosome (Fig. 48). Single fibrils pass to the opposite side of the chromosome at the sites of constriction (in the bands) and bend arc-wise between the constrictions (in the interbands). Therefore, every genofibril is zigzag or helical (the pitch of the helix is two bands), but the whole chromosome is not helical. The chromosome as a whole is still regarded as polytene.

New data on the fine structure of the dinoflagellate chromosomes were obtained by the technique of spreading whole isolated chromosomes (HAAPALA and SOYER 1973, 1974a, 1974c). Under these conditions, the greatly stretched (about five times its original size) and flattened chromosome of *Prorocentrum micans* becomes a series of "eights" formed by two mutually intersecting and oppositely directed bundles of genofibrils (Fig. 47b). The number of intersections per chromosome reaches 30–40 (15–20 "eights") in this species. Fibrils are arranged less thickly outside these two bundles, e.g. in the lumens of "eights". This pattern is consistent with the hypothesis proposed by GRASSÉ (Fig. 48). Indeed, if his model is stretched and flattened, one gets the same configuration as in Fig. 47*b*. According to HAAPALA and SOYER, GRASSÉ's model correctly describes the nature of transverse bands and arched fibrils in the middle part of any chromosome, though it does not consider their arrangement in the telomeric regions of chromosomes.

According to HAAPALA and SOYER, many fibrils at both ends of a spread chromosome form a loop and then go back, merging with the oppositely directed fibrillar bundle. Basing themselves on these findings, HAAPALA and SOYER (1973) proposed a fourth hypothesis for the structure of chromosomes in Dinoflagellida: the chromosome consists of a large number of independent circular genofibrils or "chromatids", i.e. it is a polytenic formation. All circular chromatids are twisted into a common bundle (Fig. 49). The only difference between this model and the one proposed by GRASSÉ is that chromatids (genofibrils) are circular in the former case.

Individual DNA molecules of dinoflagellates were observed with the electron microscope in a monomolecular layer after the complete lysis of chromosomes

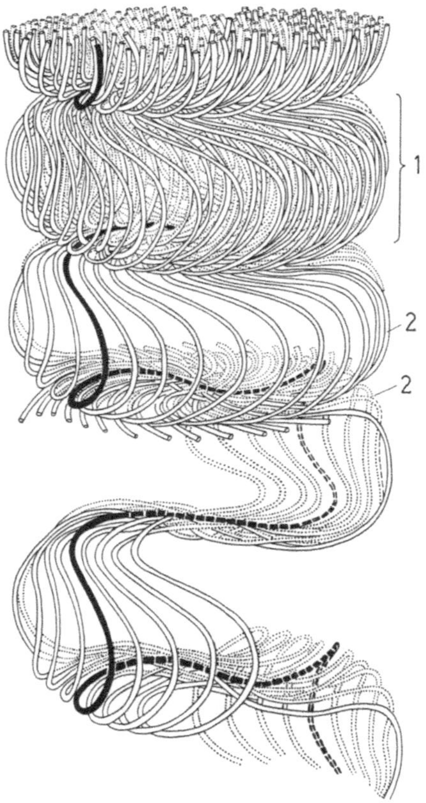

Fig. 48. Model of the chromosome structure in Dinoflagellida according to GRASSÉ *et al.* (1965a). *1* interband, *2* genofibrils.

(HAAPALA and SOYER 1974a). Some of these molecules were ring-shaped, thus corresponding to the circular chromatids postulated earlier. Their perimeter was 25–35 μm in *Crypthecodinium cohnii* and 130–160 μm in *Prorocentrum micans*. Basing themselves on these measurements and the data on the number of chromosomes and the DNA content in the nuclei of the given species, HAAPALA and SOYER have calculated that one chromosome of *C. cohnii* should comprise about 700 circular chromatids, while one chromosome of *P. micans*, about 1,300 circular chromatids, if their model of polytenic chromosomes in dinoflagellates is valid.

The polytene model of the structure of the dinoflagellate chromosomes is now debated again, however, because biochemical data on DNA renaturation have become available. When the degree of polyteny is high, one might expect considerable (and identical) repetitivity of all DNA sequences of the nucleus; this seemed to be the case in early renaturation experiments (BRITTEN and KOHNE 1968). However, ALLEN and his associates (1975) have shown that only 55%–60% of all DNA sequences in *Crypthecodinium* are repetitive, while the remaining sequences either are unique or repeat only a few times. What is even more important is that the length of DNA molecules from *Crypthecodinium* nuclei (determined by the method of viscoelastometry) is much greater than the polytene models allow. One chromosome of *C. cohnii* may contain not more than five independent DNA molecules, and it probably contains only one (ALLEN *et al.*

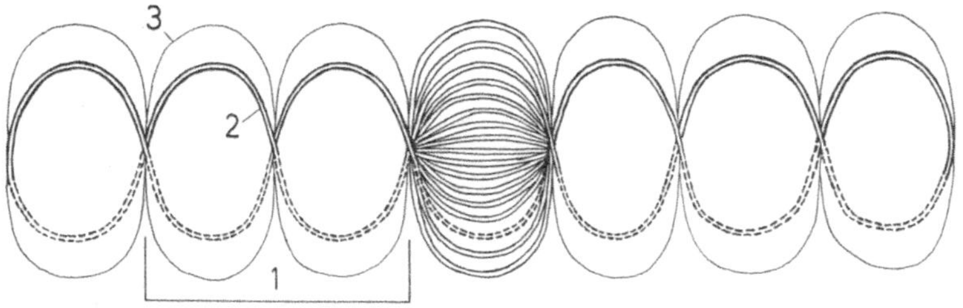

Fig. 49. Model of the chromosome structure in Dinoflagellida according to HAAPALA and SOYER (1973). *1* pitch of the helix of each circular chromatid (*2*), *3* outline of the chromosome.

1974). All these data are hardly consistent with the hypotheses of polyteny of the dinoflagellate chromosomes (GRASSÉ *et al.* 1965a, HAAPALA and SOYER 1973), but are, rather, an indication of their uninemic nature.

OAKLEY and DODGE (1979) proposed a fifth hypothesis for the structure of dinoflagellate chromosomes. To a certain extent, it revives the hypothesis of helical chromonemata. Using isolated chromosomes spread to a varying degree as well as serial sections of chromosomes *in situ*, they have demonstrated that the chromosomes of *Amphidinium* and *Glenodinium* consist of a rather thick toroidal ring (circular "chromonema") that is twisted upon itself in a very tight plectonemic helix (Fig. 50*a*). The "chromonema" forms a loop at each end of the chromosome and turns around in the opposite direction. Since the "chromonema" is circular, the chromosomal helix is double or binemic (Fig. 50*b*). But when spreading is great, the double helix becomes a series of regular "eights" (Fig. 50*c*) similar to that described by HAAPALA and SOYER (Fig. 47*b*).

The hypothesis proposed by OAKLEY and DODGE takes into account the above data on the uninemic or oligonemic nature of the dinoflagellate chromosomes. These authors presume that the chromosome contains from one to five DNA molecules that are very long and "are looped around many times to form a large torus [i.e.] the circular chromonema", rather than many discrete circular DNA fibrils, as supposed by HAAPALA and SOYER. Another feature that distinguishes the model proposed by OAKLEY and DODGE from the one proposed by HAAPALA and SOYER is that the former allows the existence of a helically packed chromonema in the chromosome. The model of OAKLEY and DODGE differs from the original hypothesis of helical chromonemata (GRASSÉ and DRAGESCO 1957) in that it acknowledges the toroidal shape of the chromonema and the uninemic nature of the chromosome as a whole.

A sixth hypothesis for the structure of the dinoflagellate chromosomes (LIVOLANT and BOULIGAND 1978, 1980) is more apt to be considered below (p. 179).

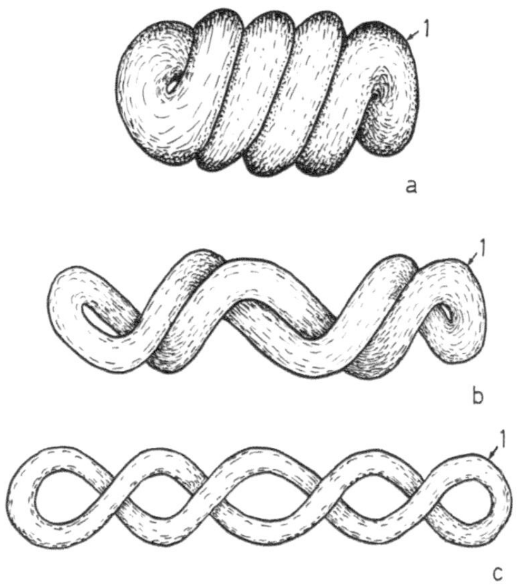

Fig. 50. Model of the chromosome structure in Dinoflagellida according to OAKLEY and DODGE (1979). *a* tightly coiled state; *b* partially uncoiled state; *c* greatly uncoiled state of the chromosome; *1* toroidal (circular) chromonema.

At present, we cannot decide which model should be accepted. However, the uninemic model of OAKLEY and DODGE appears to be consistent with most of the facts known about the structure of chromosomes in Dinoflagellida.

The degree of condensation (or hydration, to be more exact) of the dinoflagellate chromosomes may be quite different, but these differences do not involve the fundamental organization of a chromosome (i.e. the arrangement of unit fibrils in bands and interbands); they relate only to the density of fibrillar packing. For instance, the chromosomes of *Amphidinium* are loose under the optimum conditions of cultivation, but they condense if the medium is desalted or depleted (HALLER 1963, HALLER *et al.* 1964). In *Blastodinium*, *Oodinium*, and *Noctiluca*, no chromosomes can be detected in the trophozoite ("noctikarya"). The initial generations of sporocytes contain loose chromosomes without distinct boundaries between them, while the last generations have compact chromosomes (SOYER 1971,

1972, CACHON and CACHON 1974, 1977; see also p. 152). Transverse bands and arched fibrils, however, are visible in the chromosomes at any degree of hydration. Apparently, the loose state of chromosomes corresponds to their high transcriptional activity, while their condensed state corresponds to their low activity of transcription. This is confirmed by the fact that actinomycin D, an inhibitor of transcription, makes *Amphidinium* chromosomes condense (BABILLOT 1970a).

It has been discovered that dinoflagellates also have *extrachromosomal DNA fibrils*, i.e. fibrils that extend from more or less condensed chromosomes into the nuclear matrix (BOULIGAND *et al.* 1968a, 1968b, SOYER 1971, LIVOLANT and BOULIGAND 1980). These fibrils were thoroughly studied by SOYER and HAAPALA (1973, 1974a, 1974b) in *Prorocentrum*. A dense net of such fibrils occurs at the periphery of the isolated chromosomes that are spread (Fig. 47*b*). It can often be seen that some fibrils form a more or less long loop and return to the chromosome, either to the neighbouring or to a more distant band. All the extrachromosomal fibrils are probably lateral loops of ordinary chromosomal fibrils. Some of them are in close contact with the nuclear envelope, especially with the pore complexes (SOYER and HAAPALA 1973). Extrachromosomal loops are seen best of all after the digestion (with pronase) of dense proteinaceous bodies at the periphery of the chromosomes (Fig. 52*b*). It can then be observed that these sites have vast plexuses of DNA fibrils bound to one or several bands of a chromosome or are occasionally shared by two neighbouring chromosomes (Fig. 47*c*, *d*). Apparently, the lateral loops of the chromosomal fibrils that pass into the nuclear matrix are the active (transcribed) part of the genome, since labelled uridine (apart from nucleoli) is incorporated in the interchromosomal matrix of *Amphidinium* and not in the chromosomes as such (BABILLOT 1970b). SOYER and HAAPALA (1974b) believe that the extrachromosomal loops (Fig. 51 *a*) in their inactive form should condense into hypothetical spiral "hairpins", as is shown in a scheme (Fig. 51*b*) drawn in accordance with their polytenic model. Similar loops seem to exist also within the nucleolus (nucleolar organizer).

Diameter and Composition of Chromosomal Fibrils

Besides their highly ordered arrangement, chromosomal unit fibrils of dinoflagellates (GRASSÉ's "genofibrils", HAAPALA's and SOYER's "circular chromatids") are noteworthy in other respects. They are 3–4 times thinner than the DNP fibrils of eukaryotic chromosomes whose minimum thickness is about 10 nm. The thickness of chromosomal unit fibrils in dinoflagellates (*Amphidinium*, *Prorocentrum*, *Blastodinium*, etc.) is only 2.5–3 nm, occasionally 4–6 nm at the most (RIS 1962, HALLER *et al.* 1964, BOULIGAND *et al.* 1968b, SOYER 1971, HAAPALA and SOYER 1973, HAMKALO and RATTNER 1977, etc.). The fibrils are smooth and uniformly thick, i.e. they do not have globular subunits or nucleosomes, which are typical of eukaryotes (HAMKALO and RATTNER 1977, RIZZO and BURGHARDT 1980, LIVOLANT and BOULIGAND 1980, HERZOG and SOYER 1981). Apparently the fibrils of such thickness may contain only one double-helical DNA molecule (2–2.5 nm in diameter) that has either no protein sheath or (if a fibril is 5–6 nm thick) a weakly developed protein sheath (HAMKALO and RATTNER 1977).

GIESBRECHT (1961, 1962), RIS (1962), KELLENBERGER (1962), and HALLER and his associates (1964) were the first to notice that the thickness and properties of the unit fibrils of dinoflagellate chromosomes strikingly resembled those of the DNA fibrils in the nucleoids of prokaryotes (bacteria, blue-green algae, etc.). The same fibrils are found in artificial "DNA-plasms" prepared by fixing the solutions of pure DNA (KELLENBERGER 1962, SCHREIL 1964). Just as in bacteria, the unit fibrils of dinoflagellates seem to be made up of mainly pure DNA. They require the same conditions of fixation, particularly the presence of calcium ions. If the latter are bound with EDTA ("versene"), the fibrils stick together, forming a coarse net.

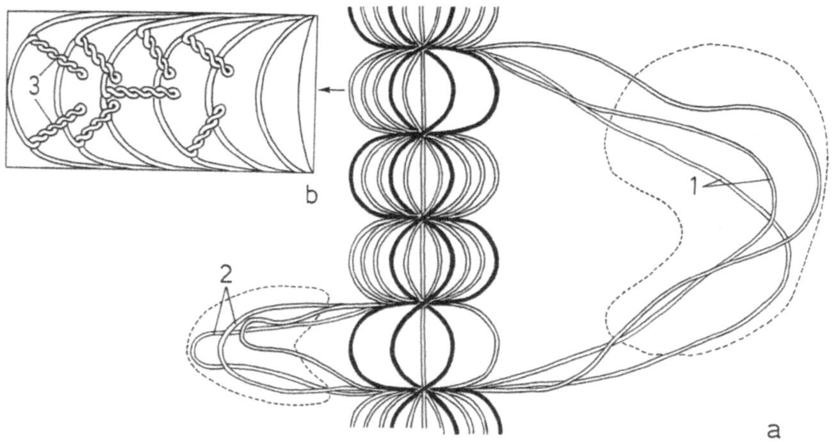

Fig. 51. Protrusion of some genofibrils from the chromosome as lateral loops (after SOYER and HAAPALA 1974b; modified). The scheme is in accordance with the polytenic model of the chromosome structure; *a* region of the chromosome with long (*1*) and short (*2*) lateral loops; *b* portion of one interband with the hypothetical "hairpins" (*3*) on genofibrils.

This "versene test" is positive for artificial "DNA-plasms", bacterial nucleoids, and the dinoflagellate chromosomes but negative for the eukaryotic chromosomes (HALLER *et al.* 1964). DNA fibrils in the nucleoids of prokaryotes also have a highly ordered spatial organization (GIESBRECHT 1962, PESCHKOFF 1966, FUHS 1969) and in some bacteria even show transverse bands connected with arched fibrils (GOURRET 1978), just as in the dinoflagellate chromosomes.

It is amazing that the same arrangement of DNA molecules can be obtained *in vitro* in concentrated solutions of pure DNA (LERMAN 1974). Indeed, the dinoflagellate chromosome may be regarded as a liquid crystal of the cholesteric type formed when DNA molecules aggregate in the absence of much basic protein (LIVOLANT and BOULIGAND 1978, 1980). This fact engendered another hypothesis of the chromosome organization in the Dinoflagellida, which may be called a "chemical" one (LIVOLANT and BOULIGAND 1978, 1980). According to it, the native unstretched chromosome consists of a "pennystack" of flat disc-shaped liquid crystals; in each the DNA is folded back on itself many times in a plane perpendicular to the chromosome axis (Bouligand *et al.* 1968b). Each disc corresponds to a band on the chromosome. The chromosome is thought to be uninemic, so that only one DNA fibril connects the neighbouring discs. All patterns seen in isolated and spread chromosomes, especially the double helical toroidal bundle (cf. p. 176), are considered to be artifacts since the structure of the dinoflagellate chromosome is supposed to be profoundly changed when isolated in the presence of water (Livolant and Bouligand 1980).

The DNA of dinoflagellates is characterized by an extremely high content of 5-hydroxymethyluracil (an analogue of thymine), which occasionally replaces thymine by 40% (Rae 1976).

The similarity between the dinoflagellate chromosomes and the prokaryotic nucleoids was confirmed also in cytochemical and biochemical studies. The nucleoids of bacteria (Belozersky 1947, Zubay and Watson 1959) and blue-green algae (De and Ghosh 1965, Makino and Tsuzuki 1971) are known to contain little or no basic proteins (of the histone type). Also, no basic proteins can be detected cytochemically in typical dinoflagellate chromosomes, e.g. by staining them with fast green in *Amphidinium* (Ris 1962) and *Prorocentrum* (Dodge 1964a). Moreover, Dodge could not find even a non-histone (acid) protein in the *Prorocentrum* chromosomes. Glutaraldehyde-fixed *Glenodinium* chromosomes are completely digested with DNase, i.e. they are likely to consist of pure DNA (Dodge 1971a). Unfixed dinoflagellate chromosomes, however, require predigestion with proteinase to make them DNase-sensitive, indicating the presence of non-histone protein(s) in them (Soyer and Herzog 1980).

Small quantities of basic protein were, nevertheless, found by means of biochemical techniques in the isolated nuclei of *Crypthecodinium*, *Peridinium*, and *Gymnodinium* (Rizzo and Noodén 1972, 1973, 1974a, Allen *et al.* 1975), but the ratio between its content and the amount of DNA varied from 0.08 to 0.13 (Rizzo and Noodén 1973). It was less than 0.1 also in *Prorocentrum* (Herzog and Soyer 1981). Since this ratio approaches unity in eukaryotic chromatin, it follows that the relative content of basic protein in dinoflagellate chromosomes must be lower than in eukaryotes by at least one order of magnitude.

The basic protein of the nuclei was studied further by electrophoretic fractionation (Rizzo and Noodén 1974b, Herzog and Soyer 1981). It proved to consist of only one or two major components (typical eukaryotes have five histone fractions). This protein differs from any of the typical histone fractions in its amino acid composition and is weakly basic; therefore, it cannot be regarded as a typical histone. The content of the basic protein is much higher at the logarithmic phase of culture growth, i.e. when the chromosomes are loose and RNA synthesis is active. It should be noted in this respect that Dodge (1964a) did not find any basic protein in chromosomes by cytochemical techniques, but he detected it in the interchromosomal nuclear matrix. We now know that the lateral loops of chromosomal fibrils pass into the matrix and that RNA synthesis occurs there. Therefore, one may suppose that the association of basic proteins with the DNA molecules of the dinoflagellate chromosomes is temporary rather than permanent and that it occurs mainly on extended DNA threads that are transcribed.

Chromosomes of such parasitic forms as *Syndinium*, *Solenodinium*, and *Haplozoon* (but not *Blastodinium* or *Oodinium*!) differ in their ultrastructure and cytochemical properties from the chromosomes of most dinoflagellates. Being permanently condensed, they have neither transverse bands nor an ordered arched arrangement of the chromosomal fibrils (Fig. 41 *a*, *d*). Basic proteins can easily be demonstrated in the chromosomes of *Syndinium* and *Solenodinium* by cytochemical techniques (RIS and KUBAI 1974, HOLLANDE 1974a). Furthermore, basic proteins were detected cytochemically in the chromosomes of the free-living dinoflagellate *Oxyrrhis* (HOLLANDE 1974a) in which the chromosomes have a less ordered ultrastructure than in typical cases (DODGE and CRAWFORD 1971, CACHON *et al.* 1979). Therefore, the content of basic proteins in dinoflagellate chromosomes seems to be inversely proportional to the extent of the ordered packing of the chromosomal fibrils. It is still unknown to which protein class the basic proteins of *Syndinium* and the related forms belong.

Chromosome Replication

Cell Cycle. It is known that DNA replication occurs continuously in prokaryote nucleoids, i.e. the cell cycle periods (G_1, S, and G_2) are absent. Continuous DNA synthesis was found autoradiographically in the chromosomes of the asynchronously dividing dinoflagellates *Prorocentrum* (DODGE 1965, 1966, FILFILAN and SIGEE 1977) and *Blastodinium* (SOYER 1971). It has been discovered biochemically, however, that the cell cycle in the cultures of *Crypthecodinium* (FRANKER *et al.* 1974) and *Cachonina* (LOEBLICH 1976), which are synchronized by photoperiodicity, includes a period when DNA synthesis is either absent or very slow (in the "light" phase of the cycle) and a period when DNA synthesis is intensive (in the "dark" phase of the cycle directly preceding mitosis). But it is still unknown whether these periods correspond to the actual phases G_1 and S or simply reflect changes in the rate of continuous DNA replication induced by synchronization.

Chromosome Splitting. As electron microscopy has shown, chromosome doubling in Dinoflagellida is very peculiar and greatly differs from that in eukaryotes. Chromosomes virtually split lengthwise while remaining condensed. Splitting starts at one end of the chromosome and gradually proceeds to the other end. The splitting chromosome thus becomess typically Y-shaped (Fig. 52*a*, *b*). Transverse bands and fibrillar arcs do not disappear, but each band subdivides in two at an angle to the original one (Fig. 52*a*, *b*). GRASSÉ and his associates (1965b), who described this phenomenon, interpret it as the "untwisting" of an integral polytenic bundle of chromosomal fibrils (Fig. 48) into two thinner bundles.

SOYER and HAAPALA (1974a) thoroughly studied the splitting of chromosomes in *Prorocentrum* and *Blastodinium*. To suit their theory of circular chromatids, they modified the concept proposed by GRASSÉ by adding that the splitting of a polytenic bundle of circular chromatids into two bundles is mediated by the nuclear envelope to which the ends of both daughter chromosomes are attached (Fig. 53). Indeed, one can often see that the ends of the dividing and nondividing chromosomes are in contact with the nuclear envelope (Figs. 47*a*, 52*a*). Apparently, the chromosomes are permanently attached to the nuclear envelope

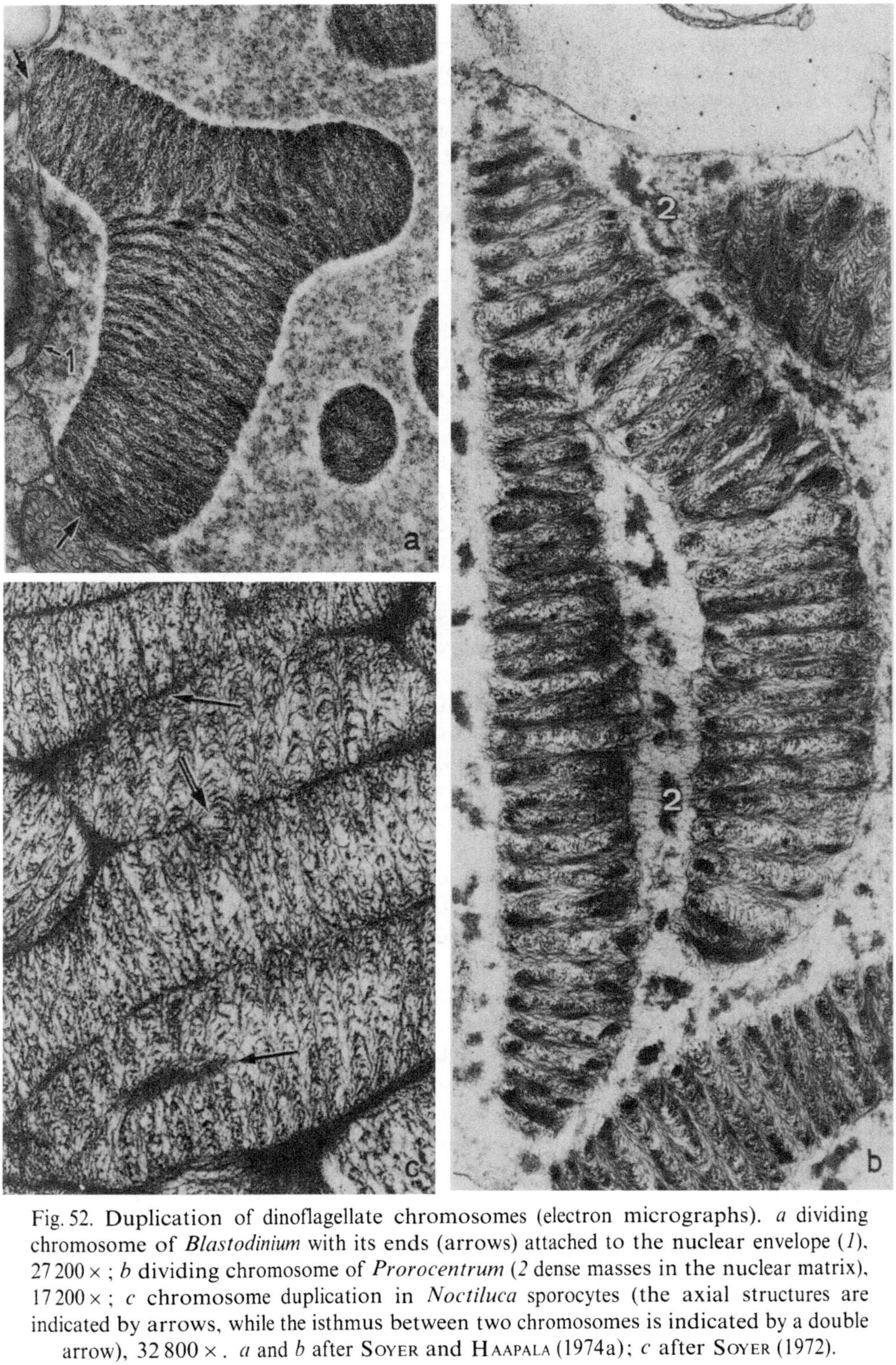

Fig. 52. Duplication of dinoflagellate chromosomes (electron micrographs). *a* dividing chromosome of *Blastodinium* with its ends (arrows) attached to the nuclear envelope (*1*), 27200 × ; *b* dividing chromosome of *Prorocentrum* (*2* dense masses in the nuclear matrix), 17200 × ; *c* chromosome duplication in *Noctiluca* sporocytes (the axial structures are indicated by arrows, while the isthmus between two chromosomes is indicated by a double arrow), 32800 × . *a* and *b* after Soyer and Haapala (1974a); *c* after Soyer (1972).

by means of extrachromosomal DNA fibrils that extend from them (Fig. 47*a*). The fibrils are very long in the interphase (each chromosome can then be rather far away from the respective site of the nuclear envelope), but they shorten when the chromosome doubles, "pulling" it to the nuclear envelope. However, the chromosome is now attached to the membrane by its telomere and not by its narrowed centromere region (Fig. 47*a*). Any structures resembling kinetochores are absent in this case. Telomeric and even lateral attachments of the chromosomes to the nuclear envelope, which occur only in the dividing nucleus,

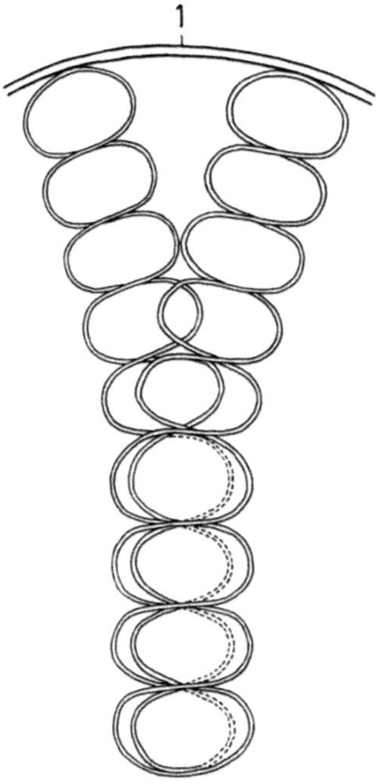

Fig. 53. Diagram of how a chromosome splits into two bundles of circular chromatids (after SOYER and HAAPALA 1974a). Only two circular chromatids are shown; *1* nuclear envelope.

were revealed in a freeze-fracture study of the *Prorocentrum* nucleus (SOYER and ESCAIG 1980). The telomeric attachment of the chromosomes to the nuclear envelope has also been observed (together with the centromeric one) in *Amphidinium* in the anaphase as well as in the daughter nuclei (MATTHYS-ROCHON 1979, cf. p. 166). This type of chromosome attachment to the nuclear envelope seems to be nonspecific, since the *Prorocentrum* chromosomes can even become attached to it with both telomeres at once (SOYER and HAAPALA 1974a). Therefore, even if such contact between the chromosomes and the nuclear envelope does play a certain role in chromosome splitting, it distinctly differs from mitosis as such, during which (as we have seen) *Prorocentrum* has cytoplasmic tunnels in the nucleus (DODGE and BIBBY 1973). In any case, the Y-shaped splitting of chromosomes has been observed also in *Woloszynskia* (LEADBEATER and DODGE 1967), *Crypthecodinium* (KUBAI and RIS 1969), and *Oxyrrhis* (CACHON *et al.* 1979). It is typical of the dinoflagellates (DODGE 1971a) and is usually confined to the early prophase of mitosis.

The splitting of the chromosomes is somewhat peculiar during the initial divisions of sporogony in *Noctiluca* and *Blastodinium* when the chromosomes are very loose and occupy almost the whole volume of the nucleus (SOYER 1970a, 1971, 1972). In this case, the lengthwise splitting of a chromosome is preceded by the appearance of the *axial structure*, an amorphous ribbon or septum structurally similar to the nuclear matrix. The axial structure appears more frequently in the middle of a chromosome than at its end (Fig. 52*c* at the arrow) and grows across its bands, gradually dividing the chromosome lengthwise in two. At the last stages of splitting, sister chromosomes remain connected with one another by a narrow fibrillar bridge somewhere in their middle (Fig. 52*c*, at the double arrow) so that together they look H-shaped. These figures are replaced by ordinary Y-shaped ones at the late stages of sporogenesis.

Problem of the Mesokaryotic Cell

Among protozoans and unicellular organisms in general, only Dinoflagellida are known to have a distinct nucleus, though with "proteinless" chromosomes. Some years ago, Euglenida were also suspected of having these features (cf. RAIKOV 1967a). But this was refuted since their nuclei are typically eukaryotic; they contain several histone fractions (JARDINE and LEAVER 1977) and have a nucleosomal organization of chromatin (HAAPALA and SOYER 1975). Can the noted features of nuclear and chromosomal organization in dinoflagellates be regarded as primitive? This question was first raised by RIS (1961, 1962) who held that the genetic systems of prokaryotes ("genophores", nucleoids) were more primitive than the eukaryotic chromosomes since the former lacked any distinct longitudinal differentiation (including kinetochores and nucleolar organizers), had thinner unit DNA fibrils, and were devoid of histones. Basing on his data about the structure of the dinoflagellate chromosomes, RIS concluded that they were similar to bacterial nucleoids. He suggested that they be regarded as an intermediate stage in the evolution of genetic systems between the prokaryotic and eukaryotic levels. From this standpoint, a bacterial nucleoid corresponds to one chromosome in Dinoflagellida. GIESBRECHT (1961, 1962), HALLER and his associates (1964), and DODGE (1963b, 1964a) also shared this point of view.

DODGE (1965, 1966) later coined the term *mesokaryotic* to designate the level of the cellular organization of the dinoflagellates, which is a combination of prokaryotic and eukaryotic features. In his opinion, the following five features of the dinoflagellate chromosomes are typically "prokaryotic": the absence of longitudinal differentiation, the lack of histones, the small thickness of the unit fibrils, continuous DNA replication, and the ordered arrangement of the chromosomal unit fibrils. Conversely, the "eukaryotic" features involve the general organization of a dinoflagellate cell: the presence of a nucleus with a typical nuclear envelope as well as of mitochondria, the Golgi apparatus, kinetosomes, flagella, and other organelles characteristic of eukaryotes.

The concept of the mesokaryotic cells was evolved further by KUBAI and RIS (1969), SOYER (1969c, 1971, 1972), and SOYER and HAAPALA (1974a), who listed three more "prokaryotic" features of the organization of dinoflagellates: the attachment of chromosomes to the nuclear envelope, the membranous mechanism of chromosome separation in mitosis, and the circular shape of DNA fibrils (chromatids) in the chromosomes. However, some corrections were later made in the list of "prokaryotic" features since kinetochores and nucleolar organizers were found on the dinoflagellate chromosomes, while basic proteins (though not typical histones) were detected in small quantities in the chromosomes of free-living dinoflagellates. Continuous DNA synthesis was then questioned. A mitotic apparatus was discovered in the form of an extranuclear spindle with chromosomal microtubules (this does not mean, however, that the membranes of the nuclear envelope are not involved in the separation of chromosomes). Moreover, the presence of numerous small circular "chromatids" in the chromosomes (the polytenic model) became doubtful. The DNA molecules of the dinoflagellate chromosomes may be circular, but the perimeter of the ring must then be very great and its packing very intricate (see the models proposed by OAKLEY and DODGE, p. 176, as well as by LIVOLANT and BOULIGAND, p. 179).

At present, the "prokaryotic" features of the chromosomal organization in Dinoflagellida may include the small thickness of "genofibrils", the low content of basic protein in them (and possibly even its absence in condensed chromosomes), the absence of nucleosomes on the "genofibrils", the ordered packing of fibrils in the chromosomes, and the attachment of the chromosomes via the kinetochores (homologues of bacterial mesosomes?) to the membrane (in this case, to the nuclear envelope). The presently accepted "eukaryotic" features of the nuclear and chromosomal organization (in addition to the aforementioned "eukaryotic" features of the cytoplasm) are the presence of a typical nuclear envelope, a nucleolus and an extranuclear mitotic apparatus made of microtubules, the existence of the repetitive sequences in the DNA that cannot be attributed to polyteny alone, and the presence of RNA polymerase of the eukaryotic type in *Crypthecodinium* (RIZZO 1979). LOEBLICH (1976) has recently analyzed the prokaryotic and eukaryotic features of organization in Dinoflagellida and concluded that the nuclei really preserved many of the prokaryotic features (though not exactly those that were supposed earlier) and that the dinoflagellates most probably branched off the evolutionary trunk of the eukaryotic organisms somewhere near their root.

Hence, what we learned about the karyology of dinoflagellates in the past 15 years has, in the main, confirmed that a special mesokaryotic level of the organization of living organisms actually exists. Even within the Dinoflagellida, however, we come across evolutionary lines that develop toward purely eukaryotic organization, such as the family Syndiniidae in which the chromosomal unit fibrils are no longer orderly packed in the chromosomes, the content of basic protein rises in them, and the chromosomal spindle microtubules become more clearly defined. Apparently the eukaryotic level of organization was reached several times by the prokaryotes, and one such parallel evolutionary line passes through the extant Dinoflagellida.

5. Meiosis in Protozoa

Biologically, meiosis consists of two fundamental events that are realized in higher organisms during an integrated meiotic process with two nuclear divisions. These events are (1) the separation of the two sets of homologous chromosomes, i.e. the *reduction of the number of chromosomes* proper and (2) the *recombination of the gene pool* at the levels of both whole chromosomes (as a result of their independent distribution with respect to every pair of homologues) and individual genes or groups of genes (as a result of crossing-over). Meiotic processes are very diverse among protozoans, while the mechanism of meiosis undoubtedly is more simple among some protozoans than among higher eukaryotes.

Alternation of Nuclear Phases

The sexual process has not been found in many groups of protozoans (for reviews, see WENRICH 1954a, GRELL 1967), including Cryptomonadida, Euglenida, Choanoflagellida, Retortamonadida, Kinetoplastida, Trichomonadida, Diplomonadida, Amoebida, Schizopyrenida, and Arcellinida. All these protozoans are either really or presumably permanent haplonts, as is evident from the frequently occurring odd chromosome numbers and occasionally from the individuality of some chromosomes (see Table 3). Polyploids (euglenids, large amoebae?) may possibly be among them, but no diplonts have been found yet. Apparently, there is an innate link between the origin of the sexual process and the formation of diploidy.

The life cycles of protozoans that undergo the sexual process include both phases of the nuclear cycle: the haploid phase (from meiosis to karyogamy) and the diploid phase (from karyogamy to meiosis). Their relative duration may differ (Fig. 54). In some forms, meiosis occurs during the first division(s) of the zygotic nucleus. These forms are *haplonts* with *zygotic meiosis* (Fig. 54*a*) and include Volvocida, Dinoflagellida, a part of Oxymonadida, most of Hypermastigida, and all Sporozoea (Table 3). In other protozoans, all the vegetative stages of the cycle are diploid, while meiosis precedes the formation of gametes or pronuclei. These are *diplonts* with *gametic meiosis* (Fig. 54*b*) and include a part of Oxymonadida and Hypermastigida, all Opalinata, Heliozoea, Labyrinthomorpha, Myxosporea, and Ciliophora. All metazoans also belong to this type. Finally, meiosis may hold an intermediate place in the life cycle; it is preceded by a diploid generation and is followed by a haploid one (*heterophasic alternation of generations*, Fig. 54*c*). This type of the nuclear cycle is characteristic of Foraminiferida (LE CALVEZ 1950, 1953, GRELL 1967, etc.). In them, a diploid *agamont* develops from the zygote. The agamont reproduces asexually, forming several *agametes*. Meiosis occurs before the formation of agametes, which are already haploid. Each agamete grows to become a specimen of the haploid generation, a *gamont*. The gamont nucleus gives rise to the nuclei of *gametes* by ordinary mitoses. Besides protozoans, the heterophasic alternation of generations is found among spore plants (the alternation of the haploid gametophyte and the diploid sporophyte).

The most primitive position of meiosis is probably the zygotic one. In other words, protozoans are primarily haplonts. The haplophase is gradually supplanted by the diplophase only in the higher protozoans.

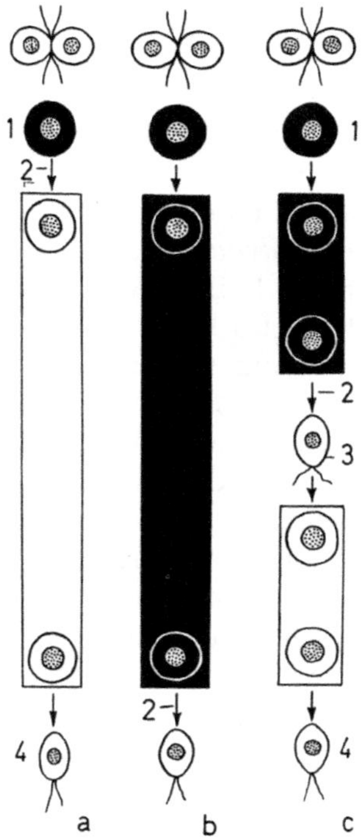

Fig. 54. Diagram of the alternation of the haplophase (*white*) and diplophase (*black*) in the life cycles of protozoans (after GRELL 1956a). *a* haplont with zygotic meiosis; *b* diplont with gametic meiosis; *c* heterophasic alternation of generations with intermediate meiosis; *1* zygote, *2* position of meiosis, *3* agamete, *4* gamete.

Meiosis with Two Nuclear Divisions

General Scheme

Classical meiosis of the metazoan type is a two-step process, i.e. it includes two nuclear divisions. In it, the homologous chromosomes of the diploid set at first unite pairwise (conjugate), then they part into different nuclei after having preliminarily exchanged parts of chromatids (crossing-over) in most cases. Only one chromosome replication occurs in two nuclear divisions.

Chromosome conjugation occurs in the prophase of the first meiotic division (or simply in the *meiotic prophase*). The chromosomes are still not paired at the leptotene stage, but they are almost completely duplicated at the molecular level, although two chromatids are usually still not seen in them at the microscopic level. At the zygotene stage, the chromosomes are oriented in the form of a "bouquet", while homologous chromosomes conjugate along their whole length, forming *bivalents* that are present in the nucleus in a haploid number. This is when crossing-over occurs. At the pachytene stage, the chromatids become microscopically visible in every chromosome, and, consequently, bivalents become four-stranded; at the same time, they shorten and thicken. At the zygotene and pachytene stages, homologous chromosomes are united into a bivalent by an axial structure known as the *synaptonemal complex*. The homologues begin to disunite at the next (diplotene) stage, but they remain linked at points where two of four chromatids cross, forming *chiasmata* that result from earlier crossing-over. During diakinesis, the chromosomes condense still more, while chiasmata gradually slip toward their ends (terminalization of chiasmata). As a result, the homologues can separate in the metaphase of the first division (metaphase I). A haploid number of bichromatidal chromosomes, or *dyads*, and not single chromatids (as in mitosis), depart to the poles in anaphase I. But since crossing-over had occurred in the meiotic prophase, the dyads are no longer identical with the chromosomes that conjugated with one another during zygotene. The chromosomal centromeres are not doubled in the first meiotic division.

In the interval between the first and second divisions (*interkinesis*), two chromatids of one dyad are connected with one another by a common centromere (characteristic four-armed figures are seen in the case of metacentric chromosomes). New chromosome replication does not occur during interkinesis. In metaphase II, the centromeres finally double and the dyad chromatids part in anaphase II. Consequently, each of the four nuclei, being the progeny of a single nucleus that underwent meiosis, receives a haploid number of unichromatidal chromosomes. In old works, the first meiotic division was often called a "reductional" one, while the second division was called an "equational" one and was considered to be similar to mitosis. However, the second meiotic division fundamentally differs from ordinary mitosis if crossing-over occurs. During this division, not only identical regions of chromosomes (unaffected by crossing-over) but also homologous ones (affected by crossing-over) undergo segregation. Consequently, two nuclei that are the progeny of the second meiotic division figure can differ genetically from one another.

Meiosis in Sarcodina

Typical two-step meiosis has been studied in the heliozoan *Actinophrys sol* (Bělař 1922), which is a diplont with gametic meiosis ($2n = 44$, $n = 22$). In this case, a "bouquet" stage is clearly expressed in the meiotic prophase and parallel conjugation of chromosomes occurs (Fig. 55*a*–*c*). During pachytene (Fig. 55*d*), the homologous chromosomes are still closely paired, while during diplotene, they begin to disunite; then chiasmata are clearly seen on the bivalents (Fig. 55*e*). During diakinesis and in metaphase I (Fig. 55*f*–*g*), the bivalents continue to condense, while in anaphase I (Fig. 55*h*), they separate into dyads. During interkinesis, the dyads have a characteristic four-arm shape (Fig. 55*i*, *k*). The chromatids of the dyads separate in metaphase II (Fig. 55*l*). Thus, the meiosis of *Actinophrys* completely fits into the classical scheme.

All the main features of Bělař's description have been confirmed in a recent electron microscopic study of meiosis in *Actinophrys* (Mignot 1980a). In addition, the chromosomal axial structures, in the form of filaments attached with both ends (telomeres) to the nuclear envelope, have been clearly observed during leptotene (Fig. 56*a*). These axial structures unite pairwise in zygotene and become the lateral elements of typical synaptonemal complexes (Fig. 56*b*). In meiosis as well as mitosis (cf. Fig. 17), the spindle is of the semiopen orthomitotic type. The

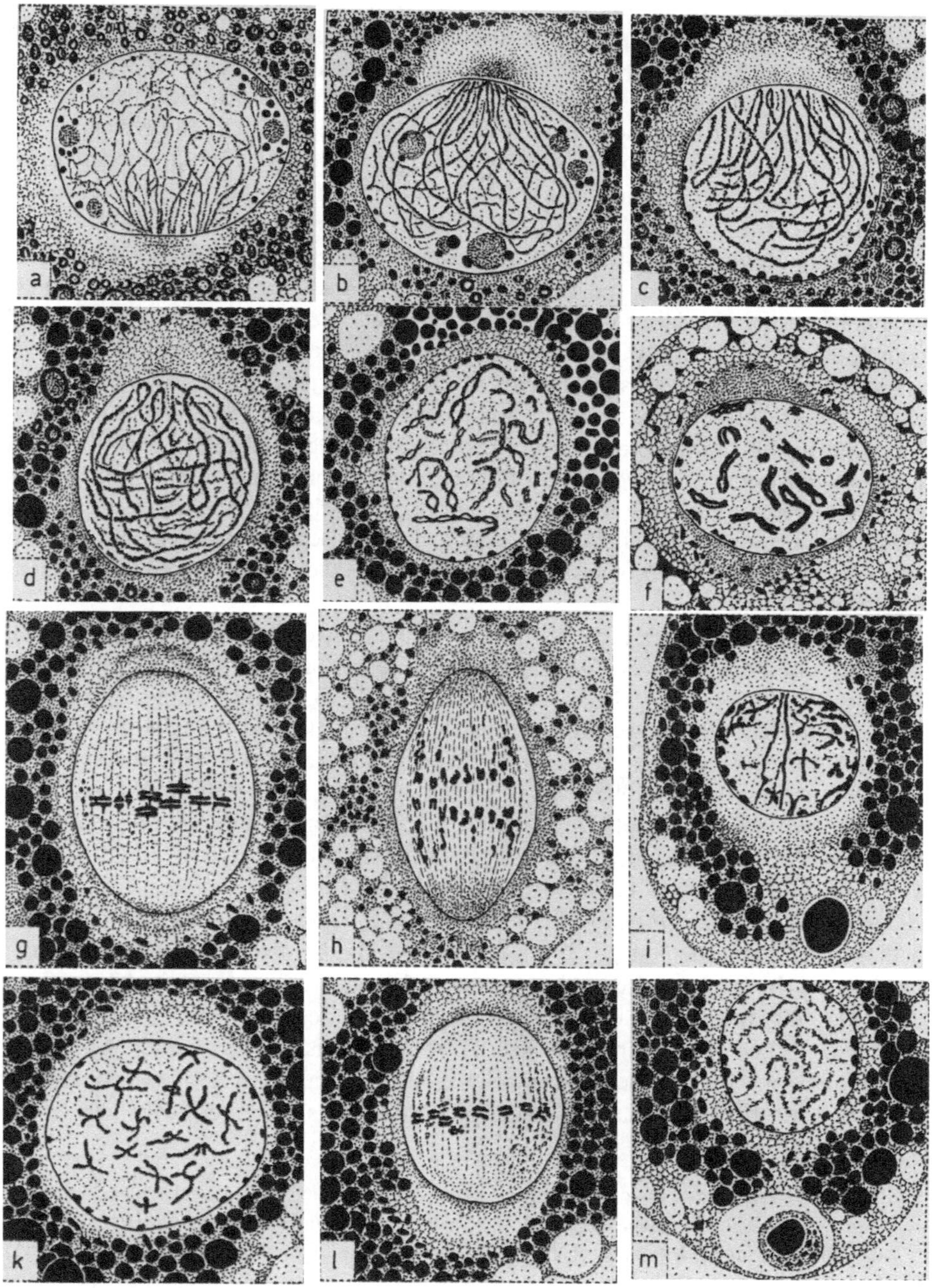

Fig. 55. Meiosis in *Actinophrys*, 2200 × (after BĚLAŘ 1922). *a*-*c* "bouquet" stage (*a* leptotene, *b* zygotene, *c* early pachytene); *d* pachytene; *e* diplotene; *f* diakinesis; *g* metaphase I; *h* anaphase I; *i* and *k* interkinesis; *l* metaphase II; *m* products of meiosis (the nucleus at the bottom is pycnotic).

microtubules of the spindle, which grow from the cytoplasmic "polar caps" (extranuclear MTOCs), penetrate into the prometaphasic nucleus through numerous polar gaps in the otherwise intact nuclear envelope (Fig. 56*c*). The polar gaps, however, are much wider in the first and second meiotic divisions than in mitosis during both the metaphase and the anaphase, while in the telophase the gaps unite and the spindle poles become completely open. The envelopes of the daughter nuclei are formed inside the remnants of the old nuclear envelope according to the "indirect" way of telophasic reorganization (cf. p. 84).

Two-step meiosis has also been discovered during sporulation in the diplont *Labyrinthula* sp. (Moens and Perkins 1969, Perkins and Amon 1969, Perkins 1970). In this case, electron microscopy has also revealed nine synaptonemal complexes (n = 9) in the meiotic prophase. The reconstruction of their arrangement in the nucleus by a series of ultrathin sections has shown that all the complexes are attached with both ends to the inner side of the nuclear envelope. Just as in higher eukaryotes, the synaptonemal complex of both *Actinophrys* and *Labyrinthula* consists of two *lateral elements* in the form of electron dense bands separated by central (synaptic) space. The axis of this space is formed by a less dense thread—the *central element*. Each lateral element belongs to its own homologous chromosome within the bivalent while they share the central element (Fig. 56*d*).

Meiosis with two nuclear divisions is characteristic of foraminiferans, in which it occurs when agametes are being formed (intermediate meiosis). In the meiotic prophase of *Myxotheca* (Grell 1956a), a typical "bouquet" stage is observed, while at its diplotene stage, bivalents with chiasma-like figures can be seen; the chromosomes are greatly condensed during both diakinesis and metaphase I. Similar patterns occur in *Allogromia* (Arnold 1955), *Rubratella* (Grell 1958a), and *Cibicides* (Voronova 1978a), which also have chiasmata in their bivalents. Meiosis with two nuclear divisions has also been described in *Discorbis* (Le Calvez 1950), *Rotaliella* (Grell 1957) and especially *Patellina* (Le Calvez 1950, Grell 1959). In the meiotic prophase of *Patellina*, homologous chromosomes conjugate without the "bouquet" stage (Fig. 57*a, b*). During diplotene and diakinesis, the bivalents at the microscopic level continue to be unichromatidal (Fig. 57*b, c*). Two diverging intranuclear half-spindles are clearly seen in the prometaphase (Fig. 57*d*). They are characteristic of both mitosis (p. 98) and meiosis of foraminiferans. In the metaphase (Fig. 57*e*), the half-spindles are completely diverged and the bivalents are arranged near the spindle equator, while the

---

Fig. 56. Ultrastructure of meiosis. *a* unpaired chromosomal axes (*1*) attached to the nuclear envelope (*2*) at the leptotene stage of the meiotic prophase in *Actinophrys*, 18 000 × ; *b* synaptonemal complexes (*3*) of the bivalents at the zygotene stage in *Actinophrys*, 13 700 × ; *c* metaphase I in *Actinophrys* that shows the developing polar gaps (*4*) in the nuclear envelope and numerous microtubules (*5*), 12 200 × ; *d* synaptonemal complex in *Labyrinthula*, 60 000 × ; *e* polycomplexes in the diplotene of micronuclear meiosis in *Dileptus anser*, 64 000 × ; *6* chromosomes, *7* nucleolar material, *8* central element and *9* lateral elements of the synaptonemal complexes or polycomplexes. *a* and *b* after Mignot (1980a); *c* unpublished micrograph, courtesy of J.-P. Mignot; *d* after Perkins and Amon (1969); *e* after Vinnikova (1976).

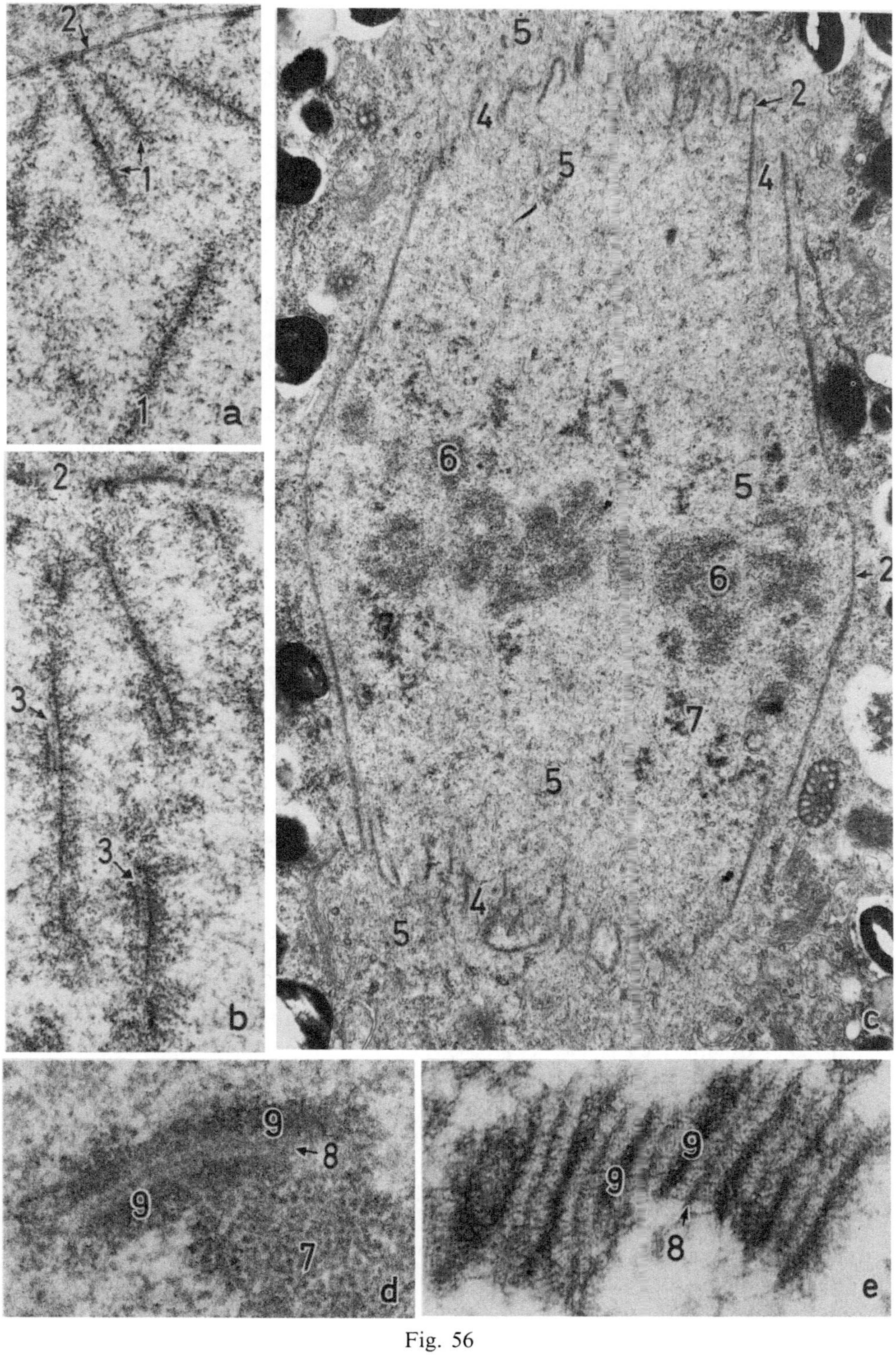

Fig. 56

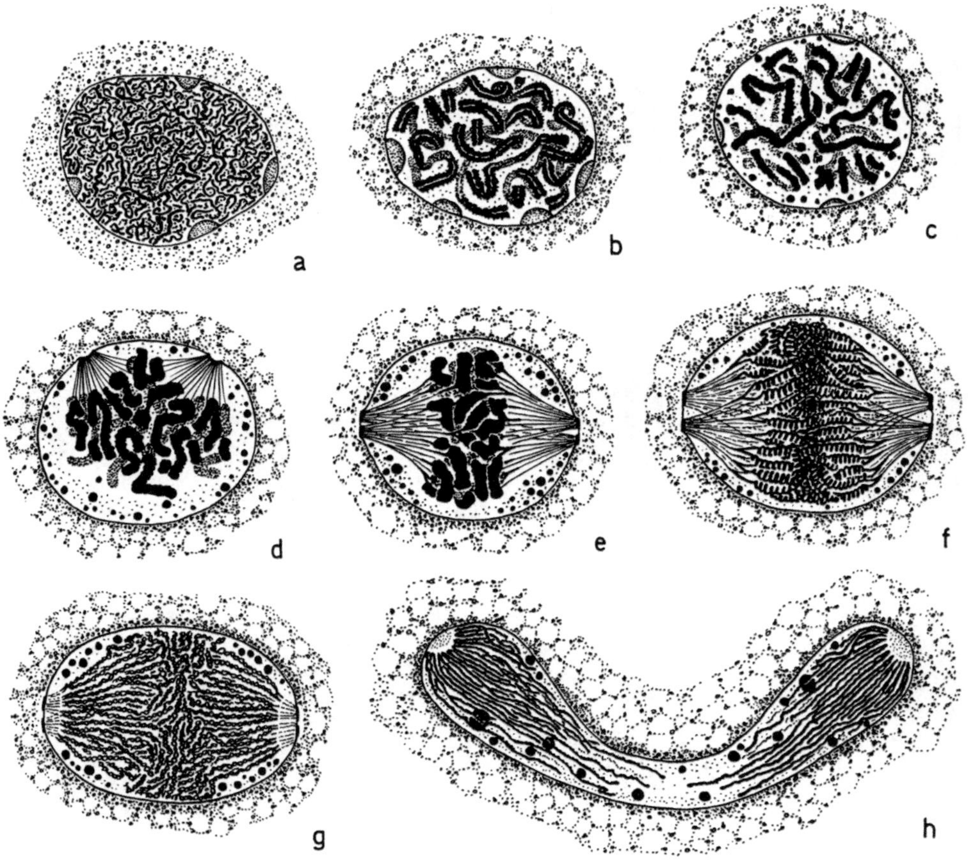

Fig. 57. First meiotic division of the nucleus in *Patellina corrugata*, 1100 × (after LE CALVEZ 1938). *a* leptotene; *b* pachytene; *c* diplotene; *d* prometaphase (two intranuclear half-spindles with MTOCs on the inner side of the nuclear envelope); *e* metaphase; *f–h* anaphase (*f* uncoiling of chromosomes, *g* their splitting into chromatids).

homologous centromeres are not on the equatorial plane but are symmetrically arranged on both sides of it (the "co-orientation of chromosomes", which is characteristic of meiosis in general). The chromosomes begin to despiralize already in anaphase I (Fig. 57*f*). They are visibly duplicated and become bichromatidal only later during the anaphase (Fig. 57*g*).

However, it has been shown cytophotometrically that there is premeiotic DNA replication in *Patellina*, *Rotaliella*, and *Cibicides*: the DNA contents of the nuclei of gametes, zygotes, and premeiotic agamonts are in a proportion of 1 : 2 : 4 (ZECH 1964, VORONOVA 1978a). These nuclei are in the G_1 period of the haplophase and the G_1 and G_2 periods of the diplophase, respectively. Consequently the chromosomes at the molecular level should already be two-stranded at the moment of their conjugation. Recently, terminalized chiasmata were discovered on the bivalents of *Patellina*, while synaptonemal complexes were revealed with the electron microscope at an earlier stage of meiosis of both *Patellina* (BERTHOLD 1977) and *Rotaliella* (GRELL and BARDELE 1977).

Two-Step Meiosis in Mastigophora

Genetic (tetrad) analysis shows that two-step meiosis occurs in *Chlamydomonas* and other Volvocida, which are haplonts. Meiosis occurs in the encysted zygote. Four haploid cells (a tetrad) are formed in the cyst as a result of two divisions of the nucleus and the cytoplasm. In a certain number of cases, all four clones produced by the four cells of the tetrad in dihybrid crosses are genetically different. This is possible only if there is crossing-over and the second meiotic division is nonequational due to it (data reviewed by PREER 1969). Synaptonemal complexes were recently discovered in the meiotic prophase of *Chlamydomonas reinhardi*; however, they are somewhat atypical (STORMS and HASTINGS 1977).

Typical two-step meiosis (with chiasmata on bivalents) has been found in *Opalina ranarum* (KACZANOWSKI 1971). In this case, meiosis is gametic and the vegetative stages of the cycle are diploid.

The most thorough study of meiosis with two nuclear divisions has been made by CLEVELAND with respect to some Hypermastigida that inhabit the intestine of the cockroach *Cryptocercus* and go through the sexual process when the host moults (see the review by CLEVELAND 1956b). It should be noted that meiosis in other Hypermastigida as well as Oxymonadida, which also live there, includes only one nuclear division and will be considered below. The forms with two-step meiosis include both haplonts with zygotic meiosis (*Trichonympha*, CLEVELAND 1949b, *Eucomonympha*, CLEVELAND 1950d, *Barbulanympha*, CLEVELAND 1953a, 1954) and diplonts with gametic meiosis (*Rhynchonympha*, CLEVELAND 1952, *Macrospironympha*, CLEVELAND 1956a). The chromosome numbers of these forms are given in Table 3.

A detailed study made by CLEVELAND has shown that meiosis in the previously mentioned flagellates markedly differs from the classic scheme. What immediately attracts our attention is the rather late conjugation of homologous chromosomes in the meiotic prophase. Accordingly, the "bouquet" stage is absent; there is no typical conjugation of despiralized chromosomes and no strict pairing of homologous loci. In *Trichonympha* (CLEVELAND 1949b), for instance, the homologues conjugate only after the visible duplication of chromosomes has occurred. Late chromosome conjugation apparently excludes the possibility of typical crossing-over, since the stage of despiralized four-strand bivalents is absent in the meiotic prophase.

Another distinguishing feature of meiosis in Hypermastigida concerns the way in which chiasmata are formed. Let us preliminarily consider two main hypotheses of the crossing-over and chiasma formation (SWANSON 1960). According to the first or "classic" hypothesis (SAX 1932), chiasmata originate before crossing-over occurs. Crossing-over occurs rather late (during diplotene). The chiasma formation is the cause of crossing-over, though not every chiasma leads to crossing-over. According to the second hypothesis, i.e. the hypothesis of chiasmatypy proposed by JANSSENS and DARLINGTON (see DARLINGTON 1937), crossing-over precedes chiasma formation and occurs early, during zygotene or pachytene; moreover, chiasma formation is the consequence of crossing-over. At present, the hypothesis of chiasmatypy is generally accepted with respect to higher eukaryotes, while SAX's hypothesis is virtually abandoned.

Cleveland (1954), however, has shown that in *Barbulanympha* (and possibly other Hypermastigida), chiasmata are formed according to Sax's hypothesis. All the chromosomes in *B. ufalula* (a haplont with n = 26) are two-armed with clearly expressed individuality. Meiosis in *Barbulanympha* may follow either karyogamy (the fusion of two pronuclei, which is often autogamous) or endomitosis. In the latter case meiosis is, in essence, somatic reduction since the sexual process is absent.

During leptotene, the chromosomes are still unichromatidal (Fig. 58*a*), and then they become bichromatidal (Fig. 58*b*). Thus, the visible duplication of the chromosomes in *Barbulanympha* (just as in *Trichonympha*) occurs before the homologues conjugate. Later the chromosomes shorten, and only at this stage do the first signs of their pairing appear (Fig. 58*c*). There are no typical stages of zygotene and pachytene. Chromosome conjugation occurs late and is never very close. It immediately produces four-chromatid bivalents that morphologically correspond to the stage of diplotene or even diakinesis (Fig. 58*d*).

Owing to the large dimensions of chromosomes (up to 20 µm in diakinesis), Cleveland succeeded in directly observing (partly *in vivo*) the formation of chiasmata on the bivalents and their further behaviour. Naturally, there could be no "early" crossing-over in conformity with the chiasmatypy hypothesis because of late chromosome conjugation. Nevertheless, chiasmata are formed because the forces of pairing between the sister chromatids prevail in some parts of the bivalent (and homologous chromatids disunite at these sites), while the pairing forces between the homologues prevail in other parts of the same bivalent (in this case, the sister chromatids disunite). The bivalent represented in Fig. 58*e* has only the first type of link and no chiasmata. This also applies to the bivalents 1, 7, 11, 13, 14, 18, and others in Fig. 58*d*. In the bivalent represented in Fig. 58*f*, the first type of interchromatidal link prevails on the right-hand side, while the second type prevails on the left-hand side; accordingly, there is one chiasma on the left-hand side of the bivalent. The bivalents 3 and 5 in Fig. 58*d* also relate to this type. Some achiasmatic bivalents completely disunite to form two homologous chromosomes already during diakinesis (the chromosome pairs 8, 21, and 22). The manner of chiasma formation is thus in full accord with that predicted by Sax's hypothesis. Moreover, Cleveland observed two variants of the behaviour of a tetrad with a chiasma. In one case, the chiasma terminalizes without crossing-over, while in the other, the stretching of the bivalent in the early anaphase causes the rupture and recombination of the two crossed chromatids, i.e. "late" crossing-over occurs according to Sax.

Fig. 58. Meiosis in *Barbulanympha ufalula* (after Cleveland 1954). *a* early leptotene; *b* duplication of chromosomes; *c* homologues beginning to pair (opposite arrows); *d* diakinesis with 26 four-chromatid bivalents (numbered); *e* bivalent where the pairing of sister chromatids predominates throughout the length; *f* bivalent with a chiasma (the pairing of nonsister chromatids prevails on the left arm); *g* separation of a bivalent into dyads in anaphase I; *h* telophase I; *i* interkinesis; *k* separation of a dyad into unichromatidal chromosomes in anaphase II; *l* telophase II. *a* and *c* 900 ×; *b* and *h* 800 ×; *d* 1600 ×; *i* and *l* 1200 ×.

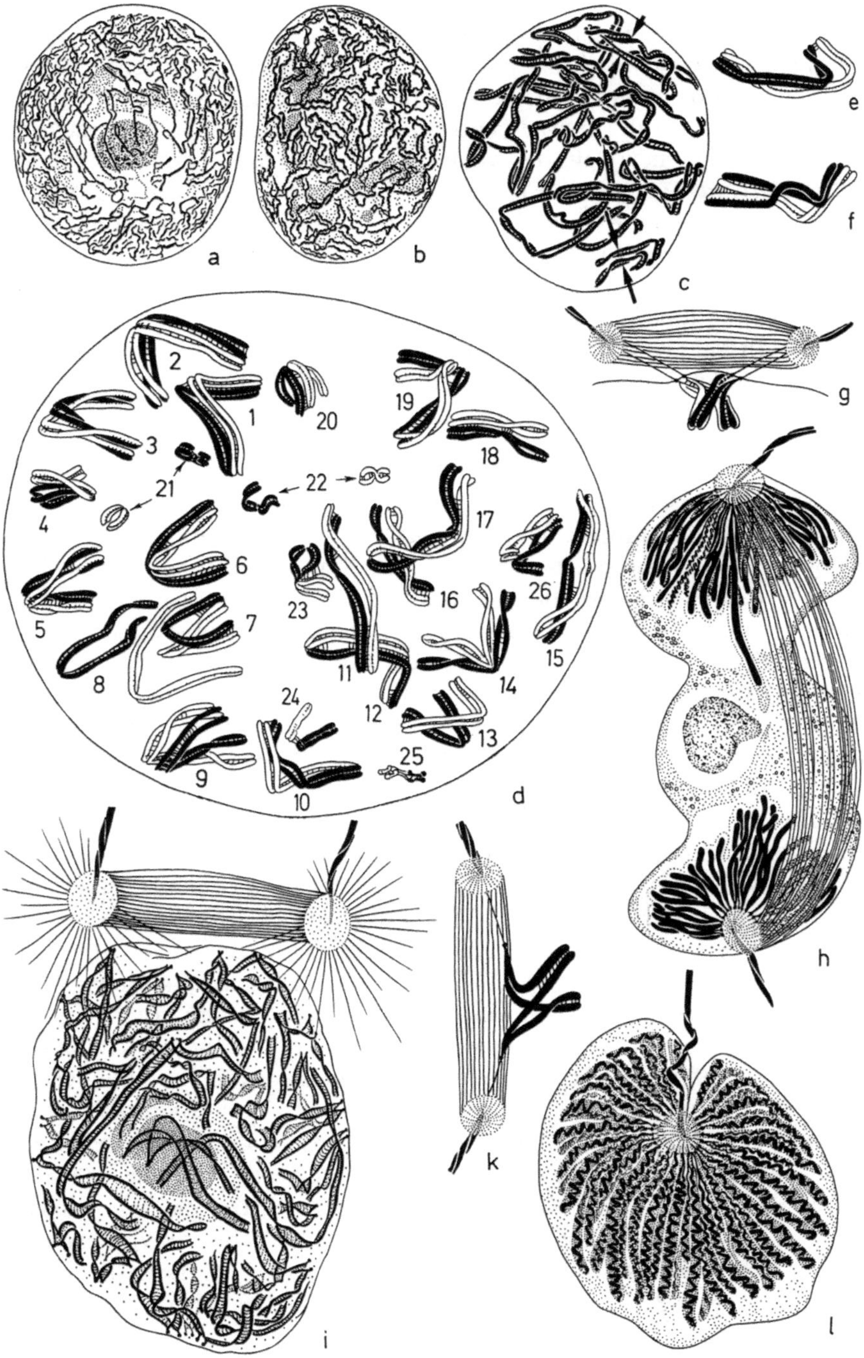
a
b
c
e
f
g
d
1
2
3
4
5
6
7
8
9
10
11
12
13
14
15
16
17
18
19
20
21
22
23
24
25
26
h
i
k
l

Fig. 58

During the further course of meiosis in *Barbulanympha*, a very peculiar circumstance is that the centromeres of the homologous chromosomes are already duplicated in the first meiotic division and not in the second, as is usually the case. In anaphase I, every dyad that goes to the pole thus has two centromeres and, accordingly, two chromosomal fibres of the spindle are attached to it (Fig. 58*g*). In telophase I (Fig. 58*h*), the dyad centromeres approach the polar atractospheres and the central spindle elongates while the nucleus begins to constrict (cf. mitosis in *Barbulanympha*, p. 118 and Fig. 32). It pinches in three parts (Fig. 58*h*), and the central part of the nucleus with the nucleolus is not incorporated into the daughter nuclei but is resorbed in the cytoplasm. During interkinesis and in prophase II, the chromatids of the dyads are prevented from completely disuniting only by the previously mentioned pairing forces since their centromeres are already separated (Fig. 58*i*). In anaphase II, the dyads divide into individual two-arm chromosomes (Fig. 58*k*), which are despiralized in telephase II (Fig. 58*l*).

Meiosis in Ciliophora

Two-step meiosis is characteristic of the *micronuclei* of all ciliates. It occurs during the sexual process, i.e. *conjugation*, when the micronucleus usually undergoes three "maturation divisions" that lead to the formation of haploid pronuclei. It is now known that the first two such divisions are meiotic ones.

Table 3 shows that ciliates are diplonts with gametic meiosis and rather large numbers of usually minute chromosomes. For a long time, their meiosis was considered to be atypical, consisting in the separation of whole chromosomes without the preliminary pairing of the homologues. The first or, more frequently, the second maturation division was considered to be a "reductional" one (data reviewed by RAIKOV 1972d). However, it has been shown (DEVIDÉ and GEITLER 1947, DEVIDÉ 1951) that these observations are erroneous due to the combination of the chromosomes into aggregates. In many species, the aggregation involves not only mitosis (see p. 111) but also meiosis, especially the second division (*Euplotes*, *Oxytricha*) and occasionally even the first division (*Chilodonella*). It is now known that the meiosis of ciliates necessarily includes the pairing of the homologous chromosomes and the formation of bivalents with chiasmata; consequently, their chromosome reduction occurs during two divisions of the micronucleus.

The meiotic prophase of the micronuclei of many ciliates includes the characteristic "parachute" stage, which is apparently homologous to the "bouquet" stage (zygotene) of classical meiosis (Fig. 59*b*). That is when homologous chromosomes conjugate, since bivalents are observed at the later stages, for instance, in *Nassula ornata* (Fig. 59*c*, *d*). In this case, the micronucleus greatly stretches at the end of the meiotic prophase (Fig. 59*e*). During metaphase I, the meiotic spindles are much larger than the mitotic ones (cf. Fig. 28), and they have characteristic pointed poles; bivalents with one or two chiasmata are seen along the equator of the spindle (Fig. 59*f*). The spindles of the second meiotic division (Fig. 59*h*–*i*) are much smaller than those of the first division.

The "parachute" stage has been observed in the meiotic prophase of many ciliates, particularly the gymnostomes *Loxodes* (BOGDANOWICZ 1930b), *Tracheloraphis* (RAIKOV 1958, DRAGESCO and RAIKOV 1966), *Trachelocerca*

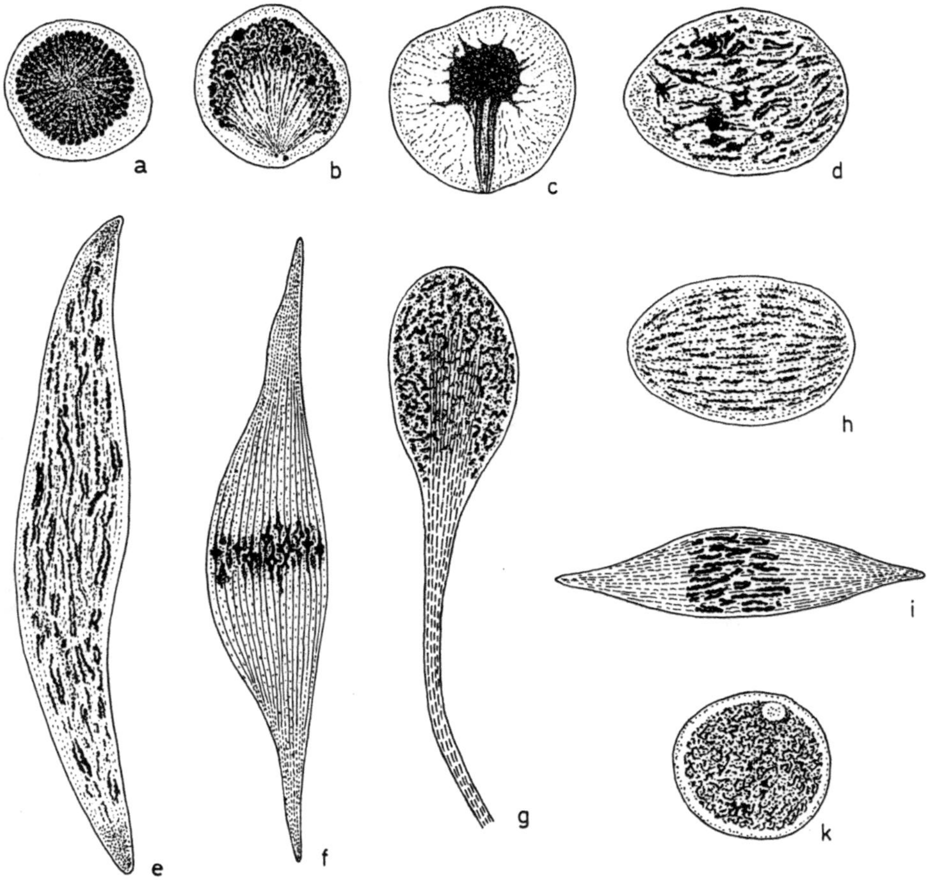

Fig. 59. Micronuclear meiosis in the ciliate *Nassula ornata*, 2070× (after RAIKOV 1967a). *a* early prophase; *b* "parachute" stage; *c* condensation of a zygotene skein of chromosomes; *d* diakinesis; *e* prometaphase; *f* metaphase; *g* telophase I; *h* prophase II; *i* metaphase II; *k* product of meiosis.

(RAIKOV 1963a), *Trachelonema* (KOVALEVA 1972), and *Didinium* (PRANDTL 1906); the vestibuliferan *Bursaria* (POLJANSKY 1934); the hypostomes *Chilodonella* (MACDOUGALL 1925) and *Nassula*; and the heterotrichids *Fabrea* (ELLIS 1937), *Spirostomum* (RAO 1968a), *Blepharisma* (MIYAKE *et al.* 1979), and *Stentor* (SKARLATO 1979a). It is especially characteristic of meiosis in the Hypotrichida, e.g. *Uroleptus* (CALKINS 1919), *Stylonychia* (PESHKOVSKAYA 1948, DEVIDÉ 1951, AMMERMANN 1965, ALONSO and PÉREZ-SILVA 1965a, etc.), *Oxytricha* (GREGORY 1923, KAY 1946), *Kahlia* (RAO 1966), *Onychodromus* (MAUPAS 1889), *Pleurotricha* (MANWELL 1928), *Euplotes* (TURNER 1930, RAO 1964), and some others. In many of these works, bivalents (whose number is indicated, as the haploid one, in Table 3), have been observed at the stages that follow the "parachute" stage.

Among ciliates that have the "parachute" stage, the most thorough analysis of the behaviour of the chromosomes in the first meiotic division was made by

KOVALEVA (1972) in *Trachelonema sulcata* (order Karyorelictida), which has rather large chromosomes (n = 28). After the typical "parachute" (Fig. 60*a*, *b*) and pachytene (Fig. 60*c*) stages, when the dual nature of bivalents is still indistinct, the diplotene stage (Fig. 60*d*) sets in. All the 28 bivalents are then arranged along the periphery of the micronucleus directly under its envelope, and the cleft between the homologues (Fig. 60*d*) is clearly seen. During diakinesis, the homologues disunite, remaining linked only by chiasmata (Fig. 60*e*,*f*). The number of the latter per

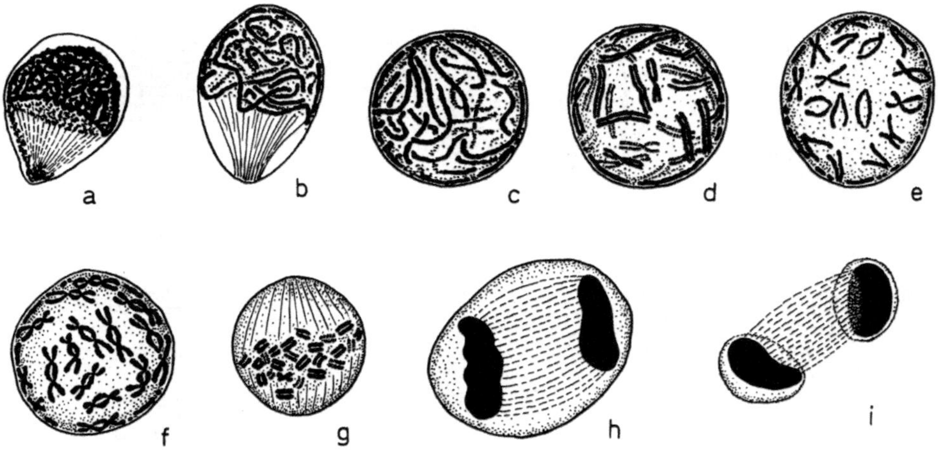

Fig. 60. First meiotic division of the micronucleus in *Trachelonema sulcata*, 1875 × (after KOVALEVA 1972). *a* and *b* "parachute" stage; *c* pachytene; *d* diplotene; *e* and *f* diakinesis; *g* metaphase; *h* and *i* telophase.

bivalent is one (Fig. 60*e*) or two (Fig. 60*f*) but rarely three. By metaphase I, all the chiasmata terminalize and disappear (Fig. 60*g*). Although four chromatids in the bivalent were not directly observed in *Trachelonema*, the existence of chiasmata makes their presence very probable.

Bivalents that separate into dyads in the first division and into individual chromosomes in the second division have also been observed in the meiosis of micronuclei of other karyorelictids: *Tracheloraphis phoenicopterus* (RAIKOV 1958) and *Trachelocerca coluber* (RAIKOV 1963a). In this case, bivalents are minute and mostly cruciform (i.e. probably with one chiasma).

Among other ciliates, mainly of the Oligohymenophorea class, the "parachute" stage is absent, but the "crescent" stage, which is just as characteristic, occurs in the meiotic prophase. The micronucleus strongly elongates and often bends, while the chromosomes become oriented along its axis. The "crescent" stage exists in *Paramecium* (MAUPAS 1889, HERTWIG 1889, HAMBURGER 1904, CALKINS and CULL 1907, DEHORNE 1920, CHEN 1940b, 1946a, 1946b, PESHKOVSKAYA 1941, EGELHAAF 1955, KOŚCIUSZKO 1965, etc.), *Colpidium*, *Tetrahymena* (RAY 1956), *Glaucoma* (MAUPAS 1889), *Frontonia* (DEVI 1961), and some other forms.

The clearest pictures of meiosis with the "crescent" stage were obtained by DEVIDÉ and GEITLER (DEVIDÉ and GEITLER 1947, DEVIDÉ 1951) for *Colpidium campylum*. At the "crescent" stage, the micronucleus becomes almost filiform. It could be that the chromosomes are arranged in tandem in it, since only one chromatin strand is seen in the "crescent". When the crescent shortens, 21 typical bivalents with one, two, or three chiasmata appear in the nucleus. In metaphase I, the bivalents are divided into dyads, some of which are typically four-armed.

Meiosis with the "crescent" stage and a small number of chromosomes (n = 5) has also been clearly observed in *Tetrahymena thermophila* (RAY 1956). However, according to RAY as well as SUGAI and HIWATASHI (1974), chromosomes arranged in tandem in the "crescent" are already paired into bivalents, i.e. the "crescent" would correspond to the pachytene stage. Later, the ends of the "crescent" are drawn in and from one to three chiasmata appear on the bivalents. The two-step nature of meiosis in *Tetrahymena*, just as in *Colpidium*, is accentuated by the presence of four-armed dyads, which exist from anaphase I to metaphase II (RAY 1956).

Thus, the "crescent" stage clearly precedes diakinesis, but it does not necessarily represent zygotene, i.e. the stage at which the chromosomes pair. In *Paramecium caudatum* (CALKINS and CULL 1907, DEHORNE 1920, PESHKOVSKAYA 1941, etc.), the "crescent" is much wider than in *Tetrahymena* and *Colpidium* and the chromosomes are arranged in it parallel to one another rather than in tandem. At the later stages of the meiotic prophase, they already have the form of bivalents, occasionally with chiasma-like figures (CALKINS and CULL 1907). Therefore, the "crescent" stage in *Paramecium* was usually regarded as a homologue of the "parachute" stage in other ciliates, i.e. a homologue of zygotene. However, the data recently obtained by electron microscopy on meiosis in the *Paramecium* species of the *aurelia* group show that it is more likely the pachytene or even the diplotene stage, since the synaptonemal complexes (though rather indistinct ones) appear in the micronucleus even before it stretches into a "crescent" (STEVENSON 1972). The chromosomes are rather dense at the beginning of "crescent" formation, but they decondense so much in the developed "crescent" that they can no longer be detected on electron micrographs; synaptonemal complexes are also absent in the "crescent" (STEVENSON 1972, JURAND 1976).

STEVENSON discovered longitudinal microtubules in the meiotic micronucleus of *P. primaurelia* as it stretched into a "crescent". These microtubules are situated only on the inner side of its envelope, which has many pores probably due to the intense entry of tubulin into the micronucleus. In *P. tetraurelia*, according to JURAND (1976), the microtubules in the developed "crescent" (which can be as much as 20 μm long) are packed strictly into one layer, while the envelope of the micronucleus forms a system of longitudinal folds. Apparently, it is the growth of the microtubules that causes the micronucleus to stretch into a "crescent". As soon as the reverse shortening of the "crescent" begins, microtubules in it are depolymerized (STEVENSON 1972). It is possible that paired intranuclear dense bodies, which according to JURAND (1976) are in one of the ends of the "crescent", participate in the assembly of microtubules. If this is confirmed, they would be regarded as intranuclear microtubule-organizing centres (MTOCs).

All traces of the layer of the submembranous microtubules disappear in metaphase I of *P. primaurelia*. The bivalents are attached to an ordinary spindle that consists of both continuous and chromosomal microtubules. However, the whole division figure is larger than in mitosis and, unlike in mitosis, the spindle poles become pointed in the anaphase (cf. Fig. 59*f*). The nuclear envelope is preserved in meiosis, while the nucleus simply pinches in two in the telophase (STEVENSON 1972). The spindles of the second meiotic division are ultrastructurally similar to the mitotic ones (STEVENSON 1972, JURAND 1976).

The "crescent" stage of *Tetrahymena thermophila* also has a layer of microtubules that underlie the envelope of the micronucleus and are apparently responsible for its elongation (DAVIDSON and LA FOUNTAIN 1975, WOLFE *et al.* 1976). At the beginning of "crescent" formation, these microtubules grow from the intranuclear electron dense MTOC that lies in one of the ends of the stretching micronucleus (WOLFE *et al.* 1976) and is similar to such bodies in *Paramecium*. The above mentioned researchers admit that two sets of submembranous microtubules are pushed apart as a result of their mutual sliding, since microtubules are more sparse in the fully developed "crescent" than in the growing one. At the "crescent" stage, the chromosomes of *T. thermophila* are greatly decondensed, and there are no synaptonemal complexes (i.e. this is not zygotene and probably not even pachytene). The micronucleus of *Paramecium bursaria*, however, contains not only microtubules, but also many microfilaments at the "crescent" stage (LEWIS 1977).

Thus, the mechanism for stretching the micronucleus into a "crescent" has now become clear, although the homology of this stage and its biological significance are still ambiguous. This may be a stage at which the chromosomes are temporarily decondensed during diplotene (e.g. for transcribing certain information that is coded in the micronucleus and is needed for the further course of conjugation). This assumption is supported by the fact that although the interphase micronucleus of *Tetrahymena* practically does not synthesize RNA, labelled uridine is intensely incorporated into the micronucleus at the "crescent" stage (SUGAI and HIWATASHI 1974). This is all the more interesting because the lampbrush chromosomes, which actively function during the oogenesis of metazoans, are also diplotene chromosomes.

The above mentioned data on the meiosis of *Nassula* also support the nonhomology of the "crescent" and "parachute" (zygotene) stages. The elongation of the micronucleus at the end of the meiotic prophase (Fig. 59*e*) greatly resembles a "crescent". *Nassula* also has the "parachute" stage, but it occurs much earlier. Hence, if the "parachute" stage corresponds to zygotene, the "crescent" stage corresponds to diplotene in this case, too.

Two nucleolus-like bodies connected with the chromosomes appear in the leptotene stage (?) of the micronucleus of *Tetrahymena* (MURTI 1972). This also indirectly shows that the chromosomes of the micronucleus can function during the meiotic prophase. Unfortunately, the chemical nature of these bodies is still unknown. They have no granular component and do not resemble real nucleoli. Similar fibrillar "spheres", which are linked with chromosomes, have been observed in the micronuclei of *Didinium* (KARADZHAN 1977) and *Stentor* (SKARLATO 1979a) at the leptotene stage of meiosis.

Besides *Paramecium*, the complete course of meiosis has been studied with the electron microscope in the gymnostome *Dileptus anser* (VINNIKOVA 1975, 1976). In this case, neither the "parachute" stage nor the "crescent" stage is clearly defined (VINNIKOVA 1974b). At the zygotene stage, threadlike chromosomes conjugate pairwise, as is evident from the appearance of the synaptonemal complexes. Somewhat later (pachytene?), all the chromosomes condense into an eccentric mass (cf. Fig. 59*c*) in which the synaptonemal complexes are preserved. At the end of this stage, the bivalents begin to disunite (transition to diplotene) and a large amount of fragments of the synaptonemal complexes are released from the chromatin cluster into the karyolymph where they form typical *polycomplexes* (Fig. 56*e*). These are stacks of densely packed parallel segments of the synaptonemal complexes. The fact that the synaptonemal complexes can separate from the chromosomes shows that the former are not an integral part of the latter but are additional structures that ensure their conjugation.

At the later stages of the meiotic prophase of *Dileptus*, the chromosomal mass breaks up into condensed bivalents that usually have one chiasma. In both meiotic divisions of the micronucleus, the nuclear envelope is preserved, while in the telophase, it is reorganized in the same way as in mitosis, i.e. by the "indirect" ("intranuclear") method (cf. p. 111).

Synaptonemal complexes of a typical structure have been found also in the meiotic prophase of *Didinium* (KARADZHAN 1977) and *Stentor* (SKARLATO 1979a). No formation of polycomplexes has been observed, however. The spindle of the first meiotic division in *Didinium* is much larger than that of mitosis; in the anaphase, it does not stretch into a narrow pointed figure but remains broad and blunt. Accordingly, the daughter micronuclei immediately receive a large amount of karyolymph, while the chromosomes (dyads) remain condensed and the nucleus enters metaphase II directly after telophase I.

The two-step nature of meiosis in cilates has been confirmed cytophotometrically and autoradiographically. The amounts of DNA in the micronucleus in the meiotic prophase and in the products of the first and second nuclear divisions are in a ratio of 4 : 2 : 1, as in classic meiosis with two nuclear divisions. This fact was demonstrated in *Paramecium* (WOODARD *et al.* 1966, PIERI *et al.* 1968, DUPY-BLANC 1969a, GOLIKOVA 1974), *Tetrahymena* (MCDONALD 1973, DOERDER and DE BAULT 1975), *Stylonychia* (PIERI 1965, AMMERMANN 1971b, CHADHA *et al.* 1978), and *Euplotes* (SALVANO 1974a). In *Bursaria*, ^{3}H-thymidine is incorporated into the micronuclei of the conjugants before the first meiotic division but not before the second one (POLJANSKY and SERGEJEVA 1977). Consequently, the ciliates undergo the *premeiotic replication* of DNA, and their bivalents must be four-stranded. Conversely, DNA replication is absent between the first and second meiotic divisions. As for the third (postmeiotic) division of the micronucleus, the previously mentioned works show that it is ordinary haploid mitosis that is preceded by DNA replication.

In most cases, ciliates conjugate when their micronuclei are in the G_1 period. This applies to the previously mentioned *Paramecium*, *Stylonychia*, and *Bursaria* as well as to *Diophrys* and *Oxytricha* (LUPORINI and DINI 1975). Premeiotic DNA replication occurs in the conjugants. Since it is known that this replication differs by its nature from the premitotic one, it can be assumed that DNA synthesis is switched over to the premeiotic course as a result of the union of partners. The hypotrich *Euplotes crassus* mates when its micronucleus is in G_2, but then meiosis is preceded by one mitotic (preliminary) division of the micronucleus, while premeiotic DNA synthesis occurs only after this mitosis, namely, also in the conjugants (SALVANO 1974a, LUPORINI and DINI 1975). Things are more complicated with *Tetrahymena*, which also mate with their micronuclei in G_2, since the micronucleus directly begins meiosis in this case, and there is no new DNA replication in it (WOLFE 1973, SUGAI and HIWATASHI 1974, DOERDER and DE BAULT 1975). It turns out that ordinary premitotic DNA synthesis, which occurs in the micronucleus before the union of partners, can replace specific premeiotic replication in *Tetrahymena*. Apparently, further investigations are needed in this respect. In *Paramecium caudatum*, the micronucleus that undergoes premeiotic DNA replication (in the conjugants) is not yet irreversibly differentiated for meiosis; it can return to the normal S phase of the mitotic cycle if it is transplanted from the conjugating cell into a vegetative one (FUJISHIMA and HIWATASHI 1978).

Meiosis with a Single Nuclear Division

As we have already seen, the duplication of chromosomes and that of centromeres in two-step meiosis are *temporally separated:* the former occurs in the prophase of the first meiotic division, while the latter occurs in the metaphase of the second division. This unites the two nuclear divisions into a single meiotic process: crossing-over occurs in the meiotic prophase, while the homologues are separated in two stages (in some loci, in the first division, and in other loci, in the second one). Besides such meiosis, however, a simpler type of chromosomal reduction occurs in protozoans, namely, meiosis with one nuclear division, or *one-step meiosis*. CLEVELAND (1947), who proposed this concept, understands it as the pairing and subsequent separation of the homologous chromosomes without their preliminary duplication. In such meiosis, the bivalents are not four-chromatidal but bichromatidal, and chiasmata do not occur in them. Crossing-over is impossible, since it can occur only at the stage of four-strand bivalents, which is absent in one-step meiosis. The whole process of chromosomal reduction is completed in one nuclear division. Even if it is followed by a second division, the latter will be purely mitotic and will be preceded by ordinary chromosome replication.

One-Step Meiosis in Mastigophora

One-step meiosis was first described by CLEVELAND in flagellates that are intestinal symbionts of the cockroach *Cryptocercus*, namely, all the representatives of Oxymonadida that have been studied up to now, i.e. *Oxymonas*, *Saccinobaculus*, and *Notila* (CLEVELAND 1950a, 1950b, 1950c) as well as some Hypermastigida, i.e. *Leptospironympha* and *Urinympha* (CLEVELAND 1951a, 1951b); other Hypermastigida undergo two-step meiosis. The group of forms with one-step meiosis includes both haplonts and diplonts.

Among haplonts (*Oxymonas*, *Saccinobaculus*, *Leptospironympha*), the first division of the zygotic nucleus is a meiotic one. It is accompanied by cytoplasmic division and produces two vegetative cells that continue to multiply by mitotic

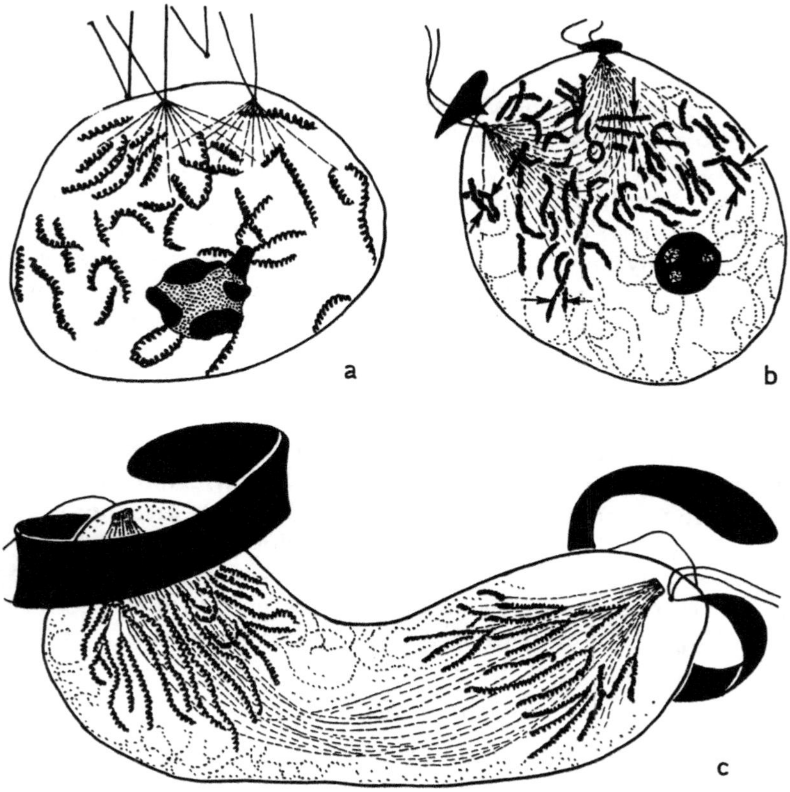

Fig. 61. One-division meiosis in *Oxymonas*, 2200 × (after CLEVELAND 1950a). *a* early prophase (unichromatidal chromosomes); *b* homologues beginning to pair (opposite arrows); *c* anaphase. Ribbon-shaped structures are axostyles.

divisions. Among diplonts (*Notila*, *Urinympha*), the one-step nature of meiosis is especially obvious since meiosis is gametic and directly precedes the formation of pronuclei. The latter originate from the nucleus of the diploid trophozoite as a result of a *single* nuclear division; the second nuclear division does not occur, and meiosis cannot be two-step.

Let us consider, for instance, the cytology of meiosis in *Oxymonas* (CLEVELAND 1950a). In the early prophase, the chromosomes of the zygote ($2\,n = 50$) spiralize without preliminary homologue conjugation and, accordingly, without the "bouquet" stage (Fig. 61 *a*). The condensing chromosomes are also not duplicated (unlike in *Trichonympha* or *Barbulanympha*): they invariably remain unichromatidal, at least at the microscopic level. Almost all the stages of the meiotic prophase are thus omitted. The momentary, poorly expressed pairing of the homologous chromosomes which are already coiled occurs only in the late prophase (Fig. 61 *b*); no precise conjugation of homologous loci is possible, and no chiasmata are formed. Such pairs of homologous chromosomes can hardly be even called real bivalents. In the metaphase, the homologues easily separate, while in the anaphase, they go to the poles as unichromatidal two-armed chromosomes (Fig. 61 *c*).

Apparently, the most significant feature of one-step meiosis is the absence of chromosome replication in the meiotic prophase. Consequently, there cannot be any chiasmata and crossovers. Likewise, premeiotic DNA replication is also unlikely to occur in one-step meiosis. However, this has not been verified yet with respect to the above-mentioned flagellates.

Recently, data have appeared on the existence of the sexual process in mesokaryotic Dinoflagellida (reviewed by LOEBLICH 1976). As a rule, isogamic (more rarely, anisogamic) copulation of the free-swimming cells occurs in this case. The zygote usually encysts and divides twice, and four vegetative cells come out of the cyst. The assumption that zygotic meiosis should occur in the cyst is confirmed by the recombination of genes, which takes place at this moment (BEAM and HIMES 1974, TUTTLE and LOEBLICH 1974). The nature of meiosis in *Crypthecodinium cohnii* was genetically ascertained (HIMES and BEAM 1975, BEAM *et al.* 1977) by tetrad analysis (compare the previously mentioned data on two-step meiosis in *Chlamydomonas*). In *Crypthecodinium*, unlike in *Chlamydomonas*, all the four descendants of one zygote are *never* genotypically different. A zygote always produces two identical clones of one type and two clones of another type (six genes were analyzed). Apparently, this means that homologous loci are wholly segregated in the first division of the zygotic nucleus, while the second division is equational, i.e. it represents ordinary mitosis. Thus, according to genetic data, the meiosis of Dinoflagellida is most probably a one-step process (or, in any case, it is achiasmatic).

Meiosis in Sporozoea

Besides Oxymonadida and some Hypermastigida, one-step meiosis is apparently characteristic of all Sporozoea (GRELL 1956a). Neither four-stranded bivalents nor chiasmata have been found yet in any sporozoan.

In Coccidia, e.g. *Karyolysus* (REICHENOW 1921), *Aggregata* (DOBELL 1925, NAVILLE 1925, BĚLAŘ 1926a, 1926b), *Klossia* (NAVILLE 1927), *Eucoccidium* (GRELL 1953a), *Coelotropha* (VIVIER and HENNERÉ 1964, PORCHET-HENNERÉ 1971), and *Eimeria* (CANNING and ANWAR 1968, 1969), meiosis begins immediately after karyogamy (coccidians are haplonts with zygotic meiosis). In the meiotic prophase, the zygote nucleus greatly swells and stretches out (the "fertilization spindle", which probably corresponds to zygotene). Longitudinal threadlike chromosomes are seen in it, and they combine pairwise into bivalents. Later, bivalents condense, and then it can be seen that they consist of two unichromatidal chromosomes and do not have chiasmata. Homologous chromosomes disunite in the anaphase of the first division of the zygote nucleus. Apparently, this division is a meiotic one, while the subsequent divisions of sporogony are ordinary mitoses.

The occurrence of the parallel conjugation of despiralized chromosomes, however, left some doubt about the one-step nature of meiosis in coccidians: since the chromosomes are minute, their splitting into chromatids could have simply been overlooked. But CANNING and MORGAN (1975) have cytophotometrically confirmed that premeiotic DNA replication was absent in *Eimeria* and *Hepatozoon*. They showed that the fusing pronuclei were in the G_1 period of the haplophase (the amount of DNA in them was 1c). Accordingly, the zygote nucleus received 2c of DNA (G_1 period of the diplophase). During the meiotic prophase, this amount of DNA did not change, remaining at the 2c level. The 4c level, being typical of the prophase of two-step meiosis, was not found in this case. This fact indicates that chromosome replication is absent. The products of the first division of the zygote nucleus initially have only 1 c of DNA and are in the G_1 period of the haplophase. Before the second division of the zygote nucleus, their DNA replicates (1c → 2c). Hence, the second division is ordinary haploid mitosis. Thus, meiosis in coccidians is really completed during the first division of the zygote nucleus, i.e. it is one-step meiosis.

In Gregarinia, meiosis occurs in approximately the same way as in coccidians, but the zygotes (spores) are much smaller and the "fertilization spindle" is not expressed. Apparently, homologous chromosomes conjugate later and in a more spiralized state, while the link between the homologues in bivalents is expressed far more faintly in gregarines than in coccidians. In many cases, the bivalents completely separate into two homologous chromosomes long before the metaphase, indicating that no chiasmata exist on them, e.g. in *Actinocephalus* (WESCHENFELDER 1938) or *Stylocephalus* (GRELL 1940, UTTANGI and DESAI 1963, DESAI 1965a, 1965b, 1966).

In the gregarine *Grebnickiella*, the relatively weak union of the homologues seems to depend on the fact that the synaptonemal complexes connect the homologous chromosomes at only one point per bivalent, and not along their entire length. The synaptonemal complexes are very short (0.16–0.18 μm) and exist solely between the kinetochores of the two homologues (MOLON-NOBLOT and DESPORTES 1977). The kinetochores, which display a typical trilaminar structure, actually become the lateral elements of the synaptonemal complex. This is also the first record of synaptonemal complexes in sporozoan meiosis.

Meiosis in Microspora and Myxosporea

It was unknown for a long time whether microsporidians underwent a sexual process, an alternation of the haplophase and the diplophase, and, accordingly, meiosis. Synaptonemal complexes were discovered only recently in the nuclei of sporonts and young pansporoblasts of several microsporidians (LOUBÈS *et al.* 1976b, LOUBÈS 1979, DESPORTES and THÉODORIDÈS 1979, HAZARD *et al.* 1979, VIVARÈS 1980). They are found in both uninucleate sporonts (*Gurleya*, *Glugea*, *Baculea*) and paired sporont nuclei, the "diplokarya" (*Thelohania*, *Pleistophora*, *Tuzetia*). Consequently, both nuclei that constitute the diplokaryon are diploid and cannot be pronuclei, as was often believed. Moreover, it was discovered that these nuclei do not fuse before the beginning of sporogenesis but separate, and each one undergoes at least two divisions. Thus, the divisions of sporont nuclei can be regarded as meiotic, while the nuclei of sporoblasts and spores, which are formed due to them, can be regarded as haploid (LOUBÈS 1979). However, it is still unknown whether one-step or two-step meiosis occurs in this case. The sexual process proper (karyogamy) in microsporidians has also not been studied yet, and it is unknown how diploidy is restored in the life cycle. There are no synaptonemal complexes in the nuclei of *Nosema*. LOUBÈS (1979) believes that this genus has secondarily lost the sexual process and meiosis, i.e. it has become apogamic.

As for myxosporidians, karyogamy occurs in their spore. It has been shown cytophotometrically that in *Sphaeromyxa* and *Myxidium*, the two nuclei of the amoeboid sporoplasm (as well as the nuclei of other spore cells) are haploid, while the nuclei of sporoblasts are diploid. Apparently, meiosis occurs in microsporidians during sporogenesis after a series of mitoses (USPENSKAJA 1976, 1979, 1981). Synaptonemal complexes were discovered in the nuclei of myxosporidians (SIAU 1979). However, they occur in the schizonts of *Myxobolus* and *Ceratomyxa* and not in the sporoblasts, as might be expected. The site and nature of myxosporidian meiosis thus await further investigation.

Problem of the Origin and Evolution of Meiosis

There are still very few works on meiosis in protozoans, especially those employing modern methods. Therefore, all the following reasonings on the possible ways of the evolution of meiosis are highly hypothetical.

Origin of Meiosis

Apparently, the sexual process and meiosis, which is coupled with it, had repeatedly and independently originated in the evolution of protozoans. In other words, the agamic life cycles, being observed in all the lower groups of protozoans (amoebae, many orders of flagellates), are primary and the haploid state of the nucleus corresponds to them. A momentary diploid phase and zygotic meiosis (which originated as a compensatory mechanism for the return to the haplophase) had appeared in every line of development together with the sexual process. Such views, which uphold the polyphyletism of meiosis and the sexual process in protozoans, are presently shared by most specialists (WENRICH 1954b, DOGIEL 1965, IVANOV 1968, 1970, etc.).

According to CLEVELAND (1947) and GRELL (1953b), one-step meiosis is the most primitive type of meiosis. It is commonly found among protozoans, especially in the lower groups: Mastigophora and Sporozoea. A particularly important fact is that it occurs in very primitive mesokaryotic protozoans, the Dinoflagellida. Conversely, one-step meiosis is not found in higher protozoans (the ciliates), just as it is not found in metazoans. Genetically, one-step meiosis is a mechanism of gene recombination that is less perfect than two-step meiosis. Since crossing-over is absent, recombination does not occur inside linkage groups (chromosomes) in this case.

It can be imagined that one-step meiosis really originated as a compensatory mechanism for the return to the haplophase from the diplophase, which appeared in one way or another in the life cycle of a species. In this case, homologous chromosomes in the diploid nucleus united pairwise at first, then parted into daughter nuclei in one division. Apparently, the ability of chromosomes to undergo parallel conjugation is their fundamental property, which is expressed not only in meiosis. CLEVELAND's observations with respect to *Holomastigotoides* are interesting in this respect (CLEVELAND 1949a, 1953b, 1961b). No sexual process is known in this flagellate. Normally, it is a haplont with two chromosomes. In ordinary mitosis, the chromatids separate very early and most of the time the nucleus contains four free chromatids; namely, two chromatids of one chromosome and two of the other (Fig. 37*d*). But the sister chromatids often pair again, forming bivalent-like structures (somatic conjugation). The somatic conjugation of homologous chromosomes widely occurs not only among protozoans. This ability of the chromosomes to undergo conjugation was a prerequisite of the transformation of mitosis into meiosis (CLEVELAND 1953b, STACK and BROWN 1969). If we suppose that the ordinary replication of chromosomes is omitted before one of the divisions of the diploid nucleus in which homologous chromosomes are somatically paired, homologous chromosomes instead of sister chromatids will part into the daughter nuclei. The result is one-step meiosis.

Origin of Two-Step Meiosis

If we suppose that one-step meiosis is primitive, the possible way of its transformation into two-step meiosis is a problem. In principle, this transformation can occur by drawing the next mitosis (which originally followed one-step meiosis) into the general process of chromosome reduction. It can be assumed that in classic two-step meiosis, chromosome replication that occurs in the premeiotic interphase relates by its origin not to the first but to the second meiotic division; however, it is shifted one cell generation forward. This is evident from the fact that centromere reproduction, which corresponds to the given chromosome replication, occurs only in the second meiotic division. Such uncoupling of chromosome reproduction from centromere reproduction as well as a shift of chromosome reproduction to an earlier time causes the appearance of the four-stranded stage in the meiotic prophase, making crossing-over possible and thus combining both nuclear divisions into a single process of two-step meiosis.

The evolutionary advantages of two-step meiosis are evident: it ensures a much higher frequency of hereditary recombination in comparison with one-step meiosis. Apparently, two-step meiosis had repeatedly evolved in protozoans: in Sarcodina (Heliozoea, Foraminiferida), Labyrinthomorpha, Mastigophora (Volvocida, Hypermastigida), Opalinata, and Ciliophora.

Evolution of Crossing-Over

After seeing that the formation of chiasmata "according to SAX" and the "late" crossing-over associated with it in *Barbulanympha* were feasible, CLEVELAND (1953b, 1954) tried to apply the "classical" hypothesis of crossing-over also to higher eukaryotes and completely rejected the hypothesis of chiasmatypy. However, a large amount of experimental material now shows that crossing-over occurs very early in higher animals and plants. Therefore, we cannot share CLEVELAND's extreme standpoint. It follows that both types of crossing-over, i.e. the "early" and "late" ones, should exist in the organic world. If two-step meiosis had often arisen in evolution, the mechanism that ensures crossing-over should also have been formed many times. When early and precise conjugation of the homologues had appeared, "early" crossing-over could have evolved in accordance with the hypothesis of chiasmatypy. When chromosome pairing was late and incomplete, there could have been only "late" crossing-over.

Apparently, "early" crossing-over turned out to be more advantageous evolutionarily since it ensures a much higher frequency of genetic recombination inside the linkage groups. Indeed, in spite of the large dimensions of chromosomes, more than two chiasmata per bivalent are rare in *Barbulanympha*, in which chiasmata are formed "according to SAX". The frequency of crossing-over in it must be even less since not every chiasma, in this case, leads to an exchange of chromatid parts. There may be as many as ten chromatid exchanges per bivalent in higher organisms in which chiasmata are formed "according to DARLINGTON". "Early" crossing-over was apparently retained in the evolution of all higher organisms, while "late" crossing-over could have been preserved only in some individual blind-end branches of the evolution of lower eukaryotes, including some protozoans.

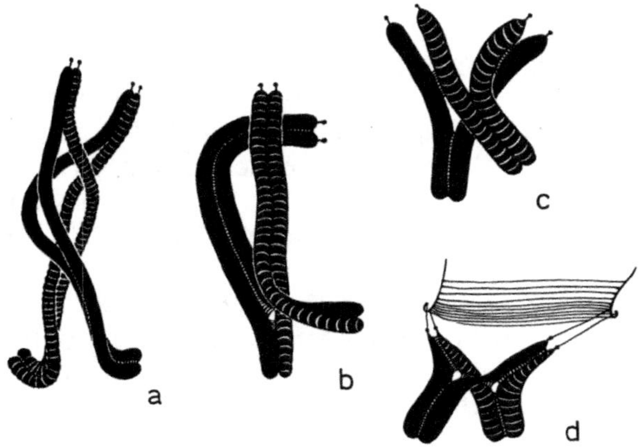

Fig. 62. "Somatic bivalents" in the diplonts of *Holomastigotoides* (after CLEVELAND 1953b). *a–c* prophase four-chromatid bivalents with a chiasma (pairing of sister chromatids is replaced by the pairing of nonsister chromatids); *d* separation of a bivalent with a chiasma into two dyads in the anaphase (chromatid parts are likely to be recombined).

The answer to the question of the origin of "early" crossing-over is to be found in certain very general regularities in the recombination of genetic material, while CLEVELAND's previously mentioned data on the somatic pairing of spiralized chromatids in the agamic flagellate *Holomastigotoides* (CLEVELAND 1949a, 1953b, 1961b) could apparently shed light on the origin of "late" crossing-over. The fact is that, besides haploid races, *Holomastigotoides* has diploid races in which the "resting" nucleus has four separate chromatids of every pair of homologous chromosomes (there are eight chromatids in all, since n = 2). In such races, all four chromatids combine in a group that is similar to the four-chromatidal meiotic bivalent; even chiasmata whose type is the same as in *Barbulanympha* can be formed (Fig. 62*a–c*). In mitosis, such bivalents are divided into two "dyads". Since all four chromatids have separate centromeres, two variants of the distribution of chromatids between dyads exist. In some cases, the dyads are pairs of homologous chromatids, while the sister chromatids separate just as in mitosis. In others, the dyads consist of pairs of sister chromatids, while the homologues are segregated as in the first meiotic division. Moreover, CLEVELAND even observed a rupture of two chromatids and the exchange of their parts, i.e. somatic crossing-over, in "somatic bivalents" with a chiasma (Fig. 62*d*).

Thus, mitosis in the diplonts of *Holomastigotoides* acquires many meiotic features: chromosome pairing, occasionally the segregation of homologues and crossing-over. The chromosome number, however, is not reduced in this case, since every nuclear division is accompanied by chromosome replication. Nevertheless, the homologous chromosomes in diploid races may genetically recombine with one another.

Hence, meiosis unquestionably has a polyphyletic origin in protozoans. It also underwent an evolution within their limits. This evolution involved especially the mechanisms of genetic recombination and consisted in the transformation of one-step meiosis into the more complicated two-step meiosis as well as in the retention of a more effective "early" crossing-over in the higher groups of protozoans.

6. Polygenomic Nuclei of Radiolaria

Many radiolarians have highly polyploid or *polygenomic* nuclei whose division involves the *segregation of genomes*. Such nuclei are characteristic of only some stages of the life cycle, i.e. their polyploidy is *cyclic*. Consequently, the life cycle has stages at which the nucleus undergoes polyploidization as well as other stages at which it undergoes depolyploidization.

Polygenomic nuclei of the Phaeodarea have been studied in the greatest detail. In this case, the sole[1] *primary nucleus* of the vegetative stage of the life cycle becomes polyploid. During the transition to sporogenesis, the primary nucleus breaks up into many *secondary nuclei* that multiply by mitoses and give rise to the nuclei of the flagellated zoospores. The zoospores apparently grow again to become vegetative cells. Besides Phaeodarea, the nuclear cycle has been studied in some Polycystinea (the order Spumellarida). In this case, the primary nucleus likewise exists, and it later breaks up into secondary ones. It is likely, however, that the primary nucleus of Spumellarida becomes polyploid just for a short time.

Only the nuclei of Phaeodarea and Spumellarida will be considered in this chapter since the karyology of Nassellarida and Acantharea has scarcely been studied.

Nuclei of Phaeodarea

The karyology of Phaeodarea has been studied best of all with respect to the Mediterranean species *Aulacantha scolymantha*. KARAWAIEW (1895) and BORGERT (1900, 1909) were among the first to study the nucleus of *Aulacantha*. This radiolarian reproduces not only by zoospores but also by binary fission, during which the primary nucleus divides in two. In this nucleus, BORGERT discovered more than 1000 rather large chromosomes that were subdivided by constrictions into segments. According to him, the chromosomes longitudinally split *twice*

---

[1] The central capsule of some Phaeodarea, however, always has two primary nuclei (RESHETNYAK 1966).

before each nuclear division. An equatorial plate is formed during this division, but the sister chromosomes, which are formed during the second splitting, get into the same daughter nucleus. In this respect, nuclear division in *Aulacantha* substantially differs from mitosis, although that is what BORGERT calls it. BĚLAŘ (1926a) tried to interpret BORGERT's observations differently, reducing the nuclear division in *Aulacantha* to real mitosis: he believed that BORGERT's description of the dual splitting of the chromosomes was a mistake. BĚLAŘ'S criticism was not very convincing, however.

According to BORGERT, the double cleavage of the chromosomes causes an increase in the number of chromosomes in the primary nucleus (to use modern terminology, at least one out of two chromosome duplications was endomitosis). He believed that amitosis was the mechanism regulating an increased number of chromosomes and described three varieties of "amitosis" in *Aulacantha*. As for sporogenesis, he held that the primary nucleus at first broke up into many (more than 1000) secondary nuclei, each of which received only one large segmented chromosome of the primary nucleus. This chromosome then fragmented into 10-12 small ones, and the secondary nuclei started to divide mitotically to form the nuclei of the flagellated zoospores (BORGERT called them "gametes")[2].

HARTMANN (1952) critically reviewed the data on the karyology of *Aulacantha*, concluding that the primary nucleus of this radiolarian was highly polyploid. Later, a number of researchers (GRELL 1953c, GRELL and RUTHMANN 1964, CACHON-ENJUMET 1961a, 1964, CACHON *et al.* 1973, LÉCHER 1978) carefully reexamined the karyology of *Aulacantha*.

The *nondividing primary nucleus* of *Aulacantha* is spherical, about 100 μm in diameter, and enclosed in the central capsule (Fig. 63*a*). The nucleus is very Feulgen-positive, and its chromatin structures resemble spongy masses oriented more or less radially. Nucleoli are absent[3]. Ultrastructural data show that the primary nucleus is surrounded by a two-membrane nuclear envelope with typical pores, while chromatin consists of densely packed fibrils about 8 nm thick and display no orderly packing (GRELL and RUTHMANN 1964, RUTHMANN and GRELL 1964).

The *stages of endomitosis* of the primary nucleus are found in *Aulacantha* together with the ordinary nondividing nuclei (in about 9% of the specimens;

[2] This refers to *isosporogenesis*, which leads to the formation of zoospores with a peculiar crystalloid in the cytoplasm. As for "anisosporogenesis", which has often been described in *Aulacantha* and other radiolarians, it has nothing to do with the life cycle of radiolarians and represents sporogenesis of the parasitic Dinoflagellida (*Syndinium*, *Merodinium*, etc.) that inhabit the cytoplasm and the nucleus of radiolarians (CHATTON 1920b, HOLLANDE *et al.* 1953, etc.).

[3] Other phaeodarians, however, have nucleoli (CACHON-ENJUMET 1961a, RESHETNYAK 1966); they may be either massive spheres (*Miracella*, *Tuscaridium*) or convoluted, occassionally branched strands (*Phaeodina*, *Gymnocella*).

Fig. 63. *Aulacantha* nucleus and its division (photomicrographs; after GRELL 1953c). *a* central capsule with a primary nucleus (*1*) *in vivo*, 650 × ; *b* stage of daughter plates *in vivo*, 500 × ; *c* young daughter nucleus, a Feulgen-stained section, 1200 ×.

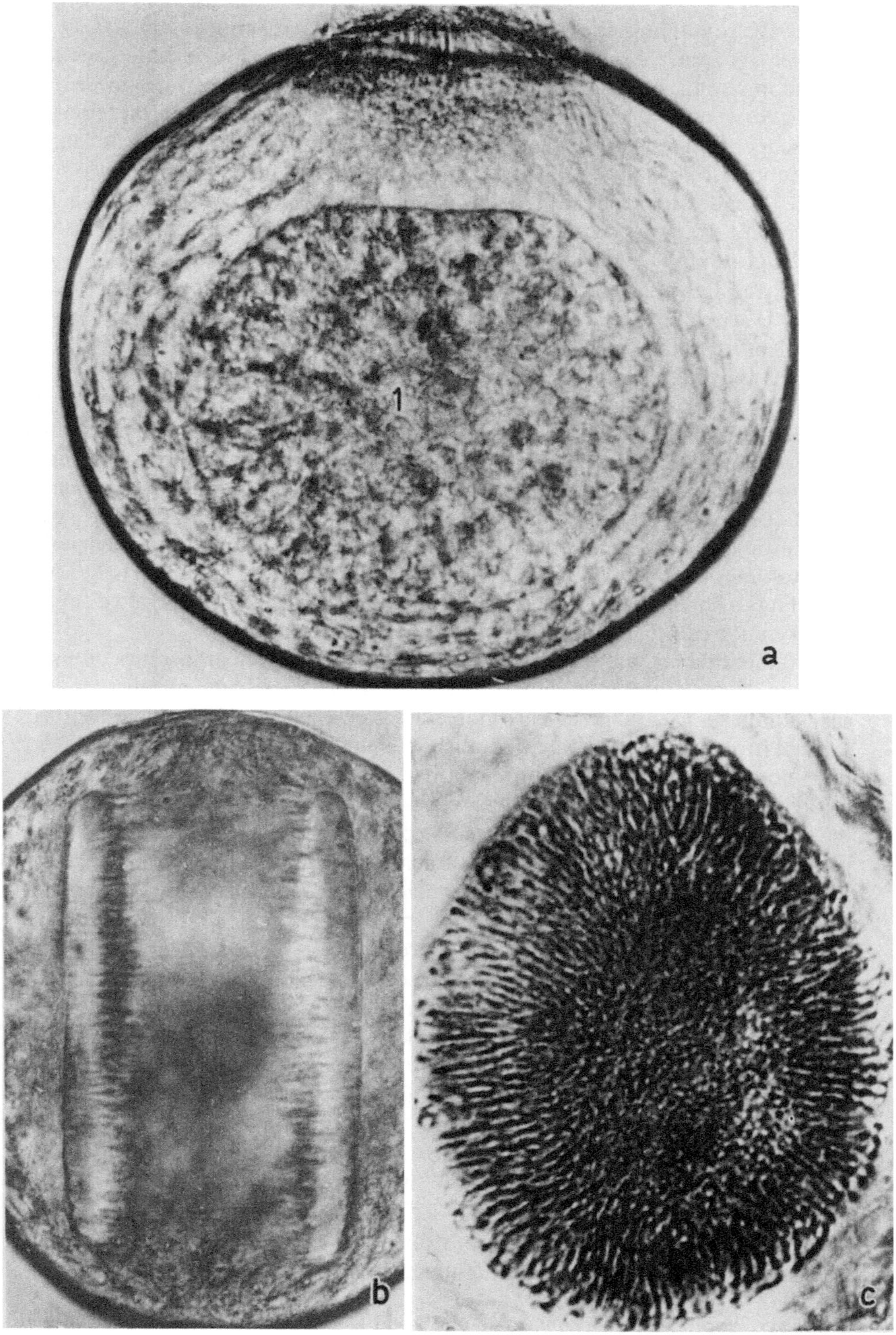

Fig. 63

GRELL 1953c). The nucleus swells at these stages, reaching 200–250 μm in diameter. At first, threadlike chromosomes appear in it (endoprophase stage). In the endometaphase, the chromosomes greatly shorten and become divided by constrictions into a number of segments (GRELL 1953c). Using the electron microscope, GRELL and RUTHMANN (1964) discovered that every segment contained a tubular *axial complex* about 150 nm in diameter, whose axis is formed by a central thread with thin radial fibrils. Such a complex exists not only in the chromosome segments but also in the bridges between them. Every segment soon splits longitudinally along the axial complex into two chromatids.

When the light microscope is used, many double chains made of rod-shaped segments, which are interconnected by very thin fibres (Fig. 64*a–d*), can be seen in the nucleus after the segmented chromosomes split longitudinally (GRELL 1953c). Segments may differ from one another in length and stainability. According to GRELL, segments are the true chromosomes of *Aulacantha*, while chains are whole genomes or composite chromosomes (Sammelchromosomen) in which individual chromosomes are connected end-to-end. This is consistent with BORGERT's data (1900, 1909), according to which secondary nuclei receive only one chromatin strand (composite chromosome) each, that later breaks up into 10–12 true chromosomes. It also follows from GRELL's hypothesis that all the composite chromosomes are homologous to one another[4] and that their number (1000–2700, according to various authors' estimates) is equal to the degree of polyploidy of the primary nucleus.

At the endoanaphase stage, the two sister composite chromosomes disunite, although they continue to be more or less parallel to one another. In the endotelophase, the composite chromosomes stretch out into threads, while their segmentation vanishes; they are obviously paired (Fig. 64*e*). According to GRELL, this is not the result of another duplication of chromosomes; it is simply due to the fact that the sister composite chromosomes remain parallel to one another.

The *division of the primary nucleus in two* precedes the division of the radiolarian itself. In this case, the whole primary nucleus flattens, forming a massive disc-shaped "equatorial plate", while the nuclear envelope disintegrates. Chromosomes do not lie on the plane of this plate, as in typical mitosis, but are perpendicular to it. The chromosomes are at first single, but later they split longitudinally into two chromatids. In the equatorial plate, however, the chromosomes do not condense as much as in endomitosis, and they are not visibly segmented. Later, the equatorial plate divides into two daughter plates that begin to move apart (Fig. 63*b*). Light microscopy does not show any spindle (BORGERT 1900, 1909, GRELL 1953c), but electron microscopy has revealed spindle microtubules in the interzonal region (GRELL and RUTHMANN 1964) and near the spindle poles. The latter are attached to chromosomes. Chromosomal kinetochores are described as being "diffuse" (LÉCHER 1972).

GRELL (1953c) has confirmed BORGERT's data according to which two sister chromatids, formed by chromosome splitting at the equatorial plate stage, are not segregated but get into the same daughter nucleus. Thus, chromosomes continue

[4] Hence, composite chromosomes should consist of the same number of segments. However, this has not been shown yet because of technical imperfections.

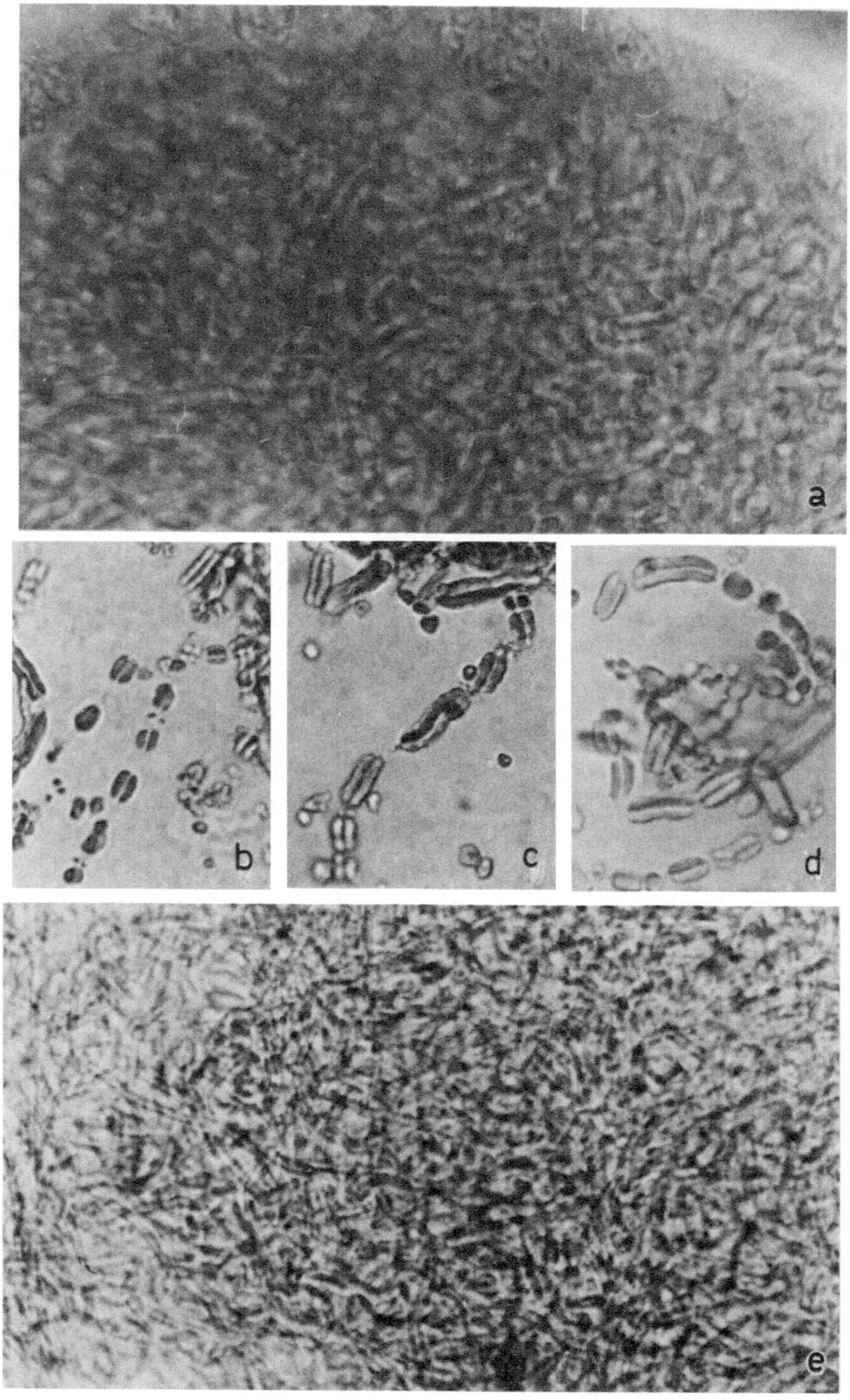

Fig. 64. Composite chromosomes and endomitosis in *Aulacantha* (photomicrographs, after GRELL 1953c). *a* endometaphase *in vivo*, 1700 × ; *b*–*d* endometaphase composite chromosomes, acetocarmine staining (*b* 1200 ×, *c* 1300 ×, *d* 1600 ×); *e* endotelophase, acetocarmine staining, 820 ×.

to be bichromatidal in the daughter plates. This is what clearly distinguishes the nuclear division in *Aulacantha* from mitosis, which is characterized by the regular segregation of chromatids. HARTMANN (1952) proposed to call this type of nuclear division *polyploid pseudomitosis*. However, GRELL (1953c) indicated that "chromosomes" corresponded in *Aulacantha* to whole genomes and called this type of nuclear division the *segregation of genomes*. Since all the genomes (composite chromosomes) are homologous, it does not matter whether two copies of every composite chromosome get into the same nucleus or into different nuclei during such division. It is only later in the daughter plates that chromatids disunite and form two independent composite chromosomes. This is apparently endomitosis, which is timed to the division of the nucleus. Thus, the nuclear cycle of *Aulacantha* is a combination of endomitosis and genome segregation. Moreover, endomitosis also occurs without any connection with nuclear division.

Daughter nuclei are reconstructed at the late stages of the division. The daughter plates of the chromosomes become cupped (with the polar sides inside), then form hollow spheres. Chromosomes are still seen in the young daughter nuclei. These chromosomes are arranged radially due to the bending of the daughter plates (Fig. 63*c*). The nuclear envelopes of the daughter nuclei are restored due to annulate lamellae, which exist in the cytoplasm of *Aulacantha* and adjoin the surface of the chromatin masses; later the lamellae fuse with one another (FEBVRE-CHEVALIER and LÉCHER 1971).

In her early works, CACHON-ENJUMET (1961a, 1961b) did not agree with GRELL's conception and tried to reduce nuclear division in *Aulacantha* to ordinary mitosis. She also doubted the existence of sporogenesis in phaeodarians. In 1964, however, she acknowledged the polyploidy of the primary nucleus of *Aulacantha* because she discovered sporogenesis in *Aulacantha* and *Coelodendron*, thus confirming BORGERT's data.

According to CACHON-ENJUMET (1964), sporogenesis begins with the appearance of "centrosomes" in the primary nucleus of *Aulacantha*; these are granules that lie in small outpocketings of the nuclear envelope. A bundle of thread-like chromosomes converge on every granule (Fig. 65*a*). Later the envelope of the primary nucleus disappears (at least, it cannot be seen with the light microscope), while every "centrosome" with a bundle of chromosomes initiates a small secondary nucleus. The chromosomes noticeably shorten. The wall of the central capsule is then resorbed, and the whole cytoplasm breaks up into several multinucleate spheres (Fig. 65*b*). Secondary nuclei (Fig. 65*c*) continue to undergo division in these spheres, forming at first multipolar (Fig. 65*d*) and then bipolar spindles (Fig. 65*e*). Chromosomes greatly condense after these divisions of secondary nuclei, and they number 8 (Fig. 65*f*). The cytoplasm then breaks up into uninucleate prespores that undergo division again and produce biflagellated zoospores with a peculiar crystalloid in the cytoplasm (Fig. 65*g*).

That process should apparently be regarded as the fragmentation of the polygenomic primary nucleus into separate genomes. The multipolar division figures of secondary nuclei also indicate the continuing depolyploidization of the nucleus, which, consequently, occurs in several stages. However, the existence of sporogenesis in phaeodarians is now being questioned again, because electron microscopy indicates that the secondary nuclei and the crystalloid-bearing zoospoores of *Aulacantha* and *Coelodendron* may in fact belong to an intracellular parasite, possibly a phycomycete (HOLLANDE 1981). As for BORGERT's three varieties of "amitosis", they are likely to be the pathologically changed forms of the binary division of the nucleus when chromosomes stick together (GRELL 1953c).

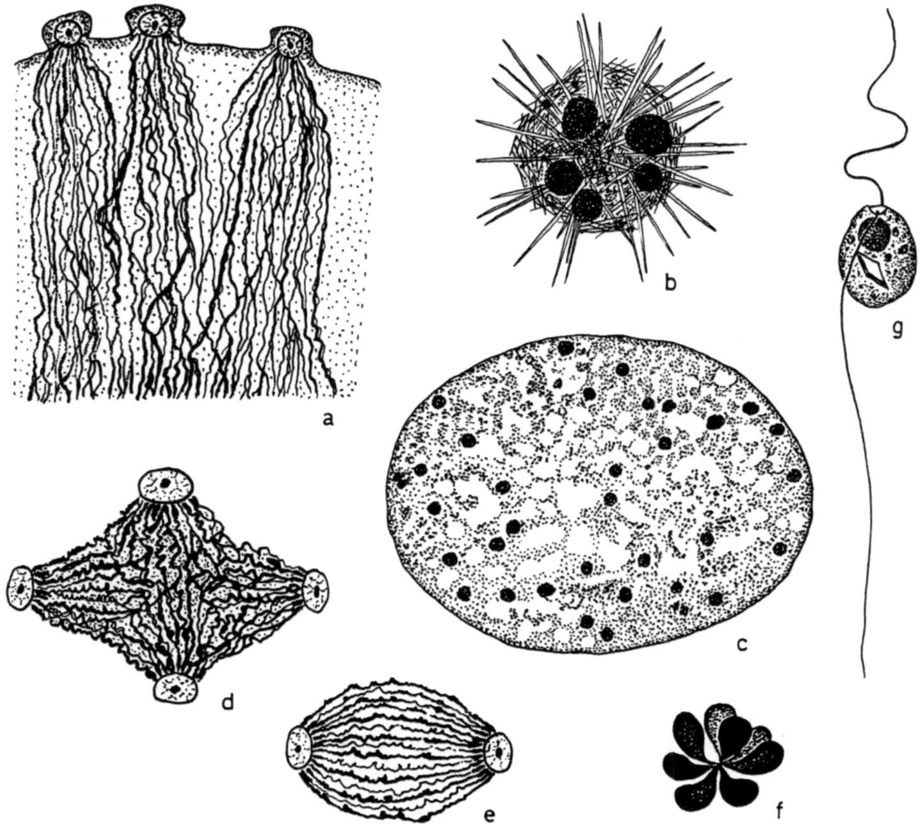

Fig. 65. Sporogenesis in *Aulacantha* (after CACHON-ENJUMET 1964). *a* polarization of chromosomes toward intranuclear "centrosomes"; *b* fragmentation of the cytoplasm of the radiolarian into multinucleate spheres; *c* multinucleate sphere with secondary nuclei; *d* quadripolar division of a secondary nucleus; *e* bipolar division of a secondary nucleus; *f* condensed chromosomes of a prespore; *g* zoospore.

It should be admitted, however, that the generally harmonious conception of GRELL's has one weak point: the interpretation of the "axial complex" in the chromosome as the site of its future cleavage into chromatids. Therefore, it was not clear why these chromatids were paired again in the endotelophase (Fig. 64*e*). This aspect was reexamined by CACHON and other colleagues (1973) and by LÉCHER (1978). They showed that the axial complex of a chromosome became the lateral element of a typical synaptonemal complex when it linked up with the axial complex of another chromosome. Consequently, the synaptonemal complex originates not when the chromosome splits into two chromatids, but when two initially independent chromosomes conjugate in a parallel manner. In this respect, CACHON and coworkers (1973) and LÉCHER (1978) proposed a new interpretation of the nuclear cycle of *Aulacantha*. According to them, the cycle GRELL regarded as endomitosis is that of the *somatic pairing of homologous chromosomes* in the primary polyploid nucleus. Axial complexes are the future lateral elements of the synaptonemal complexes. Consequently, single chromosomes in the "endoprophase" are those that have not paired yet, while double chromosomes in the "endotelophase" are somatic bivalents (Fig. 64*e*). The whole cycle does not double the chromosome number (in this case, it does not matter whether the chromosomes are composite or not). On the contrary, it prepares for the halving of this number, i.e. for *somatic reduction*, which can be compared with that in the oxymonadid *Pyrsonympha* (see Chapter 3, p. 147).

According to CACHON and LÉCHER, somatic reduction proper occurs during nuclear division in *Aulacantha* when homologous chromosomes disunite. They regard the paired state of the chromosomes in the daughter plates as the outcome of a new endomitotic replication of every homologue in the early anaphase (BORGERT's "second cleavage" of chromosomes). Thus, according to them, the nuclear cycle of *Aulacantha* is an alternation of endomitosis and somatic reduction, processes that balance one another in the nucleus of an adult trophozoite.

Obviously, further research is needed to interpret all the phases of the nuclear cycle of *Aulacantha*. However, the polygenomic nature of the primary nucleus can be considered proven regardless of the true sequence of chromosome behaviour.

The chromosomes of other phaeodarians have been scarcely studied. The primary nuclei of some forms are very large, e.g. they reach 650 μm in the Tuscaroridae (RESHETNYAK 1966). In the deep-sea species of *Planktonetta atlantica*, the nucleus is lenticular, 400 μm in diameter, while the nuclear envelope has tubular outgrowths that go into the cytoplasm (CACHON and CACHON-ENJUMET 1965b). In the nondividing nucleus, the chromosomes have many lateral loops on which RNP globules are formed, thus resembling lampbrush chromosomes. When the nucleus begins to divide, the RNP globules separate from the loops of the lampbrush chromosomes and dissolve in the karyolymph, while the loops are gradually drawn in.

Nuclei of Spumellarida

The order Spumellarida contains two suborders: Sphaerellarina (solitary forms with a skeleton that has the shape of reticular spheres) and Sphaerocollina (large solitary or colonial forms with a reduced skeleton). The karyology of Sphaerocollina has been studied in greater detail.

Sphaerocollina

It has long been known that the large primary nucleus of such forms as *Thalassicolla* and *Thalassophysa* can break up into many secondary nuclei

(HERTWIG 1879, BRANDT 1902, 1905, HUTH 1913, etc.). This allowed HARTMANN (1909) to conclude that the primary nucleus of sphaerocollids was *polyenergid*. The above works are now outdated, however. For instance, as is the case with phaeodarians, the stages of development of parasitic Dinoflagellida were regarded by BRANDT and HUTH as a form of sporogenesis (anisosporogenesis) of the radiolarians themselves (see the footnote on p. 210). Later, the karyology of Sphaerocollina was thoroughly studied by HOLLANDE and CACHON-ENJUMET

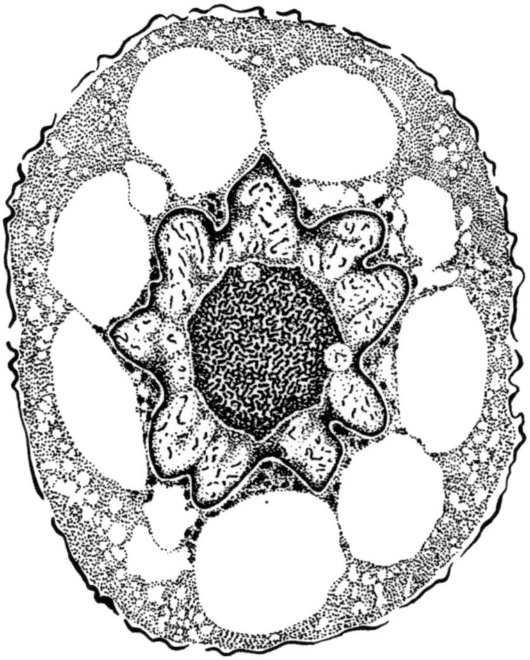

Fig. 66. Section through the central capsule of *Thalassophysa cladococcus* with a lobed primary nucleus, 150 × (after HOLLANDE and ENJUMET 1953).

(HOLLANDE and ENJUMET 1953, 1960, HOLLANDE and CACHON-ENJUMET 1959a, 1959b) with the light microscope, but less thoroughly with the electron microscope.

The sphaerocollid primary nucleus attains gigantic dimensions — 300 μm and occasionally even 1 mm in diameter. This nucleus is spherical in *Thalassicolla* (Fig. 68*a*), *Thalassoxanthium*, and *Bathysphaera*, while it is lobed with vast saccular pockets in *Thalassophysa* (Fig. 66). Many threadlike chromosomes can be seen with the light microscope in the primary nucleus, which always also contains nucleoli of diverse shape.

Unlike the phaeodarian nucleus, the sphaerocollid primary nucleus is virtually unstained by the Feulgen method. Therefore, HOLLANDE and ENJUMET (1953) at first assumed that it was not polyploid and was comparable with the "ovular" nuclei of some foraminiferans (cf. Fig. 11*a*). According to this viewpoint, the nucleus becomes polyploid only for a short time at the beginning of sporogenesis.

Later these authors changed their mind and acknowledged that the primary nucleus of Sphaerocollina was polyploid, since they discovered in it radially arranged bundles of parallel "chromonemata" that apparently resulted from endomitosis (HOLLANDE and CACHON-ENJUMET 1959a, HOLLANDE and ENJUMET 1960). However, these bundles only stained with iron haematoxylin. According to HOLLANDE and ENJUMET, there are two or three thousand such "chromonemata" in the primary nucleus but only about 50 chromosomes in the nucleus of the zoospore. If the latter number is haploid, the degree of polyploidy of the primary nucleus would be 40–60 n, with five or six endomitoses occurring during its development. But HOLLANDE (1972) has recently been tending to return to his initial viewpoint.

The morphology of the chromosomes of the primary nucleus seems to be species-specific. According to PÄTAU (1937a), the distal segments of some chromosomes in the primary nucleus of *Thalassophysa* and *Collosphaera* are heterochromatic and form two chromocentres. However, according to HOLLANDE and CACHON-ENJUMET, all the chromosomes in *Thalassophysa* (Fig. 66), *Thalassoxanthium*, and *Bathysphaera* are uniformly stained by iron haematoxylin. *Thalassicolla* (Fig. 68*a*) and *Thalassolampe*, however, have heterochromatic segments on the chromosomes that are linked with nucleoli, while the bulk of the chromosome threads is hardly discernible.

The nucleoli of the primary nucleus of Sphaerocollina are very peculiar (HOLLANDE and ENJUMET 1953, 1960, HOLLANDE and CACHON-ENJUMET 1959b). They are usually located at the periphery of the nucleus and are formed by the distal regions of the nucleolus-organizing chromosomes. These regions are always long (up to 600 μm in *Thalassophysa*), suggesting that they have many linearly repeating genes for rRNA. This has recently been confirmed in *Thalassicolla* by the spreading technique, which makes it possible to visualize the transcription of individual genes. The transcribing nucleolar genes for rRNA (monomers) proved to be 3.75 μm long and were arranged in tandemic arrays, while the transcribing non-nucleolar genes were of varying size (HOLLANDE *et al.* 1980).

The nucleoli themselves are either sausage-shaped with the nucleolus-organizing segment of a chromosome passing along them (*Thalassicolla*, *Thalassoxanthium spumidum*) or spherical with the appropriate chromosome segment wound into a skein in every nucleolus (*Thalassoxanthium sphaeronucleolatum*, *Bathysphaera pelagica*). The nucleoli of *Thalassophysa* are located in the saccular pockets of the primary nucleus (Fig. 66) whose centre contains chromosomal material. Nucleoli have the form of cords more than 0.5 mm long and up to 20 μm thick and folded in the pockets of the nuclear envelope. The nucleolar cord (Fig. 67) consists of both a spiral chromonema with a pyroninophilic "sheath" around it (the fibrillar zone of the nucleolus?) and the peripheral "matrix", which also contains RNA (granular zone?).

Unlike the nucleus of Phaeodarea, the primary nucleus of Spumellarida is apparently unable to divide in two. The only way in which these radiolarians reproduce themselves is by sporogenesis, which is accompanied by the fragmentation of the primary nucleus into many small secondary nuclei.

The *sporogenesis* of Sphaerocollina leads to the formation of biflagellate zoospores with a crystalloid in the cytoplasm. HOLLANDE and ENJUMET (1953) have established that sporogenesis is either direct (*Thalassicolla*, *Thalassoxanthium*) or indirect (*Thalassophysa*).

In direct sporogenesis, the primary nucleus (Fig. 68*a*) at first breaks up into rather large fragments (Fig. 68*b*) that continue to split up and initiate many secondary nuclei (Fig. 68*c*). The latter form zoospore nuclei after more mitotic divisions. This is when the cytoplasm of the radiolarian directly breaks up into zoospores. The stage of numerous secondary nuclei, which are in the common cytoplasm, is brief in this case.

In indirect sporogenesis, the stage of secondary nuclei is very long. The central capsule also divides into secondary central capsules, each of which has several secondary nuclei. The overall mass of the extracapsular cytoplasm greatly

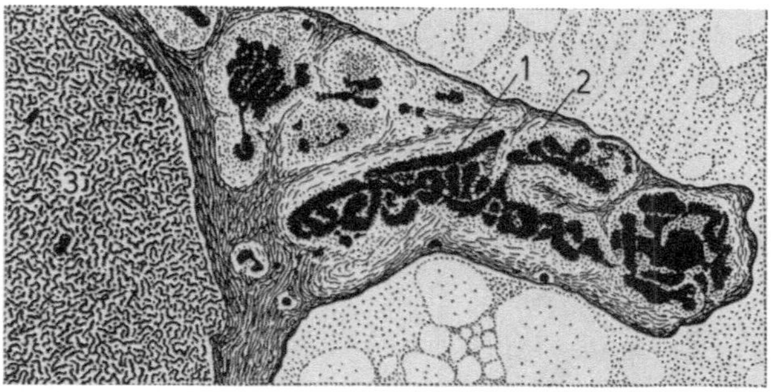

Fig. 67. Sacculiform outpocketing of the nucleus with a nucleolar cord within it in *Thalassophysa sanguinolenta* (redrawn from a photomicrograph, 785 × ; after HOLLANDE and CACHON-ENJUMET 1959a). *1* pyroninophilic sheath around the chromonema, *2* "matrix" of the nucleolus, *3* central part of the nucleus.

increases, producing temporary multinucleate and multicapsular colonies that occasionally grow to several centimetres (polycyttaric stages). It is they that were formerly included in a special suborder Polycyttarina (the genera *Collozoum*, *Sphaerozoum*, etc.). It could be that some "*Polycyttarina*" are really independent forms. However, "*Collozoum pelagicum*" is known to be the polycyttaric stage of *Thalassophysa sanguinolenta* (BRANDT 1902, HOLLANDE and ENJUMET 1953).

The structure of the primary nucleus changes before the beginning of sporogenesis. A region of a homogeneous karyoplasm, which HOLLANDE and ENJUMET call the "intranuclear centrosphere" (or archoplasmic sphere), appears in the centre of that nucleus. All the chromosomes of the primary nucleus radially converge on this sphere. This is when all the chromosomes are greatly despiralized, and they almost cease to be stained by haematoxylin. Much earlier, authors (HUTH 1913, etc.) erroneously interpreted this "homogeneous" stage as the extrusion of all "chromatin" in the form of "generative chromidia" from the primary nucleus into the cytoplasm. In reality, only nucleolar RNP is extruded into the cytoplasm, while all the chromosomal material remains in the nucleus.

Nucleoli break down before the beginning of sporogenesis, while the basophilia of the cytoplasm grows. Apparently, RNA is actually released from the primary

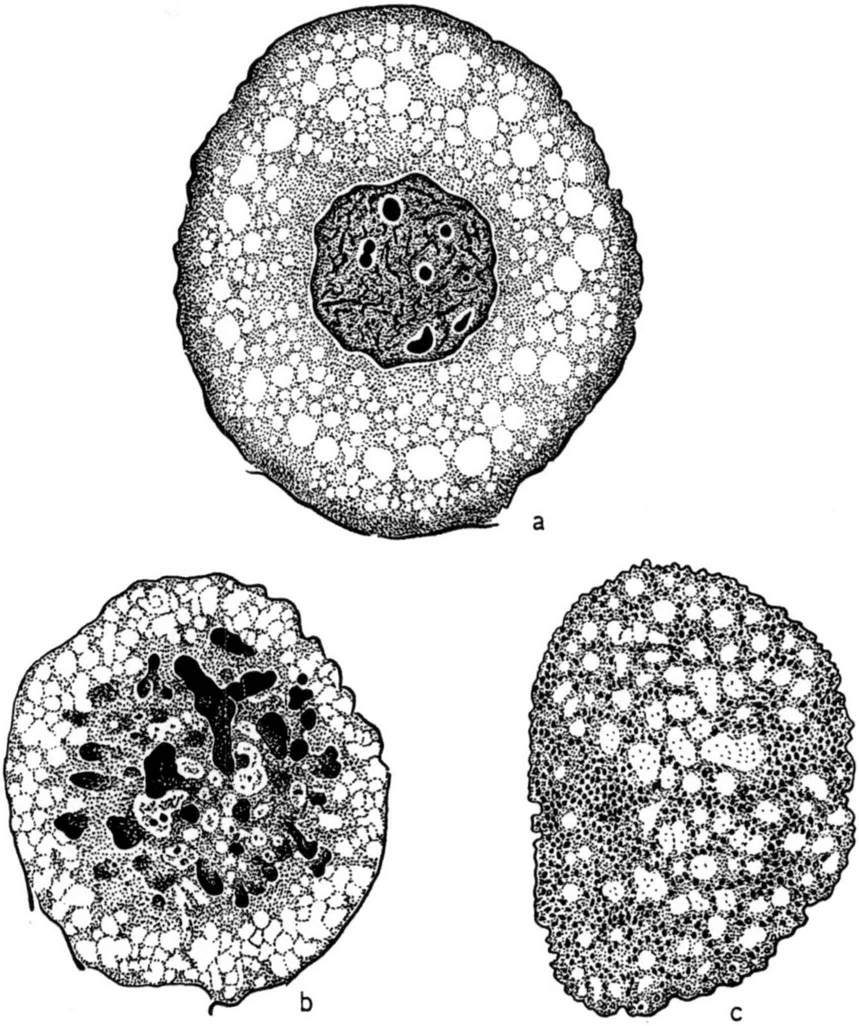

Fig. 68. Sporogenesis in *Thalassicolla spumida* (after HOLLANDE and ENJUMET 1953). *a* central capsule with a primary nucleus, 160 × ; *b* fragmentation of the primary nucleus, 85 × ; *c* central capsule with numerous secondary nuclei, 100 × .

nucleus into the cytoplasm (HOLLANDE and ENJUMET 1953, 1960, HOLLANDE and CACHON-ENJUMET 1959b). The primary nucleus of *Thalassophysa* undergoes the greatest changes. Before the beginning of sporogenesis, the nucleolus-forming chromosomes are shortened and drawn into the central chromatin zone of the nucleus where they are seen as radially arranged fibres that are thicker than the other chromosomes. Consequently, the peripheral nucleoli lose their axial chromonema, and only the nonchromosomal material remains in the pockets of the nucleus. The peripheral pockets then simply pinch off the nucleus and undergo lysis in the cytoplasm (Fig. 69). What is left of the primary nucleus is its central part, which contains chromosomes. All the chromosomes, including the nucleolar ones, are despiralized by then (homogeneous stage). In this case, a comparison can certainly be made with the extrusion of the nonchromosomal material during the first division of the large nonpolyploid "ovular" nuclei of the gamonts of gregarines and foraminiferans (Fig. 14).

The central "archoplasmic sphere" in the primary nucleus disappears later, while many small "secondary centrospheres" (intranuclear MTOCs on the inner

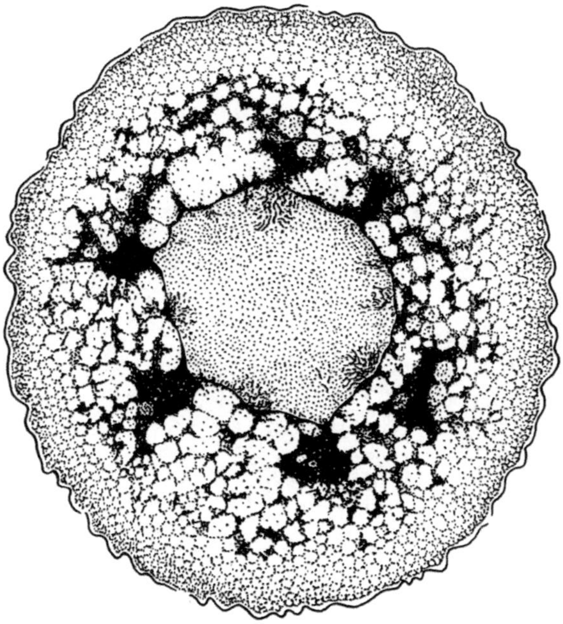

Fig. 69. Central capsule of *Thalassophysa sanguinolenta* at the beginning of sporogenesis with the nucleolar sacs of the primary nucleus being resorbed, 135 × (after Hollande and Enjumet 1953).

side of the nuclear envelope?) appear at its periphery. Rosettes of threadlike chromosomes converge on these "centrospheres". Their centromeres are apparently turned toward the "centrospheres". These chromosomal rosettes are the future secondary nuclei.

According to the hypothesis initially proposed by Hollande and Enjumet (1953), the primary nucleus becomes polyploid only at this stage due to the repeated "intranuclear mitoses" (duplications of both "centrospheres" and chromosome rosettes without the division of the nucleus itself). Later these researchers (1960) assumed that the primary nucleus was polyploid during the vegetative period, too. It has already been mentioned that Hollande (1972) tends to return to his initial hypothesis. It should be recalled, here, that such a type of momentary polyploidization of the nucleus prior to its fragmentation occurs during the gamogony of Haemosporina (see Chapter 2). In the latter, every set of chromosomes is permanently linked with a MTOC (which is built into the nuclear envelope) by means of a half-spindle of microtubules. In haemosporidians, genomes rapidly reproduce themselves in the nucleus as a result of the repeated duplication of MTOCs, half-spindles, and chromosomes. In this case, the integrity of the genomes is ensured by their permanent link with the MTOCs. A similar situation may exist in Sphaerocollina. At least, the morphological pattern of

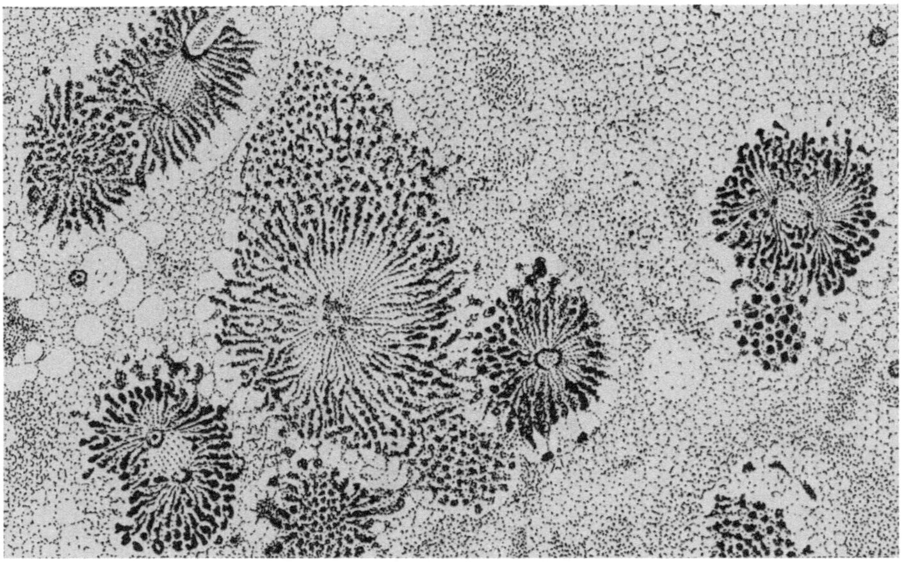

Fig. 70. Division of the secondary nuclei inside the central capsule of *Thalassophysa sanguinolenta* (redrawn from a photomicrograph, 800×; after HOLLANDE and CACHON-ENJUMET 1959a).

sporogenesis allows this. To prove or refute such a hypothesis, however, it is necessary to have quantitative data on the changes in the DNA content of the primary nucleus during its preparation for sporogenesis.

Later the primary nucleus breaks up into large parts (Fig. 68*b*), which continue to fragment further. During this time, every anlage of a secondary nucleus has its own "centrosphere" (Fig. 70). The chromosomes considerably spiralize and finally begin to be stained by the Feulgen method. There are about 40–50 chromosomes in each rosette. The anlagen of the secondary nuclei continue to reproduce by mitoses. These mitoses have been studied with the electron microscope in *Collozoum* (the polycyttaric stage of *Thalassophysa*) and *Sphaerozoum* and have been described earlier (Fig. 26). They belong to the closed type of intranuclear pleuromitosis with half-spindles and MTOCs in the form of plaques on the inner side of the intact nuclear envelope. The equatorial plate is absent (Fig. 70), as is usually the case in this type of mitosis, while the chromosome arms disunite very gradually.

In the last generations of the secondary nuclei, the chromosomes greatly condense and are brilliantly stained by the Feulgen reaction. They become the nuclei of biflagellated zoospores (there are still no grounds for regarding them as gametes). HOLLANDE (1974b) has studied the fine structure of the zoospore of *Sphaerozoum*. The chromosomes in its nucleus merge into a spongy mass. The remnant of the half-spindle, which has the shape of a cone of microtubules, can be seen at the front pole of the nucleus. The nucleolus is absent. It has also been established that the zoospore chromosomes are rich in basic proteins.

Regardless of whether the *vegetative* primary nucleus of Sphaerocollina is polyploid or not, the *presporogenic* primary nucleus of these radiolarians is certainly polyploid (polygenomic), while the fragmentation of the primary nucleus into secondary ones undoubtedly represents the segregation of genomes (depolyploidization). The permanent link between all the chromosomes of every genome and the intranuclear "centrosphere" (by analogy with later mitoses, this link is probably effected by half-spindles made of microtubules) apparently plays an extremely important role in the mechanism of such segregation. This mechanism does not resemble that of the formation of composite chromosomes in phaeodarians, but is functionally equivalent to the latter since it also ensures the integrity of genomes during the fragmentation of the primary nucleus and prevents the chromosomes from being randomly distributed and aneuploid secondary nuclei from being formed. However, it is still unknown how a transition is made from the single "centrosphere", which is in the centre of the presporogenic nucleus, to the numerous peripheral "centrospheres". If there is a series of duplications of both "centrospheres" and chromosomes (and DNA synthesis, accordingly) at this stage, this would indicate the nonpolyploidy of the vegetative primary nucleus of Sphaerocollina and its polyploidization only before the very beginning of sporogenesis. If, however, there is only the segregation of the existing chromosomes (without DNA synthesis), this would testify to the polyploidy of the primary nucleus in the vegetative phase of the cycle, too.

Sphaerellarina

The primary nuclei of Sphaerellarina are usually smaller than those of Sphaerocollina (50–70 μm in diameter) and have very diverse shapes (HOLLANDE and ENJUMET 1960). Just as in sphaerocollids, these nuclei are Feulgen-negative. This fact requires caution when the idea of their polyploidy is considered.

In the group Anaxoplastidia, the primary nucleus of the trophozoite is either spherical or lobed and has large pyroninophilic nucleoli (Fig. 71 *a*). The chromosomes are arranged radially in it, forming bundles that converge on the centre of the nucleus (the result of endomitoses?). The nucleoli usually consist of a pyroninophilic mass with a region of the nucleolar chromosome that is wound into a skein in it. Sometimes, nucleoli have the shape of long cords. The nuclear envelope of some species is very thick, forming peculiar papillae to which the bases of axopodia are attached (Fig. 71 *a*). It has been shown with the electron microscope that the envelope thickens due to the deposition of lamellar material in the perinuclear space (CACHON and CACHON 1972b).

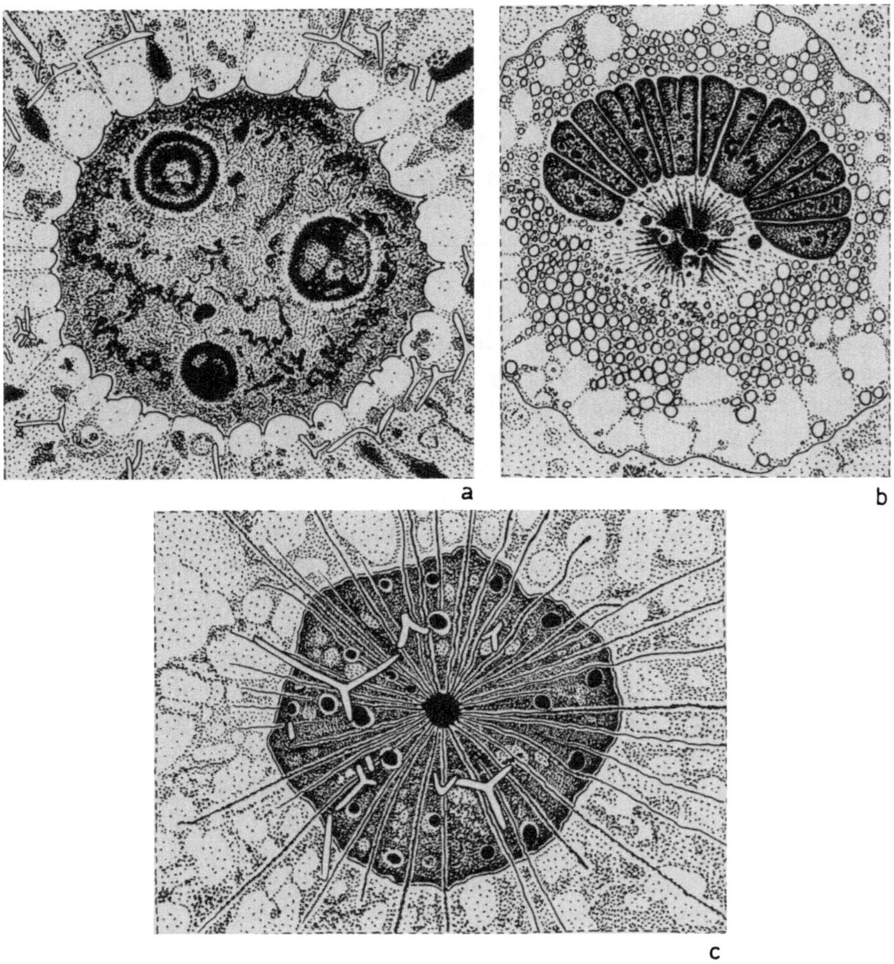

Fig. 71. Primary nuclei of Sphaerellarina (redrawn from photomicrographs; after Hollande and Enjumet 1960). *a Arachnosphaera* nucleus (3 nucleoli are visible), 1000 × ; *b* cupped nucleus of *Arachnostylus* (with an axoplast in the centre), 800 × ; *c Actinosphaera* nucleus (an axoplast with radial axonemes of the axopodia in the centre of the nucleus), 900 × .

In the group Periaxoplastidia, axopodial axonemes (bundles of cytoplasmic microtubules) converge on a special central body, the *axoplast*, that is adjacent to the primary nucleus. In this case, the nucleus frequently becomes cupped and embraces the axoplast from three sides. The axonemes of many axopodia then pierce the nucleus, passing through it in narrow cytoplasmic tunnels, e.g. in *Arachnostylus* (Fig. 71*b*). This fact has been confirmed by electron microscopy (Cachon and Cachon 1972b). The extent of "immersion" of the axoplast into the nucleus differs from one species to another.

Finally, in the group Centroaxoplastidia, the nucleus forms a hollow sphere and embraces the axoplast from all sides (Fig. 71*c*). The axonemes of all axopodia pass through the radial cytoplasmic tunnels that pierce the nucleus. Electron microscopy has shown that the axoplast consists of densely interwoven microfilaments. Bundles of such microfilaments also form the initial parts of the axonemes (which are in the tunnels of the nucleus). It is only at the level of the nuclear surface that the microfilaments are replaced by bundles of regularly packed microtubules (CACHON and CACHON 1972a). The nuclear envelope of the tunnels displays numerous pore complexes (about 100 per square μm).

Sporogenesis in Sphaerellarina has been studied only in its general features (HOLLANDE and ENJUMET 1960). Just as the sporogenesis of Sphaerocollina, it includes the lysis of nucleoli, the "homogeneous" stage, the reorientation of chromosomes with their centromeres toward the periphery of the nucleus, the fragmentation of the primary nucleus, and additional mitoses of secondary nuclei. Thus, the primary nucleus of Sphaerellarina should become polygenomic at least by the time it undergoes fragmentation. It is still unknown whether the primary nucleus is polyploid in trophozoites.

Thus, radiolarians apparently differ from one another in the organization of their primary nucleus. In Spumellarida, the primary nucleus is most probably not polyploid for a long time and resembles "ovular" nuclei (see Chapter 1). In this case, the amplification of the ribosomal gene could possibly be connected with the formation of the nucleoli (especially in such forms as Thalassophysa), but this has not been verified yet. Such nuclei probably become polygenomic only for a short time before the very beginning of sporogenesis, which involves the segregation of genomes (depolyploidization). In Phaeodarea, the primary nucleus is polyploidized early and remains polygenomic during the whole vegetative phase of the life cycle. It is also depolyploidized in sporogenesis. The group to which Nassellarida belong is still unknown. On the whole, the karyology of radiolarians offers a wide field for further research.

7. Nuclear Dualism in Protozoa

The phenomena of *nuclear differentiation* are widespread among Protozoa. They consist in the appearance of morphological and/or functional differences between originally identical, often sister nuclei. In the light of modern knowledge, such phenomena should be regarded as being based on the differential activation of certain genes.

GRELL (1964) distinguishes between *functional* and *structural* differentiations of nuclei in Protozoa. In functional differentiations, the differences between nuclei are morphologically indistinct but are expressed in different effects on the phenotype, e.g. during modificatory (or phenotypic) sex determination in some Hypermastigida (CLEVELAND 1949b, 1951a), Foraminiferida (GRELL 1958c, 1959, etc.), Coccidia (GRELL 1953a), and other protozoans. In structural differentiation, the differences between nuclei are expressed both functionally and morphologically. The structural differentiation of nuclei can further be divided into *sequential* (or *cyclic*) and *simultaneous* ones. The former means that genotypically identical nuclei (and sometimes even the same nucleus) can greatly differ at different stages of the life cycle (e.g. during the development of trophozoites from sporozoites in gregarines and coccidians).

The simultaneous structural differentiation of nuclei localized in a common cytoplasm is called *nuclear dualism*, and protozoans displaying this dualism are termed *heterokaryotic*. Hence, nuclear dualism can be defined as the simultaneous presence in the cell of two different nuclei of common origin (or of many nuclei belonging to two types). Generally, the nuclei of one type are metabolically more or less inert and are called *generative nuclei*, while those of the other type are very active and are called *somatic* (or vegetative) *nuclei*. Nuclear dualism in Protozoa is somewhat analogous to the differentiation of metazoan cell nuclei into those of the germ line (or generative) cells and those of specific tissue (or somatic) cells. In Protozoa, however, both types of nuclei are in a common cytoplasm, and somatic nuclei are not differentiated further into several types (consequently, there are no analogies to the tissue differentiation of the metazoan somatic cells).

Typical nuclear dualism is known to exist in two protozoan groups: in some Foraminiferida and in Ciliophora. In the former, nuclear differentiation is usually not accompanied by the polyploidization of the somatic nucleus, while in the latter, the somatic nucleus most often becomes polyploid and its genetic apparatus is greatly reorganized. Such complex somatic nuclei of the ciliates will be considered in Chapter 8. In some lower ciliates, however, nuclear dualism is still not complicated by the polyploidy of somatic nuclei; such forms are discussed in this chapter.

Nuclear Dualism in Foraminiferida

Morphology and Cytochemistry of Nuclei

As we have seen in Chapter 5, the life cycle of foraminiferans consists in the heterophasic alternation of a haploid sexual generation (gamonts) and of a diploid asexual generation (agamonts). Nuclear dualism has been observed so far in the agamonts of only some polythalamous foraminiferans (for reviews, see GRELL 1958c, 1964, 1973, 1979); it is unknown in monothalamous forms. It was discovered and thoroughly studied by GRELL and his students in the family Rotaliidae, namely, in *Rotaliella heterocaryotica*, *R. roscoffensis* (GRELL 1954, 1957, CZIHAK and GRELL 1960, CZIHAK 1964, ZECH 1964), *Metarotaliella parva* (GRELL 1962, WEBER 1965), *M. simplex* (GRELL 1979), *Rubratella intermedia* (GRELL 1958a), and *Glabratella sulcata* (GRELL 1958b). Later, nuclear dualism was found in the agamonts of the rotaliid *Rosalina* sp. (VORONOVA 1971), the anomaliid *Cibicides lobatulus* (VORONOVA 1976), the globigerinids *Globigerina bulloides* and *Globigerinoides ruber* (LEE *et al.* 1965), and the soritid *Sorites marginalis* (MÜLLER-MERZ and LEE 1976).

The sexual process in heterokaryotic foraminiferans is either autogamous (paedogamous) or gamontogamous (plastogamous). In *Rotaliella roscoffensis* (Fig. 72), which displays autogamy, the gamont is uninucleate (*1*); the number of gametes produced is small as compared with the gametogamous forms. The gametes are amoeboid and copulate pairwise with each other inside the gamont's shell (*2*, *3*). The nucleus of every zygote (*4*) divides twice, forming four genotypical-

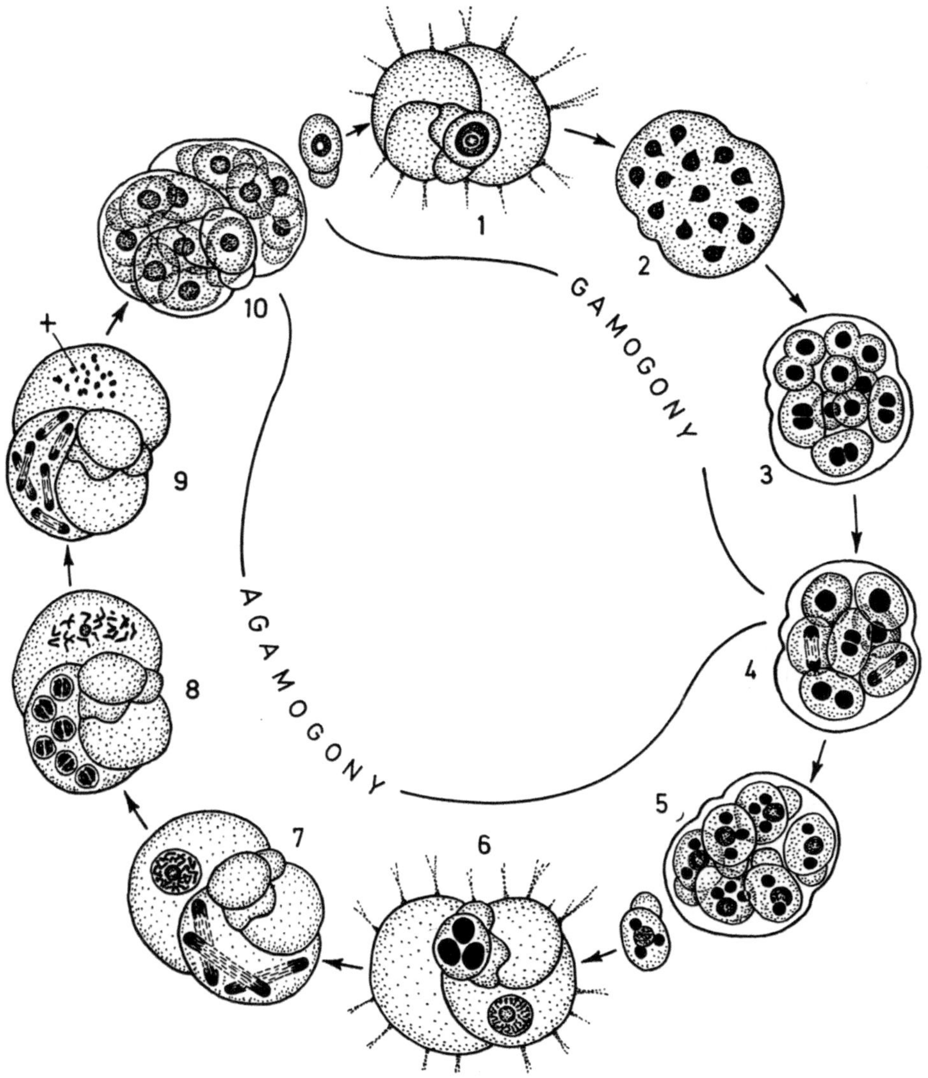

Fig. 72. Life cycle of *Rotaliella roscoffensis* (after GRELL 1957). *1* gamont; *2* formation of gamete nuclei; *3* paedogamous copulation of gametes; *4* zygotes and the first division of their nuclei; *5* young agamonts with one somatic and three (occasionally two) generative nuclei; *6* adult agamont; *7* first meiotic division of the three generative nuclei; *8* disintegration of the somatic nucleus; *9* second division of the generative nuclei, resorption of the chromosomes of the somatic nucleus (+); *10* formation of agametes.

ly identical nuclei of the young agamont. At first, these nuclei are similar morphologically, too, but soon they differentiate into two types: a single nucleus grows and forms a nucleolus, while three nuclei remain small and compact. Such agamonts leave the parent's empty shell (*5*). In growing agamonts, the large *somatic* nucleus moves into one of the younger chambers from the initial chamber, where three small *generative* nuclei remain (*6*). When the growth period is completed, the generative nuclei undergo meiosis (*7*). As a result of two meiotic nuclear divisions (*7–9*), three generative nuclei produce 12 haploid nuclei of agametes around which regions of the cytoplasm are separated (*10*). Amoeboid agametes grow into gamonts. The somatic nucleus of the agamont degenerates during meiosis (*7–9*). *Rotaliella heterocaryotica* has a similar life cycle (GRELL 1954).

In gamontogamous species, e.g. in *Metarotaliella parva* (WEBER 1965), the life cycle includes the pairing of two gamonts. Each of them produces amoeboid gametes, and only gametes coming from different gamonts copulate (unlike in autogamy). The zygotes are in a common cavity that was formed when the shells of two gamonts stuck together. The further development of agamonts (including nuclear differentiation and meiosis) and agamete formation are similar to those in *R. roscoffensis*. Gamontogamy is characteristic of most heterokaryotic foraminiferans, and it has been observed also in *Rubratella*, *Glabratella* (GRELL 1958a, 1958b) and *Rosalina* (VORONOVA 1971).

The *gamonts* of almost all foraminiferans, which are heterokaryotic in the agamont phase, remain uninucleate (Fig. 72, *1*). Only the gamonts of *Cibicides lobatulus* are known to display nuclear dualism (VORONOVA 1978b). The haploid nucleus of a young gamont divides in two in *Cibicides*, and the daughter nuclei immediately differentiate into a generative one and a somatic one. The former remains small, haploid, and gives rise to gamete nuclei after a series of mitoses. The latter grows greatly (to 40 μm in diameter) and develops peripheral nucleoli; its DNA content increases about 20-fold as a result of either polyploidy or more likely gene amplification. The somatic nucleus of the *Cibicides* gamont is thus similar to the type of "ovular" nuclei. During gamogony, the somatic nucleus of the gamont disintegrates.

The *agamonts* of heterokaryotic forms have from four to a few hundred nuclei, the number of generative nuclei usually being several times greater than that of somatic nuclei. Only the agamonts of *Metarotaliella simplex* frequently have a single generative nucleus and a single somatic nucleus (GRELL 1979). *Rotaliella heterocaryotica*, *R. roscoffensis*, and *Metarotaliella parva* normally have a single somatic and three generative nuclei (Fig. 72, *6*). In this case, nuclear differentiation occurs after the second division of the zygote nucleus. In *Rubratella intermedia*, it occurs after the third postzygotic mitosis, but at this stage there are six instead of eight nuclei since one derivative of the second mitosis degenerates (GRELL 1958a). Let us consider the nuclear dualism in *Rubratella* in greater detail (Fig. 73).

Soon after the formation of six diploid nuclei in a young agamont, one nucleus becomes larger, produces peripheral nucleoli, and migrates from the initial chamber into the next one (Fig. 73*a*). This is the somatic nucleus, and it resembles the unique nucleus of the gamont. The other five nuclei are generative; they remain dense, form no nucleoli, and lie in the initial chamber. This state of the nuclear apparatus is conserved throughout the agamont's growth period. Later on, the generative nuclei of the adult agamont undergo meiosis. Bivalents are formed in them during the meiotic prophase (Fig. 73*b*), and homologous chromosomes separate in anaphase I (Fig. 73*c*). As a result of the second meiotic division, 20 haploid nuclei of the future agametes (Fig. 73*d*) are formed from the five

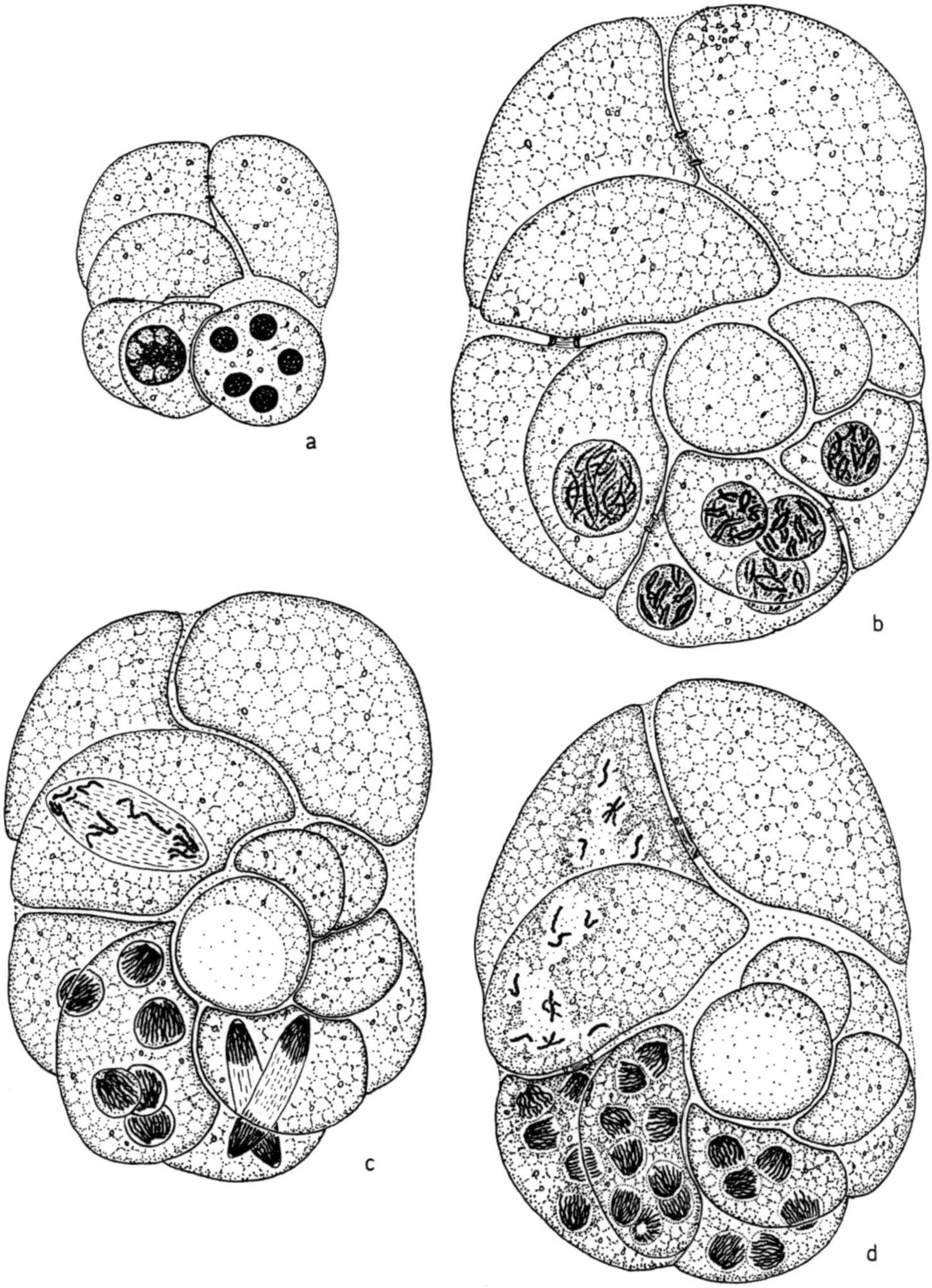

Fig. 73. Agamonts of *Rubratella intermedia*, 1050 × (after GRELL 1958a). *a* young agamont with one somatic and five generative nuclei; *b* adult agamont (generative nuclei contain meiotic bivalents, while the somatic nucleus contains univalents); *c* anaphase and telophase of the first meiotic division of generative nuclei, irregular distribution of the univalents along the spindle in the somatic nucleus; *d* generative nuclei in the telophase of the second meiotic division, the chromosomes of the somatic nucleus being resorbed in the cytoplasm.

generative nuclei. The somatic nucleus degenerates in *Rubratella* as in the other forms. An interesting fact, however, is that it makes an "attempt" to undergo meiosis; its chromosomes condense during the meiotic prophase, but homologues do not pair and the chromosomes remain "univalents" (Fig. 73*b*). During anaphase I of the generative nuclei, a spindle also appears in the somatic nucleus, but the univalents scatter throughout the nucleus and no anaphase groups are formed (Fig. 73*c*). During the second division of the generative nuclei, the envelope of the somatic nucleus disintegrates, the spindle disappears, and the univalent chromosomes are resorbed in the cytoplasm (Fig. 73*d*). The somatic differentiation of a nucleus thus seems to be coupled with the blocking of both the mechanism of chromosome conjugation and the mechanism of chromosome attachment to spindle fibres. A significant fact is that, in *Rotaliella*, synaptonemal complexes are formed only in generative nuclei and not in the somatic nucleus (Grell and Bardele 1977).

The mode of degeneration of somatic nuclei differs in various species of heterokaryotic foraminiferans (Grell 1956b). In *Rotaliella roscoffensis* and *Glabratella sulcata*, they behave as in *Rubratella*, i.e. they undergo an abortive asynaptic meiosis (Grell 1957, 1958b). In *Rotaliella heterocaryotica* and *Metarotaliella parva*, the somatic nucleus simply undergoes pycnosis (Grell 1954, Weber 1965). In *Cibicides lobatulus*, it fragments and then degenerates by pycnosis (Voronova 1976).

A single somatic nucleus is formed in *Rotaliella*, *Metarotaliella*, *Rubratella*, and *Cibicides* irrespective of the number of generative nuclei, which may be a few hundred in *Cibicides* (Voronova and Selivanova 1976). But in *Glabratella sulcata*, several (usually three) somatic nuclei differentiate. Initially, a single somatic nucleus appears in a young agamont, but later, a second and a third, and sometimes a fourth and a fifth, somatic nucleus is subsequently formed (Grell 1958b). The large rotaliid *Rosalina* sp. can have up to 10–15 somatic nuclei, the number of generative nuclei being 40–70 (Voronova 1971). The agamonts of *Globigerina* and *Globigerinoides* also seem to have several somatic nuclei (Lee *et al.* 1965). Finally, the huge *Sorites marginalis* produces dozens of somatic nuclei, the number of generative nuclei being several hundred (Müller-Merz and Lee 1976).

The somatic nucleus of *Rotaliella heterocaryotica* is not polyploid: it contains as much DNA as the generative diploid nuclei (Zech 1964). Single DNA replication takes place in both the somatic nucleus and generative nuclei as the agamont grows; this means, the two types of nuclei are in the G_1 period in young agamonts and in the G_2 period in adult ones. In generative nuclei, this DNA replication is premeiotic. The curves of the growth of the DNA content are parallel in both types of nuclei. The inability of the somatic nucleus to undergo meiosis thus seems to be due to causes other than the blocking of DNA replication.

The conclusion that the agamont's somatic nucleus is diploid, however, apparently cannot be applied to all heterokaryotic foraminiferans. The large agamont of *Cibicides* has a single big (up to 40 μm) somatic nucleus whose DNA content is 6–15 times higher than that of the diploid generative nucleus or 12–30 times higher than that of the haploid postmeiotic nucleus (Voronova 1976, Voronova and Selivanova 1976). The somatic nucleus may thus be polyploid to a degree of 12 to 30n. The DNA content of the somatic nucleus of *Cibicides* increases with the agamont's age: 11–12 n, 12–24 n, and 20–30 n, respectively, in three age groups (VORONOVA 1976). Such an increase in the DNA content, however, may also be entirely or partially connected with gene amplification. As we have seen, there is a similar problem with respect to the somatic nucleus of the gamont of the same species.

The RNA and protein contents of the two types of agamont nuclei have, as yet, only been investigated in *Rotaliella* (ZECH 1964). The protein content approximately doubles in generative nuclei as the agamont grows, while it increases five or six times in the somatic nucleus. The RNA content of generative nuclei remains virtually unchanged during the same period, while it sharply increases in the somatic nucleus. By the end of the growth period, the somatic nucleus contains approximately three times more protein and several times more RNA than the generative nucleus. This is in accord with the fact that no nucleoli are formed in the generative nuclei of *Rotaliella* (also *Rubratella* and *Glabratella*).

Cytochemical differences between generative and somatic nuclei, however, are not so clearly defined in *Metarotaliella parva*. Though its three generative nuclei remain compact and lack nucleoli at the beginning of the growth period of the agamont, they swell and form pyroninophilic nucleoli by the end of this period; then they differ from somatic nuclei only by their smaller size. However, the ultimate fate of the nuclei is the same: generative nuclei undergo meiosis and give rise to agamete nuclei, while the somatic nucleus perishes (WEBER 1965). It seems that transcription is activated in generative nuclei for a short time by the end of the growth period. Nucleoli have been observed also in the generative nuclei of the agamonts of *Rosalina* (VORONOVA 1971) and *Sorites* (MÜLLER-MERZ and LEE 1976).

Mechanisms of Nuclear Differentiation

The abundance of RNA and proteins in the somatic nucleus already indicates that the metabolic activity of this nucleus is far more intensive than that of generative nuclei. Moreover, direct evidence shows that this nucleus supports the growth of the agamont and all types of its physiological activity, while generative nuclei are inactive during this period. Thus, agamonts of *Rotaliella roscoffensis*, which lack generative nuclei as a result of their spontaneous degeneration, grow quite normally. Such agamonts, however, perish later because their somatic nucleus ultimately degenerates and the agamont becomes anucleate (GRELL 1957). These observations also show that the differentiation of the somatic nucleus is irreversible: it cannot undergo normal meiosis and give rise to agamete nuclei even when generative nuclei are absent. If the activation of certain genes is really the basis of the differentiation of somatic nuclei, this activation is coupled with the stable repression of some other genes that ensure normal meiosis. Conversely, some parts of the genome may be physically lost in somatic nuclei, making meiosis impossible.

These observations have been confirmed in other species, too. If the initial number of nuclei in a young agamont of *Glabratella sulcata* is abnormally small (e.g. two or four), they may all become somatic during the agamont's growth. In this case, the agamonts likewise normally grow and then perish. The same result is produced when all generative nuclei are experimentally destroyed, leaving only the somatic nucleus. This has been carried out in *Rotaliella heterocaryotica* by means of ultraviolet microbeam irradiation (CZIHAK 1964).

Unlike somatic nuclei, generative nuclei are not irreversibly differentiated. CZIHAK and GRELL (1960) destroyed the somatic nucleus of *R. heterocaryotica* with an ultraviolet microbeam, while ZECH (1964) also used its microsurgical excision in the same species. A new somatic nucleus was always formed (to replace the inactivated one) in both experiments; it developed as a result of the differentiation of one of the generative nuclei. This phenomenon can be induced twice in the same agamont. The "secondary" (or "tertiary") somatic nucleus later degenerates in exactly the same way as the primary nucleus. Such agamonts produce normal agametes but in reduced quantity since they have less generative nuclei left over. Somatic and generative nuclei thus cannot be determined during the divisions of the zygote nucleus (e.g. by differential mitosis). Every generative nucleus can become somatic. In other words, generative nuclei are omnipotent, while somatic nuclei are irreversibly differentiated.

If that is so, a mechanism that prevents the differentiation of all generative nuclei into somatic ones should exist. GRELL (1958b, 1958c) believes that the number of somatic nuclei in the cell can be regulated by a "feedback" mechanism. According to his hypothesis, the differentiation of somatic nuclei is induced by a factor that accumulates in the cytoplasm of the agamont. The generative nucleus that has the lowest threshold sensitivity to the factor differentiates first. As soon as it becomes somatic, it begins to inhibit the differentiation of other nuclei (e.g. by binding or inactivating the differentiating factor). However, the inhibitory action of the somatic nucleus is removed as a result of its destruction or excision, and then the differentiating factor can induce another generative nucleus to become somatic.

This hypothesis explains also the sequential formation of several somatic nuclei in *Glabratella* (GRELL 1958b). While the agamont is small, it forms one somatic nucleus. But adult agamonts of *Glabratella* are very large, and perhaps a single somatic nucleus is unable to inactivate or bind the differentiating factor in a large cell. Then, a second and later a third nucleus differentiates, etc. It would be very interesting to reveal the chemical nature of the differentiating factor and the molecular mechanism of its action.

Nuclear Dualism of Lower Ciliates

The second protozoan group with clear differentiation of diploid nuclei into generative nuclei (or *micronuclei*) and somatic nuclei (or *macronuclei*) consists of the lower ciliates that belong to the order Karyorelictida. In most of the other ciliates, macronuclei are secondarily modified as a result of their polyploidy and/or other processes involving the reorganization of their genetic apparatus (see Chapt. 8). In Karyorelictida, however, macronuclei remain diploid or nearly diploid and resemble the somatic nuclei of Foraminiferida.

Besides karyorelictids, there are probably lower ciliates without nuclear dualism, i.e. the *homokaryotic* ciliates. We now know of a single genus of homokaryotic ciliates, *Stephanopogon*, that comprises three species: *S. mesnili*, *S. colpoda*, and *S. apogon*. They are all small free-living marine forms. Because he

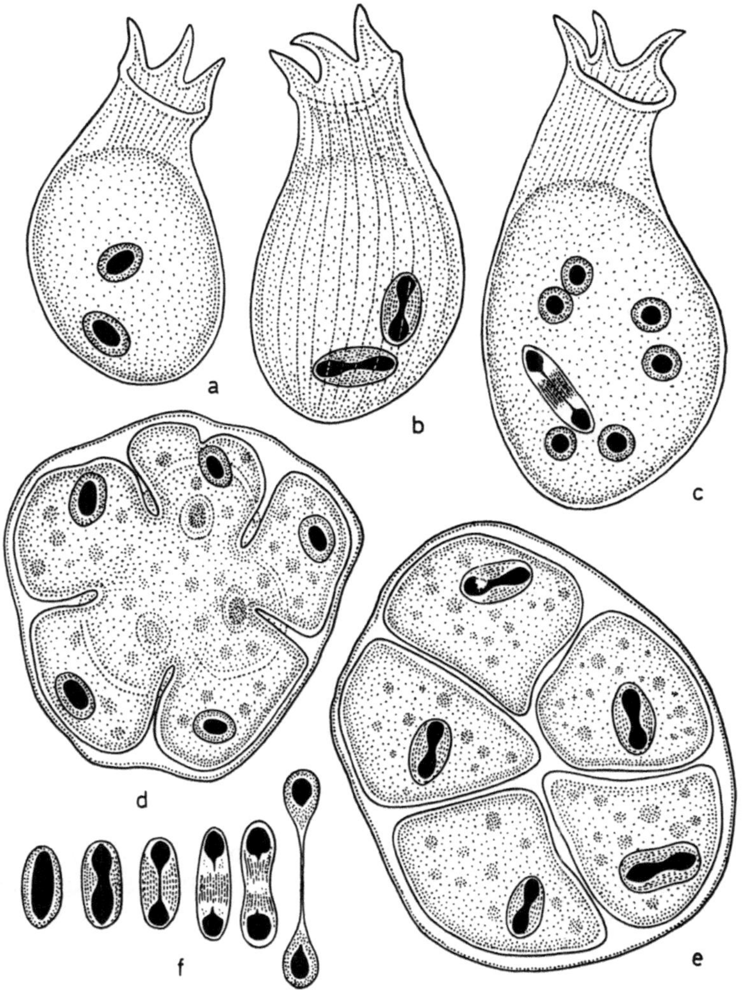

Fig. 74. Life cycle of *Stephanopogon colpoda* (haemalum staining, 1850 × ; after RAIKOV 1967a). *a* young binucleate trophont; *b* division of its nuclei; *c* multinucleate trophont with asynchronous mitosis of one of the nuclei; *d* plasmotomy inside a cyst; *e* mitosis of the nucleus of each tomite; *f* phases of mitosis.

considered the absence of nuclear dualism in *Stephanopogon* to be of paramount taxonomic importance, CORLISS (1975a, 1979a) created a separate, new order for this genus – Primociliatida.

LWOFF (1923) discovered that nuclear dualism did not exist in *S. mesnili*. Later, he described the nuclear phenomena accompanying the life cycle in greater detail (LWOFF 1936). The accuracy of LWOFF's observations was later confirmed with respect to *S. mesnili*, *S. colpoda* (DRAGESCO 1963, RAIKOV 1969), and *S. apogon* (BORROR 1965).

The life cycle of *S. colpoda* (Fig. 74) includes a stage of hypertrophic growth without cell division and a stage of multiple cell division inside a cyst. The smallest active cells (trophonts) have two identical nuclei (Fig. 74*a*). As the trophont grows, the number of nuclei increases as a result of mitoses (Fig. 74*b*), which are often asynchronous (Fig. 74*c*). Multinucleate trophonts may have up to 16 nuclei. After encystment, the trophont "dedifferentiates" and becomes a tomont, which simultaneously divides into a number of cytoplasmic portions, or tomites, corresponding to the number of nuclei present (Fig. 74*d*). But even before tomites excyst, the sole nucleus of each of them divides mitotically in two (Fig. 74*e*). The tomites then redifferentiate their buccal and ciliary structures and leave the cyst in a binucleate state (Fig. 74*a*). The number of tomites is usually about eight.

The nuclei of *Stephanopogon* are vesicular; they have a central pyroninophilic nucleolus and peripheral, faintly Feulgen-positive chromatin (Dragesco 1963, Borror 1965). Mitosis in *Stephanopogon* seems to be closed intranuclear orthomitosis with a dumbbell-like division of the nucleolus and an acentric spindle (Fig. 74*f*). It resembles the mitoses of schizopyrenid amoebae (*Vahlkampfia*, *Naegleria*, etc.).

Phylogenetically, the nuclear apparatus of *Stephanopogon* can be interpreted in various ways. Lwoff (1923, 1936), Raikov (1957b, 1969), and Corliss (1974, 1975a, 1975b, Corliss and Hartwig 1977) consider this organism a *primitive* relict ciliate that maintained the type of organization of the nuclear apparatus characteristic of the remote homokaryotic ancestors of ciliates. Another interpretation is that one may consider *Stephanopogon* a descendant of some heterokaryotic ciliates that *secondarily* lost nuclear dualism. The problem, however, has recently shifted to another side, i.e. whether *Stephanopogon* is really a ciliate. New data obtained with the electron microscope show that *Stephanopogon apogon* has many mastigophoran features in its ciliary (flagellar?) and mitochondrial structure and that its division is longitudinal rather than transverse (Corliss 1979b, Lipscomb 1980). If *Stephanopogon* turns out to be a flagellate, its primary homokaryotic condition will be unquestionable, but then this organism may be unrelated to the ancestors of ciliates and have nothing to do with the origin of their nuclear dualism. *Stephanopogon* should be studied further to clear this up.

Morphology and Fine Structure of the Nuclear Apparatus of Karyorelictida

More than 120 species of Karyorelictida are now known. They belong to the families Trachelocercidae (genera *Trachelocerca*, *Tracheloraphis*, and *Trachelonema*), Loxodidae (genera *Loxodes*, *Remanella*, and *Kentrophoros*), and Geleiidae (genera *Geleia* and *Avelia*), and to the taxonomically enigmatic genus *Ciliofaurea*. Almost all these ciliates belong to the interstitial fauna of marine sand (the psammon), except the freshwater genus *Loxodes*.

The investigation of the karyorelictid nuclear apparatus began with *Loxodes*. Much earlier, authors (Bütschli 1876, Balbiani 1890, Joseph 1907, Kasanzeff 1910, Rossolimo 1916) not only noticed the unusual morphology of the macronuclei of *Loxodes*, but also succeeded in correctly interpreting the nuclear reorganization during the division of these ciliates. The nuclei of some Trachelocercidae (it is now difficult to say which species) were studied by Lebedeff (1909), but he did not understand either the structure of nuclei or the sequence of nuclear phenomena during cell division. Then for almost 40 years, no new studies were made of karyorelictid nuclei. The nuclei of *Loxodes* were regarded as an isolated aberrant case, while the nuclei of Trachelocercidae were regarded as a "protistological enigma". Karyorelictid nuclei began to be studied again only in the 1950s. The similarity between the karyological features of Trachelocercidae and those of *Loxodes* was soon revealed, and it was discovered that the given type of nuclear apparatus was widespread among lower ciliates.

The light microscopic morphology and cytochemistry of nuclei have now been studied in detail in four species of *Tracheloraphis* (RAIKOV 1957a, 1958, DRAGESCO and RAIKOV 1966), one species of *Trachelonema* (KOVALEVA and RAIKOV 1970), three species of *Loxodes* (FAURÉ-FREMIET 1954, RAIKOV 1959a), three species of *Remanella* (FAURÉ-FREMIET 1954, RAIKOV 1963c, 1963d), two species of *Kentrophoros* (FAURÉ-FREMIET 1954, RAIKOV 1972b, 1972c), three species of *Geleia* (RAIKOV 1959b, 1963e, NOUZARÈDE 1976), and one species of *Avelia* (NOUZARÈDE 1976). There are less detailed data on the nuclei of many other karyorelictid species. Electron microscopic data on the structure of the macronuclei and micronuclei of the karyorelictid species that have been investigated thus far are summarized in Table 5.

Number and Arrangement of Nuclei. All karyorelictids have more than one macronucleus[1]. The minimum number of nuclei encountered among them is two Ma and one Mi, e.g. in *Loxodes rostrum* (Fig. 75*a*), *Remanella granulosa*, *R. rugosa*, and many Geleiidae (DRAGESCO 1960, and others). In this case, the nuclei always form a single group, the Mi being intercalated between the two Ma. Such a nuclear group has been studied by electron microscopy in *Geleia orbis* and *Avelia martinicense* (NOUZARÈDE 1975, 1976). The group is surrounded by a zone of dense vesiculate cytoplasm that seems to hold the nuclei together.

Two stable nuclear groups, each with one Ma and one Mi, are characteristic of *Loxodes striatus* (Fig. 75*b*). Almost every genus of Karyorelictida also comprises true multinucleate species with dozens of Ma and Mi, the number of nuclei usually varying greatly. In the genus *Loxodes*, that applies to *L. magnus* (Fig. 75*c*), which has up to 30 Ma, and the giant tropical species *L. rex*, which has up to 180 Ma (DRAGESCO 1970).

The nuclei of multinucleate karyorelictids can be arranged very differently. In *L. magnus*, many (but not all) nuclei form groups of two Ma and one Mi each (Fig. 75*c*); the nuclear apparatus of *Remanella multinucleata* and *Geleia murmanica* (RAIKOV 1963d, 1963e) is basically similar. Electron microscopy shows that Ma and Mi lie freely in the cytoplasm in both *L. magnus* and *R. multinucleata* (RAIKOV 1975a, 1979). Their union into groups is unstable: due to the vacuolization of the cytoplasm, several nuclei often occupy a common cytoplasmic "islet". Among Trachelocercidae, species with many freely scattered Ma and Mi also exist, e.g. *Tracheloraphis dogieli* (Fig. 85), *T. margaritatus* (RAIKOV 1957a, 1958, 1962b), and *T. crassus* (RAIKOV 1963b). Electron microscopy shows that the nuclei of *T. dogieli* lie freely in the highly vacuolized cytoplasm (Fig. 76*a*). In other species, nuclear groups exist, e.g. in *Trachelonema oligostriata*, which displays a longitudinal row of groups of two Ma and one Mi each (Fig. 77).

[1] The following abbreviations are used in the text: Ma, macronucleus; Mi, micronucleus.

Table 5. *Fine structural data on the nuclear apparatus of Karyorelictida**

Species	Number of Ma	Number of Mi	Arrangement of nuclei	Envelopes of nuclear complexes	Structure of Ma: Chromocentres	Structure of Ma: Nucleoli	Structure of Ma: Intranuclear inclusions	References
Tracheloraphis dogieli	10–30	5–15	free nuclei	absent	up to 15, large	10–15, medium sized	protein sphere	Raikov 1972a, 1974, Raikov and Kovaleva 1980
T. caudatus	28–200	14–100	groups of 4 Ma, 2 Mi	additional membrane	up to 4, small	1–5, medium sized	crystalloids	Raikov and Dragesco 1969
T. phoenicopterus	12	6	single complex	additional membrane	several, composite	1–3, medium sized	crystalloids	Raikov and Kovaleva 1978
T. totevi	16–22	2	single complex	additional membrane	several, composite	3–6, medium sized	crystalloids	Kovaleva *et al.* 1979
Trachelocerca geopetiti	4–8	2–4	single group	dense cytoplasm	up to 10, large	5–10, medium sized	protein sphere	Raikov and Kovaleva 1981

Trachelonema sulcata	9 60	3 18	unstable groups	absent	1 3, large, composite	up to 4, medium, or 1, large	sphere	Kovaleva and Raikov 1970, 1973
Loxodes magnus	8 31	5 32	unstable groups	absent	many small bodies	single, composite	absent	Mashansky 1963, Puytorac and Njiné 1970, Raikov 1975a
Remanella multinucleata	7 35	2 16	unstable groups	absent	several composite and many small bodies	up to 10, medium sized	sphere	Raikov 1979
Kentrophoros fistulosum	15 165	5 80	groups of 4 Ma, 2 Mi	absent	single, large	2 4, small, fibrillar	several spheres	Raikov 1972b
K. latum	6 42	2 14	1 7 complexes	multimembranous capsule	several, medium sized	1 4, medium sized	large sphere	Raikov 1972c
Geleia orbis	2	1	single group	dense cytoplasm	absent or small	many, small	large sphere	Nouzarède 1976, Raikov 1981
Avelia martinicense	2	1	single group	dense cytoplasm	absent or small	many, small	large sphere	Nouzarède 1975, 1976

* For details of the fine structure of micronuclei, see Table 1.

Stable groups of nuclei are characteristic of *Tracheloraphis caudatus* (Dragesco and Raikov 1966). In this case, nuclei form several flat "rosettes", each of which consists of four Ma and two Mi. A fine structural investigation of *T. caudatus* (Raikov and Dragesco 1969) revealed that a special cytoplasmic (additional) membrane surrounds each nuclear group outside the usual two-membraned envelopes of individual macronuclei. This membrane passes from one Ma to another at their junctions. It seems to be the proximal (in respect to nuclei) membrane of a special perinuclear lacuna, comparable to that in some foraminiferans (cf. p. 44).

A single rosette-like nuclear group is typical of Trachelocercidae that have only four Ma, e.g. *Trachelocerca coluber* (Fig. 89*a*). When there are 6–16 Ma (and 2–8 Mi), the nuclear group is usually also single, but it is dense and morula-like and not rosette-shaped. Macronuclei are then always peripheral, while micronuclei occupy the centre of the nuclear complex or cluster (Fig. 78*a*, *b*). A single morula-like nuclear cluster exists in *Tracheloraphis phoenicopterus*, *T. drachi*, and other forms; two such clusters exist in *T. dicaryon*, and 3–11 clusters exist in *T. kahli* (Raikov 1962b, 1963b).

The fusion of all the macronuclei, which make up the periphery of the morula-like nuclear complex, into a "composite nucleus" has been reported by means of light microscopy for some species. The impression was that micronuclei got inside the fused Ma. Such "composite nuclei" have been described, for instance, with respect to *Tracheloraphis phoenicopterus* (Raikov 1958, 1962b). The problem of "composite nuclei" has recently been investigated by electron microscopy in *T. phoenicopterus* (Raikov and Kovaleva 1978). No true fusion of Ma has been observed however (Fig. 79*a*). Individual macronuclei are closely adjacent to each other within the single nuclear cluster, which contains 12 Ma and 6 Mi. The envelopes of adjacent Ma remain intact, however, and there are systems of cytoplasmic membranes that intercalate between them (Fig. 79*a*). Similar data are available for *T. totevi* (Kovaleva *et al.* 1979). It is highly probable that no fused "composite nuclei" exist at all; they correspond in reality to very compact morula-like nuclear complexes, the boundaries of individual Ma being beyond the resolving power of the light microscope.

In the genus *Kentrophoros* (mouthless, ribbon-like ciliates carrying symbiotic sulphur bacteria on the dorsal, non-ciliated side), nuclear morphology is also very diverse. The "minimal" nuclear apparatus, consisting of 2 Ma and 1 Mi, exists in *K. fistulosum*. There are several groups of both Ma and Mi in some other species (Dragesco 1960). A row of free Ma and Mi is typical of *K. gracile* (Raikov 1963b). Finally, species with one (*K. uninucleatum*) or several (*K. latum*) "composite nuclei" were described within this genus (Raikov 1962b). The fusion of Ma

---

Fig. 75. Nuclear apparatus of the *Loxodes* species (photomicrographs). *a Loxodes rostrum*, 480 × ; *b L. striatus*, 480 × ; *c L. magnus*, 230 × (*a–c* haemalum staining); *d–f* cytochemistry of *L. magnus* nuclei, 1200 × ; *d* Feulgen reaction for DNA, *e* toluidine blue staining for DNA plus RNA, *f* mercuric bromophenol blue staining for protein; *1* Ma, *2* Mi, *3* macronuclear anlage. *a* and *f* after Raikov (1967a); *b* and *e* after Raikov (1969); *c* and *d* after Raikov and others (1963).

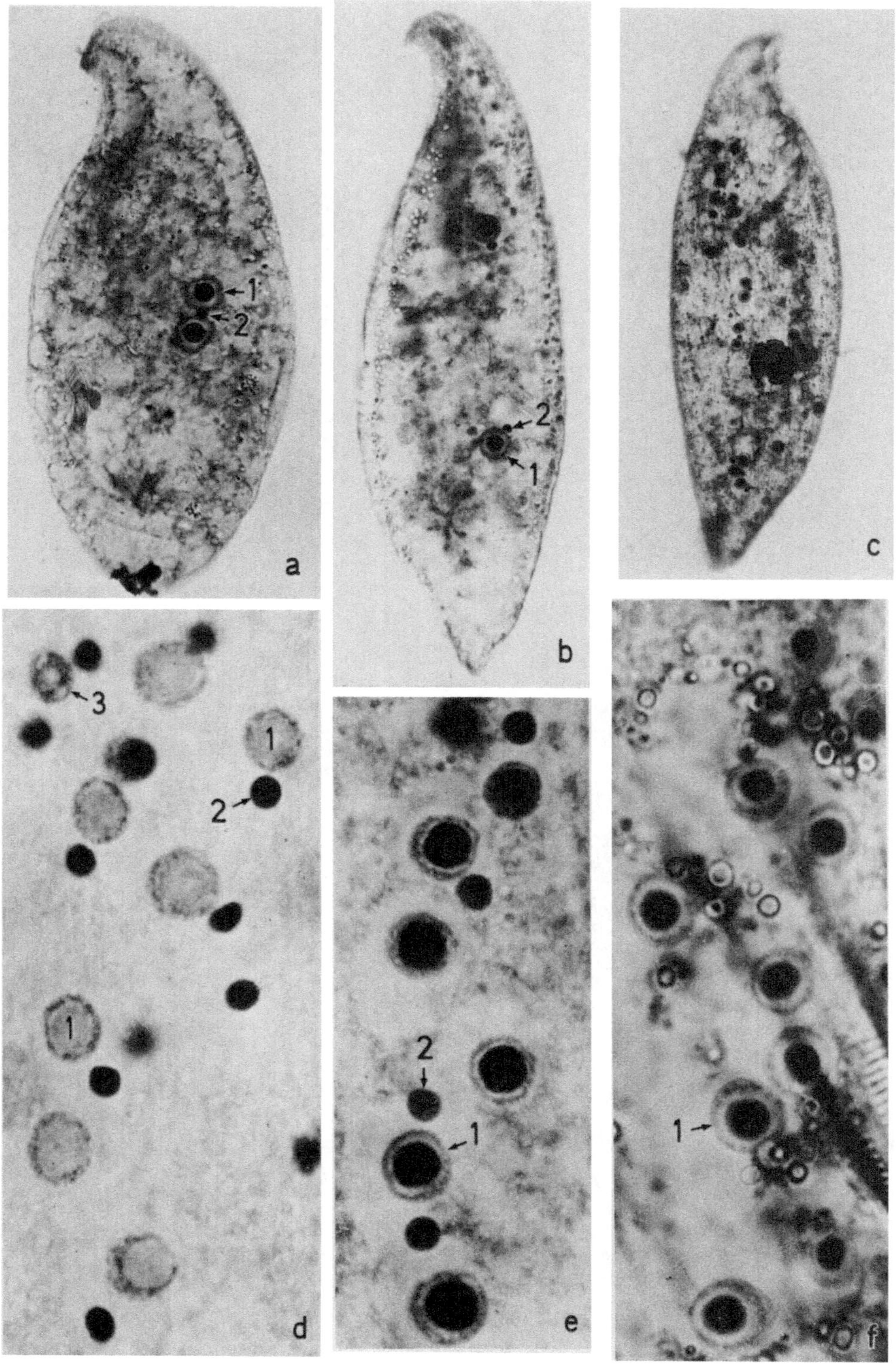

Fig. 75

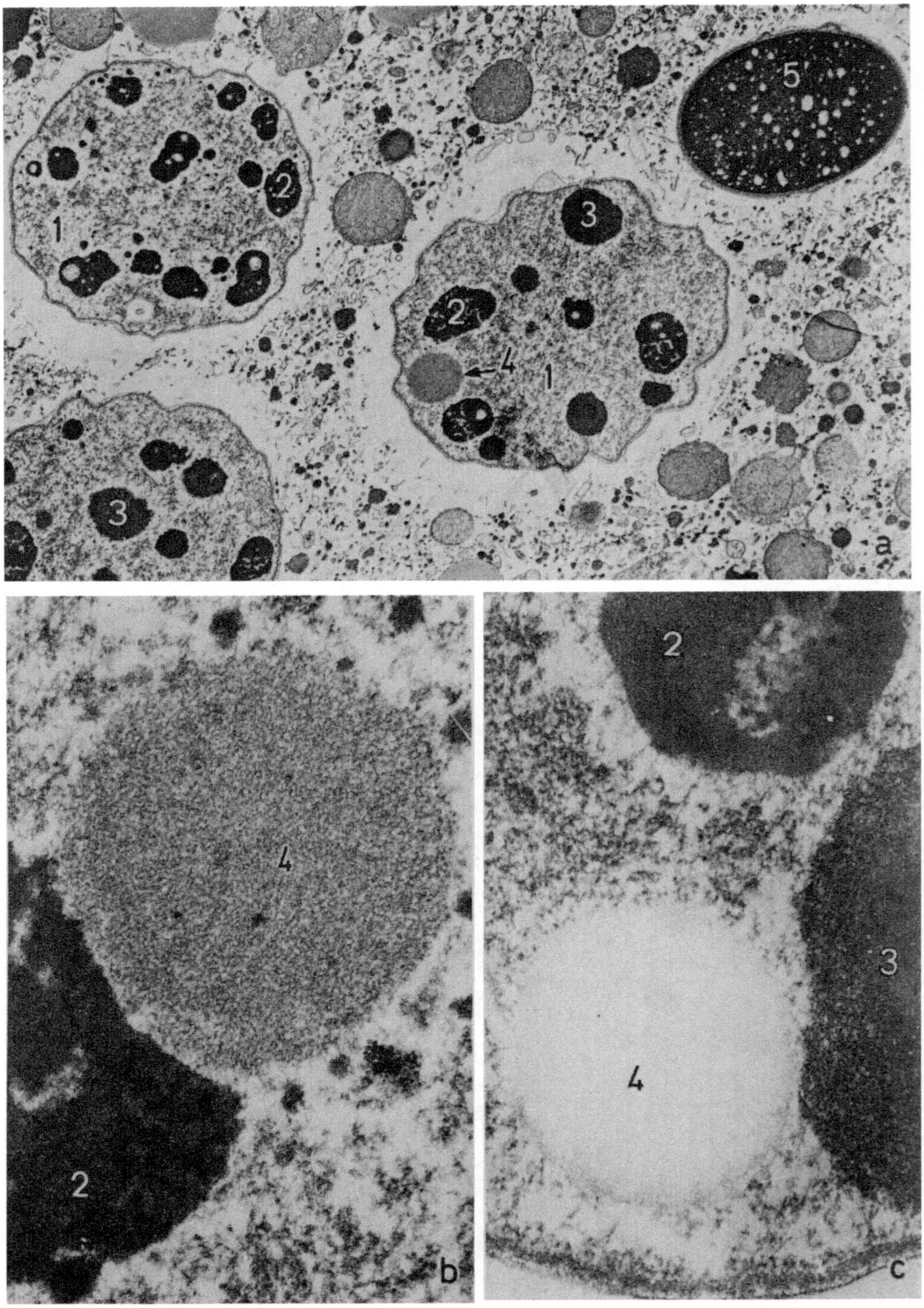

Fig. 76. Ultrastructure of the nuclei of *Tracheloraphis dogieli*. *a* 3 Ma and 1 Mi, 4500 ×; *b* sphere and chromocentre in a Ma, 40 000 ×; *c* pronase digestion of a sphere, the chromocentre and nucleolus remaining intact, 35 000 ×; *1* Ma, *2* its chromocentre, *3* its nucleolus, *4* its sphere, *5* Mi. *a* and *b* after RAIKOV (1974); *c* after RAIKOV and KOVALEVA (1980).

into "composite nuclei" in *K. latum*, however, proved to be an optical illusion, just as in *Tracheloraphis phoenicopterus*. It has been shown by electron microscopy that the macronuclei of *K. latum* do not fuse inside each nuclear complex, but they retain their independent envelopes (RAIKOV 1972c). Unlike *T. phoenicopterus*, the entire complex of nuclei in *K. latum* is additionally surrounded by a thick

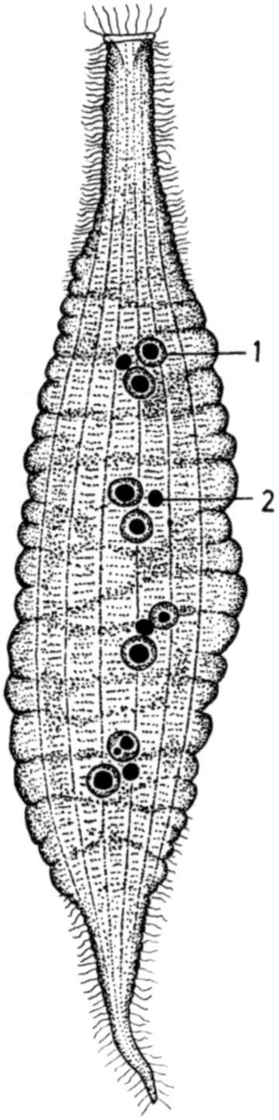

Fig. 77. *Trachelonema oligostriata*. Haemalum staining, 1100 × (after RAIKOV 1962b). *1* Ma, *2* Mi.

multimembranous *nuclear capsule* that consists of several concentric layers of flattened cisternae (Fig. 80*a*, *b*). The cisternae of the nuclear capsule lack pore complexes and thus differ from the nuclear envelopes proper (Fig. 80*b*). No ribosomes are attached to their membranes. The functional significance of the encapsulation of the nuclear complexes of *K. latum* is still unknown.

Structure and Cytochemistry of Micronuclei. The micronuclei of all Karyorelictida investigated until now have rather uniform morphology, differing primarily in size in various species. These nuclei are always small (seldom exceeding 3 μm in diameter), compact, and sharply Feulgen-positive (Figs. 75, 77, 78, 85). They generally resemble the Mi of higher ciliates (see Chapt. 1 and Table 1).

Micronuclear chromatin positively stains for DNA (Fig. 75*d*) and for both basic and total protein. No RNA was detected in the Mi of any species at the light microscopic level, but it was found in the persisting kinetochores of the Mi of *Trachelocerca geopetiti* (Fig. 10*f*) by ultrastructural cytochemistry (Raikov and Kovaleva 1981). The absence of RNA synthesis in karyorelictid Mi has been demonstrated autoradiographically in *Tracheloraphis* sp. (Torch 1964) and *Loxodes striatus* (Ron and Urieli 1977). As for the replication of DNA, it seems to

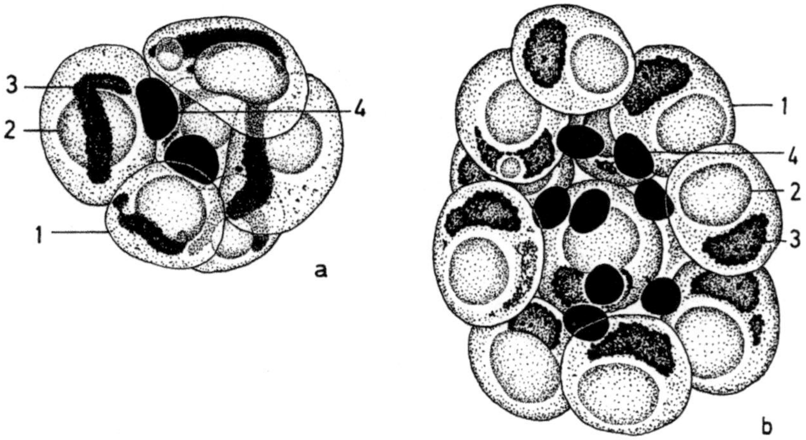

Fig. 78. Nuclear groups of *Tracheloraphis drachi*. Feulgen, light green staining. 2300 × (after Raikov 1962b). *a* group of 6 Ma and 2 Mi; *b* group with 8 Mi; *1* Ma, *2* its nucleolus, *3* its chromocentre, *4* Mi.

occur in *L. magnus* soon after mitosis (Raikov and Morat 1977, Bobyleva *et al.* 1980).

Some fine structural features of karyorelictid Mi have already been considered in Chapter 1 (see also Table 1). To sum up, the micronuclei of all species that have been investigated belong to the "spermal" type of nuclei. Their chromatin consists of densely packed fibrillar material that forms either a spongy mass (Figs. 76*a*, 80*a*) or a compact electron-dense body (Figs. 10*e*, 79*a*). Nucleoli are always absent, but the micronuclei of *Loxodes magnus* contain peculiar "spheres", that are apparently proteinaceous (Raikov 1975a). The micronuclear envelope always has a classical ultrastructure and many pores.

Fig. 79. Ultrastructure of the nuclear complex of *Tracheloraphis phoenicopterus* (after Raikov and Kovaleva 1978). *a* longitudinal section through the nuclear cluster, 7300 × ; *b* Ma with two crystalloids, 11 000 × ; *c* part of a crystalloid showing lengthwise packing of the component fibres, 114 000 × ; *1* Ma, *2* their nucleoli, *3* their chromatin bodies, *4* their envelopes, *5* Mi, *6* cytoplasmic membranes between the nuclei, *7* crystalloids.

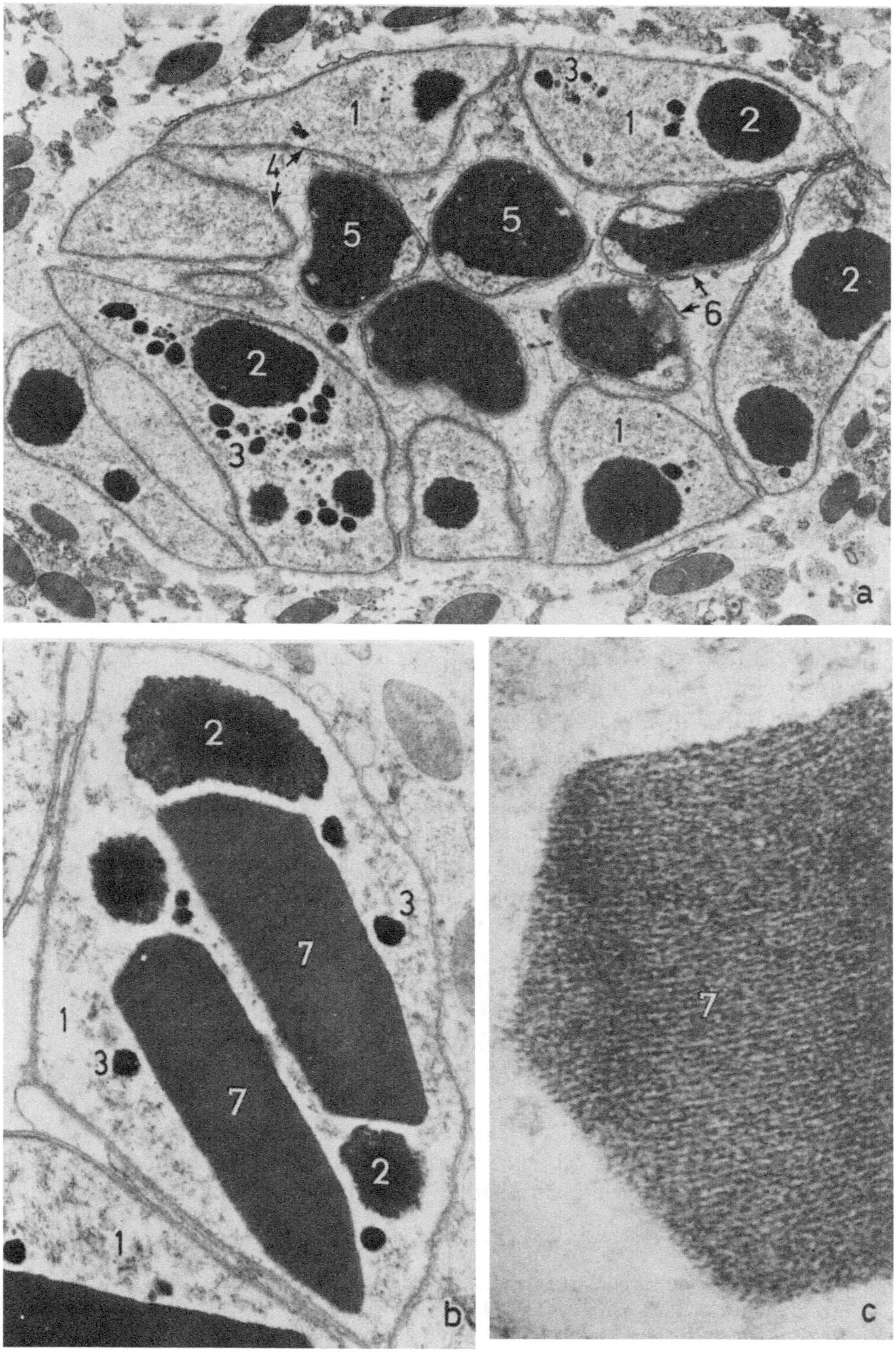
1
3
1
2
4
5
5
2
6
2
3
1
a
2
3
7
1
3
7
2
1
b
7
c

Structure of Macronuclei. The macronuclei of all Karyorelictida belong to the vesicular type of nuclei. They are comparatively small (5–10 μm, rarely more than 15 μm in diameter) and usually have a simple shape, i.e. they are spherical, ellipsoid, etc. Their internal structure, however, is more diverse and depends on varying degrees of chromatin condensation, nucleolar development, the presence of intramacronuclear protein inclusions, and on their type (Table 5). Nevertheless, all karyorelictid macronuclei have an important feature in common: they are comparatively poor in DNA and thus differ clearly from the "polyploid" Ma of other ciliates (see Chapter 8).

The *envelope* of karyorelictid Ma is typically two-membraned and carries pore complexes. In most species, nuclear pores lack typical annuli, but they usually have central granules. Annuli are prominent, however, in *Loxodes magnus* (RAIKOV 1975a) and *Geleia orbis* (NOUZARÈDE 1976). The number of pores per square micrometre of the macronuclear surface is about 24 in *Tracheloraphis dogieli* and *Kentrophoros latum*, 40 in *T. caudatus*, 60 in *Loxodes magnus*, 70 in *Geleia orbis*, and 85 in *T. phoenicopterus* (for references, see Table 5). In the species in which Ma are united into groups or complexes (*T. caudatus*, *T. phoenicopterus*, *K. latum*), pores are present on the outer surfaces of these nuclei as well as on the surfaces turned inside the complex (Fig. 80*b*). The union of Ma into complexes, even if they are encapsulated (as in *K. latum*), seems to have no effect on either pore frequency or morphology.

The *chromatin structures* of karyorelictid Ma usually appear to be small Feulgen-positive granules (Fig. 75*d*) and/or larger *chromocentres*. Large chromocentres are often single (Fig. 78*b*) and are sometimes sausage-shaped (Fig. 78*a*). A very large micronucleus-like spherical chromocentre is typical of every Ma in *Kentrophoros fistulosum* (RAIKOV 1972b). This chromocentre can even be called a *karyosome*. But in the *Geleia* species, chromatin is finely dispersed in the peripheral zone of each Ma, which is faintly and diffusely stained by the Feulgen reaction (RAIKOV 1959b).

The cytochemical peculiarities of the macronuclei of *Loxodes magnus* (RAIKOV 1959a) may be taken as an example. These macronuclei are rather small (about 7 μm in diameter) spherical nuclei with a large central nucleolus. Only small granules at the macronuclear periphery stain for DNA (Fig. 75*d*) as well as basic protein and thus consist of DNP. The macronuclear periphery also has some basophilic material that contains RNA (Fig. 75*e*). DNP granules and RNA-containing material produce positive reactions for total protein (Fig. 75*f*). The nucleolus is Feulgen-negative, but it is rich in RNA and non-histone proteins (Fig. 75*e*,*f*).

The fine structure of the Ma of *Loxodes*, first described by MASHANSKY (1963), proved to be rather typical of vesicular nuclei (Fig. 81*a*). Chromatin elements are represented by a moderate number of *small bodies* that are electron-dense,

---

Fig. 80. Nuclear complex of *Kentrophoros latum* (after RAIKOV 1972c). *a* survey micrograph, 6500 × ; *b* detail of the nuclear capsule, 44 000 × ; *1* Ma, *2* its nucleolus, *3* its chromocentre (patches of fibrils at the arrowheads), *4* its sphere, *5* Mi, *6* nuclear capsule, *7* macronuclear envelope (pores indicated by arrows), *8* membrane systems between the nuclei, *9* cytoplasm.

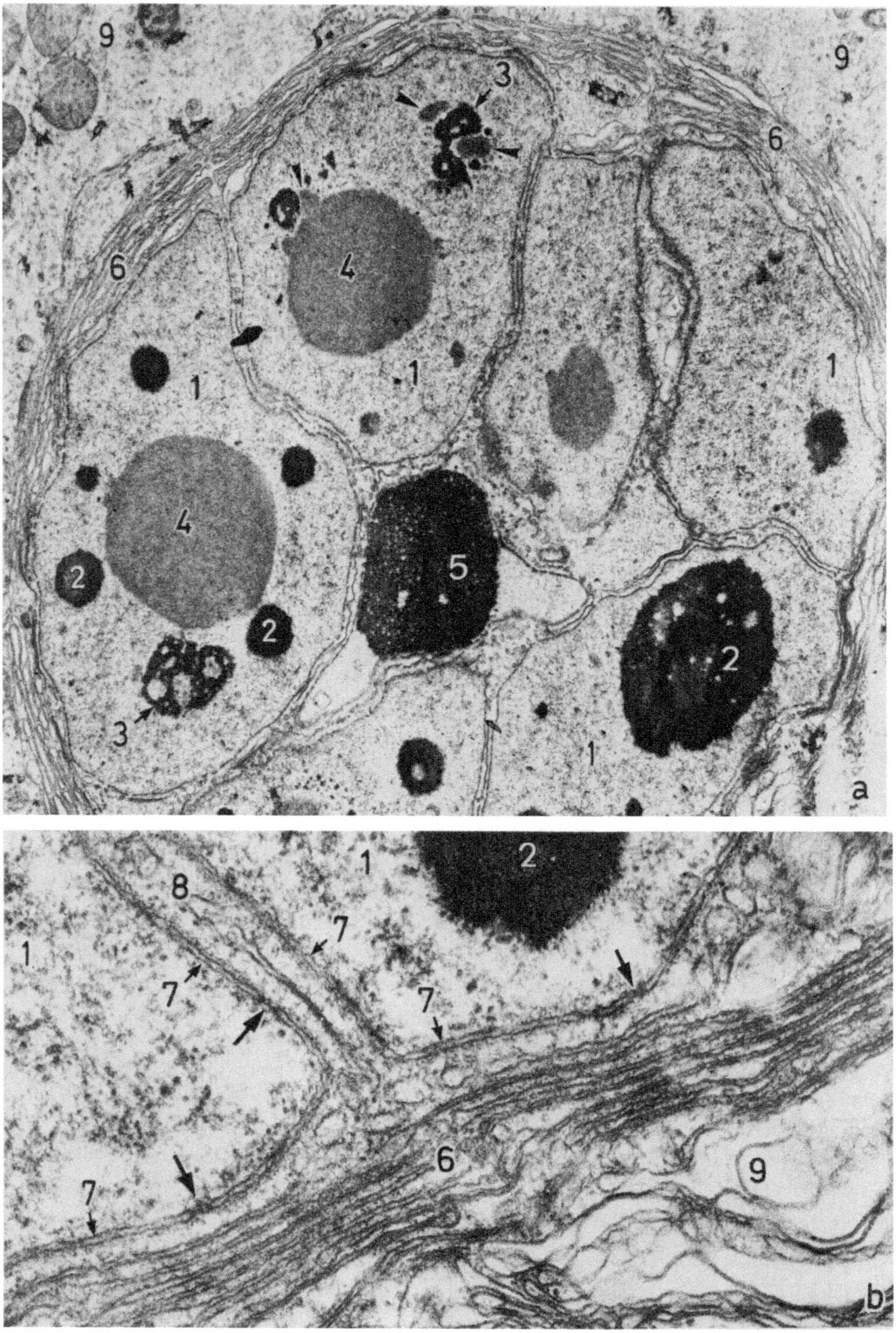

Fig. 80

measure about 0.1–0.3 μm, and occur in the peripheral zone of every Ma. The greater part of chromatin, however, seems to be dispersed. Local granular condensations of the macronuclear matrix correspond to the peripheral RNA-containing material. Some chromatin bodies are embedded into the nucleolus and seem to be nucleolar organizers (Fig. 81 *a*). The fine structure of the nucleolus will be considered below.

The ultrastructure of the chromatin elements of karyorelictid Ma proved to be rather diverse (see Table 5). The macronuclei of *Remanella multinucleata*, like those of *Loxodes*, contain numerous small chromatin bodies that correspond to the Feulgen-positive granules of the light microscope studies (Fig. 81 *b*). But they also contain several larger spherical chromocentres that are often vacuolized and measure about 1 μm in diameter. They seem to result from the fusion of a group of small bodies. The macronuclei of *Tracheloraphis caudatus* contain a few rather small (less than 1 μm) chromocentres that are usually adjacent to nucleoli. The rest of the chromatin appears to be decondensed.

Two types of chromatin structures exist in the Ma of *T. phoenicopterus*. Young macronuclei contain many small (0.1–0.3 μm) electron-dense chromatin bodies scattered in the karyolymph (Fig. 79*a*). In older Ma, these bodies assemble into several groups, or composite chromocentres, that measure 1–2 μm and appear to be annular in sections because their chromatin bodies are surrounding a mass of less dense material. Adult macronuclei of *Trachelonema sulcata* also contain several composite chromocentres that consist of aggregated dense chromatin bodies or strands; in young Ma, only separate small bodies of chromatin are present.

The macronuclei of some karyorelictids contain rather large chromocentres that do not seem to be composite. They often have a spongy structure, and their cavities are filled with karyolymph, e.g. in *Kentrophoros latum* (Fig. 80*a*). Many large polymorphic chromocentres of this type, reaching 3 μm in size, exist in each Ma of *Tracheloraphis dogieli* (Fig. 76*a*) and *Trachelocerca geopetiti* (Table 5). The chromocentres are very electron-dense and consist of a mass of fibrils about 10 nm thick. They also show narrow cleft-like lacunae that communicate with the karyolymph, and it seems likely that a chromocentre is really a single cord of condensed chromatin folded several times upon itself (Fig. 76*b*). The chromocentres of both species can be completely bleached with EDTA according to the method of W. Bernhard (1969), thus indicating their DNP nature. However, a rim of perichromatin fibrils and granules can be seen around the bleached chromocentres and in some of their lacunae (Fig. 82*a*). This shows that the periphery of the chromocentres is active in transcription (Raikov and Kovaleva 1980, 1981).

The *nucleolar apparatus* of karyorelictid Ma is usually well developed, suggesting that the rRNA-synthesizing activity of these nuclei is intense. Macronuclei often have a large single central nucleolus (Figs. 75, 77, 78, 81*a*). In other

---

Fig. 81. Ultrastructure of macronuclei in Loxodidae (after Raikov 1975a, 1979). *a* Ma of *Loxodes magnus*, 11 000 ×; *b* Ma of *Remanella multinucleata*, 19 000 ×; *1* nucleolus, *2* its fibrillar zones, *3* nucleolar chromatin, *4* chromatin bodies, *5* larger chromocentre, *6* sphere.

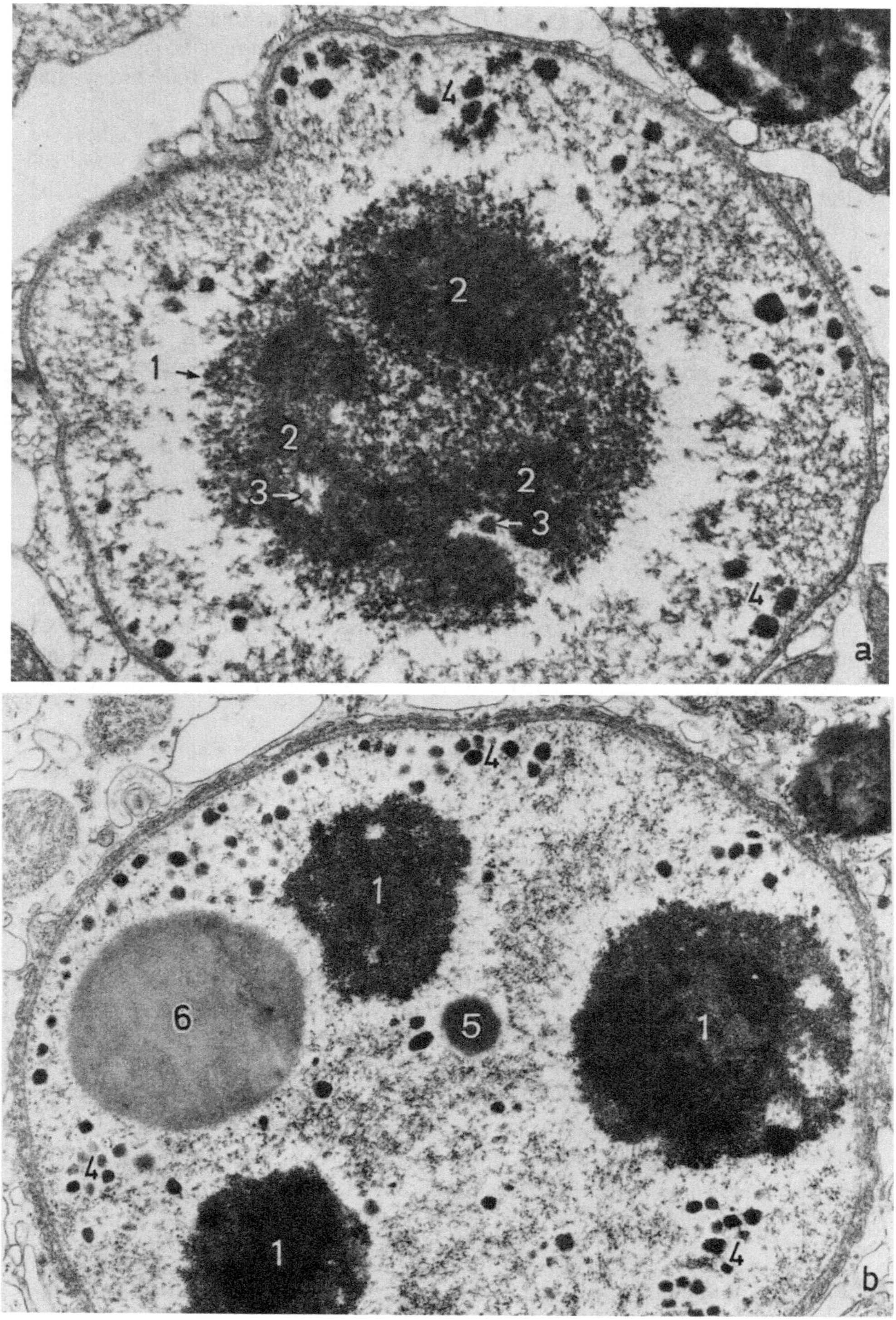

Fig. 81

species, macronuclei contain several (2–5) comparatively large nucleoli (Figs. 79*a*, 80*a*). The number of nucleoli per Ma reaches 10 in *Remanella multinucleata* (Fig. 81*b*) and 15 in *Tracheloraphis dogieli* (Fig. 76*a*). Many small nucleoli are scattered in the karyolymph of the Ma of *Geleia orbis* (Fig. 83).

The nucleoli of various Karyorelictida are cytochemically similar: they are always rich in RNA (Fig. 75*e*) and non-histone proteins (Fig. 75*f*). It is not yet clear whether nucleoli contain basic protein; they usually stain moderately or weakly with fast green at high pH (Raikov 1959a, 1963c) and may display even a strong reaction for arginine, such as in *Geleia nigriceps* (Raikov 1959b).

Cyclical changes of the nucleolar apparatus, which can be interpreted as the release of nucleolar RNP into the cytoplasm, have been observed with the light microscope in the species of *Geleia* and *Remanella* (Raikov 1959b, 1963c, 1963d, 1963e). Such changes, however, seem to be absent in *Trachelocercidae* and *Loxodes*. For example, a single large RNA- and protein-rich sphere appears among numerous small nucleoli in every Ma of *Geleia orbis* (Fig. 83*a*–*c*). The macronuclei of other individuals contain a large basophilic "RNP-vacuole" (Fig. 83*d*) instead of a typical sphere. Stages at which the vacuole contents "pour out" into the cytoplasm through an opening in the macronuclear envelope have been observed. Anyway, the vacuoles may have different sizes, and the smallest ones show almost no basophilia (Fig. 83*e*,*f*). These observations have been interpreted as the accumulation of RNP in macronuclear spheres with the subsequent release of RNP into the cytoplasm (Raikov 1963e). Similar observations have been made with respect to *Remanella multinucleata* in which a single sphere or a "RNP-vacuole" is also formed among nucleoli in every Ma (Raikov 1963d).

It has been shown by means of the electron microscope that the RNP-containing macronuclear sphere of *R. multinucleata* has medium electron density (Fig. 81*b*) and a finely granular structure, the diameter of its granules being less than 2.5 nm (Raikov 1979). The macronuclear spheres of the geleiids *Avelia martinicense* and *Geleia orbis*, however, have a homogeneous structure and high electron density (Nouzarède 1975, 1976).

Basophilic spheres also exist in the Ma of *Geleia nigriceps*, but they are formed at first inside the large central nucleolus, and only later do they pass into the karyolymph (Raikov 1959b). The spheres are rich in RNA and non-histone proteins with much tyrosine and tryptophan.

The ultrastructure of nucleoli is classic in most karyorelictids studied thus far (Table 5). They usually have a concentric organization with a dense fibrillar core and a looser granular cortex. Aggregates of granules and even spherical "micro-

---

Fig. 82. Ultrastructure of karyorelictid macronuclei. *a* Ma of *Trachelocerca geopetiti*, stained after Bernhard (the DNP of the chromocentres is bleached, while the nucleoli and the sphere retain the contrast), 15 500 × ; *b* part of the Ma of *Kentrophoros fistulosum* with three spheres, 30 000 × ; *c* macronuclear anlage of *Trachelonema sulcata*, 18 500 × ; *1* chromocentre, *2* rim of perichromatin fibrils (the arrows indicate perichromatin granules), *3* nucleolus, *4* sphere, *5* chromatin strands, *6* fibrogranular bodies extruded from the macronuclear anlage. *a* after Raikov and Kovaleva (1981); *b* after Raikov (1972b); *c* after Kovaleva and Raikov (1978).

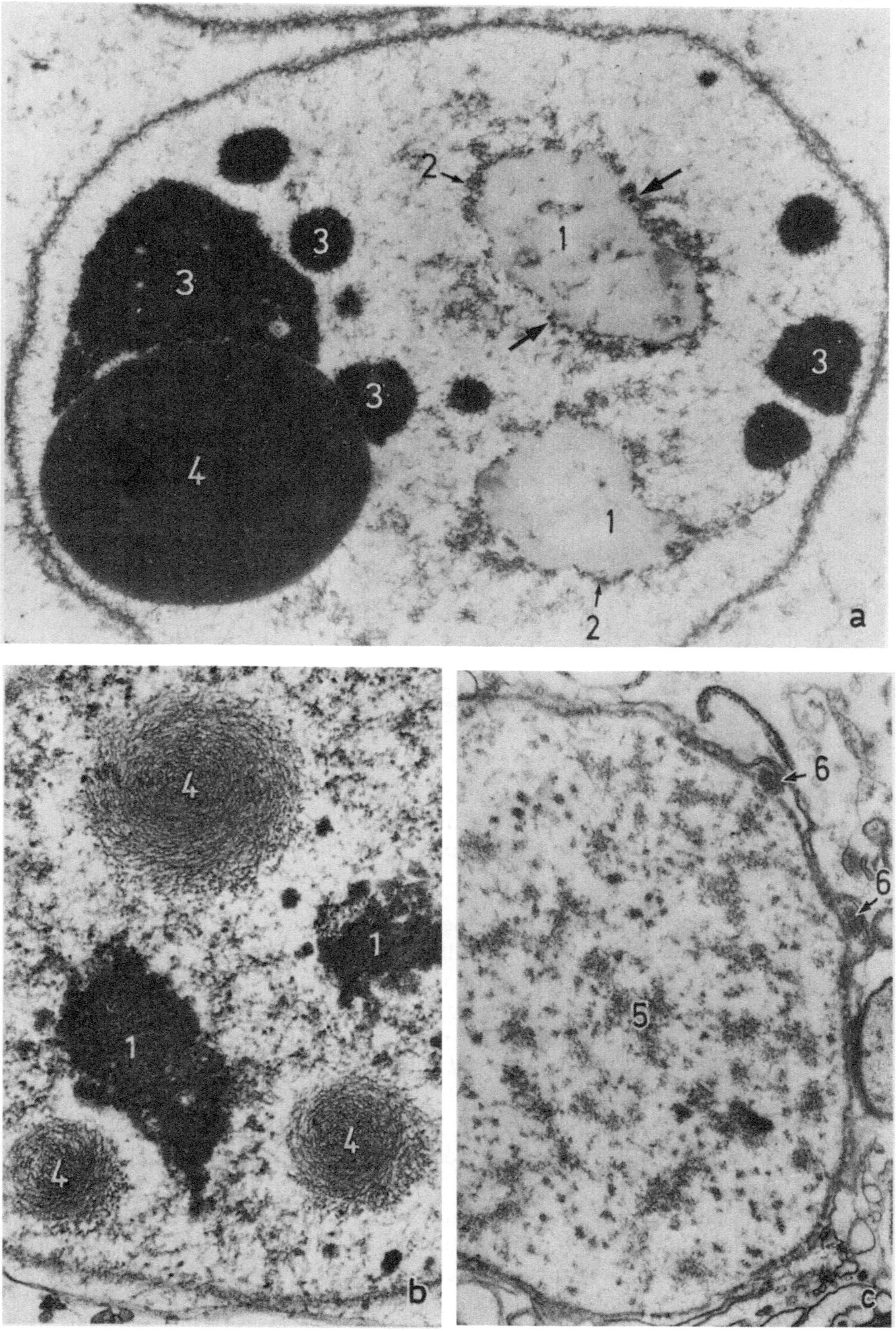

Fig. 82

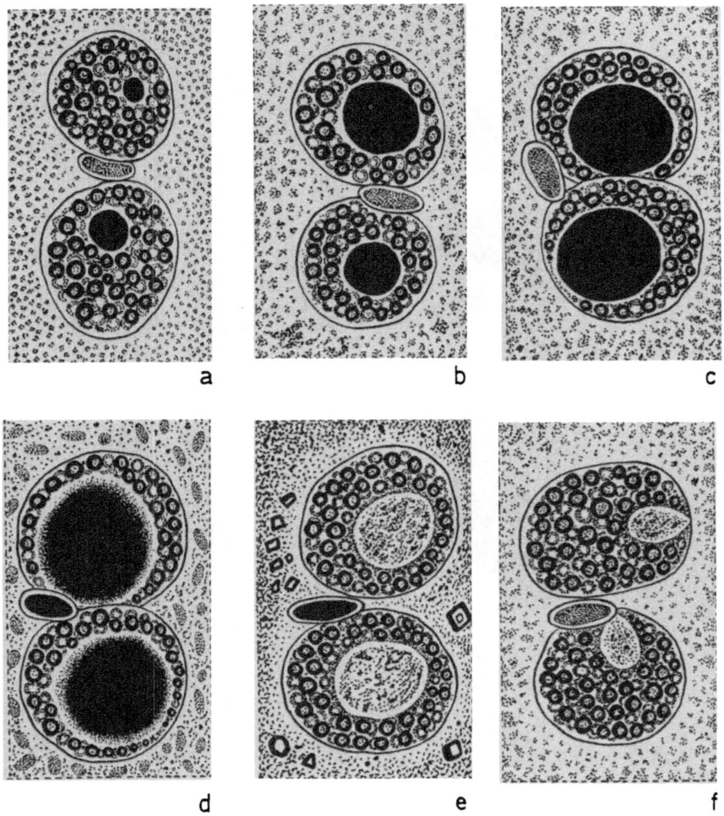

Fig. 83. Nucleolar cycle in *Geleia orbis*, 2200 × (after RAIKOV 1963e). *a–c* formation of spheres; *d* basophilic vacuoles in Ma; *e* and *f* their involution. Staining: methyl green – pyronine (*a–c*, *f*), mercuric bromophenol blue (*d*), and gallocyanin (*e*).

nucleoli" may become detached from the nucleolar surface, such as in *Tracheloraphis caudatus* (RAIKOV and DRAGESCO 1969). *Nucleolar organizers* that have the form of either fibrillar strands or dense bodies that penetrate deep into the nucleolus and contact its fibrillar core were observed in almost every species studied. Such a strand is occasionally in a light channel embedded in the fibrous part of the nucleolus, such as in *Tracheloraphis caudatus*, *T. dogieli*, *T. phoenicopterus*, and *Remanella multinucleata*. The presence of DNA in the intranucleolar fibrillar strands of *T. dogieli* and *Trachelocerca geopetiti* has recently been confirmed by electron microscopic cytochemistry (RAIKOV and KOVALEVA 1980, 1981).

The large central nucleolus of a *Loxodes* macronucleus is clearly composite: it is formed during macronuclear development by the fusion of three or four unit nucleoli (RAIKOV 1959a). Accordingly, it has three or four fibrillar cores (each with its own nucleolar organizer) inside a common granular cortex (Fig. 81*a*). A composite nucleolus with several organizers is also formed in adult Ma of *Trachelonema sulcata* by the fusion of several unit nucleoli (Fig. 84*b–d*).

Protein inclusions in macronuclei exist in all marine species of Karyorelictida studied thus far with the electron microscope. But they are absent in the freshwater *Loxodes* species. Protein inclusions are either *spheres* or *crystalloids*. Unlike chromocentres and nucleoli, these inclusions are facultative structures and occur in only some Ma.

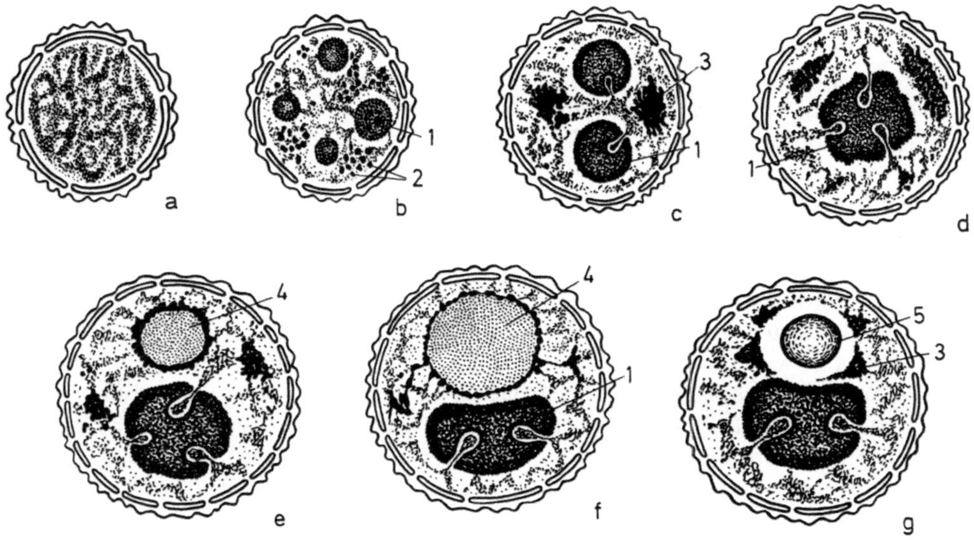

Fig. 84. Diagram of age-related changes in the ultrastructure of the Ma of *Trachelonema sulcata* (after KOVALEVA and RAIKOV 1970). *a* macronuclear anlage; *b* and *c* young Ma; *d g* adult Ma; *1* nucleolus, *2* chromatin bodies, *3* chromocentre, *4* fibrous material inside the chromocentre, *5* sphere.

A single sphere of regular shape and high electron density is formed in each Ma of *Trachelonema sulcata*. According to light microscopic cytochemistry, it consists mainly of protein (KOVALEVA and RAIKOV 1970). No sphere exists in young Ma (Fig. 84*b d*); its primordium appears only in adult Ma in the form of a mass of loose fibrillar material inside a composite chromocentre (Fig. 84*e*). As this mass grows, the chromocentre becomes distended into a thin layer of chromatin bodies around it (Fig. 84*f*). Finally, the fibrillar material condenses into a compact sphere 1 3 μm large (Fig. 84*g*). The sphere consists of very tightly packed sinuous fibres about 7 nm thick (KOVALEVA and RAIKOV 1973, 1978).

The spheres in the Ma of *Tracheloraphis dogieli* reach 3 μm in size and differ from both nucleoli and chromocentres by their lower electron density (Fig. 76*a*, *b*). They also consist of sinuous fibres 8 nm thick. The spheres are always in close contact with one or several chromocentres, which form slight depressions at the junction surface (Fig. 76*b*). Hence, it is likely that condensed chromatin contributes to the formation of the spheres. They possibly represent intranuclear stocks of chromatin proteins. The proteinaceous nature of the spheres of *T. dogieli* is confirmed by the fact that they can be completely digested with pronase in ultrathin sections (Fig. 76*c*).

The formation of protein spheres has been studied in greater detail in *Kentrophoros latum* (RAIKOV 1972c). A single central sphere of a fibrous structure that reaches 4 μm in diameter is gradually formed in each Ma of the nuclear cluster (Fig. 80*a*). The sphere itself usually does not directly contact the chromocentres; however, small patches of fibres 8 nm thick and identical to the fibres making up the sphere appear in the lacunae of the spongy chromocentres or in the karyolymph adjacent to them (Fig. 80*a* at the arrowheads). These fibrillar patches later fuse with the main sphere, as is seen in Fig. 80*a*. Thus, condensed chromatin probably participates in the formation of sphere material in *K. latum*, too.

The intramacronuclear spheres of *Kentrophoros fistulosum* have a rather unusual structure (RAIKOV 1972b). Several spheres are formed in each aging Ma. They consist of spirally arranged fibres 8 nm thick (Fib. 82*b*). Masses of loose microfibrillar material that are formed inside vacuoles of a large chromocentre seem to be the precursors of the spheres. These masses pass into the karyolymph where microfibrils are transformed into thicker (8 nm) fibres that gradually become spirally arranged.

A single large sphere is formed in each Ma of *Trachelocerca geopetiti* (RAIKOV and KOVALEVA 1981). Unlike *Tracheloraphis dogieli*, *K. latum*, and *K. fistulosum*, the sphere in *T. geopetiti* is closely adjacent to several nucleoli; it is not bleached with BERNHARD's EDTA and thus seems to have no DNA (Fig. 82*a*). Just as in *T. dogieli*, the sphere can be completely digested with pronase and thus appears to be purely proteinaceous. However, its association with nucleoli suggests that it represents a stock of nucleolar, and not chromatinic, proteins in *T. geopetiti*.

Protein crystalloids are found in the Ma of many Trachelocercidae. Within the single genus of *Tracheloraphis*, some species have spheres while others have crystalloids. The two inclusions seem to be homologous, and both are intranuclear stocks of some proteins.

Crystalloids are usually prismatic and lie freely in the macronuclear matrix. They have been observed with the light microscope in the Ma of *Trachelocerca coluber*, *Tracheloraphis incaudatus*, *T. prenanti*, and several other species. Protein in a high concentration is always cytochemically detected in them. The question of whether crystalloids contain any RNA remains an open one: they are often basophilic, but their basophilia is ribonuclease-resistant, and they do not incorporate labelled RNA precursors (TORCH 1964). Most probably, they consist of pure crystalline protein, but its type is still unknown. The functions of crystalloids as well as those of the protein spheres are also obscure. There is no evidence that crystalloids might eventually pass into the cytoplasm.

Intramacronuclear crystalloids have been studied with the electron microscope in *Tracheloraphis caudatus* (RAIKOV and DRAGESCO 1969) and *T. phoenicopterus* (RAIKOV and KOVALEVA 1978). A crystalloid per Ma, about 3–5 μm in length, is formed in the former species, and up to three crystalloids having the shape of hexahedral prisms and reaching 5 μm in length are formed in the latter (Fig. 79*b*). In both species, crystalloids display definite longitudinal striation with a period of 13 nm; the transverse period of the crystal lattice is less pronounced (8 nm). After some fixations, it becomes obvious that the crystalloid consists of an array of parallel fibres about 8 nm thick that resemble the component fibres of the spheres (Fig. 79*c*). Both the spheres and the crystalloids thus seem to be different forms of storage of the same protein. Crystalloids are formed in *T. phoenicopterus* due to the condensation (crystallization) of the loose microfibrillar material that again accumulates in contact with condensed chromatin, particularly inside "annular" composite chromocentres.

The interrelations of the protein spheres and crystalloids, described mainly in the Trachelocercidae and *Kentrophoros*, and of the RNP-containing spheres (or "RNP-vacuoles") of *Remanella* and *Geleia* are still unknown. In spite of the doubtless presence of RNA in the intramacronuclear spheres of *Remanella multinucleata*, survey electron micrographs (Fig. 81*b*) show them to be rather similar to the spheres of Trachelocercidae and *Kentrophoros latum*. Higher resolution, however, shows that instead of fibres 7–8 nm thick, they consist of much smaller granules (Raikov 1979). The spheres in the macronuclei of *Geleia* and *Avelia*, which were studied by Nouzarède (1975, 1976), also resemble those in Trachelocercidae, although they are much more dense. It is not yet known why spheres sometimes contain a large amount of RNA and sometimes either little or no RNA.

Nuclear Reorganization During Cell Division in Karyorelictida

The most remarkable peculiarity of the macronuclei of all karyorelictids is that they are *unable to undergo any kind of division*. During cytokinesis, they are simply distributed among daughter individuals, their number being approximately (or exactly) halved. The original number of Ma is then restored as some Mi are differentiated into Ma in the cycle of every cell division. Before that, micronuclei multiply by mitoses. Thus, the *development of macronuclear anlagen*, a phenomenon that occurs in other ciliates only during sexual processes (conjugation or autogamy), is observed in the cycle of every agamic division in karyorelictids.

Both the absence of the division of Ma and the transformation of some Mi into Ma during cell division were first observed in *Loxodes* a century ago by Bütschli (1876). Later, they were confirmed and studied in greater detail by Balbiani (1890), Joseph (1907), Kasanzeff (1910), and Rossolimo (1916). It is now firmly established that these peculiarities are common to all Karyorelictida.

In multinucleate species, nuclei are randomly segregated during cytokinesis, and daughter specimens often receive unequal numbers of Ma and Mi. This is the case, for instance, in *Tracheloraphis dogieli* (Fig. 85*a*), *T. margaritatus*, *Loxodes magnus*, *Remanella multinucleata*, *Geleia murmanica*, and other species (Fauré-Fremiet 1954, Raikov 1957a, 1958a, 1959a, 1963d, 1963e). The random segregation of nuclei results in the wide variability in the number of nuclei in most multinucleate species (see Table 5).

The reorganization of the nuclear apparatus, which supplements the number of Ma, is observed in most multinucleate karyorelictids during the interdivision period. Micronuclei reproduce by mitoses that are usually asynchronous (*Tracheloraphis margaritatus*, *Loxodes magnus*) and occur during both cytokinesis and the interdivision period. In other species, mitoses are more synchronous, but they are still not concomitant with cytokinesis, such as in *T. dogieli* (Fig. 85*b*).

The *morphology of micronuclear mitosis* has been investigated with the light microscope in *Loxodes striatus*, *L. magnus* (Rossolimo 1916, Raikov 1959a),

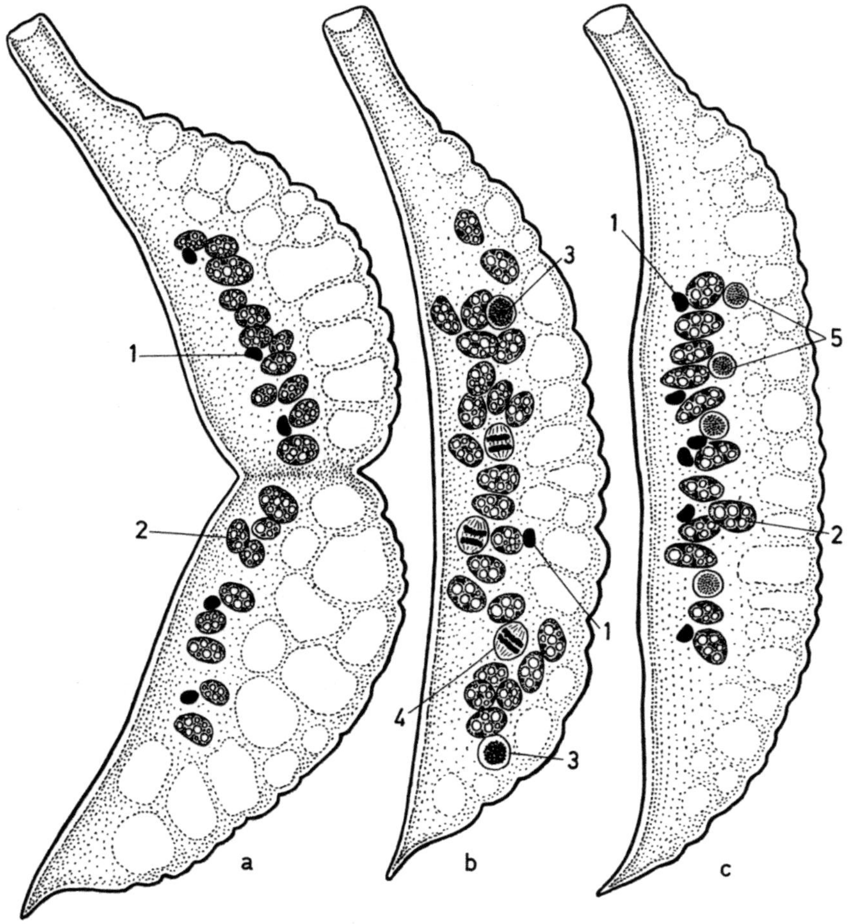

Fig. 85. Division and nuclear reorganization in *Tracheloraphis dogieli*, 430 × (after RAIKOV 1957a). *a* cytokinesis with the segregation of nuclei; *b* mitoses of most Mi; *c* formation of macronuclear anlagen; *1* Mi, *2* Ma, *3* prophase, *4* anaphase of mitosis, *5* macronuclear anlagen.

Tracheloraphis margaritatus, *T. dogieli*, *T. phoenicopterus* (RAIKOV 1957a, 1958), *Remanella rugosa*, *R. granulosa* (RAIKOV 1963c), and *Geleia nigriceps* (RAIKOV 1959b). Mitoses are always acentric, being accompanied by the preservation of the nuclear envelope and the appearance of an obtuse intranuclear spindle (Figs. 85*b*, 89*b*, *e*), i.e. they belong to the type of closed intranuclear orthomitoses. A fine structural description of micronuclear mitosis in *Loxodes magnus* has been presented in Chapter 2 (see also Fig. 29), and the chromosome numbers of various karyorelictids are included in Table 3.

As a result of mitoses, the number of micronuclei increases, then some of them are transformed into macronuclear anlagen. The formation of anlagen in most multinucleate species is also asynchronous (*Loxodes magnus*, *Tracheloraphis margaritatus*) or only partially synchronous (*T. dogieli*, Fig. 85*c*).

The *development of macronuclear anlagen* has been investigated with the light microscope in three species of *Tracheloraphis* (RAIKOV 1957a, 1958), two species of *Loxodes* (FAURÉ-FREMIET 1954, RAIKOV 1959a), three species of *Remanella* (RAIKOV 1963c, 1963d), and three species of *Geleia* (RAIKOV 1959b, 1963e). It occurs in a more or less similar way in all these forms. The development in *Loxodes magnus* may be taken as an example (Fig. 86). The process begins with the swelling of the micronucleus and the vacuolization of its chromatin (Fig. 86*a*). A "spireme" stage, i.e. a threadball of thin chromosomes resembling the mitotic

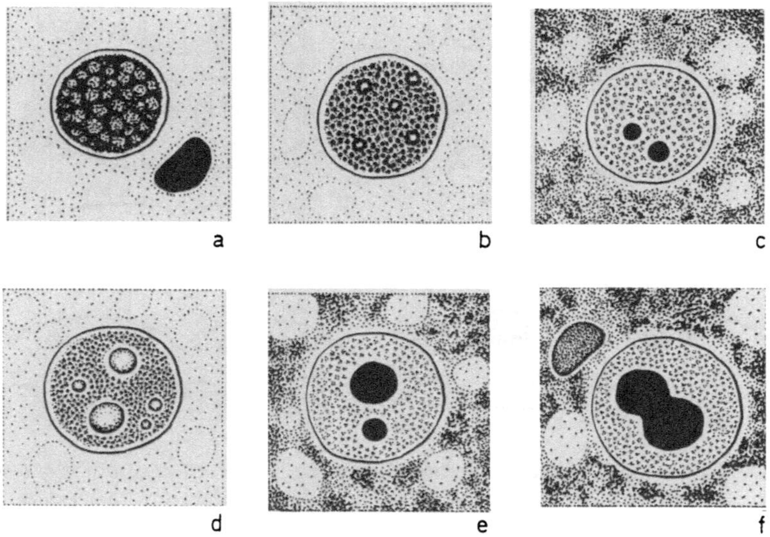

Fig. 86. Development of the macronuclear anlage in *Loxodes magnus*, 2500 × (after RAIKOV 1959a). *a* vacuolization of the young anlage (Mi at bottom right); *b* granular stage; *c*–*e* development of nucleoli; *f* fusion of nucleoli (Mi at top left). Staining: Feulgen (*a*, *b*, *d*), toluidine blue (*c*), and methyl green – pyronin (*e* and *f*).

prophase, sometimes ensues (in *Tracheloraphis phoenicopterus*, *Geleia nigriceps*, and *Remanella multinucleata*). Only uniform Feulgen-positive granules are seen in the anlage at the next stage (Fig. 86*b*); there are no signs of the endomitotic reproduction of chromosomes, and RNA cannot be revealed cytochemically. Then, chromatin granules become smaller and 2–4 nucleoli appear in the anlagen (Fig. 86*c*, *d*). These nucleoli have RNA from the very beginning (Fig. 86*c*, *e*). Nucleoli rapidly grow and then fuse with each other into a composite central nucleolus (Fig. 86*f*), which is typical of the adult Ma of *Loxodes* (Figs. 75*e*, *f*, 81*a*). Nucleoli do not fuse, although multiple Feulgen-positive chromocentres are formed in the late macronuclear anlagen of some other karyorelictids (e.g. *Tracheloraphis dogieli*).

Some stages of macronuclear development have been studied with the electron microscope in *Trachelonema sulcata*, *Tracheloraphis dogieli*, *Loxodes magnus*, and *Kentrophoros fistulosum* (KOVALEVA and RAIKOV 1970, 1978, RAIKOV 1972b, 1974,

1975a). In *T. sulcata* and *L. magnus*, the chromosomes of the anlage considerably decondense before the first nucleoli are formed, and they have the appearance of fluffy irregular strands; the small bodies of condensed chromatin are still absent (Fig. 82*c*; cf. Fig. 84*a*). When nucleoli appear, the chromosomes remain decondensed, but some small chromatin bodies are simultaneously formed (Fig. 84*b*). The number of these bodies increases as the anlage develops. Macronuclear development is similar in other species that have been investigated.

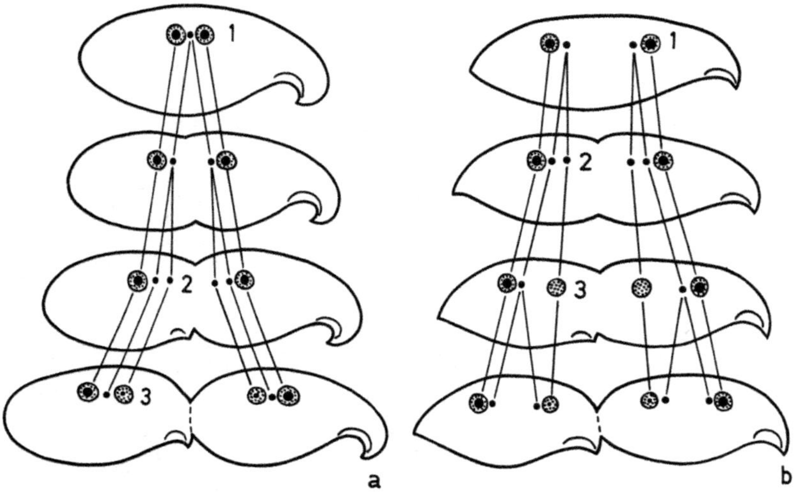

Fig. 87. Diagrams of nuclear reorganization during the division of *Loxodes rostrum* (*a*) and *L. striatus* (*b*) (after RAIKOV 1957b). *1* Ma, *2* Mi, *3* macronuclear anlage.

It can easily be calculated that at least a part of the micronuclei of a multinucleate ciliate must divide more than once per cell cycle in order to provide an "excess" of Mi, which can then be transformed into Ma. If the number of Ma is twice that of Mi (and such numerical relations are common in *Loxodes magnus* and other species), all Mi should, on the average, divide twice per cell cycle.

This repeated mitosis of Mi can easily be observed in species with a small and constant number of nuclei. In this case, mitoses and the formation of macronuclear anlagen are strictly synchronous and occur during cytokinesis, such as in *Loxodes rostrum* and *L. striatus* (RAIKOV 1959a).

In *L. rostrum* (Fig. 87*a*), two Ma of the mother cell segregate during cytokinesis and each daughter cell receives one of them. The sole Mi divides mitotically twice in succession and each daughter cell receives two Mi. During late cytokinesis, one of them differentiates into a macronuclear anlage, and the original number of nuclei is restored after its development is completed. Exactly the same pattern of

Fig. 88. Division of *Loxodes striatus*, 600 × (after RAIKOV 1959a). *a* nondividing specimen; *b* first mitosis of both Mi; *c* beginning of cytokinesis; *d* formation of macronuclear anlagen; *e* stage after second micronuclear mitosis; *f* young daughter cell; *1* Ma, *2* Mi, *3* macronuclear anlage.

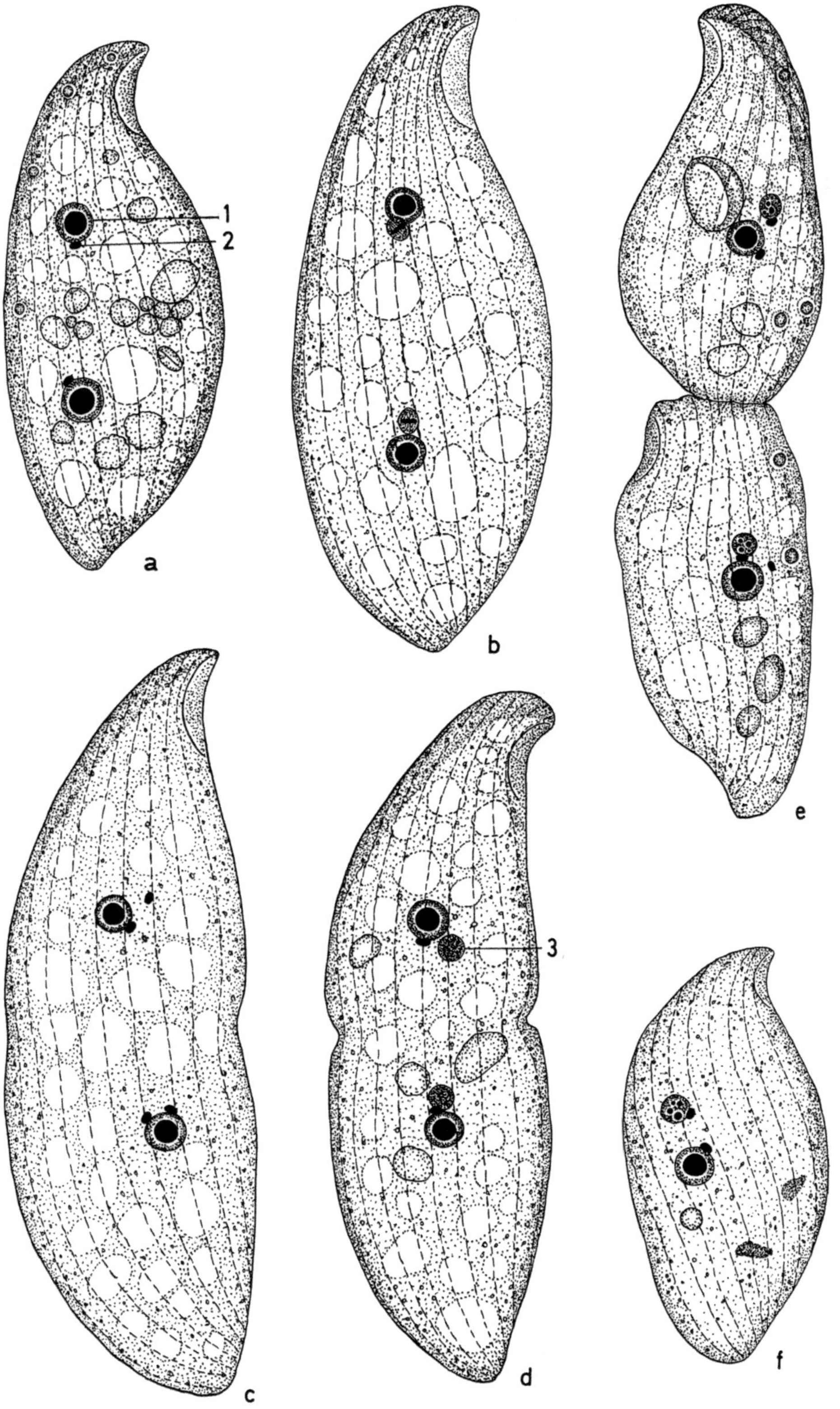

Fig. 88

nuclear reorganization has been observed in *Geleia nigriceps*, *G. orbis*, *Remanella rugosa*, and *R. granulosa* (RAIKOV 1959b, 1963c, 1963e).

In *Loxodes striatus*, which has two Ma and two Mi (Figs. 75*b*, 88*a*), the sequence of nuclear phenomena is somewhat different, as can be seen in the diagram (Fig. 87*b*). In this case, too, each daughter cell receives one mother Ma. Both Mi divide mitotically for the first time (Fig. 88*b*), and the future daughter cells at first receive two Mi each (Fig. 88*c*). One of them gives rise to the anlage of the second Ma (Fig. 88*d*), while the other divides once more (Fig. 88*e*).

These observations imply that the two Ma of every individual of *Loxodes rostrum*, *L. striatus*, *Geleia nigriceps*, and others always differ in age. One of them is old and has been received from the mother cell, while the other is new and has been formed during the last cell division (Fig. 88*f*). It should be noted that our diagrams (Fig. 87*a, b*) are somewhat simplified: in the proters of both species, the new Ma is shown behind the old one. In reality, the new Ma in all bimacronucleate species investigated is always the anterior one, both in the proter and the opisthe, since the entire nuclear group of the proter turns with its macronuclear anlage anteriad (Fig. 88*c–e*). The macronuclei of every individual of *Loxodes magnus* and other multinucleate species obviously should have different ages as well, a fact first emphasized by FAURÉ-FREMIET (1954).

In the species in which the entire nuclear apparatus is united into a single group or cluster of nuclei, reorganization phenomena are basically similar and usually occur synchronously during cell division. Thus, in *Trachelocerca coluber* (Fig. 89), which has a single group of four Ma and two Mi (Fig. 89*a*), both Mi divide by mitosis during early cytokinesis (Fig. 89*b, c*). Then, the nuclear group segregates into two daughter groups (d and d_1), each of which at first receives two mother Ma and two Mi. Both Mi later divide for the second time (Fig. 89*e*). Among four Mi (Fig. 89*f*), those two that lie farther from old Ma differentiate into macronuclear anlagen (Fig. 89*g*). The development of anlagen (Fig. 89*h*) leads to the restoration of normal nuclear groups in both daughter cells. Nuclear reorganization in *Tracheloraphis* sp. is similar (TORCH 1961). In *T. phoenicopterus*, a species with 12 Ma and 6 Mi that form a compact cluster (Fig. 79*a*), nuclear reorganization also includes twofold mitosis of all Mi, the subdivision of the nuclear complex into two groups of nuclei, the formation of six macronuclear anlagen in each daughter group, and the re-union of the nuclei of each daughter group into a compact cluster (RAIKOV 1958).

Finally, in species with several or many nuclear groups, reorganization processes are also synchronous, such as in *Tracheloraphis caudatus* (DRAGESCO and RAIKOV 1966) and *Kentrophoros fistulosum* (RAIKOV 1972b). They lead to the formation of two nuclear groups from each original one.

Since macronuclei pass without division from one generation cells into the next, the longevity of individual Ma is an interesting question. This longevity is apparently restricted to several cell generations, after which old Ma become pycnotic and are finally resorbed (FAURÉ-FREMIET 1954, RAIKOV 1959a, BOBYLEVA *et al.* 1980). The calculated longevity of individual Ma of *Loxodes magnus* ranges from four to seven cell generations. A special reorganization process leading to the replacement of the posterior (i.e. the older) Ma has been observed in some specimens of *Loxodes striatus*: the posterior Mi undergoes asynchronous mitosis,

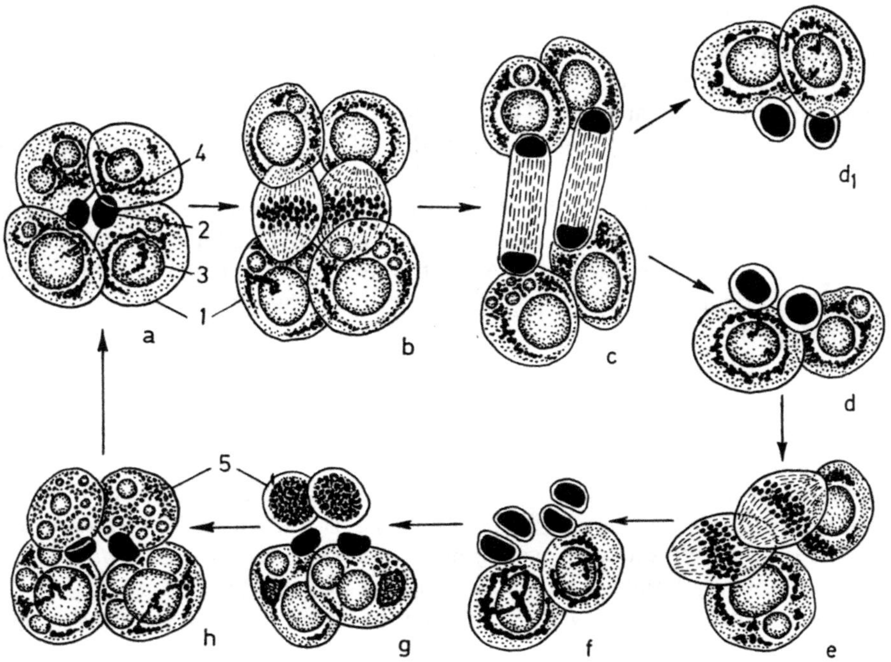

Fig. 89. Nuclear reorganization during division in *Trachelocerca coluber*, 2300× (after RAIKOV 1969). *a* nuclear group of a nondividing specimen; *b* metaphase, *c* telophase of the first mitosis of both Mi; *d* and d_1 separation of the nuclear group in two; *e* metaphase of second micronuclear mitosis; *f* stage after second mitosis; *g* and *h* formation and development of macronuclear anlagen; *1* Ma, *2* Mi, *3* nucleolus, *4* chromatin, *5* macronuclear anlagen.

and one of its derivatives differentiates into a macronuclear anlage (RAIKOV 1959a).

Content and Synthesis of DNA and RNA in Karyorelictid Macronuclei

Content and Synthesis of DNA. Definite parallelism exists between the quantities of DNA in the Mi and Ma of Karyorelictida. Species with large micronuclei that are brightly Feulgen-stained also have more chromatin (especially condensed chromatin) in Ma, e.g. *Tracheloraphis dogieli* with multiple large chromocentres in Ma (Fig. 76*a*). Such parallelism is obvious also in *Kentrophoros fistulosum* and in the species of Trachelocercidae in which each Ma contains a single large chromocentre. This chromocentre is sometimes similar to a Mi in size as well as in the visually evaluated DNA content (Fig. 78*a, b*). The species of *Loxodes* that have small Mi (*L. striatus*, *L. rostrum*) have less DNA in their Ma in comparison with *L. magnus* (RAIKOV 1959a).

These observations suggested (RAIKOV 1957b, 1963f, etc.) that unlike the macronuclei of higher ciliates, karyorelictid macronuclei were *diploid.* On the whole, this supposition has been confirmed by the cytophotometry of the Feulgen-stained nuclei of *Loxodes magnus* (RAIKOV *et al.* 1963). Adult Ma proved to contain, on the average, only a little more DNA than the diploid postsynthetic Mi, the difference averaging 11% but varying greatly from nucleus to nucleus. A study recently carried out with a more sensitive cytofluorimetric method, however, has shown that the macronuclear anlagen of *L. magnus* differentiate from presynthetic micronuclei (which are still in G_1) and that a single replication of DNA occurs in them during their early development (BOBYLEVA *et al.* 1980). This resembles the single DNA replication in the foraminiferan somatic nuclei and brings the DNA content of young Ma up to the diploid postsynthetic level (4c). But in adult and especially aging macronuclei, there is also some minor DNA synthesis that results in a gradual increase of their DNA content to 5-6c, rarely to 10c. This increase is likely to reflect some type of partial replication of macronuclear DNA, e.g. gene amplification.

The adult macronuclei of *L. magnus* show in their DNA content cyclical changes that are correlated with cytokinesis, although macronuclei do not undergo division in this case. The average DNA content of Ma proved to be lower in dividing animals than in nondividing ones. This can be attributed to the destruction of some of the overreplicated DNA copies in the Ma of the dividers. The macronuclei of *Loxodes* thus seem to be basically *diploid postsynthetic* ones (4c), but are subject to some overreplication of DNA (BOBYLEVA *et al.* 1980). They can be properly termed *paradiploid* or *hyperdiploid* nuclei.

DNA synthesis in *L. magnus* nuclei has also been studied by means of autoradiography (RAIKOV and MORAT 1977). Developing macronuclear anlagen were labelled with ^{3}H-thymidine. This is obviously consistent with the data on DNA replication in them. Most adult macronuclei remained unlabelled, however, and the rest displayed only slight incorporation of ^{3}H-thymidine. Therefore, the full replication of the bulk DNA in adult Ma is hardly possible, while minor DNA synthesis in them seems to correspond to the previously mentioned gradual increase of the DNA content in adult and aging Ma of *Loxodes*. Highly DNA-labelled Ma have been obtained only when ciliates were kept in a nonradioactive medium for at least 20 hours after washing them off ^{3}H-thymidine. Probably they were new Ma formed from Mi or macronuclear anlagen that had already been labelled.

RON and URIELI (1977) have also caused the macronuclear DNA of *Loxodes striatus* to be intensely labelled mainly by a long-term incubation of ciliates with ^{3}H-thymidine (1-5 days). When labelling was carried out for 2-6 hours, only 10% of Ma contained the label. This labelling seems to either reflect partial DNA replication in some Ma or occur in macronuclei that have completed their development from labelled macronuclear anlagen during incubation. On the other hand, TORCH (1964) failed to detect any incorporation of labelled thymidine in the Ma of *Tracheloraphis* sp.

The changes in the DNA content of the macronuclei of *Trachelonema sulcata*, which occur at various stages of their development (see Fig. 84), are more complicated. These changes have been investigated cytophotometrically by KOVALEVA and RAIKOV (1978). A clear *decrease in the DNA content* to about one-half the original value, or to 1c, has been revealed in the developing macronuclear anlagen as compared with the DNA content of presynthetic (2c) Mi from which they differentiate. This can be attributed only to the elimination of some DNA sequences from the macronuclear anlagen at the time of their differentiation. The question of what these sequences are remains open. In principle, these sequences could be genic, and then their elimination would mean *genetic diminution* during macronuclear development. Small fibrogranular bodies are extruded from the young macronuclear anlagen of *T. sulcata* into the surrounding cytoplasm (Fig. 82*c*). These bodies may contain the eliminated chromatin or its degradation products.

Young macronuclei (Fig. 84*b*) also contain only about 1c of DNA. But when they are transformed into adult Ma (Fig. 84*c*), one full replication of the DNA that remained after diminution is likely to occur. At any rate, the DNA content of adult Ma (Fig. 84*d*) returns to the nearly diploid level, or 2c. Later, only some DNA fractions are replicated in aging Ma. Consequently, old macronuclei (Fig. 84*g*) contain DNA in quantities considerably higher than the diploid level, quantities that vary greatly from nucleus to nucleus. About half of them contain 2-4c of DNA, roughly 40% contain 4-6c, and about 10% contain more than 6c (up to 12c at the most). One can suppose that this partial DNA replication in aging Ma is gene amplification. The polyploidy of aging macronuclei is excluded in this case since DNA accumulates in them gradually and not stepwise (KOVALEVA and RAIKOV 1978).

It has thus been shown that the development of paradiploid macronuclei in *T. sulcata* involves the reorganization of the genetic apparatus of these nuclei in the form of chromatin diminution and partial DNA replication. The loss of some genes during the development of macronuclei could, for instance, make these nuclei unable to divide. In other words, nuclear differentiation in Karyorelictida may well have a genetic basis (differences in the gene complement between the two types of nuclei) instead of merely an epigenetic basis, as was believed earlier.

NOUZARÈDE (1976) demonstrated that the DNA content of Ma is also rather variable in *Geleia orbis*. The minimum DNA content of Ma approximately equals 4c, but the maximum DNA content of individual Ma can exceed this level by about seven times. As NOUZARÈDE himself said, it seems likely that DNA synthesis in the Ma of *G. orbis* is not polyploidization, but rather the partial replication of repetitive DNA sequences. The amplification of ribosomal genes seems most probable in this case since the macronuclei of *G. orbis* contain many nucleoli (Fig. 83). DNA synthesis in macronuclei (especially in their peripheral area, which contains nucleoli) also has been revealed by NOUZARÈDE (1976) autoradiographically after ^{3}H-thymidine incorporation.

Content and Synthesis of RNA. No quantitative data on the RNA content of karyorelictid nuclei are available so far. According to qualitative cytochemical reactions, RNA is almost absent in Mi (see p. 242), while it is found in a high concentration in macronuclear nucleoli (and sometimes spheres and "RNP-vacuoles", such as in *Remanella* and *Geleia*).

According to autoradiographic data, the micronuclei of Karyorelictida incorporate no labelled uridine (TORCH 1964, RON and URIELI 1977). RNA synthesis occurs only in Ma. Newly synthesized RNA then migrates into the cytoplasm. TORCH (1964) revealed the incorporation of labelled precursors into the macronuclear RNA of *Tracheloraphis* sp. with incubation periods varying from 20 minutes to 6 hours; after 6 hours, labelled RNA appears in the cytoplasm, too. Thus, at least a part of cytoplasmic RNA of *Tracheloraphis* sp. has a nuclear origin

(TORCH 1965). RNA synthesis in Ma is inhibited by actinomycin D, i.e. it is normal DNA-primed transcription. However, the cytoplasm continues to slowly incorporate ^{3}H-uridine in the presence of actinomycin D. According to TORCH, this means that *Tracheloraphis* also has a nucleus-independent cytoplasmic system of RNA synthesis. Its nature is still unknown, but it may depend on the activity of endozoic bacteria, which are common in the cytoplasm of many Karyorelictida.

RON and URIELI (1977) observed that ^{3}H-uridine is intensively incorporated into nucleoli and less intensively into the peripheral zone of the Ma of *Loxodes striatus*. In *L. magnus*, both adult and developing Ma synthesize RNA (RAIKOV and MORAT 1977). If ciliates are kept in a nonradioactive medium after incubation with a labelled precursor prior to fixation, the labelling of all Ma gradually becomes less intensive and, in about 24 hours, even weak. This seems to be due to the release of RNA from Ma into the cytoplasm. The intensive incorporation of ^{3}H-uridine into Ma has also been observed in *Geleia orbis* by NOUZARÈDE (1976), who used both light and electron microscopic autoradiography. The peripheral zone of every Ma, which contains numerous nucleoli, is preferentially labelled. Moreover, the perinuclear accumulation of the endoplasmic reticulum becomes labelled, and its RNA seems to have a nuclear origin.

Thus, cytochemical and autoradiographic results show that the Ma of Karyorelictida, unlike Mi, are active in transcription.

Primitiveness of Karyorelictid Macronuclei

Do paradiploid Ma of karyorelictids correspond to a primitive stage of evolutionary development of nuclear dualism in ciliates, or are these nuclei secondarily simplified and derived from DNA-rich Ma of higher ciliates? FAURÉ-FREMIET (1954, 1961, 1970) repeatedly supported the latter hypothesis. Macronuclei that cannot undergo division are found mainly in psammobiotic ciliates, i.e. the representatives of the interstitial fauna of marine sand; therefore FAURÉ-FREMIET considered these nuclei to be highly specialized in the course of adaptation to this particular mode of life. However, the adaptive significance (if any) of macronuclei with respect to the deficiency of DNA and/or their inability to undergo division remained unclear. Another argument of FAURÉ-FREMIET (1970) was that both the buccal ciliature and the stomatogenesis of *Loxodes* cannot be considered primitive among ciliates, and thus their nuclei cannot be primitive either. It should be noted, however, that evolutionary specialization is known to proceed rather independently with respect to different organs (or organelles) of the same animals (mosaic evolution); therefore, forms with a primitive nuclear apparatus and a specialized ciliature, and vice versa, are likely to exist (CORLISS 1975b, CORLISS and HARTWIG 1977).

The author of this book develops an alternative point of view, i.e. paradiploid karyorelictid macronuclei are primitive and phylogenetically more ancient than the Ma of the "polyploid" type (RAIKOV 1957b, 1963f, 1969, 1976b). Similarities in the structure and development of the Ma of various karyorelictids thus seem to be due to a common origin and not to convergent evolution. The surviving species of ciliates with paradiploid Ma are thus peculiar "karyological relicts". This hypothesis has recently been supported by CORLISS (1974, 1975b, HUTNER and CORLISS 1976, CORLISS and HARTWIG 1977), who also made the relevant changes in the system of Ciliophora, introducing a new order, i.e. Karyorelictida (CORLISS 1975a, 1979a).

The following considerations speak in favour of the primitiveness of karyorelictid Ma and against their secondary origin from "polyploid" Ma:

1. Polyploidy (and other modes of accumulaton of large quantities of DNA, see Chap. 8) seems to be irreversible in the evolutionary sense.

2. Karyorelictids belong to the lowest major ciliate taxon, the class Kinetofragminophorea. At the same time, paradiploid macronuclei are not restricted to a single highly specialized minor taxon since three karyorelictid families with rather similar somatic infraciliatures differ greatly from one another with respect to their buccal structure.

3. Higher ciliate taxa (especially the Polyhymenophorea) also contain many typical psammobionts, but their nuclear apparatus does not have the characteristic form peculiar of Karyorelictida.

4. The psammon, or the interstitial fauna of marine sands, generally contains many primitive relict animals, including metazoans (Turbellaria, Gnathostomulida, Crustacea, etc.). The interstitial sand biotope seems to be an ecological niche that favours the survival of ancient primitive forms, which were supplanted in other biotopes by more advanced forms.

5. Karyorelictids are mostly (except the genus *Kentrophoros*) predatory motile forms that are unlikely to undergo morphophysiological regress. Also, they can secondarily pass from the psammon into other biotopes; for instance, *Loxodes* occur in both the freshwater psammon (SASSUCHIN 1931, DRAGESCO 1960) and the plankton of ponds and lakes (RYLOV 1923, GOULDER 1980).

6. The paradiploid macronuclei of Karyorelictida have many features in common with the somatic nuclei of heterokaryotic Foraminiferida. Both originate agamically from generative nuclei and, for the most part, remain nearly diploid. In both groups, the differentiation of somatic nuclei is coupled with the activation of transcription and the loss of the ability to undergo division, thus ultimately dooming these nuclei to degeneration after functioning for a certain length of time. This clear parallelism in the structure and development of the somatic nuclei of Foraminiferida and of the Ma of Karyorelictida suggests that they originated in phylogenesis in a similar, though independent way. But since homokaryotic forms predominate among foraminiferans, while the polyploidy of somatic nuclei (if it exists at all) is an exception, it is unlikely that diploid foraminiferan somatic nuclei originated as a result of the "simplification" of polyploid ones. Undoubtedly these nuclei originated from the ordinary diploid nuclei of homokaryotic foraminiferans by nuclear differentiation. Hence, the paradiploid, never-dividing karyorelictid Ma probably have the same origin.

Therefore, it can be concluded that in spite of the substantial specialization of the ciliary and especially buccal structures of karyorelictids, their paradiploid macronuclei originated from the normal diploid nuclei of some ancient homokaryotic ciliates. The presently living *Stephanopogon* may give us an idea of what their nuclei might have been like. Nuclear differentiation was apparently coupled with some irreversible change of the genome (loss of certain genes by genetic diminution?). As a result, such nuclei lost the ability to undergo division and became physiologically active somatic nuclei, or macronuclei. This change has been termed the *primary differentiation* of somatic nuclei (RAIKOV 1963f) to distinguish it from the next step of the phylogenetic development of nuclear dualism, i.e. the polyploidization of Ma.

It is quite possible that the "feedback mechanism" that regulates the differentiation of somatic nuclei in Foraminiferida is active also in Karyorelictida. This is likely because macronuclear anlagen in many karyorelictids differentiate from micronuclei that are the most distant from old Ma (Fig. 89*g*). It looks as if every old Ma inhibited the formation of new Ma inside a certain zone around itself.

Other Descriptions of Nuclear Dualism

False Nuclear Dualism

Cases of nuclear dualism have often been described in other protozoan groups, also. Many of them, however, are cases of false nuclear dualism. For example, the *kinetoplast* of Trypanosomidae and Bodonidae (members of the flagellate order Kinetoplastida) has long been considered a second nucleus and even called a "kinetonucleus". But it is now firmly established that this Feulgen-positive body contains circular mitochondrial DNA and is a specialized part of a single giant mitochondrion (for reviews, see SIMPSON 1972, KALLINIKOVA 1981). No continuity exists between the nucleus and the kinetoplast: they are not formed from one another.

Another example of false nuclear dualism is provided by marine amoebae of the family Paramoebidae, which have, in addition to the nucleus, at least one Feulgen-positive associated body called "Nebenkörper" or *paranucleus* (GRELL 1961). Electron microscopy, however, showed that the associated body was a whole endosymbiotic organism with its own cytoplasm and a large, highly Feulgen-positive prokaryotic-type nucleoid (which is the paranucleus proper) and that it had from one to three small eukaryotic-type nuclei of unknown origin (GRELL and BENWITZ 1970, PERKINS and CASTAGNA 1971, HOLLANDE 1980).

Finally, false nuclear dualism is characteristic of Myxosporea. It has long been believed that two categories of nuclei that often differ in size, i.e. generative and vegetative ones, exist in the common cytoplasm of their plasmodia. But electron microscope studies have shown that only "vegetative" nuclei belong to the plasmodium proper, while the "generative" ones are localized in separate generative cells (GRASSÉ 1960, LOM and PUYTORAC 1965, LOM 1969, USPENSKAJA 1969). It has been shown cytophotometrically that the nuclei of the generative cells of *Sphaeromyxa* are diploid, while plasmodial (vegetative) nuclei are mostly tetraploid (USPENSKAJA 1976). GRASSÉ and LAVETTE (1978) have shown that the generative cells of *Sphaeromyxa* are in fact differentiated into further categories: the *sporogenic* cells (or generative cells proper), the *pericytes* (cells that envelop and carry the sporogenic ones), and the *lobocytes* (phagocytizing cells). The last two categories of separate cells are, in essence, not generative, but somatic. The sporogenic cells divide and give rise to pansporoblasts and later to sporoblasts and spores. "Vegetative" nuclei (and probably pericytes and lobocytes) remain in the plasmodium and degenerate with it. Consequently, *cell differentiation* analogous to that in Metazoa, and not nuclear dualism, is found in Myxosporea (GRASSÉ and LAVETTE 1978).

"Nuclear Dualism" in Dinoflagellida

The cells of some species of free-living Dinoflagellida always contain two morphologically different nuclei. Such "nuclear dualism" was discovered and investigated with the electron microscope in *Glenodinium foliaceum* by DODGE (1971b). In this case, one nucleus is a typical dinokaryon that contains permanently condensed chromosomes, while the other is an "eukaryotic" nucleus that lacks visible chromosomes and has an oval, elongated, or branching form. The fine structure of the dinokaryon (or the mesokaryotic nucleus) is quite typical: its chromosomes display characteristic transverse bands and interbands that consist of highly ordered DNA molecules, and its few nucleoli contain the same nucleolar organizers as in other dinoflagellates (see Chapter 4). During mitosis, this nucleus becomes pierced with cytoplasmic tunnels (multichannelled dinomitosis). Hence, the dinokaryon is doubtlessly *Glenodinium*'s own nucleus.

The eukaryotic nucleus is also Feulgen-positive; it is surrounded with a typical porous nuclear envelope and contains several nucleoli. The nuclear matrix is granular; chromosomes were not observed. The division of the eukaryotic nucleus of *Glenodinium* has been investigated so far only with the light microscope (BLANCHARD-BABILLOT 1972). It divides later than the dinokaryon by apparent amitosis.

Similar "nuclear dualism" has also been observed in *Peridinium balticum* (TOMAS *et al.* 1973), with the eukaryotic nucleus also dividing by apparent amitosis. Neither microtubules (inside or outside the nucleus) nor condensed chromosomes have been found with the electron microscope in the dividing eukaryotic nucleus of *P. balticum* (TIPPIT and PICKETT-HEAPS 1976).

There are two hypotheses concerning the nature of nuclear dualism in dinoflagellates (DODGE 1971b). According to one of them, nuclear dualism is true and results from the differentiation of identical, basically mesokaryotic nuclei into a generative one (the dinokaryon) and a somatic one (the "eukaryotic" nucleus). According to the other hypothesis, nuclear dualism is false and the eukaryotic nucleus has a symbiotic origin.

Most of the presently available evidence favors the symbiotic hypothesis. First, no formation of the eukaryotic nucleus from the dinokaryon has ever been observed. Second, other species of the same two genera (*Glenodinium* and *Peridinium*) have no nuclear dualism at all, suggesting that the eukaryotic nucleus originated rather recently. This is more compatible with symbiogenesis. Third, the eukaryotic nucleus of *P. balticum* displays, on its chromatin fibres, typical nucleosomes that never occur in dinokarya (RIZZO and BURGHARDT 1980). Fourth, and this is most important, the eukaryotic nucleus of both *P. balticum* (TOMAS and COX 1973) and *G. foliaceum* (JEFFREY and VESK 1976, FINE and LOEBLICH 1976) proved to be surrounded by a membrane that encloses a large compartment of the cytoplasm that is likely to be the cytoplasm of a well integrated endosymbiont. It contains all the chloroplasts that differ in type from the typical dinoflagellate chloroplasts and at least some mitochondria. The cytoplasm of the host cell lacks chloroplasts (and, according to FINE and LOEBLICH, also mitochondria), so that the symbiotic complex uses only the respective organelles of the endosymbiont. According to the array of chloroplast pigments, the latter is either a chrysomonad

(Tomas and Cox 1973, Jeffrey and Vesk 1976) or a chloromonad (Fine and Loeblich 1976).

Hence, we have one more case of false nuclear dualism in the Dinoflagellida, the two nuclei being of independent origin. Thus, no cases of true nuclear dualism have been discovered so far outside Foraminiferida and Ciliophora. Nevertheless, an investigation of the metabolic relationships between the two nuclei of the symbiotic complex seems to be very promising.

8. Polyploid Macronuclei of Ciliates

In this chapter, we will consider somatic nuclei, i.e. *macronuclei* (Ma) of all Ciliophora except the orders Primociliatida and Karyorelictida, which were described in Chapter 7. All the ciliates considered here are heterokaryotic, i.e. they possess nuclear dualism. Their generative nuclei, or *micronuclei* (Mi), belong to small vesicular, "spermal", or "chromosomal" nuclei (see Chapter 1) and divide by closed intranuclear orthomitosis (see Chapter 2). Their ultrastructure is shown in Figs. 10*c*, *d*, 91*a*, 93*a*. Conversely, Ma are usually very large and extremely rich in DNA. During the sexual process (conjugation or autogamy), the old Ma degenerates and a new Ma develops from certain products of the synkaryon division. The *synkaryon* is formed by the pairwise fusion of *pronuclei*, which are haploid products of micronuclear meiosis. Therefore, the sexual process of ciliates not only fulfils its usual function (recombination of the gene pool), but also plays another important role, that of replacement of the old Ma with a new one that eventually develops from the micronucleus.

Morphology, Cytochemistry, and Fine Structure of the Macronucleus

The morphology of polyploid macronuclei is highly diverse, and is also species-specific. The number of Ma is, in most cases, one, occasionally two (particularly in Hypotrichida), and more rarely higher – up to 500 in *Dileptus anser* (Studitsky 1930, Vinnikova 1974c) and *Helicoprorodon orientalis* (Raikov 1967b). Macronuclei may be spherical, oval, or even C-shaped (*Didinium*), ribbon-shaped (*Bursaria*), moniliform (*Spirostomum*), branched (various Suctoria), and so on (Fig. 90).

In ciliates with a complex life cycle, the shape of Ma changes regularly at its different stages. For instance, the Ma is reticular in the adult form of *Chromidina elegans* (Apostomatida), but its regions condense, producing compact elongated macronuclei of the separating buds (Wermel 1928). The adult sessile specimens of *Magnifolliculina* (Heterotrichida) have spherical Ma, while vermiform swarmers have moniliform Ma (Uhlig 1964). Such cases may be regarded as recapitulation of the ancestral shape of Ma at the dispersive stages of the cycle (buds, swarmers).

The size of Ma is always rather large and more or less proportional to the mass of the cytoplasm. The length of Ma reaches 2 mm in large ciliates (*Spirostomum ambiguum*, *Bursaria truncatella*). Almost all large Ma have a complicated shape: they are moniliform, ribbon-shaped, and so forth. In this case, the surface of Ma increases proportionally to their volume, thus creating favourable conditions for the nucleocytoplasmic transport of substances.

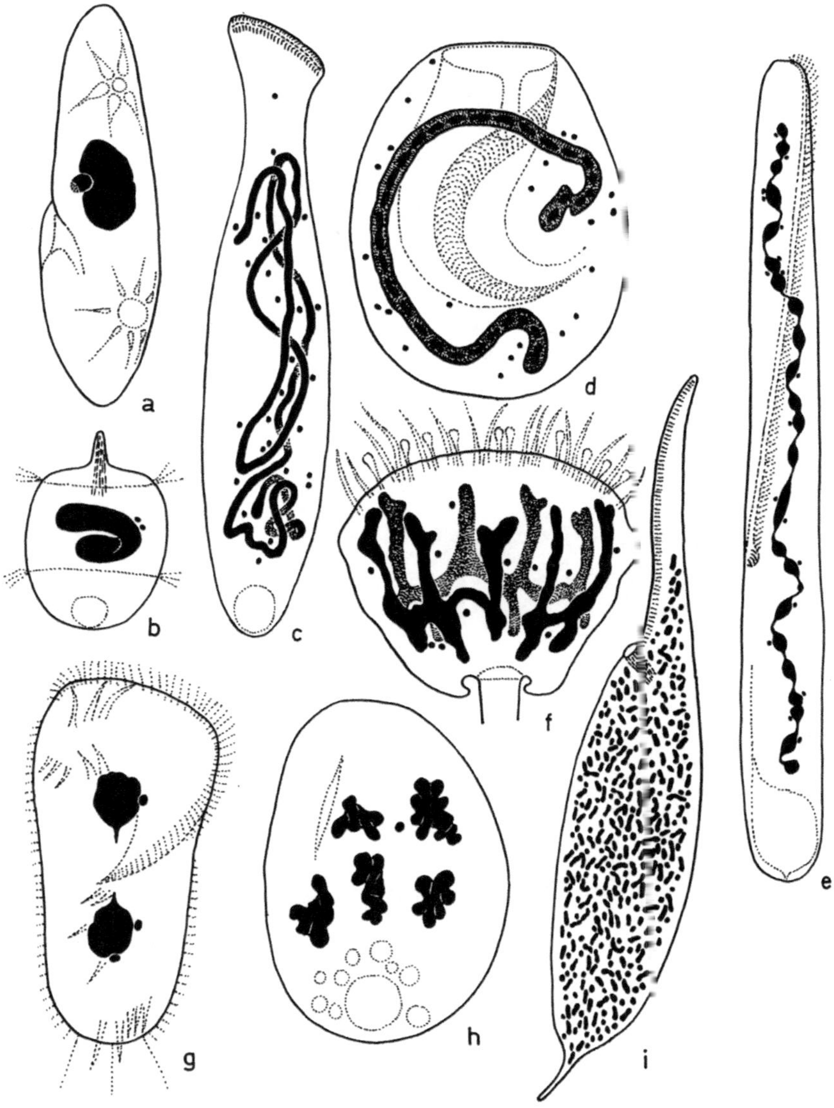

Fig. 90. Various forms of polyploid macronuclei (after RAIKOV 1969). *a Paramecium caudatum; b Didinium nasutum; c Spathidium spathula; d Bursaria truncatella; e Spirostomum minus; f Ephelota gemmipara; g Stylonychia mytilus; h Schizokaryum dogieli; i Dileptus anser.*

The structural components of Ma are the same as in the nuclei of other protozoans (see Chapter 1): they are the nuclear envelope, the karyolymph, the chromatin, and the nucleoli. However, macronuclei usually differ from most nonpolyploid nuclei in that they have abundant compact chromatin and numerous nucleoli, while the content of "free" karyolymph in them is relatively small; therefore, they are often called *massive nuclei*. Much recent electron microscopic data on Ma are available (see Table 6).

Table 6. *Ultrastructural characterization of ciliate macronuclei (except the order Karyorelictida)*

Species	Number of Ma	Chromatin elements	Nucleoli	Intra-nucleolar chromatin bodies	Inclusions	Nuclear envelope	References
Class Kinetofragminophorea							
Subclass Gymnostomatia							
Order Prostomatida							
Coleps hirtus	1	sm. bodies	many, subm.	?	–	cl.	Rodrigues de Santa Rosa 1976
Prorodon palustris	1	chr. str., chr. kar.	many, small	?	–	cl.	Puytorac and Savoie 1968
Helicoprorodon gigas	1	chr. str.	many, large	+	cyt. incl.	cl.	Kovaleva *et al.* 1977
H. multinucleatum	many	chr. str.	many, small	?	(bact.)	cl.	Puytorac and Kattar 1969
Sorogena stoianovitchae	1	sm. bodies	many, small	?	–	cl.	Bradbury and Olive 1980
Chaenea vorax	many	disp.	1, ret.	?	–	cl. + ER	Fauré-Fremiet and Ganier 1969
Lacrymaria olor	1	sm. bodies	many, large	large	fib. zones	cl.	Bohatier 1972, 1975
Lagynophrya fusidens	1	chr. str.	many, small	?	–	cl.	Grain 1970
Bryophyllum tegularum	1	sm. bodies	many, small	?	–	cl.	Bohatier *et al.* 1978
Spathidium (3 species)	1	sm. bodies	many, small	+	–	cl.	Bohatier *et al.* 1978, Ivashchenko 1980

Homalozoon vermiculare	1	sm. bodies	many, small and large	?	–	cl.	Saxena *et al.* 1978
Didinium nasutum	1	sm. bodies	many, large	+	–	cl. + ER	Rieder 1971, Karadzhan and Raikov 1977
Dileptus anser	many	sev. sm. bodies	1, cent.	?	–	cl. + ER	Vinnikova 1974c, 1974d, Bohatier 1977a, 1977b
Subclass Vestibuliferia							
Order Trichostomatida							
Isotricha sp.	1	chr. str.	sev.	?	bact.	cl.	Grain 1966
Balantidium coli	1	sm. bodies	many, small	?	–	cl.	Kraschenínnikow and Scholtyseck 1975, Avakyan 1976
B. caviae	1	sm. bodies	sev., small	?	–	cl.	Paulin and Kraschenínnikow 1973
Order Colpodida							
Colpoda cucullus	1	sm. bodies	1, comp.	?	–	cl.	Kawakami and Yagiu 1963
C. maupasi	1	sm. bodies	1, comp.	+	–	cl. + ER	Rudzinska *et al.* 1966, Lynn 1976a
C. steini	1	sm. bodies	1, comp.	+	–	cl. + ER	Lynn 1976b, Frenkel 1980
C. simulans	1	sm. bodies	1, comp.	+	–	cl.	Lynn 1976d
Tillina magna	1	sm. bodies	1, ret.	?	–	cl. + out. l., ER	Lynn 1976c, Frenkel 1979
Order Platyophryida							
Woodruffia metabolica	1	sm. bodies	many, small	?	–	cl. + Mi	Golder 1976
Platyophrya spumacola	1	sm. bodies	many, small	?	–	cl. + Mi	Dragesco *et al.* 1977
Cyrtolophosis mucicola	1	sm. bodies	sev., per.	?	–	cl. + Mi, bact.	Detcheva 1976

Table 6 (continued)

Species	Number of Ma	Chromatin elements	Nucleoli	Intra-nucleolar chromatin bodies	Inclusions	Nuclear envelope	References
Aristerostoma marinum	1	sm. bodies	sev., large	?	–	cl. + Mi	DETCHEVA and PUYTORAC 1979
Order Bursariida							
Bursaria truncatella	1	sm. bodies	many, large	+	–	cl.	MASHANSKY 1963, SERGEJEVA 1976, 1977
Order Entodiniomorphida							
Diplodinium sp.	1	sm. bodies	many, small	?	–	cl.	NOIROT-TIMOTHÉE 1960, ROTH and SHIGENAKA 1964
Epidinium ecaudatum	1	large bodies	many, small	?	–	cl.	GERASSIMOVA and SERAVIN 1978
Subclass Hypostomatia							
Order Nassulida							
Nassula ornata	1	ret.	many, small	?	–	cl.	RAIKOV 1966
Nassula sp.	1	sm. bodies	many, small	?	–	cl. + ER	TUCKER 1967
Order Cyrtophorida							
Chilodonella cucullulus	1	orthomere: large bodies; paramere: disp. + endosome	sev., in orthomere	+	–	cl. + ER	RADZIKOWSKI 1973a
Ch. uncinata	1	as above	as above	?	–	cl. + ER	PYNE 1978

Chlamydodon pedarius	1	orthomere: chr. str.; paramere: disp. + endosome	sev., per., in ortho- and paramere	?	–	cl.	Kaneda 1961b
Order Chonotrichida							
Spirochona gemmipara	1	as above	as above	?	–	cl.	Guénin and Fahrni 1971
Cavichona elegans	1	as above	as above	?	–	cl.	Karadzhan 1976
Order Rhynchodida							
Ancistrocoma pelseneeri	1	large bodies	many, small	?	–	cl. + ER	Khan 1969
Order Apostomatida							
Hyalophysa chattoni							Bradbury 1966, 1974
tomite	1	large bodies	sev., small, per.	?	–	cl.	
phoront	1	as above	as above	?	mtb. bundle	cl.	
Subclass Suctoria							
Podophrya parameciorum	1	large bodies	1, lob.	+	–	cl.	Jurand and Bomford 1965
Phalacrocleptes verruciformis	1	large bodies	sev., per.	?	–	cl.	Lom and Kozloff 1967
Acineta tuberosa	1	anast. th. str.	none	–	mtb. bundle	cl.	Bardele 1968, 1969
Paracineta limbata	1	as above	none	–	as above	cl.	Hauser 1970
Tokophrya infusionum							Rudzinska and Porter 1955, Rudzinska 1956, 1961, Millecchia and Rudzinska 1971
trophont	1	large bodies	none	–	as above	cl.	
swarmer	1	as above	none	–	–	cl.	

Table 6 (continued)

Species	Number of Ma	Chromatin elements	Nucleoli	Intra-nucleolar chromatin bodies	Inclusions	Nuclear envelope	References
Trematosoma bocqueti							BATISSE 1972
trophont	1	anast. th. str.	none or small	–	mtb. bundles	cl.	
swarmer	1	spon. mass	none	–	sep. mtb.	cl.	
Class Oligohymenophorea							
Subclass Hymenostomatia							
Order Hymenostomatida							
Tetrahymena, species of the *pyriformis* group	1	sm. bodies	many, per. cup-shaped, or sev. large aggregates	+	(RNA bodies)	cl.	ROTH and MINICK 1961, ELLIOTT *et al.* 1962, ELLIOTT 1963, FLICKINGER 1965, CAMERON and GUILE 1965, CAMERON *et al.* 1966, CHARRET 1969, NILSSON and LEICK 1970, NILSSON 1970, 1976a, 1976b, SATIR and DIRKSEN 1971
Paramecium, species of the *aurelia* group	1	sm. bodies	many, small	+	PF, PG, IG (bact.)	cl. + ER	DIPPELL and SONNEBORN 1957, JURAND *et al.* 1962, JURAND and JACOB 1969, JURAND and SELMAN 1969, STEVENSON and LLOYD 1971b, KRAWCZYNSKA 1979, 1980
P. caudatum	1	sm. bodies	many, small	+	(bact., fib. bundles)	cl. + ER	TSUJITA *et al.* 1954, VIVIER and ANDRÉ 1961, VIVIER 1963b, MASHANSKY 1963, OSSIPOV *et al.* 1976, OSSIPOV 1981

P. multimicronucleatum	1	sm. bodies	as above	+ (?)	–	cl. (+ bact.)	Jenkins 1970, Inaba and Kudo 1972
Neobursaridium gigas	1	sm. bodies	as above	?	bact.	cl. + ER	Nilsson 1969, Kattar 1975
Ophryoglena bacterocaryon	1	sm. bodies	as above	?	mtb. at periphery	cl.	Roque *et al.* 1965
Ichthyophthirius multifiliis							Peschkoff and Tikhomirova 1967, Mosevich 1968
trophont	1	sm. bodies	as above	?	mtb. bundles	cl.	
tomite	1	spon. mass	none	–	–	cl.	
Order Scuticociliatida							
Thigmophrya macomae	1	sm. bodies	sev., small	?	–	cl.	Khan 1970
Uronema sp.	1	sm. bodies	sev., per.	?	–	cl. + ER	Kaneshiro and Holz 1976
Cardiostomatella vermiforme	1	sm. bodies	sev., small	?	–	cl.	Puytorac *et al.* 1981
Order Astomatida							
Haptophrya (2 species)	1	sm. bodies	many, small	?	–	cl.	Puytorac 1959, 1963a
Anoplophrya commune	1	sm. bodies	sev., ret.	?	–	cl.	Puytorac 1961
Mesnilella trispiculata	1	chr. str.	many, small	?	(bact.)	cl. + gran.	Puytorac 1963b
M. fastigata	1	sm. bodies	none (?)	–	–	cl.	Mashansky 1963
Class Polyhymenophorea							
Subclass Spirotrichia							
Order Heterotrichida							
Protocruzia (2 species)	~ 10	sm. bodies	1, cent.	?	–	cl.	Ruthmann and Hauser 1974, Grolière *et al.* 1980
Blepharisma intermedium	1	chr. str.	many, small	?	–	cl.	Seshachar 1964

Table 6 (continued)

Species	Number of Ma	Chromatin elements	Nucleoli	Intra-nucleolar chromatin bodies	Inclusions	Nuclear envelope	References
B. americanum	1	anast. str.	as above	?	ds. spheres	cl.	Kennedy 1965, Jenkins 1977
B. wardsi	1	chr. str.	as above	+	as above	cl.	Inaba and Sotokawa 1968
Spirostomum ambiguum	1	sm. bodies	as above	?	–	cl.	Seshachar 1965, Astakhova *et al.* 1975, Zazulina and Polyakov 1976
Stentor coeruleus	1	sm. bodies	as above	+	PG, fib. zones	cl.	Paulin and Brooks 1975, Pelvat and Haller 1976, Skarlato 1977, 1979b
Climacostomum virens	1	sm. bodies	as above	?	–	cl.	Peck *et al.* 1975
Condylostoma magnum	1	chr. str.	as above	?	–	cl.	Bohatier 1978
Nyctotherus cordiformis	1	sm. bodies	as above	+	PF, PG	cl. + out. l.	Vinnikova and Golikova 1979
Order Oligotrichida							
Halteria grandinella	1	sm. bodies	sev., small	?	ds. spheres	cl.	Grain 1972
Favella ehrenbergii	2	chr. str.	many, small	?	–	cl.	Hedin 1975
Parafavella gigantea	2	chr. str.	many, small	?	–	cl.	Hedin 1975
Order Hypotrichida							
Urostyla grandis	40 80	anast. str.	sev., small	?	–	cl.	Inaba and Suganuma 1966
Keronopsis rubra	~ 100	chr. str.	1, cent.	?	–	cl.	Ruthmann 1973
Oxytricha platystoma	2	anast. str.	many, large	?	–	cl.	Chakraborty 1967

O. fallax	2	as above	many, large	?	–	cl.	Grimes 1973
Gastrostyla steini	4	as above	sev., large	?	–	cl.	Walker and Goode 1976
Stylonychia mytilus	2	as above	sev., large	?	–	cl.	Walker and Goode 1976
Euplotes patella	1	sm. bodies	many, large	?	–	cl.	Roth 1957
E. eurystomus	1	chr. bodies united in chains	many, large	?	RNP helices	cl.	Fauré-Fremiet *et al.* 1957, Kluss 1962, Ringertz *et al.* 1967b, Fauré-Fremiet and André 1968, Murti 1976a, Ruffolo 1978b, 1979

Abbreviations:

In all columns: sm. = small; sev. = several; th. = thick; ret. = reticular; sep. = separate; ds. = dense.

Chromatin elements: chr. = chromatin(ic); kar. = karyosome; disp. = dispersed; anast. = anastomosing (strands); spon. = spongy; str. = strands.

Nucleoli: subm. = submembranous; cent. = central; comp. = composite; per. = peripheral; lob. = lobulated.

Inclusions: PF = perichromatin fibrils; PG = perichromatin granules; IG = interchromatin granules; cyt. incl. = cytoplasmic inclusion; bact. = bacteria; fib. = fibrillar; mtb. = microtubules; parentheses indicate the facultative occurrence of an inclusion.

Nuclear envelope: cl. = classic (two-membraned with pores); + ER = connections of the outer membrane with endoplasmic reticulum; + Mi = common outer membrane of the macro- and micronucleus; + bact. = bacteria in perinuclear space; + gran. = granules in perinuclear space; + out. l. = outer additional layer.

Macronuclear Envelope

The envelope of Ma has a classical two-membrane structure and contains many pore complexes; it is frequently connected with the endoplasmic reticulum through membrane continuities (Table 6). It has already been described in Chapter 1. In spite of the great size of Ma, their envelope rarely has any additional layers. *Nyctotherus* has a fibrillar layer rich in ribosomes located on the outside of the macronuclear envelope (VINNIKOVA and GOLIKOVA 1979); a similar layer (though without ribosomes) is found in *Tillina* (LYNN 1976c). In some cases, symbiotic bacteria are located in special "pockets" of the perinuclear space of Ma in *Paramecium multimicronucleatum* (JENKINS 1970) and *Cyrtolophosis* (DETCHEVA 1976).

In *Cyrtolophosis* (DETCHEVA 1976, DIDIER *et al.* 1980), *Platyophrya* (DRAGESCO *et al.* 1977), *Woodruffia* (GOLDER 1976), and *Aristerostoma* (DETCHEVA and PUYTORAC 1979), members of the Platyophryida, the envelopes of the macronucleus and the micronucleus have the outer membrane in common, i.e. that membrane passes from one nucleus onto the other; the inner membranes of these two closely adjacent nuclei are separate (Fig. 91*a*). Transitions between the outer membranes of macro- and micronuclei were found also in *Uronema* (KANESHIRO and HOLZ 1976) and *Metacystis* (DETCHEVA *et al.* 1979). In these cases, however, the two nuclei are not so close to one another, and fragments of their outer membranes as well as islets of the cytoplasm can be seen between them.

Each node of the moniliform Ma in *Helicoprorodon gigas* sometimes contains an inner cavity, but only in some specimens of the population (RAIKOV 1967b). Electron microscopy has shown (KOVALEVA *et al.* 1977) that the cavity is filled with cytoplasm rich in mitochondria and is surrounded by a normal nuclear envelope (Fig. 91*b*). It is likely that such cytoplasmic inclusions appear in the macronuclear nodes only at certain stages of the cell cycle. Their functional role is still obscure.

Chromatin

The Ma of almost all ciliates (except Karyorelictida, certain Colpodida, and some enigmatic forms, such as *Protocruzia*) are very rich in DNA. The Feulgen reaction stains them heavily; only nucleoli remain unstained. It was demonstrated in cytochemical studies that the macronuclear chromatin contains not only DNA but also RNA (MOSES 1950, etc.) and basic proteins (ROSKIN and STRUVE 1953, ALFERT and GOLDSTEIN 1955, FAURÉ-FREMIET 1957, RAIKOV 1962a, etc.). Biochemical studies have shown that these basic proteins are true histones consisting of the usual five fractions (GOROVSKY 1970b, 1973, ISAACKS and SANTOS 1973, LIPPS and HANTKE 1975, etc.). Macronuclear chromatin, just as the

---

Fig. 91. Peculiarities of the envelope of some macronuclei (electron micrographs). *a* part of Ma (*1*) and Mi (*2*) sharing a common outer nuclear membrane in *Cyrtolophosis* (*3* pair of inner membranes which separate the two nuclei), 28000 ×; *b* section through a node of the moniliform Ma of *Helicoprorodon gigas* with an inner cavity (*7*) containing cytoplasm and mitochondria (*8*), 8500 ×; *4* nucleoli, *5* chromatin elements, *6* macronuclear envelope. *a* unpublished electron micrograph, courtesy of P. DIDIER; *b* after KOVALEVA and others (1977).

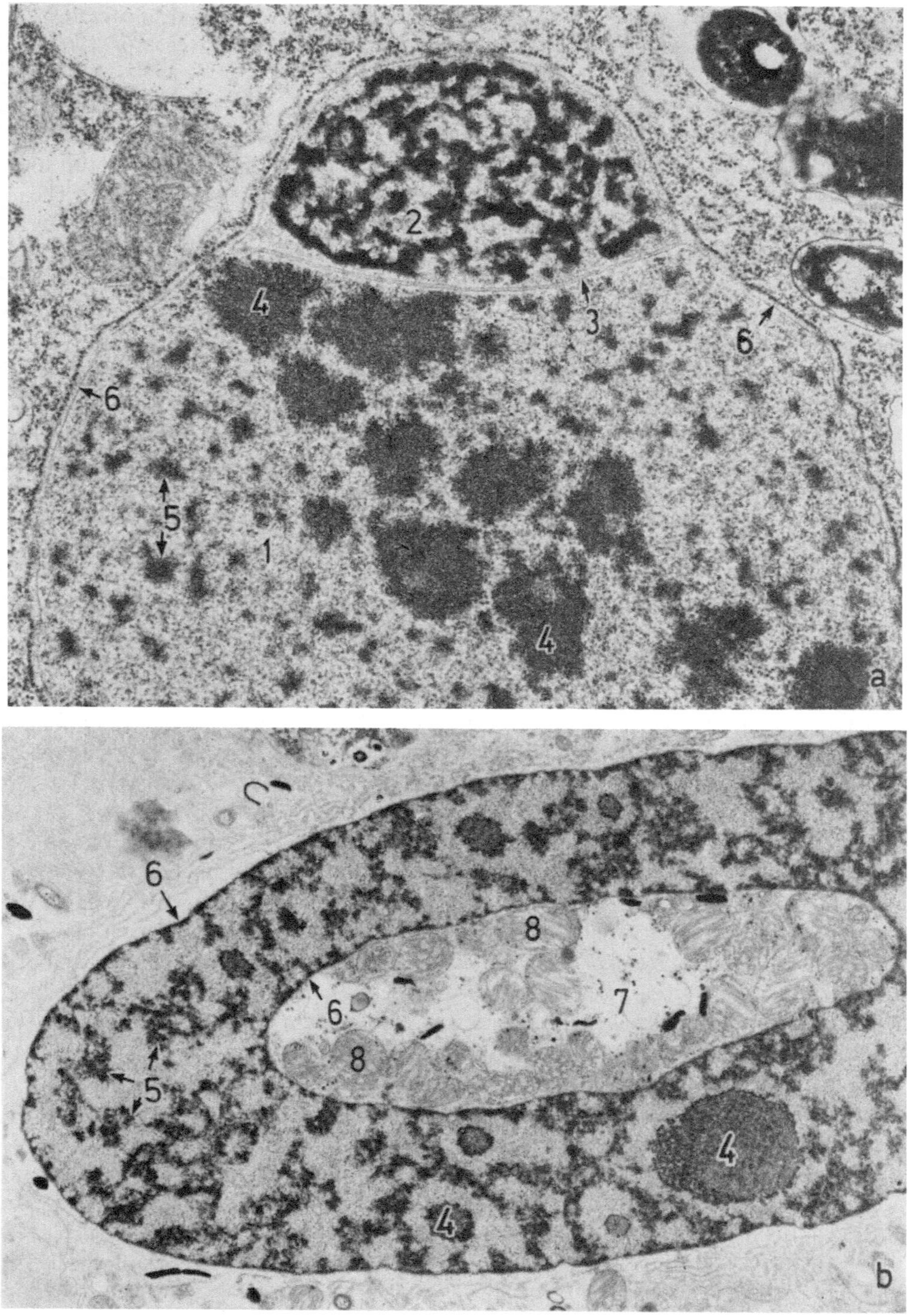

Fig. 91

chromatin of other eukaryotes, has a nucleosomal organization (LIPPS and MORRIS 1977, GOROVSKY *et al.* 1978, WADA and SPEAR 1980, etc.). Moreover, it contains a large amount of non-histone proteins (STEVENSON 1967, RINGERTZ *et al.* 1967a, CUMMINGS 1972, etc.).

Macronuclear chromatin elements are most frequently seen under the electron microscope as numerous discrete lumps of condensed chromatin known as "small bodies" (Table 6). They are usually of irregular shape, about 0.1–0.2 μm (up to 0.5 μm) in size, and have high electron density (Figs. 92, 93*a*, 94, 95*c*, 102*c*). The location of DNA in the "small bodies" was corroborated ultracytochemically by treatment with nucleases (GOROVSKY 1965), the elective "staining" of DNA (BOHATIER 1975), and the selective bleaching of chromatin in ultrathin sections with BERNHARD's (1969) solution of EDTA (BOHATIER 1975, VINNIKOVA and GOLIKOVA 1979, SKARLATO 1979b, KRAWCZYNSKA 1979, 1980, FRENKEL 1980). This location was also confirmed by ultrastructural autoradiography, which revealed ^{3}H-thymidine incorporation in the "small bodies" (JURAND and JACOB 1969, CHARRET 1969, NILSSON 1973). The bodies consist of densely packed chromatin fibres about 12 nm thick, which are chains of nucleosomes. Using the spreading technique, SAMUEL *et al.* (1981) revealed that sections of nucleosomal fibres wind up to form supranucleosome globules (30–40 nm), which aggregate into thicker (40–60 nm) knotty threads; in their turn, the latter wind up forming the "small bodies" of the *Paramecium* macronucleus.

It can occasionally be seen that the "small bodies" are interconnected as chains (MASHANSKY 1963, MORAT 1973, SERGEJEVA 1977, RUFFOLO 1978). The interconnection is visible in sections and especially in whole mounts of chromatin isolated from Ma and spread on the water surface. These bodies then form long chains, e.g. in *Paramecium tetraurelia* (WOLFE 1967) and *Colpidium campylum* (MORAT 1973). In the latter case, the chromatin bodies proved to be connected with one another by DNA and not proteins since treatment with pronase did not destroy the links between them. The interconnection of the "small bodies" can easily be seen also with the scanning electron microscope in the intact isolated Ma of *Paramecium primaurelia*, whose nuclear envelope has been removed (CUMMINGS and KOTASEK 1978). Sections of the Ma of *Bursaria truncatella* show that chromatin fibres pass parallel to one another through several chromatin bodies (SERGEJEVA 1977). In *Euplotes eurystomus*, some chromatin bodies in the chains are fixed at the inner membrane of the macronuclear envelope via special attachment sites. It is likely that all the chains are bound to the nuclear envelope in the same manner (RUFFOLO 1978b).

The organization of macronuclear chromatin in the form of numerous "small bodies" is typical of many Kinetofragminophorea, e.g. *Didinium* (Fig. 92*a*),

---

Fig. 92. Ultrastructure of the macronucleus in the prostomatid ciliates. *a* part of the Ma of *Didinium nasutum*, 10 000 × ; *b* a nucleolus of this Ma, 46 500 × ; *c* nucleolus of the Ma of *Lacrymaria olor* with a large body of intranucleolar chromatin (*4*) and fibrogranular lamellae (*7*), 27 000 × ; *d* fibrous zone (*8*) in the Ma of *L. olor*, 56 000 × ; *1* chromatin bodies, *2* fibrillar and *3* granular zones of the nucleolus, *4* intranucleolar chromatin body, *5* perinucleolar chromatin body, *6* macronuclear envelope. *a* and *b* unpublished electron micrographs, courtesy of B. P. KARADZHAN; *c* and *d* after BOHATIER (1975).

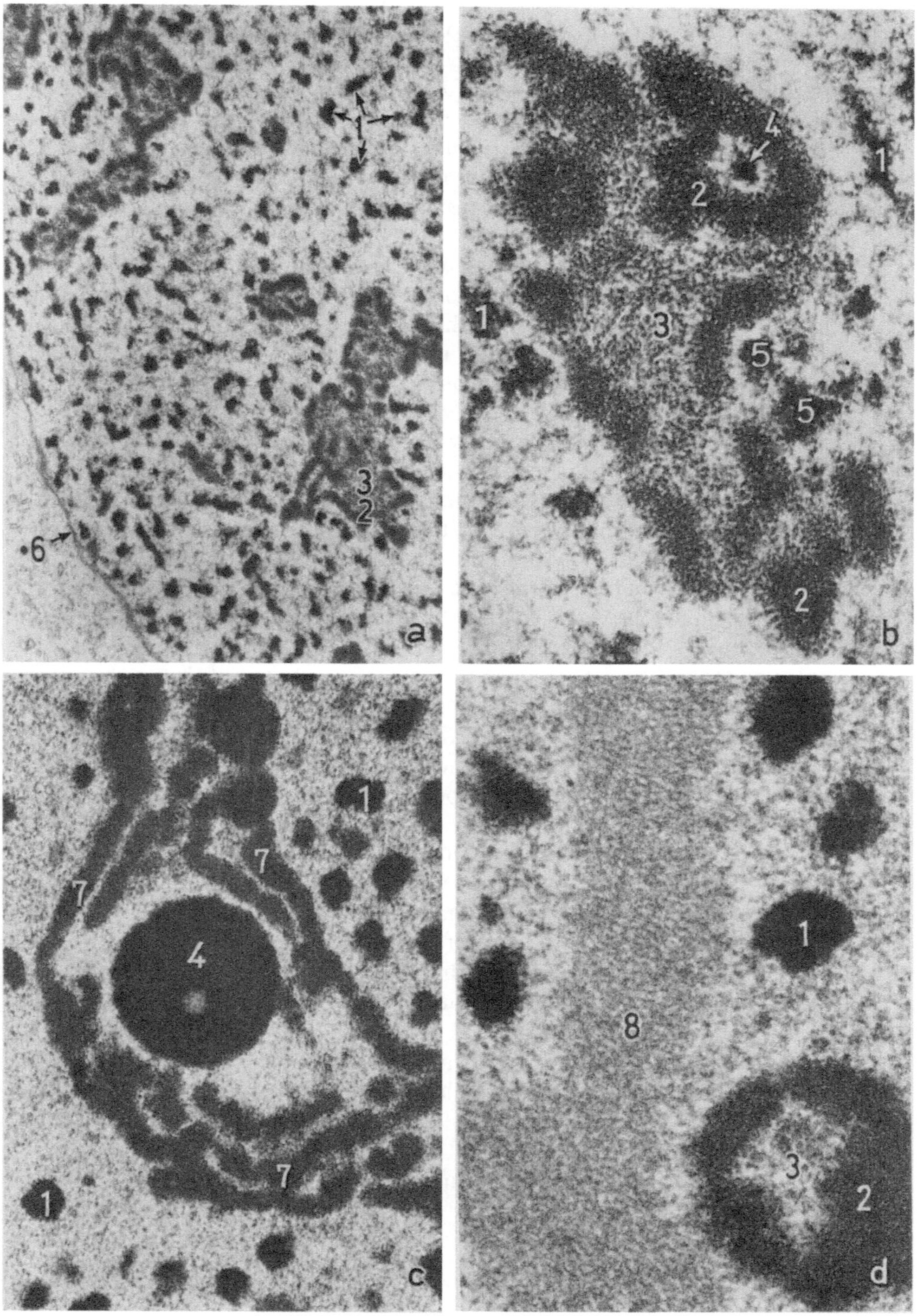

Fig. 92

Coleps, *Lacrymaria* (Fig. 92*c*, *d*), *Tillina*, *Colpoda* (Fig. 93*a*), *Bursaria* and *Balantidium*, and of Oligohymenophorea, e.g. *Tetrahymena* (Fig. 94*d*), *Paramecium*, and many others. This type of chromatin structure is also common among Polyhymenophorea, such as *Spirostomum*, *Stentor*, and *Euplotes* (Table 6).

In some ciliates, e.g. in *Blepharisma intermedium* and *B. wardsi* (SESCHACHAR 1964, INABA and SOTOKAWA 1968), the macronuclear chromatin is represented by convoluted threads whose cross-sections imitate isolated lumps. Such threads have probably originated as a result of the fusion of the "small bodies" interconnected in chains.

Chromatin threads may then anastomose and combine with one another, forming a reticulum. In *Prorodon palustris* (PUYTORAC and SAVOIE 1968), the chromatin reticulum is rather loose at the macronuclear periphery but condenses to form a spongy chromatin karyosome in the centre of the Ma. This was observed also by light microscopy in *Prorodon teres* (KASANZEFF 1928) and in the oligotrichid *Strombidium sulcatum* (SALVANO 1974b).

In *Blepharisma americanum* (KENNEDY 1965), the spongy mass of chromatin fills up the entire Ma (Fig. 95*b*). Such hypotrichids as *Urostyla*, *Keronopsis*, *Oxytricha*, *Gastrostyla*, and *Stylonychia* have the same chromatin structure (Table 6). Reticular chromatin is typical also of Ma in some Kinetofragminophorea, such as *Helicoprorodon* (Fig. 91*b*). Moreover, condensed chromatin in this case closely contacts the inner membrane of the macronuclear envelope: the chromatin forms a continuous submembranous layer with openings opposite the nuclear pores (KOVALEVA *et al.* 1977). In the hypotrichids *Gastrostyla* and *Stylonychia*, chromatin cords are likewise attached to the inner membrane of the macronuclear envelope via numerous electron-dense "plaques" (WALKER and GOODE 1976). A network of anastomosing chromatin cords is found also in Ma of several suctorian trophonts, e.g. in *Acineta*, *Paracineta*, and *Trematosoma* (Table 6). The chromatin condenses even more and becomes spongy in the swarmer of *Trematosoma* (BATISSE 1972).

SESHACHAR (1960, 1964, 1965) has confirmed that the Ma of *Blepharisma* and *Spirostomum* contain long chromatin strands: he isolated bundles of thin Feulgen-positive threads from Ma by centrifugation and treatment with KCN. If the moniliform Ma of *Spirostomum* is stretched with two needles, the nodes on it disappear. The chromatin threads, which pass in general along the macronucleus, are apparently folded in each node, but they straighten out when Ma is being stretched (SESHACHAR 1965). SESHACHAR believes that the chromatin threads are identical with the chromatin cords that are visible under the electron microscope and are chromosomes of the macronucleus. Long Feulgen-positive threads that spiral along the whole Ma were also detected in *Bursaria* (RUTHMANN and HECKMANN 1961).

A plexus of thin beaded chromatin threads was observed with the light microscope in the Ma of *Paramecium* "*aurelia*" (KIMBALL 1953), *P. bursaria* (SCHWARTZ 1957, 1958), *P. caudatum* (SATO and SAITO 1958, SAITO and SATO 1961a), *Tetrahymena* (SATO and SAITO 1959), *Vorticella* (MÜGGE 1957, SAITO 1961b), and several other ciliates. These threads apparently correspond to the chains of "small bodies" revealed by electron microscopy. In the aforementioned works, they were interpreted as despiralized macronuclear chromosomes.

SCHWARTZ (1978), however, has recently isolated such beaded threads from the unfixed Ma of *P. bursaria* and has shown that they were 4–6 times shorter than isolated despiralized micronuclear chromosomes. SCHWARTZ has concluded that Ma contain chromosome fragments and not entire chromosomes. Rod-shaped, chromosome-like formations interconnected in chains were revealed by light microscopy in the Ma of *Loxophyllum* (RUTHMANN 1963, Fig. 100*a*). The exact nature of these rods, however, is unknown.

Another variety of the structure of macronuclear chromatin is found among Suctoria and Apostomatida. Their Ma contain large discrete chromatin bodies. In *Tokophrya infusionum*, these bodies are spherical, with a diameter of about 0.5 μm, and are distributed throughout the nucleus (Fig. 95*d*); they number 100–300 per Ma (RUDZINSKA and PORTER 1955, RUDZINSKA 1956, 1961, MILLECCHIA and RUDZINSKA 1971). In *Podophrya parameciorum* (JURAND and BOMFORD 1965), a lobed nucleolus is in the centre of the macronucleus, while large (up to 3 μm) chromatin bodies lie at its periphery attached to the nuclear envelope from the inside and elongated radially. In *Tokophrya* and *Podophrya*, the chromatin bodies are much larger than the micronuclear chromosomes; they visually correspond more to the whole micronuclear chromosome set. Therefore, it could be that they are whole *genomes* that are spatially separated in Ma. But this possibility has not been checked yet by cytophotometric studies. Large (up to 2 μm) oval chromatin bodies have been found in the Ma of tomites and phoronts of the apostomatid *Hyalophysa* (BRADBURY 1966, 1974).

Nucleoli

Polyploid Ma usually contain many nucleoli that are either distributed throughout the nucleus (Fig. 92*a*) or, less frequently, occupy a submembranous position, e.g. in *Tetrahymena* (Fig. 94*d*). The number and size of nucleoli in Ma vary from small and multiple ones to relatively large and few (KASANZEFF 1928; see Table 6). Only one central nucleolus is found in each of the numerous Ma in *Dileptus anser* as well as in the Ma of the small parasitic suctorian *Podophrya parameciorum* (JURAND and BOMFORD 1965). One central nucleolus is seen with the light microscope also in the *Colpoda* species; it is spherical in *C. steini* or lobed in other forms (KIDDER and CLAFF 1938, BURT *et al.* 1941). Electron microscopy, however, has shown that this nucleolus is composite, i.e. it consists of several fused unit nucleoli (KAWAKAMI and YAGIU 1963, LYNN 1976a, 1976b, 1976d, FRENKEL 1980). For example, the nucleolus of *C. steini* has dozens of fibrillar cores embedded in a common granular mass (Fig. 93*a*). The larger nucleolar aggregate of *Tillina* is reticular and contains hundreds of fibrillar cores (LYNN 1976c, FRENKEL 1979). Cytochemical analysis has shown that macronuclear nucleoli contain RNA and proteins, mainly non-histone ones (ROSKIN and GINSBURG 1944, EGELHAAF 1955, SCHWARTZ 1956, FAURÉ-FREMIET 1957, RAIKOV 1962a, etc.).

In general, macronuclear nucleoli have a rather typical ultrastructure (see Chapter 1). They usually consist of fibrillar and granular zones. The former contains densely packed fibrils, about 5 nm thick, that consist of preribosomal RNA bound to proteins; the latter displays RNP granules about 14 nm in diameter, the precursors of ribosome subunits (LEICK 1969, NILSSON and LEICK

1970). Moreover, the macronuclear nucleoli of many ciliates contain chromatin bodies, i.e. nucleolar organizers (Table 6).

The spatial arrangement of the fibrillar and granular zones may differ. In most cases, fibrils form a dense core of the nucleolus, while more loosely packed granules are located at the periphery and constitute its cortex. Nucleolar chromatin always contacts the fibrillar core of the nucleolus and, therefore, usually occupies the central position in the nucleoli of this type. In *Tetrahymena*, for instance, all macronuclear nucleoli (in the specimens of a growing culture) are cupped, with their concave side facing the centre of the nucleus and the convex side adjacent to the macronuclear envelope (for references, see Table 6). The dense body of nucleolar chromatin is in the depression of the cupped nucleolus, bordering on its fibrillar core (CAMERON *et al.* 1966, CHARRET 1969, NILSSON and LEICK 1970, SATIR and DIRKSEN 1971, NILSSON 1976a, 1976b, etc.). The granular zone forms the convex outer surface of the nucleolus (Fig. 94*a*). In *Paramecium*, macronuclear nucleoli are not connected with the membrane; they have an even more distinct concentric structure with a DNA-containing centre, a fibrillar intermediate zone, and a granular cortex (JURAND and SELMAN 1969).

The presence of DNA in the intranucleolar chromatin bodies of *Tetrahymena* was demonstrated both ultracytochemically (GOROVSKY 1965, NILSSON 1976b) and autoradiographically (CHARRET 1969, Fig. 94*d*). This DNA has also been extracted from isolated nucleoli and its *in vitro* transcription visualized by the spreading technique (GALL *et al.* 1977, BORKHARDT and NIELSEN 1981). In *Paramecium* (KRAWCZYŃSKA 1980) and *Stentor* (SKARLATO 1979b), these bodies can be bleached with EDTA by BERNARD's method; this is typical of DNP. In *Colpoda steini*, each of the multiple fibrillar cores in the composite nucleolus has its own DNA-containing body bleached with EDTA (FRENKEL 1980).

A fibrillar core and a granular cortex are typical also of relatively large nucleoli in the macronucleus of *Helicoprorodon*. In this case, intranucleolar chromatin looks like a rather loose cord embedded in the peripheral part of the fibrillar zone of the nucleolus. The granular cortex is developed mainly at one side of the nucleolus, where it forms a "cap" (Fig. 93*b*). Such nucleoli resemble the amphinucleoli of some oocytes (KOVALEVA *et al.* 1977).

The inverted arrangement of the nucleolar components is less frequent in ciliates. In this case, the nucleolar fibrils usually form clumps or trabeculae at the periphery of the nucleolus, while its centre is filled with nucleolar granules. Nucleolar chromatin bodies are on the outside, bordering only on fibrillar material, e.g. in *Didinium* (Fig. 92*b*), *Dileptus* (VINNIKOVA 1974c), and *Nyctotherus* (VINNIKOVA and GOLIKOVA 1979).

The macronuclear nucleoli of *Lacrymaria* (BOHATIER 1975) contain an extremely large (up to 1 μm) spherical body of nucleolar chromatin (Fig. 92*c*). The

Fig. 93. Macronuclear nucleoli in some ciliates (electron micrographs). *a* Ma (*1*) and Mi (*2*) of *Colpoda steini* (*3* composite nucleolus of the Ma, *4* its fibrous cores, *5* its granular component, *6* small chromatin bodies), 27000×; *b* amphinucleolus in the Ma of *Helicoprorodon gigas* (*1* fibrous core, *2* granular cortex, *3* intranucleolar chromatin, *4* chromatin strands of Ma), 24000×. *a* after LYNN (1976b); *b* after KOVALEVA and others (1977).

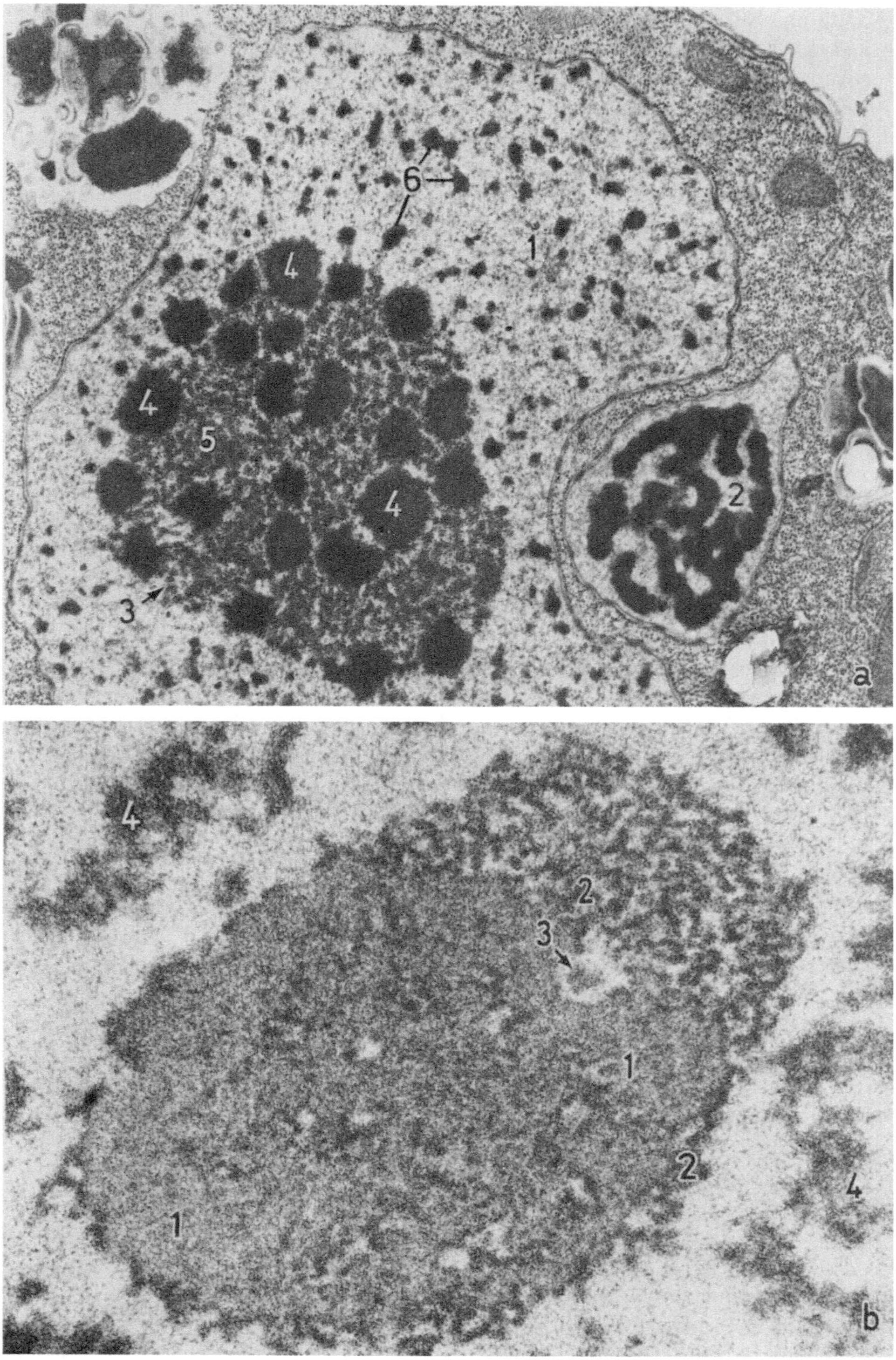

Fig. 93

body produces ultracytochemical reactions for DNA: it is bleached with EDTA and is electively stained with thallium ethylate. One or several fibrillar lamellae covered with granules on both sides protrude from the body (Fig. 92*c*). The lamellae can be superposed as scales, anastomose, and so on; they form the nucleolus proper. The nucleolar chromatin bodies of *Lacrymaria* are much larger than the free chromatin "small bodies" and have another ultrastructure, which is distinctly granular. One may assume, therefore, that these bodies contain amplified extrachromosomal DNA that represents ribosomal genes (rDNA). The presence of such DNA in the Ma was confirmed in biochemical studies involving *Tetrahymena* and *Paramecium* (see p. 328); an analysis of isolated nucleoli in *Tetrahymena* has shown that nucleolar chromatin consists of rDNA (LEER *et al.* 1978, HIGASHINAKAGAWA *et al.* 1979).

The macronuclear nucleoli of *Chilodonella* have a peculiar structure. The nucleolar cortex is granular, but a long, convoluted channel is found in the fibrillar core. A filament passes along the channel axis; RADZIKOWSKI (1973a) believes that it contains DNA and is a nucleolar organizer. However, BOHATIER (personal communication) has shown that the "filament" is in fact a thin membranous tubule, probably a deep invagination of the inner membrane of the macronuclear envelope, which enters each nucleolus and forms loops there (Fig. 97*c*). The molecules of nucleolar DNA (which are possibly extrachromosomal) may be attached radially to this tubule. However, the precise location of nucleolar DNA in *Chilodonella* is still unknown.

The relative content of fibrils and granules in macronuclear nucleoli may also vary within a wide range; it seems to depend on the ratio between the rate at which preribosomal RNA is processed and that at which the ribosome subunits are released into the cytoplasm. For instance, the nucleoli of *Bursaria* are completely fibrillar (SERGEJEVA 1977), while those of *Euplotes* are mainly granular (KLUSS 1962). Many suctorians have no typical nucleoli at all (Fig. 95*d*, Table 6).

The morphology of nucleoli may change drastically, depending on the physiological state of a ciliate. In *Paramecium*, multiple small nucleoli fuse when the ciliates are starved, forming several vacuolized masses (GROMOVA 1941, EGELHAAF 1955). Even when ciliates are fed daily, the ultrastructure of their nucleoli undergoes cyclic changes due to the temporary fusion of small nucleoli into composite reticular ones (VIVIER 1960a, 1963b).

These changes were studied especially in *Tetrahymena*. The small cupped macronuclear nucleoli, which are typical of the exponentially growing cells (Fig. 94*a*, *d*), fuse into large aggregates during the stationary phase of growth (CAMERON and GUILE 1965, CAMERON *et al.* 1966, SATIR and DIRKSEN 1971,

---

Fig. 94. Macronuclear nucleoli of *Tetrahymena* (electron micrographs). *a* nucleoli (*1*), nucleolar chromatin (*2*) and macronuclear chromatin bodies (*3*) after ribonuclease treatment; the contrast of the fibrous (*4*) and granular (*5*) RNP components of the nucleolus has decreased, 84000 × ; *b* nucleolus fused from two unit ones: it has two nucleolar chromatin bodies (*2*), 60000 × ; *c* nucleolar aggregate from which granular trabeculae (*5*) separate, 60000 × ; *d* incorporation of ^{3}H-thymidine only into nucleolar chromatin (a radioautograph; *6* silver grains), 21000 × ; *7* macronuclear envelope, *8* cytoplasm with ribosomes. *a*, *b*, and *d* after CHARRET (1969); *c* after NILSSON (1976a).

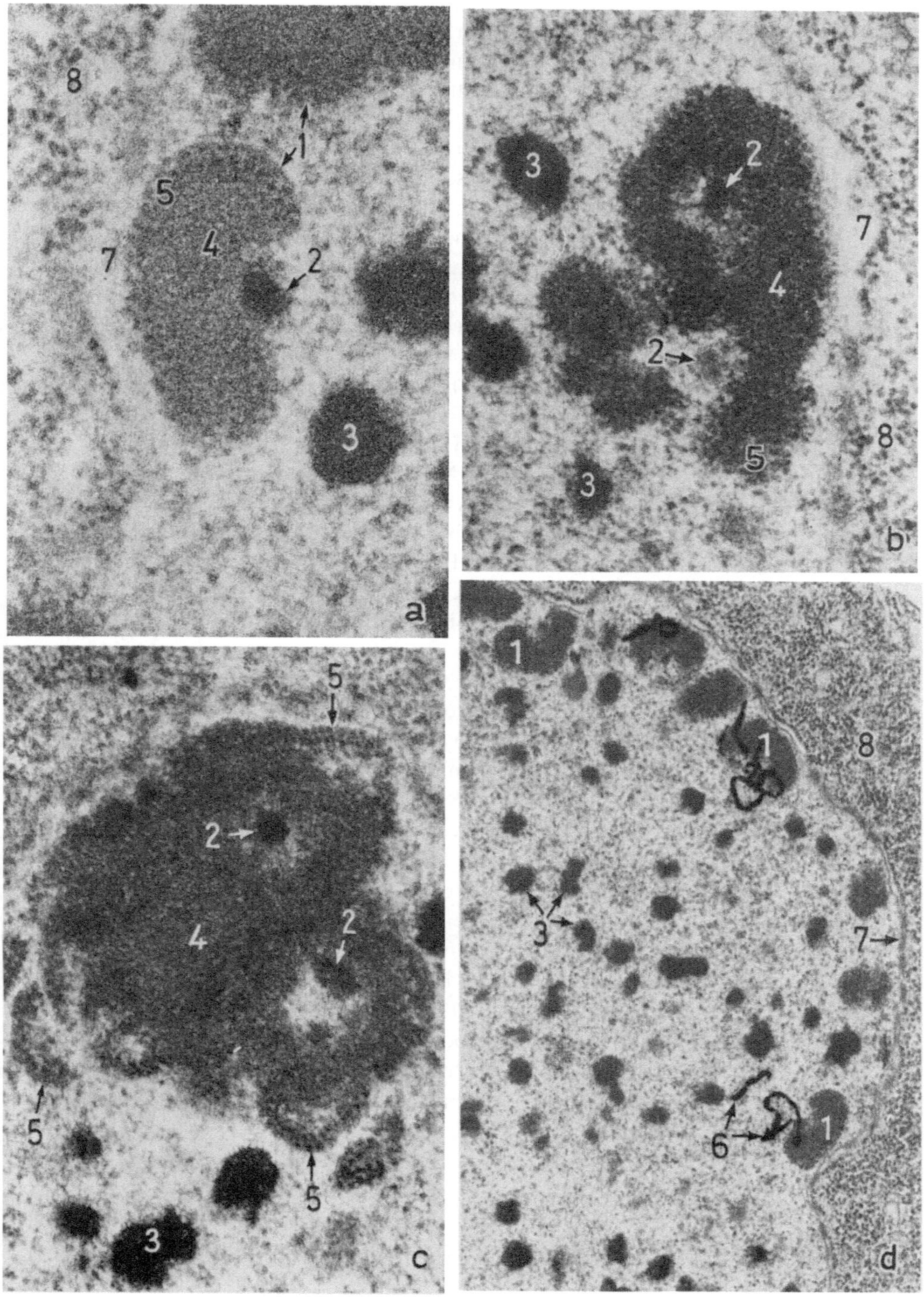

Fig. 94

NILSSON 1976a). This occurs also when the ciliates are completely starved even for a short time (LEVY and ELLIOTT 1969, NILSSON and LEICK 1970, NILSSON 1976a, 1976b), when the cultivation temperature is elevated to 34 °C (LEVY *et al.* 1969, NILSSON 1976a), and when the medium contains dimethyl sulfoxide (NILSSON 1977), which reversibly inhibits RNA synthesis and other functions of the cell, etc. When the nucleoli fuse into aggregates, they lose their cupped shape and usually separate from the macronuclear envelope. The content of granules in them decreases while that of fibrils increases; several lumps of nucleolar chromatin can be seen in them (Fig. 94*b*, *c*). The synthesis of rRNA in such nucleoli is strongly inhibited (CAMERON and GUILE 1965, LEICK 1969, NILSSON and LEICK 1970, LEICK *et al.* 1976). Once the ciliates are transferred to a fresh medium at optimal temperature (28 °C), the aggregates dissociate into typical cupped nucleoli, and rRNA synthesis in them is resumed even earlier (KAFFENBERGER and ECKERT 1980).

Similar morphological changes of macronuclear nucleoli are induced in *Tetrahymena* by actinomycin D and cycloheximide, which inhibit RNA synthesis (SATIR and DIRKSEN 1971, ECKERT and FRANKE 1975). Both compounds also inhibit the transport of rRNA from the Ma to the cytoplasm. Another inhibitor, thioacetamide, does not inhibit RNA synthesis in the nucleoli, but it blocks the processing of preribosomal RNA, which then accumulates as peculiar, coarsely granular masses (ECKERT and FRANKE 1975). In *Dileptus anser*, actinomycin D makes the only nucleolus in each Ma spherical and devoid of granules; at the same time, it blocks ^{3}H-uridine incorporation into the nucleolus (BOHATIER 1977a, 1977b).

Nucleolus-like bodies of a peculiar type appear in the macronucleus of *Tetrahymena* when cell cycles are synchronized by a series of heat shocks (34 °C). These formations are called "RNA bodies" (Fig. 95*a*). They have a purely fibrillar ultrastructure and contain no chromatin elements (ELLIOTT *et al.* 1962, NILSSON and ZEUTHEN 1974, NILSSON 1976a). "RNA bodies" remain in Ma until the end of the first synchronous division of the cells and then disappear. Their chemical nature, origin, and the mode of disappearance are obscure. It is likely that they are produced by the complete segregation of the fibrillar component of the fused nucleoli. "RNA bodies" are said to be released into the cytoplasm as nuclear fragments bud off Ma (ELLIOTT *et al.* 1962, LEVY *et al.* 1969). According to other data, however, they are resorbed in the Ma (NILSSON and ZEUTHEN 1974).

Regular changes in the morphology of macronuclear nucleoli were found during the regeneration of the anterior body part in *Lacrymaria* (BOHATIER 1972). A few minutes after the operation, nucleoli become annular, and in 1–1.5 hour, they become loose and reticular, comprising a large number of fibrogranular lamellae (as in Fig. 92*c*).

Karyolymph and Inclusions in the Macronucleus

The macronuclear karyolymph is seen under the electron microscope as a general fibrillar matrix (Figs. 91–95). The macronuclear matrix can be obtained in pure form by extracting nucleic acids and soluble proteins from isolated Ma; in this case, it can be seen that the matrix is a common skeletal framework of the nucleus that interconnects all its structural components. Electron microscopy of such extracted Ma of *Tetrahymena* shows that the matrix comprises not only a common fibrillar network but also residues of chromatin bodies, nucleoli, and the nuclear envelope, particularly its pore complexes (HERLAN and WUNDERLICH 1976, WUNDERLICH and HERLAN 1977, HERLAN *et al.* 1979). A biochemical analysis of the nuclear matrix has shown that almost 97% of it is protein, which corresponds to about 15% of the total protein of Ma. The matrix also contains small quantities of phospholipids as well as firmly bound DNA and RNA (HERLAN and WUNDERLICH 1976). The proteins of the matrix include structural proteins with a molecular weight of 50000–70000 and a contractile protein with a molecular weight about 18000, as a result of which the whole macronuclear matrix can considerably contract (WUNDERLICH and HERLAN 1977). The contractile protein does not belong to the actinomyosin system, however, since the contraction of the nuclear matrix does not depend on ATP. The protein matrix firmly binds certain high molecular weight RNA's (preribosomal ones and others) as well as some low molecular weight "small nuclear RNA's" (HERLAN *et al.* 1979).

The macronuclear karyolymph always contains various RNP particles: nucleolar granules that migrate to the cytoplasm and special RNP-containing structures. The latter are abundant near the surface of chromatin bodies. They include *perichromatin fibrils*, which seem to be new gene transcripts that form a rim around macronuclear chromatin bodies; this rim is easily discernible once the chromatin is bleached with EDTA. They were found in *Paramecium* (KRAWCZYNSKA 1979, 1980) and *Nyctotherus* (VINNIKOVA and GOLIKOVA 1979). Perichromatin fibrils are apparently transformed into *perichromatin granules* with a diameter of 40–50 nm (the storage and/or transport form of messenger RNA; see BOUTEILLE *et al.* 1974). In the Ma of ciliates, they are also arranged in the vicinity of chromatin bodies. They have been detected in *Paramecium*, *Nyctotherus*, and *Stentor* (SKARLATO 1979b).

The macronucleus of *Euplotes* contains "nuclear helices" instead of perichromatin granules. The "helices" resemble those of amoebae (see Chapter 1) and seem to be a transport form of mRNA that occurs only in protozoans (MURTI 1976a, RUFFOLO 1978b). Such "helices" are found at the periphery of both chromatin bodies and nucleoli; they can migrate to the cytoplasm through pores in the macronuclear envelope (RUFFOLO 1979).

Groups of *interchromatin granules* were also observed in the macronuclear karyolymph (NILSSON 1976a, FRENKEL 1980, KRAWCZYNSKA 1980). The chemical nature and functions of these granules are obscure, however.

Unlike the paradiploid Ma of Karyorelictida (see Chapter 7), polyploid Ma rarely have formed protein inclusions. Electron-dense homogeneous spheres about 1 μm in diameter were found in meshes of the macronuclear chromatin network of *Blepharisma americanum* (Fig. 95*b*). The spheres contain protein and a polysaccharide that can be digested with amylase (KENNEDY 1965). INABA and SOTOKAWA (1968) detected purely proteinaceous and less dense spheres in the Ma of *B. intermedium*. In the Ma of *Halteria*, such spheres develop in nucleoli and not in the karyolymph (GRAIN 1972). Their chemical nature is unknown.

Fibrillar zones that look like branched ribbons 0.5–0.8 μm wide and up to 10 μm long were observed in the Ma of certain *Lacrymaria* cells. These zones are formed by a mass of fibrils with low electron density and a diameter of 5–6 nm (Fig. 92*d*). Ultracytochemical data suggest that the fibrils have neither DNA nor RNA. Apparently, they are proteinaceous (BOHATIER 1975). SKARLATO (1977) found similar inclusions of loose fibrillar material in the form of "nuclear vacuoles" with a diameter up to 4 μm in the Ma of *Stentor*. This material cannot be DNP since it is not bleached with EDTA (SKARLATO 1979b). A bundle of microfilaments 8–9 nm thick fills the narrow bridges between Ma in the hypotrichids *Gastrostyla* and *Stylonychia*. It could be that the filaments are responsible for the rigidity of the bridges, which have neither chromatin nor nucleoli (WALKER and GOODE 1976).

The nature of fibrillar bundles in the Ma of *Paramecium caudatum*, which are hereditary in some clones, is obscure (VIVIER and ANDRÉ 1961). These bundles are Feulgen-negative, rich in protein, and possibly contain some RNA. Electron microscopy has shown that they are bundles of straight, nontubular fibrils 25 nm thick. It is likely that these fibrils are intranuclear RNA-containing viruses (Fig. 95*c*).

The nondividing Ma of some ciliates include *bundles of microtubules* that are particularly characteristic of suctorian trophonts (Table 6). One thick bundle of microtubules traverses the nondividing Ma of *Tokophrya* (Fig. 95*d*). The microtubules are not in contact with the chromatin bodies, but they are often surrounded by clusters of RNP granules (MILLECCHIA and RUDZINSKA 1971). The Ma of the adult *Acineta* and *Trematosoma* trophonts have many microtubular bundles. The microtubules are in close contact with granular material that resembles nucleolar granules (BARDELE 1968, 1969, BATISSE 1972).

Besides suctorians, a thick axial bundle of parallel microtubules was detected in the Ma of the encysted stage (phoront) of *Hyalophysa* (BRADBURY 1974) but not in the actively swimming tomite (BRADBURY 1966). In the suborder Ophryoglenina (order Hymenostomatida), bundles of microtubules occur at the periphery of Ma in the *Ichthyophthirius* trophont (PESCHKOFF and TIKHOMIROVA 1967, MOSEVICH 1968) where they develop especially during encystment that precedes a series of divisions. The role of microtubules in the nondividing Ma will be discussed further (p. 318).

Heteromeric Macronuclei

So far, we have considered the *homeomeric* Ma that are not divided into segments. Conversely, *heteromeric* Ma consist of two distinct parts (karyomeres): a *paramere* poor in DNA and an *orthomere* rich in DNA. Such Ma are typical of the orders Cyrtophorida and Chonotrichida (subclass Hypostomatia).

Fig. 95. Inclusions in the macronucleus (electron micrographs). *a* "RNA body" (*1*) in the Ma of *Tetrahymena pyriformis* after the synchronous division induced by heat shocks, 32000 ×; *b* glycoprotein sphere (*2*) surrounded with chromatin strands (*3*) in the Ma of *Blepharisma americanum*, 39000 ×; *c* bundle of fibres (*4*) in the Ma of *Paramecium caudatum*, 30000 ×; *d* Ma of nondividing *Tokophrya infusionum* with a thick bundle of microtubules (*5*) among large chromatin bodies (*6*), 13700 ×; *7* chromatin bodies, *8* nucleoli, *9* macronuclear envelope. *a* after NILSSON and ZEUTHEN (1974); *b* after KENNEDY (1965); *c* after VIVIER and ANDRÉ (1961); *d* after MILLECCHIA and RUDZINSKA (1971).

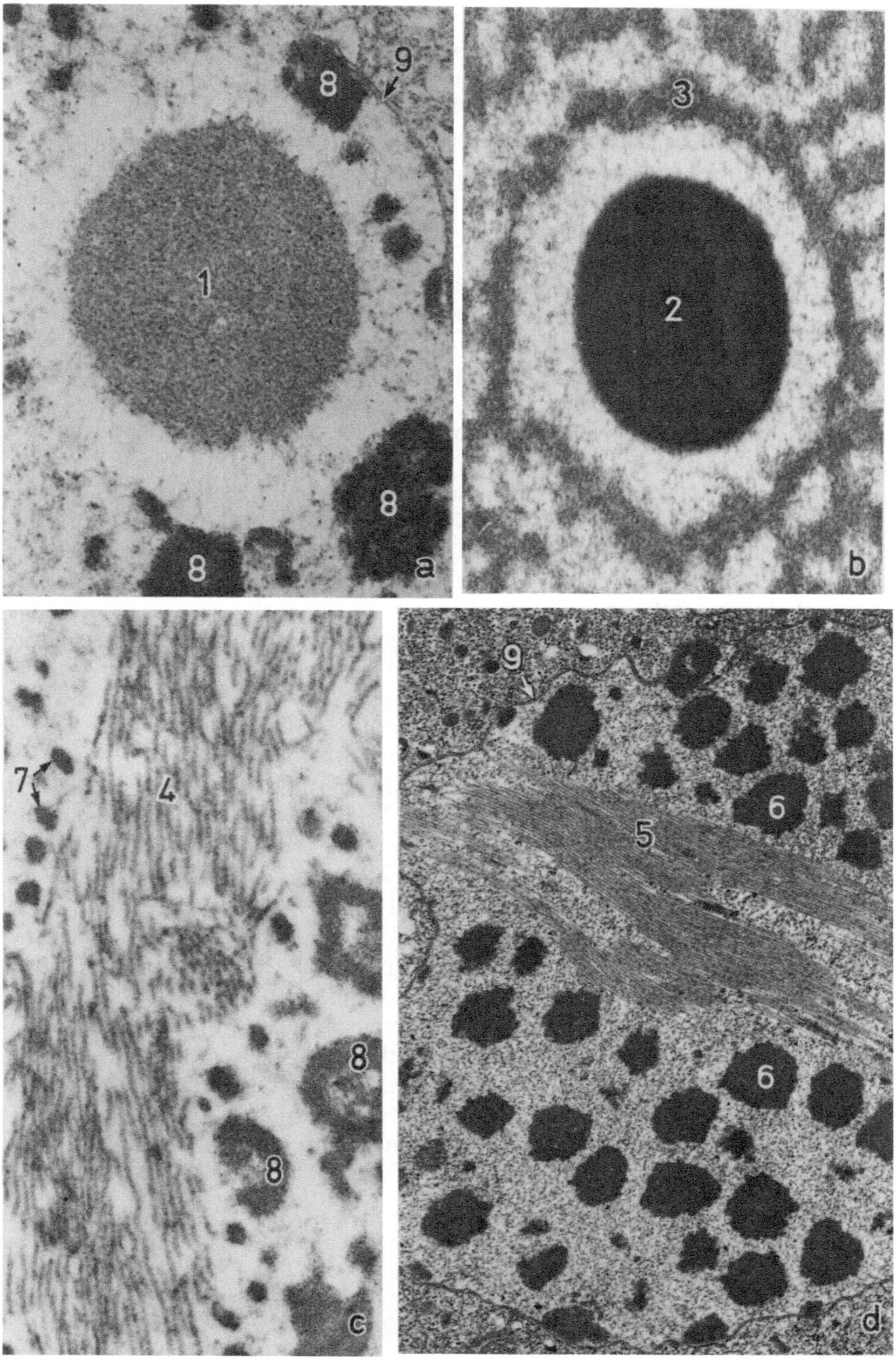

Fig. 95

There are two types of heteromeric Ma: one type has a *juxtaposed* arrangement of karyomeres when one of them is located behind the other (Fig. 96*a*), and the other has a *concentric* arrangement of karyomeres when the paramere is surrounded with the orthomere (Fig. 97*a*).

The heteromeric Ma of the juxtaposed type were studied by light microscopy in the cyrtophorids *Dysteria* and *Chlamydodon* (FAURÉ-FREMIET 1957, KANEDA 1960, 1961a) as well as in the chonotrichids *Spirochona*, *Heliochona*, *Lobochona*, and some other forms (GUILCHER 1951, TUFFRAU 1953, DOBRZAŃSKA-KACZANOWSKA 1963, MATSUDO 1966, JANKOWSKI 1973). The Ma of this type were examined by electron microscopy in *Chlamydodon* (KANEDA 1961 b), *Spirochona* (GUÉNIN and FAHRNI 1971) and *Cavichona* (KARADZHAN 1976). The paramere is always clearly separated from the orthomere, although there is no membrane between them (Fig. 96*a*); a continuous layer of condensed chromatin of the orthomere often serves as a boundary between the karyomeres (GUÉNIN and FAHRNI 1971, KARADZHAN 1976). A nuclear envelope of usual structure surrounds the two karyomeres together.

The heteromeric Ma of the concentric type were studied most thoroughly in *Chilodonella* with the use of both light microscopy (MACDOUGALL 1925, 1936, REICHENOW 1928, SESHACHAR 1950b, 1953, RADZIKOWSKI 1965) and electron microscopy (RADZIKOWSKI 1973a, PYNE 1978). The boundary between the peripheral orthomere and the central paramere is less distinct in this case (Fig. 97*a*) and is not marked with a chromatin layer. Cytochemical properties and the ultrastructure of karyomeres are rather similar in both types of heteromeric Ma.

The *orthomere* is always definitely Feulgen-positive, thus resembling the whole homeomeric Ma. In the Ma of the juxtaposed type, the chromatin is usually a network of anastomosing electron-dense threads 0.1–0.3 μm thick (Fig. 96*a*); they were called "chromonemata" in light microscopic studies (KANEDA 1960, 1961a). A lighter band appears in the orthomere of these Ma shortly before division; the band may be connected with the replication of the orthomere DNA (Fig. 96*a*).

In the concentric macronucleus of *Chilodonella cucullulus*, the ultrastructure of chromatin in the orthomere is different: it consists of discrete electron-dense clumps 0.1–0.5 μm in size that resemble the "small bodies" of many homeomeric Ma. The chromatin bodies of the orthomere in *Ch. uncinata* are even larger (up to 1 μm) and spherical (PYNE 1978, Fig. 97*a*). No replication bands have been found. Therefore, both types of orthomeres in heteromeric Ma fit the same varieties of chromatin elements (reticular and discrete) that occur in homeomeric Ma.

As do homeomeric macronuclei, the orthomere always contains nucleoli located mainly at the periphery of Ma in *Chlamydodon*, *Spirochona* (Fig. 96*a*), *Cavichona*, and *Chilodonella* (Fig. 97*a*). In all species studied, the nucleoli are compact and

---

Fig. 96. Heteromeric macronucleus of *Spirochona gemmipara* and its division (electron micrographs). *a* longitudinal section through Ma in the later half of the interphase, 5300×; *b* late stage of macronuclear division, 3900× (cf. Fig. 106*i*); *1* orthomere, *2* paramere, *3* nucleoli, *4* endosome, *5* replication band of the orthomere, *6* macronuclear envelope, *7* material of the future orthomere, *8* material of the future paramere, *9* material of the future epimere, *10* median thickening, *11* micronucleus. *a* after FAHRNI (1977); *b* after GUÉNIN and FAHRNI (1971).

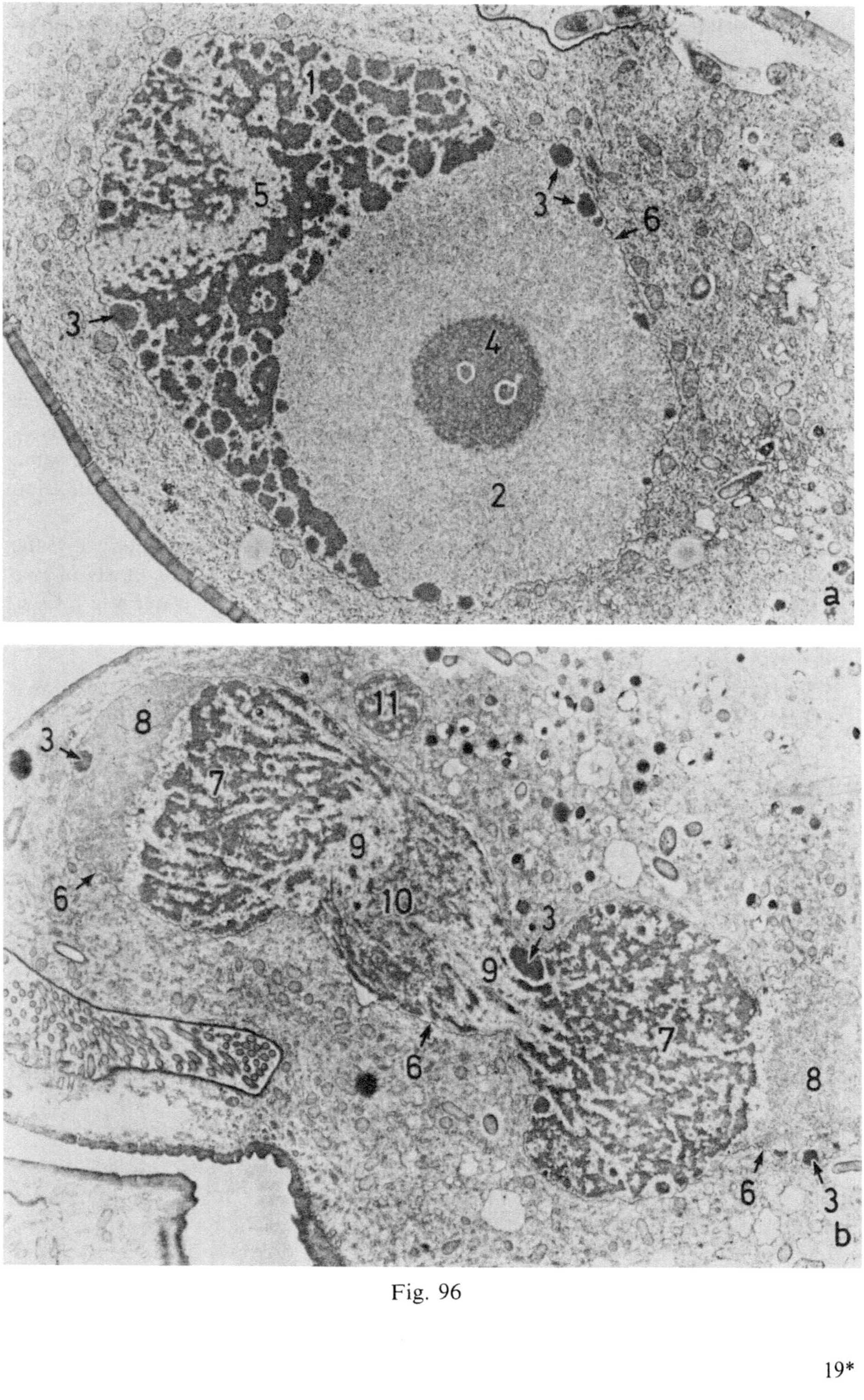

Fig. 96

have a fibrillar core and a granular cortex; the nucleoli of *Chilodonella* also have nucleolar organizers in the form of twisted channels with a tubular membranous axis (Fig. 97*c*; cf. p. 284).

In contrast to the orthomere, the *paramere* has neither clumps nor threads of condensed chromatin (Figs. 96*a*, 97*a*). Its "matrix" in both types of heteromeric Ma is faintly Feulgen-positive (FAURÉ-FREMIET 1957). It is slightly denser than the "matrix" of the orthomere (Fig. 97*a*) and has an indistinct fibrillar ultrastructure (Fig. 97*b*). In juxtaposed macronuclei, nucleoli (the same as in the orthomere) are arranged at the periphery of the paramere that borders on the nuclear envelope. In concentric macronuclei, the paramere does not come into contact with the nuclear envelope and has no nucleoli (Fig. 97*a*).

In both types of heteromeric Ma, a distinctly Feulgen-positive body (*endosome*) develops in the centre of the paramere during the interphase (REICHENOW 1928, SESHACHAR 1950b, 1953, GUILCHER 1951, TUFFRAU 1953, FAURÉ-FREMIET 1957, RADZIKOWSKI 1969). Ultrastructural data show that the endosome of *Cavichona* is a plexus of condensed chromatin threads thinner than the chromatin threads of the orthomere (KARADZHAN 1976). The endosome of *Chilodonella* has the same structure (RADZIKOWSKI 1973a, PYNE 1978). It is formed by anastomosing chromatin threads about 70 nm thick (Fig. 97*b*).

At the beginning of the interphase, the endosome is usually absent, while the paramere (particularly in the Ma of the juxtaposed type) often consists of two unequal parts located on both sides of the orthomere (GUÉNIN and FAHRNI 1971, KARADZHAN 1976). In these cases, the lesser segment of the paramere is sometimes called the *epimere* (JANKOWSKI 1973).

The functional role of macronuclear heteromery remains obscure. It is known that the DNA of the orthomere in *Chilodonella* replicates later than that of the paramere, while endosomal DNA does not replicate at all and is eliminated in the course of division. RNA synthesis is the most intensive in the orthomere, and less so in the paramere, while it is absent in the endosome (RADZIKOWSKI 1976). The types of RNA synthesized in the two different karyomeres have not been established yet.

DNA Content of the Macronucleus

The DNA content of Ma is always much higher (occasionally hundreds and even thousands of times) than that of Mi. The respective cytophotometric data are presented in Table 7. In the fifties and early sixties, such data served as one of the foundations of the theory of macronuclear polyploidy. The degree of macronuclear ploidy (n) was determined at the time simply as a double ratio between the

---

Fig. 97. Ultrastructure of the heteromeric macronucleus of *Chilodonella* (electron micrographs). *a* section through the Ma of *Chilodonella uncinata*, 17 000 × ; *b* endosome of the paramere of *Ch. cucullulus*, 17 000 × ; *c* part of the nucleolus of the *Ch. cucullulus* orthomere with a tube penetrating into it and cut tubes (arrows) inside the nucleolus, 21 000 × ; *1* macronuclear envelope, *2* chromatin bodies of the orthomere, *3* nucleoli of the orthomere, *4* paramere, *5* endosome. *a* after PYNE (1978); *b* and *c* unpublished micrographs, courtesy of J. BOHATIER and P. DIDIER.

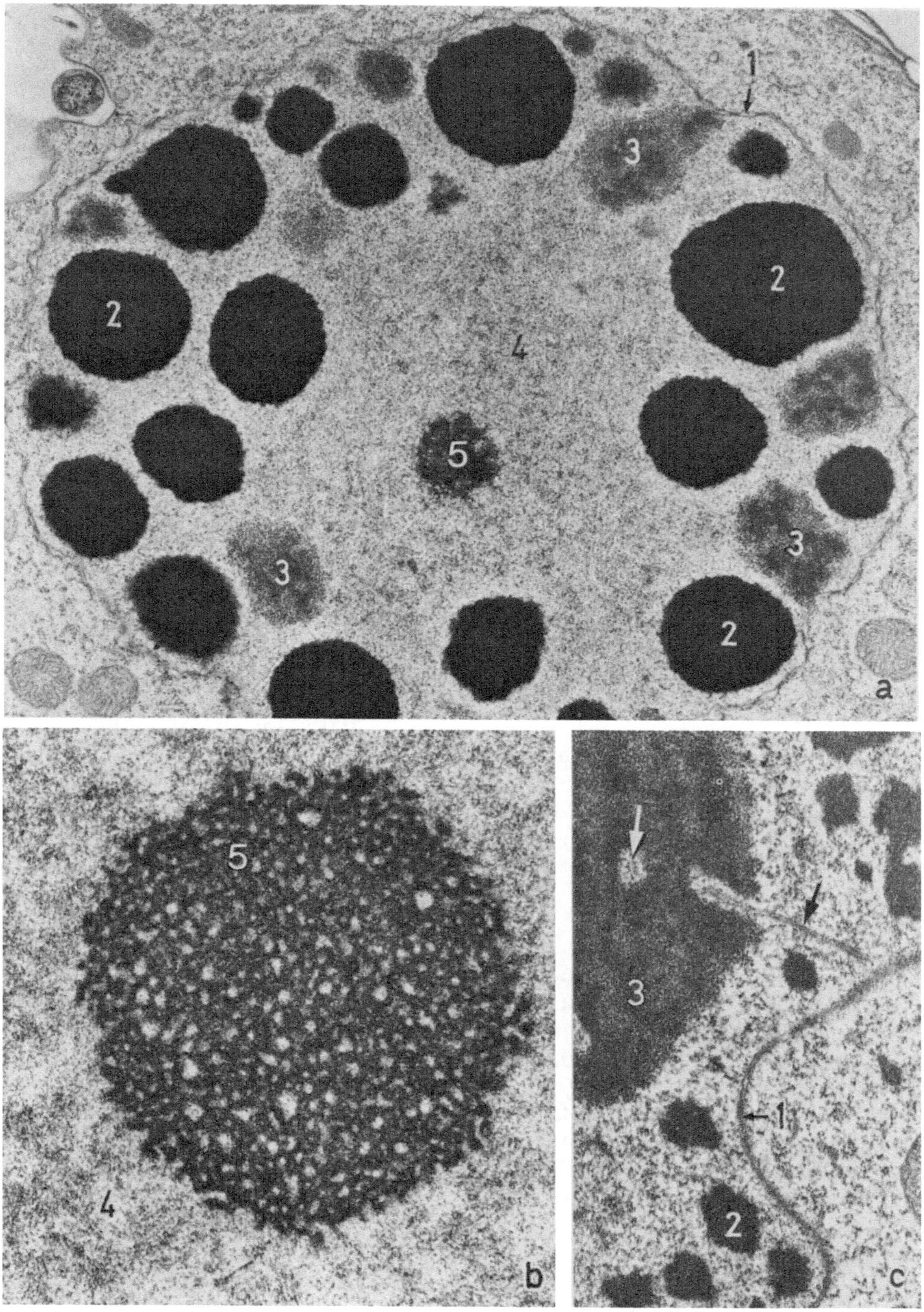

Fig. 97

Table 7. *Mean ratio between the DNA contents of Ma and presynthetic Mi in different ciliates*

Species	Q_{Ma}/Q_{Mi}	References
Subclass Gymnostomatia		
Dileptus anser		
single Ma	6	VINNIKOVA 1977
set of Ma	1000–2500	
Subclass Vestibuliferia		
Colpoda steini		
mature trophont	8	FRENKEL *et al.* 1973
tomite	2	
Tillina magna, tomites[a]	30, 135	FRENKEL 1969, 1975
Bursaria truncatella	2620	RUTHMANN 1964
Subclass Hypostomatia		
Chilodonella uncinata	32	SESHACHAR 1950b, PYNE *et al.* 1974
Nassula ornata	115	RAIKOV *et al.* 1963
Nassulopsis elegans	125.5	RAIKOV 1964
Subclass Hymenostomatia		
Tetrahymena thermophila[b]	21–25	GIBSON and MARTIN 1971, WOODARD *et al.* 1972, DOERDER and DE BAULT 1975, SEYFERT and PREPARATA 1979
Tetrahymena thermophila[c]	17.5	SEYFERT 1979
T. canadensis	23	GIBSON and MARTIN 1971
T. pyriformis, strain HSM[c]	35–36	CLEFFMANN 1968, SEYFERT 1979
T. limacis[d]	7.5, 15, 22.5, 30, 45	DYSART 1963
T. rostrata[d]	15, 22.5, 30, 45	Ibid.
T. patula[d]	75, 150, 225, 300, 375, 450	Ibid.
Colpidium campylum	44	MORAT 1970
Paramecium primaurelia	430	GIBSON and MARTIN 1971
P. biaurelia	450	Ibid.
P. tetraurelia	467	Ibid.
P. tetraurelia[d]	430, 860	WOODARD *et al.* 1961, 1966
P. septaurelia	430	GIBSON and MARTIN 1971
P. octaurelia[a]	474, 530, 540	Ibid.
P. jenningsi	244	DUPY-BLANC 1969b
P. calkinsi	280	CHEISSIN and OVCHINNIKOVA 1964
P. woodruffi	441	Ibid.

Species	Q_{Ma}/Q_{Mi}	References
P. caudatum [a, e]	25 80	Moses 1950, Walker and Mitchison 1957, Blanc 1963, Cheissin *et al.* 1964, Dupy-Blanc 1969b
P. bursaria [e]	8	Cheissin and Ovchinnikova 1964
	7	Dupy-Blanc 1969b
	20	Cullis 1973
P. putrinum [e]	11.5	Cheissin and Ovchinnikova 1964
Ichthyophthirius multifiliis		
mature trophont	6300	Uspenskaja and Ovchinnikova 1966
tomite	24	
Metaradiophrya varians	290	Morat 1965
M. gigas	330	Ibid.
M. lumbrici	660	Ibid.
Subclass Spirotrichia		
Protocruzia sp.		
single Ma	~ 1	Ruthmann and Hauser 1974
set of Ma	6-10	
Spirostomum ambiguum	6575	Ovchinnikova *et al.* 1965
Laurentia acuminata		
single Ma	38.5	Torres *et al.* 1979
set of 4 Ma	154	
Keronopsis rubra		
single Ma	~ 1	Ruthmann 1972
set of Ma	~ 100	
Stylonychia mytilis	32	Ammermann 1965
Oxytricha sp.		
single Ma	45	Lauth *et al.* 1976
set of 2 Ma	90	
Gastrostyla steini	80	Torres *et al.* 1980
Euplotes eurystomus	200	Gall 1959
E. eurystomus [a]	111, 117, 127	Ringertz *et al.* 1967b

[a] Figures stand for several different clones or strains.
[b] Micronuclear diploidy was checked by counting the chromosomes.
[c] It could be that the micronucleus of this strain is aneuploid.
[d] Several classes of the DNA content of Ma within a clone or strain.
[e] Micronuclear polyploidy is likely.

DNA content (Q) of Ma and that of Mi, which was assumed to be a diploid nucleus:

$$n_{Ma} = 2\,(Q_{Ma}/Q_{Mi})$$

However, it is now known that this method of calculating the degree of macronuclear polyploidy is far from accurate.

First, the micronuclei of ciliates are often polyploid themselves, e.g. in some *Paramecium* species (see Chapter 3). Accordingly, the calculated degree of macronuclear polyploidy for *P. bursaria*, *P. caudatum*, *P. putrinum*, and some other species would be greatly underestimated. It is more reliable to determine the degree of macronuclear polyploidy by comparing the content of DNA in the macronucleus not with the overall DNA content of the micronucleus but with the true size of the macronuclear unit haploid genome (i.e. its kinetic complexity) calculated from the rate of macronuclear DNA renaturation. Unfortunately, there is not enough data in this respect. The first way of calculating the degree of macronuclear polyploidy in *P. bursaria* gives values ranging from 14 n to 40 n (Table 7), whereas the second way produces a value of about 560 n! (CULLIS 1973). However, micronuclear polyploidy is rather rare in *P. aurelia*, *P. woodruffi*, *P. calkinsi*, and outside the genus *Paramecium*. In *Tetrahymena thermophila*, for instance, micronuclei are diploid (2 n = 10) in most strains; in these strains, the ratio between the DNA contents of Ma and Mi varies very slightly, i.e. 21–25 times (Table 7). Other clones of this species have abnormal micronuclear karyotypes: the DNA content of their Mi varies accordingly while that of their Ma is almost constant (apparently because the DNA content is adjusted during macronuclear development). Consequently, the ratio Q_{Ma}/Q_{Mi} is 23 in the diploid clone, 34 in the haploid clone (n = 5), and only 14 in the triploid clone (3 n = 15; SEYFERT and PREPARATA 1979). It is likely that Mi is triploid (or hyperdiploid) also in the clone of the same species studied by SEYFERT (1979; see Table 7).

Second, only a part of and not the whole micronuclear genome is preserved and multiplied in the Ma of at least some ciliates, particularly Hypotrichida. Therefore, a direct comparison of the DNA quantities in Ma and Mi will tend to underestimate the repetitiveness of the genome parts preserved in Ma in this case, too, even if Mi were diploid. For instance, photometric data might suggest that the degree of macronuclear polyploidy is 64 n in *Stylonychia mytilus* and 180 n in *Oxytricha* sp. (Table 7); actually, the multiplicity of the genetic elements in Ma is nearly 4000 in *Stylonychia* and 2000 in *Oxytricha*. Taking these two sources of error into consideration, we include only the actual ratios between the DNA contents of the Ma and Mi of different ciliates in Table 7; we do not interpret these values in terms of macronuclear ploidy, as we used to earlier (see, for example, RAIKOV 1969).

The data in Table 7 show that the ratios between the DNA quantities in Ma and Mi vary greatly in different ciliates: from a twofold quantity in the *Colpoda steini* tomites to more than a 6000-fold quantity in *Spirostomum* and the *Ichthyophthirius* trophonts. No distinct correlation can be established between the relative macronuclear DNA content and the taxonomic position of a ciliate. This content, however, seems to depend on the size of the ciliate's body. The relative (and usually absolute) content of DNA in Ma is always higher in large ciliates than in the smaller related forms (*Tillina magna* and *Colpoda steini; Tetrahymena patula* and *T. thermophila*). This correlation is even more obvious in species with a complicated life cycle, one that includes a stage of hypertrophic growth (trophont) and small dispersive stages (tomites). The difference between the DNA contents of Ma and Mi comes to thousands in the *Ichthyophthirius* trophonts, although it is only 24 in the tomites of the same species, the DNA content of Mi being constant (USPENSKAJA and OVCHINNIKOVA 1966). Changes in the DNA content of Ma are proportional to the changes in the body dimensions when the microstome form of *Tetrahymena paravorax*, which feeds on bacteria, is transformed into a predatory macrostome form and vice versa (DUPY-BLANC and MÉTENIER 1975). In this case, the macrostome form has 4–16 times more DNA in Ma than the microstome form.

Several classes of Ma occur in three *Tetrahymena* species (*T. limacis*, *T. rostrata*, and *T. patula*) and in *Paramecium tetraurelia;* the DNA contents of these classes are multiples of one another, while that of Mi is the same in all the classes (Table 7). The DNA content of Ma decreases in *P. tetraurelia*, *P. bursaria*, and *Oxytricha* sp. as the clones age (SCHWARTZ and MEISTER 1975b, KLASS and SMITH-SONNEBORN 1976, LIEDER and RUTHMANN 1976).

In ciliates with multiple Ma, the DNA content of single Ma usually varies even within a specimen. In *Dileptus anser*, for instance, the ratio of the DNA contents of individual macronuclei to that of the micronucleus varies from 4 : 1 to 30 : 1 (about 6 : 1, on the average), i.e. the differences between individual macronuclei may be sevenfold (VINNIKOVA 1977). In *Keronopsis rubra*, these values range from 0.25 : 1 to 3.75 : 1, i.e. the maximum differences between individual macronuclei are even 15-fold, while the mean DNA content of a Ma is close to that of a Mi (RUTHMANN 1972).

Reproduction of the Macronucleus

DNA Replication

Periods of Macronuclear Interphase. Just as in the nuclei of other protozoans (see Chapter 3), macronuclear DNA is replicated in the S period, which may have different positions in the interval between two divisions of Ma (Table 8). In some species, the S period begins as soon as Ma has divided; then, the G_1 period is either very short or absent, while the G_2 period is well expressed (*Nassulopsis*, *Paramecium bursaria*). In *Tetrahymena* and *Colpidium*, the S period is in the middle of the macronuclear interphase. *Chilodonella*, most species of *Paramecium*, *Blepharisma*, *Stentor*, and all Hypotrichida have the S period either in the second half or at the end of the macronuclear interphase. In the latter case, the Ma divides as soon as DNA replication is completed.

The relative duration of the periods G_1, S, and G_2 may vary even among strains of the same species (*Tetrahymena pyriformis*). It also depends very much on the cultivation conditions. The data in Table 8 mostly refer to optimum cultivation conditions when generation time was the shortest. In *T. pyriformis*, all three periods are prolonged more or less proportionally if the cultivation temperature is below or above the optimum (29 °C; MACKENZIE *et al.* 1966, CLEFFMANN 1967). Conversely, if the growth medium is depleted, the periods G_1 and G_2 lengthen more than the S period (SUHR-JESSEN *et al.* 1977). In *Euplotes*, the G_1 period

Table 8. *Position of the period of DNA synthesis (S) in the macronuclear division cycle (Ma, solid line) and the micronuclear mitotic cycle (Mi, double line) in ciliates*[a]

Species	Type of nucleus	Generation time, %	References
Subclass Gymnostomatia			
Homalozoon vermiculare	Ma		SAXENA and SAXENA 1978b
Subclass Vestibuliferia			
Bursaria truncatella	Ma		RUTHMANN 1964
B. truncatella	Ma		NILOVA 1968
Subclass Hypostomatia			
Chilodonella cucullulus	Ma		RADZIKOWSKI 1969, 1976
	Mi		
Nassulopsis elegans	Ma		RAIKOV 1964
	Mi		
Subclass Hymenostomatia			
Tetrahymena pyriformis			
strain W	Ma		WALKER and MITCHISON 1957
strain H	Ma		McDONALD 1958
strain HS	Ma		PRESCOTT 1960
strain HSM	Ma		CAMERON and STONE 1964, FLICKINGER 1965, PRESCOTT and STONE, 1966, CLEFFMANN 1967
	Mi		
T. thermophila	Ma		McDONALD 1962, 1973, WOODARD *et al.* 1972
	Mi		
Colpidium campylum	Ma		MORAT 1970
	Mi		
Paramecium tetraurelia	Ma		KIMBALL and BARKA 1959, KIMBALL and PERDUE 1962, BERGER and KIMBALL 1964, BERGER 1971
P. tetraurelia	Ma		WOODARD *et al.* 1961
	Mi		
P. caudatum	Ma		WALKER and MITCHISON 1957, CHEISSIN *et al.* 1963
P. caudatum	Ma		BLANC 1965
	Mi		
P. caudatum	Ma		RAO and PRESCOTT 1967
	Mi		
P. putrinum	Ma		KUDRJAVTSEV 1966
	Mi		
P. bursaria	Ma		JUST 1973
	Mi		

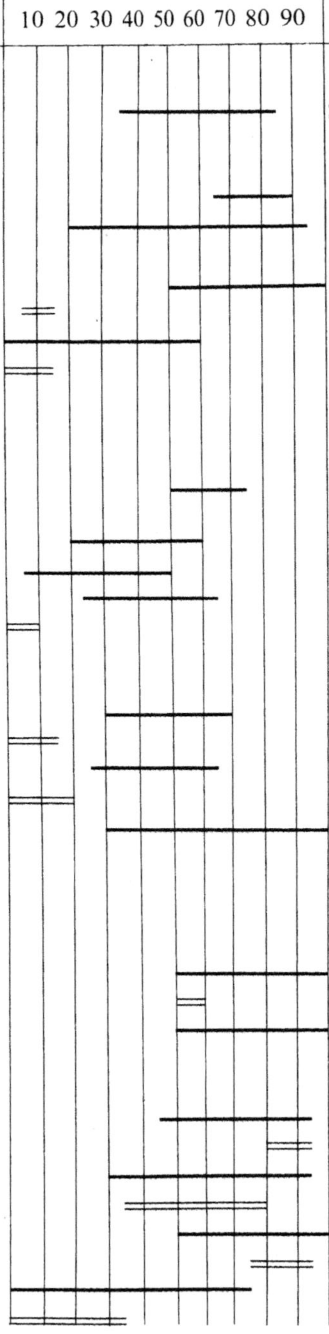

Species	Type of nucleus	Generation time, % (10 20 30 40 50 60 70 80 90)	References
Subclass Spirotrichia			
Blepharisma americanum	Ma		MUNITOLI and HIRSHFIELD 1968
B. intermedium	Ma		SHIVAJI *et al.* 1978b
Stentor coeruleus	Ma		GUTTES and GUTTES 1960, DE TERRA 1967, PELVAT *et al.* 1973
Spirostomum ambiguum	Ma		OVCHINNIKOVA *et al.* 1965
Urostyla weissei	Ma		JERKA-DZIADOSZ and
	Mi		FRANKEL 1970
Laurentia acuminata	Ma		TORRES *et al.* 1979
	Mi		
Oxytricha bifaria	Ma		DINI *et al.* 1975
	Mi		
Oxytricha sp.	Ma		LIEDER and RUTHMANN 1976
Stylonychia mytilus	Ma		AMMERMANN 1970,
	Mi		CHADHA *et al.* 1978, CHADHA 1980
Onychodromus grandis	Ma		SAXENA and SAXENA
	Mi		1978a
Gastrostyla steini	Ma		TORRES *et al.* 1980
	Mi		
Euplotes eurystomus	Ma		FAURÉ-FREMIET *et al.* 1957, RINGERTZ and HOSKINS 1965, WISE 1965
E. eurystomus	Ma		PRESCOTT *et al.* 1962
	Mi		
Diophrys scutum	Ma		DINI *et al.* 1975
	Mi		
Aspidisca aculeata	Ma		DINI and BRACCHI
	Mi		1976

[a] The duration of the periods of the division cycle is expressed in percentage of generation time under optimum cultivation conditions; the interval from zero to the beginning of the S period corresponds to the G_1 period; the interval from the end of the S period to 100% makes up the periods G_2 and D (nuclear division proper).

increases under similar conditions much more than the S period; in the stationary phase, Ma are blocked in the G_1 period (RINGERTZ and HOSKINS 1965). The generation time of *Paramecium aurelia* grows as the clones age; this growth is completely due to the elongation of the S period, i.e. to a decrease in the speed of DNA replication (SMITH-SONNEBORN and KLASS 1974).

The S period in macronuclei never coincides exactly with that in micronuclei. In some species (*Nassulopsis*, *Paramecium tetraurelia*, *P. bursaria*), DNA synthesis commences almost simultaneously in Ma and Mi but ends in Mi much earlier than in Ma. Occasionally, DNA synthesis in Mi starts later than in Ma and coincides with the middle or, more frequently, with the end of the S period in Ma (*Paramecium caudatum*, *P. putrinum*, *Urostyla*, *Oxytricha*, *Stylonychia*, *Diophrys*). DNA replication in *Chilodonella*, *Tetrahymena*, *Colpidium*, and *Euplotes* is especially noteworthy: it begins in Mi much earlier than in Ma and ends while Ma are still in G_1. Apparently, the ciliates have independent systems of DNA synthesis initiation for Mi and for Ma (PRESCOTT and STONE 1966).

The period of DNA synthesis is always longer in Ma than in Mi (Table 8). It is often longer than 50% of generation time and on an absolute scale may take more than a day, e.g. in *Stentor* and *Spirostomum*. This fact indicates that DNA replication in macronuclear subunits is asynchronous. Biochemical studies have shown that the DNA molecules in the Ma of *Tetrahymena*, which replicated simultaneously in one cell cycle, would replicate asynchronously (throughout the whole S period) in the next cycle (ANDERSEN and ZEUTHEN 1971). Therefore, the sequence of replication of macronuclear subunits is determined anew in each cell cycle.

Replication Bands. In all Hypotrichida, DNA is replicated locally and sequentially along the macronucleus. DNA is synthesized only in the "replication" (or reorganization) bands, i.e. clear discs that gradually move through the entire Ma. In *Euplotes*, two replication bands appear at the ends of a C-shaped Ma (where, apparently, the initiation sites for DNA synthesis are located), move toward each other, meet in the middle of the Ma, and disappear (TURNER 1930). Conversely, the bands of *Aspidisca* move from the middle of a C-shaped Ma toward its ends (SUMMERS 1935). In *Stylonychia*, one replication band passes in each of the two Ma from its distal end toward the proximal end (Fig. 98*a–c*); however, electron microscopy has shown that both Ma are interconnected by a narrow isthmus and that actually the replication bands move from the ends toward the middle of a sole dumbbell-shaped Ma (WALKER and GOODE 1976). In Hypotrichida with several or many isolated Ma, replication bands pass through each Ma separately, e.g. in *Uroleptus* (CALKINS 1919), *Urostyla* (RAABE 1947, INABA and SUGANUMA 1966), or *Keronopsis* (RUTHMANN 1972).

A replication band consists of two zones: a front one that deeply stains by the Feulgen method and a rear one that stains very faintly (TURNER 1930, etc.). GALL (1959) has discovered by means of autoradiography that ^{3}H-thymidine was electively incorporated into the macronuclear replication bands of *Euplotes*. It has been demonstrated (PRESCOTT and KIMBALL 1961, PRESCOTT 1962) that DNA is synthesized only at the front edge of the rear zone of the band (Fig. 99*a*). The newly synthesized DNA immediately combines with histones, the amount of which doubles in the replicated part of Ma (GALL 1959). The concentration of

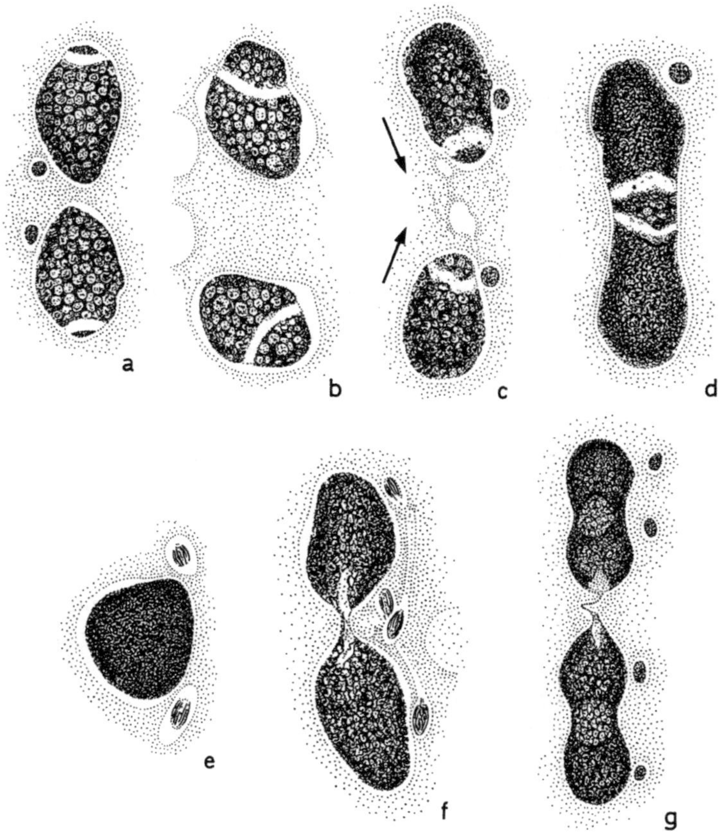

Fig. 98. Macronuclear division in *Stylonychia pustulata*, 875× (after SUMMERS 1935). *a* appearance of replication bands; *b* and *c* their movement through both Ma (arrows); *d* fusion of Ma; *e* maximal condensation of Ma, mitosis of Mi; *f* first division of Ma; *g* second division of the daughter Ma.

histones increases even prior to DNA replication in the front zone of the replication band (PIERI and VAUGIEN 1967a, SALVANO 1975). Histones can be detected also by preferential incorporation of labelled histidine (Fig. 99*b*), arginine, and lysine into the replication band of *Euplotes* (PRESCOTT and KIMBALL 1961, PRESCOTT 1962, 1966, RINGERTZ and HOSKINS 1965).

Unlike DNA, no RNA is synthesized in the replication band (Fig. 99*c*): labelled uridine is incorporated throughout the Ma except in the band. In most cases, macronuclear nucleoli vanish when the replication band reaches them. Autoradiography has shown that all preexisting, prelabelled RNA then disappears from the Ma (possibly by being released into the cytoplasm) and new RNA is synthesized behind the band (PRESCOTT and KIMBALL 1961, PRESCOTT 1962, STEVENS 1963). Some nucleoli pass through the replication band (RUFFOLO 1978b), however.

Replication bands were examined with the electron microscope in *Euplotes* (FAURÉ-FREMIET *et al.* 1957, ROTH 1957, KLUSS 1962, FAURÉ-FREMIET and ANDRÉ 1968, RUFFOLO 1978b) and other Hypotrichida (INABA and SUGANUMA 1966,

Chakraborty 1967, Walker and Goode 1976, Chadha 1980). The bodies of condensed macronuclear chromatin disintegrate in the front zone, producing a plexus of chromatin fibrils 10–11 nm in diameter (Fig. 99*d*). These structures abruptly disappear at the edge of the rear zone, probably because protein is stripped off DNA molecules. Only very thin filaments (apparently replicating DNA molecules) are visible in this zone; this is confirmed by ultrastructural autoradiography (Stevens 1963). At the end of the rear zone, DNA again assembles into chromatin bodies that gradually grow as the replication band moves away (Fig. 99*d*). Nucleoli reappear in this zone as soon as transcription recommences.

We still do not know what mechanism is responsible for the local DNA synthesis in the replication bands and for the ordered movement of the bands through the hypotrichid macronucleus. Apparently, a Ma should have either long chromatin threads (chromosomes or genomes) that are arranged along it and contain definite initiation sites for DNA synthesis (in *Euplotes*, for instance, at both ends of the Ma) or some other information that controls the sequential replication of the chromatin bodies along Ma. The nuclear protein matrix or the macronuclear envelope may serve as a scaffold that fixes the position of DNA molecules and determines the sequence of their replication. Indeed, threads and bodies of macronuclear condensed chromatin are locally (and, possibly, specifically) attached to the inner membrane of the nuclear envelope of *Gastrostyla*, *Stylonychia*, and *Euplotes* (Walker and Goode 1979, Ruffolo 1978b; cf. p. 309).

In this respect, it would be interesting to consider how newly synthesized DNA molecules behave in the subsequent cell cycles of Hypotrichida. In *Euplotes* (Kimball and Prescott 1962) and *Stylonychia* (Ammermann 1971a, Chadha *et al.* 1979), the macronuclear contents are known to be mixed before macronuclear division when the S period is over and the replication bands disappear. Pulse-labelled macronuclear DNA that was located in short segments of the nucleus (Fig. 99*a*) now spreads throughout the Ma (Fig. 99*e*). If Ma had long DNP fibrils reaching from one of its ends to the other, the fibrils would occupy the same position in the S period of the next cell cycle; then the labelled sections of the fibrils would automatically re-form labelled transverse bands on Ma. If the DNP fibrils in the macronucleus are shorter than Ma itself, the label may not return to the former position in the subsequent cell cycle (although it may return if some extra allowances are made). Such a return of the label was not detected in *Stylonychia*: the label remained dispersed throughout the Ma in the next cell

Fig. 99. Replication bands in the macronucleus of *Euplotes eurystomus*. *a*–*c* radioautographs: *a* after brief incorporation of ^{3}H-thymidine (DNA synthesis), *b* after incorporation of ^{3}H-histidine for 40 min (*de novo* synthesized protein), *c* after incorporation of ^{3}H-uridine for 30 min, RNA is synthesized throughout the Ma except in the replication band (*1*); *d* electron micrograph of the replication band (*1* chromatin bodies, *2* nucleoli, *3* front zone, *4* rear zone, *5* macronuclear envelope), 30000 × ; *e* dispersion of DNA labelled with ^{3}H-thymidine throughout the Ma during its condensation (radioautograph). The arrows indicate the direction in which the replication bands move. *a* unpublished micrograph, courtesy of D. M. Prescott; *b* and *c* after Prescott (1962); *d* after Fauré-Fremiet and André (1968); *e* after Kimball and Prescott (1962).

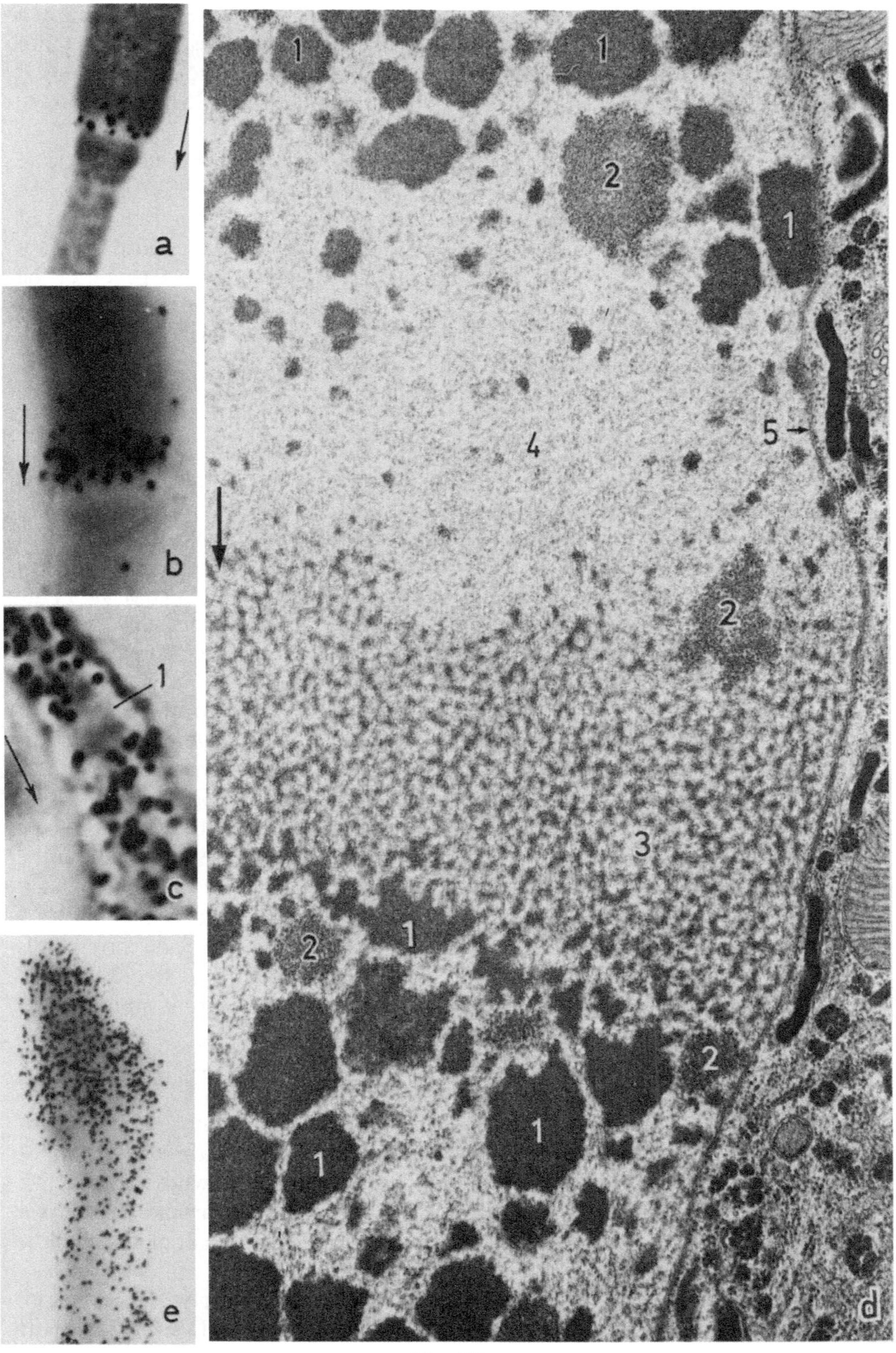

Fig. 99

cycle (AMMERMANN 1971a). Consequently, the DNA molecules that replicated simultaneously as they were on the same transverse plane in one cell cycle get into different parts of a Ma in the next cycle and, therefore, replicate asynchronously. The label also did not return to its former position in the Ma of *Stylonychia* in the subsequent (up to the sixth) cell cycles (CHADHA *et al.* 1979).

The behaviour of labelled macronuclear DNA in *Euplotes* is more intricate (BONIFAZ and PLAUT 1974). This DNA remains dispersed in the G_1 period of the next (second) cell cycle but concentrates at both ends of a Ma in the S period of this cycle (irrespective of where the replication bands were located during the actual labelling). The labelled DNA then disperses again. In the S period of the third cell cycle, however, the label returns to the same positions the replication bands held two generations earlier! Thus, the Ma of *Euplotes* has a two-generation cycle of translocations of DNA molecules. Hence, there should be a "scaffold" that fixes the positions of these molecules. This fact, however, cannot be interpreted yet.

Replication bands are also found in the Ma of ciliates other than Hypotrichida. Such bands occur in the orthomere of the heteromeric Ma of the juxtaposed type in some Cyrtophorida and Chonotrichida. Only one replication band appears at the boundary between the orthomere and the paramere and moves toward the free end of the orthomere (Fig. 96*a*). It was seen with the light microscope in *Dysteria* (FAURÉ-FREMIET 1957), *Chlamydodon* (KANEDA 1960, 1961a), *Spirochona* (TUFFRAU 1953), *Lobochona* (MATSUDO 1966), and others, and was examined with the electron microscope in *Chlamydodon* (KANEDA 1961b) and *Spirochona* (GUÉNIN and FAHRNI 1971). The fine structure of the replication band in the Ma of *Spirochona* resembles the one in Hypotrichida in that it consists of a denser front zone and a looser rear zone. Nevertheless, there is still no direct (e.g. autoradiographic) evidence which confirms that the DNA of the orthomere replicates in this band in heteromeric Ma. No replication bands exist in the paramere.

No replication bands were found in the heteromeric Ma of the concentric type (e.g. in *Chilodonella*). RADZIKOWSKI (1969, 1976) studied DNA replication in the Ma of *Chilodonella* by photometry and autoradiography. He discovered that it consists of two phases. The first phase (slow synthesis) corresponds to DNA replication in the paramere, while the second phase (rapid synthesis) corresponds to DNA replication in the orthomere. No DNA synthesis was recorded in the endosome, which apparently contains defective nonreplicating macronuclear DNA molecules.

Replication bands were detected also in the Ma of Oligotrichida. They have a typical fine structure in *Halteria* (GRAIN 1972) as well as in the tintinnids *Petalotricha* (LAVAL 1972), *Favella*, and *Parafavella* (HEDIN 1975). Cytophotometry has shown that when the replication band passes through the Ma of *Strombidium*, the content of DNA and protein in it doubles, while that of RNA decreases (SALVANO 1975). Only one replication band moves through the sole Ma of *Halteria* and *Strombidium* as well as through each of the two or more Ma in the tintinnids.

Control of DNA Synthesis in Macronuclei. Unlike the nonpolyploid cell nuclei, the Ma of ciliates can also divide with incompletely duplicated DNA, e.g. when its DNA replication is interrupted by an inhibitor. Such a phenomenon was described with respect to *Tetrahymena* (ANDERSEN 1972a) and *Urostyla* (JERKA-DZIADOSZ and FRANKEL 1970). The DNA content of a Ma may vary considerably owing to such processes. Moreover, the division of a Ma is frequently unequal (RUTHMANN 1964, ZECH 1966b, CLEFFMANN 1968, EIDE 1968). The DNA content of the Ma of *Bursaria truncatella* (ZECH 1966b) and *Paramecium tetraurelia* (BERGER 1973c) is adjusted after such divisions. The adjustment seems to occur during DNA synthesis due to the underreplication or overreplication of certain DNA sequences.

It has been shown recently that the quantity of DNA synthesized in the Ma of *Paramecium tetraurelia* during one cell cycle does not depend on the initial DNA content of Ma: it is identical in cells with a small Ma and in cells with a large Ma (BERGER and SCHMIDT 1978, BERGER 1979). Even the cells that got the whole mother Ma after division (this happens in certain mutants) synthesize the same clone-specific quantity of DNA. Consequently, the DNA content of the Ma of such cells increases only by 50% instead of doubling (BERGER 1978). With such a control mechanism, some parts of the macronuclear genome would replicate more than once in the cells with a small Ma but would not replicate at all in the cells with a large Ma (BERGER 1979). As the process is repeated in a series of generations, the quantity of DNA would gradually increase in the Ma of the progeny of the first-type cells, but it would decrease in those of the second-type cells until both quantities reach the "norm" typical of the clone.

CLEFFMANN (1968) discovered another mechanism controlling the DNA content of Ma that operates in *Tetrahymena pyriformis*. In this case, besides the unequal divisions of Ma, chromatin clumps are extruded from Ma during each division. However, no regulatory overreplication or underreplication of macronuclear DNA occurs, but it only exactly doubles per cell cycle. Since the extrusion of chromatin from Ma is not compensated, the DNA content of each cell line steadily decreases in the following generations. Nevertheless, when it reaches a lower critical level, the control mechanism switches on an additional round of replication of all DNA, and the DNA content of Ma quadruples during one interval between divisions. Then, the amount of DNA in Ma gradually decreases again.

The same control mechanism operates in *Tetrahymena thermophila*. In this case, cells with a Ma poor in DNA may have an extra S period, and DNA replication may be skipped completely in cells that originally contained a large Ma (as a result of unequal division). Thus, the mechanism of the "stepwise" adjustment of the macronuclear DNA content that is typical of *Tetrahymena* may either double or halve the content of DNA within a cell cycle (DOERDER and DE BAULT 1978, DOERDER 1979). This mechanism is active also in *Colpidium campylum* (MORAT 1980) and in many species of the genus *Tetrahymena* (DOERDER *et al.* 1981), including the polymorphic species *Tetrahymena paravorax* (MÉTÉNIER 1979). CLEFFMANN (1975) has detected additional rounds of DNA synthesis in the Ma of *Tetrahymena* when cell division was synchronized with heat shocks, particularly in cells with an originally low DNA content of Ma.

Apparently, the two types of adjustment of the DNA level in a Ma, i.e. the gradual one (during replication) and the stepwise one (additional rounds of DNA synthesis), may coexist. For instance, FRENKEL (1975) has found both mechanisms in *Tillina magna*, which reproduces itself by two successive palintomic divisions in a cyst. DNA does not replicate in Ma between these two divisions, while two cycles of DNA replication occur normally at the trophont stage (FRENKEL 1969). Both divisions of *Tillina* in a cyst are accompanied by the extrusion of chromatin bodies from Ma into the cytoplasm. In spite of this, the DNA level in the Ma of these cell lines does not decrease regularly but only fluctuates owing to the unequal divisions of Ma (FRENKEL 1975). Apparently, DNA is overreplicated in the trophont so that the mature trophont has almost five (and not four) times more DNA in its Ma than the tomite. This overreplication compensates for a loss of DNA caused by chromatin extrusions during macronuclear divisions. Moreover, the DNA content of Ma is occasionally adjusted stepwise toward either its increase or decrease. When it increases, the Ma of the trophont undergoes three and not two cycles of DNA replication, and when it decreases, only one replication cycle occurs before the two divisions in a cyst.

Doubling of Chromatin Elements

The duplication of chromatin bodies or threads, which follows DNA replication and precedes macronuclear division, has been observed in the Ma of only a few ciliates. Nevertheless, such duplication must regularly occur; otherwise, the quantity of chromatin elements in a Ma would decrease from generation to generation.

Light microscopic studies of Ma in *Tetrahymena*, *Paramecium*, and *Vorticella* have shown that their "chromatin threads" double shortly before macronuclear division (SATO and SAITO 1959, SAITO and SATO 1961a, SAITO 1961b). "Chromonemata" in the orthomere of the heteromeric Ma in *Chlamydodon* also double lengthwise during the G_2 period (KANEDA 1960, 1961a). "Chromatin threads" in the Ma of *Bursaria* (RUTHMANN and HECKMANN 1961), chromatin bodies in the Ma of *Tokophrya* (RUDZINSKA 1961), and rod-like "chromosomes" in the Ma of *Loxophyllum* (RUTHMANN 1963) are usually paired, too (Fig. 100*a*). But it is difficult to correlate these data with ultrastructural observations.

The duplication of the chromatin elements was clearly seen with the light microscope (RAIKOV 1960, 1962a) in the Ma of *Nassula ornata* (subclass Hypostomatia). Its nondividing Ma contains elongated chromatin aggregates 5–10 μm long and 1–2 μm thick (Fig. 100*b*) that join one another, forming a network. Shortly before division, the Ma swells and the chromatin aggregates stretch into long Feulgen-positive threads that soon double. The sections reveal only fragments of these paired threads (Fig. 100*c*). This stage was interpreted as endoanaphase in the endomitosis of macronuclear chromosomes. Then the daughter chromatin threads

---

Fig. 100. Duplication of chromatin elements in the macronucleus (photomicrographs). *a* paired chains of chromosome-like elements (arrows) in nodes of the moniliform Ma of *Loxophyllum meleagris*, phase contrast, 1760 ×; *b–d* reproduction of the Ma of *Nassula ornata*: *b* nondividing Ma with chromatin aggregates and the adjacent Mi, 1800 ×; *c* duplication of the aggregates drawn into threads (opposite arrows), 2200 ×; *d* macronuclear division with segregation of compact daughter aggregates (arrows), 1400 ×; Feulgen-stained sections. *a* after RUTHMANN (1963); *b–d* after RAIKOV (1962a, 1967a, 1976b).

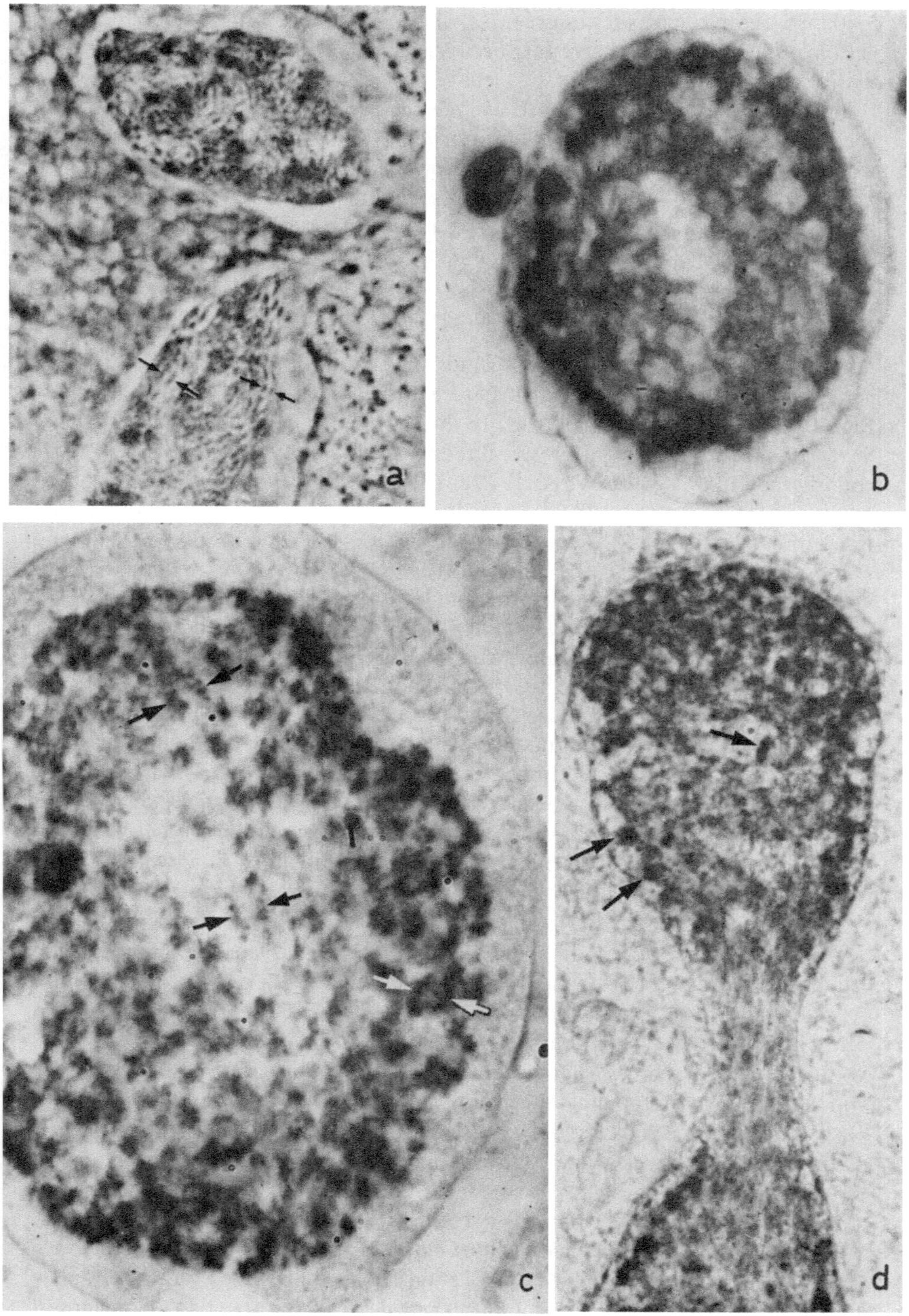

Fig. 100

condense again in compact aggregates whose number is consequently doubled. The chromatin aggregates segregate between daughter nuclei during macronuclear division (Fig. 100*d*).

The macronucleus of *Nassulopsis elegans* has a similar structure (RAIKOV 1964). The chromatin aggregates of the nondividing Ma elongate prior to division, producing threads that duplicate and become paired. The daughter threads then condense again into aggregates. All this occurs in the G_2 period, which is followed by macronuclear division and the segregation of aggregates. The nature of the chromatin aggregates in the Ma of *Nassula* and *Nassulopsis* will be discussed further (p. 315).

Macronuclear Condensation

Macronuclei with an intricate shape (e.g. ribbon-like, dumbbell-shaped, moniliform) usually contract just before division, forming a compact body (Fig. 101*a–c*); only then do they start to divide (Fig. 101*d*, *e*). The typical shape of the Ma is restored only after cytokinesis. Such behaviour is typical of the moniliform Ma of *Stentor* (WEISZ 1949a, TARTAR 1961), *Blepharisma* (McLOUGHLIN 1957, HELSON *et al.* 1959), *Spirostomum* (EBERHARDT 1962), *Loxophyllum* (RUTHMANN 1963), *Homalozoon* (DILLER 1964, SAXENA and SAXENA 1978b), and *Dileptus cygnus* (GOLINSKA 1965), and of the sausage-shaped or ribbon-like Ma of *Bursaria* (PESCHKOFF 1929, RUTHMANN and HECKMANN 1961), *Fabrea* (ELLIS 1937), *Euplotes* (TURNER 1930), *Didinium* (RIEDER 1971), and many other ciliates.

In ciliates with several or many Ma, the latter usually fuse into one compact "division macronucleus" (Fig. 98*d*, *e*). In Hypotrichida, fusion always occurs after the passage of replication bands through individual Ma, i.e. in the G_2 period. The fused Ma divides in two simultaneously with cytokinesis; then, the daughter nuclei divide again once or several times until the typical number of Ma is restored (Fig. 98*f*, *g*). Macronuclei fuse even if their number is very large, e.g. over 100 in *Urostyla grandis* (RAABE 1947) or around 200 in *Swedmarkia arenicola* (LUPORINI and MAGAGNINI 1970). Numerous cytoplasmic microtubules interconnecting the nuclear envelopes of individual nuclei develop in *Swedmarkia* before Ma fuse. The microtubules apparently help to draw individual Ma together.

Apart from the species displaying macronuclear fusion, the Hypotrichida include some rare species whose numerous Ma divide one by one without preliminary fusion, e.g. *Keronopsis rubra* (RUTHMANN 1972). Such behaviour also occurs in some Gymnostomatia, e.g. *Dileptus anser* (STUDITSKY 1930, JONES 1951, VINNIKOVA 1974c) and *Chaenea vorax* (FAURÉ-FREMIET and GANIER 1969), as well as in some Hymenostomatia, e.g. *Conchophthirius steenstrupii* with seven Ma (ROSSOLIMO and JAKIMOWITSCH 1929). Thus, the division of Ma without their preliminary fusion is occasionally found in different groups of ciliates and is not characteristic of any particular taxon.

Macronuclei always condense or fuse in the G_2 period, i.e. when DNA replication is over. The condensed Ma of *Bursaria* (RUTHMANN 1964) or *Blepharisma* (PARKER and GIESE 1966, MUNITOLI and HIRSHFIELD 1968) no longer incorporates ^{3}H-thymidine. The Ma of *Euplotes* condenses from its ends, i.e. regions in which DNA replication terminates first. When stretched, the condensed Ma of *Spirostomum* is unable to produce the long chromatin threads that are easily obtained from nondividing Ma. It is likely that the filamentous chromatin structures are extended in the nondividing Ma, but are coiled in the condensed Ma (SESHACHAR 1966). In *Bursaria*, small chromatin bodies fuse into long strands at the macronuclear condensation phase; membrane-bound vesicles separate from the nuclear envelope into both the Ma (these vesicles have a simple or double membrane without pores) and the cytoplasm (these vesicles have a double membrane with pores) (SERGEJEVA 1979).

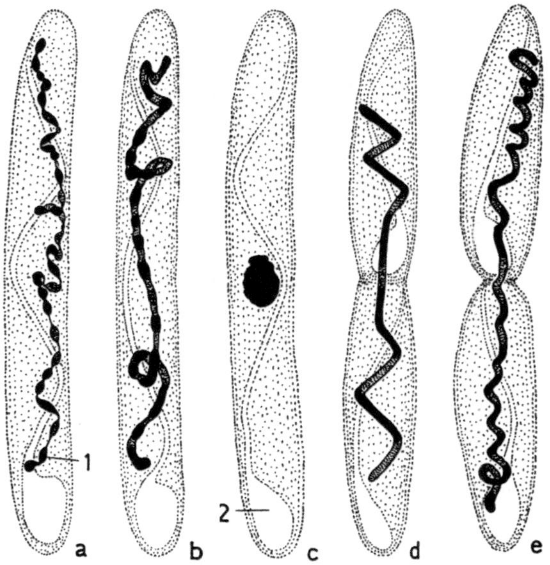

Fig. 101. Condensation (*a–c*) and division (*d* and *e*) of the macronucleus in *Spirostomum ambiguum*, 63 × (after DOFLEIN and REICHENOW 1953). *1* Ma, *2* contractile vacuole.

Chromatin strands in the condensed Ma of the hypotrichids *Gastrostyla* and *Stylonychia* are arranged either concentrically or as vortices, suggesting that the macronuclear contents rotate; nonoriented microtubules are located between the chromatin strands (WALKER and GOODE 1976). A noteworthy fact is that the strands then lose contact with the inner membrane of the macronuclear envelope to which they are attached at all the other stages of the cycle. However, "plaques" of condensed chromatin remain on the inner nuclear membrane and the chromatin strands are reattached to them as soon as the macronuclear condensation phase is over. All these data suggest that macronuclear chromatin is redistributed and that the configuration of chromatin elements changes when Ma condense. It is then that the label from the pulse-labelled sections of Ma disperses throughout the hypotrichid macronucleus (cf. p. 302).

In some ciliates, the structural components of Ma mix so vigorously at the condensation stage that this is seen *in vivo* as the rapid rotation of the macronuclear contents, e.g. in the suctorians *Tokophrya* (HECKMANN 1966),

Acineta (BARDELE 1969), and *Paracineta* (HAUSER 1973) as well as in the hymenostomatid *Ichthyophthirius* (HAUSER 1973). Electron micrographs show that the bundles of microtubules typical of the nondividing Ma of *Acineta* then disappear almost entirely, while macronuclear chromatin strands become thin and arched (Fig. 102*a*). In *Tokophrya*, large spherical chromatin bodies that are characteristic of the nondividing Ma (cf. Fig. 95*d*) fuse with one another when the macronuclear contents rotate, producing irregular strands that bend as vortical arcs (Fig. 102*b*). The large bundle of microtubules of the nondividing nucleus also disappears. It should be noted that the nondividing Ma of *Acineta* and *Tokophrya* are already compact (they are oval); therefore, the stage of macronuclear condensation can be recognized more by the rotation of the macronuclear contents and the concomitant changes in the fine structure of the macronucleus than by its shortening, which is rather faint.

HAUSER (1973) studied the *mechanism of the rotation* of the macronuclear contents in *Ichthyophthirius*. This rotation lasts only 10–30 seconds and coincides with the maximum condensation of the horseshoe-shaped Ma. The speed of rotation increases noticeably when ATP is added, indicating that contractile proteins are involved in the process. A branched system of channels free of chromatin bodies can be detected by electron microscopy in the rotating Ma. HAUSER discovered that these zones contain a network of microfilaments that may be either thick (10–20 nm) or thin (about 5 nm). The filaments are either arranged loosely, forming only some acicular aggregates (Fig. 102*c*), or the filaments are closer to one another and the number of aggregates is greater; moreover, filaments may form tight balls in the macronuclear channels (Fig. 102*d*). In the acicular aggregates and balls, thick and thin filaments are parallel to one another and are connected with cross-bridges, thus resembling fibrils of myosin and actin, respectively. Such a configuration of filaments is especially typical of Ma treated with an ATP-containing "contraction solution" before fixation. Long bundles of parallel thick and thin filaments were isolated also from glycerinized Ma. The actinomyosin nature of filaments found in the Ma of *Ichthyophthirius* at the rotation stage has been confirmed by an immunofluorescent technique: zones with bundles of filaments preferentially bound fluorescent antibodies against myosin (HAUSER *et al.* 1975). The nondividing Ma of *Ichthyophthirius* likewise contain myosin that is uniformly distributed throughout the nucleus. Myosin was found also in the predivision Ma of *Paramecium bursaria*. Therefore, the macronuclear contents are apparently mixed in ciliates with the participation of contractile proteins of the actinomyosin type during macronuclear condensation.

Fig. 102. Macronuclei during the phase of the rotation of their contents before division (electron micrographs). *a* and *b* vortical torsion of chromatin strands and disappearance of microtubule bundles in Suctoria (*a* part of the Ma of *Acineta*, 8750 × ; *b* Ma of *Tokophrya*, 9500 ×); *c* and *d* islets containing actin and myosin filaments in the Ma of *Ichthyophthirius*: *c* loose filaments with single needle-shaped aggregates (arrows), 13 000 × ; *d* compact skein of myosin filaments (thick arrows) and actin filaments (thin arrows), 48 000 × ; *1* chromatin strands, *2* macronuclear envelope, *3* small chromatin bodies, *4* islet without chromatin. *a* after BARDELE (1969); *b* after MILLECCHIA and RUDZINSKA (1971); *c* and *d* after HAUSER (1973).

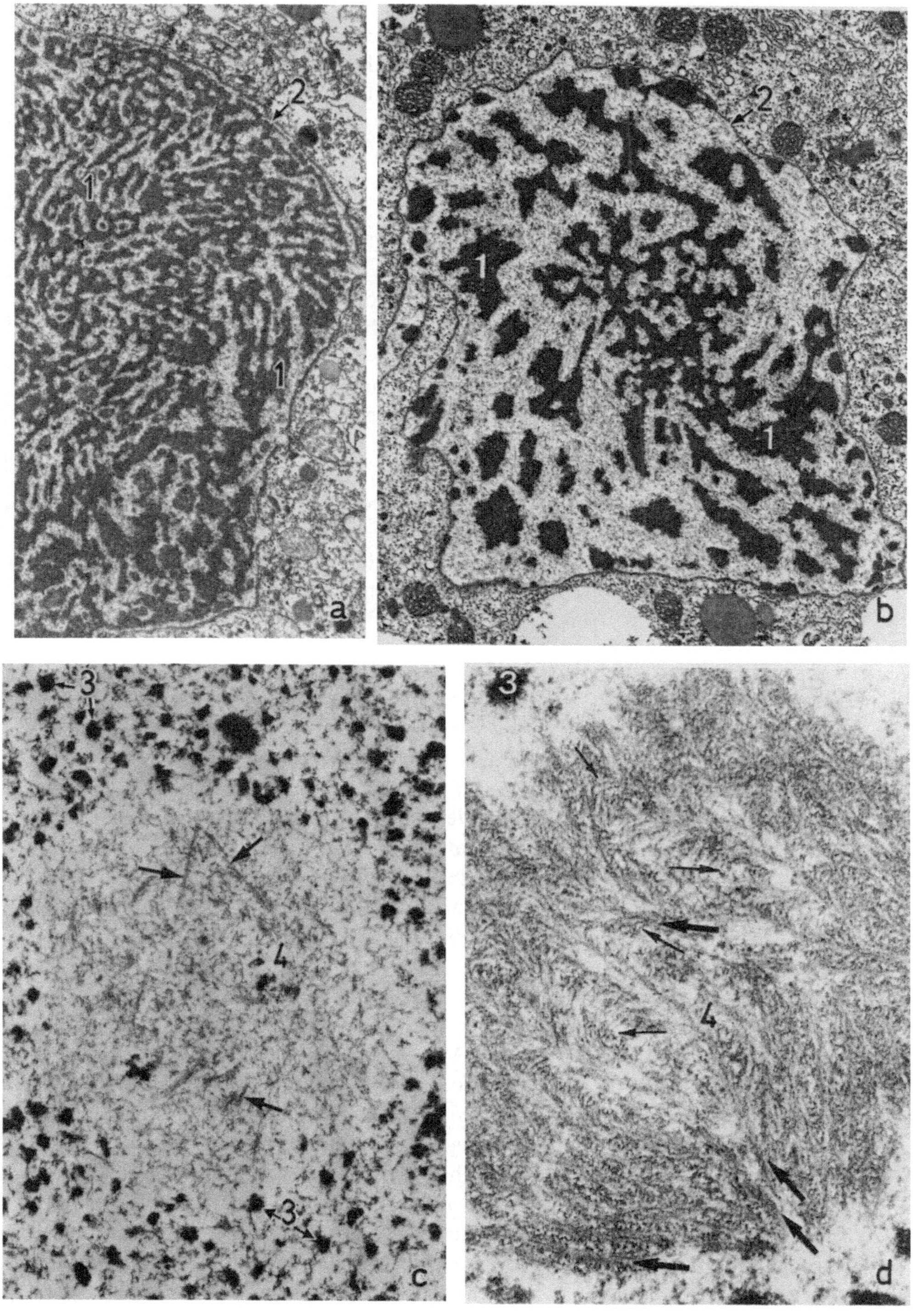

Fig. 102

It is still unclear why the macronucleus should condense and its contents be mixed. During the interphase, the rear nodes of the moniliform Ma of *Stentor* and *Blepharisma* were reported to lose their ability to support regeneration and their DNA was reported to be stained less with methyl green (WEISZ 1949b, 1950). In WEISZ's opinion, macronuclei condense in order to eliminate this morphogenetic gradient. However, TARTAR (1957, 1961) has not confirmed the fact that the front and rear nodes of the Ma in *Stentor* and *Condylostoma* are morphogenetically unequal. Indeed, the DNA content of different macronuclear nodes in *Stentor* may vary (SCHWEIKHARDT 1965, ZECH 1966a, MEISTER 1970), but this is related only to the functional activity of different parts of Ma. The anterior nodes, which lie in a greater amount of the cytoplasm near the peristome, are normally much richer in DNA than the posterior nodes (SCHWEIKHARDT 1965). The content of DNA increases in the posterior nodes as a result of its partial local replication when the rear end of the body regenerates (MEISTER 1970). These differences are smoothed out when Ma condense.

However, since the condensation and rotation of the nuclear contents are also typical of Ma with a relatively simple shape, it could be that the main purpose of macronuclear condensation is something else. Most likely, it is needed for compacting the macronuclear subunits that had to be maximally expanded earlier, e.g. during their replication. The compact subunits may then segregate when the nucleus divides.

Macronuclear Division

Types of Division. The equal division of a Ma in two (which looks like an amitotic one) is most common. Two phases, i.e. the stretching of a Ma and its constriction, can usually be distinguished in this process. Some Ma, however, are capable of undergoing multiple and unequal division as well as budding; the resultant daughter Ma are much smaller than half the mother nucleus, and only later do they grow to the size of the latter (Fig. 103). This process resembles the *regeneration* of Ma. Indeed, macronuclei are known to be able to regenerate when a ciliate is cut into parts, e.g. a normal Ma of *Stentor* can regenerate from one nodule, i.e. from 1/10 to 1/20 of the original nucleus (SCHWARTZ 1935, TARTAR 1961).

In ciliates that reproduce themselves by palintomy, the repeated macronuclear divisions in two usually do not involve the intermediate replication of DNA, i.e. they bring about *depolyploidization* of Ma. The content of DNA per Ma is halved in every division. In *Ichthyophthirius*, for example, the relative content of DNA in Ma (compared with Mi) decreases during palintomy from 6300 : 1 in a mature trophont to 24 : 1 in a swarmer (USPENSKAYA and OVCHINNIKOVA 1966), while DNA is distributed between the daughter cells as a variable number of haploid subunits (EIDE 1968). The reverse process of DNA accumulation occurs in a trophont whose Ma grows and undergoes several rounds of DNA replication but does not divide. Such reproduction cycles are typical of the order Colpodida as well as of the suborder Ophryoglenina and certain Tetrahymenina (order Hymenostomatida). The lack of DNA synthesis in Ma between two palintomic divisions was demonstrated cytophotometrically in *Tillina magna* (FRENKEL 1969) and *Colpoda steini* (FRENKEL *et al.* 1973), and it was corroborated both cytofluorimetrically and autoradiographically in *Tetrahymena patula*, which undergoes two or three such divisions (GABE and DE BAULT 1973), and in the macrostome form of *T. paravorax* (MÉTENIER 1981).

Behaviour of Chromatin Elements. The alignment of chromatin granules in longitudinal rows or even their fusion into longitudinal threads was often observed under the light microscope in the dividing Ma of different ciliates (CALKINS 1919, WEYER 1930, RAABE 1947, GRELL 1950a, 1953d, GOLIKOVA 1965, DWORAKOWSKA

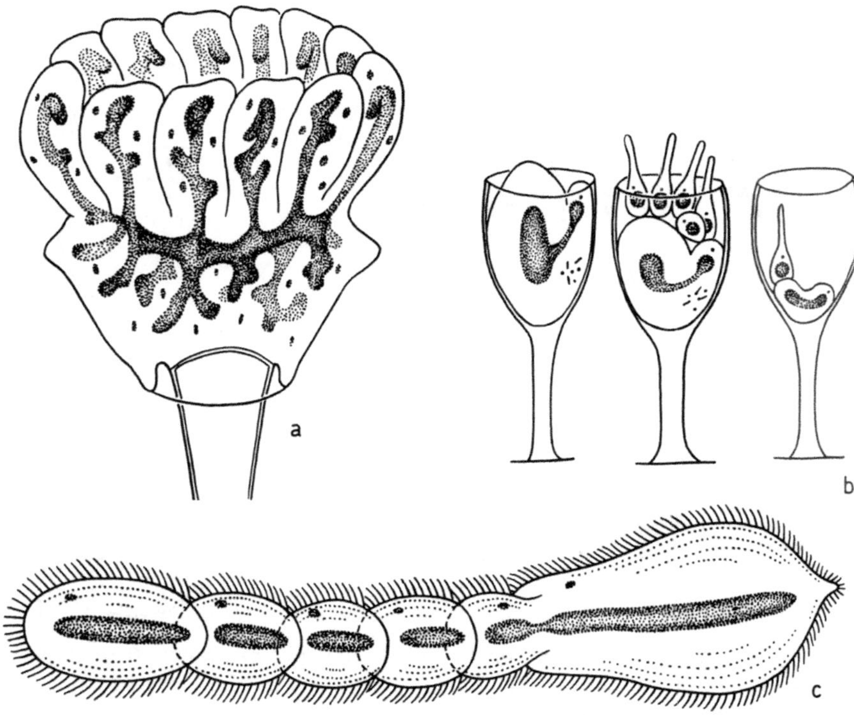

Fig. 103. Unequal and multiple macronuclear division (after GRELL 1953b). *a* simultaneous multiple division in *Ephelota gemmipara*; *b* sequential multiple division in the "*Dactylophrya*-stage" of the parasitic suctorian *Tachyblaston ephelotensis*; *c* unequal division in *Anoplophrya*.

1966, LIEDER and RUTHMANN 1976). It could be that in many cases, the effect is imaginary and is caused by bundles of microtubules that develop in Ma and force the chromatin elements back into the spaces between them. At the fine structural level, the "small bodies" of chromatin do not undergo any essential changes in the dividing Ma of most ciliates, e.g. in *Paramecium* (JURAND and SELMAN 1970, INABA and KUDO 1972).

But when macronuclear chromatin permanently consists of anastomosing threads (*Acineta*, many Hypotrichida), or when discrete chromatin bodies are transformed into threads during macronuclear condensation (*Tokophrya*), real threads are oriented along the stretching nucleus and apparently divide transversely during its division. In *Stylonychia* and *Gastrostyla*, the chromatin strands of the stretching Ma reattach themselves to the chromatin "plaques" that remain on the inner membrane of the nuclear envelope during macronuclear condensation. Thus, the nuclear envelope in these forms may be involved in the segregation of the macronuclear chromatin elements. In some Heterotrichida, such as *Blepharisma* (Inaba and Sotokawa 1968) and *Spirostomum* (Zazulina and Polyakov 1976), the macronuclear chromatin elements decondense during nuclear division. Condensed chromatin bodies or strands reappear only at the end of macronuclear division.

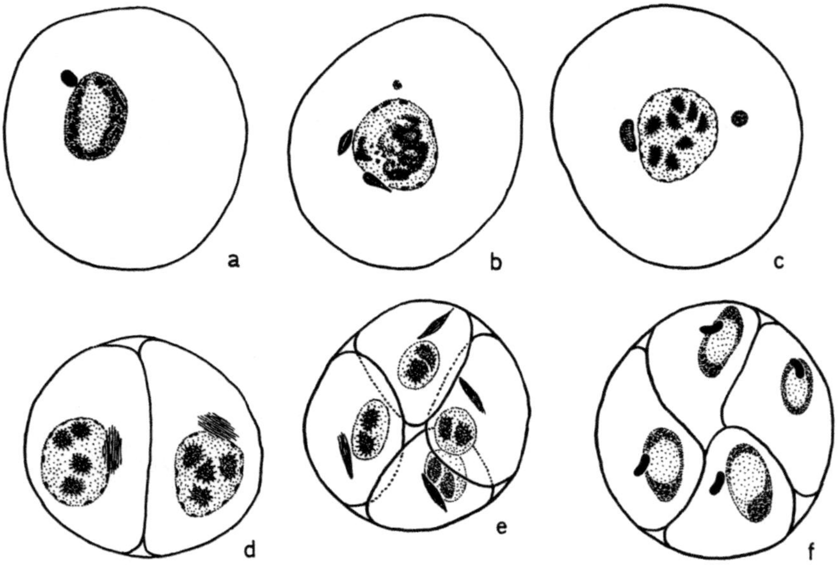

Fig. 104. Appearance (*a*–*c*), segregation (*d* and *e*), and disappearance (*f*) of chromatin aggregates during two consecutive macronuclear divisions in the reproduction cysts of *Colpoda steini* (Feulgen staining, 1000 ×; after Piekarski 1939).

The composite central nucleolus of the Colpodida disappears before division, and macronuclear chromatin contracts into peculiar stellate clumps. Eight such aggregates are formed in *Colpoda steini* (Fig. 104). When the ciliate divides twice (in a cyst), the chromatin aggregates segregate between daughter Ma, first in fours and then in twos (Piekarski 1939). In its DNA content (2c), each aggregate is equal to the presynthetic micronucleus, i.e. it is a diploid macronuclear subunit (Frenkel *et al.* 1973, Frenkel 1978). Therefore, the Ma of a *C. steini* trophont (with eight chromatin aggregates) may be polyploid to 16 n, while the Ma of a tomite that leaves the cyst (with two aggregates) is only tetraploid. Electron microscopy shows that chromatin aggregates in the dividing Ma of *Colpoda steini* consist of ordinary "small bodies" that are slightly drawn together. The composite nucleolus disintegrates and its fibrillar cores scatter throughout the Ma in the spaces between the chromatin aggregates; they have never been found inside the aggregates (FRENKEL 1980). It is very likely that the chromatin aggregates, being segregation units in macronuclear division, correspond to diploid chromosome sets (i.e. pairs of haploid genomes). The macronuclear division of Colpodida would then be the separation of whole chromosome sets, i.e. the *segregation of genomes* (GRELL 1950b, 1953b).

The number of chromatin aggregates differs in other Colpodida (BURT *et al.* 1941). *Colpoda duodenaria* has only four chromatin aggregates that segregate in twos during the single macronuclear division (TAYLOR and GARNJOBST 1941). *C. maupasi* has about 70 such aggregates (PADNOS 1962), and *Tillina magna* has over 100 of them (BEERS 1946). Accordingly, the relative content of DNA in the Ma of *T. magna* is greater (as compared with the Mi), e.g. up to 135:1 in tomites of certain clones (FRENKEL 1975). DNA segregation between two sister Ma in this species is often unequal, indicating that the chromatin aggregates can segregate randomly if they are numerous.

The chromatin aggregates in the Ma of *Nassula* and *Nassulopsis* (cf. p. 306) behave in a similar way. The replicated aggregates segregate between daughter nuclei almost equally (Fig. 100*d*). The number of aggregates in *Nassulopsis elegans* comes close to 130 before duplication and 260 after duplication. The DNA content of the Ma of *N. elegans* in G_1 is almost 125 times as high as that of the presynthetic micronucleus, while in G_2, it is 250 times as high (RAIKOV 1964). A comparison of these numbers shows that just as in *Colpoda steini*, the chromatin aggregates in the Ma of *Nassulopsis* (and probably *Nassula*) are diploid subunits, while macronuclear division seems to be the segregation of whole genomes.

About 80 discrete chromatin aggregates appear before division in the Ma of the GL strain of *Tetrahymena pyriformis* (NILSSON 1970, 1976a). Autoradiography has shown that the aggregates are formed in the G_2 period (NILSSON 1973); they contain equal amounts of DNA. Electron microscopy has revealed that the aggregates are groups of "small bodies" of chromatin drawn together; they may be coiled chains of such bodies. The aggregates are also segregation units in macronuclear division. But as regards the DNA content, they are smaller than diploid subunits and come close to haploid ones (NILSSON 1970, 1976a). SEYFERT (1979) has found similar aggregates, though in smaller quantities (35–60) in the predivision Ma of the strain HSM of *T. pyriformis*, but his interpretation is different (see p. 334).

Even ordinary macronuclear divisions in two are frequently unequal. The following ratios of DNA distribution between daughter nuclei were observed: from 1:1 to 1.14 in *Tetrahymena pyriformis*, strain H (MCDONALD 1958), and *Oxytricha* sp. (LIEDER and RUTHMANN 1976); up to 1:1.35 in *T. pyriformis*, strain HSM (CLEFFMANN 1968) and *Euplotes eurystomus* (WITT 1977); up to 1:1.5 in *Paramecium putrinum* (KUDRJAVTSEV 1966); up to 1:1.62 in *Spirostomum ambiguum* (OVCHINNIKOVA *et al.* 1965); and up to 1:1.88 in *Tillina magna* (FRENKEL 1975) and *Keronopsis rubra* (RUTHMANN 1972). In these cases, deviations from equal macronuclear division seem to be random. But in *Bursaria truncatella*, unequal macronuclear division is a regular occurrence: the Ma of the rear daughter cell is always larger and contains, on the average, 1.4 time more DNA than the Ma of the front cell (ZECH 1966b); individual variations of the ratio reach 1:2.4 (RUTHMANN 1964). Very sharp variations in the DNA content between sister

Ma are found in the *Paramecium tetraurelia* cells, which are homozygous with respect to the mutant gene *am* (BERGER 1973c).

Unequal DNA distribution in the macronuclear division of many ciliates can be explained if it is assumed that whole genomes randomly segregate between daughter Ma, just as in Colpodida and *Nassulopsis*. According to the cytophotometric data provided by EIDE (1968), DNA is distributed as haploid subunits during the repeated macronuclear divisions in *Ichthyophthirius*. But this assumption cannot be applied to small *Keronopsis* macronuclei which, on the average, contain diploid DNA quantities in the G_1 period and divide without preliminary fusion with one another. Apparently, only macronuclear subunits, which are smaller than even haploid ones, can segregate in this case (RUTHMANN 1972).

Microtubules in the Dividing Macronucleus. SCHWARTZ (1957) was the first to show that the Ma of *Paramecium bursaria* becomes birefringent during its division; this suggested that longitudinal fibrils of a protein nature appeared in Ma (birefringence was eliminated when the nucleus was treated with proteases). In *Nassula ornata*, a bundle of protein fibrils is seen in the constriction zone of the dividing Ma. The bundle was interpreted as a "pushing body" of the dividing Ma; its extension apparently elongates the entire nucleus (RAIKOV 1960, 1962a).

Longitudinal bundles of microtubules can, as a rule, be detected in the dividing Ma, especially during its stretching. Such bundles were found in the gymnostomes *Didinium* (RIEDER 1971), *Spathidium* (IVASHCHENKO 1980) and *Dileptus* (VINNIKOVA 1974c, 1974d, BOHATIER 1977b; Fig. 105*a*), the vestibuliferans *Bursaria* (SERGEJEVA 1979), *Diplodinium* (ROTH and SHIGENAKA 1964) and *Isotricha* (GRAIN 1966), the hypostome *Nassula* (RAIKOV 1966, TUCKER 1967), the hymenostomes *Tetrahymena* (ROTH and MINICK 1961, FALK *et al.* 1968, ITO *et al.* 1968, TAMURA *et al.* 1969, DAVIDSON and LA FONTAIN 1975, etc.), *Paramecium* (JURAND and SELMAN 1969, STEVENSON and LLOYD 1971b, INABA and KUDO 1972, TUCKER *et al.* 1980; Fig. 105*b*), *Ophryoglena* (ROQUE *et al.* 1965), and *Ichthyophthirius* (MOSEVICH 1968). Bundles comprising 20–25 microtubules were found in the stretching Ma of the peritrich *Campanella* (CARASSO and FAVARD 1965; Fig. 105*c*). Among spirotrichs, intranuclear microtubules seem to be absent in Heterotrichida but present in Hypotrichida, e.g. *Stylonychia*, *Gastrostyla* (ROTH 1960, WALKER and GOODE 1976) and *Euplotes* (RUFFOLO 1978b).

The spatial arrangement of microtubules may vary in the dividing Ma. Usually, they form small bundles that are spread throughout the Ma (Fig. 105*b*). Each

---

Fig. 105. Microtubules in macronuclear division (electron micrographs). *a* part of the dividing Ma of *Dileptus anser* with a bundle of microtubules (*1*) surrounded by nucleolar granules (*2*); intranuclear vesicles (*3*) are seen at the pole, 54 000 ×; *b* longitudinal section through the constriction of the dividing Ma of *Paramecium aurelia* showing microtubules (*1*), 36 000 ×; *c* bundles of microtubules (*1*) in longitudinal and cross (inset) sections in the dividing Ma of *Campanella*, 65 000 ×; *d* tangential section through the dividing Ma of *Blepharisma japonicum* during the stretching phase, with parallel microtubules (*1*) on the outer surface of the nuclear envelope, 26 500 ×; *4* macronuclear envelope, *5* its pore complexes, *6* chromatin bodies, *7* nucleoli, *8* Ma. *a* after VINNIKOVA (1974c); *b* after JURAND and SELMAN (1970); *c* after CARASSO and FAVARD (1965); *d* after JENKINS (1977).

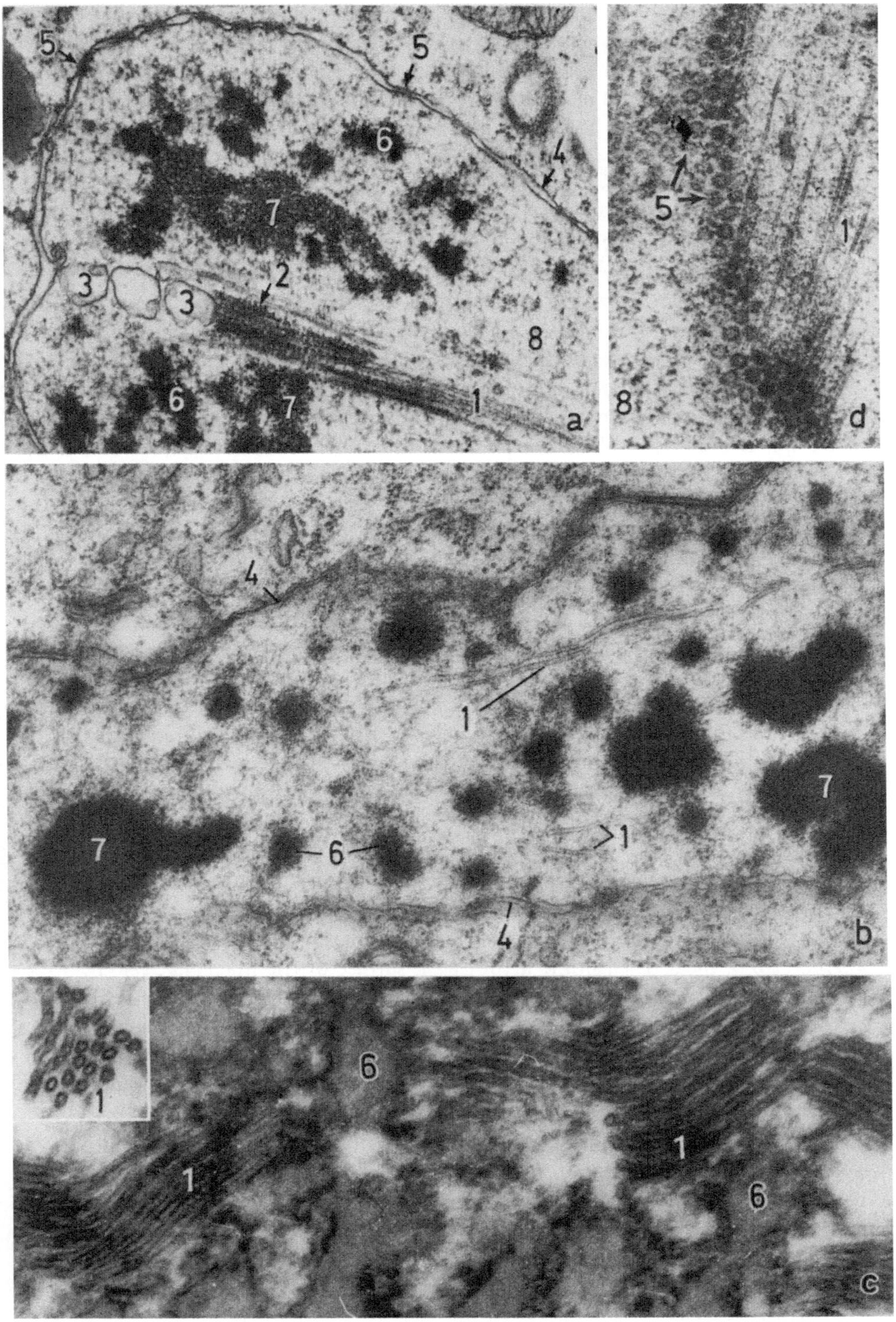

Fig. 105

small Ma of *Dileptus* usually has only one bundle of microtubules that passes along the axis of the stretching nucleus and through the material of the loosened nucleolus; its individual microtubules are often coated with nucleolar granules (Fig. 105*a*). At the poles of the division figure, the bundle of microtubules often butts against membrane-bound vesicles that bud from the inner membrane of the macronuclear envelope (Fig. 105*a*). The microtubules are also coated with nucleolar granules in *Tetrahymena* (Falk *et al.* 1968), *Didinium* (Rieder 1971), and *Paramecium* (Tucker *et al.* 1980). In *Stylonychia* and *Gastrostyla*, the microtubules butt against the macronuclear envelope from the inside at the poles of the division figure (Walker and Goode 1976).

In other cases, the bundles or ribbons of microtubules are mainly at the periphery of a Ma. When the Ma of *Tetrahymena* elongates, all the microtubules form a single layer just beneath the macronuclear envelope. But when Ma pinches in two, another set of microtubules appears in the constriction zone; this set is known as the "separation spindle" (Davidson and La Fountain 1975). A similar pattern is characteristic of *Paramecium primaurelia* (Stevenson and Lloyd 1971 b).

A more sophisticated relay of different populations of microtubules occurs during macronuclear division in *Paramecium tetraurelia* (Tucker *et al.* 1980). Before stretching, the Ma flattens dorsoventrally against the dorsal cell cortex and then contains an elliptical marginal band of microtubules that "spans" the Ma to a flat shape. During the stretching phase, these microtubules disappear and are replaced by a population of straight longitudinal microtubules that are preferentially peripheral, arranged in bundles, and interconnected with cross-bridges. They seem to cause macronuclear elongation due to their mutual sliding with a velocity of about 40 μm per minute. In a mutant in which intramacronuclear microtubules fail to be regularly arranged, the stretching of Ma is delayed and reduced (Cohen *et al.* 1980). Most microtubules are disassembled when the stretching phase is over. and only a few microtubules are present in macronuclei that pinch in two (Tucker *et al.* 1980).

Microtubules involved in macronuclear division may be found not only in the nucleus but also in the cytoplasm, where they retain their general longitudinal orientation and remain in close contact with the outer membrane of the macronuclear envelope. In Heterotrichida, all microtubules are extranuclear. In *Blepharisma*, they form a single layer on the surface of the macronuclear envelope (Fig. 105*d*) and pass parallel to it at the sides of the stretching Ma and perpendicular to it at the poles (Inaba and Sotokawa 1968, Jenkins 1977). In *Stentor*, all microtubular bundles are enclosed in special cytoplasmic channels that penetrate the Ma at the condensation stage and at the beginning of the stretching stage; the Ma has no microtubules inside it (Paulin and Brooks 1975).

Microtubules behave in a peculiar manner during macronuclear budding in Suctoria. As we have already seen (Fig. 95*d*), bundles of microtubules are well developed in the nondividing Ma, but they disappear almost completely during the condensation and rotation of the macronuclear contents (Fig. 102*a*, *b*). When the Ma of *Tokophrya* divides, small longitudinal bundles of microtubules reappear between the chromatin strands (Millecchia and Rudzinska 1976); only single microtubules develop in the dividing Ma of *Acineta* (Bardele 1969) and *Paracineta* (Hauser 1970). Therefore, Bardele and Hauser believe that the microtubules are a storage form of proteins that participate in macronuclear division and not a kinetic element of the dividing Ma.

However, experimental data that are available for other ciliates suggest that the microtubules are the locomotory component needed at least for macronuclear stretching; in this respect, they are similar to the mitotic spindle. Mercaptoethanol, a spindle inhibitor, blocks or seriously impairs the macronuclear division in *Tetrahymena* (HOLZ 1958, GAVIN and FRANKEL 1966) and *Stylonychia* (DASS and SAPRA 1966). The Ma does not stretch in a normal way and, in cytokinesis, one daughter cell receives no Ma or an abnormally small Ma. Using colchicine treatment together with electron microscopy, TAMURA and others (1969), as well as WUNDERLICH and SPETH (1970), have shown that depolymerization of the intranuclear microtubules in the Ma of *Tetrahymena* coincides with an arrest of its division, while reformation of the microtubules coincides with the resumption of the division. N. E. WILLIAMS and R. J. WILLIAMS (1976) studied the effect of colchicine on macronuclear division in *T. pyriformis* in greater detail; they showed that microtubules may not be involved in the initial phases of macronuclear stretching but that they are necessary for the terminal phases of stretching as well as for the regular pinching of Ma in two. When there are no microtubules, a partially stretched Ma can only be cleaved in a mechanical way by the cytoplasmic fission furrow, usually in two unequal parts. Dimethylsulphoxide, another compound inhibiting the assembly of microtubules, produces a similar effect in *Tetrahymena* (NILSSON 1974).

A noteworthy fact is that treatment with colchicine, podophyllotoxin or mercaptoethanol blocks not only macronuclear division with intranuclear microtubules but also such division with cytoplasmic (perinuclear) microtubules, e.g. in *Blepharisma* (JENKINS 1977) and *Spirostomum* (SUNDARARAMAN and DASS 1973). Therefore, the function of the latter microtubules is at least partially similar to that of the former. An opposite effect (the stabilization of microtubules) is induced by heavy water. If the dividing Ma of *Stylonychia* and *Gastrostyla* are treated with D_2O, abnormally thick and long bundles of microtubules develop in them and the whole Ma stretches greatly (WALKER and GOODE 1976).

Thus, intranuclear and probably also extranuclear microtubules are responsible for the stretching and regular division of the Ma. The mechanism of their participation in macronuclear division may involve assembly of microtubules at the ends and/or their mutual sliding. Different populations of microtubules that regularly replace one another are probably needed for macronuclear stretching and pinching in two. Moreover, some stages of macronuclear division apparently need other structures (possibly, contractile microfilaments) in addition to or instead of microtubules. This should be studied further, especially in Suctoria.

Behaviour of Nucleoli and the Macronuclear Envelope. The fragmentation, size diminution, and even the disappearance of nucleoli were often observed under the light microscope during macronuclear division, e.g. in *Fabrea* (ELLIS 1937), *Paramecium bursaria* (SCHWARTZ 1956), *Loxophyllum* (RUTHMANN 1963), *Dileptus cygnus* (GOLINSKA 1965), and *Paramecium putrinum* (JANKOWSKI 1966). But in other species, the morphology of nucleoli does not undergo any significant changes during macronuclear division.

Electron microscopy has revealed that, in most cases, nucleoli change only slightly during macronuclear division, e.g. in *Tetrahymena* (FLICKINGER 1965). In *Didinium*, the granular component of nucleoli loosens and they stretch along the dividing Ma (RIEDER 1971, KARADZHAN and RAIKOV 1977). Nucleolar granules in *Spirostomum* almost completely disappear during macronuclear division (ZAZULINA and POLYAKOV 1976). The single nucleolus of small Ma in *Dileptus* becomes very loose during division and is permeated with bundles of microtubules

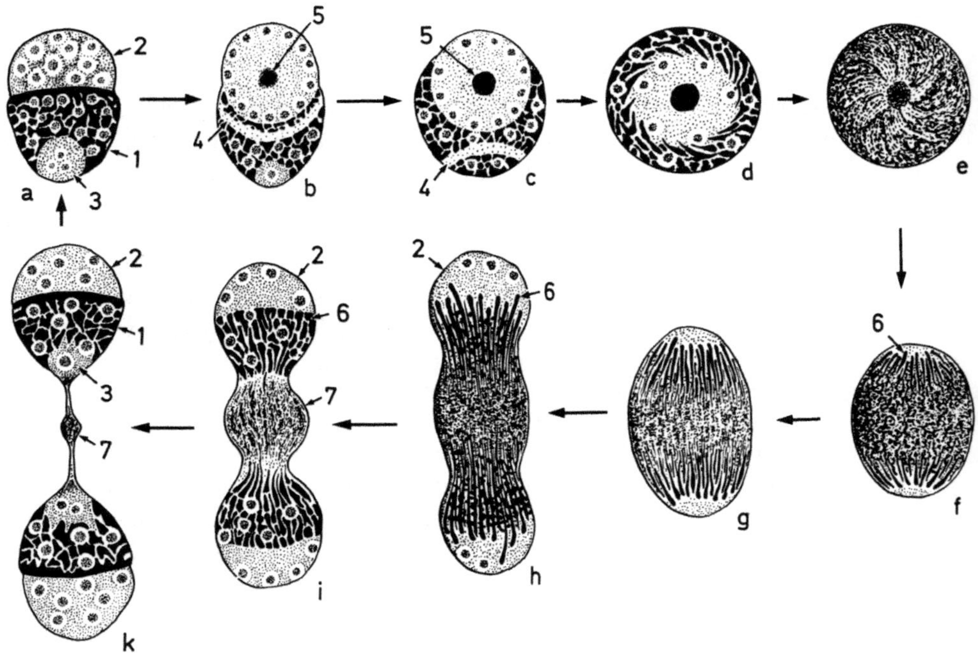

Fig. 106. Division cycle (schematic) of the heteromeric macronucleus in *Spirochona gemmipara* (after FAHRNI 1977). *a–c* interphase; *d* and *e* predivisional reorganization: *f–k* division; *1* orthomere, *2* paramere, *3* epimere, *4* replication band, *5* endosome. *6* chromatin strands, *7* median thickening.

(VINNIKOVA 1974c, 1974d; Fig. 105*a*). Many nucleoli are attached to microtubules during the phase of macronuclear stretching in *Paramecium tetraurelia;* since it is likely that these microtubules may mutually slide, the nucleoli may be actively propulsed into the future daughter nuclei (TUCKER *et al.* 1980).

The nuclear envelope is always preserved in macronuclear division. In some cases, membrane-bound vesicles separate either from its inner membrane and move inward the Ma (Fig. 105*a*) or from its outer membrane and move into the cytoplasm, e.g. in *Nassula* (RAIKOV 1966) and *Didinium* (RIEDER 1971).

Division of Heteromeric Macronuclei. The division of the "juxtaposed" heteromeric Ma was studied by light microscopy in *Cavichona* (SWARCZEWSKY 1928), *Spirochona* (GUILCHER 1951, TUFFRAU 1953, FAHRNI 1977), *Dysteria* (FAURÉ-FREMIET 1957), *Chlamydodon* (KANEDA 1960, 1961a), *Allosphaerium*

(Dobrzanska-Kaczanowska 1963), and *Lobochona* (Matsudo 1966), and by electron microscopy only in *Spirochona gemmipara* (Guénin and Fahrni 1971, Fahrni 1977). Macronuclear division is rather similar in all these forms. A chromatin endosome develops gradually in the paramere during the interphase and a replication band passes through the orthomere (Figs. 96*a*, 106*a–c*). The orthomere surrounds the paramere just before division, and the Ma becomes concentric for a short time (Fig. 106*d*). The chromatin of the orthomere decondenses and is apparently mixed (Fig. 106*e*). Later, it condenses again and forms lengthwise strands at the nuclear periphery (Fig. 106*f*, *g*). The paramere material gradually migrates to the poles of the division figure (Fig. 106*g*, *h*). The chromatin strands of the orthomere then subdivide into two daughter groups and also migrate to the poles (Figs. 96*b*, 106*h*, *i*), while a part of the chromatin remains in a median thickening at the isthmus between the daughter Ma (Fig. 106*i*, *k*). This chromatin possibly originates from the endosome. Immediately after the division, the daughter Ma has, besides a paramere, an epimere structurally similar to the paramere (Fig. 106*k*, *a*).

As for the chromatin of the median thickening, most authors believe that it is extruded into the cytoplasm as a residual body (Tuffrau 1953, Kaneda 1960, Dobrzanska-Kaczanowska 1963, Matsudo 1966). Fahrni (1977), however, believes that this material is either resorbed in situ or is incorporated in the daughter Ma.

Many longitudinal bundles of microtubules appear between the chromatin strands of the orthomere at the late division stages. Thus, the dividing orthomere is similar to a dividing homeomeric Ma. Microtubules are absent from achromatic polar caps (the future parameres of daughter Ma). The macronuclear envelope remains intact throughout the division (Guénin and Fahrni 1971).

Heteromeric Ma of the "concentric" type, such as in *Chilodonella*, divide in a simpler manner (MacDougall 1925, 1936, Radzikowski 1965). The paramere remains surrounded by the orthomere during division. The endosome seems to be extruded into the cytoplasm (Radzikowski 1965, 1976).

Morphology and Reproduction of the Macronuclei of Protocruzia

Marine ciliates of the genus *Protocruzia*, which belong to the lower Heterotrichida (Grolière *et al.* 1980), have a very peculiar nuclear apparatus. In *P. depressa*, all macronuclei, whose number varies from 5 to 15–20 in different races, are arranged as a rosette around a unique micronucleus or around a group of 2–6 micronuclei (Ammermann 1968a). Hence, there is some similarity between the nuclear apparatus of *Protocruzia* and that of certain Karyorelictida where Ma and Mi are assembled in a nuclear group (see Chapter 7). But unlike Ma in Karyorelictida, all Ma in *Protocruzia* are able to divide without preliminary fusion (Ammermann 1968a).

Macronuclear division was studied in *Protocruzia* sp. (Ruthmann and Hauser 1974). The nuclear apparatus of this species comprises a tight cluster of ten small vesicular Ma around a unique Mi (Fig. 107*a*). Electron microscopy has shown that each Ma contains a nucleolus and is surrounded by a typical bimembranous envelope. The chromatin material is dispersed in the peripheral zone of each Ma.

It has been demonstrated cytophotometrically that the complex of ten Ma contains about ten times (at least six times) as much DNA as the Mi. The division of the nuclear apparatus in *Protocruzia* has been studied so far only by light microscopy. The micronucleus divides by mitosis with the chromosome number not less than 5, possibly around 10 (Fig. 107*c*–*e*). Two parallel chromatin strands then develop in each Ma, while macronuclear envelopes seem to dissolve (Fig. 107*b*). The threads are much larger than the micronuclear chromosomes and they differ in length. Later, the paired strands gradually shorten (Fig. 107*c*, *d*),

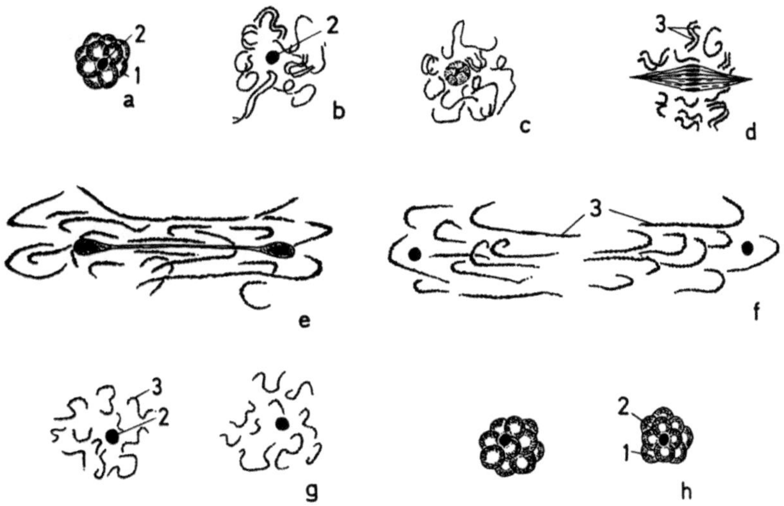

Fig. 107. Division of the nuclear apparatus in *Protocruzia* sp., 1400 × (after Ruthmann and Hauser 1974). *a* cluster of ten Ma (*1*) and one Mi (*2*); *b*–*d* formation of a pair of chromatin threads (*3*) in each Ma, mitosis of Mi; *e* and *f* separation of the sister chromatin threads (*3*); *g* and *h* two daughter nuclear clusters before (*g*) and after (*h*) the formation of new Ma.

surrounding the micronuclear spindle. During the telophase of Mi, two strands of each pair separate just as chromatids do in the anaphase of mitosis, migrating in opposite directions (Fig. 107*e*, *f*). Consequently, each daughter nuclear group gets one Mi and ten macronuclear strands (Fig. 107*g*), which are then transformed again into ten vesicular Ma (Fig. 107*h*). The mechanism of the separation and migration of daughter chromatin strands remains obscure; no structures resembling a "macronuclear spindle" have been described so far.

Ruthmann and Hauser (1974) proposed the following interpretation of the above pattern. Individual Ma of *Protocruzia* are *karyomeres*, each of which is formed by an individual macronuclear chromosome. The latter is homologous to one of the micronuclear chromosomes, but it is larger and contains several times more DNA owing to oligoteny (from six- to tenfold). The individuality of the respective micronuclear chromosomes would account for differences in the length of the chromatin strands that originate from various karyomeres (individual Ma).

The unique oligotene chromosome of each karyomere replicates in the interphase and gives rise to two oligotenic chromatids that segregate during the subsequent division of the nuclear cluster. Thus, according to RUTHMANN and HAUSER, the division of the macronuclear apparatus in *Protocruzia* is basically *mitosis*, although it is complicated with karyomery and chromosome oligoteny.

Development and Genetic Organization of the Macronucleus

The macronucleus develops from a diploid synkaryon derivative after conjugation or autogamy. Chromosomal transformations that occur in a developing macronuclear anlage obviously determine the organization of the genetic apparatus of mature Ma. Since the course of macronuclear development is quite different in various ciliates, the organization of mature Ma also cannot be the same in all ciliates. Therefore, data on macronuclear development as well as on the organization of the genome of adult Ma will be considered separately for different groups of ciliates. The most detailed information of this kind is available for Hymenostomatida (*Tetrahymena*, *Paramecium*, etc.) on the one hand, and for Hypotrichida (*Stylonychia*, *Oxytricha*, *Euplotes*, etc.) on the other.

Macronuclei of Hymenostomatida

Development of Ma. The development of macronuclear anlagen in hymenostomatids was studied morphologically and cytochemically in *Paramecium* and *Tetrahymena*.

A skein of despiralized chromosomes appears in a young macronuclear anlage of *Paramecium caudatum* or *P. bursaria;* the chromosomes later condense and partly fuse, producing a *karyosome*, i.e. an intranuclear chromatin body (KLITZKE 1915, DEHORNE 1920, PESHKOVSKAYA 1941, GROMOVA 1948, EGELHAAF 1955, SAITO and SATO 1961 b). Most authors believe that only the heterochromatin chromosome segments fuse with one another, while the euchromatin regions redecondense. The karyosomal stage is short, and the karyosome soon disappears. SCHWARTZ (1978) discovered that only the thickened pericentromeric regions of the chromosomes are preserved in the macronuclear anlage of *P. bursaria*, while the chromosome arms detach and disintegrate (in some strains, they preliminarily fuse into the karyosome). These observations are consistent with SCHWARTZ's own findings to the effect that the adult Ma of *P. bursaria* contains shortened chromosome fragments rather than whole chromosomes (cf. p. 281). The next stage is an "achromatic" one, when the chromatin of the anlage is fully decondensed and the anlage is faintly stained by the Feulgen reaction. The macronuclear anlagen of *Paramecium trichium* (DILLER 1948), *P. putrinum* (JANKOWSKI 1966), and *P. tetraurelia* (BERGER 1973a) also display this stage. Then 1–3 RNA-containing nucleoli appear in the anlage. These "primary" nucleoli are probably linked with definite chromosomes. Later, however, many "secondary" nucleoli appear in the anlage; at the same time, the anlage begins to accumulate fine granular chromatin (PESHKOVSKAYA 1941, GROMOVA 1948, EGELHAAF 1955). Three generations of nucleoli develop consecutively in the macronuclear anlage of *P. bursaria* (EHRET and POWERS 1955). Some of them are extruded into the cytoplasm; this has been observed also in *P. putrinum* (JANKOWSKI 1966).

Most cytophotometric studies indicate that DNA accumulates progressively (without any phase of its partial loss) in the developing macronuclear anlagen of *P. caudatum* (DUPY-BLANC 1969a) and *P. tetraurelia* (BERGER 1973a). More than ten cycles of DNA replication occur in the anlagen of the latter species. The DNA breakdown products from the old Ma may be reutilized as DNA precursors, especially in the starving exconjugants (BERGER 1974). Consequently, the young Ma of *P. tetraurelia* that is ready for the first division contains almost 1700 times more DNA than its presynthetic micronucleus and twice as much DNA as the postsynthetic vegetative Ma (WOODARD *et al.* 1966). It is still unknown how this DNA excess is normalized in the initial generations after conjugation.

However, the DNA content is almost halved temporarily in the macronuclear anlage of *Paramecium bursaria* at the "achromatic" stage of its development. This is preceded by an initial sixfold increase in the DNA content of the anlage above the diploid level (SCHWARTZ and MEISTER 1975a). Then, DNA synthesis is resumed in the anlage. It could be that a similar decrease in the DNA content occurs also in *Paramecium caudatum* and the members of the *aurelia* group since the initial stages of the development of their Ma have not been sufficiently studied photometrically. The loss of DNA from the macronuclear anlagen of *P. bursaria* apparently reflects the disintegration of the chromosome arms described by SCHWARTZ (1978); it is *genetic diminution*, i.e. the elimination of a part of the micronuclear genome.

The development of the macronuclear anlagen was studied by electron microscopy in the *Paramecium* species of the *aurelia* group (JURAND *et al.* 1964, JURAND and SELMAN 1970) and in *P. caudatum* (PUYTORAC and BLANC 1967), but mainly at the late stages, with attention given particularly to the nucleoli. The anlagen are filled with a loose network of fibres 10–30 nm thick at the "achromatic" stage; no chromosomes can be discerned at this stage as well as at the later ones. Nucleoli then appear in the anlagen, growing into large spongy or reticular aggregates. At the later stages of development, the nucleolar aggregates break up into small definitive nucleoli and many chromatin "small bodies" appear in the anlage.

MURTI (1973) used the electron microscope to investigate the development of the much smaller macronuclear anlagen in *Tetrahymena thermophila*. The chromosomes of the anlage at first decondense without forming a chromatin karyosome. Then, clumps of condensed chromatin appear in the matrix of the anlage and small fragments, surrounded by a nuclear envelope and containing some of the clumps, bud off the anlage. Therefore, a part of the DNA may be eliminated from the macronuclear anlage of *Tetrahymena*. The anlage grows, its chromatin clumps condense and are transformed into "small bodies" whose number gradually increases. Nucleoli appear in the Ma-anlage of *Tetrahymena* relatively late; they have a typical submembranous localization and are cupped from the very beginning (cf. Fig. 94*a*). At the same time, synthesis of rRNA begins in the anlage (LEICK *et al.* 1979).

It has been shown fluorimetrically that DNA accumulates progressively in the developing macronuclear anlage of *T. thermophila* (DOERDER and DEBAULT 1975). The anlage differentiates in the G_1 period of the diploid synkaryon derivative. The initial diploid DNA amount undergoes a series of consecutive duplications in it. No significant diminution of the DNA content, which would correspond to the chromatin extrusion reported by MURTI, was detected by fluorimetry. However, the elimination of certain DNA sequences does occur at the medium stages of development of the macronuclear anlage; that was demonstrated in *T. thermophila* by molecular hybridization methods (YOKOYAMA and YAO 1980). Just as in *Paramecium tetraurelia*, the young Ma of *Tetrahymena* ready for the first division contains 50% more DNA than the vegetative postsynthetic Ma (DOERDER and DE BAULT 1975).

Properties of DNA in Adult Macronuclei. This field has recently been reviewed by GOROVSKY (1980). The DNAs prepared from Ma and Mi of *Tetrahymena thermophila* are almost identical in their melting (heat denaturation) patterns and buoyant densities in the CsCl gradient (YAO and GOROVSKY 1974). Both DNAs consist of a single major fraction of the same density. The renaturation rates of the two DNAs are also nearly identical. However, the shapes of the renaturation curves indicate that micronuclei have some moderately repetitive DNA sequences that are absent in macronuclei. But 80%–90% of the DNA sequences present in the micronuclear genome are preserved in the Ma (YAO and GOROVSKY 1974, IWAMURA *et al.* 1979). There are also certain biochemical differences between the two DNAs of *T. thermophila:* macronuclear DNA contains much methyladenine, while micronuclear DNA has almost none (GOROVSKY 1973, 1980).

Table 9 shows that the size of the macronuclear unit genome of *T. thermophila*, calculated from the rate of DNA renaturation (ALLEN and GIBSON 1972a), approximates the size of the micronuclear genome determined by cytophotometry (GIBSON and MARTIN 1971, WOODARD *et al.* 1972, ALLEN and GIBSON 1972a). Corrected data indicate that the genomes of both nuclei have a size close to $130 \cdot 10^9$ daltons, i.e. to 70 mm of the DNA double helix (YAO and GOROVSKY 1974). Thus, if genetic diminution does occur during macronuclear development, it involves only a small part of the genome, possibly the repetitive DNA sequences present in the Mi but absent in the Ma. Indeed, one such sequence has been isolated recently from the Mi of *T. thermophila* and investigated by the gene cloning technique (YAO and GALL 1979). During macronuclear development, this sequence is at first replicated together with the bulk DNA, and only then is it eliminated (YOKOYAMA and YAO 1980). Thus, certain genes may be lost in the course of macronuclear development in *Tetrahymena* (GOROVSKY 1980).

Using DNA renaturation data, S. BORCHSENIUS and others (1978) estimated that the unit genome size of the Ma was $120 \cdot 10^9$ daltons (Table 9) or $2 \cdot 10^8$ base pairs in the amicronucleate strain GL of *T. pyriformis*. In this case, over 90% of total DNA is represented by unique sequences, and the remaining part, by moderate repeats. A noteworthy fact is that both varieties contain a large amount (3%–4% of total DNA) of inverted repeat (palindromic) sequences that are self-complementary and therefore instantly reassociate, producing "hairpins" ("foldback DNA"). This category includes amplified rDNA among moderate repeats as well as many small palindromes integrated in the bulk DNA among unique sequences (BORCHSENIUS *et al.* 1978, BORCHSENIUS and MERKULOVA 1980). The initiation sites of DNA replication in the Ma are frequently palindromic (BORCHSENIUS *et al.* 1981b). Both palindromic and nonpalindromic unique sequences are strongly methylated, while moderately repetitive DNA sequences contain somewhat less methyladenine (KIRNOS *et al.* 1980).

Table 9. *Molecular characterization of the macronuclear genome as compared with the micronuclear genome*

Species	Micronucleus		Macronucleus						References (numbers refer to data marked with the same number in the same entry)
	genome size (analytical complexity), daltons × 10^9	chromosome number (n)	size of unit genome (kinetic complexity), daltons × 10^9	number of unit genomes in Ma	size of bulk DNA molecules: daltons × 10^6	size of bulk DNA molecules: μm	size and form[a] of rDNA molecules: daltons × 10^6	size and form[a] of rDNA molecules: μm	
Tetrahymena thermophila	126–136[1] ~ 130[2]	5	135–145[1] ~ 130[2]	45	400–1000[6] 20 000–30 000[4b]	220–550 11 000 16 000	13.5[3,5] PD	7.3	Allen and Gibson 1972a (1), Yao and Gorovsky 1974 (2), Karrer and Gall 1976 (3), Williams *et al.* 1978 (4), Din and Engberg 1979 (5), Preer and Preer 1979 (6)
T. canadensis	156[1]	5	162[1]	45	?	?	12.0[2] PD	6.4	Allen and Gibson 1972a (1), Din and Engberg 1979 (2)
T. pyriformis, strain GL	—	—	120[6]	45	≥ 400[5] 20 000–30 000[7b]	≥ 220 11 000 16 000	12.6[1–4] PD	6.7	Gall 1974 (1), Engberg, Christiansen, and Leick 1974 (2), Karrer and Gall 1976 (3), Engberg *et al.* 1976 (4), Merkulova and Borchsenius 1976 (5), Borchsenius *et al.* 1978 (6), Williams *et al.* 1978 (7)
Glaucoma chattoni	?	?	?	?	1.26 60	0.7 32	5.6 (M)	3	Katzen *et al.* 1981
Paramecium primaurelia	183 189	43 63	184 189	~ 800	?	?	?	?	Allen and Gibson 1972a
Same	?	43 63	30	1400	≥ 200 8000 b	≥ 105 4400	?	?	Cummings 1975

Same	?	43 63	52	840	0.35 42	0.2 23	?	?	McTavish and Sommerville 1980
P. biaurelia	185	?	196	818	?	?	?	?	Allen and Gibson 1972a
P. tetraurelia	?	?	?	?	200 300[2]	110 160	5.5 · x[1] (x = 1 13) TR	3 · x	Findly and Gall 1978 (1), Preer and Preer 1979 (2)
P. octoaurelia	177 194	?	257	~ 600	?	?	?	?	Allen and Gibson 1972a
Same	?	?	38	1680	?	?	14.5	8	Soldo and Godoy 1972
P. bursaria, syngen 4	2250*	?	160	560	?	?	?	?	Cullis 1972, 1973
Stentor coeruleus	?	?	60	?	5 80	2.5 40	?	?	Pelvat and De Haller 1976
Stylonychia mytilis	4300[1]	125 150[1]	65[1]	4000[1]	1 70	0.5 38[1]	4.5[3]	2.4	Ammermann *et al.* 1974 (1),
	5000[2]	~ 150[2]	31[2,4]	15 000[2,4]	0.3 14[3,4] [3][2]	0.16 8 [1.6]	M		Ammermann *et al.* 1978 (2), Lipps and Steinbrück 1978 (3), Steinbrück *et al.* 1981 (4)
Oxytricha sp.	400[3]	~ 120	36[3]	2000[3]	1 1.95[1] [1.15]	0.2 1.7[1] [0.8]	4.0	2.2[1]	Prescott *et al.* 1971 (1), Wesley 1975 (2),
					0.17 4.2[2] [2.1]	0.1 2.3 [1.1]	4.9[5] M	2.7	Lauth *et al.* 1976 (3), Lawn *et al.* 1978 (4), Swanton *et al.* 1980a (5), Swanton *et al.* 1980b (6)
					0.24 12[4,6] [1.3]	0.13 6.6 [0.7]			
O. fallax	?	?	?	?	0.3 13[1]	0.16 7	4.0[1]	2.2	Rae and Spear 1978 (1),
							4.4[2] M	2.4	Spear 1980 (2)

[a] Abbreviations for the form of extrachromosomal rDNA molecules: M = monomers; PD = palindromic dimers; TR = tandem repeats.
[b] Data obtained by viscoelastometry.
* Mi is likely to be polyploid; thus, real genome size can be several times smaller.
Figures in square brackets are mean values.

The bulk of macronuclear DNA in *T. pyriformis* separates in sucrose gradients during the G_1 period as a fraction of large molecules with a sedimentation constant of about 90 S (MERKULOVA and BORCHSENIUS 1976, BORCHSENIUS *et al.* 1976). The value corresponds to a molecular size of about $400 \cdot 10^6$ daltons or 220 μm of the DNA double helix. However, the DNA molecules, particularly long ones, are likely to be sheared under these conditions. That explains why WILLIAMS and others (1978) found much larger DNA molecules ($20–30 \cdot 10^9$ daltons, i.e. 11–16 mm of the double helix) in the macronuclear lysates of *T. thermophila* and *T. pyriformis* when they used viscoelastometry. An interesting fact is that the DNA content of such a molecule approximates that of one micronuclear chromosome of *T. thermophila* ($24–27 \cdot 10^9$ daltons). Hence, at least some of the DNA molecules in the Ma of *Tetrahymena* are chromosome-sized, just as in the nuclei of other eukaryotes. Viscoelastic measurements show that the number of such molecules is 200–400 per Ma; this is consistent with the expected figure of about 225 chromosomes (five chromosomes of the micronuclear haploid set repeated 45 times in Ma; see Table 9). Apparently, either intact or slightly shortened (due to genetic diminution) micronuclear chromosomes are preserved in the Ma of *Tetrahymena*. The macronuclear chromosomes do not consist of smaller molecules connected by protein linkers since pronase causes no fragmentation of the chromosome-sized DNA molecules in Ma (WILLIAMS *et al.* 1978).

Just as in other eukaryotes, macronuclear DNA molecules in *Tetrahymena* replicate as separate sections called *replicons*. The recently synthesized single-strand DNA molecules, whose size is only $10–30 \cdot 10^6$ daltons (or 11–33 μm of a single nucleotide chain), have been detected by biochemical techniques in the Ma of *T. pyriformis* in the S period. They seem to be copies of single replicons. In about 30 minutes, they assemble in long molecules, while old DNA chains do not fragment in the course of replication (MERKULOVA and BORCHSENIUS 1976). JOHNSON and WESTERGAARD (1976) have reported that the size of replicons in the Ma of the same species varies from 17 to $35 \cdot 10^6$ daltons. According to the autoradiographic data of ANDREJEVA and others (1979), the average length of replicons is 20–30 μm.

Biochemical studies have revealed that the ribosomal gene (i.e. the nucleolar organizer gene) is amplified in the Ma of *Tetrahymena*. A Ma contains many free (extrachromosomal) copies of the ribosomal gene (rDNA). They number 170–200 per haploid macronuclear genome (ENGBERG and PEARLMAN 1972, WILLIE 1972a, YAO *et al.* 1974), while a micronucleus has only one rDNA copy per haploid genome, and it is integrated into a chromosome (YAO *et al.* 1974, YAO and GALL 1977). The number of copies of the ribosomal gene in a Ma depends on the culture conditions; their synthesis is largely independent of the replication of the bulk DNA and can be induced by the re-feeding of starved cells, heat shocks, and other factors (ENGBERG *et al.* 1972, WILLIE 1972b, ENGBERG, NILSSON *et al.* 1974, LEICK *et al.* 1976). In this case, rDNA molecules (as well as some other repetitive DNA sequences) replicate more than once per cell cycle (BORCHSENIUS *et al.* 1977). During a normal cell cycle, rDNA molecules replicate mainly at the beginning of the S period (ANDERSEN and ENGBERG 1975, ANDERSEN 1977); however, autoradiography has shown that their replication proceeds throughout the interphase, though at a lower rate (NILSSON and ZEUTHEN 1980).

GALL (1974) as well as ENGBERG, CHRISTIANSEN, and LEICK (1974) isolated the extrachromosomal rDNA molecules from the Ma of *T. pyriformis*. Their size was measured in electron micrographs (Table 9) and found to be $12.6 \cdot 10^6$ daltons or 6.7 μm (GALL 1974, KARRER and GALL 1976, ENGBERG *et al.* 1976). These molecules are somewhat larger in *T. thermophila:* about $13.5 \cdot 10^6$ daltons or 7.3 μm (KARRER and GALL 1976, GALL *et al.* 1977, 1979). The size of the rDNA molecules varies from $12 \cdot 10^6$ to $14 \cdot 10^6$ daltons in other closely related *Tetrahymena* species (DIN and ENGBERG 1979).

According to GALL (1974), most of the free rDNA molecules are linear, but some of them are rings or rings with "tails" (lariats). A thorough analysis has shown that the rDNA molecules of the Ma of *T. pyriformis* and *T. thermophila* are palindromes, i.e. they are inverted repeats that have the same nucleotide sequence when "read" from either end (KARRER and GALL 1976, ENGBERG *et al.* 1976). Such a molecule has two symmetrically arranged structural genes for 17 S rRNA (near its centre), two genes for 25 S rRNA (closer to the ends), and both central and distal regions of spacer DNA. The unique rDNA copy integrated in a micronuclear chromosome proved to be half as long as the free palindromic rDNA molecules in Ma; in other words, the latter are dimers[1] in which one of the unit rDNA molecules is inverted (YAO and GALL 1977). No ribosomal genes that are integrated into the bulk DNA exist in Ma. During macronuclear development, the rDNA copy is apparently excised from the chromosome and is duplicated, forming a dimer. The free dimer is then amplified. In Ma, the free rDNA molecules are located in the nucleolar chromatin (cf. p. 282). Using the spreading technique, their transcription has been visualized in form of two symmetrical arrays of nascent rRNA molecules on every palindromic rDNA molecule (BORKHARDT and NIELSEN 1981).

Palindromic rDNA molecules in the Ma of *Tetrahymena* can replicate. Electron microscopy has shown that this replication is initiated in the centre of the palindrome and propagates bidirectionally toward both its ends (TRUETT and GALL 1977). The rDNA molecules of *Tetrahymena* are now being thoroughly studied and their genic and spacer regions are being mapped (BLACKBURN and GALL 1978, WILD and GALL 1979, DIN and ENGBERG 1979, CECH and RIO 1979, CECH and KARRER 1980, ENGBERG *et al.* 1980, BORCHSENIUS *et al.* 1981a; for reviews, see GALL *et al.* 1977, 1979, YAO *et al.* 1979, GOROVSKY 1980).

It is still unknown when the rDNA copy is excised from the chromosome and primarily amplified during conjugation in *Tetrahymena*. Most probably, these processes occur in the anlage of the new Ma at the early stages of its development (YAO *et al.* 1979). PEARLMAN and others (1979), however, date them back to the meiotic micronucleus. At any rate, an intermediate (extrachromosomal and replicating, but still monomeric) form of rDNA has recently been discovered in the conjugating *T. thermophila* (PAN and BLACKBURN 1981). This is apparently a

[1] This is not, however, a universal phenomenon, even within the suborder Tetrahymenina: the free rDNA molecules in the Ma of the tetrahymenine ciliate *Glaucoma chattoni* proved to be nonpalindromic monomers $5.6 \cdot 10^6$ daltons or 3 μm in size (KATZEN *et al.* 1981a).

precursor of the palindromic rDNA dimer; it first appears when macronuclear anlagen start to develop.

It is remarkable that only the ribosomal gene is extrachromosomal and amplified in the Ma of *Tetrahymena*. Other repeated genes involved in protein synthesis (the gene for tRNA and the gene for 5 S rRNA) are present in the Ma in the same specific quantities as in the Mi: about 800 copies per haploid genome for the former and 300–350 copies for the latter. There are no grounds for believing that they are extrachromosomal in the Ma (KIMMEL and GOROVSKY 1976, 1978, TØNNESEN and ANDERSEN 1977, TØNNESEN 1978).

Data on the size of the macronuclear unit genome in the *Paramecium* species of the *aurelia* group are contradictory (Table 9). According to the DNA renaturation data presented by ALLEN and GIBSON (1972a), the unit genome of Ma in *Paramecium primaurelia* and *P. biaurelia* has a size of about $190 \cdot 10^9$ daltons. In other words, it is not reduced in comparison with the micronuclear genome whose size approximates $185 \cdot 10^9$ daltons according to photometric data (GIBSON and MARTIN 1971, ALLEN and GIBSON 1972a). Other authors, however, reported much lower values for the size of the macronuclear unit genome in the same species on the basis of reassociation tests: $30 \cdot 10^9$ daltons (CUMMINGS 1975) or $52 \cdot 10^9$ daltons (MCTAVISH and SOMMERVILLE 1980) for *P. primaurelia* and $38 \cdot 10^9$ daltons for *P. octoaurelia* (SOLDO and GODOY 1972). These authors have not assayed the DNA content in the Mi of their strains. Hence, it is still difficult to say whether the macronuclear genome is reduced in comparison with the micronuclear genome in the *Paramecium* species of the *aurelia* group. The size of the macronuclear unit genome of about $15–40 \cdot 10^9$ daltons was reported also for *Paramecium multimicronucleatum* and *P. caudatum* as well as for some Scuticociliatida, such as *Uronema*, *Parauronema*, and *Miamiensis* (SOLDO *et al.* 1981).

The unit genome of the Ma of *Paramecium bursaria* is rather large ($160 \cdot 10^9$ daltons) according to CULLIS (1972), but small ($11 \cdot 10^9$ daltons) according to STEINBRÜCK *et al.* (1981). Anyway, it size must be reduced in comparison with the micronuclear genome since a diminution of the chromosome parts was discovered during macronuclear development in this species (SCHWARTZ 1978). The macronuclear unit genome cannot be directly compared with the true haploid genome of the Mi, i.e. the degree of its diminution cannot be determined, since the Mi of *P. bursaria* is polyploid and photometric measurements of the size of the micronuclear genome would give overestimated values (CULLIS 1973; see Table 9).

Just as in *Tetrahymena*, the bulk DNA of the Ma of *P. primaurelia* seems to consist of very long molecules. Cummings (1975) has shown by electron microscopy that his preparations of macronuclear DNA contained molecules at least 105 μm long, i.e. their molecular weight was at least $200 \cdot 10^6$ daltons (Table 9). Many shorter DNA pieces were also observed, but CUMMINGS regards them as artifacts, i.e. fragments of longer molecules. DNA molecules 0.2–5 (up to 23) μm long predominated in the preparations of macronuclear DNA of the same species made by MCTAVISH and SOMMERVILLE (1980). These authors also believe that "nonspecific endogenous nuclease activity is responsible for at least some DNA scission during isolation". But preliminary viscoelastometric data suggest that the intact macronuclear DNA molecules from *P. primaurelia* might be at least chromosome-sized since they reach the size of $8 \cdot 10^9$ daltons (about 4 mm of the

DNA double helix), whereas the calculated average DNA content per micronuclear chromosome is only about $4 \cdot 10^9$ daltons (CUMMINGS 1975).

SOLDO and GODOY (1972) have biochemically isolated a fraction of rapidly reassociating DNA molecules from the Ma of *P. octoaurelia;* their size was $14.5 \cdot 10^6$ daltons (about 8 μm) and their content in Ma reached about 100 copies per haploid genome. This DNA fraction was found also in *P. primaurelia* by CUMMINGS (1975); such molecules hybridize selectively with rRNA. Therefore, just as in *Tetrahymena*, the Ma of *Paramecium* contains free copies of amplified rDNA (Table 9) that seem to be located in the nucleolar chromatin. Extrachromosomal rDNA molecules from the Ma of *P. tetraurelia* were characterized by the techniques of molecular biology (FINDLY and GALL 1978, 1980). They vary in size from $5.5 \cdot 10^6$ to $77 \cdot 10^6$ daltons and may be either linear or circular. The molecules are tandem repeats of a basic unit whose size is $5.5 \cdot 10^6$ daltons (ca. 3 μm). Unlike in *Tetrahymena*, no inverted repeats (palindromes) were found in the rDNA of *Paramecium*. The rDNA monomer is repeated most frequently 2, 4, or 6 times and at the most 13 times in the longer molecules (FINDLY and GALL 1978). Each monomer has one gene for 25 S rRNA, one gene for 17 S rRNA, and one gene for 5.8 S rRNA; it also contains some spacers (FINDLY and GALL 1980).

Hypotheses of the Macronuclear Structure. The problem of the genetic organization of Ma is debated mainly with respect to *Tetrahymena*. Besides the data on macronuclear development and macronuclear DNA, account is being taken of the data on the relative DNA content of Ma (Table 7), on the chromatin aggregates that appear in the dividing Ma, and on the phenotypic assortment in clones according to the characters controlled by macronuclear genes.

Many concepts of macronuclear organization in *Tetrahymena* stem from the genetic data on the heterogeneity of the Ma (see the reviews by NANNEY 1964, ALLEN and GIBSON 1972b, McCOY 1978, NANNEY and PREPARATA 1979). Macronuclear heterogeneity is expressed in the phenotypic instability of some exconjugant clones with respect to the mating type (in both homo- and heterozygous clones) as well as the serotype, the esterase isozyme array, and some other characters (only in heterozygotes). A constant fraction of stable cells is assorted in the phenotypically unstable clones, beginning with a certain generation after conjugation. All the progeny of the stable cells is irreversibly differentiated according to the mating type; in heterozygotes, it phenotypically corresponds to only one of the two alleles that control a particular character (for instance, an isozyme array). Stable cell lines behave as if the other allele was either physically lost or irreversibly repressed.

Differentiation according to various genes occurs after different numbers of generations following conjugation. This number is specific for each gene. However, the rate at which stable cells are produced was found to be the same for all the characters investigated and is 0.0113 cell per generation (NANNEY 1964, ORIAS and FLACKS 1975). These data indicate that the Ma of *Tetrahymena* contains a certain quantity of nondifferentiated genetic subunits after conjugation; some subunits then differentiate in one direction, while the others differentiate in the opposite direction. The differentiated subunits later randomly segregate during macronuclear divisions; consequently, stable Ma, in which all subunits are of the same type, appear at a constant rate. The same assortment rate is also typical of the somatic (macronuclear) cycloheximide resistance mutations, which originate and are expressed regardless of conjugation (ORIAS and NEWBY 1975). SCHENSTED (1958) has demonstrated that the assortment rate (0.0113) depends only on the number of subunits, corresponding to 45 subunits in a recently divided Ma or 90 subunits in a Ma that is about to divide.

There are now many hypotheses of the macronuclear organization of *Tetrahymena*. They may be subdivided into three groups.

1. *Subnuclear Hypotheses*. The first subnuclear hypothesis was formulated by NANNEY (1964), who assumed that the genetic subunits in the Ma of *Tetrahymena* are *diploid subnuclei*, each of which equals a Mi by its DNA content and genic complement. According to the hypothesis, phenotypic assortment in clones is due to "allelic repression", i.e. the stable epigenetic inactivation of one allele of a given gene in a part of the subnuclei and of the other allele of the same gene in the remaining subnuclei of a Ma. The differentiated subnuclei are then randomly assorted in a series of divisions, replicating normally in each cell cycle. Since the subnuclei are always diploid, allelic repression may occur at different times for different genes.

If the genetic subunits are diploid subnuclei, however, the latter should number about 45 in the G_1 period, i.e. the Ma of *T. thermophila* should be about 90-ploid. At the same time, photometric data (Table 7) show that the Ma of normal *T. thermophila* (which have diploid Mi) contains enough DNA for only 21–25 diploid subnuclei, i.e. a Ma in G_1 can be merely 45-ploid. This is the main shortcoming of the hypothesis of diploid subnuclei.

Some authors tried to overcome that shortcoming while remaining within the framework of the given hypothesis. NANNEY (quoted by ALLEN and GIBSON 1972a) believed that roughly one-half of the micronuclear genome disintegrates when macronuclear anlagen develop; in other words, although macronuclear subnuclei are diploid, their genome is half as large as that of a Mi. This interpretation is, however, inconsistent with the above DNA renaturation data, which indicate that the macronuclear genome of *T. thermophila* is not reduced even nearly so much. DOERDER and DE BAULT (1975) supposed that the Ma in G_1 contains only 22–23 diploid subnuclei, but they do not randomly segregate during division (two sister subnuclei go into different Ma more often than they remain in the same Ma). Then stable cells would be assorted at a lower rate, simulating a larger number of subnuclei. However, ORIAS and FLACKS (1975) have confirmed that the macronuclear genetic subunits are randomly segregated in *Tetrahymena*, i.e. their number in G_1 cannot be less than 45. Thus, the hypothesis of the diploid subnuclei seems to be in an impasse.

Another subnuclear hypothesis suggests that the Ma of *Tetrahymena* contains not diploid but *haploid subnuclei* (i.e. spatially separated genomes), which are the macronuclear genetic subunits (NILSSON 1970, WOODARD *et al.* 1972, ORIAS and FLACKS 1975, DOERDER *et al.* 1977). The hypothesis is based on the fact that the amount of DNA in a Ma is enough for exactly 45 haploid genomes, in accord with the estimated number of genetic subunits in G_1 (WOODARD *et al.* 1972) and with the number of chromatin aggregates received by each daughter Ma during division (NILSSON 1970). According to the hypothesis, each haploid subnucleus contains only one allele from every pair of genes. The subnuclei randomly segregate in macronuclear division; consequently, a part of the progeny may possess Ma, all the subnuclei of which have the same allele. These cells are stable ones. Thus, according to the hypothesis, phenotypic assortment in clones is due to the physical loss of alleles by some of the daughter Ma (*allelic assortment*) rather than to the mutual repression of the alleles (ORIAS and FLACKS 1975).

For quite some time it was believed that the hypothesis of haploid subnuclei contradicted some direct genetic evidence. However, new research data show that this is not so (DOERDER *et al.* 1977).

This does not mean, however, that the haploid subnuclei in the Ma of *Tetrahymena* are merely haploid sets of five invariable micronuclear chromosomes interconnected end-to-end or in some other manner. As we have seen, the rDNA copy is excised and several other DNA sequences are eliminated from the chromosomes in the course of macronuclear development. Apparently, as the Ma develops, the chromosomes first fragment at specific breakage sites in the micronuclear DNA, which seem to be tandem repeats of a simple hexanucleotide (YAO and GALL 1979, YAO *et al.* 1981). Then, some fragments are eliminated, rDNA is amplified as free molecules, and the remaining fragments reassemble, producing modified chromosomes. The latter correspond to the chromosome-sized DNA molecules detected in Ma (WILLIAMS *et al.* 1978). The modified chromosomes combine with one another, producing the haploid subnuclei (DOERDER *et al.* 1977). It is also possible that the haploid subnuclei in the adult Ma exchange chromosomes or their parts, undergoing somatic recombination, which is likely, as is evident from the genetic data (MCCOY 1978, 1979, NANNEY and PREPARATA 1979).

It is still unknown how macronuclear chromosomes may combine so as to form the haploid (or diploid) subnuclei. Most probably, they assemble end-to-end in a chain that is a composite chromosome (GRELL 1953b, RAIKOV 1969). This is in accord with the fact that the Ma of *Tetrahymena* contains long chains of chromatin "small bodies" (WOLFE 1967) that contract into compact aggregates prior to division (NILSSON 1970). WILLIAMS and others (1978), however, have shown that chromosomes cannot combine with one another either through the DNA itself or through protein linkers since viscoelastometry has not revealed any DNA molecules of superchromosomal length in the Ma of *Tetrahymena*, even when pronase treatment was omitted. Only such linkers as membrane bonds, which are disrupted by detergents alone, may be present.

2. *Chromosomal Hypothesis*. The idea that the Ma of ciliates is highly polyploid, i.e. that it contains numerous free copies of chromosomes, dates back to PIEKARSKI (1941) and GEITLER (1941). However, these authors as well as GRELL (1950b, 1953b) saw the main difficulty with this hypothesis, namely, that random segregation of a mass of separate free chromosomes during macronuclear divisions would quickly result in aneuploidy and the death of the clone, at least when there are no regulation phenomena. Hence, the hypothesis envisaging many free chromosomes in a Ma was soon replaced by different varieties of the subnuclear hypothesis (SONNEBORN 1947, GRELL 1953b) and was not seriously considered for a long time.

However, the chromosomal hypothesis was recently revived and updated by SEYFERT (1979) with respect to *Tetrahymena*. He detected rod-shaped, chromosome-like elements in the dividing Ma and cytophotometrically showed that, in their DNA content, they fall into five basic classes and a number of classes that are multiples of the basic ones. SEYFERT interpreted the five basic classes of elements as macronuclear chromosomes and interpreted the higher classes as the polytenic variants of the basic classes. The chromosome-like elements aggregate into chromatin clumps resembling those described by NILSSON (1970). SEYFERT, however, regards these units as groups of homologous chromosomes and not as genomes. In the interphasic Ma, the chromosomes are highly decondensed and, therefore, invisible.

SEYFERT's findings are consistent with the existence of chromosome-sized DNA molecules in the Ma of *Tetrahymena*. In this case, separate chromosomes should be regarded as the genetic subunits sorted out during macronuclear division; each chromosome is repeated in a Ma either in 45 copies (in G_1) or in 90 copies (in G_2). Some of these copies contain only one allele of each particular gene, while the others have only the other allele of the same gene. As did the haploid subnuclear hypothesis, the chromosomal hypothesis explains the phenotypic assortment in clones by a loss, from some Ma, of all chromosomes that carried an allele and the replacement of that loss with an equivalent number of homologous chromosomes that carry the other allele (i.e. by allelic assortment).

Many daughter Ma, however, would quickly lose some types of chromosomes when free chromosomes are randomly segregated; in other words, they would lose *both* alleles of many genes. This would lead to the early and high mortality of the subclones which, however, has not been observed. Hence, SEYFERT believes that the number of each of the five chromosome types is under the control of a separate regulatory mechanism. For instance, it could be that when the content of any of the five chromosome elements in a Ma falls below a certain critical level, the mechanism of additional replication of all macronuclear DNA (an extra S phase), which was described in *Tetrahymena* (cf. p. 305), is switched on. After DNA quadruples, all the chromosomes now in excess are eliminated from the Ma via chromatin extrusion bodies (SEYFERT 1979).

3. *Chromosome Fragmentation Hypotheses*. According to all these hypotheses, the chromosomes in Ma fragment into many free DNA molecules, each of which contains one or several genes. The molecules are genetic subunits that segregate in macronuclear divisions. These hypotheses easily explain such phenomena as "linkage disruption", i.e. why genes located in one micronuclear chromosome behave in Ma independently of one another (NANNEY and PREPARATA 1979). To interpret this phenomenon in terms of subnuclear or chromosomal hypotheses, one has to admit that homologous chromosomes or genomes are intensively recombined in Ma (MCCOY 1978).

The first fragmentation hypothesis for the Ma of *Tetrahymena*, the *"master-slave" hypothesis*, was formulated by ALLEN and GIBSON (1972a) to offset the diploid subnuclear hypothesis. According to them, only two fragments, which originate from two homologous chromosomes and can replicate (they are known as "masters"), are available for each macronuclear locus in G_1. But in the course of macronuclear development, the "masters" produce many copies that cannot replicate (they are known as "slaves"). In the S period, the "masters" replicate once and also produce a number of additional "slaves". When Ma divides, the two "master" copies regularly separate into two nuclei, while the "slaves" randomly segregate between them.

According to this hypothesis, genetic determination occurs when one "master" of a pair is lost from certain Ma because it failed to replicate. The "slaves" of the lost "master" will gradually disappear from the progeny of such cells and will be replaced by the "slaves" of its homologue, leading to phenotypic assortment in the respective character.

However, the "master-slave" hypothesis was checked and refuted by direct biochemical evidence (VOROB'EV *et al.* 1975, BORCHSENIUS *et al.* 1976). Using the double labelling of macronuclear DNA in *Tetrahymena*, the authors concerned have demonstrated that all DNA molecules replicate once and semiconservatively in the Ma. If the "master-slave" hypothesis were true, many DNA molecules would be conserved without replication (old "slaves") and many molecules would be synthesized as a result of the repeated replication of the "masters" (new "slaves").

Another fragmentation hypothesis was proposed by VOROB'EV and others (1975). It was called the "nucleosomal" hypothesis. However, this name should be changed since the term "nucleosome" now has another meaning (it is used to designate the chromatin subunits). According to the hypothesis, all the DNA fragments may replicate in Ma. When the primary chromosomal fragments replicate in the developing Ma, their two homologues remain somatically paired ("diploid fragments") so that each fragment contains both alleles of a given gene. A young Ma has 22 23 diploid fragments of each type in the G_1 period. At an instant that is specific for every gene, each fragment is split into two haploid chromosome fragments, each containing only one allele (for this reason, the hypothesis may be renamed the *allelic splitting hypothesis*). As a result, the genetic subunits that are open to assortment reach the predicted number (45 in G_1), although the DNA content and the genic complement of Ma remain the same. Then the haploid fragments replicate semiconservatively in each cell cycle and randomly segregate in each macronuclear division, producing a certain fraction of "pure" Ma which lost all fragments that carry one of the two alleles.

There are other varieties of the fragmentation hypotheses. According to some of them, the macronuclear chromosomes of *Tetrahymena* progressively fragment as the number of generations grows after conjugation. According to others, chromosome fragmentation occurs in each interphase and the fragments reassemble into transient haploid subnuclei prior to each macronuclear division (McCOY 1978, 1979). These hypotheses, however, are not enough elaborated in the details.

The stumbling block of any fragmentation hypothesis is the discovery that macronuclear DNA in *Tetrahymena* is chromosome-sized and therefore chromosomes cannot substantially fragment in Ma (WILLIAMS *et al.* 1978). Moreover, any concept other than the subnuclear one (including the chromosome fragmentation hypotheses) should explain how both alleles of some macronuclear genes are not lost in a long series of generations; otherwise, many genetically defective lines would appear and the clone would die. Nevertheless, *Tetrahymena* clones can live for decades without their Ma being replaced.

Just as in the case of the chromosomal hypothesis, an explanation can be found if it is assumed that there are control systems that prevent the content of a genetic element in the Ma from falling below a critical level. For instance, such systems may stimulate either the selective overreplication of the rare elements or the elimination of the excessive ones. The computer simulation of such a system provides satisfactory results for both whole chromosomes and their fragments (PREER and PREER 1979). But while the chromosomal hypothesis requires the existence of only five control systems (one per each type of chromosomes), the fragmentation hypotheses require the existence of hundreds of such systems.

Hence, the haploid subnuclear hypothesis and the chromosomal hypothesis with replication control seem to offer the most plausible explanations for the macronuclear genetic organization in *Tetrahymena*. But neither of them has been proved yet.

The macronuclear structure of *Paramecium* is less controversial. In a series of classic works, SONNEBORN (1947) proposed that a paramecian Ma consists of diploid, relatively autonomous subnuclei. This supposition is based mainly on the ability of the macronuclear fragments to regenerate into normal Ma during conjugation. Mathematical models (PREER 1976) and experiments (NYBERG 1976) involving *P. tetraurelia* have shown that the inheritance of the resistance to copper salts, controlled by a pair of alleles, may be interpreted satisfactorily only in terms of the subnuclear hypothesis (the number of subnuclei in the Ma of *P. tetraurelia* must reach hundreds; see Table 7). The hypotheses that admit genome fragmentation are compatible with the facts only if the genetic subunits are repeated in a Ma 2000 times and even more. In turn, such a high degree of repetition requires a reduction of the unit genome size during macronuclear development, i.e. genetic diminution, which has not been proved so far in paramecia belonging to the *aurelia* group. The presence of superchromosome-sized DNA molecules in the Ma of *P. primaurelia* (CUMMINGS 1975; Table 9) also shows that it may contain subnuclei or composite chromosomes.

Different data have been obtained for *P. bursaria*, whose Ma seems to contain free chromosomes, although they are shortened as a result of genetic diminution (SCHWARTZ 1978). It is unknown whether they can assemble into subnuclei.

Therefore, the genetic organization of the Ma is still an open question in both *Tetrahymena* and *Paramecium*.

Macronuclei of Hypotrichida

The development and genetic organization of Ma in Hypotrichida have been thoroughly studied during the past decade (see the reviews by AMMERMANN 1973, PRESCOTT *et al.* 1973, PRESCOTT and MURTI 1974, RAIKOV and AMMERMANN 1974, RAIKOV 1976b, MURTI 1976b, LAWN *et al.* 1978, PRESCOTT *et al.* 1979). The hypotrichid Ma is similar both morphologically and ultrastructurally to the Ma of other ciliates. Its distinguishing feature, however, is DNA synthesis that involves replication bands (Fig. 99).

Macronuclear Development. Unlike in the Ma of Hymenostomatida and many other ciliates, distinct polytene chromosomes appear in the developing macronuclear anlagen of Hypotrichida. They were first discovered in *Stylonychia mytilus* by AMMERMANN (1965) and simultaneously, in other species as well (ALONSO and PÉREZ-SILVA 1965b, 1966, PÉREZ-SILVA and ALONSO 1966, RAO 1966, SAPRA and

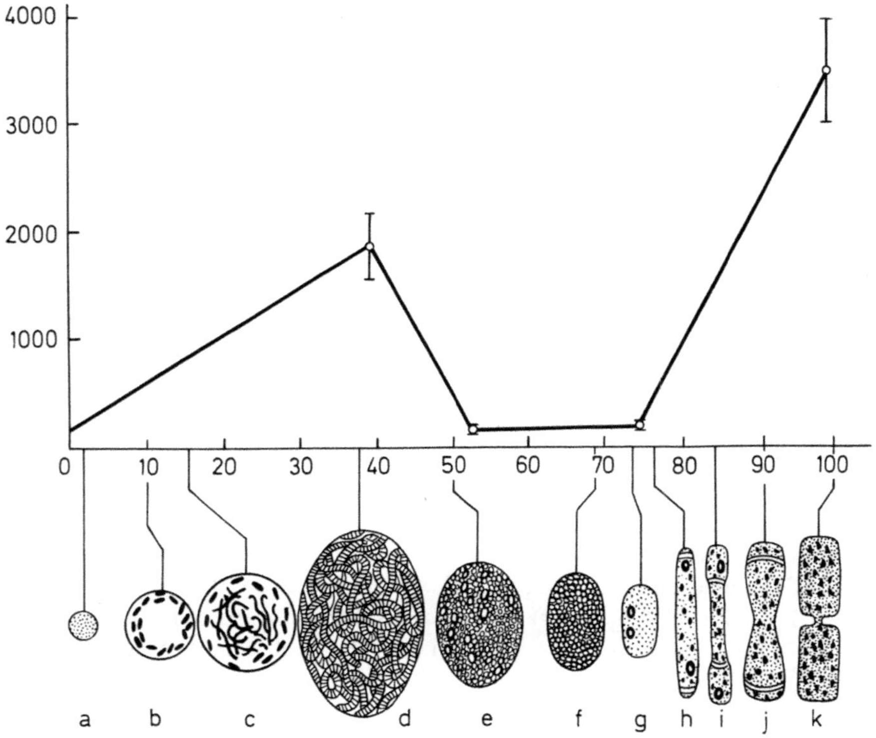

Fig. 108. Changes in the DNA content during the development of the macronuclear anlage in *Stylonychia mytilus* (after AMMERMANN *et al.* 1974). The abscissa represents the age of the anlage (hours), while the DNA content (arbitrary units) is shown on the ordinate. Stages in the development of the anlage are at the bottom: *a* synkaryon derivative; *b* anlage with condensed chromosomes; *c* decondensation of some chromosomes and their polytenization; *d* maximal development of polytene chromosomes; *e* and *f* vesicle stage after the fragmentation of polytene chromosomes; *g* "achromatic" stage, appearance of the first nucleoli; *h*–*j* second phase of DNA synthesis (with replication bands); *k* division of the anlage into two definitive Ma.

DASS 1970, ALONSO 1978). Their development was thoroughly studied by light microscopy in *S. mytilus* (AMMERMANN 1965, 1971a, JAREÑO *et al.* 1972, JAREÑO 1974, AMMERMANN *et al.* 1974), *Euplotes aediculatus* (RAO and AMMERMANN 1970, AMMERMANN 1971a), and *Oxytricha* sp. (SPEAR and LAUTH 1976).

A macronuclear anlage begins to develop in *S. mytilus* (Fig. 108) with the condensation of all the 250–300 chromosomes in the diploid synkaryon derivative (the number of chromosomes is half as much in the anlagen that develop from the haploid hemikaryon). The condensed chromosomes are arranged at the periphery of the anlage. Despite the constant chromosome number, the DNA content is four times higher in the anlagen at this stage in comparison with the initial level, indicating that chromosomal polytenization has started (Fig. 108*b*). The chromosomes then differentiate: less than a third of them decondense again into threads, while the others remain compact and later degenerate (Figs. 108*c*, 109*a*). Thus, a considerable part of the micronuclear genome is eliminated as a result of *chromosome diminution* at the very beginning of the macronuclear development in *S. mytilus*[2]. Such a phenomenon has not been found in *Euplotes* and *Oxytricha*.

The polytenizing chromosomes then become ribbon-shaped and form distinct transverse bands (Fig. 109*b, h*). The number of polytene chromosomes in the macronuclear anlage of *Stylonychia mytilus* that develops from the haploid hemikaryon is 25–50[3], whereas the original haploid chromosome number is about 125 (AMMERMANN 1971a). The number of polytene chromosomes in the macronuclear anlage that develops from a diploid nucleus is 50–70 in *S. muscorum* (ALONSO and PÉREZ-SILVA 1965b, PÉREZ-SILVA and ALONSO 1966), about 120 in *Oxytricha* sp. (SPEAR and LAUTH 1976), and 130–155 in *Euplotes aediculatus* (AMMERMANN 1971a). It is very likely that the homologous polytene chromosomes are paired band-to-band, i.e. they undergo *somatic conjugation*, since asynaptic regions can be discerned on the chromosomes of *S. mytilus* (JAREÑO 1974: Fig. 109*h*) and *Oxytricha* sp. (SPEAR and LAUTH 1976). Also, the number of polytene chromosomes in *Oxytricha* (in which no chromosomes degenerate in the early anlage) is haploid, not diploid.

When the polytene chromosomes reach maximum development, the DNA content of the macronuclear anlage of *S. mytilus* increases 16-fold in comparison with the diploid level (Fig. 108*d*). But since less than a third of the chromosomes are involved in macronuclear development, the real extent of chromosomal polyteny must be much higher: it approaches 64-fold on the average, corresponding to six duplications of the chromonema (AMMERMANN *et al.* 1974). Moreover, different chromosomal loci do not replicate to the same extent: most bands of the polytene chromosomes remain thin, but some of them grow into large heterochromatin blocks (Figs. 109*c*, 110*a*). Judging from ^{3}H-thymidine incorporation, DNA synthesis in these blocks lasts much longer than in the ordinary bands (Fig. 109*d, e*). The DNA content of the macronuclear anlage of *Euplotes aedicula-*

[2] In the ultrastructural study carried out by GIL and others (1972), the degenerating compact chromosomes were apparently taken for nucleoli (ALONSO 1978).

[3] AMMERMANN (1965) at first believed, in this case, that all the polytene chromosomes assembled end-to-end into one or two giant composite chromosomes. But this has not been confirmed.

Fig. 109. Polytene chromosomes in the macronuclear anlage of *Stylonychia mytilus* (photomicrographs). *a* decondensation and the beginning of polytenization of only some chromosomes in the early anlage (whole mount, Feulgen reaction, 840 ×); *b* fully developed polytene chromosomes in the anlage (phase contrast, 1660 ×); *c* isolated polytene chromosome with large heterochromatin blocks (acetoorcein, 1340 ×); *d* and *e* incorporation of ^{3}H-thymidine in isolated polytene chromosomes: uniform during the middle stage (*d*) and selective (in the heterochromatin blocks) at the late stage of polytenization (*e*) (radioautographs, 1340 ×); *f* and *g* molecular hybridization *in situ* of the polytene chromosome DNA with macronuclear labelled RNA (radioautographs, 1440 ×); *h* isolated polytene chromosomes showing somatic conjugation (the zones of asynapsis are indicated by arrows; phase contrast, 820 ×). *a* and *c*–*g* after AMMERMANN and others (1974); *b* after AMMERMANN (1965); *h* after JAREÑO (1974).

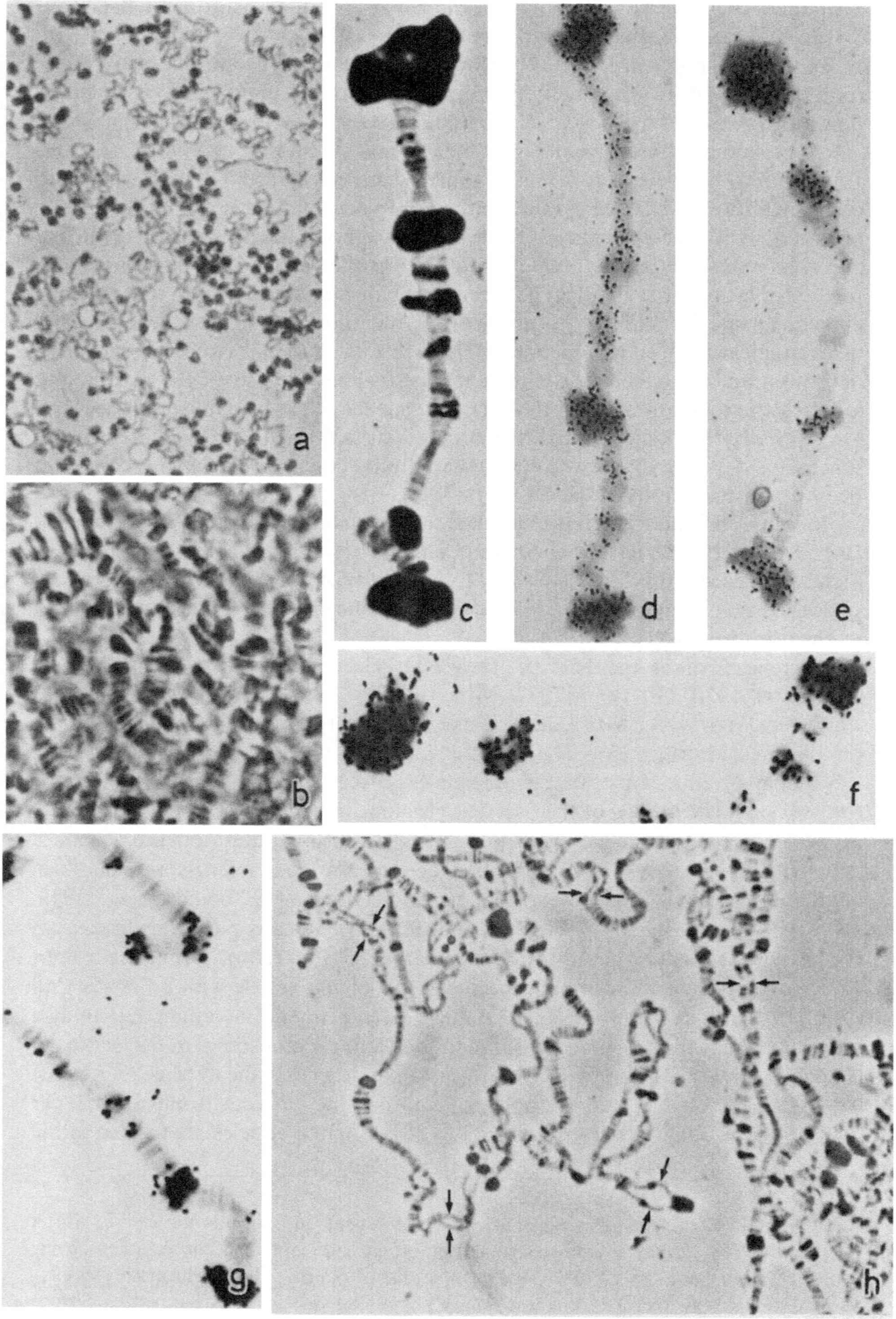

Fig. 109

tus increases 50 times during polytenization, and some heterochromatin bands are "overreplicated" in this case, too (AMMERMANN 1971a).

As for *transcription*, it is rather slow in the polytene chromosomes of Hypotrichida (unlike in those of the dipteran salivary glands). The polytene chromosomes of *Stylonychia*, *Oxytricha*, and *Euplotes* incorporate almost no RNA precursors (AMMERMANN 1968b, RAO and AMMERMANN 1970, SPEAR and LAUTH 1976). However, weak RNA synthesis has been detected in them when high concentrations of labelled uridine were used (ALONSO and JAREÑO 1974). This is consistent with the presence of RNP particles (about 10 nm in diameter) in the chromosomes and the karyolymph; it seems that the particles can be released into the cytoplasm (GIL 1976). The occurrence of transcription in the polytene chromosomes is indirectly supported by the morphogenetic activity of the macronuclear anlagen in both *Stylonychia* and *Euplotes*: they issue information at definite moments of their development, first for the pycnosis of the old Ma and later for the reorganization of the ciliary organelles of the exconjugant (SAPRA and DASS 1978, KLOETZEL 1981a). Moreover, squashed polytene chromosomes can synthesize RNA in vitro, i.e. they have RNA polymerase, which belongs to the α-amanitin-sensitive type (GAUDE 1981).

Polytene chromosomes have loosened regions that resemble puffs (JAREÑO *et al.* 1972, MURTI 1976b). However, it is still unknown whether RNA is synthesized in them. They are more likely sites of DNA replication than sites of transcription, since the puffs seem to be the precursors of the condensed heterochromatin blocks in chromosomes (ALONSO 1978).

Polytene chromosomes have been studied by electron microscopy in *Stylonychia* (KLOETZEL 1970, GIL *et al.* 1972, MURTI 1973), *Oxytricha* (MURTI 1976b) and *Euplotes* (KLOETZEL 1970). Both ordinary and heterochromatin bands can easily be discerned on them (Fig. 110*a*).

When polytenization of the chromosomes is over, a striking phenomenon of the *transverse fragmentation* of the polytene chromosomes into separate bands occurs in the macronuclear anlage of Hypotrichida. The disintegration of chromosomes, which AMMERMANN (1965, 1971a) first detected by light microscopy, has been confirmed in electron microscopic studies by KLOETZEL (1970) and MURTI (1973, 1976b). The process starts with the appearance of fibrous partitions that cross the polytene chromosomes through the interbands (Fig. 110*a*). Then the chromosomes disappear, and each band becomes an isolated vesicle with a fibrous wall (Fig. 110*b*). There are thousands of such vesicles in the macronuclear anlage. However, the transverse fragmentation of polytene chromosomes in *Oxytricha* sp. is preceded by the fusion of several adjacent bands so that the average number of bands drops from 80 to 20 per polytene chromosome, or from 10000 to 2400 per macronuclear anlage. Accordingly, about 2500 isolated vesicles are formed in the

---

Fig. 110. Ultrastructure of macronuclear development in *Stylonychia mytilus* (after KLOETZEL 1970). *a* part of the macronuclear anlage with polytene chromosomes being transected into bands by fibrous lamellae (*1* ordinary bands, *2* heterochromatin blocks), 10000 × ; *b* part of the macronuclear anlage containing vesicles (*3*) separated by fibrous material (*4*), 10000 × ; *5* envelope of the anlage.

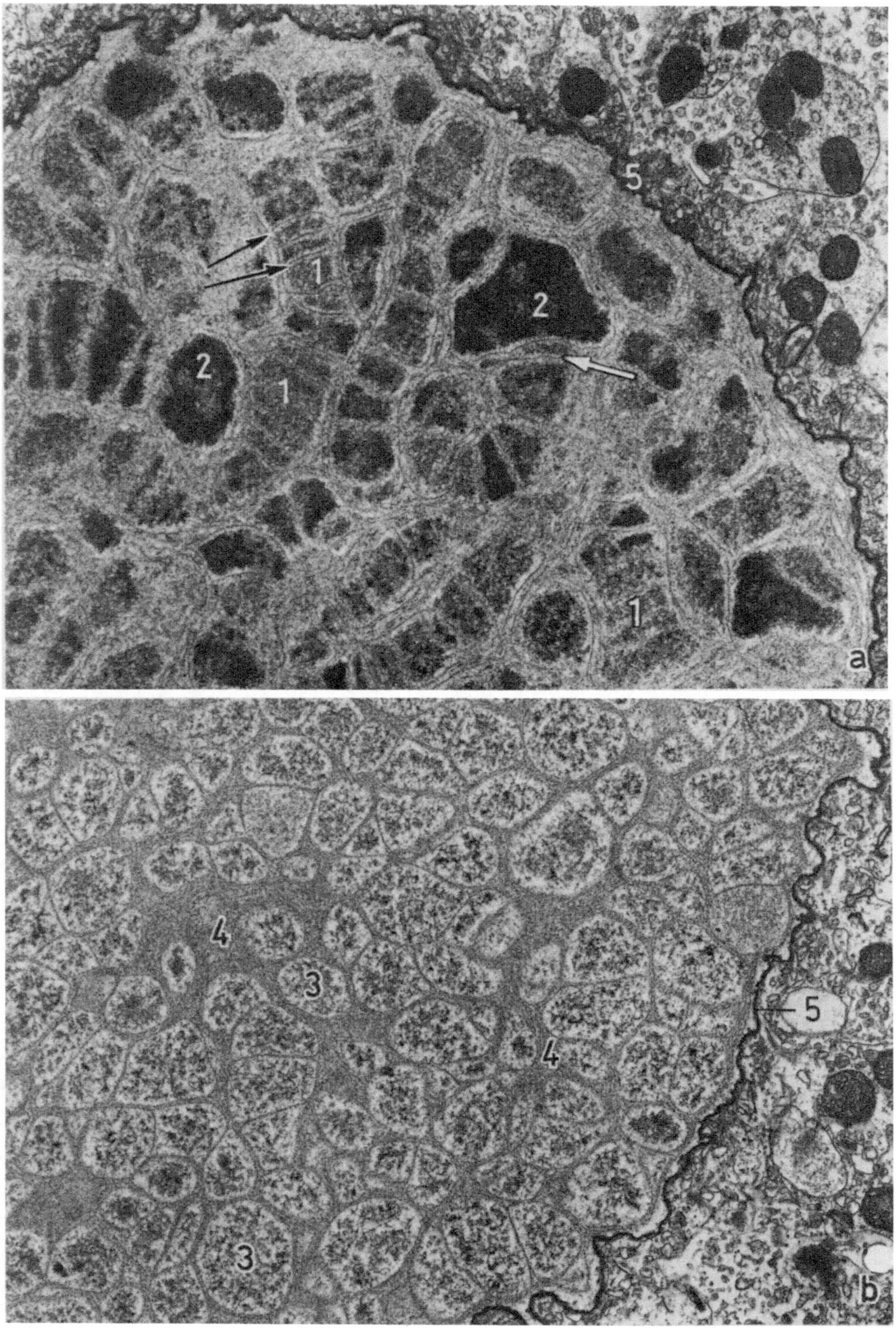

Fig. 110

anlage of *Oxytricha* (SPEAR and LAUTH 1976). Thus, the macronuclear anlage contains, at the vesicle stage, only isolated groups of homologous chromomeres that earlier constituted either one band or several adjacent bands.

The DNA content of the macronuclear anlage diminishes sharply at this stage (Fig. 108*e*). It almost returns to the diploid level in *Stylonychia* in which about 93% of the DNA of the polytene chromosomes is eliminated (AMMERMANN 1965, 1971a). About 80% of DNA disappears from the macronuclear anlage of *Euplotes aediculatus* (RAO and AMMERMANN 1970, AMMERMANN 1971a). The eliminated DNA is degraded to low molecular weight products, which are released from the exconjugant and are no longer reutilized (AMMERMANN 1968b, 1969).

The elimination of DNA from the polytene chromosomes of *Stylonychia* has recently been studied by the technique of chromatin microspreading combined with electron microscopy (MEYER and LIPPS 1980, 1981). Two types of DNP fibrils were discovered in these chromosomes: those 30 nm thick (in bands and heterochromatin blocks) and those 12 nm thick ("axial" fibrils). During chromatin diminution, only fibrils of the first type disintegrate: at first, they assume a circular shape, then they fragment into supranucleosomal elements ("super-beads").

The amount of electron-dense material in the vesicles of the macronuclear anlage also decreases (KLOETZEL 1970, MURTI 1973, 1976b), and the whole anlage visibly shrinks (Fig. 108*f*). PRESCOTT and MURTI (1974) have established autoradiographically that DNA is lost more or less proportionally from all the vesicles of the anlage. AMMERMANN and others (1974), however, hybridized *in situ* the DNA of isolated polytene chromosomes with labelled RNA that was complementary to the DNA of the adult Ma of *Stylonychia;* they showed that the DNA loss was selective and varied for different chromosome bands (and, consequently, vesicles). Only about 10% of the bands of the polytene chromosomes (including all the heterochromatic ones) contain DNA sequences preserved in the adult Ma; these bands bind complementary labelled RNA (Fig. 109*f*, *g*). The remaining 90% of the bands do not bind labelled RNA; therefore, either their DNA is entirely degraded at the vesicle stage or its content in the adult Ma is very low. Apparently, the gene complement of the macronuclear anlage is reduced as a result of the selective disintegration of some genes, i.e. *gene diminution* occurs. This is the second diminution in *Stylonychia* (the first one involved the degeneration of many whole chromosomes).

The walls of the vesicles disappear in the macronuclear anlage at the end of the DNA destruction phase; this is the beginning of the "achromatic" phase (Fig. 108*g*). Only finely dispersed bodies of electron-dense chromatin are now seen in the anlage (Fig. 111*a*). But RNA synthesis soon starts in it, and the first nucleoli

---

Fig. 111. Ultrastructure of macronuclear anlagen and DNA molecules in the adult macronuclei of the chromomeric type. *a* macronuclear anlage of *Stylonychia mytilus* at the "achromatic" stage, 5700 ×; *b* the same during the second phase of DNA synthesis, showing a replication band (*1*), nucleoli (*2*), and chromatin bodies (*3*), 7200 ×; *c* electron micrograph of short DNA molecules isolated from the Ma of *Oxytricha* sp., 41 500 ×; *d* part of the macronuclear anlage in *Chilodonella uncinata*, showing oligotene chromosomes, 54 000 ×. *a* and *b* after MURTI (1973); *c* after PRESCOTT and others (1971); *d* after PYNE (1978).

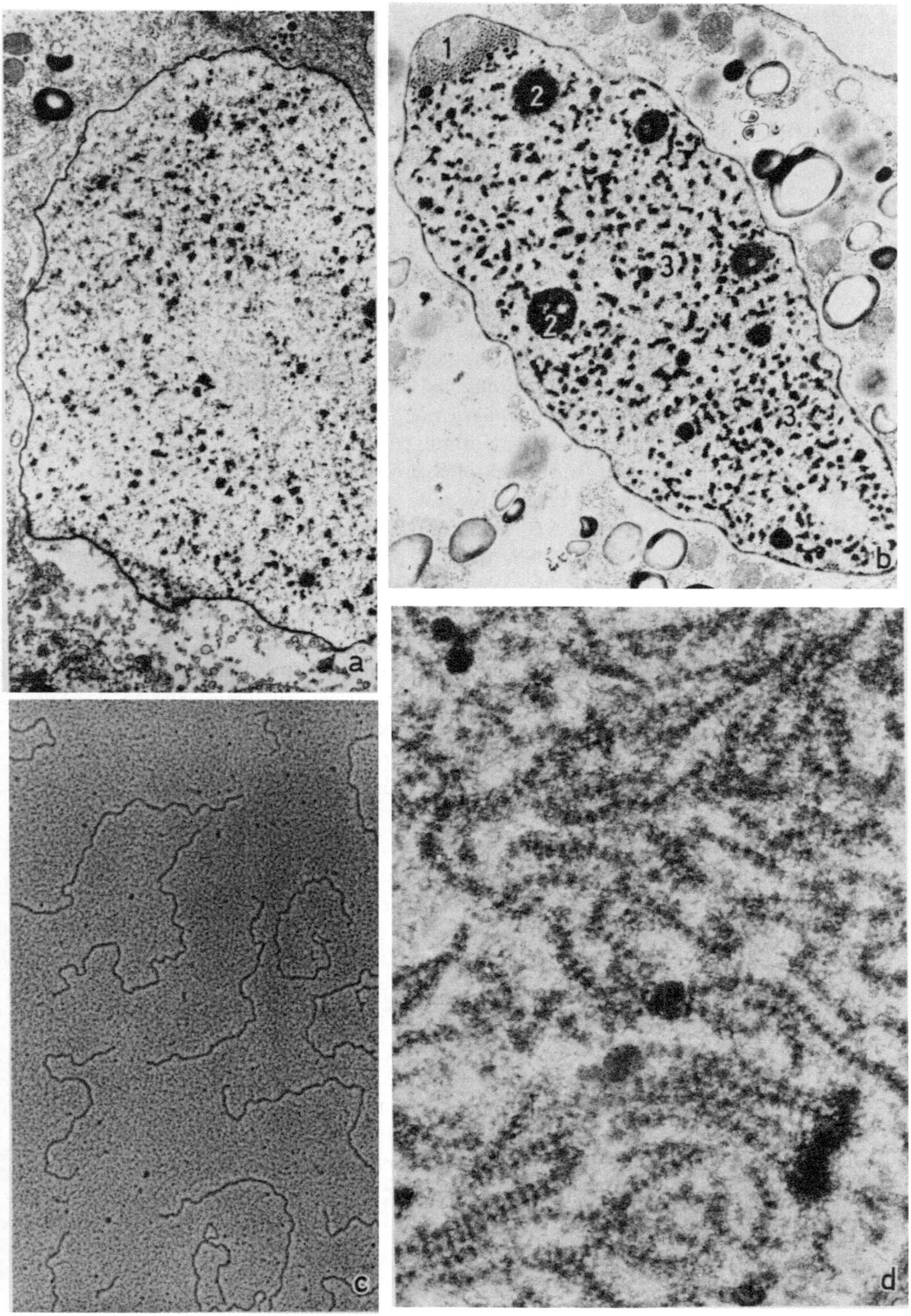

Fig. 111

appear (Fig. 108*g*). At the same time, acidic proteins accumulate in the anlage of *Euplotes* (ALONSO 1974). DNA synthesis is soon resumed. The DNA replicates in several discrete cycles, and each cycle involves a pair of replication bands that pass through the macronuclear anlage (Fig. 111*b*). Five pairs of replication bands pass through the anlage of *S. mytilus*. Consequently, its DNA content increases 32 times in comparison with the achromatic stage (Fig. 108*h*–*k*). DNA replication includes four rounds in *Euplotes aediculatus* (AMMERMANN 1971a). During this period, the DNA disintegration products of the old Ma are reutilized and incorporated into the macronuclear anlage of *E. woodruffi* (RAO 1968b). After the second phase of DNA synthesis, the new Ma assumes its final shape: it divides in two in *Stylonychia* and *Oxytricha* (Fig. 108*k*) or elongates and bends in *Euplotes*.

Since polytene chromosomes fragment into bands, the adult Ma of Hypotrichida cannot contain chromosomes or whole genomes but only short free fragments of chromosomes (chromomeres). Accordingly, this type of macronuclear organization is called a *chromomeric* one (AMMERMANN 1971a). According to AMMERMANN and others (1974), only about 1.6% of the micronuclear genome is preserved as free fragments in the Ma of *S. mytilus*. This is due to two consecutive genetic diminutions: at the level of whole chromosomes and at the level of chromosomal loci. The preserved genome fragments, however, have a very high degree of multiplicity: each DNA piece is present in the Ma of *S. mytilus* in thousands of copies. Such a high degree of multiplicity is reached after 11–13 cycles of DNA replication, including 6–8 cycles during chromosome polytenization and 5 cycles during the second phase of DNA synthesis.

Properties of DNA in Adult Macronuclei. PRESCOTT and others (1971) isolated DNA separately from the Ma and Mi of *Oxytricha* sp.[4] and studied it both biochemically and by electron microscopy (Table 9). They discovered that all macronuclear DNA is represented by *short molecules* (Fig. 111*c*), whose size is $1–2 \cdot 10^6$ daltons (biochemical determinations) or 0.2–2.2 μm (electron microscopic measurements), which averages 0.8 μm, i.e. $1.5 \cdot 10^6$ daltons or 2500 base pairs. A separate fraction contains DNA molecules 2.0–2.2 μm long (about $4 \cdot 10^6$ daltons). The micronuclear DNA molecules proved to be longer by several orders of magnitude. According to more recent data (WESLEY 1975, SWANTON *et al.* 1980b), the mean size of the DNA molecules in the Ma of *Oxytricha* sp. is within the range of 0.7 to 1.1 μm (or 1.3 to $2.1 \cdot 10^6$ daltons), while the longest molecules may reach 7 μm (SWANTON *et al.* 1980b). RAE and SPEAR (1978) obtained similar values for the macronuclear DNA of *Oxytricha fallax* (see Table 9).

PRESCOTT and others (1971, 1973) as well as LAWN and others (1978) interpret these short molecules of macronuclear DNA as *free genes*. Indeed, a DNA length of 0.7–1.0 μm is enough to code for a medium-sized protein. As for the DNA molecules whose length is 2.0–2.5 μm, i.e. about $4 \cdot 10^6$ daltons (occasionally, up to $4.9 \cdot 10^6$ daltons; see Table 9), it has been shown that they selectively hybridize with ribosomal RNA. In other words, they are free ribosomal genes (PRESCOTT *et al.* 1973). A thorough study of these molecules in various hypotrichs has shown that,

[4] The ciliate was mistaken for *Stylonychia mytilus* in this and other studies made in PRESCOTT's laboratory; later, the ciliate turned out to be *Oxytricha* sp. (MURTI 1976b, LAUTH *et al.* 1976).

unlike the macronuclear rDNA of *Tetrahymena* and *Paramecium*, they are monomers, each containing only one coding region for 25 S rRNA and one coding region for 19 S rRNA (SPEAR 1980, SWANTON *et al.* 1980a, SPEAR *et al.* 1980).

The interpretation of macronuclear DNA molecules in *Oxytricha* as free genes is supported by data on their heteropolar nature (one end of each molecule is richer in adenine and thymine than the other) as well as by the ability of each molecule to bind a RNA polymerase molecule at one of its ends. Therefore, each gene-sized DNA molecule is capable of undergoing transcription (MURTI *et al.* 1972).

Identical inverted repetitive sequences 26 base pairs long were found at the ends of the short DNA molecules in the Ma of *Oxytricha*, *Euplotes*, and *Paraurostyla* (WESLEY 1975, LAWN 1977, HERRICK and WESLEY 1978, PRESCOTT *et al.* 1979). Due to them, denatured molecules can form one-strand circles with a short "neck". Such sequences may be important in the transverse fragmentation of chromosomes during macronuclear development because they can be recognition sites for specific endonucleases.

Contrary to the above data, AMMERMANN and others (1974) found not only short (less than 2.5 μm) DNA molecules in the Ma of *Stylonychia mytilus* but also medium-sized and long (up to 38 μm) molecules (Table 9). Later, however, STEINBRÜCK *et al.* (1981) reported values of the same order as in *Oxytricha*, i.e. from $0.3 \cdot 10^6$ to $14 \cdot 10^6$ daltons (500 23000 base pairs) or from 0.16 to 8 μm; the average length of the molecules was about 1.6 μm (AMMERMANN *et al.* 1978). Thus, the interpretation of macronuclear DNA molecules as single free genes applies to *Stylonychia*, too. Moreover, certain gene-sized DNA classes have been identified as particular genes, e.g. those coding for the five known types of histones (ELSEVIER *et al.* 1978). Just as in *Oxytricha*, the macronuclear DNA molecules in *Stylonychia* have identical terminal repeat sequences (LIPPS 1980). The Ma of *Stylonychia* also contains a class of rDNA molecules (free ribosomal genes) whose length is $4.5 \cdot 10^6$ daltons (7500 base pairs or 2.4 μm) and a class of free genes that code for 5S RNA and whose length is about $0.2 \cdot 10^6$ daltons (LIPPS and STEINBRÜCK 1978). All these molecules exist in the Ma in the form of short nucleosomal filaments, which comprise from 3 to 70 nucleosomes (LIPPS *et al.* 1978).

Gene-sized DNA molecules of almost the same length as those in *Oxytricha* and *S. mytilus* (0.24 14 $\cdot 10^6$ daltons) were found also in the Ma of *Stylonychia pustulata*, *Paraurostyla weissei*, and *Euplotes aediculatus* (SWANTON *et al.* 1980b, STEINBRÜCK *et al.* 1981).

After the destruction of the polytene chromosomes, a large amount of DNA may, in principle, be degraded in three different ways. First, most of the copies of a particular region of the chromonema, which are enclosed in a vesicle, may disintegrate and only some of the copies may be preserved. In this case, DNA degradation will not qualitatively change the genome of the developing Ma, and no genes from the polytene chromosomes will be lost. Second, all the copies of the chromonema in each vesicle may shorten even more by the destruction of some specific parts of each copy. Then the biochemical composition of DNA will change, but the gene complement of the nucleus may remain unaltered if the destroyed regions are noncoding spacer DNA. Third, the DNA may be completely degraded in some vesicles and completely or partly preserved in others. In this case, not only the DNA sequences will change, but also the gene complement of the Ma will be reduced (gene diminution). Moreover, every two and even all three models may combine. The right model can be chosen when the biochemical properties of the DNA at different phases of macronuclear development are known.

The qualitative composition of macronuclear DNA in Hypotrichida undoubtedly undergoes drastic changes in comparison with that of micronuclear DNA. Thus, the first model can be discarded at once. The macronuclear DNA of *Oxytricha* and *Stylonychia* turned out to be quite homogeneous as regards buoyant density (1.701 g/cm^3) and the melting profile, while micronuclear DNA proved to be heterogeneous, consisting of four of five fractions (Bostock and Prescott 1972, Ammermann *et al.* 1974, Lauth *et al.* 1976). The behaviour of these fractions during macronuclear development was studied in both *Stylonychia* (Ammermann *et al.* 1974) and *Oxytricha* (Spear and Lauth 1976). In *Stylonychia*, some satellite (highly repetitive) DNA fractions, present in micronuclei, disappear as early as the stage of polytene chromosomes. They were possibly located in the chromosomes that degenerate without undergoing polytenization. Three more DNA fractions disappear as polytene chromosomes disintegrate and much DNA is degraded. That is when nearly all the repetitive DNA sequences are lost. In *Oxytricha*, where all the chromosomes become polytenic, the fraction of satellite DNA with high buoyant density disappears, all the same, during polytenization possibly due to its selective nonreplication. A DNA with low buoyant density (1.695 g/cm^3) is synthesized at a high rate and constitutes the bulk DNA of the polytene chromosomes. But this fraction is selectively destroyed after the chromosomes fragment, and only the DNA which is typical of adult Ma (with a buoyant density of 1.701 g/cm^3) replicates in the second phase of DNA synthesis. Thus, both DNA degradation and DNA replication are selective processes in *Stylonychia* and *Oxytricha*. As a result, macronuclear DNA qualitatively differs from micronuclear DNA.

Prescott, Murti, and others (Bostock and Prescott 1972, Prescott *et al.* 1973, 1979, Murti 1976b) interpret the transformations of the genetic apparatus during macronuclear development in terms of the second model of DNA degradation (Fig. 112). These authors believe that all the DNA pieces produced as polytene chromosome fragments are shortened even more due to the degradation of their component nongenic (spacer) DNA. Only structural genes with the control regions are preserved, multiplying again during the second phase of DNA synthesis. Accordingly, Prescott and others (1973) regard the adult Ma of Hypotrichida as "a bag of free floating genes". Ammermann and others (1974), however, hold that true gene diminution occurs during macronuclear development, as predicted by the third model of DNA degradation. They drew this conclusion on the basis of their data, which show that the material of many bands of the polytene chromosomes is absent from the adult Ma (Fig. 109*f*, *g*).

Almost all the repetitive DNA sequences are eliminated at one stage or another of macronuclear development. This was demonstrated by DNA renaturation tests involving both *Stylonychia* (Ammermann *et al.* 1974) and *Oxytricha* (Lauth *et al.* 1976). An exception is ribosomal DNA, but its relative content in Ma is low.

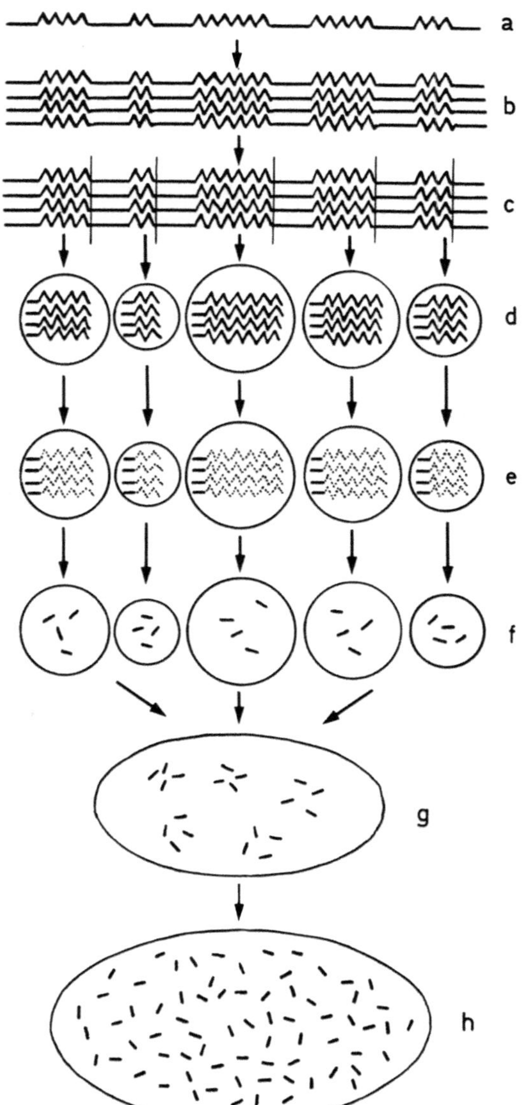

Fig. 112. Diagram explaining the derivation of gene-sized DNA molecules in the macronucleus of *Oxytricha* (after MURTI 1976b). *a* high molecular weight DNA of a micronuclear chromosome; *b* DNA in a polytene chromosome; *c* transection of the polytene chromosome at the interbands; *d* formation of vesicles; *e* destruction of the spacer DNA; *f* gene-sized molecules of DNA in the vesicles; *g* the macronuclear anlage after the disappearance of the vesicles; *h* multiple replication of the gene-sized DNA molecules in the adult macronucleus.

Otherwise, only unique DNA sequences are preserved in the adult Ma of Hypotrichida. Consequently, the renaturation curve of the macronuclear DNA is similar to the theoretical second-order reaction curve and can be used to calculate the *kinetic complexity* of the DNA, i.e. the true size of the macronuclear unit genome (Table 9). This is very small for an eukaryotic nucleus, merely $36 \cdot 10^9$ daltons in *Oxytricha* (LAUTH *et al.* 1976) and exceeding only 14 times the genome of *Escherichia coli*. The size of the macronuclear unit genome in *Stylonychia mytilus* was first reported to be $65 \cdot 10^9$ daltons (AMMERMANN *et al.* 1974), but after corrections were made, it came to only $31 \cdot 10^9$ daltons (STEINBRÜCK *et al.* 1981). When the total DNA content of Ma is known from photometric data, the true degree of multiplicity of this unit genome can be determined; for instance, it is about 15000 in *S. mytilus* and 2000 in *Oxytricha* sp.

Calculations show that the macronuclear unit genome of *Oxytricha* should comprise from 17000 to 25000 qualitatively different DNA molecules, each being about 1 μm long. Then, with a genome multiplicity of 2000 times, the whole Ma should contain 34–50 million free genes (LAUTH *et al.* 1976, SWANTON *et al.* 1980b). Similar calculations for *Stylonychia* give about 10000 different gene-sized molecules in the unit genome and 150 million genes in the whole Ma (AMMERMANN *et al.* 1978).

The macronuclear unit genome is greatly reduced in both ciliates in comparison with the haploid genome of their micronuclei (Table 9), namely, by 11 times in *Oxytricha* and by 160 times in *Stylonychia*. Therefore, a very small part of the eukaryotic genome is enough for ensuring the vegetative functions of the cell. In *Stylonychia*, however, the size of the haploid genome of a Mi, calculated with respect to the DNA content of this nucleus, is very large (10–12 times larger than in *Oxytricha;* see Table 9). This indicates the latent polyploidy of Mi (polyteny of its chromosomes?). If this is confirmed, the true degree of genome reduction in the Ma of *Stylonychia* will possibly be of the same order as in *Oxytricha*.

Problem of Division of Chromomeric Macronuclei. Both AMMERMANN and PRESCOTT believe that when the Ma of a hypotrichid divides, its DNA molecules (free genes) randomly segregate between the daughter Ma. Consequently, the genetic imbalance will increase in a long series of generations, and eventually certain genes will be completely lost. In other words, inviable cells will appear. The higher the degree of multiplicity of each gene, the greater is the number of viable generations. AMMERMANN (1971a) calculated, for instance, that if the degree of gene multiplicity is 60×, all cells will die out after 100 generations, and if it is 240×, they will do so only after 400 generations. *Stylonychia* clones actually age and begin to die out after 400–500 generations. Therefore, gene multiplicity ranging from 2000 to 15000 times can ensure even greater longevity of the clones than that observed. No hypotrichids capable of unlimited agamic reproduction are known yet. Therefore, the hypothesis of random segregation of the free DNA molecules in macronuclear division seems to hold true in this case.

Macronuclei of Keronopsis. Macronuclear development in the multinucleate ciliate *Keronopsis rubra* apparently differs from that in other hypotrichids. Its macronuclear anlage contains paired filamentous chromosomes (possibly in a state of somatic conjugation), but they are not clearly polytenic (RUTHMANN 1972). Neither the transverse fragmentation of chromosomes nor the "achromatic" phase in macronuclear development have been found. RUTHMANN (1973) has shown by electron microscopy that the chromatin of the macronuclear anlage gradually condenses into compact bodies that are separated by layers of lamellar material. The bodies are then successively extruded from the anlage into the cytoplasm and become small definitive Ma, each of which contains a paradiploid amount of DNA. RUTHMANN (1972) believes that the chromatin bodies preformed in the anlage are *diploid subnuclei* that later become individual Ma.

These numerous Ma divide without preliminary fusion with one another. Often, they divide into parts with an unequal DNA content. This means that the genome of the *Keronopsis* macronuclei fragments into subunits smaller than even the haploid genome (RUTHMANN 1972). Thus, the macronuclear apparatus of *Keronopsis* has features of both the subnuclear and the chromomeric types. It certainly has to be studied further.

Macronucleus of Chilodonella

Polytene chromosomes are formed and then fragment during macronuclear development not only in Hypotrichida, but also in *Chilodonella cucullulus* (class Kinetofragminophorea, order Cyrtophorida). The vegetative Ma of this ciliate has a heteromeric structure (see p. 288). RADZIKOWSKI (1967, 1973b) found typical ribbon-shaped polytene chromosomes with distinct transverse bands in the macronuclear anlage of *Ch. cucullulus*. At the beginning of its development, the anlage becomes polyploid: the number of chromosomes in it increases from 8 to about 128, corresponding to four duplications. Only 60–80 of these chromosomes become polytene, while the remaining ones degenerate (RADZIKOWSKI 1979). It could be that certain individual types of chromosomes (four of eight at the most) are completely eliminated, but this is only a supposition. DNA puffs, i.e. thickened and widened bands with a granular structure that are very rich in DNA, develop on the chromosomes at the end of their polytenization. Differential DNA overreplication of some chromosomal loci thus occurs in *Chilodonella* (just as in *Stylonychia*). This has been confirmed by autoradiography (RADZIKOWSKI 1979).

As the chromosomes of *Ch. cucullulus* become polytene, the DNA content increases almost 60-fold in the macronuclear anlage. Then all the polytene chromosomes transversely fragment into sections that have several bands, not into single bands. Moreover, sections with DNA puffs split longitudinally into short chains of chromomeres that then wind up to form chromatin bodies of the orthomere in the adult Ma (RADZIKOWSKI 1973b, 1979). Hence, these bodies should contain chromosome fragments with a length of several genes. Other polytene chromosome fragments (those containing no DNA puffs) fuse with one another and with the chromosomes not involved in polytenization, forming a large dense pycnotic body. The body gets into the constriction of a young Ma during its first division and is extruded into the cytoplasm where it is resorbed, carrying away about 35% of the total DNA of the young Ma. Therefore, macronuclear development in *Ch. cucullulus* must involve the differential replication of genes and genetic diminution. There is no additional replication of the conserved chromosome fragments, however, since no second phase of DNA synthesis exists in the macronuclear anlage. The Ma of *Chilodonella* should be assigned to the chromomeric type because it contains only chromosome fragments. It is still unknown how the genetic material is distributed when the Ma divides.

The macronuclear anlage in *Chilodonella uncinata*, another species of the same genus, develops in a somewhat different way. This development was studied by light microscopy and cytophotometry (PYNE *et al.* 1974), electron microscopy (PYNE 1978), and ultrastructural autoradiography (PYNE and GACHE 1979). Just as in *Ch. cucullulus*, four cycles of chromosomal duplication and DNA replication occur at the first stage of development, bringing the polyploidy of the anlage to 32 n. The chromosomes are seen as dense bodies under the electron microscope. The anlage actively incorporates ^{3}H-thymidine. At the second stage of development, the anlage becomes much larger, its chromosomes elongate into filaments only 25–50 nm thick, and transverse bands develop on them (Fig. 111 *d*). The chromosomes thus seem to become polytene, though much less than those in *Ch. cucullulus*. Such chromosomes are said to be *oligotene* ones (PYNE 1978). No DNA puffs develop on them. The DNA content of the anlage increases throughout the second stage (PYNE *et al.* 1974). The incorporation of ^{3}H-thymidine into the anlage stops for a short time when the chromosomes elongate, and then it goes on again (PYNE and GACHE 1979). Photometric data show that the chromonema may duplicate two or three times in the oligotene chromosomes.

At the third stage of development, the macronuclear anlage condenses and shrinks. However, its DNA content does not decrease, but even slightly increases (PYNE *et al.* 1974). The oligotene chromosomes gradually disappear at this stage and are replaced by dense chromatin bodies typical of the adult Ma. We still do not know how this happens since no transverse fragmentation of the oligotene chromosomes was detected (PYNE 1978). The incorporation of ^{3}H-thymidine slows down, being confined to the loose chromatin of the anlage rather than to the dense bodies. Thus, the DNA content in the developing Ma of *Ch. uncinata* rises progressively without a phase of DNA degradation, as has been shown also in the earlier works (SESHACHAR 1950b, 1953, PIERI and VAUGHIEN 1967b). No chromatin body is extruded during the first division of a young Ma in *Ch. uncinata*. However, at the second stage of development, the macronuclear anlage contains not only the oligotene chromosomes, but also blocks of condensed chromatin (PYNE 1978), which incorporate no ^{3}H-thymidine (PYNE and GACHE 1979) and disappear later. It could be that the blocks are eliminated chromosomes or parts of chromosomes that are resorbed *in situ*.

Therefore, even if there is chromosome diminution in *Ch. uncinata*, the process is less pronounced than in *Ch. cucullulus*.

Macronucleus of Bursaria

The development of the macronuclear anlage in *Bursaria truncatella* (subclass Vestibuliferia, order Bursariida[5]) was thoroughly studied with the light microscope by POLJANSKY (1934) and with ^{3}H-thymidine autoradiography by POLJANSKY and SERGEJEVA (1981).

[5] Until recently, *Bursaria* was assigned to the order Heterotrichida, subclass Spirotrichia.

Young macronuclear anlagen contain a threadball of filamentous chromosomes that gradually thicken and possibly become oligotenic (Fig. 113*b*); such anlagen incorporate ^{3}H-thymidine at a high rate. The filamentous chromosomes are then transformed into compact elements apparently by transverse fragmentation (Fig. 113*c*). The DNA synthesis slows down for a short time just before fragmentation but resumes at a high rate as the compact elements duplicate twice, forming "dyads" (Fig. 113*d*) and then "tetrads" (Fig. 113*e*). The process seems to

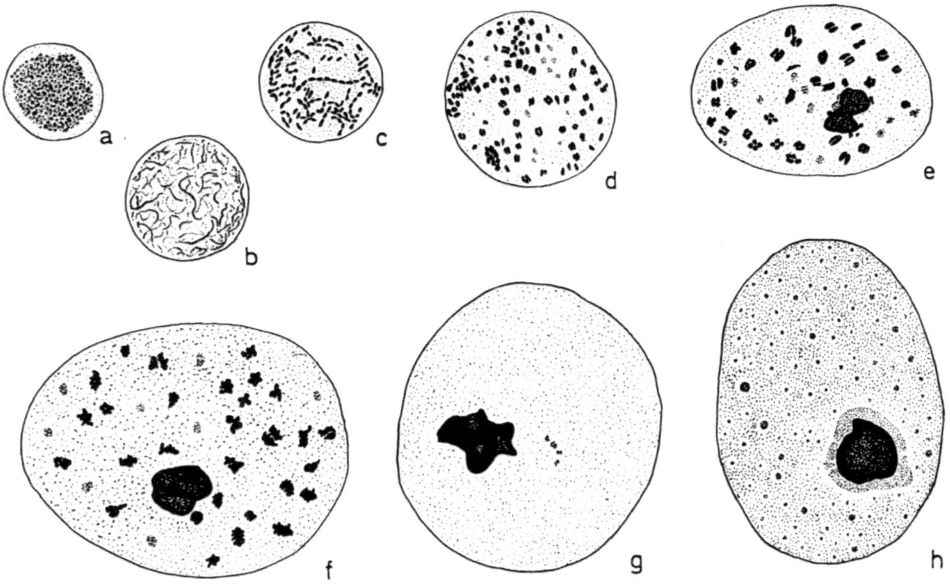

Fig. 113. Development of the macronuclear anlage in *Bursaria truncatella* (sections, iron haematoxylin, 830 ×, after POLJANSKY 1934). *a* young anlage; *b* threadlike chromosomes; *c* fragmentation of chromosome threads; *d* duplication of condensed elements, forming "dyads"; *e* formation of "tetrads" and the appearance of a karyosome; *f* distortion of "tetrads"; *g* karyosomal stage; *h* beginning of the accumulation of diffuse chromatin and nucleoli.

be the endomitotic reproduction of either whole chromosomes or their fragments. But endomitoses stop at this point; the chromosomal "tetrads" become distorted (Fig. 113*f*) and gradually dissolve in the karyolymph (Fig. 113*g*). A chromatin karyosome simultaneously appears in the anlage (Fig. 113*e*, *f*), possibly due to the fusion of some "tetrads". DNA synthesis in "dyads" and "tetrads" gradually ceases; it goes on a little longer in the karyosome and then stops, too. Labelled thymidine is no longer incorporated by the anlage once the karyosome is fully formed; the macronuclear anlage remains at the karyosome stage for at least a day (Fig. 113*g*).

Fine-granular chromatin then begins to accumulate in the anlage (Fig. 113*h*). DNA synthesis starts again, this time throughout the anlage except the karyosome (POLJANSKY and SERGEJEVA 1981). The macronuclear anlage of *Bursaria* displays no replication bands, which are found in Hypotrichida. The chromatin karyosome is gradually resorbed and multiple nucleoli appear (Fig. 113*h*). At the end of this stage, the anlage elongates to form the adult Ma. Four macronuclear anlagen develop in each exconjugant of *Bursaria;* they do not fuse but segregate in the first two cell divisions (POLJANSKY 1934).

Therefore, DNA synthesis in *Bursaria* comprises two phases: the first one involves the oligotenization and/or the endomitotic duplication of chromosomes, while "diffuse" chromatin accumulates during the second phase. The two phases are separated by a pause, i.e. the karyosome stage. But it is still unknown how the genetic apparatus of the anlage is reorganized at the karyosome stage. Nor is it known whether true chromosomes fragment at the stage when "dyads" and "tetrads" are formed or composite chromosomes are disassembled to form single chromosomes.

S. BORCHSENIUS and SERGEJEVA (1979) have reported some properties of the DNA of the adult Ma in *Bursaria*. The DNA is homogeneous in buoyant density (1.684 g/cm^3); however, DNA renaturation studies have shown that it contains up to 50% of repetitive sequences. According to its sedimentation rate, the DNA consists of two classes of molecules: large ones (about $100 \cdot 10^6$ daltons) and small ones (about $10 \cdot 10^6$ daltons). It is likely that the latter are rDNA molecules. These data are naturally not enough to determine how the genetic apparatus of the *Bursaria* Ma is organized. However, they at least indicate that the Ma contains DNA molecules much longer than the gene size.

Macronuclei of Heterotrichida

Macronuclear Development. The development of the macronuclear anlagen in the free-living Heterotrichida resembles that in *Bursaria*. It was studied by light microscopy in *Stentor* (MULSOW 1913), *Climacostomum* (PESHKOVSKAYA 1936), *Fabrea* (ELLIS 1937), and *Spirostomum* (RAO 1968a, SESHACHAR and BAI 1968) as well as by electron microscopy in *Stentor* (SKARLATO 1978a).

The young macronuclear anlage of *Stentor* contains filamentous chromosomes that later condense and duplicate once or twice (MULSOW 1913). Similar stages were observed in *Climacostomum*, *Fabrea*, and *Spirostomum*. The anlagen apparently become polyploid at this stage. Paired chromosomes are visible under the electron microscope as well. In addition to chromosomes the anlagen contain electron-dense fibrillar globules (Fig. 114*a*). Later, the chromosomes vanish and a Feulgen-positive karyosome appears in the macronuclear anlage of *Stentor*. Under the electron microscope, the karyosome looks like a group of loose chromatin aggregates (Fig. 114*b*). Karyosomes were also found in *Climacostomum* and *Fabrea*.

Fig. 114. Development of macronuclear anlagen in Heterotrichida. *a* and *b* macronuclear anlagen of *Stentor coeruleus*, electron micrographs: *a* during the doubling of chromosomes (*1*), a fibrous globule (*2*) is visible, 7200 × ; *b* at the stage of diffuse chromatin aggregates (*3*), 7200 × ; *c* and *d* polytene chromosomes in the macronuclear anlage of *Nyctotherus cordiformis: c* photomicrograph of a section stained with iron haematoxylin, 3750 × ; *d* electron micrograph, the arrows indicate the chromomeres that constitute the bands, 9000 × . *a* and *b* after SKARLATO (1978a); *c* after GOLIKOVA (1964); *d* after VINNIKOVA and GOLIKOVA (1978).

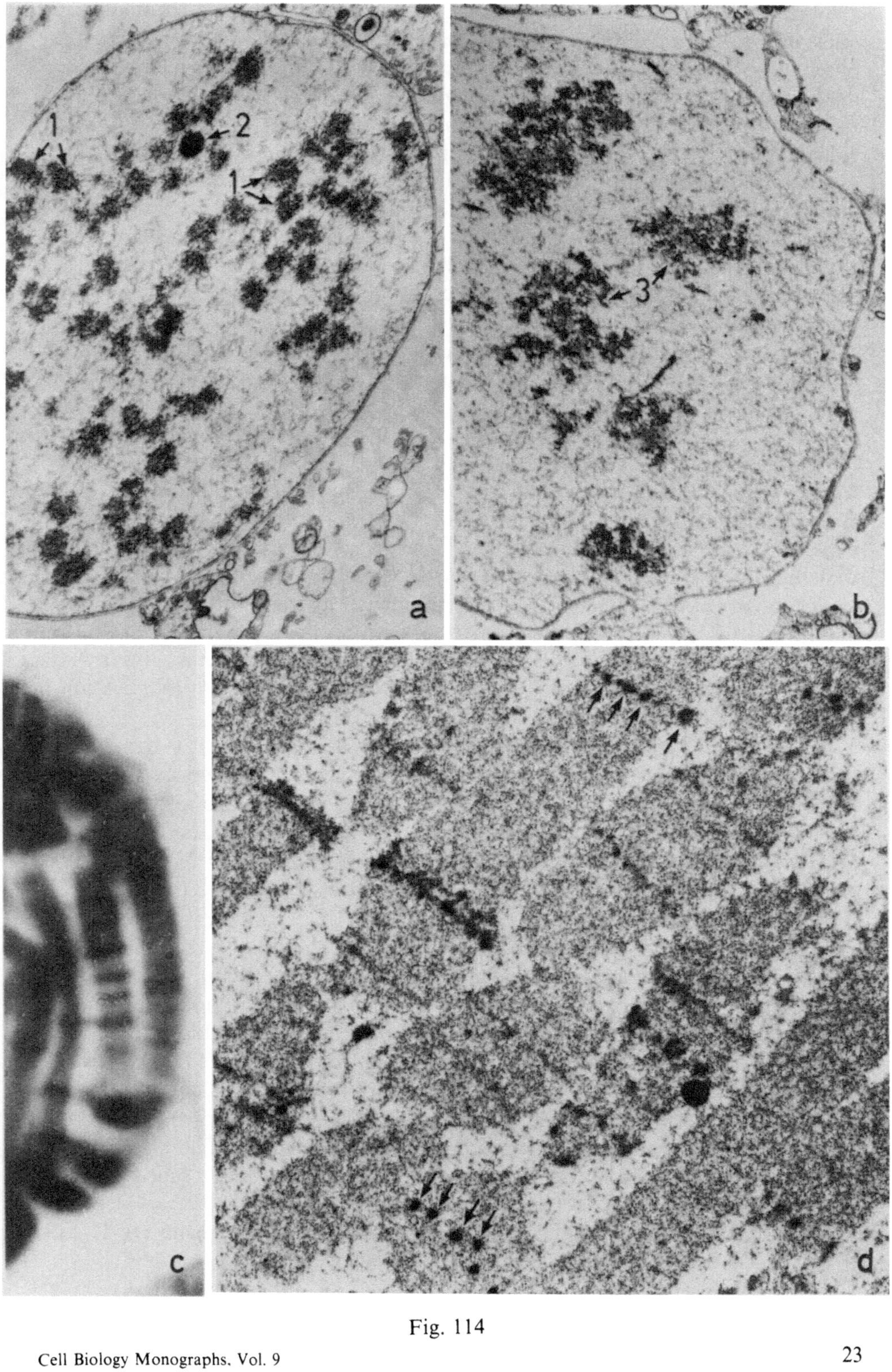

Fig. 114

When the karyosome disintegrates, small bodies of condensed chromatin and nucleoli appear in the macronuclear anlage of *Stentor*. At first, the nucleoli are purely fibrillar, but they acquire their typical structure at the later stages of macronuclear development when chromatin bodies intensively accumulate (SKARLATO 1978a). In *Stentor*, *Climacostomum*, *Fabrea*, and *Spirostomum*, several anlagen fuse, producing one adult macronucleus which then becomes moniliform.

Cytophotometric and autoradiographic studies have shown that the macronuclear anlage of *Blepharisma* synthesizes DNA continuously, i.e. without a phase during which the synthesized DNA is degraded (DASS *et al.* 1980).

Unlike the free-living forms, polytene chromosomes develop in the macronuclear anlage of the parasitic heterotrichid *Nyctotherus cordiformis* (GOLIKOVA 1964, 1965)[6]. The Ma begins to develop with a skein of filamentous chromosomes that are possibly interconnected into a composite chromosome. The anlage grows and the chromosomes become polytenic, turning into thick cords with distinct transverse bands (Fig. 114*c*). Electron microscopy (VINNIKOVA and GOLIKOVA 1978) shows that the bands consist of chromomeres on a single plane: the spaces between the bands are wide and filled with fibrillar material (Fig. 114*d*). There is no granular material in the polytene chromosomes or around them. Cytochemical tests for RNA in the macronuclear anlage of *Nyctotherus* are negative at the stage of the polytene chromosomes. Some observations indicate that the polytene chromosomes of *Nyctotherus* are somatically paired.

The macronuclear anlage of *Nyctotherus* intensively incorporates ^{3}H-thymidine at the early stages of polytenization (GOLIKOVA and NILOVA 1967); the rate of DNA synthesis diminishes toward the end of the process. Unlike in Hypotrichida, the polytene chromosomes of *Nyctotherus* show neither DNA overreplication in some loci nor the formation of large heterochromatin blocks.

It is not quite clear how the Ma of *Nyctotherus* develops further. Neither transverse fragmentation of the polytene chromosomes nor a definite "achromatic" stage was found in its development. A dense chromatin body, however, is extruded from the late macronuclear anlage into the cytoplasm. Nevertheless, we do not know the exact nature of the genome transformations in *Nyctotherus* that occur after the stage of the polytene chromosomes. It could be that they differ from those in Hypotrichida.

DNA Properties of the Adult Ma. The DNA of the adult Ma was examined biochemically in *Stentor coeruleus* (PELVAT and DE HALLER 1976). It turned out to be homogeneous in buoyant density (1.691 g/cm^3) and had no noticeable satellite fractions. Its molecular weight, calculated from the sedimentation rate, ranges from $5 \cdot 10^6$ to $40 \cdot 10^6$ daltons. Electron microscopy of the isolated DNA molecules shows that their length varies from 2–3 μm to over 40 μm. This corresponds to a molecular weight of $5 \cdot 10^6$ to $80 \cdot 10^6$ daltons (Table 9). Many of the short DNA pieces seem to be artifacts, i.e. fragments of longer molecules. Therefore, most of the DNA molecules in the Ma of *Stentor* are certainly not single gene-sized ones.

The macronuclear DNA of *Stentor* contains less than 15% of repetitive sequences. According to the DNA renaturation data (PELVAT and DE HALLER

---

[6] Polytene chromosomes were observed in *Nyctotherus* earlier by WICHTERMAN (1937) and GRASSÉ (1952).

1976), the size of the macronuclear unit genome in *Stentor* is about $60 \cdot 10^9$ daltons (Table 9). Since no data about the size of the micronuclear genome in *Stentor* are available, it is too early to conclude whether the macronuclear unit genome of this ciliate is reduced in comparison with the micronuclear genome.

Macronuclear Development in Other Ciliates

Gymnostomatia. The gymnostomes are the most primitive ciliates with polyploid Ma. Their Ma have a typical structure (see Fig. 92). Macronuclear development was studied by electron microscopy in *Didinium nasutum* (KARADZHAN 1975, KARADZHAN and RAIKOV 1979). The chromatin in the macronuclear anlage at first assumes the form of loose strands; apparently, they are despiralized chromosomes. The chromatin then condenses into loose aggregates scattered throughout the anlage. No polytenization of the chromosomes was detected, and no chromatin karyosome is formed. The loose chromatin aggregates disappear, while small chromatin bodies and numerous nucleoli develop throughout the macronuclear anlage. Peculiar bodies with a concentric lamellar structure are then formed in the anlage and are apparently extruded from it later via small vesicles that bud off the anlage. Ultrastructural autoradiography has shown that DNA synthesis commences in the anlagen immediately after their differentiation and continues without any interruption until they fully develop (KARADZHAN and RAIKOV 1979).

Therefore, macronuclear development in *Didinium* gives no indication that the genome of its Ma may be fragmented. The DNA of the adult Ma has not been investigated yet in any gymnostome.

Suctoria. GRELL (1949, 1953d) has thoroughly studied the development of the macronuclear anlage in *Ephelota gemmipara.* Its thread-like chromosomes condense and undergo at least three endomitotic replications; accordingly, about 30 bundles, each comprising 8 parallel chromosomes, appear in the anlage. Just as in *Bursaria*, the bundles then partially fuse, producing a chromatin karyosome, and partially decondense. Finally, the karyosome is resorbed, DNA synthesis starts throughout the nucleus, and the anlage becomes branched like the adult Ma. The genome transformations involved in macronuclear development in Suctoria are still unknown.

Peritrichia. Macronuclear development in Peritrichia resembles that in Hymenostomatida. Chromosomes cannot be discerned at any stage of macronuclear development in *Vorticella*, and no karyosome is formed. The accumulating chromatin is diffuse from the very beginning (MÜGGE 1957). Photometric techniques have shown that DNA accumulates in the macronuclear anlage of *Epistylis* without either a pause or a phase during which DNA is partially lost (SESHACHAR and DASS 1953, 1954a, 1954b).

Conclusion. Macronuclear development in ciliates varies greatly. On the whole, it comprises two phases of DNA synthesis that are occasionally separated by a pause (the "achromatic" or "karyosome" stage). The first phase is especially distinct in *Ephelota* and *Bursaria* where it takes the form of endomitosis, in Heterotrichida where it may be represented by either endomitosis or the polytenization of chromosomes, and in Hypotrichida and *Chilodonella* where it takes the form of polytenization. This phase is absent in Gymnostomatia, Hymenostomatia, and Peritrichia where only the second phase, i.e. DNA synthesis without visible chromosomes, is observed. The second phase of DNA synthesis also exists (after a pause) in *Bursaria*, *Ephelota*, Heterotrichida, and Hypotrichida, but not in *Chilodonella*. A noteworthy fact is that transcription (RNA synthesis) is, as a rule, weak during the first phase of DNA synthesis but strong during the second phase. The pause prior to the second phase of DNA synthesis is the most enigmatic step in the development of the macronuclear anlage, although it is also the most significant one in understanding the macronuclear structure. Chromosome fragmentation and gene diminution are known to occur during that pause in Hypotrichida and *Chilodonella*. But in Hymenostomatida, such transformations (if any) are expressed rather faintly. It is necessary to see whether they exist in *Bursaria*, Heterotrichida, Suctoria, and other forms with a distinct pause between the phases of DNA replication.

Reorganization Processes in the Macronucleus

Chromatin Extrusion

The extrusion of a part of the chromatin from the macronucleus into the cytoplasm is a common phenomenon among ciliates (for reviews of early publications, see Woodruff 1941 and Raikov 1969). It usually occurs during macronuclear division: a Feulgen-positive *residual body* is formed at the constriction site of the dividing Ma and is later resorbed in the cytoplasm of one of the daughter cells (Fig. 115). This type of chromatin extrusion occurs in many Hymenostomatia and has recently been studied in *Tetrahymena* (Cleffmann 1968, 1980, Nilsson 1976a) and *Colpidium* (Morat 1970). In other species, chromatin is extruded via large Feulgen-positive nuclear fragments that bud off both daughter Ma after division (Fig. 116). This type of extrusion is observed, for instance, within the reproduction cysts of many Colpodida (Burt *et al.* 1941, Beers 1946, Padnos 1962, Frenkel 1975). But in *Colpoda steini*, the extrusion body is poor in chromatin and can therefore be revealed only by electron microscopy (Frenkel 1980).

The quantity of eliminated DNA may be as much as from one-fourth to one-sixth the overall DNA content in the Ma of *Tetrahymena* (Dysart *et al.* 1962), although it usually constitutes only about 4% of the macronuclear DNA in *T. thermophila* (Cleffmann 1980). Over 50% of the cells in the exponentially growing culture of *T. pyriformis* divide with chromatin extrusion (Salamone and Nachtwey 1979), while the number of such cells in *T. thermophila* reaches 80%–90% (Cleffmann 1980). DNA is not synthesized in the extrusion body (Dysart 1963, Shepard 1965, Salamone and Nachtwey 1979). Most of the extrusion bodies in *Tetrahymena* are rapidly resorbed in the cytoplasm, especially in starved cells, but some are preserved for more than one cell cycle. Therefore, the cell treats them as food inclusions (Salamone and Nachtwey 1979).

The specific functioning of the DNA in extrusion bodies, however, has also been reported. In *Tetrahymena limacis*, for instance, they incorporate labelled thymidine (Dysart 1960). Chromatin extrusion in *T. pyriformis* can be induced by blocking DNA replication in Ma with inhibitors in the last third of the S period

(ANDERSEN 1972b, 1977). Then incompletely replicated DNA is extruded into the cytoplasm; this DNA is not degraded, and it even completes its replication in the G_2 phase of the next cell cycle (ANDERSEN 1974, 1977).

In all the cases studied by electron microscopy, the extruded DNA is surrounded by a nuclear envelope (i.e. it lies within macronuclear fragments), at least, soon after its extrusion (MORAT 1973, NILSSON 1976a).

In *Nassulopsis lagenula*, the residual body formed during the division of the Ma regenerates in one of the two daughter cells to form a new Ma, and the old Ma of this cell is then resorbed (TUFFRAU 1962). The phenomenon is known as *parahemixis*.

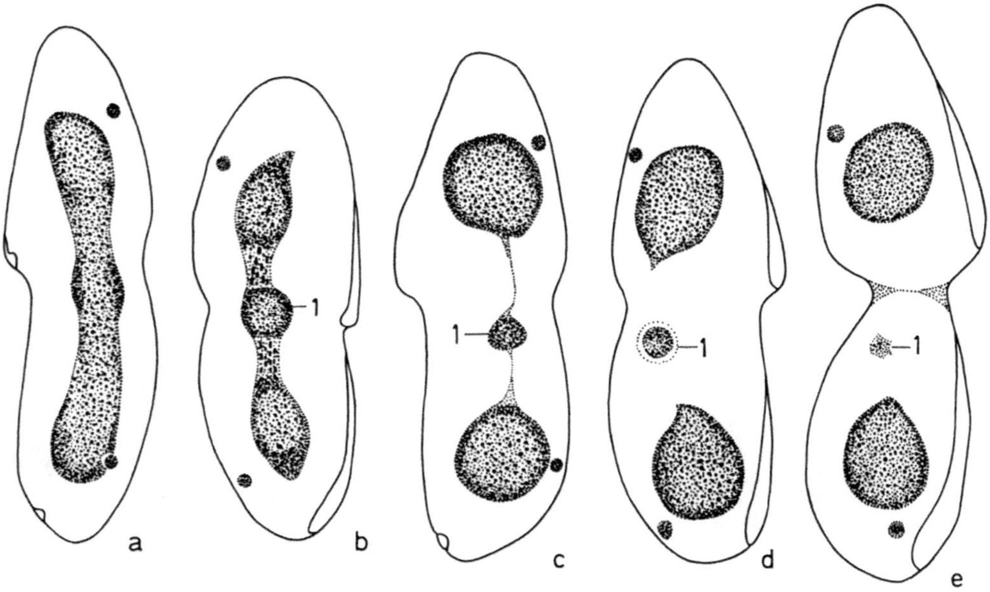

Fig. 115. Macronuclear division with chromatin extrusion in *Ancistruma isseli*. 750 × (after KIDDER 1933). *a*–*c* formation of a residual body (*1*) at the constriction site of Ma; *d* and *e* resorption of the residual body.

Chromatin can also be extruded from Ma without any relation to cell division, particularly in aging clones. This process, known as *hemixis*, occurs in *Paramecium* (DILLER 1936, 1940, 1958, VIVIER 1960b) and some other forms. Both large and small fragments can bud off the Ma in hemixis; the large fragments usually regenerate into a whole Ma, while the small fragments are resorbed.

Chromatin extrusion occurs also in the resting cysts of many Colpodida. Just as before a division, four chromatin aggregates appear in the Ma of *Colpoda duodenaria* (TAYLOR and GARNJOBST 1941). Two of them get into the extrusion body, which undergoes pycnosis. In this case, chromatin extrusion may be interpreted as abortive division. But in other Colpodida, the extrusion body is less than half of Ma in size (KIDDER and CLAFF 1938, BURT *et al.* 1941, BEERS 1946, etc.). Chromatin is repeatedly extruded from the Ma of *Tillina* when resting cysts are formed so that the mean DNA content in Ma decreases almost to one-third (FRENKEL 1977). In this case, chromatin extrusion in the resting cysts resembles hemixis.

The significance of chromatin extrusion remains obscure. FAURÉ-FREMIET (1953) has presumed that this process regulates either the degree of macronuclear polyploidy (by eliminating the extra genomes) or the chromosome equilibrium in Ma (by extruding the "extra" chromosomes). Indeed, chromatin extrusion is expressed most clearly after DNA overreplication in Ma (i.e. after DNA has replicated more than once), e.g. during the first divisions in *Tetrahymena* after

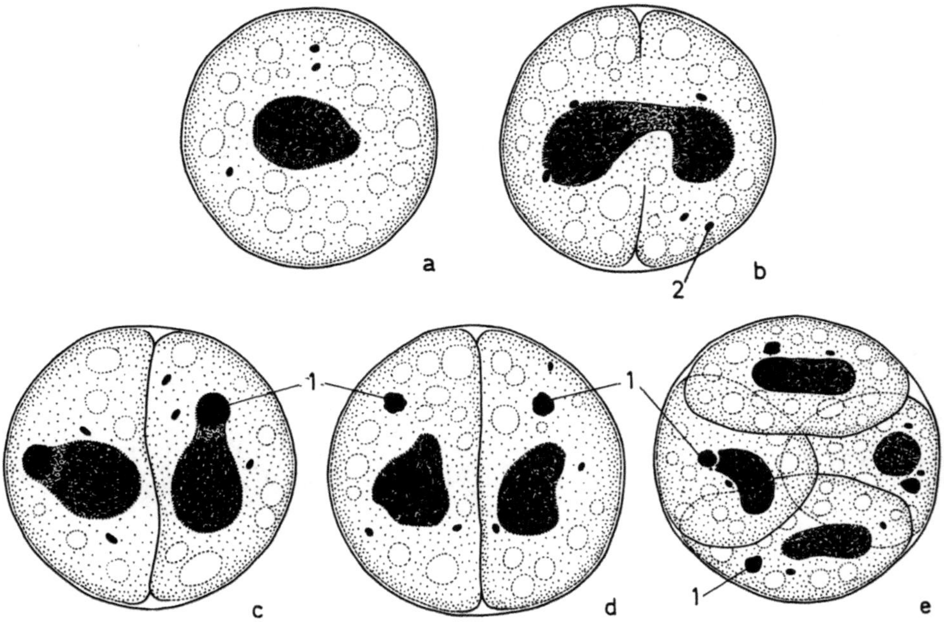

Fig. 116. Chromatin extrusion from the daughter macronuclei after each division in the reproduction cysts of *Tillina magna*. Feulgen reaction, 350 × (unpublished drawing, courtesy of M. A. FRENKEL). *a* encysted cell; *b* first division; *c* and *d* chromatin extrusion after the first division; *e* chromatin extrusion after the second division; *1* extrusion bodies, *2* micronuclei.

either UV irradiation (SHEPARD 1965) or treatment with a thymidine analogue (LYKKESFELDT 1979). It is prominent also in cells with a spontaneously high DNA content in Ma (CLEFFMANN 1968, MORAT 1970). Biochemical and autoradiographic data indicate that DNA in the extrusion bodies does not differ from the bulk DNA of the Ma (LYKKESFELDT 1979); consequently, it cannot be a specific DNA fraction, e.g. amplified rDNA. The DNA content of the extrusion bodies of *T. thermophila* is not a multiple of the DNA content in the micronucleus, i.e. chromatin extrusion cannot be the elimination of whole genomes from Ma (CLEFFMANN 1980). If the Ma has a chromosomal organization, extra chromosomes can be eliminated via the extrusion bodies (SEYFERT 1979). This, however, would necessitate the existence of a device that can recognize such supernumerary elements and collect them into the extrusion body.

According to another hypothesis, chromatin extrusion is not selective in itself but is only a means of regulating the nucleocytoplasmic ratio in a *Tetrahymena* cell (SALAMONE and NACHTWEY 1979). In that case, however, elements that are present in insufficient numbers should differentially replicate in Ma (CLEFFMANN 1980). To put it differently, either DNA extrusion or DNA replication should be a selective process.

A particular category consists of chromatin extrusion bodies that separate from Ma in amicronucleate ciliates, especially after the experimental elimination of Mi; they can, to a certain extent, play the functional role of the absent Mi. Such "pseudomicronuclei" divide by means of a mitosis-like process. They were found in *Paramecium bursaria* (SCHWARTZ 1958, GOLIKOVA 1978), *P. putrinum* (FOKIN 1979), and *Stylonychia mytilus* (AMMERMANN 1970).

Macronuclear Condensation Without Division

The macronucleus with a complicated shape may condense into an oval body not only before division but also during the *physiological reorganization* of the cell, which involves the dedifferentiation of the old buccal ciliary apparatus and the development of a new one, e.g. in *Stentor* (SCHWARTZ 1935, WEISZ 1949a, TARTAR 1961, DE TERRA 1967, PAULIN and BROOKS 1975), *Spirostomum* (EBERHARDT 1962), and other species. Macronuclear condensation also usually accompanies the posttraumatic regeneration of ciliates (TARTAR 1961, PARKER and GIESE 1966, MEISTER 1970).

In Hypotrichida, two or more Ma usually fuse into one in resting cysts (WEYER 1930, POMRYASKINSKAYA 1940, GRIMES 1973, WALKER *et al.* 1975, 1980, GUTIERREZ *et al.* 1981). Apparently, this phenomenon is similar to macronuclear condensation during the physiological reorganization. The DNA content in the Ma of *Oxytricha* nearly halves at this time, but then it reaches the original level during excystment (LIEDER and RUTHMANN 1976). Hence, the process seems to be coupled with the reorganization of the macronuclear genetic apparatus.

Changes in the Macronucleus During Sexual Processes

The macronuclear fragments produced when the old Ma disintegrates during conjugation or autogamy are normally resorbed. But they can also regenerate, e.g. in *Paramecium*, though only under specific conditions. A Ma can regenerate from its fragments only in the exconjugants (or their progeny) that had either a synkaryon or a hemikaryon but later lost their macronuclear anlagen owing to certain factors (SONNEBORN 1947, OSSIPOV and SKOBLO 1968, 1973, BERGER 1973b). The anlagen normally seem to inhibit macronuclear regeneration. In the *Paramecium* species of the *aurelia* group, it can be produced experimentally by heating exconjugants to 38 C, thus suppressing the development of the macronuclear anlagen (SONNEBORN 1947). Then a part of the exconjugant progeny does not get any more anlagen of the new Ma, while the old macronuclear fragments regenerate in them. In *Euplotes*, macronuclear regeneration can be induced by an UV damage of the macronuclear anlage, but only after the latter has issued information for the cortical morphogenesis (KLOETZEL 1981 b).

The ability of the macronuclear fragments to regenerate means they do not undergo irreversible degenerative changes for a long time. But DNA synthesis is absent in the fragments of most ciliates, e.g. *Tetrahymena* (SUGAI and HIWATASHI 1974, DOERDER and DE BAULT 1975), *Chilodonella* (RADZIKOWSKI 1973b, PYNE *et al.* 1974, PYNE and GACHE 1979), *Nyctotherus* (GOLIKOVA and NILOVA 1967), *Stylonychia* (CHADHA *et al.* 1978), and *Euplotes* (DINI and LUPORINI 1979). DNA synthesis continues for some time in the fragments of the old Ma of *Paramecium tetraurelia* (BERGER 1973a, 1973b), *P. caudatum* (MIKAMI 1979), and *Bursaria truncatella* (POLJANSKY and SERGEJEVA 1977). In *P. caudatum*, it goes on until the development of a new Ma ends; the influence of this Ma eventually represses DNA replication in the fragments of the old Ma. DNA synthesis in the macronuclear fragments of *P. tetraurelia* sharply accelerates when the fragments begin to regenerate.

The rate of RNA synthesis in the fragments of old Ma remains, as a rule, very high; most probably, they supply the cell with various RNA species throughout conjugation. Experiments involving labelled uridine incorporation into the macronuclear fragments show that this is so in *Paramecium* (KIMBALL and PERDUE 1962, BERGER 1973a), *Tetrahymena* (SUGAI and HIWATASHI 1974), *Didinium* (KARADZHAN and RAIKOV 1979), *Chilodonella* (RADZIKOWSKI 1973b), *Bursaria* (SERGEJEVA 1976), and *Euplotes* (RAO 1968b). In *Stylonychia*, however, RNA synthesis in the old Ma occurs only during a short time at the beginning of conjugation (between the fifth and sixth hour) and then stops. SAPRA and AMMERMANN (1973, 1974) have shown that long-lived messenger RNA is synthesized during this time, programming all the events of further conjugation until a new Ma begins to function. After that, SAPRA and AMMERMANN (1974) failed to reactivate RNA synthesis in the macronuclear fragments or to produce macronuclear regeneration even when the anlage of the new Ma had been eliminated. Apparently, the old Ma functions differently during conjugation in different ciliates.

Electron microscopy has revealed the following occurrences in the macronuclear fragments of *Didinium*: the chromatin bodies fuse into a spongy mass, the nucleoli segregate into granular and fibrillar zones, the nuclear envelope "peels off" in the form of separate cisternae, and the nucleolar material is extruded into the cytoplasm. Ultrastructural autoradiography, however, indicates that RNA is still synthesized in the chromatin and residual nucleoli until the macronuclear fragments are enclosed in autophagic vacuoles (KARADZHAN and RAIKOV 1979). The envelopes of numerous Ma in *Dileptus* form vesicles during conjugation (VINNIKOVA 1974d). In *Stentor*, typical nucleoli disappear from the fragments of the old Ma, being replaced by fibrillar RNP globules, which are probably produced by the transcription of the chromatin strands (SKARLATO 1978b, 1979b). In *Bursaria*, the chromatin of the macronuclear fragments decondenses and vesicles separate from the nuclear envelope in the middle of conjugation. This possibly indicates intensive transcription (SERGEJEVA 1976).

The old Ma begins to degenerate only at the late stages of conjugation. Its DNA is degraded. The process involves acid phosphatase (SERGEJEVA 1964, SAXENA and

JURAND 1973, SERGEJEVA and POLJANSKY 1974). The products of DNA degradation may be reutilized by the developing macronuclear anlage, particularly in the starved exconjugants (RAO 1968b, SAPRA and DASS 1970, BERGER 1974). In other cases, there is no such reutilization (BERECH and WAGTENDONK 1962). Electron microscopy has revealed that the old macronuclei or their fragments in *Dileptus*, *Bursaria*, *Stentor*, and *Didinium* get into autophagic vacuoles and are digested there (VINNIKOVA 1974d, SERGEJEVA 1976, SKARLATO 1978b, KARADZHAN and RAIKOV 1979).

Transcription in the Macronucleus

Intensive RNA synthesis in the Ma of *Tetrahymena* and *Paramecium* was detected by autoradiography (PRESCOTT 1960, 1962, KIMBALL and PERDUE 1962). Labelled RNA is released from Ma into the cytoplasm after one or two hours (ECKERT *et al.* 1975). RNA synthesis was registered also by biochemical techniques in the isolated Ma of *Tetrahymena* (MITA *et al.* 1966), *Paramecium* (STEVENSON 1967), and *Blepharisma* (SESHACHAR and SAXENA 1968). Electron microscopic autoradiography shows that the RNA precursors are incorporated into both the chromatin bodies and the nucleoli of Ma (STEVENS 1963, CHARETT 1969, NILSSON 1973, BOHATIER 1977b). Transcription of the macronuclear chromatin has recently been visualized in *Paramecium* using the spreading technique and electron microscopy (SAMUEL *et al.* 1981).

Ribosomal RNA synthesis and processing in the Ma of *Tetrahymena* have been studied by both biochemical and spreading techniques. The rate of rRNA synthesis in the Ma proved to be very high – ten times higher than in the HeLa cells (LEICK 1969). The palindromic extrachromosomal rDNA molecules are transcribed bidirectionally from the centre of the molecule toward its ends; the terminal spacers are not transcribed (GALL *et al.* 1977, NILES 1978, CECH and RIO 1979, DIN *et al.* 1979, ENGBERG *et al.* 1980, GOROVSKY 1980). The nontranscribed spacers have a typical nucleosome structure, while the transcribed ribosomal genes have no such structure (S. BORCHSENIUS *et al.* 1981a). Both structural ribosomal genes of each half of the rDNA molecule are transcribed together to form a molecule of preribosomal RNA with a molecular weight of $2.2 \cdot 10^6$ daltons and a sedimentation constant of 35S (LEICK and PLESNER 1968, LEICK 1969, KUMAR 1970, ECKERT and FRANKE 1975, HERMOLIN and ZIMMERMAN 1976, LEICK *et al.* 1976, NILES 1978). The molecule is broken in two and, after a series of intermediate steps, produces 17S and 25S rRNA molecules ($0.7 \cdot 10^6$ and $1.3 \cdot 10^6$ daltons, respectively). Similar values were reported for the preribosomal and two ribosomal RNAs in *Paramecium* (CUMMINGS 1972), *Stylonychia* (NOCK 1981), and *Euplotes* (LIPPS 1974). At the beginning, primary rRNA transcripts seem to be closely associated with the macronuclear matrix, whose proteins could be active enzymatically and could participate in rRNA processing (HERLAN *et al.* 1979).

RNA is synthesized in Ma during all the phases of the cell cycle (RAO and PRESCOTT 1967, PELVAT *et al.* 1973, etc.). The rate of total RNA synthesis is the highest in the G_2 period in *Paramecium bursaria* (JUST 1973), at the end of the G_1 period in *Euplotes eurystomus* and *Blepharisma intermedium* (EVENSON and PRESCOTT 1970, SHIVAHI *et al.* 1978c), and in late S in *Stylonychia notophora*

(SHIVAHI *et al.* 1978a). The rate of rRNA synthesis in the Ma of *Tetrahymena* clearly increases in early S, i.e. immediately after the replication of ribosomal genes, which may be attributed to the gene dose effect (KEIDING and ANDERSEN 1978).

RNA synthesis in Ma clearly depends on the culture conditions. The rate of RNA synthesis in *Tetrahymena* drops during starvation mainly because of a decrease in its synthesis on repetitive DNA sequences, i.e. due to the decelerated synthesis of rRNA, tRNA, 5 S RNA, and some other types of RNA (MERKULOVA and IRLINA 1974). In particular, the rate of preribosomal RNA synthesis drops to 3%-5% of that which occurs under the optimal growth conditions, rRNA processing decelerates, and the rate of rRNA transport into the cytoplasm diminishes to 1% 2% of the optimal rate (ECKERT and KAFFENBERGER 1980). rRNA synthesis, processing, and transport resume as soon as the starved *Tetrahymena* are fed again (KAFFENBERGER and ECKERT 1980). The rate of RNA synthesis in the Ma of *Paramecium* diminishes also with clonal age (KLASS and SMITH-SONNEBORN 1976).

It is a well known fact that the Ma is needed for regeneration in ciliates and that its effect on regeneration is species-specific (SCHWARZ 1935, TARTAR 1953, 1961). The effect seems to be determined, first of all, by the transcriptional activity of the Ma.

There are at least two types of dependence of posttraumatic regeneration on the Ma in ciliates (BOHATIER 1981). In some species, the lost parts of the body will not regenerate unless messenger RNA is synthesized *de novo* in the Ma. This synthesis is induced by the trauma itself. Such species have no stock of long-lived mRNA in their cells. Therefore, actinomycin D and other inhibitors of transcription block regeneration, too. This is typical of *Stentor* (JAMES 1967), *Blepharisma* (GIESE 1970), *Condylostoma* (BOHATIER 1981), and some other ciliates. In other species, regeneration does not require instantaneous RNA synthesis but proceeds owing to a large intracellular stock of stable messenger RNA (which is synthesized in the Ma independently of the eventual trauma) or, occasionally, even to stocked proteins. Naturally, actinomycin D does not block regeneration in such species of which *Dileptus* is an example (BOHATIER 1977a, 1977b).

Long-lived messenger RNA is synthesized in the Ma of *Stylonychia* not only at the beginning of conjugation but also prior to the cell division and during the regeneration of cortical organelles. This RNA is needed also for micronuclear mitosis during regeneration (SAPRA and AMMERMANN 1974, SAPRA *et al.* 1976). Therefore, just as in Metazoa, long-lived mRNA in ciliates participates in many cell differentiation phenomena.

The mechanism of RNA transport from a Ma into the cytoplasm varies among ciliates. In *Tetrahymena*, electron-dense material (tentatively, RNP) is extruded in vesicles that separate from the macronuclear surface (ELLIOTT *et al.* 1962). RNA transport through the pores in the macronuclear envelope is a more universal mechanism (KANEDA 1961c, WUNDERLICH *et al.* 1974). The rate of transport of total RNA as well as ribosome subunits through the pores in the macronuclear envelope of *Tetrahymena* clearly depends on temperature (NAGEL and WUNDERLICH 1977); actinomycin D and cycloheximide inhibit RNA transport from a Ma (ECKERT *et al.* 1975).

Theories of the Macronuclear Structure

Development of Ideas About the Macronuclear Structure

It was long held (almost till the late thirties) that the Ma of ciliates had no chromosome structure, contained no genes, and was filled with "trophochromatin". In the 1940s and 1950s, this concept was succeeded by a theory proposed by PIEKARSKI (1941), GEITLER (1941), SONNEBORN (1947), and GRELL (1950b, 1953b). According to the theory, the Ma is polyploid, i.e. it has a manifold unaltered micronuclear genome and contains either many separate chromosomes or a large number of subunits. Each subunit or subnucleus is a complete diploid set of chromosomes separated from other such sets (SONNEBORN 1947). The "subnuclear" variety of the theory of macronuclear polyploidy gave an especially good explanation of the "amitotic" division of the Ma: it was maintained that pairs of genomes (diploid subnuclei) segregated between the daughter nuclei and, since they were identical, even their random segregation would not lead to aneuploidy in a series of cell generations.

Various views were expressed on the nature of subnuclei. SONNEBORN's initial assumption, that the subnuclei were membrane-bound units that reproduced themselves by mitosis within a Ma, has not been confirmed in cytological studies. Then GRELL (1953b) assumed that each subnucleus corresponded to a pair of composite chromosomes, while each composite chromosome resulted from the end-to-end union of all the chromosomes of a single genome. As we have seen, this concept was somewhat confirmed, for instance, by RUTHMANN's (1963) cytological studies involving *Loxophyllum* (see Fig. 100*a*). The pairing of two haploid composite chromosomes (similar to somatic conjugation) would then produce structures equivalent to diploid subnuclei. The composite chromosomes would reproduce themselves by endomitosis without fragmenting into single chromosomes and thus would remain integral during all the stages of the cell cycle. Other authors (NILSSON 1970, WOODARD *et al.* 1972) allowed for the existence of *haploid* rather than diploid subnuclei in the Ma. This did not run counter to the hypothesis of composite chromosomes (provided the latter were not paired).

However, the universal subnuclear hypothesis formulated with respect to the macronuclear structure of all ciliates (except karyorelictids) in the 1960s (POLJANSKY and RAIKOV 1960, 1961, RAIKOV 1963f, 1967a, 1969) should be reconsidered now. First, it does not account for chromosome fragmentation and genetic diminution during macronuclear development in Hypotrichida and *Chilodonella*. In addition, more and more data show that macronuclear development in other ciliates may also involve genetic diminution; consequently, the gene complement of a Ma (even if it contains subnuclei) would not fully correspond to that of a Mi. At the same time, morphological data on the diploid subunits in the macronuclei of *Nassula*, *Nassulopsis*, and Colpodida gained support from cytophotometric studies. Therefore, we can no longer maintain that a *universal* type of genetic organization of a "polyploid" Ma exists. The Ma of various ciliates must fundamentally differ in their organization. This raises the question of the evolution of macronuclear organization.

Evolution of the Types of "Polyploid" Macronuclei

Our ideas about the evolution of macronuclei are rather speculative. Only two types of organization of "polyploid" Ma can be clearly defined now.

1. *Polyploid subnuclear macronuclei*, which are typical of certain lower ciliates (orders Colpodida, Synhymeniida, Nassulida): Apparently, there is a minimum number of rearrangements in the genome of a subnuclear Ma, and they probably correspond to the most primitive form of a polyploid Ma. Such Ma may be called *polygenomic* ones. We once believed that such an organization was typical of all polyploid Ma (RAIKOV 1967a, 1969). It is now necessary, however, to narrow down the range of forms with polygenomic Ma.

2. *"Polyploid" macronuclei of the chromomeric type*, which are characteristic of several higher ciliates (Hypotrichida) and of at least one form of lower ciliates (*Chilodonella*): Both the genomes and the chromosomes of such Ma are fragmented into pieces one or several genes long. Free fragments of chromosomes randomly segregate during macronuclear division, while macronuclear development involves drastic genetic diminution. Consequently, the macronuclear genome is greatly reduced in comparison with the micronuclear genome. The chromomeric Ma may be termed "polyploid" only conventionally; instead, SCHWARTZ (1978) proposed the term *ampliploid*. Ampliploid (chromomeric) macronuclei seem to have originated more than once from polygenomic Ma of the subnuclear type in the course of evolution, since Hypotrichida and *Chilodonella* are not related forms.

It is still impossible to specify all ciliate taxa that have subnuclear and chromomeric types of macronuclei. It is still unknown to what types most Kinetofragminophorea (especially Gymnostomatia, Bursariida, and Suctoria) and Heterotrichida belong. The large and generally well known group of Oligohymenophorea (*Tetrahymena*, *Paramecium*, etc.) is also a problem. According to morphological, cytochemical, biochemical, and genetic data, the Ma of the above organisms fit best into either the category of subnuclear Ma (particularly its haploid variety) or the category of chromosomal Ma (with a mechanism that controls the quantity of different chromosomal types). If the chromosomal model proposed for the Ma of *Tetrahymena* is corroborated, it will be necessary to recognize a third independent type of macronuclear organization, i.e. the *chromosomal* (or simply polyploid) type. The chromosomal model does not rule out the possibility of substantial genetic diminution (the shortening of chromosomes) during macronuclear development, as was demonstrated by SCHWARTZ (1978) with respect to *Paramecium bursaria*.

Detailed studies of the nuclear apparatus in such aberrant forms as *Protocruzia* are likely to throw light on other types of macronuclear organization.

The major ways of evolution of the ciliate macronucleus are consistent with the theory proposed by DOGIEL (1929, 1954). According to it, the polymerization of nuclei and organelles is one of the ways of the progressive evolution of protozoans (reviewed by POLJANSKY and RAIKOV 1976). Only identical nuclei polymerize in homokaryotic ciliates (*Stephanopogon*), while both Ma and Mi are capable of undergoing polymerization in karyorelictids. In higher ciliates, polymerization switches over to the intranuclear level (POLJANSKY and RAIKOV 1961) and involves whole genomes (in the Ma of the subnuclear type), free chromosomes (in the Ma of the chromosomal type), or only active chromosome fragments (in the Ma of the chromomeric type). In the last case, polymerization occurs at the molecular (genic) level. Polymerization always results in both a greater "gene dose" in the cell and a higher yield of the primary gene products. This is done most efficiently in the Ma of the chromomeric type in which only actively functioning parts of the genome are preserved and multiplied.

Conclusion

Much progress has been made in the comparative karyology of Protozoa in the last 10–15 years. The main achievements will now be briefly reviewed and the most important unresolved problems will be mentioned.

Morphology and Ultrastructure of the Nondividing Nucleus. The fine structure of the nuclei has now been studied in most protozoan taxa. The comparative karyology of Protozoa has presently shifted to the ultrastructural level. It has been proved that all the protozoans, except dinoflagellates, are typical eukaryotic organisms.

A wide field for further research has been opened up in the areas of the ultracytochemistry and functional morphology of the nuclear structures (electron microscopy combined with autoradiography, spreading techniques, the use of inhibitors, etc.). The unresolved problems in this respect include the functions of the pore complex components, of the additional layers of the nuclear envelope, and of various categories of intranuclear structures that differ from chromatin and the nucleolus, especially the protein spheres. It is likely that the forms of rDNA amplification in protozoans differ from those in the higher eukaryotes, and free rDNA somehow substitutes the linear repeats integrated in DNA (GALL 1974). There is still no molecular characterization of the RNP helices of amoebae and some ciliates, i.e. structures that are possibly analogous to informosomes.

Various types of nondividing nuclei are hardly of much phylogenetic significance, their morphology being connected to a far greater extent with their functional (especially transcriptional) activity than with the phylogeny of the appropriate taxon. Conversely, the experimental investigation of the mechanism of the interchange of nuclear forms in the life cycle, i.e. an analysis of cyclic nuclear differentiation (by nuclear transplantation, the use of inhibitors, etc.), is promising.

Mitosis. The ultrastructural morphology of almost all the types of protozoan mitosis has been studied in the past 15–20 years. For instance, the mechanism of semiopen and closed intranuclear pleuromitoses (in coccidians, haemosporidians, microsporidians, etc.) has been understood for the first time, and the structure of both the mitotic apparatus and kinetochores has been revealed in closed extranuclear pleuromitosis (in hypermastigids). Many errors and ambiguities have been eliminated in the descriptions of mitosis based on light microscopy, and a new, more natural classification of the types of nuclear division has been created on the basis of work by A. HOLLANDE, H. RIS, D. KUBAI, and others. These authors have proposed that the most primitive type of mitosis is closed intranuclear pleuromitosis (i.e. mitosis with intranuclear half-spindles whose poles are on the inner side of the nuclear envelope). This type of mitosis is most commonly found among the lower eukaryotes, including protozoans (HEATH 1980).

Another important achievement is the elaboration of the concept of the microtubule-organizing centres (MTOCs), which can be localized in different sites in the cell. This concept, mainly developed by J. D. PICKETT-HEAPS, is connected with the view that the centrioles in the lower eukaryotes are, in most cases, not the inducers of the mitotic spindle.

The great diversity of the types of mitosis in Protozoa apparently reflects the "period of search" in evolution when the most rational type of mitotic apparatus had still not been consolidated. Some phylogenetic links between various types of mitosis are still obscure. For instance, it is still unknown how closed extranuclear pleuromitosis, which resembles the dinomitosis of dinoflagellates, had originated. It could be that these two types of mitoses had originated independently of the other types.

The mitotic apparatus in various types of protozoan mitosis should be thoroughly studied and biochemical methods should also be used for this. This will probably show how the mechanism of the chromosome movement has evolved, i.e. we will learn about the role of the microtubules, their assembly, disassembly, and mutual sliding, and the role of contractile microfilaments at various stages of evolution.

Chromosomes. Only modest advances were made in studying the morphology of the chromosomes of eukaryotic protozoans in the past 15 years. This is probably because the main method used during this time was electron microscopy of ultrathin sections, which is hardly suitable for studying the three-dimensional organization of chromosomes. Progress has been made, however, in studying the structure of kinetochores, their links with the spindle microtubules, and, occasionally (in hypermastigids), their incorporation into the nuclear envelope (HOLLANDE and CARRUETTE-VALENTIN 1971, KUBAI 1973). Much new data have been obtained on the processes of DNA replication, mitotic cycles, and their control in various protozoans. It is necessary, however, to thoroughly study the chromosomes of hypermastigids with their unusually distinct spiral structure and their clearly defined matrix and also the chromosome aggregates observed in micronuclear mitosis of many ciliates.

Nuclei of Mesokaryotic Protozoans. On the whole, it has been ascertained that Dinoflagellida belong to a special mesokaryotic level of organization of the living cell (DODGE 1965). The organization of their unusual chromosomes has been thoroughly studied. In particular, it has been shown that they virtually have no histones and no subunits of the nucleosome type. The mechanism of dinomitosis has also been revealed. In spite of its seeming dissimilarity, this mechanism turned out to be common to the free-living and parasitic dinoflagellates and to be related to the closed extranuclear pleuromitosis of the eukaryotic protozoans (KUBAI and RIS 1969, OAKLEY and DODGE 1974, and others). The list of the prokaryotic and eukaryotic features of the organization of dinoflagellate nuclei has been specified and supplemented.

Unresolved questions include the possible polyteny of the dinoflagellate chromosomes, the way they function, and the way the complexly packed "genofibrils" segregate when chromosomes are replicated. It is unknown whether dinoflagellates have a typical mitotic cycle.

Meiosis. New data underpin the hypothesis that the forms of meiosis are more primitive in protozoans than in the higher eukaryotes; this applies particularly to one-step meiosis, which was originally recognized in L. R. CLEVELAND's and K. G. GRELL's work. The existence of one-division meiosis has been proved by cytophotometric methods (premeiotic DNA replication was absent) in sporozoans and by genetic methods in Dinoflagellida. However, the one-step nature of meiosis in foraminiferans has not been corroborated. Typical synaptonemal complexes have been detected in some protozoans.

The peculiarities of crossing-over and chiasma formation, particularly in hypermastigids, should be studied further by modern methods. The homology of some stages of meiosis in ciliates, especially the "crescent" stage, and the way the micronuclear chromosomes function in meiosis are still unclear. It would be interesting to learn from the evolutionary standpoint how one-division meiosis was established when the sexual process originated and how it was transformed into two-division meiosis.

The *primary nuclei of radiolarians* still await investigation by modern methods. Although the primary nucleus of Phaeodarea is undoubtedly polyploid and polygenomic, the organization and replication of its chromosomes are still disputed. As for Spumellarida, it is very likely that their primary nuclei become polyploid only before sporogenesis. Precise data on the mechanism of genome reproduction in the nucleus are still unavailable, however.

Nuclear Dualism of Foraminiferans and Karyorelictid Ciliates. This dualism in foraminiferans was discovered and thoroughly studied by K. G. GRELL in the 1950s and 1960s. It has been shown in the past decade that the somatic nuclei of some foraminiferans can accumulate much more DNA than the diploid amount. It remains to be ascertained, however, whether this is caused by polyploidy or gene amplification.

Nuclear differentiation has been studied much in the ciliates of the order Karyorelictida, mainly by I. B. RAIKOV and his colleagues. Comparative data have been obtained on the nuclear fine structure of several representatives of that order. It has been confirmed that the macronuclei of karyorelictids never divide and that DNA is not completely replicated in them, although minor DNA synthesis may occur. The ultrastructural pattern of the development of macronuclei from micronuclei has been studied in several species. Moreover, it has been shown that some DNA is destroyed and then DNA is resynthesized when macronuclei develop in at least some karyorelictids. It is important to learn which DNA fractions break down and are synthesized and to ascertain whether such DNA synthesis is gene amplification. Thus, data indicate that the nuclear differentiation in karyorelictids may be due not only to epigenetic phenomena (such as the activation or repression of definite genes) but also to some structural reorganization of the macronuclear genetic apparatus.

The factors that induce macronuclear differentiation, the causes of the inability of the macronuclei to undergo division, the nature and functions of protein inclusions in the macronuclei of karyorelictids, and many other aspects are still unknown. It is important to learn about the place of karyorelictids and homokaryotic ciliates in the phylogeny of Ciliophora. The nuclear apparatus of *Stephanopogon* should be studied by modern methods. Moreover, the somatic nuclei of foraminiferans, especially their ultrastructure, should be investigated further.

"Polyploid" Macronuclei of Ciliates. The polygenomic theory of the macronuclear structure has been radically revised since 1965. It has become clear that "polyploid" Ma of ciliates cannot belong to a single type. New morphological and biochemical methods have shown that the macronuclear genome breaks up into subchromosomal sections (chromomeres) and that drastic genetic diminution occurs when macronuclei develop in both Hypotrichida and *Chilodonella*. This was discovered by D. AMMERMANN and his colleagues, D. M. PRESCOTT and his colleagues, S. RADZIKOWSKI, and some others. These Ma can be regarded as amplified. However, the polygenomic (subnuclear) nature of the Ma of Colpodida and Nassulida has also been confirmed cytophotometrically. The amplification of the ribosomal gene with the formation of free molecules of rDNA has been discovered in various ciliates by J. G. GALL and his colleagues, J. ENGBERG and his coworkers, and other investigators. These molecules differ structurally in different ciliates. The Ma of various ciliates has been thoroughly studied with the electron microscope in the interphase and during division. It has been confirmed that intranuclear microtubules directly participate in macronuclear division. Moreover, it has been discovered that extranuclear (cytoplasmic) microtubules and contractile intranuclear microfilaments occasionally participate in that process.

Many questions remain unresolved as regards the macronucleus. It is still unknown how the macronuclear genome is organized in Hymenostomatia (*Paramecium*, *Tetrahymena*, etc.), as well as in Gymnostomatia, Vestibuliferia (*Bursaria*), Suctoria, Peritrichia, Spirotrichia, and other ciliates. The structure and division of the Ma of *Protocruzia* and other related forms are still obscure (probably they are very important in evolution). It is necessary to learn how genetic material is segregated during the division of the Ma of all types except the polygenomic one. The kinetic elements that participate in macronuclear division in Suctoria are unknown, and the significance of chromatin extrusion from Ma is obscure. The ways in which "polyploid" Ma evolve and the phylogenetic relationships between various types of their organization have not yet been ascertained.

Hence, the karyology of Protozoa has recently turned from a purely descriptive discipline into an experimental one and has become closely associated with cellular and molecular biology. Unicellular organisms (*Euglena*, *Chlamydomonas*, *Amoeba*, *Tetrahymena*) are increasingly used as convenient models of the eukaryotic cell in general, partly because molecular biology is passing over from prokaryotic organisms to eukaryotic cells. In using protozoans simply as eukaryotic cells, however, it is often forgotten that they are also organisms whose nuclear and especially cytoplasmic structures and functions are differentiated to a far greater extent than in the metazoan tissue cells.

It has now become clear that the true eukaryotic state of the cell first evolved in unicellular organisms. It is there that chromosomes originated and became longitudinally differentiated, histones appeared in them, the nucleosomal substructure of chromatin was produced, and repetitive sequences occurred in DNA. Most stages of the evolution of mitosis, meiosis, polyploidy, polyteny, gene amplification, and some other phenomena also occur in these organisms. Therefore, protozoans and other lower eukaryotes should become the preferential models for studying the evolution of these important biological phenomena.

References

Afzelius, B. A., 1955: The ultrastructure of the nuclear membrane of the sea urchin oocyte as studied with the electron microscope. Exper. Cell Res. **8**, 147–158.

– 1963: The nucleus of *Noctiluca scintillans*. Aspects of nucleocytoplasmic exchanges and the formation of nuclear membrane. J. Cell Biol. **19**, 229–238.

Aikawa, M., 1966: The fine structure of the erythrocytic stages of three avian malarial parasites. *Plasmodium fallax, P. lophurae*, and *P. cathemerium*. Amer. J. Trop. Med. Hyg. **15**, 449–471.

– 1971: *Plasmodium:* the fine structure of malarial parasites. Exper. Parasitol. **30**, 284–320.

– Beaudoin, R. L., 1968: Studies on nuclear division of a malarial parasite under pyrimethamine treatment. J. Cell Biol. **39**, 749–754.

– – 1969: Effects of chloroquine on the morphology of the erythrocytic stages of *Plasmodium gallinaceum*. Amer. J. Trop. Med. Hyg. **18**, 166–181.

– Cook, R. T., Sakoda, J. J., Sprinz, H., 1969: Fine structure of the erythrocytic stages of *Plasmodium knowlesi*. A comparison between intracellular and free forms. Z. Zellforsch. **100**, 271–284.

– Huff, C. G., Sprinz, H., 1967: Fine structure of the asexual stages of *Plasmodium elongatum*. J. Cell Biol. **34**, 229–249.

– – – 1968: Exoerythrocytic stages of *Plasmodium gallinaceum* in chick-embryo liver as observed electron microscopically. Amer. J. Trop. Med. Hyg. **17**, 156–169.

– – – 1969: Comparative fine structure study of the gametocytes of avian, reptilian and mammalian malarial parasites. J. Ultrastr. Res. **26**, 316–331.

– – Strome, C. P. A., 1970: Morphological study of microgametogenesis of *Leucocytozoon simondi*. J. Ultrastr. Res. **32**, 43–68.

– Jordan, H. B., 1968: Fine structure of a reptilian malarial parasite. J. Parasitol. **54**, 1023–1033.

– Sterling, C. R., 1974: Intracellular parasitic protozoa. New York-San Francisco-London: Academic Press.

Akinshina, G. T., Bykovsky, A. F., 1964: Submicroscopic structure of *Toxoplasma gondii*. Zool. Zhurnal (Moscow) **43**, 1391–1394 (in Russian; English summary).

Alexeieff, A., 1913: Systématisation de la mitose dite "primitive". Sur la question du centriole. Arch. Protistenk. **29**, 345–363.

Alfert, M., Goldstein, N., 1955: Cytochemical properties of nucleoproteins in *Tetrahymena pyriformis;* a difference in protein composition between macro- and micronuclei. J. Exper. Zool. **130**, 403–421.

Allen, J. R., Roberts, T. M., Loeblich, A. R., Klotz, L. C., 1975: Characterization of the DNA from the dinoflagellate *Crypthecodinium cohnii* and implications for nuclear organization. Cell **6**, 161–169.

– – Tuttle, R. C., Klotz, L. C., Loeblich, A. R., 1974: Dinoflagellate DNA organization and chromosome structure. J. Protozool. **21**, 426.

Allen, S., Gibson, I., 1972a: Genome amplification and gene expression in the ciliate macronucleus. Biochem. Genet. **6**, 293–313.
– – 1972b: Genetics of *Tetrahymena*. In: Biology of *Tetrahymena* (Elliott, A. M., ed.), pp. 307–373. Stroudsburg, Penn.: Dowden, Hutchinson, and Ross.

Allis, C. D., Bowen, J. K., Abraham, G. N., Glover, C. V. C., Gorovsky, M. A., 1980: Proteolytic processing of histone H3 in chromatin: a physiologically regulated event in *Tetrahymena* micronuclei. Cell **20**, 55–64.
– Glover, C. V. C., Gorovsky, M. A., 1979: Micronuclei of *Tetrahymena* contain two types of histone H3. Proc. Nat. Acad. Sci. USA **76**, 4857–4861.

Alonso, P., 1974: Differential nuclear coloration in ciliates. J. Protozool. **21**, 751–754.
– 1978: Les chromosomes polytènes des Ciliés. Année Biol. **17**, 369–386.
– Jareño, M. A., 1974: Incorporación de uridina-H^3 en el esbozo macronuclear de *Stylonychia mytilus*. Microbiol. Esp. **27**, 199–211.
– Pérez-Silva, J., 1965a: Conjugation in *Stylonychia muscorum* Kahl. J. Protozool. **12**, 253–258.
– – 1965b: Giant chromosomes in Protozoa. Nature **205**, 313–314.
– – 1966: Cromosomas politénicos en ciliados hipotricos. Bol. Real Soc. Esp. Hist. Nat. (Biol.) **64**, 171–176.

Alov, I. A., Aspiz, M. E., Starosvetskaya, N. A., 1977: Evolution of the mitotic apparatus and of the chromosome distribution processes in a dividing cell. Zhurnal Obsch. Biol. **38**, 204–217 (in Russian; English summary).

Ammermann, D., 1965: Cytologische und genetische Untersuchungen an dem Ciliaten *Stylonychia mytilus* Ehrenberg. Arch. Protistenk. **108**, 109–152.
– 1968a: Die Kernverhältnisse des Ciliaten *Protocrucia depressa* n. sp. Arch. Protistenk. **110**, 434–438.
– 1968b: Synthese und Abbau der Nucleinsäuren während der Entwicklung des Makronucleus von *Stylonychia mytilus* (Protozoa, Ciliata). Chromosoma **25**, 107–120.
– 1969: Release of DNA breakdown products into the culture medium of *Stylonychia mytilus* exconjugants (Protozoa, Ciliata) during the destruction of the polytene chromosomes. J. Cell Biol. **40**, 576–577.
– 1970: The micronucleus of the ciliate *Stylonychia mytilus*: its nuclei acid synthesis and its function. Exper. Cell. Res. **61**, 6–12.
– 1971a: Morphology and development of the macronuclei of the ciliates *Stylonychia mytilus* and *Euplotes aediculatus*. Chromosoma **33**, 209–238.
– 1971b: Micronuclear DNA synthesis in *Stylonychia mytilus*. Naturwissenschaften **58**, 149.
– 1973: Cell development and differentiation in the ciliate *Stylonychia mytilus*. In: Cell cycle in development and differentiation, pp. 51–60. Cambridge Univ. Press.
– Lipps, H. J., Steinbrück, G., 1978: The nuclei of the ciliate *Stylonychia mytilus*. Cytobiologie **18**, 201–202.
– Steinbrück, G., Berger, L. von, Hennig, W., 1974: The development of the macronucleus in the ciliated protozoan *Stylonychia mytilus*. Chromosoma **45**, 401–429.

Andersen, H. A., 1972a: Requirements for DNA replication preceding cell division in *Tetrahymena pyriformis*. Exper. Cell Res. **75**, 89–94.
– 1972b: Induced elimination of DNA from macronucleus of *Tetrahymena pyriformis*. Exper. Cell Res. **74**, 610–613.
– 1974: Replication of macronuclear DNA in the cytoplasm of *Tetrahymena pyriformis*. J. Cell Sci. **14**, 289–300.
– 1977: Replication and functions of macronuclear DNA in synchronously growing populations in *Tetrahymena pyriformis*. Carlsberg Res. Commun. **42**, 225–248.

ANDERSEN, H. A., ENGBERG, J., 1975: Timing of the ribosomal gene replication in *Tetrahymena pyriformis*. Exper. Cell Res. **92**, 159–163.

– ZEUTHEN, E., 1971: DNA replication sequence in *Tetrahymena* is not repeated from generation to generation. Exper. Cell Res. **68**, 309–314.

ANDERSON, E., 1955: The electron microscopy of *Trichomonas muris*. J. Protozool. **2**, 114–124.

– BEAMS, H. W., 1959: The cytology of *Tritrichomonas* as revealed by the electron microscope. J. Morphol. **104**, 205–219.

ANDERSON, O. R., 1976a: Ultrastructure of a colonial radiolarian *Collozoum inerme* and a cytochemical determination of the role of its zooxanthellae. Tissue and Cell **8**, 195–208.

– 1976b: Fine structure of a collodarian radiolarian (*Sphaerozoum punctatum* MÜLLER 1858) and cytoplasmic changes during reproduction. Marine Micropaleontol. **1**, 287–297.

ANDREADIS, T. G., HALL, D. W., 1979: Development, ultrastructure, and mode of transmission of *Amblyospora sp.* (Microspora) in the mosquito. J. Protozool. **26**, 444–452.

ANDREJEVA, L. F., ANDREJEVA, T. F., MERKULOVA, N. A., BORCHSENIUS, S. N., 1979: Size of replicons and rate of replication of the macronuclear DNA of the ciliate *Tetrahymena pyriformis*. Tsitologiya (Leningrad) **21**, 318–326 (in Russian; English summary).

ANDRESEN, N., CHAPMAN-ANDRESEN, C., NILSSON, J. R., 1968: The fine structure of *Pelomyxa palustris*. C. r. Trav. Lab. Carlsberg **36**, 285–317.

ANGELL, R. W., 1976: Observations on *Trichosphaerium platyxyrum* sp. n. J. Protozool. **23**, 357–364.

ANTUKHAEV, I. K., 1977: A study of the ultrafine organization of the goose coccidian, *Eimeria kotlani*. I. Intranuclear development of macrogametes. Tsitologiya (Leningrad) **19**, 288–292 (in Russian; English summary).

ARNDT, A., 1924: Rhizopodenstudien. Arch. Protistenk. **49**, 1–83.

ARNOLD, Z. M., 1955: Life history and cytology of the foraminiferan *Allogromia laticollaris*. Univ. Calif. Publs Zool. **61**, 167–252.

– 1972: Observations on the biology of the protozoan *Gromia oviformis* Dujardin. Univ. Calif. Publs Zool. **100**, 1–168.

ASTAKHOVA, I. N., DORONIN, YU. K., TROFIMOV, S. S., TUSHMALOVA, N. A., 1975: Electron microscopic study of the ciliate *Spirostomum ambiguum*. Vestnik Moskovsk. Univ. **30**, No. 3, 37–42 (in Russian; English summary).

AVAKYAN, A. A., 1976: Atlas of the anatomy of Protozoa pathogenic for man and animals. Moscow: Medicina Press (in Russian; English summary).

– MANUKYAN, I. A., 1967: Electron microscopic study of the intracellular stage of *Leishmania tropica*. Biol. Zhurnal Armenii **20**, No. 6, 40–43 (in Russian).

– SARKISYAN, S. A., 1965: Submicroscopic organization of *Entamoeba histolytica*. Izvest. Akad. Nauk Arm. SSR, Biol. Nauki **18**, 12, 73–76 (in Russian).

BABILLOT, C., 1970a: Étude des effets de l'actinomycine D sur le noyau du dinoflagellé *Amphidinium carteri*. J. Microscopie **9**, 485–502.

– 1970b: Étude de l'incorporation d'uridine-H^3 dans le noyau chez *Amphidinium carteri*, Dinoflagellé. C. r. Acad. Sci. Paris **D 271**, 828–831.

BAJER, A. S., MOLÈ-BAJER, J., 1972: Spindle dynamics and chromosome movements. Internat. Rev. Cytol., Suppl. 3, 1–271.

BAKER, J. R., BIRD, R. G., 1968: *Trypanosoma avium:* fine structure of all developmental stages. J. Protozool. **15**, 298–308.

BAKER, W., 1926: Studies in the life history of *Euglena*. I. *Euglena agilis* Carter. Biol. Bull. **51**, 321–362.

BALBIANI, E. G., 1890: Étude sur le Loxode. Ann. Micrograph. **2**, 401–431.

BAND, R. N., MOHRLOK, S., 1973: The cell cycle and induced amitosis in *Acanthamoeba*. J. Protozool. **20**, 654–657.

BARDELE, C. F., 1966: Elektronenmikroskopische Untersuchung an dem Sporozoon *Eucoccidium dinophili* Grell. Z. Zellforsch. **74**, 559–595.

– 1968: *Acineta tuberosa*. I. Der Feinbau des adulten Suktors. Arch. Protistenk. **110**, 403–421.

– 1969: *Acineta tuberosa*. II. Die Verteilung der Mikrotubuli im Makronucleus während der ungeschlechtlichen Fortpflanzung. Z. Zellforsch. **93**, 93–104.

– 1975: The fine structure of the centrohelidian heliozoan *Heterophrys marina*. Cell and Tissue Res. **161**, 85–102.

BATISSE, A., 1972: Premières observations sur l'ultrastructure de *Trematosoma bocqueti* (GUILCHER) Batisse (Ciliata, Suctorida). Protistologica **8**, 477–495.

BAUDOIN, J., ORMIÈRES, R., 1973: Sur quelques particularités de l'ultrastructure de *Didymophyes chaudefouri* ORMIÈRES, Eugregarine, Didymophyidae. C. r. Acad. Sci. Paris **D 277**, 73–75.

BAYER, M., DENNIG, H. K., 1961: Elektronenmikroskopische Untersuchungen an *Babesia canis*. Z. Tropenmed. Parasitol. **12**, 28–35.

BEAM, C. A., HIMES, M. H., 1974: Evidence for sexual fusion and recombination in the dinoflagellate *Crypthecodinium (Gyrodinium) cohnii*. Nature **250**, 435–436.

– – HIMMELFARB, J., LINK, C., SHAW, K., 1977: Genetic evidence of unusual meiosis in the dinoflagellate *Crypthecodinium cohnii*. Genetics **87**, 19–32.

BEAMS, H. W., SEKHON, S. S., 1969: Further studies on the fine structure of *Lophomonas blattarum* with special reference to the so-called calyx, axial filament and parabasal body. J. Ultrastr. Res. **26**, 296–315.

– TAHMISIAN, T. N., ANDERSON, E., WRIGHT, B., 1961: Studies on the fine structure of *Lophomonas blattarum* with special reference to the so-called parabasal apparatus. J. Ultrastr. Res. **5**, 166–183.

– – DEVINE, R., ANDERSON, E., 1957: Ultrastructure of the nuclear membrane of a gregarine parasitic in grasshoppers. Exper. Cell Res. **13**, 200–204.

– – – – 1959: The fine structure of the nuclear envelope of *Endamoeba blattae*. Exper. Cell Res. **18**, 366–369.

BEERS, C. D., 1946: History of the nuclei of *Tillina magna* during division and encystment. J. Morphol. **78**, 181–200.

– 1964: Cytochemical observations on the interphase nucleus of the parasitic ameba *Hydramoeba hydroxena* (ENTZ). J. Parasitol. **50**, 630–633.

BĚLAŘ, K., 1922: Untersuchungen an *Actinophrys sol* Ehrenberg. I. Die Morphologie des Formwechsels. Arch. Protistenk. **46**, 1–96.

– 1926a: Der Formwechsel der Protistenkerne. Eine vergleichend-morphologische Studie. In: Ergebnisse und Fortschritte der Zoologie, pp. 235–654. Jena: G. Fischer-Verlag.

– 1926b: Zur Cytologie von *Aggregata eberthi*. Arch. Protistenk. **53**, 312–325.

BELCHER, J. H., 1975: The fine structure of the loricate colourless flagellate *Bicoeca planctonica* Kisselew. Arch. Protistenk. **117**, 78–84.

– SWALE, E. M. F., 1972: The morphology and fine structure of the colourless colonial flagellate *Anthophysa vegetans* (O. F. MÜLLER) Stein. Brit. Phycol. J. **7**, 335–346.

– – 1976: *Spumella elongata* (Stokes) nov. comb., a colourless flagellate from soil. Arch. Protistenk. **118**, 215–220.

BELNAT, P., RIOU, G., 1979: The organization of chromatin from *Trypanosoma cruzi*. J. Protozool. **26**, No. 3 (part 1), 52A.

BELOZERSKY, A. N., 1947: On the nucleoproteins and polynucleotides of certain bacteria. Cold Spring Harbor Sympos. Quant. Biol. **12**, 1–14.

BENWITZ, G., GRELL, K. G., 1971a: Ultrastruktur mariner Amöben. II. *Stereomyxa ramosa*. Arch. Protistenk. **113**, 51–67.

– – 1971b: Ultrastruktur mariner Amöben. III. *Stereomyxa angulosa*. Arch. Protistenk. **113**, 68–79.

BERECH, J., WAGTENDONK, W. J. VAN, 1962: An autoradiographic study of the macronuclear changes occurring in *Paramecium aurelia* during autogamy. Exper. Cell Res. **26**, 360–372.

BERGER, J. D., 1971: Kinetics of incorporation of DNA precursors from ingested bacteria into macronuclear DNA of *Paramecium aurelia*. J. Protozool. **18**, 419–429.

– 1973a: Nuclear differentiation and nucleic acid synthesis in well-fed exconjugants of *Paramecium aurelia*. Chromosoma **42**, 247–268.

– 1973b: Selective inhibition of DNA synthesis in macronuclear fragments in *Paramecium aurelia* exconjugants and its reversal during macronuclear regeneration. Chromosoma **44**, 33–48.

– 1973c: Regulation of DNA content and length of the cell cycle in *Paramecium aurelia*. In: Progress in protozoology (Abstr. IV Internat. Congr. Protozool.), p. 39. Clermont-Ferrand: Université de Clermont.

– 1974: Selective autolysis of the nuclei as a source of DNA precursors in *Paramecium aurelia* exconjugants. J. Protozool. **21**, 145–151.

– 1978: Downward regulation of macronuclear DNA content in *Paramecium tetraurelia*. Effects of excess DNA on the subsequent DNA and protein content and the cell growth rate. Exp. Cell Res. **114**, 253–262.

– 1979: Regulation of macronuclear DNA content in *Paramecium tetraurelia*. J. Protozool. **26**, 18–28.

– KIMBALL, R. F., 1964: Specific incorporation of precursors into DNA by feeding labelled bacteria to *Paramecium aurelia*. J. Protozool. **11**, 534–537.

– SCHMIDT, H. J., 1978: Regulation of macronuclear DNA content in *Paramecium tetraurelia*. J. Cell Biol. **76**, 116–126.

BERNHARD, W., 1969: A new staining procedure for electron microscopical cytology. J. Ultrastr. Res. **27**, 250–265.

BERTAUX, O., MOYNE, G., LAFARGE-FRAYSSINET, C., VALENCIA, R., 1978: The nucleus of *Euglena*. II. Ultrastructural modifications of the nucleus of B_{12}-deprived *Euglena gracilis* Z. J. Ultrastr. Res. **62**, 251–269.

BERTHOLD, W.-U., 1977: Synaptinemale Komplexe bei *Patellina corrugata* Williamson (Protozoa, Foraminifera). Cytobiologie **14**, 253–258.

BESSE, C., SCHRÉVEL, J., 1973: Étude ultrastructurale de la gamétogenèse de la Grégarine *Lecudina tuzetae*. In: Progress in protozoology (Abstr. IV Internat. Congr. Protozool.), p. 40. Clermont-Ferrand: Université de Clermont.

BEYER, T. V., 1977: Electron microscope study of *Karyolysus* sp. (Sporozoa: Adeleida: Haemogregarinidae) and of changes induced in the infected host cell. Protistologica **13**, 57–66.

– SHIBALOVA, T. A., KOSTENKO, L. A., 1978: Cytology of the Coccidia. 188 pp. Leningrad: Nauka Press (in Russian).

BHOWMICK, D. K., 1967: Electron microscopy of *Trichamoeba villosa* and amoeboid movement. Exper. Cell Res. **45**, 570–589.

BIANCHI, L., RONDANELLI, E. G., CAROSI, G., GERNA, G., 1969: Endonuclear mitotic spindle in the leptomonad of *Leishmania tropica*. J. Parasitol. **55**, 1091–1092.

BLACKBURN, E. H., GALL, J. G., 1978: A tandemly repeated sequence at the termini of the extrachromosomal ribosomal RNA genes in *Tetrahymena*. J. Molec. Biol. **120**, 33–53.

Blanc, J., 1963: Troubles de la teneur en ADN dans le macronucléus de certains mating types de paramécies ayant perdu leur réactivité sexuelle. Arch. Zool. Expér. Génér. **102**, 173–176.
– 1965: Étude cytophotométrique des périodes de duplication de l'acide désoxyribonucléique dans l'appareil nucléaire de *Paramecium caudatum*. Protistologica **1**, No. 2, 11–15.
Blanchard-Babillot, C., 1972: Recherches sur le mode de division du Péridinien *Glenodinium foliaceum*. C. r. Acad. Sci. Paris **D 275**, 2635–2638.
Bloodgood, R. A., Miller, K. R., Fitzharris, T. P., McIntosh, J. R., 1974: The ultrastructure of *Pyrsonympha* and its associated microorganisms. J. Morphol. **143**, 77–105.
Blum, J. J., Buetow, D. E., 1963: Biochemical changes during acetate deprivation and repletion in *Euglena*. Exper. Cell Res. **29**, 407–421.
– Padilla, G. M., 1962: Studies on synchronized cells: the time course of DNA, RNA and protein synthesis in *Astasia longa*. Exper. Cell Res. **28**, 512–523.
– Sommer, J. R., Kahn, V., 1965: Some biochemical, cytological and morphogenetic comparisons between *Astasia longa* and a bleached *Euglena gracilis*. J. Protozool. **12**, 202–209.
Bobyleva, N. N., Kudrjavtsev, B. N., Raikov, I. B., 1980: Changes of the DNA content of differentiating and adult macronuclei of the ciliate *Loxodes magnus* (Karyorelictida). J. Cell Sci. **44**, 375–394.
Bogdanowicz, A., 1930a: Ergebnisse mit der Nuclealfärbung bei einigen Rhizopoden. Z. Zellforsch. **10**, 471–480.
– 1930b: Über die Konjugation von *Loxodes striatus* (Engelm.) Penard und *Loxodes rostrum* (O. F. M.) Ehrenb. Zool. Anz. **87**, 209–222.
Bohatier, J., 1972: Étude cytologique de la régénération de la région buccale chez le Cilié *Lacrymaria olor* (O. F. M., 1786). Protistologica **8**, 439–459.
– 1975: Structure fine et caractérisation cytochimique des composants macronucléaires chez le Cilié *Lacrymaria olor*. Protistologica **11**, 251–262.
– 1977a: A propos des effets de l'actinomycine D, au niveau des ultrastructures, chez le Cilié *Dileptus anser*, pendant la régénération et la division. Protistologica **13**, 77–88.
– 1977b: Incorporation d'actinomycine D-H^3 et d'uridine-H^3 pendant la régénération et la division du Cilié *Dileptus anser*. Étude autoradiographique en microscopie électronique. Protistologica **13**, 89–99.
– 1978: Morphologie ultrastructurale de *Condylostoma*, Cilié Polyhymenophora. Protistologica **14**, 433–450.
– 1981: Synthèses macromoléculaires au cours de la morphogenèse de régénération chez le Cilié *Condylostoma magnum*: Action de différents inhibiteurs. I. Synthèses d'acides nucléiques. Arch. Protistenk. **124**, 144–172.
– Iftode, F., Didier, P., Fryd-Versavel, G., 1978: Sur l'ultrastructure des genres *Spathidium* et *Bryophyllum*, Ciliés Kinetofragmophora (De Puytorac *et al.* 1974). Protistologica **14**, 189–200.
Bonifaz, E. L., Plaut, W., 1974: The synthesis and distribution of DNA in the macronucleus of *Euplotes eurystomus*. Chromosoma **46**, 261–278.
Bonnet, L., Brabet, J., Comoy, N., Guitard, J., 1981: Nouvelles données sur le thécamoebien Filosia *Amphitrema flavum* (Archer, 1877) Penard, 1902. Protistologica **17**, 225–233.
Borchsenius, O. N., 1975: On the activity of the generative nucleus in ciliates. Vestnik Leningrad. Univ. No. 21, 7–18 (in Russian; English summary).
– Ossipov, D. V., 1971: Polymorphism of the micronuclei of *Paramecium caudatum*. III. Cytofluorimetric study of the DNA content. Tsitologiya (Leningrad) **13**, 1041–1043 (in Russian; English summary).

BORCHSENIUS, O. N., SKOBLO, I. I., OSSIPOV, D. V., 1968: Polymorphism of the micronuclei of *Paramecium caudatum*. I. Morphological types of the interphasic nuclei. Tsitologiya (Leningrad) **10**, 227–235 (in Russian: English summary).

BORCHSENIUS, S. N., BELOZERSKAYA, N. A., MERKULOVA, N. A., IRLINA, I. S., VOROB'EV, V. I., 1976: Peculiarities of replication of *Tetrahymena* DNA in relation to hypotheses of macronuclear structure. In: Karyologiya i genetika prostejshikh (Series "Protozoology", No. 1), pp. 91–100. Leningrad: Nauka Press (in Russian; English summary).

– – – – – 1977: Possible repeated replication of a fraction of nuclear DNA in *Tetrahymena* during a single S phase. Molek. Biol. (Moscow) **11**, 171–180 (in Russian; English summary).

– – – WOLFSON, V. G., VOROB'EV, V. I., 1978: Genome structure of *Tetrahymena pyriformis*. Chromosoma **69**, 275–289.

– BONVEN, B., LEER, J. C., WESTERGAARD, O., 1981a: Nuclease sensitive regions on the extrachromosomal r-chromatin from *Tetrahymena pyriformis*. Europ. J. Biochem. **117**, 245–250.

– MERKULOVA, N. A., 1980: Discrete length of palindromic segments from *Tetrahymena* DNA. FEBS Letters **119**, 90–92.

– – ROMANOV, G. A., VANYUSHIN, B. F., 1981b: Palindromic and repetitive nucleotide sequences at the DNA synthesis initiation sites in the ciliate *Tetrahymena*. Molek. Biol. (Moscow) **15**, 816–823.

– SERGEJEVA, G. I., 1979: Characterization of the DNA of vegetative individuals of the ciliate *Bursaria truncatella*. Tsitologiya (Leningrad) **21**, 327–333 (in Russian; English summary).

BORGERT, A., 1900: Untersuchungen über die Fortpflanzung der tripyleen Radiolarien, speziell von *Aulacantha scolymantha* H. Teil I. Zool. Jahrb. (Abt. Anat. und Ontog.) **14**, 203–276.

– 1909: Untersuchungen über die Fortpflanzung der tripyleen Radiolarien, speziell von *Aulacantha scolymantha* H. Teil II. Arch. Protistenk. **14**, 134–261.

BORKHARDT, B., NIELSEN, O. F., 1981: An electron microscopic analysis of transcription of nucleolar chromatin isolated from *Tetrahymena pyriformis*. Chromosoma **84**, 131–143.

BORROR, A. C., 1965: New and little known tidal marsh ciliates. Trans. Amer. Microscop. Soc. **84**, 550–565.

BOSTOCK, C. J., PRESCOTT, D. M., 1972: Evidence of gene diminution during the formation of the macronucleus in the protozoan, *Stylonychia*. Proc. Nat. Acad. Sci. USA **69**, 139–142.

BOUCK, G. B., 1971: The structure, origin, isolation and composition of the tubular mastigonemes of the *Ochromonas* flagellum. J. Cell Biol. **50**, 362–384.

– BROWN, D. L., 1973: Microtubule biogenesis and cell shape in *Ochromonas*. I. The distribution of cytoplasmic and mitotic microtubules. J. Cell Biol. **56**, 340–359.

BOULIGAND, Y., PUISEUX-DAO, S., SOYER, M.-O., 1968a: Liaisons morphologiquement définies entre chromosomes et membrane nucléaire chez certains Péridiniens. C. r. Acad. Sci. Paris **D 266**, 1287–1289.

– SOYER, M.-O., PUISEUX-DAO, S., 1968b: La structure fibrillaire et l'orientation des chromosomes chez les Dinoflagellés. Chromosoma **24**, 251–287.

BOUTEILLE, A., LAVAL, M., DUPUY-COIN, A. M., 1974: Localization of nuclear functions as revealed by ultrastructural autoradiography and cytochemistry. In: The cell nucleus (BUSCH, H., ed.), Vol. 1, pp. 3–71. New York-London: Academic Press.

BOWERS, B., KORN, E. D., 1968: The fine structure of *Acanthamoeba castellanii*. I. The trophozoite. J. Cell Biol. **39**, 95–111.

– – 1969: The fine structure of *Acanthamoeba castellanii* (Neff strain). J. Cell Biol. **41**, 786–805.

BRADBURY, P. C., 1966: The fine structure of the mature tomite of *Hyalophysa chattoni*. J. Protozool. **13**, 591–607.
– 1974: The fine structure of the phoront of the apostomatous ciliate, *Hyalophysa chattoni*. J. Protozool. **21**, 112–120.
– GALLUCCI, B. B., 1971: The fine structure of differentiating merozoites of *Haemoproteus columbae* Kruse. J. Protozool. **18**, 679–686.
– – 1972: Observations on the fine structure of the schizonts of *Haemoproteus columbae* Kruse. J. Protozool. **19**, 43–49.
– OLIVE, L. S., 1980: Fine structure of the feeding stage of a sorogenic ciliate, *Sorogena stoianovitchae* gen. n., sp. n. J. Protozool. **27**, 267–277.
– ROBERTS, J. F., 1970: Early stages in the differentiation of gametocytes of *Haemoproteus columbae* Kruse. J. Protozool. **17**, 405–414.
– TRAGER, W., 1968a: The fine structure of the mature gametes of *Haemoproteus columbae* Kruse. J. Protozool. **15**, 89–102.
– – 1968b: The fine structure of microgametogenesis in *Haemoproteus columbae* Kruse. J. Protozool. **15**, 700–712.

BRADLEY, D. M., GOLDIN, H. H., CLAYBROOK, J. R., 1974: Histone analysis in *Volvox*. FEBS Letters **41**, 219–222.

BRANDT, K., 1902: Beiträge zur Kenntnis der Colliden. I. Über Bau und Fortpflanzung von Thalassophysiden. II. Die Einteilung der Colliden und ihre Stellung im System der Radiolarien. Arch. Protistenk. **1**, 59–88.
– 1905: Beiträge zur Kenntnis der Colliden. III. Über den Bau, die multiple Kernteilung und den Generationswechsel von Thalassicollen. Arch. Protistenk. **6**, 245–271.

BRANDT, P. W., PAPPAS, G. D., 1963: The fine structure of helices in *Amoeba* nuclei. In: Progress in protozoology (Proc. I Internat. Congr. Protozool.), pp. 408–409. Prague: Publ. House ČSAV.

BRÉ, M.-H., DELPECH, S., CHAMPAGNE, M., MAZEN, A., LEFORT-TRAN, M., 1980: Analyse des histones et de l'ADN nucléosomal de l'Euglène normale et carencée en vitamine B_{12}. C. r. Acad. Sci. Paris **D 290**, 93–96.

BRITTEN, R. J., KOHNE, D. E., 1968: Repeated sequences in DNA. Science **161**, 529–540.

BROOKER, B. E., 1971: Fine structure of *Bodo saltans* and *Bodo caudatus* (Zoomastigophora: Protozoa) and their affinities with the Trypanosomatidae. Bull. Brit. Museum Natur. Hist., Zool. **22**, 89–102.
– OGDEN, C. G., 1972: Encystment of *Bodo caudatus*. Protoplasma **74**, 397–409.

BROWN, D. L., MASSALSKI, A., PATENAUDE, R., 1976: Organization of the flagellar apparatus and associated cytoplasmic microtubules in the quadriflagellate alga *Polytomella agilis*. J. Cell Biol. **69**, 106–125.

BROWN, V. E., 1930: Cytology and binary fission of *Peranema*. Quart. J. Microscop. Sci. **73**, 403–419.

BRUGEROLLE, G., 1971: Ultrastructure du genre *Trichomitus* Swezy 1915, Zooflagellata, Trichomonadida. Protistologica **7**, 171–176.
– 1973: Étude ultrastructurale du trophozoite et du kyste chez le genre *Chilomastix* ALEXEIEFF 1910 (Zoomastigophorea, Retortamonadida GRASSÉ 1952). J. Protozool. **20**, 574–585.
– 1974: Contribution à l'étude cytologique et phylétique des Diplozoaires (Zoomastigophorea, Diplozoa DANGEARD 1910). III. Étude ultrastructurale du genre *Hexamita* (DUJARDIN 1836). Protistologica **10**, 83–90.
– 1975a: Étude ultrastructurale du genre *Enteromonas* da Fonseca (Zoomastigophorea) et revision de l'ordre des Diplomonadida Wenyon. J. Protozool. **22**, 468–475.

Brugerolle, G., 1975b: Étude de la cryptopleuromitose et de la morphogenèse de division chez *Trichomonas vaginalis* et chez plusieurs genres de Trichomonadines primitives. Protistologica **11**, 457–468.

– 1977: Ultrastructure du genre *Retortamonas* Grassi 1879 (Zoomastigophorea, Retortamonadida Wenrich 1931). Protistologica **13**, 233–240.

– Joyon, L., 1973: Sur la structure et la position systématique du genre *Monocercomonoides* (Travis 1932). Protistologica **9**, 71–80.

– – 1975: Étude cytologique ultrastructurale des genres *Proteromonas* et *Karotomorpha* (Zoomastigophorea, Proteromonadida Grassé 1952). Protistologica **11**, 531–546.

– – Oktem, N., 1973: Contribution à l'étude cytologique et phylétique des Diplozoaires (Zoomastigophorea, Diplozoa Dangeard 1910). I. Étude ultrastructurale du genre *Trepomonas* (Dujardin). Protistologica **9**, 339–348.

– – – 1974: Contribution à l'étude cytologique et phylétique des Diplozoaires (Zoomastigophorea, Diplozoa Dangeard 1910). IV. Étude ultrastructurale du genre *Octomitus* (Prowazek 1904). Protistologica **10**, 457–463.

– Lom, J., Nohýnková, E., Joyon, L., 1979: Comparaison et évolution des structures cellulaires chez plusieurs espèces de Bodonidés et Cryptobiidés appartenant aux genres *Bodo, Cryptobia* et *Trypanoplasma* (Kinetoplastida, Mastigophora). Protistologica **15**, 197–221.

– Mignot, J.-P., 1979: Observations sur le cycle, l'ultrastructure et la position systématique de *Spiromonas perforans* (*Bodo perforans* Hollande 1938), Flagellé parasite de *Chilomonas paramaecium*: ses relations avec les Dinoflagellés et Sporozoaires. Protistologica **15**, 183–196.

Burt, R. L., Kidder, G. W., Claff, C. L., 1941: Nuclear reorganization in the family Colpodidae. J. Morphol. **69**, 537–561.

Burzell, L. A., 1973: Observations on the proboscis–cytopharynx complex and flagella of *Rhynchomonas metabolita* Pshenin 1964 (Zoomastigophorea: Bodonidae). J. Protozool. **20**, 385–393.

– 1975: Fine structure of *Bodo curvifilus* Griessman (Kinetoplastida: Bodonidae). J. Protozool. **22**, 35–39.

Busch, H., Smetana, K., 1970: The nucleolus. New York-London: Academic Press.

Bütschli, O., 1876: Studien über die ersten Entwicklungsvorgänge der Eizelle, die Zellteilung und die Conjugation der Infusorien. Abhandl. Senckenberg. Naturforsch. Ges. **10**, 213–464.

Büttner, D. W., 1966: Über die Feinstruktur der erythrozytären Formen von *Theileria mutans*. Z. Tropenmed. Parasitol. **17**, 397–406.

– 1967a: Die Feinstruktur der Merozoiten von *Theileria parva*. Z. Tropenmed. Parasitol. **18**, 224–244.

– 1967b: Elektronenmikroskopische Studien der Vermehrung von *Theileria parva* im Rind. Z. Tropenmed. Parasitol. **18**, 245–268.

– 1968: Vergleichende Untersuchungen der Feinstruktur von *Babesia gibsoni* und *Babesia canis*. Z. Tropenmed. Parasitol. **19**, 330–342.

Cachon, J., 1953: Morphologie et cycle évolutif de *Diplomorpha paradoxa*, Péridinien parasite des Siphonophores. Bull. Soc. Zool. France **78**, 408–414.

– 1964: Contribution à l'étude des Péridiniens parasites. Cytologie, cycles évolutifs. Ann. Sci. Natur., Zool. sér. 12, **6**, 1–158.

– Cachon, M., 1966: Ultrastructure d'un Péridinien parasite d'Appendiculaires, *Neresheimeria catenata* (Neresheimer). Protistologica **2**, No. 4, 17–25.

– – 1967: Contribution à l'étude des Noctilucidae Saville-Kent. I. Les Kofoidininae Cachon J. et M. Evolution morphologique et systématique. Protistologica **3**, 427–444.

Cachon, J., Cachon, M., 1969a: Contribution à l'etude des Noctilucidae Saville-Kent. Evolution morphologique, cytologie, systématique. II. Les Leptodiscinae Cachon J. et M. Protistologica **5**, 11–33.

– – 1969b: Les processus sporogénétiques du Radiolaire *Sticholonche zanclea* Hertwig. Arch. Protistenk. **111**, 87–99.

– – 1970: Ultrastructure des Amoebophryidae (Péridiniens Duboscquodinida). II. Systèmes atractophoriens et microtubulaires; leur intervention dans la mitose. Protistologica **6**, 57–70.

– – 1972a: Le système axopodial des Radiolaires sphaeroïdés. I. Les Centroaxoplastidiés. Arch. Protistenk. **114**, 51–64.

– – 1972b: Le système axopodial des Radiolaires sphaeroïdés. II. Les Periaxoplastidiés. III. Les Cryptoaxoplastidiés (Anaxoplastidiés). IV. Les fusules et le système rhéoplasmique. Arch. Protistenk. **114**, 291–307.

– – 1974: Comparaison de la mitose des Péridiniens libres et parasites à propos de celle des *Oodinium*. C. r. Acad. Sci. Paris **D 278**, 1735–1737.

– – 1977: Observations on the mitosis and on the chromosome evolution during the life cycle of *Oodinium*, a parasitic dinoflagellate. Chromosoma **60**, 237–251.

– – 1978: *Sticholonche zanclea* Hertwig: a reinterpretation of its phylogenetic position based upon new observations on its ultrastructure. Arch. Protistenk. **120**, 148–168.

– – 1979: Singular kinetochore structure in a parasitic dinoflagellate. Arch. Protistenk. **122**, 267–274.

– – Bouquaheux, F., 1969: *Myxodinium pipiens* gen. nov. sp. nov., Péridinien parasite d'*Halosphaera*. Phycologia **8**, 157–164.

– – Lécher, P., 1973: Nouvelle interprétation de la division nucléaire des Phaeodariés (Actìnopodes). C. r. Acad. Sci. Paris **D 276**, 3311–3314.

– – Salvano, P., 1979: The nuclear division of *Oxyrrhis marina*: an example of the role played by the nuclear envelope in chromosome segregation. Arch. Protistenk. **122**, 43–54.

– Cachon-Enjumet, M., 1963: Ultrastructure de la membrane nucléaire du Foraminifère *Pontomyxa flava* (Topsent). J. Microscopie **2**, 103–106.

– – 1964a: Cytologie et cycle évolutif de *Cinetidomyxa chattoni* nov. sp., Héliozoaire Protéomyxée. In: Groupement des protistologues de langue française, pp. 47–60. Paris: Masson et Cie.

– – 1964b: *Leptospathium navicula* nov. gen. nov. sp. et *Leptophyllus dasypus* nov. gen. nov. sp., Péridiniens Noctilucidae (Hertwig) du plancton néritique de Villefranche-sur-Mer. Bull. Inst. Océanogr. Monaco **62**, No. 1292, 1–12.

– – 1964c: Cycle évolutif et cytologie de *Neresheimeria catenata* Neresheimer, Péridinien parasite d'Appendiculaires. Rapports de l'hôte et du parasite. Ann. Sci. Natur., Zool. sér. 12, **6**, 779–800.

– – 1965a: *Atlanticellodinium tregouboffi* nov. gen. nov. sp., Péridinien Blastuloidae Neresheimer, parasite de *Planktonetta atlantica* Borgert, Phaeodarié Atlanticellide. Cytologie, cycle biologique, évolution nucléaire au cours de la sporogenèse. Arch. Zool. Expér. génér. **105**, 369–379.

– – 1965b: Étude cytologique et caryologique d'un Phaeodarié bathypélagique *Planktonetta atlantica* Borgert. Bull. Inst. Océanogr. Monaco **64**, No. 1330, 1–23.

Cachon-Enjumet, M., 1961a: Contribution à l'étude des Radiolaires Phaeodariés. Arch. Zool. Expér. Génér. **100**, 151–237.

– 1961b: Le mécanisme caryocinétique chez les Phaeodariés. Bull. Inst. Océanogr. Monaco **58**, No. 1214, 1–8.

– 1964: L'évolution sporogénétique des Phaeodariés (Radiolaires). C. r. Acad. Sci. Paris **259**, 2677–2679.

CALKINS, G. N., 1899: Mitosis in *Noctiluca* and its bearing on the nuclear relations of the Protozoa and the Metazoa. J. Morphol. **15**, 711–772.

– 1919: *Uroleptus mobilis* Engelm. I. History of the nuclei during division and conjugation. J. Exper. Zool. **27**, 293–357.

– CULL, S. W., 1907: The conjugation of *Paramecium aurelia* (*caudatum*). Arch. Protistenk. **10**, 375–415.

CAMERON, I. L., GUILE, E. E., JR., 1965: Nucleolar and biochemical changes during unbalanced growth of *Tetrahymena pyriformis*. J. Cell Biol. **26**, 845–855.

– PADILLA, G. M., MILLER, O. L., JR., 1966: Macronuclear cytology of synchronized *Tetrahymena pyriformis*. J. Protozool. **13**, 336–341.

– STONE, G. E., 1964: Relation between the amount of DNA per cell and the duration of DNA synthesis in three strains of *Tetrahymena pyriformis*. Exper. Cell Res. **36**, 510–514.

CAMP, R. R., MATTERN, C. F. T., HONIGBERG, B. M., 1974: A study of *Dientamoeba fragilis* JEPPS and DOBELL. I. Electron microscopic observations of the binucleate stages. II. Taxonomic position and revision of the genus. J. Protozool. **21**, 69–82.

CANN, J. P., PAGE, F. C., 1979: *Nucleosphaerium tuckeri* nov. gen. nov. spec.—a new freshwater filose amoeba without motile form in a new family Nucleariidae (Filosea: Aconchulida) feeding by ingestion only. Arch. Protistenk. **122**, 226–240.

CANNING, E. U., ANWAR, M., 1968: Studies on meiotic division in coccidial and malarial parasites. J. Protozool. **15**, 290–298.

– – 1969: Nuclear studies on sporozoan oocysts. In: Progress in protozoology (Abstr. III Internat. Congr. Protozool.), p. 23. Leningrad: Nauka Press.

– MORGAN, K., 1975: DNA synthesis, reduction, and elimination during life cycles of the eimeriine coccidian, *Eimeria tenella*, and the haemogregarine, *Hepatozoon domerguei*. Exper. Parasitol. **38**, 217–227.

– SINDEN, R. E., 1973a: The organization of the ookinete and observations on nuclear division in oocysts of *Plasmodium berghei*. Parasitology **67**, 29–40.

– – 1973b: Ultrastructural observations on the development of *Nosema algerae* VÁVRA and UNDEEN (Microsporidia, Nosematidae) in the mosquito *Anopheles stephensi* Liston. Protistologica **9**, 405–415.

– – 1975: Development of *Farinocystis tribolii* WEISER (Sporozoa, Neogregarinida) in *Palembus ocularis* CASEY (Coleoptera, Tenebrionidae) and some observations on its fine structure. Protistologica **11**, 221–231.

CARASSO, N., FAVARD, P., 1965: Microtubules fusoriaux dans les micro- et macronucléus de Ciliés péritriches en division. J. Microscopie **4**, 395–402.

CAROSI, G., SCAGLIA, M., FILICE, G., WILLAERT, E., 1976: An electron microscope study of *Naegleria jadini* nov. sp. (WILLAERT LE RAY 1973) in axenic "medium". The ameboid stage. Protistologica **12**, 31–36.

– – – – 1977: A comparative electron microscope study of axenically cultivated trophozoites of free-living amoebae of the genus *Acanthamoeba* and *Naegleria* with special reference to the species *N. gruberi* (SCHARDINGER 1899), *N. fowleri* (CARTER 1970), and *N. jadini* (WILLAERT et LE RAY 1973). Arch. Protistenk. **119**, 264–273.

CAVE, S., POCOCK, M. A., 1951: Karyological studies in the Volvocaceae. Amer. J. Bot. **38**, 800–811.

CECH, T. R., KARRER, K. M., 1980: Chromatin structure of the ribosomal RNA genes of *Tetrahymena thermophila* as analyzed by trimethylpsoralen crosslinking *in vivo*. J. Molec. Biol. **136**, 395–416.

– RIO, D. C., 1979: Localization of transcribed regions on extrachromosomal ribosomal RNA genes of *Tetrahymena thermophila* by R-loop mapping. Proc. Nat. Acad. Sci. U.S.A. **76**, 5051–5055.

Césana, D., 1971: Le mécanisme de la mitose gamogonique chez le Foraminifère *Iridia lucida*. C. r. Acad. Sci. Paris **D 272**, 3057–3060.
— 1972: Ultrastructure des gamètes chez un Foraminifère: *Iridia lucida* Le Calvez. C. r. Acad. Sci. Paris **D 274**, 1044–1047.
— 1975: Les stades initiaux de la gamogonie chez *Iridia lucida* Le Calvez (Foraminifera, Lagynidae). C. r. Acad. Sci. Paris **D 281**, 263–266.
— 1978: La mitose gamogonique chez *Iridia lucida* (Foraminifera, Lagynidae). Ann. Sci. Nat., Zool. sér. **12**, **20**, 287–320.
Chadha, R., 1980: Fine structure of the chromatin during the micronuclear mitotic cycle in *Stylonychia mytilus* Ehrenberg. Arch. Protistenk. **123**, 406–421.
— Sapra, G. R., Dass, C. M. S., 1978: Cell cycle stages and their relationship with conjugation in *Stylonychia mytilus* Ehrenberg. Ind. J. Exp. Biol. **16**, 5–9.
— — — 1979: Distribution of macronuclear DNA in *Stylonychia mytilus*—an autoradiographic study. Ind. J. Exp. Biol. **17**, 539–545.
Chakraborty, J., 1967: Electron microscopy of the hypotrich ciliate *Oxytricha platystoma*, with consideration of its macronuclear organization. J. Protozool. **14**, 59–64.
Chaly, N., Lord, A., Lafontaine, J. G., 1977: A light- and electron-microscope study of nuclear structure throughout the cell cycle in the euglenoid *Astasia longa* (Jahn). J. Cell Sci. **27**, 23–45.
— — — 1979: Nucleolar organization in the euglenoid *Astasia longa* as disclosed by selective staining, actinomycin D treatment, and cold shock. Canad. J. Botany **57**, 2031–2043.
Charret, R., 1967: Caractères cytologiques du Thécamoebien *Arcella polypora*. Protistologica **3**, 73–78.
— 1969: L'ADN nucléolaire chez *Tetrahymena pyriformis*: chronologie de sa réplication. Exper. Cell Res. **54**, 353–361.
— Vivier, E., 1964: Sur l'ultrastructure *d'Arcella polypora* (Rhizopode testacé). In: Electron microscopy (Proc. 3rd Europ. Reg. Conf. Electron Microscopy). Vol. B, pp. 187–188. Prague: Publ. House ČSAV.
Chatton, E., 1910: Essai sur la structure du noyau et la mitose chez les Amoebiens. Faits et théories. Arch. Zool. Expér. Génér. **5**, 267–337.
— 1920a: Les péridiniens parasites. Morphologie, reproduction, éthologie. Arch. Zool. Expér. Génér. **59**, 1–475.
— 1920b: Existence chez les Radiolaires de Péridiniens parasites considérés comme formes de reproduction de leurs hôtes. C. r. Acad. Sci. Paris **170**, 413–415.
— 1921: Sur un mécanisme cinétique nouveau: la mitose syndinienne chez les Péridiniens parasites plasmodiaux. C. r. Acad. Sci. Paris **173**, 859–862.
— 1952: Classe des Dinoflagellés. In: Traité de Zoologie (P.-P. Grassé, ed.), Vol. 1, fasc. 1, pp. 309–406. Paris: Masson et Cie.
Cheissin, E., 1958: Cytologische Untersuchungen verschiedener Stadien des Lebenszyklus der Kaninchencoccidien. I. *Eimeria intestinalis* E. Cheissin 1948. Arch. Protistenk. **102**, 265–290.
— 1964: Electron microscopic study of the microgametes of *Eimeria intestinalis* (Sporozoa, Coccidia). Zool. Zhurnal (Moscow) **43**, 647–651 (in Russian: English summary).
— 1965: Electron microscopic study of microgametogenesis in two species of coccidia from rabbit (*Eimeria magna* and *E. intestinalis*). Acta Protozool. **3**, 215–224.
— Ovchinnikova, L. P., 1964: A photometric study of DNA content in macronuclei and micronuclei of different species of *Paramecium*. Acta Protozool. **2**, 225–236.
— — Kudriavtsev, B. N., 1964: A photometric study of DNA content in macronuclei and micronuclei of different strains of *Paramecium caudatum*. Acta protozool. **2**, 237–245.

CHEISSIN, E., OVCHINNIKOVA, L. P., SELIVANOVA, G. V., BUZE, E. G., 1963: Changes of the DNA content of the macronucleus of *Paramecium caudatum* in the interdivision period. Acta Protozool. **1**, 63–69 (in Russian; English summary).
– ROZANOV, J. M., KUDRIAVTSEV, B. N., 1968: Fluorescence microscopic study on the content of nucleic acids in merozoites and microgametes of *Eimeria intestinalis* from the intestine of rabbit. Acta Protozool. **5**, 389–394.
– SNIGIREVSKAYA, E. S., 1965: Some new data on the fine structure of the merozoites of *Eimeria intestinalis* (Sporozoa, Eimeriidae). Protistologica **1**, 121–126.
CHEN, T. T., 1936a: Observations on mitosis in Opalinids (Protozoa, Ciliata). I. The behaviour and individuality of chromosomes and their significance. Proc. Nat. Acad. Sci. U.S.A. **22**, 594–602.
– 1936b: Observations on mitosis in Opalinids (Protozoa, Ciliata). II. The association of chromosomes and nucleoli. Proc. Nat. Acad. Sci. U.S.A. **22**, 602–607.
– 1940a: Polyploidy and its origin in *Paramecium*. J. Heredity **31**, 175–184.
– 1940b: Conjugation in *Paramecium bursaria* between animals with diverse nuclear constitutions. J. Heredity **31**, 185–196.
– 1940c: A further study on polyploidy in *Paramecium*. J. Heredity **31**, 249–251.
– 1940d: Polyploidy in *Paramecium bursaria*. Proc. Nat. Acad. Sci. U.S.A. **26**, 239–240.
– 1940e: Evidences of exchange of pronuclei during conjugation in *Paramecium bursaria*. Proc. Nat. Acad. Sci. U.S.A. **26**, 241–243.
– 1940f: Conjugation in *Paramecium bursaria* between animals with very different chromosome numbers and between animals with and without micronuclei. Proc. Nat. Acad. Sci. U.S.A. **26**, 243–246.
– 1946a: Conjugation in *Paramecium bursaria*. I. Conjugation of three animals. J. Morphol. **78**, 353–395.
– 1946b: Conjugation in *Paramecium bursaria*. II. Nuclear phenomena in lethal conjugation between varieties. J. Morphol. **79**, 125–262.
– 1948: Chromosomes in Opalinidae (Protozoa, Ciliata) with special reference to their behavior, morphology, individuality, diploidy, haploidy, and association with nucleoli. J. Morphol. **83**, 281–358.
– 1972: Chromosomes and nucleoli in some opalinid Protozoa. In: Research in protozoology (T. T. CHEN, ed.), Vol. 4, pp. 351–391. Oxford-New York: Pergamon Press.
CHOBOTAR, B., SCHOLTYSECK, E., SÉNAUD, J., ERNST, J. V., 1975: A fine structural study of asexual stages of the murine coccidium *Eimeria ferrisi* LEVINE and IVENS 1965. Z. Parasitenk. **45**, 291–306.
– SÉNAUD, J., ERNST, J. V., SCHOLTYSECK, E., 1975: The ultrastructure of macrogametes of *Eimeria ferrisi* LEVINE and IVENS 1965 in *Mus musculus*. Z. Parasitenk. **48**, 111–121.
– – – – 1980: Ultrastructure of macrogametogenesis and formation of the oocyst wall of *Eimeria papillata* in *Mus musculus*. Protistologica **16**, 115–124.
CHRISTIE, E., PAPPAS, P. W., DUBEY, J. P., 1978: Ultrastructure of excystment of *Toxoplasma gondii* oocysts. J. Protozool. **25**, 438–443.
CHUNOSOFF, L., HIRSHFIELD, H. I., 1967: Nuclear structure and mitosis in the dinoflagellate *Gonyaulax monilata*. J. Protozool. **14**, 157–163.
CLEFFMANN, G., 1967: Temperaturabhängigkeit der Phasen des Teilungszyklus von *Tetrahymena pyriformis* HSM. Z. Zellforsch. **79**, 599–602.
– 1968: Regulierung der DNS-Menge im Makronucleus von *Tetrahymena*. Exper. Cell Res. **50**, 193–207.
– 1975: Amount of DNA produced during extra S phases in *Tetrahymena*. J. Cell Biol. **66**, 204–209.
– 1980: Chromatin elimination and the genetic organization of the macronucleus in *Tetrahymena thermophila*. Chromosoma **78**, 313–325.

CLEVELAND, L. R., 1935a: The intranuclear achromatic figure of *Oxymonas grandis* sp. n. Biol. Bull. **69**, 54–63.
– 1935b: The centrioles of *Pseudotrichonympha* and their role in mitosis. Biol. Bull. **69**, 46–53.
– 1938a: Mitosis in *Pyrsonympha*. Arch. Protistenk. **91**, 452–455.
– 1938b: Origin and development of the achromatic figure. Biol. Bull. **74**, 41–55.
– 1938c: Longitudinal and transverse division in two closely related flagellates. Biol. Bull. **74**, 1–24.
– 1947: The origin and evolution of meiosis. Science **105**, 287–288.
– 1949a: The whole life cycle of chromosomes and their coiling systems. Trans. Amer. Philos. Soc. **39**, 1–100.
– 1949b: Hormone-induced sexual cycles of flagellates. I. Gametogenesis, fertilization and meiosis in *Trichonympha*. J. Morphol. **85**, 197–296.
– 1950a: Hormone-induced sexual cycles of flagellates. II. Gametogenesis, fertilization and one-division meiosis in *Oxymonas*. J. Morphol. **86**, 185–214.
– 1950b: Hormone-induced sexual cycles of flagellates. III. Gametogenesis, fertilization and one-division meiosis in *Saccinobaculus*. J. Morphol. **86**, 215–228.
– 1950c: Hormone-induced sexual cycles of flagellates. IV. Meiosis after syngamy and before nuclear fusion in *Notila*. J. Morphol. **87**, 317–348.
– 1950d: Hormone-induced sexual cycles of flagellates. V. Fertilization in *Eucomonympha*. J. Morphol. **87**, 349–367.
– 1951a: Hormone-induced sexual cycles of flagellates. VI. Gametogenesis, fertilization, meiosis, oocysts and gametocysts in *Leptospironympha*. J. Morphol. **88**, 199–244.
– 1951b: Hormone-induced sexual cycles of flagellates. VII. One-division meiosis and autogamy without cell division in *Urinympha*. J. Morphol. **88**, 385–440.
– 1952: Hormone-induced sexual cycles of flagellates. VIII. Meiosis in *Rhynchonympha* in one cytoplasmic and two nuclear divisions followed by autogamy. J. Morphol. **91**, 269–324.
– 1953a: Hormone-induced sexual cycles of flagellates. IX. Haploid gametogenesis and fertilization in *Barbulanympha*. J. Morphol. **93**, 371–404.
– 1953b: Studies on chromosomes and nuclear division. III. Pairing, segregation, and crossing-over. Trans. Amer. Philos. Soc. **43**, 809–869.
– 1954: Hormone-induced sexual cycles of flagellates. XII. Meiosis in *Barbulanympha* following fertilization, autogamy, and endomitosis. J. Morphol. **95**, 557–619.
– 1956a: Hormone-induced sexual cycles of flagellates. XIV. Gametic meiosis and fertilization in *Macrospironympha*. Arch. Protistenk. **101**, 99–170.
– 1956b: Brief accounts of the sexual cycles of the flagellates of *Cryptocercus*. J. Protozool. **3**, 161–180.
– 1957: Types and life cycles of centrioles of flagellates. J. Protozool. **4**, 230–241.
– 1958a: A factual analysis of chromosomal movement in *Barbulanympha*. J. Protozool. **5**, 47–62.
– 1958b: Movement of chromosomes in *Spirotrichonympha* to centrioles instead of the ends of central spindle. J. Protozool. **5**, 63–68.
– 1960: The centrioles of *Trichonympha* from termites and their functions in reproduction. J. Protozool. **7**, 326–341.
– 1961a: The centrioles of *Trichomonas* and their function in cell reproduction. Arch. Protistenk. **105**, 149–162.
– 1961b: Pairing and segregation in haploids and diploids of *Holomastigotoides*. Arch. Protistenk. **105**, 163–172.
– 1966a: Reproduction in *Deltotrichonympha*. Arch. Protistenk. **109**, 8–14.
– 1966b: General features and reproduction in *Koruga bonita*, gen. nov. sp. nov. Arch. Protistenk. **109**, 18–23.

Cleveland, L. R., 1966c: General features and reproduction in *Mixotricha*. Arch. Protistenk. **109**, 26–36.
– 1966d: Reproduction by binary and multiple fission in *Gigantomonas*. J. Protozool. **13**, 573–585.
– Day, M., 1958: Spirotrichonymphidae of *Stolotermes*. Arch. Protistenk. **103**, 1–53.
– Grimstone, A. V., 1964: The fine structure of the flagellate *Mixotricha paradoxa* and its associated microorganisms. Proc. Roy. Soc. **B 159**, 668–686.
– Hall, S. R., Sanders, E. P., Collier, J., 1934: The wood-feeding roach *Cryptocercus*, its protozoa, and the symbiosis between protozoa and roach. Mem. Amer. Acad. Arts Sci. **17**, 185–342.
Codreanu, R., Codreanu-Balcescu, D., Porchet-Henneré, E., 1974: Caractères ultrastructuraux dans la sporogenèse d'une Microsporidie de type thélohanien parasite musculaire de la crevette *Palaemon serratus* des côtes de France. C. r. Acad. Sci. Paris **D 279**, 2047–2049.
– Vávra, J., 1970: The structure and ultrastructure of the microsporidian *Telomyxa glugeiformis* Léger and Hesse 1910, parasite of *Ephemera danica* (Müll.) nymphs. J. Protozool. **17**, 374–384.
Cohen, A. I., 1957: Electron microscopic observations of *Amoeba proteus* in growth and inanition. J. Biophys. Biochem. Cytol. **3**, 859–865.
Cohen, J., Beisson, J., Tucker, J. B., 1980: Abnormal microtubule deployment during defective macronuclear division in a *Paramecium* mutant. J. Cell Sci. **44**, 153–167.
Coleman, A. W., 1979: Feulgen microspectrophotometric studies of *Pandorina morum* and other Volvocales (Chlorophyceae). J. Phycol. **15**, 216–220.
– Zollner, J., 1977: Cytogenetic polymorphism within the species *Pandorina morum* Bory de St. Vincent (Volvocaceae). Arch. Protistenk. **119**, 224–232.
Colley, F. C., 1967a: Fine structure of sporozoites of *Eimeria nieschulzi*. J. Protozool. **14**, 217–220.
– 1967b: Fine structure of microgametocytes and macrogametes of *Eimeria nieschulzi*. J. Protozool. **14**, 663–674.
– 1968: Fine structure of schizonts and merozoites of *Eimeria nieschulzi*. J. Protozool. **15**, 374–382.
Comandon, J., Fonbrune, P. de, 1937: Caryocinèse d'une amibe (*Acanthamoeba*). Ann. Physiol. **13**, 1068–1069.
Corbel, J.-C., Desportes, I., Theodorides, J., 1979: Étude de *Gonospora beloneides* (Ming.) (= *Lobianchella beloneides* Ming.) (Grégarine, Urosporidae), parasite coelomique d'une Alciopide (Polychaeta) et remarques sur d'autres Grégarines d'Alciopidae. Protistologica **15**, 55–65.
Corff, S., Yuyama, S., 1976: Cessation of nuclear DNA synthesis in differentiating *Naegleria*. J. Protozool. **23**, 587–593.
Corliss, J. O., 1974: Remarks on the composition of the large ciliate class Kinetofragmophora De Puytorac *et al.* 1974, and recognition of several new taxa therein, with emphasis on the primitive order Primociliatida n. ord. J. Protozool. **21**, 207–220.
– 1975a: Taxonomic characterization of the suprafamilial groups in a revision of recently proposed schemes of classification for the phylum Ciliophora. Trans. Amer. Microscop. Soc. **94**, 224–267.
– 1975b: Nuclear characteristics and phylogeny in the protistan phylum Ciliophora. BioSystems **7**, 338–349.
– 1979a: The ciliated Protozoa. Characterization, classification and guide to the literature, 2nd ed. Oxford-New York: Pergamon Press.
– 1979b: L'impact de la microscopie électronique sur la systématique des protistes et le cas curieux du *Stephanopogon*: un flagellé au lieu d'un cilié? J. Protozool. **26**, No. 3 (part 1), 56A.

CORLISS, J. O., HARTWIG, E., 1977: The "primitive" interstitial ciliates: their ecology, nuclear uniqueness, and postulated place in the evolution and systematics of the phylum Ciliophora. Mikrofauna Meeresboden **61**, 65–88.

COSGROVE, W. B., SKEEN, M. J., 1970: The cell cycle in *Crithidia fasciculata*. Temporal relationships between synthesis of deoxyribonucleic acid in the nucleus and in the kinetoplast. J. Protozool. **17**, 172–178.

COSS, R. A., 1974: Mitosis in *Chlamydomonas reinhardii*. Basal bodies and the mitotic apparatus. J. Cell Biol. **63**, 325–339.

CREEMERS, J., JADIN, J. M., 1967: Étude de l'ultrastructure et de la biologie de *Leishmania mexicana* BIAGI 1953. I. Les modifications qui surviennent lors de la transformation leishmania-leptomonas. Bull. Soc. Pathol. Exot. **60**, 53–58.

CROFT, S. L., 1979: Ultrastructural study of the nucleus of *Leishmania hertigi*. Protistologica **15**, 103–110.

CULLIS, C. A., 1972: The basis of cell-to-cell transformation in *Paramecium bursaria*. II. Investigation into the molecular nature of the transforming agent. J. Cell Sci. **11**, 611–619.

– 1973: DNA amounts in the nuclei of *Paramecium bursaria*. Chromosoma **40**, 127–133.

CUMMINGS, D. J., 1972: Isolation and partial characterization of macro- and micronuclei from *Paramecium aurelia*. J. Cell Biol. **53**, 105–115.

– 1975: Studies on macronuclear DNA from *Paramecium aurelia*. Chromosoma **53**, 191–208.

– KOTASEK, W. E., 1978: Three-dimensional structure of macronuclei from *Paramecium* determined by scanning electron microscopy. J. Protozool. **25**, 334–337.

CURRENT, W. L., 1979: *Henneguya adiposa* MINCHEW (Myxosporida) in the channel catfish: ultrastructure of the plasmodium wall and sporogenesis. J. Protozool. **26**, 209–217.

– JANOVY, J., 1977: Sporogenesis in *Henneguya exilis* infecting the Channel catfish: an ultrastructural study. Protistologica **13**, 157–167.

– – KNIGHT, S. A., 1979: *Myxosoma funduli* KUDO (Myxosporida) in *Fundulus kansae*: ultrastructure of the plasmodium wall and of sporogenesis. J. Protozool. **26**, 574–583.

CZIHAK, G., 1964: Experiments on nuclear differentiation in the foraminifer *Rotaliella heterocaryotica* GRELL by means of UV-microirradiation. Exper. Cell Res. **35**, 372–380.

– GRELL, K. G., 1960: Zur Determination der Zellkerne bei der Foraminifere *Rotaliella heterocaryotica*. Naturwissenschaften **47**, 211–212.

DAHLGREN, L., 1964: On nuclear cytology and reproduction in the monothalamous foraminifer *Ovammina opaca* DAHLGREN. Zool. Bidrag Uppsala **36**, 315–341.

– 1967a: On the ultrastructure of the gamontic nucleus and the adjacent cytoplasm of the monothalamous foraminifer *Ovammina opaca* DAHLGREN. Zool. Bidrag Uppsala **37**, 77–112.

– 1967b: On the nuclear distribution of RNA and DNA and on the ultrastructure of nuclei and adjacent cytoplasm of the foraminifers *Hippocrepinella alba* HERON-ALLEN and EARLAND and *Globobulimina turgida* (BAILEY). Zool. Bidrag Uppsala **37**, 113–138.

DANFORTH, H. D., HAMMOND, D. M., 1972: Stages of merogony in multinucleate merozoites of *Eimeria magna* PÉRARD 1925. J. Protozool. **19**, 454–457.

DANIELS, E. W., 1973: Ultrastructure. In: Biology of Amoeba, pp. 125–169. New York-London: Academic Press.

– BREYER, E. P., 1966: Stratification within centrifuged amoeba nuclei. Z. Zellforsch. **70**, 449–460.

– – 1967: Ultrastructure of the giant amoeba *Pelomyxa palustris*. J. Protozool. **14**, 169–179.

– – KUDO, R. R., 1966: *Pelomyxa palustris* GREEFF. II. Its ultrastructure. Z. Zellforsch. **73**, 367–383.

DANIELS, E. W., MCNIFF, J. M., EKBERG, D. R., 1969: Nucleopores of the giant amoeba, *Pelomyxa carolinensis*. Z. Zellforsch. **98**, 357–363.

– ROTH, L. E., 1964: Electron microscopy of mitosis in a radiosensitive giant amoeba. J. Cell Biol. **20**, 75–84.

DARLINGTON, C. D., 1937: Recent advances in cytology, 2nd ed. London: Churchill.

DASS, C. M. S., SAPRA, G. R., 1966: Effect of β-mercaptoethanol on the binary fission in *Stylonychia notophora* (STOKES). Exper. Cell Res. **43**, 622–632.

– – KUMAR, R., 1980: Macronuclear development in the ciliate *Blepharisma musculus*. Europ. J. Cell Biol. **22**, 105.

DAVIDSON, L., LA FOUNTAIN, J. R., 1975: Mitosis and early meiosis in *Tetrahymena pyriformis* and the evolution of mitosis in the phylum Ciliophora. BioSystems **7**, 326–336.

– – 1977: Quantitative analysis of the changes in distribution of the three classes of spindle microtubules in mitotic micronuclei of *Tetrahymena*. Abstr. V Internat. Congr. Protozool., p. 49. New York.

DE, D. N., GHOSH, S. N., 1965: Cytochemical evidence for the apparent absence of histone in the cells of Cyanophyceae. J. Histochem. Cytochem. **13**, 298.

DEHORNE, A., 1920: Contribution à l'étude comparée de l'appareil nucléaire des Infusoires ciliés (*Paramecium caudatum* et *Colpidium truncatum*), des Euglènes et des Cyanophycées. Arch. Zool. Expér. Génér. **60**, 47–176.

DESAI, R. N., 1965a: Tetraploidy in a cephaline gregarine *Stylocephalus conoides* DEVDHAR. Naturwissenschaften **52**, 146.

– 1965b: Chromosome cycle in the gregarine *Stylocephalus elongatus* DEVDHAR. Naturwissenschaften **52**, 649.

– 1966: Chromosome cycle in the gregarine *Stylocephalus mesomorphi* DEV. Parasitology **56**, 199–208.

DE SOUZA, W., MEYER, H., 1974: On the fine structure of the nucleus in *Trypanosoma cruzi* in tissue culture forms. Spindle fibers in the dividing nucleus. J. Protozool. **21**, 48–52.

DESPORTES, I., 1966: L'ultrastructure du gamète mâle de l'Eugrégarine *Stylocephalus longicollis* F. ST. (Stylocephalidae). C. r. Acad. Sci. Paris **D 263**, 517–520.

– 1967: Ultrastructure et évolution du sporozoite de *Stylocephalus africanus* THÉODORIDÈS, DESPORTES et JOLIVET, Eugrégarine Stylocephalidae. C. r. Acad. Sci. Paris **D 265**, 423–426.

– 1969: Ultrastructure et développement des Grégarines du genre *Stylocephalus*. Ann. Sci. Natur., Zool. sér. 12, **11**, 31–96.

– 1970: Ultrastructure des Grégarines du genre *Stylocephalus*: la phase enkystée. Ann. Sci. Natur., Zool. sér. 12, **12**, 73–169.

– 1974: Ultrastructure et évolution nucleaire des trophozoïtes d'une Grégarine d'Ephémeroptère: *Enterocystis fungoides* M. CODREANU. J. Protozool. **21**, 83–94.

– 1976: Ultrastructure de *Stempellia mutabilis* LEGER et HESSE, Microsporidie parasite de l'éphémère *Ephemera vulgata* L. Protistologica **12**, 121–150.

– THÉODORIDÈS, J., 1969: Ultrastructure de la Grégarine *Callyntrochlamys phronimae* FRENZEL: étude comparée de son noyau avec celui de *Thalicola salpae* (FRENZEL) (Eugregarina). J. Protozool. **16**, 449–460.

– – 1979: Étude ultrastructurale *d'Amphiamblys laubieri* n. sp. (Microsporidie, Metchnikovellidae), parasite d'une Grégarine (*Lecudina* sp.) d'un Echiurien abyssal. Protistologica **15**, 435–457.

DESSER, S. S., 1972: Gametocyte maturation, exflagellation and fertilization in *Parahaemoproteus* (= *Haemoproteus*) *velans* (COATNEY and ROUDABUSH) (Haemosporidia: Haemoproteidae): an ultrastructural study. J. Protozool. **19**, 287–296.

Desser, S. S., Baker, J. R., Lake, P., 1970: The fine structure of *Leucocytozoon simondi*. I. Gametocytogenesis. Canad. J. Zool. **48**, 331–336.

– Paterson, W. B., 1978: Ultrastructural and cytochemical observations on sporogenesis of *Myxobolus* sp. (Myxosporida: Myxobolidae) from the common shiner *Notropis cornutus*. J. Protozool. **25**, 314–326.

– Weller, I., 1973: Structure, cytochemistry and locomotion of *Haemogregarina* sp. from *Rana berlandieri*. J. Protozool. **20**, 65–73.

Detcheva, R., 1976: Particularités ultrastructurales du Cilié *Cyrtolophosis mucicola* Stokes 1885. C. r. Soc. Biol. **170**, 112–114.

– De Puytorac, P., 1979: Un nouvel exemple de Cilié à micronoyau inclus dans l'enveloppe macronucléaire: le genre *Aristerostoma* Kahl 1926. Ann. Station Biol. Besse-en-Chandesse **13**, 247–251.

– – Grolière, C.-A., 1979: Intérêt d'une étude des particularités de l'enveloppe nucléaire de certains Ciliés. C. r. Acad. Sci. Paris **D 288**, 955–958.

Deutsch, K., Swann, M. M., 1959: An electron microscope study of a small free-living amoeba (*Hartmanella astronyxis*). Quart. J. Microscop. Sci. **100**, 13–16.

– Zaman, V., 1959: An electron microscope study of *Entamoeba invadens* Rodhain 1934. Exper. Cell Res. **17**, 310–319.

Devauchelle, G., 1968: Étude ultrastructurale du développement des Grégarines du *Tenebrio molitor*. Protistologica **4**, 313–332.

Devi, R. V., 1961: Autogamy in *Frontonia leucas* (Ehrbg.) J. Protozool. **8**, 277–283.

Devidé, Z., 1951: Chromosomes in ciliates (Euciliata and Opalinidae). Bull. Internat. Acad. Yougosl. Sci. Beaux-Arts, n. sér. **3**, 75–114.

– Geitler, L., 1947: Die Chromosomen der Ciliaten. Chromosoma **3**, 110–136.

Dickinson, A. H., 1966: Cytochemistry of the helices in the nucleus of *Amoeba proteus*. J. Cell Biol. **31**, 27A.

– 1970: An electron microscope cytochemical and radioautographic study of the nucleus of *Amoeba proteus*. J. Cell Biol. **47**, 49A.

Didier, P., Puytorac, P. de, Wilbert, N., Detcheva, R., 1980: A propos d'observations sur l'ultrastructure du Cilié *Cyrtolophosis mucicola* Stokes 1885. J. Protozool. **27**, 72–79.

Diller, W. F., 1936: Nuclear reorganization processes in *Paramecium aurelia*, with description of autogamy and "hemixis". J. Morphol. **59**, 11–67.

– 1940: Nuclear variation in *Paramecium caudatum*. J. Morphol. **66**, 605–633.

– 1948: Nuclear behavior of *Paramecium trichium* during conjugation. J. Morphol. **82**, 1–52.

– 1958: Studies on conjugation in *Paramecium polycaryum*. J. Protozool. **5**, 282–292.

– 1964: Observations on the morphology and life history of *Homalozoon vermiculare*. Arch. Protistenk. **107**, 351–362.

Din, N., Engberg, J., 1979: Extrachromosomal ribosomal RNA genes in *Tetrahymena*. Structure and evolution. J. Molec. Biol. **134**, 555–574.

– – Kaffenberger, W., Eckert, W. A., 1979: The intervening sequence in the 26S rRNA coding region of *T. thermophila* is transcribed within the largest stable precursor for rRNA. Cell **18**, 525–532.

Dingle, A. D., Fulton, C., 1966: Development of the flagellar apparatus of *Naegleria*. J. Cell Biol. **31**, 43–54.

Dini, F., Bracchi, P., 1976: Ciclo cellulare di *Aspidisca aculeata* (Ehrenberg). Atti Accad. Naz. dei Lincei, Cl. Sci. Fis., Mat. Nat., ser. 8, **60**, 64–69.

– – Luporini, P., 1975: Cellular cycle in two ciliates hypotrichs. Acta Protozool. **14**, 59–66.

– Luporini, P., 1979: Preconjugant cell interaction and cell cycle in the ciliate *Euplotes crassus*. Switching from the vegetative to the sexual stage. Devel. Biol. **69**, 506–516.

DIPPELL, R. V., 1954: A preliminary report on the chromosomal constitution of certain variety 4 races of *Paramecium aurelia*. In: Atti IX Congr. Internaz. Genet., pp. 1109–1111. Firenze.

– SONNEBORN, T. M., 1957: Structure of the *Paramecium aurelia* macronucleus as revealed by electron microscopy. Proc. Indiana Acad. Sci. **66**, 60.

DOBELL, C., 1925: The life history and chromosome cycle of *Aggregata eberthi* (Protozoa: Sporozoa: Coccidia). Parasitology **17**, 1–136.

DOBRZAŃSKA-KACZANOWSKA, J., 1963: Comparaison de la morphogenèse des Ciliés: *Chilodonella uncinata* (EHRBG.), *Allosphaerium paraconvexa* sp. n. et *Heliochona scheuteni* (STEIN). Acta Protozool. **1**, 353–394.

DODGE, J. D., 1963a: The nucleus and nuclear division in the Dinophyceae. Arch. Protistenk. **106**, 442–452.

– 1963b: Chromosome structure in the Dinophyceae. I. The spiral chromonema. Arch. Mikrobiol. **45**, 46–57.

– 1963c: Chromosome numbers in some marine dinoflagellates. Botanica Marina (Hamburg) **5**, 121–127.

– 1964a: Chromosome structure in the Dinophyceae. II. Cytochemical studies. Arch. Mikrobiol. **48**, 66–80.

– 1964b: Nuclear division in the dinoflagellate *Gonyaulax tamarensis*. J. Gen. Microbiol. **36**, 269–276.

– 1965: Chromosome structure in the dinoflagellates and the problem of the mesocaryotic cell. In: Progress in Protozoology (Abstr. II Internat. Conf. Protozool.), pp. 264–265. London: Excerpta Medica.

– 1966: The Dinophyceae. In: The chromosomes of the algae (M. B. E. GODWARD, ed.), pp. 96–115. London: E. Arnold.

– 1971a: Fine structure of the Pyrrophyta. Bot. Rev. **37**, 481–508.

– 1971b: A dinoflagellate with both a mesokaryotic and a eukaryotic nucleus. I. Fine structure of the nuclei. Protoplasma **73**, 145–157.

– 1975: The fine structure of *Trachelomonas* (Euglenophyceae). Arch. Protistenk. **117**, 65–77.

– BIBBY, B. T., 1973: The Prorocentrales (Dinophyceae). I. A comparative account of fine structure in the genera *Prorocentrum* and *Exuviaella*. Bot. J. Linnean Soc. London **67**, 175–187.

– CRAWFORD, R. M., 1968: Fine structure of the dinoflagellate *Amphidinium carteri* HULBERT. Protistologica **4**, 231–242.

– – 1969: The fine structure of *Gymnodinium fuscum* (Dinophyceae). New Phytologist **68**, 613–618.

– – 1970: The morphology and fine structure of *Ceratium hirundinella* (Dinophyceae). J. Phycol. **6**, 137–149.

– – 1971: Fine structure of the dinoflagellate *Oxyrrhis marina*. I. The general structure of the cell. Protistologica **7**, 295–304.

DOERDER, F. P., 1979: Regulation of macronuclear DNA content in *Tetrahymena thermophila*. J. Protozool. **26**, 28–35.

– DE BAULT, L. E., 1975: Cytofluorimetric analysis of nuclear DNA during meiosis, fertilization and macronuclear development in the ciliate *Tetrahymena pyriformis*, syngen 1. J. Cell Sci. **17**, 471–493.

– – 1978: Life cycle variation and regulation of macronuclear DNA content in *Tetrahymena thermophila*. Chromosoma **69**, 1–19.

– DITARANTO, J., DE BAULT, L. E., 1981: Evolutionary constraints on quantitative variation and regulation of macronuclear DNA content in the genus *Tetrahymena*. J. Cell Sci. **49**, 177–193.

DOERDER, F. P., LIEF, J. H., DE BAULT, L. E., 1977: Macronuclear subunits of *Tetrahymena thermophila* are functionally haploid. Science **198**, 946–948.

DOFLEIN, F., REICHENOW, E., 1953: Lehrbuch der Protozoenkunde, 6. Aufl. Jena: VEB G. Fischer-Verlag.

DOGIEL, V. A., 1910a: Beiträge zur Kenntnis der Gregarinen. IV. *Callyntrochlamys phronimae* FRENZ. Arch. Protistenk. **20**, 60–78.

– 1910b: Catenata. Organization of the genus *Haplozoon* and some similar forms. St. Petersburg: M. M. Stasyulevitch Publ. (in Russian).

– 1929: Polymerisation als ein Prinzip der progressiven Entwicklung bei Protozoen. Biol. Zbl. **49**, 451–469.

– 1954: Oligomerization of homologous organs. Leningrad Univ. Press (in Russian).

– 1965: General protozoology, 2nd ed. (Revised by J. I. POLJANSKIJ and E. M. CHEJSIN.) Oxford: Clarendon Press.

DRAGESCO, J., 1960: Les Ciliés mésopsammiques littoraux (systématique, morphologie, écologie). Trav. Stat. Biol. Roscoff, n. sér. **12**, 1–356.

– 1963: Compléments à la connaissance des Ciliés mésopsammiques de Roscoff. I. Holotriches. Cahiers Biol. Marine **4**, 91–119.

– 1970: Ciliés libres du Cameroun. Ann. Fac. Sci. Univ. Féd. Cameroun (Yaoundé), Numéro hors-série, 1–141.

– FRYD-VERSAVEL, G., IFTODE, F., DIDIER, P., 1977: Le Cilié *Platyophrya spumacola* KAHL 1926: Morphologie, stomatogenèse et ultrastructure. Protistologica **13**, 419–434.

– RAIKOV, I., 1966: L'appareil nucléaire, la division et quelques stades de la conjugaison de *Tracheloraphis margaritatus* (KAHL) et *T. caudatus* sp. nov. (Ciliata, Holotricha). Arch. Protistenk. **109**, 99–113.

DUBEY, J. P., SPEER, C. A., DOUGLASS, T. G., 1980: Development and ultrastructure of first-generation meronts of *Sarcocystis cruzi* in calves fed sporocysts from coyote feces. J. Protozool. **27**, 380–387.

DUBREMETZ, J.-F., 1971: L'ultrastructure du centriole et du centrocône chez la Coccidie *Eimeria necatrix*. Etude au cours de la schizogonie. J. Microscopie **12**, 453–458.

– 1973: Étude ultrastructurale de la mitose schizogonique chez la Coccidie *Eimeria necatrix* (JOHNSON 1930). J. Ultrastr. Res. **42**, 354–376.

– 1975: La genèse des mérozoites chez la Coccidie *Eimeria necatrix*. Étude ultrastructurale. J. Protozool. **22**, 71–84.

– ELSNER, Y. Y., 1979: Ultrastructural study of schizogony of *Eimeria bovis* in cell cultures. J. Protozool. **26**, 367–376.

DUPY-BLANC, J., 1969a: Étude cytophotométrique des teneurs en ADN des micronucléus de *Paramecium caudatum* au cours de la conjugaison et pendant la différenciation des "Anlages" en macronucléus. Protistologica **5**, 239–248.

– 1969b: Étude par cytophotométrie des teneurs en ADN nucléaire chez trois espèces de Paramécies, chez differentes variétés d'une même espèce et chez différents types sexuels d'une même variété. Protistologica **5**, 297–308.

– MÉTENIER, G., 1975: Étude par cytophotométrie et autoradiographie de l'activité autocatalytique de l'ADN macronucléaire au cours du cycle de polymorphisme du Cilié *Tetrahymena paravorax* CORLISS 1957. Protistologica **11**, 159–167.

DWORAKOWSKA, I., 1966: The structures of the nuclear apparatus in *Plagiotoma lumbrici* DUJARDIN and their transformations in division cycle. Acta Protozool. **4**, 109–125.

DYSART, M. P., 1960: Study of macronuclear chromatin extrusion in *Tetrahymena limacis* using tritiated thymidine. J. Protozool. **7**, suppl., 10–11.

– 1963: Cytochemical and quantitative DNA analyses of the macronucleus and its extrusion body in species of *Tetrahymena*. J. Protozool. **10**, suppl., 8.

– CORLISS, J. O., TORRE, L. DE LA, 1962: Comparative DNA measurements in two species of *Tetrahymena*. J. Protozool. **9**, suppl., 17.

EBERHARDT, R., 1962: Untersuchungen zur Morphogenese von *Blepharisma* und *Spirostomum*. Arch. Protistenk. **106**, 241–341.

ECKERT, W. A., FRANKE, W. W., 1975: Changes in fine structure and composition of macronuclei of *Tetrahymena pyriformis* induced by drugs interfering with RNA synthesis and processing. Cytobiologie **11**, 392–418.

– – SCHEER, U., 1975: Nucleocytoplasmic translocation of RNA in *Tetrahymena pyriformis* and its inhibition by actinomycin D and cycloheximide. Exp. Cell Res. **94**, 31–46.

– KAFFENBERGER, W., 1980: Regulation of rRNA metabolism in *Tetrahymena pyriformis*. I. Nutritional shift-down. Eur. J. Cell Biol. **21**, 53–62.

EGELHAAF, A., 1955: Cytologisch-entwicklungsphysiologische Untersuchungen zur Konjugation von *Paramecium bursaria* FOCKE. Arch. Protistenk. **100**, 447–514.

EHRET, C. F., POWERS, E. L., 1955: Macronuclear and nucleolar development in *Paramecium bursaria*. Exper. Cell Res. **9**, 241–257.

EIDE, A., 1968: Subunit composition of the macronucleus in the ciliate *Ichthyophthirius multifiliis*. Årbok Univ. Bergen (Mat.-Naturvit.), No. 7, 3–16.

ELLIOTT, A. M., 1963: The fine structure of *Tetrahymena pyriformis* during mitosis. In: The cell in mitosis, pp. 107–121. New York-London: Academic Press.

– KENNEDY, J. R., BAK, I. J., 1962: Macronuclear events in synchronously dividing *Tetrahymena pyriformis*. J. Cell Biol. **12**, 515–531.

ELLIS, I. M., 1937: The morphology, division and conjugation of the salt-marsh ciliate *Fabrea salina* HENNEGUY. Univ. Calif. Publs Zool. **41**, 343–388.

ELSEVIER, S. M., LIPPS, H. J., STEINBRÜCK, G., 1978: Histone genes in macronuclear DNA of the ciliate *Stylonychia mytilus*. Chromosoma **69**, 291–306.

ENGBERG, J., ANDERSSON, P., LEICK, V., COLLINS, J., 1976: Free ribosomal DNA molecules from *Tetrahymena pyriformis* GL are giant palindromes. J. Molec. Biol. **104**, 455–470.

– CHRISTIANSEN, G., LEICK, V., 1974: Autonomous rDNA molecules containing single copies of the ribosomal RNA genes in the macronucleus of *Tetrahymena pyriformis*. Biochem. Biophys. Res. Comm. **59**, 1356–1365.

– DIN, N., ECKERT, W. A., KAFFENBERGER, W., PEARLMAN, R. E., 1980: Detailed transcription map of the extrachromosomal ribosomal RNA genes in *Tetrahymena thermophila*. J. Molec. Biol. **142**, 289–313.

– MOWAT, D., PEARLMAN, R. E., 1972: Preferential replication of the ribosomal RNA genes during a nutritional shift-up in *Tetrahymena pyriformis*. Biochim. Biophys. Acta **272**, 312–320.

– NILSSON, J., PEARLMAN, R. E., LEICK, V., 1974: Induction of nucleolar and mitochondrial DNA replication in *Tetrahymena pyriformis*. Proc. Nat. Acad. Sci. U.S.A. **71**, 894–898.

– PEARLMAN, R. E., 1972: The amount of ribosomal RNA genes in *Tetrahymena pyriformis* in different physiological states. Europ. J. Biochem. **26**, 393–400.

ENTZ, G., 1921: Über mitotische Teilung von *Ceratium hirundinella*. Arch. Protistenk. **43**, 415–430.

EPSTEIN, G. V., 1941: Parasitic amoebae. Moscow-Leningrad: Medgiz Press (in Russian).

ETTL, H., MANTON, I., 1964: Die feinere Struktur von *Pedinomonas minor* KORSCHIKOFF. Nova Hedwigia **8**, 421–451.

EVENSON, D. P., PRESCOTT, D. M., 1970: RNA metabolism in the macronucleus of *Euplotes eurystomus* during the cell cycle. Exper. Cell Res. **61**, 71–74.

EYDEN, B. P., 1975: Light and electron microscope study of *Dunaliella primolecta* BUTCHER (Volvocida). J. Protozool. **22**, 336–344.

– 1977: Morphology and ultrastructure of *Bodo designis* SKUJA 1948. Protistologica **13**, 169–179.

Fahrni, J., 1977: Contribution à la connaissance de *Spirochona gemmipara* Stein (Cilié Chonotriche). Morphologie, ultrastructure, réproduction sexuée et cycle vital. Thèse de Doctorat, Fac. Sci., Univ. de Lausanne.

Falk, H., Wunderlich, F., Franke, W. W., 1968: Microtubular structures in macronuclei of synchronously dividing *Tetrahymena pyriformis*. J. Protozool. **15**, 776–780.

Fauré-Fremiet, E., 1953: L'hypothèse de la sénescence et les cycles de réorganisation nucléaire chez les Ciliés. Rev. Suisse Zool. **60**, 426–438.

– 1954: Réorganisation du type endomixique chez les Loxodidae et chez les *Centrophorella*. J. Protozool. **1**, 20–27.

– 1957: Le macronucléus hétéromère de quelques Ciliés. J. Protozool. **4**, 7–17.

– 1961: Quelques considérations sur les Ciliés mésopsammiques à propos d'un récent travail de J. Dragesco. Cahiers Biol. Marine **2**, 177–186.

– 1970a: Microtubules et mécanismes morphopoïétiques. Année Biol. **9**, 1–61.

– 1970b: A propos de la note de M. Thomas Njiné sur le Cilié *Loxodes magnus*. C. r. Acad. Sci. Paris **D 270**, 523–524.

– André, J., 1968: Structure fine de l'*Euplotes eurystomus* (Wrz.). Arch. Anat. Microscop. Morphol. Expér. **57**, 53–78.

– Ganier, M.-Cl., 1969: Morphologie et structure fine du Cilié *Chaenea vorax* Quenn. Protistologica **5**, 353–361.

– Rouiller, C., Gauchery, M., 1957: La réorganisation macronucléaire chez les *Euplotes*. Étude au microscope électronique. Exper. Cell Res. **12**, 135–144.

Febvre, J., 1977: La division nucléaire chez les Acanthaires. I. Étude ultrastructurale de la mitose. Comparison avec la caryocinèse d'autres organismes. J. Ultrastr. Res. **60**, 279–295.

Febvre-Chevalier, C., 1971: Constitution ultrastructurale de *Globigerina bulloides* d'Orbigny 1826 (Rhizopoda–Foraminifera). Protistologica **7**, 311–324.

– 1973: *Hedraiophrys hovassei*: morphologie, biologie et cytologie. Protistologica **9**, 503–520.

– 1975: Étude cytologique de *Gymnosphaera albida* Sassaki 1894 (Héliozoaire Centrohélidié). Protistologica **11**, 331–344.

– Lécher, P., 1971: Étude ultrastructurale des lamelles annelées intracytoplasmiques chez les Foraminifères et les Radiolaires Phaeodariés. C. r. Acad. Sci. Paris **D 272**, 1264–1267.

Fedoseyenko, V. M., 1979: *Besnoitia besnoiti* (Sporozoa, Eimeriidae): the cyst ultrastructure. In: Toxoplasmidy (Series "Protozoology", No. 4), pp. 100–105. Leningrad: Nauka Press (in Russian; English summary).

– Levit, A. V., 1979: Electron microscopic study of the cyst of *Sarcocystis muris* (Sporozoa, Eimeriidae) in skeletal muscles of white mice. In: Toxoplasmidy (Series "Protozoology", No. 4), pp. 106–110. Leningrad: Nauka Press (in Russian; English summary).

Feldherr, C. M., 1962: The nuclear annuli as pathways for nucleocytoplasmic exchanges. J. Cell Biol. **14**, 65–72.

– 1965: A study of the changes in the nuclear envelope during mitosis. J. Cell Biol. **27**, 28A–29A.

– 1966: Nucleocytoplasmic exchanges during cell division. J. Cell Biol. **31**, 199–203.

– 1968a: Nucleocytoplasmic exchanges during early interphase. J. Cell Biol. **39**, 49–54.

– 1968b: Changes in the nuclear envelope of *Amoeba* during mitosis. Nature **218**, 184–185.

– 1970: Nuclear permeability during different physiological states. J. Cell Biol. **47**, 60A.

Ferguson, D. J. P., Hutchison, W. M., Dunachie, J. F., Siim, J. Chr., 1974: Ultrastructural study of early stages of asexual multiplication and microgametogony of *Toxoplasma gondii* in the small intestine of the cat. Acta Pathol. Microbiol. Scand. **B 82**, 167–181.

FERGUSON, D. J. P., HUTCHISON, W. M., SIIM, J. CHR., 1975: The ultrastructural development of the macrogamete and formation of the oocyst wall of *Toxoplasma gondii*. Acta Pathol. Microbiol. Scand. **B 83**, 491–505.

FERNANDO, M. A., 1973: Fine structural changes associated with microgametogenesis of *Eimeria acervulina* in chickens. Z. Parasitenk. **43**, 33–42.

– 1974: Fine structure of the schizonts and merozoites of *Eimeria acervulina* in the chicken. J. Parasitol. **60**, 149–159.

– STOCKDALE, P. H. G., 1974: Fine structural changes associated with schizogony in *Eimeria necatrix*. Z. Parasitenk. **43**, 105–114.

FILFILAN, S. A., SIGEE, D. C., 1977: Continuous DNA replication in the nucleus of the dinoflagellate *Prorocentrum micans* (EHRENBERG). J. Cell Sci. **27**, 81–90.

– – 1978: Cell division in *Prorocentrum micans*. J. Protozool. **25**, No. 3 (part 2), 8 B.

FINDLY, R. C., GALL, J. G., 1978: Free ribosomal RNA genes in *Paramecium* are tandemly repeated. Proc. Nat. Acad. Sci. U.S.A. **75**, 3312–3316.

– – 1980: Organization of ribosomal genes in *Paramecium tetraurelia*. J. Cell Biol. **84**, 547–559.

FINE, K. E., LOEBLICH, A. R., 1976: Endosymbiosis in the marine dinoflagellate *Kryptoperidinium foliaceum*. J. Protozool. **23**, No. 2, 8 A.

FLICKINGER, C. J., 1965: The fine structure of the nuclei of *Tetrahymena pyriformis* throughout the cell cycle. J. Cell Biol. **27**, 519–529.

– 1968: The fine structure of the nucleoli of normal and actinomycin D treated *Amoeba proteus*. J. Ultrastr. Res. **23**, 260–271.

– 1974: The fine structure of four "species" of *Amoeba*. J. Protozool. **21**, 59–68.

FOKIN, S. I., 1979: On existence of internuclear interactions in ciliates. Tsitologiya (Leningrad) **21**, 1348–1352 (in Russian; English summary).

– BORCHSENIUS, O. N., 1978: Effect of local UV irradiation on the vegetative functions of uni- and bimicronucleate cells of *Paramecium caudatum*. Tsitologiya (Leningrad) **20**, 570–574 (in Russian; English summary).

– OSSIPOV, D. V., 1981: Generative nucleus control over cell vegetative functions in *Paramecium*. Acta Protozool. **20**, 51–73.

FORER, A., 1969: Chromosome movements during cell division. In: Handbook of molecular cytology (LIMA-DE-FARIA, A., ed.), pp. 554–601. Amsterdam: North-Holland.

FØYN, B., 1936: Über die Kernverhältnisse der Foraminifere *Myxotheca arenilega* SCHAUDINN. Arch. Protistenk. **87**, 272–295.

– 1937: Zur Kenntnis der asexuellen Fortpflanzung und der Entwicklung der Gamonten von *Discorbina vilardeboana* D'ORBIGNY. Årbok Bergens Museum (Naturvit.) **5**, 3–16.

FRANKE, W. W., 1967: Zur Feinstruktur isolierter Kernmembranen aus tierischen Zellen. Z. Zellforsch. **80**, 585–593.

– SCHEER, U., 1974: Structures and function of the nuclear envelope. In: The cell nucleus (BUSCH, H., ed.), Vol. 1, pp. 219–347. New York-London: Academic Press.

FRANKER, C. K., SAKHRANI, L. M., PRICHARD, C. D., LAMDEN, C. A., 1974: DNA synthesis in *Crypthecodinium cohnii*. J. Phycol. **10**, 91–94.

FRENKEL, M. A., 1969: The action of external factors on the DNA content of the macronuclei in the reproduction cysts of *Tillina magna* (Ciliata, Colpodidae). Zool. Zhurnal (Moscow) **48**, 970–979 (in Russian; English summary).

– 1975: Regulation of macronuclear DNA content in the holotrichous ciliate *Tillina magna* (Trichostomatida, Colpodidae). Protistologica **11**, 41–48.

– 1977: Morphological and photometrical investigation of chromatin extrusion from the macronucleus in resting cysts of the ciliate *Tillina magna*. Tsitologiya (Leningrad) **19**, 1029–1035 (in Russian, English summary).

Frenkel, M. A., 1978: DNA content of micronuclei and chromatin aggregates of the dividing macronucleus in reproduction cysts of the ciliate *Colpoda steini*. Tsitologiya (Leningrad) **20**, 465–469 (in Russian; English summary).

– 1979: Ultrastructure and division of the macronucleus of *Colpoda steini* and *Tillina magna* (Colpodidae). J. Protozool. 26, No. 3 (part 1), 38A.

– 1980: Fine structure of the macronucleus of the active and encysted (dividing) forms of ciliate *Colpoda steini*. Protistologica **16**, 339–351.

– Kudrjavtsev, B. N., Kudrjavtseva, M. V., 1973: DNA content of macro- and micronuclei of *Colpoda steini* before and after division in the reproduction cysts. Tsitologiya (Leningrad) **15**, 353–357 (in Russian; English summary).

Friedhoff, K., Scholtyseck, E., 1968: Feinstrukturen von *Babesia ovis* (Piroplasmida) in *Rhipicephalus bursa* (Ixodoidea): Transformation sphäroider Formen zu Vermiculaformen. Z. Parasitenk. **30**, 347–359.

– – 1969: Feinstrukturen der Merozoiten von *Babesia bigemina* im Ovar von *Boophilus microplus* und *Boophilus decoloratus*. Z. Parasitenk. **32**, 266–283.

– – 1977: Fine structural identification of erythrocytic stages of *Babesia bigemina, B. divergens* and *B. bovis*. Protistologica **13**, 195–204.

– – Weber, G., 1972: Die Feinstruktur der differenzierten Merozoiten von *Babesia ovis* in den Speicheldrüsen weiblicher Zecken. Z. Parasitenk. **38**, 132–140.

Friend, D. S., 1966: The fine structure of *Giardia muris*. J. Cell Biol. **29**, 317–332.

Fuge, H., 1974: Ultrastructure and function of the spindle apparatus microtubules and chromosomes during nuclear division. Protoplasma **82**, 289–320.

Fuhs, G. W., 1969: The nuclear structures of protocaryotic organisms (Bacteria and Cyanophyceae). Protoplasmatologia **5**, No. 4, 1–186.

Fujishima, M., Hiwatashi, K., 1978: Transplantation of germ nucleus in *Paramecium caudatum*. I. Nuclei in the pre-meiotic S phase can enter into mitotic cycle. Exp. Cell Res. **111**, 468–471.

– Watanabe, T., 1981: Transplantation of germ nuclei in *Paramecium caudatum*. III. Role of germinal micronucleus in vegetative growth. Exper. Cell Res. **132**, 47–52.

Fuller, M. S., 1976: Mitosis in Fungi. Internat. Rev. Cytol. **45**, 113–153.

Fulton, C., 1970: Amoebo-flagellates as research partners: the laboratory biology of *Naegleria* and *Tetramitus*. In: Methods in cell physiology (Prescott, D. M., ed.), Vol. 4, pp. 341–395. New York: Academic Press.

– Dingle, A. D., 1971: Basal bodies, but not centrioles, in *Naegleria*. J. Cell Biol. **51**, 826–836.

Gabe, P. R., De Bault, L. E., 1973: Macromolecular syntheses related to the reproductive cyst of *Tetrahymena patula*. J. Cell Biol. **59**, 615–623.

Gaffal, K. P., Schneider, G. J., 1978: The changes in ultrastructure during fertilization of the colourless flagellate *Polytoma papillatum* with special reference to the configural changes of their mitochondria. Cytobiologie **18**, 161–173.

Gaibova, G. D., 1972: On revealing of DNA in the nuclei of growing coccidian macrogametes. Tsitologiya (Leningrad) **14**, 1427–1429 (in Russian; English summary).

Gall, J. G., 1959: Macronuclear duplication in the ciliated protozoan *Euplotes*. J. Biophys. Biochem. Cytol. **5**, 295–308.

– 1974: Free ribosomal RNA genes in the macronucleus of *Tetrahymena*. Proc. Nat. Acad. Sci. U.S.A. **71**, 3078–3081.

– Karrer, K., Yao, M.-C., Grainger, R., 1977: The ribosomal RNA genes in *Tetrahymena*. In: The organization and expression of the eukaryotic genome (Bradbury, E. M., Javaherian, K., eds.), pp. 437–444. New York: Academic Press.

– Yao, M.-C., Blackburn, E., Findly, R. C., Wild, M., 1979: The extrachromosomal ribosomal DNA of *Tetrahymena* and *Paramecium*. In: Specific eukaryotic genes (Alfred

Benzon Symposium No. 13) (ENGBERG, J., KLENOW, H., LEICK, V., eds.), pp. 299–304. Copenhagen: Munksgaard.

GALLUCCI, B. B., 1974a: Fine structure of *Haemoproteus columbae* KRUSE during macrogametogenesis and fertilization. J. Protozool. **21**, 254–263.

– 1974b: Fine structure of *Haemoproteus columbae* KRUSE during differentiation of the ookinete. J. Protozool. **21**, 264–275.

GARBER, R. C., AIST, J. R., 1979: The ultrastructure of mitosis in *Plasmodiophora brassicae* (Plasmodiophorales). J. Cell Sci. **40**, 89–110.

GARNHAM, P. C. C., 1977: A new malaria parasite of pigeons and ducks from Venezuela. Protistologica **13**, 113–125.

– BAKER, J. R., BIRD, R. G., 1962a: The fine structure of *Lankesterella garnhami*. J. Protozool. **9**, 107–114.

– – – 1962b: Fine structure of cystic form of *Toxoplasma gondii*. Brit. Med. J. No. 1, 83–84.

– BIRD, R. G., BAKER, J. R., 1960: Electron microscope studies of motile stages of malarial parasites. I. The fine structure of the sporozoites of *Haemamoeba* (*Plasmodium*) *gallinaceum*. Trans. Roy. Soc. Trop. Med. Hyg. **54**, 274–278.

– – – 1962c: Electron microscope studies of motile stages of malaria parasites. III. The ookinetes of *Haemamoeba* and *Plasmodium*. Trans. Roy. Soc. Trop. Med. Hyg. **56**, 116–120.

– – – 1967: Electron microscope studies of motile stages of malaria parasites. V. Exflagellation in *Plasmodium*, *Hepatocystis* and *Leucocytozoon*. Trans. Roy. Soc. Trop. Med. Hyg. **61**, 58–68.

– – – BRAY, R. S., 1961: Electron microscope studies of motile stages of malaria parasites. II. The fine structure of the sporozoites of *Laverania* (*Plasmodium*) *falcipara*. Trans. Roy. Soc. Trop. Med. Hyg. **55**, 98–102.

– – – DESSER, S. S., EL-NAHAL, H. M. S., 1969: Electron microscope studies of motile stages of malaria parasites. VI. The ookinete of *Plasmodium berghei yoelii* and its transformation into the early oocyst. Trans. Roy. Soc. Trop. Med. Hyg. **63**, 187–194.

GAUDE, H., 1981: Zur Funktion der Ciliaten-Riesenchromosomen: Nachweis von RNS-Polymerase in den polytänen Chromosomen der Makronucleusanlage von *Stylonychia mytilus*. Arch. Protistenk. **124**, 252–258.

GAVIN, M. A., WANKO, T., JACOBS, L., 1962: Electron microscope studies of reproducing and interkinetic *Toxoplasma*. J. Protozool. **9**, 222–234.

GAVIN, R. H., FRANKEL, J., 1966: The effects of mercaptoethanol on cellular development in *Tetrahymena pyriformis*. J. Exper. Zool. **161**, 63–82.

GEITLER, L., 1941: Das Wachstum des Zellkerns in tierischen und pflanzlichen Geweben. Ergebn. Biol. **18**, 1–54.

– 1953: Endomitose und endomitotische Polyploidisierung. Protoplasmatologia **6**, Teil C, 1–89.

GEORGIEV, G. P., 1974: Precursor of mRNA (pre-mRNA) and ribonucleoprotein particles containing pre-mRNA. In: The cell nucleus (BUSCH, H., ed.), Vol. 3, pp. 67–108. New York-London: Academic Press.

GERASSIMOVA, Z. P., SERAVIN, L. N., 1978: Fine structure of the ciliate *Epidinium ecaudatum* (Entodiniomorpha, Ophryoscolecidae) from the rumen of the ruminants. In: Fauna i Systematika Odnokletochnych Zhivotnych (Proceed. Zool. Inst. Acad. Sci. USSR, Vol. 78) (KRYLOV, M. V., ed.), pp. 44–50. Leningrad: Zool. Institute (in Russian).

GIBBS, S. P., 1962: Nuclear envelope-chloroplast relationships in algae. J. Cell Biol. **14**, 433–444.

– 1970: Nuclear and organelle RNA synthesis in *Ochromonas*: the effects of light. J. Cell Biol. **46**, 559–604.

GIBBS, S. P., 1979: The route of entry of cytoplasmatically synthesized proteins into chloroplasts of algae possessing chloroplast ER. J. Cell Sci. **35**, 253–266.

– 1981: The chloroplast endoplasmic reticulum: structure, function, and evolutionary significance. Int. Rev. Cytol. **72**, 49–99.

GIBSON, I., MARTIN, N., 1971: DNA amounts in the nuclei of *Paramecium aurelia* and *Tetrahymena pyriformis*. Chromosoma **35**, 374–382.

GICQUAUD, C. R., 1979: Étude de l'ultrastructure du noyau et de la mitose de *Entamoeba histolytica*. Biol. Cellul. **35**, 305–312.

GIESBRECHT, P., 1961: Über das „Supercoiling"-System der Chromosomen von Bakterien und Flagellaten und seine Beziehungen zu Nucleolus und Kerngrundsubstanz. Zbl. Bakteriol., Abt. I (Orig.) **183**, 1–44.

– 1962: Vergleichende Untersuchungen an den Chromosomen des Dinoflagellaten *Amphidinium elegans* und denen der Bakterien. Zbl. Bakteriol., Abt. I (Orig.) **187**, 452–498.

– 1965: Über das Ordnungsprinzip in den Chromosomen von Dinoflagellaten und Bakterien. Zbl. Bakteriol., Abt. I (Orig.) **196**, 516–519.

GIESE, A. C., 1970: Macromolecular synthesis during regeneration in *Blepharisma* determined by specific inhibitors and incorporation of ^{14}C-tracers. Exp. Cell Res. **61**, 91–102.

GIL, R., 1976: Possible nuclear-cytoplasmic transfers in the exconjugants of *Stylonychia mytilus*: ultrastructural observations. Microbiol. Esp. **29**, 125–134.

– ALONSO, P., PÉREZ-SILVA, J., 1972: Ultrastructure of the macronuclear anlage in *Stylonychia mytilus*. Exper. Cell Res. **72**, 509–518.

GILLOTT, M. A., TRIEMER, R. E., 1978: The ultrastructure of cell division in *Euglena gracilis*. J. Cell Sci. **31**, 25–35.

GINSBURGER-VOGEL, T., DESPORTES, I., 1979: Étude ultrastructurale de la sporulation de *Paramarteilia orchestiae* gen. n. sp. n., parasite de l'amphipode *Orchestia gammarellus* (PALLAS). J. Protozool. **26**, 390–403.

GODWARD, M. B. E., 1966: The Cryptophyceae. In: The chromosomes of the algae (GODWARD, M. B. E., ed.), pp. 116–117. London: E. Arnold.

GOLDER, T. K., 1976: The macro-micronuclear complex of *Woodruffia metabolica*. J. Ultrastr. Res. **54**, 169–175.

GOLDSTEIN, L., PRESCOTT, D. M., 1967: Nucleocytoplasmic interactions in the control of nuclear reproduction and other cell cycle stages. In: The control of nuclear activity (GOLDSTEIN, L., ed.), pp. 3–17. Englewood Cliffs, N.J.: Prentice-Hall.

GOLIKOVA, M. N., 1964: Polytene chromosomes in developing macronuclei of ciliates. Tsitologiya (Leningrad) **6**, 250–253 (in Russian; English summary).

– 1965: Der Aufbau des Kernapparates und die Verteilung der Nukleinsäuren und Proteine bei *Nyctotherus cordiformis* STEIN. Arch. Protistenk. **108**, 191–216.

– 1972: Polymorphism of the micronuclei in *Paramecium bursaria* FOCKE. Tsitologiya (Leningrad) **14**, 637–646 (in Russian; English summary).

– 1974: A cytophotometric study of the DNA content in the micronucleus of *Paramecium bursaria*. Acta Protozool. **13**, 109–121 (in Russian; English summary).

– 1978: Morphology and viability of amicronucleate clones of the ciliate *Paramecium bursaria*. Acta Protozool. **17**, 89–108 (in Russian; English summary).

– NILOVA, V. K., 1967: On DNA synthesis in the macronuclear anlage of *Nyctotherus cordiformis*. Tsitologiya (Leningrad) **9**, 465–469 (in Russian; English summary).

GOLIŃSKA, K., 1965: Macronucleus in *Dileptus cygnus* and its changes in division. Acta Protozool. **3**, 143–152.

GOMEZ, M. P., ALDRICH, H. C., WALNE, P. L., 1975: Ultrastructural observations of age-related changes in the nuclear envelope of *Euglena gracilis* "Z". J. Protozool. **22**, 8A.

GOODE, D., 1965: Isolation and fine structure of the mitotic apparatus from the giant ameba. J. Cell Biol. **27**, 36A.

– 1975: Evolution of mitosis in Protozoa: the association of chromosomes, nuclear envelope, kinetochores and microtubules. BioSystems 7, 318–325.

– ROTH, L. E., 1969: The mitotic apparatus of a giant ameba. Solubility properties and induction of elongation. Exper. Cell Res. **58**, 343–352.

GOODFELLOW, L. P., BELCHER, J. H., PAGE, F. C., 1974: A light- and electron-microscopical study of *Sappinia diploidea*, a sexual amoeba. Protistologica **10**, 207–216.

GORDEYEVA, L. M., SOLOVJEV, M. M., 1967: Nucleic acids in the interphasic and dividing nucleus of *Entamoeba* species (Protozoa, Sarcodina). In: Struktura i funksiya kletochnogo yadra, pp. 165–168. Moscow: Nauka Press (in Russian).

GOROVSKY, M. A., 1965: Electron microscopic cytochemistry of isolated nuclei of *Tetrahymena*. J. Cell Biol. **27**, 37A.

– 1970a: Studies on nuclear structure and function in *Tetrahymena pyriformis*. II. Isolation of macro- and micronuclei. J. Cell Biol. **47**, 619–630.

– 1970b: Studies on nuclear structure and function in *Tetrahymena pyriformis*. III. Comparison of the histones of macro- and micronuclei by quantitative polyacrylamide gel electrophoresis. J. Cell Biol. **47**, 631–636.

– 1973: Macro- and micronuclei of *Tetrahymena pyriformis:* a model system for studying the structure and function of eukaryotic nuclei. J. Protozool. **20**, 19–25.

– 1980: Genome organization and reorganization in *Tetrahymena*. Ann. Rev. Genetics **14**, 203–239.

– GLOVER, C., JOHMANN, C. A., KEEVERT, J. B., MATHIS, D. J., SAMUELSON, M., 1978: Histones and chromatin structure in *Tetrahymena* macro- and micronuclei. Cold Spring Harbor Symp. Quant. Biol. **42**, 493–503.

– KEEVERT, J. B., 1975: Subunit structure of a naturally occurring chromatin lacking histones F_1 and F_3. Proc. Nat. Acad. Sci. U.S.A. **72**, 3536–3540.

GOULDER, R., 1980: The ecology of two species of primitive ciliated protozoa commonly found in standing freshwaters (*Loxodes magnus* STOKES and *L. striatus* PENARD). Hydrobiologia **72**, 131–158.

GOURRET, J.-P., 1978: Description et interprétation des nucléoïdes structurés observés dans des bactéroïdes de *Rhizobium*. Biol. Cellul. **32**, 299–306.

GRAIN, J., 1966: Étude cytologique de quelques Ciliés holotriches endocommensaux des Ruminants et des Equidés. Protistologica **2**, No. 2, 5–52.

– 1970: Structure et ultrastructure de *Lagynophrya fusidens* KAHL 1927. Protistologica **6**, 37–51.

– 1972: Étude ultrastructurale *d'Halteria grandinella* O.F.M. (Cilié oligotriche) et considérations phylogénétiques. Protistologica **8**, 179–197.

GRASSÉ, P.-P., 1939: Étude de mécanisme cellulaire: centromères et centrosomes dans la mitose de certains Flagellés. C. r. Soc. Biol. **131**, 1015–1018.

– 1952: Embranchement des Protozoaires. Generalités. In: Traité de Zoologie (GRASSÉ, P.-P., ed.), Vol. 1, fasc. 1, pp. 37–132. Paris: Masson et Cie.

– 1956: L'ultrastructure de *Pyrsonympha vertens* (Zooflagellata Pyrsonymphina): les flagelles et leur coaptation avec le corps, l'axostyle contractile, le paraxostyle, le cytoplasme. Arch. Biol. **67**, 595–611.

– 1960: Les Myxosporidies sont des organismes pluricellulaires. C. r. Acad. Sci. Paris **251**, 2638–2640.

– DRAGESCO, J., 1957: L'ultrastructure du chromosome des Péridiniens et ses conséquences génétiques. C. r. Acad. Sci. Paris **245**, 2447–2452.

– FAURE, A., 1939: Quelques données nouvelles sur la cytologie et la reproduction de *Trichomonas caviae* DAV. Bull. Biol. France et Belgique **73**, 1–18.

GRASSÉ, P.-P., HOLLANDE, A., 1951: Cytologie et mitose des *Pseudotrichonympha* GRASSI et FOÀ 1911. Ann. Sci. Natur., Zool., sér. 11, **13**, 237–248.
– – 1963: Les flagellés des genres *Holomastigotoides* et *Rostronympha*. Structure et cycle de spiralisation des chromosomes chez *Holomastigotoides psammotermitidis*. Ann. Sci. Natur., Zool., sér. 12, **5**, 749–792.
– – CACHON, J., CACHON-ENJUMET, M., 1965a: Nouvelle interprétation de l'ultrastructure du chromosome de certains Péridiniens (*Prorocentrum*, *Gymnodinium*, *Amphidinium*, *Plectodinium* et Xanthelles d'Anémones). C. r. Acad. Sci. Paris **260**, 1743–1747.
– – – – 1965b: Interprétation de quelques aspects infrastructuraux des chromosomes de Péridiniens en division. C. r. Acad. Sci. Paris **260**, 6575–6978.
– LAVETTE, A., 1978: La myxosporidie *Sphaeromyxa sabrazesi* et le nouvel embranchement des myxozoaires (Myxozoa). Recherches sur l'état pluricellulaire primitif et considérations phylogénétiques. Ann. Sci. Nat., Zool. et Biol. Anim. (12e sér.) **20**, 193–285.
– THÉODORIDÈS, J., 1957: L'ultrastructure de la membrane nucléaire des Grégarines. C. r. Acad. Sci. Paris **245**, 1985–1986.
– – 1959: Recherches sur l'ultrastructure de quelques Grégarines. Ann. Sci. Natur., Zool., sér. 12, **1**, 296–305.
GREGORY, L. H., 1923: The conjugation of *Oxytricha fallax*. J. Morphol. **37**, 555–581.
GREIDER, M. H., KOSTIR, W. J., FRAJOLA, W. J., 1956: Electron microscopy of the nuclear membrane of *Amoeba proteus*. J. Biophys. Biochem. Cytol. **2**, suppl., 445–448.
– – – 1958: Electron microscopy of *Amoeba proteus*. J. Protozool. **5**, 139–146.
GRELL, K. G., 1940: Der Kernphasenwechsel von *Stylocephalus* (*Stylorhynchus*) *longicollis* F. STEIN. Arch. Protistenk. **94**, 161–200.
– 1949: Die Entwicklung der Makronucleusanlage im Exkonjuganten von *Ephelota gemmipara* R. HERTWIG. Biol. Zbl. **68**, 289–312.
– 1950a: Der Generationswechsel des parasitischen Suktors *Tachyblaston ephelotensis* MARTIN. Z. Parasitenk. **14**, 499–534.
– 1950b: Der Kerndualismus der Ciliaten und Suctorien. Naturwissenschaften **37**, 347–356.
– 1953a: Entwicklung und Geschlechtsbestimmung von *Eucoccidium dinophili*. Arch. Protistenk. **99**, 156–186.
– 1953b: Der Stand unserer Kenntnisse über den Bau der Protistenkerne. Verhandl. Deutsch. Zool. Ges. (Freiburg), 212–251.
– 1953c: Die Chromosomen von *Aulacantha scolymantha* HAECKEL. Arch. Protistenk. **99**, 1–54.
– 1953d: Die Struktur des Makronucleus von *Tokophrya*. Arch. Protistenk. **98**, 466–468.
– 1953e: Die Konjugation von *Ephelota gemmipara* R. HERTWIG. Arch. Protistenk. **98**, 287–326.
– 1954: Der Generationswechsel der polythalamen Foraminifere *Rotaliella heterocaryotica*. Arch. Protistenk. **100**, 268–286.
– 1956a: Protozoologie. Berlin-Göttingen-Heidelberg: Springer.
– 1956b: Über die Elimination somatischer Kerne bei heterokaryotischen Foraminiferen. Z. Naturforsch. **11b**, 759–761.
– 1957: Untersuchungen über die Fortpflanzung und Sexualität der Foraminiferen. I. *Rotaliella roscoffensis*. Arch. Protistenk. **102**, 147–164.
– 1958a: Untersuchungen über die Fortpflanzung und Sexualität der Foraminiferen. II. *Rubratella intermedia*. Arch. Protistenk. **102**, 291–308.
– 1958b: Untersuchungen über die Fortpflanzung und Sexualität der Foraminiferen. III. *Glabratella sulcata*. Arch. Protistenk. **102**, 449–472.
– 1958c: Studien zum Differenzierungsproblem an Foraminiferen. Naturwissenschaften **45**, 25–32.

GRELL, K. G., 1959: Untersuchungen über die Fortpflanzung und Sexualität der Foraminiferen. IV. *Patellina corrugata*. Arch. Protistenk. **104**, 211–235.

– 1961: Über den „Nebenkörper" von *Paramoeba eilhardi* SCHAUDINN. Arch. Protistenk. **105**, 303–312.

– 1962: Entwicklung und Geschlechtsdifferenzierung einer neuen Foraminifere. Naturwissenschaften **49**, 214.

– 1964: The protozoan nucleus. In: The cell (BRACHET, J., MIRSKY, A., eds.), Vol. 6, pp. 1–79. New York-London: Academic Press.

– 1967: Sexual reproduction in Protozoa. In: Research in protozoology (CHEN, T. T., ed.), Vol. 2, pp. 148–213. Oxford-New York: Pergamon Press.

– 1973: Protozoology. Berlin-Heidelberg-New York: Springer.

– 1979: Cytogenetic systems and evolution in Foraminifera. J. Foram. Res. **9**, 1–13.

– BARDELE, C. F., 1977: Light and electron microscopical studies on the foraminiferan *Rotaliella heterocaryotica*. Abstr. V Internat. Congr. Protozool., p. 369. New York.

– BENWITZ, G., 1970: Ultrastruktur mariner Amöben. I. *Paramoeba eilhardi* SCHAUDINN. Arch. Protistenk. **112**, 119–137.

– – 1978: Ultrastruktur mariner Amöben. IV. *Corallomyxa chattoni* n. sp. Arch. Protistenk. **120**, 287–300.

– RUTHMANN, A., 1964: Über die Karyologie des Radiolars *Aulacantha scolymantha* und die Feinstruktur seiner Chromosomen. Chromosoma **15**, 158–211.

– SCHWALBACH, G., 1965: Elektronenmikroskopische Untersuchungen an den Chromosomen der Dinoflagellaten. Chromosoma **17**, 230–245.

– WOHLFARTH-BOTTERMANN, K. E., 1957: Licht- und elektronenmikroskopische Untersuchungen an den Dinoflagellaten *Amphidinium elegans* n. sp. Z. Zellforsch. **47**, 7–17.

GREUET, C., 1972: Intervention de lamelles annelées dans la formation de couches squelettiques au niveau de la capsule périnucléaire de péridiniens Warnowiidae. Protistologica **8**, 155–168.

GRIMES, G. W., 1973: Differentiation during encystment and excystment in *Oxytricha fallax*. J. Protozool. **20**, 92–104.

GRIMSTONE, A. V., 1959: Cytoplasmic membranes and the nuclear membrane in the flagellate *Trichonympha*. J. Biophys. Biochem. Cytol. **6**, 369–378.

GROLIÈRE, C. A., DE PUYTORAC, P., DETCHEVA, R., 1980: A propos d'observations sur la stomatogenèse et l'ultrastructure du Cilié *Protocruzia tuzeti* VILLENEUVE-BRACHON 1940. Protistologica **16**, 453–466.

GROMOVA, E. N., 1941: Effect of external factors on the macronuclear structure in *Paramecium caudatum*. Zool. Zhurnal (Moscow) **20**, 187–197 (in Russian).

– 1948: The dynamics of nucleic acids during conjugation in *Paramecium caudatum*. Doklady Akad. Nauk SSSR **63**, 73–75 (in Russian).

GRUOL, D. J., HASELKORN, R., 1976: Counting the genes for stable RNA in the nucleus and chloroplasts of *Euglena*. Biochim. Biophys. Acta **447**, 82–95.

GUÉNIN, H.-A., FAHRNI, J., 1971: La structure fine des noyaux lors du bourgeonnement chez *Spirochona gemmipara* STEIN (Cilié Chonotriche). Rev. Suisse Zool. **78**, 556–559.

GUILCHER, Y., 1951: Contribution à l'étude des Ciliés gemmipares, chonotriches et tentaculifères. Ann. Sci. Natur., Zool., sér. 11, **13**, 33–133.

GUSTAFSON, P. V., AGAR, H. D., CRAMER, D. I., 1954: An electron microscope study of *Toxoplasma*. Amer. J. Trop. Med. Hyg. **3**, 1008–1022.

GUTIERREZ, J. C., TORRES, A., PÉREZ-SILVA, J., 1981: Excystment, cortical morphogenesis, and nuclear processes during encystment in *Laurentiella acuminata* (Hypotrichida, Oxytrichidae). Acta Protozool. **20**, 145–152.

GUTTES, E., GUTTES, S., 1960: Incorporation of tritium-labeled thymidine into the macronucleus of *Stentor coeruleus*. Exper. Cell Res. **19**, 626–628.

HAAPALA, O. K., SOYER, M.-O., 1973: Structure of dinoflagellate chromosomes. Nature **244**, 195–197.

– – 1974a: Size of circular chromatids and amounts of haploid DNA in the dinoflagellates *Gyrodinium cohnii* and *Prorocentrum micans*. Hereditas **76**, 83–90.

– – 1974b: Absence of longitudinal differentiation of dinoflagellate (*Prorocentrum micans*) chromosomes. Hereditas **78**, 141–145.

– – 1974c: Electron microscopy of whole-mounted chromosomes of the dinoflagellate *Gyrodinium cohnii*. Hereditas **78**, 146–150.

– – 1974d: Effect of alkali-urea treatment on dinoflagellate chromosomes. Hereditas **78**, 295–297.

– – 1975: Organization of chromosome fibrils in *Euglena gracilis*. Hereditas **80**, 185–194.

HAFKINE, W. M., 1890: Maladies infectieuses des Paramécies. Ann. Inst. Pasteur (Paris) **4**, 148–162.

HALL, R. P., 1923: Morphology and binary fission of *Menoidium incurvum* (FRES.) Klebs. Univ. Calif. Publs Zool. **20**, 447–476.

– 1925a: Binary fission in *Oxyrrhis marina* DUJARDIN. Univ. Calif. Publs Zool. **26**, 281–324.

– 1925b: Mitosis in *Ceratium hirundinella* O.F.M., with notes on nuclear phenomena in encysted forms and the question of sexual reproduction. Univ. Calif. Publs Zool. **28**, 29–64.

– 1937: A note on behaviour of the chromosomes in *Euglena*. Trans. Amer. Microscop. Soc. **56**, 288–290.

HALLER, G. DE, 1963: Variations dans l'ultrastructure du noyau chez *Amphidinium elegans* (Gymnodiniidae). In: Progress in protozoology (Proc. I Internat. Congr. Protozool.), pp. 423–424. Prague: Publ. House ČSAV.

– KELLENBERGER, E., ROUILLER, CH., 1964: Étude au microscope électronique des plasmas contenant de l'acide désoxyribonucléique. III. Variations ultrastructurales des chromosomes *d'Amphidinium*. J. Microscopie **3**, 627–642.

HAMBURGER, C., 1904: Die Konjugation von *Paramecium bursaria* FOCKE. Arch. Protistenk. **4**, 199–239.

HAMKALO, B. A., RATTNER, J. B., 1977: The structure of a mesokaryote chromosome. Chromosoma **60**, 39–47.

HAMMOND, D. M., ROBERTS, W. L., YOUSSEF, N. N., DANFORTH, H. D., 1973: Fine structure of the intranuclear spindle poles in *Eimeria callospermophili* and *E. magna*. J. Parasitol. **59**, 581–587.

– SCHOLTYSECK, E., CHOBOTAR, B., 1969: Fine structural study of the microgametogenesis of *Eimeria auburnensis*. Z. Parasitenk. **33**, 65–84.

– – MINER, M. L., 1967: The fine structure of microgametocytes of *Eimeria perforans*, *E. stiedae*, *E. bovis*, and *E. auburnensis*. J. Parasitol. **53**, 235–247.

HARRIS, P., JAMES, T. W., 1952: Electron microscope study of the nuclear membrane of *Amoeba proteus* in thin section. Experientia **8**, 384–385.

HARRISON, F. W., DUNKELBERGER, D., WATABE, N., STUMP, A. B., 1975: The cytology of the testaceous rhizopod *Lesquereusia spiralis* (EHRENBERG) PENARD. I. Cytochemistry. Acta Histochem. **54**, 71–77.

– – – – 1976: The cytology of the testaceous rhizopod *Lesquereusia spiralis* (EHRENBERG) PENARD. II. Ultrastructure and shell formation. J. Morphol. **150**, 343–358.

HARTMANN, M., 1909: Polyenergide Kerne. Studien über multiple Kernteilungen und generative Chromidien bei Protozoen. Biol. Zbl. **29**, 481–506.

– 1952: Polyploide (polyenergide) Kerne bei Protozoen. Arch. Protistenk. **98**, 125–156.

Hauschka, T., 1943: Life history and chromosome cycle of the coccidian *Adelina deronis*. J. Morphol. **73**, 529–581.

Hauser, M., 1968: Die Bildung von Spindelfasern im Mikronucleus des Suktors *Paracineta limbata*. Z. Naturforsch. **23b**, 887.

– 1970: Elektronenmikroskopische Untersuchung an dem Suktor *Paracineta limbata* Maupas. Z. Zellforsch. **106**, 584–614.

– 1972: The intranuclear mitosis of the ciliates *Paracineta limbata* and *Ichthyophthirius multifiliis*. I. Electron microscope observations on pre-metaphase stages. Chromosoma **36**, 158–175.

– 1973: Aktomyosinartige Filamente im Teilungsmakronukleus des Ciliaten *Ichthyophthirius multifiliis*. Chromosoma **44**, 49–71.

– Beinbrech, G., 1973: Deuteriumoxid-induzierte Filamentaggregation im Mikronucleus eines Ciliaten. Z. Naturforsch. **28c**, 339–341.

– – Groschel-Stewart, U., Jockusch, B. M., 1975: Localization by immunological techniques of myosin in nuclei of lower eukaryotes. Exper. Cell Res. **95**, 127–135.

Hawes, R. S. J., 1963: On *Rosculus ithacus* gen. n. sp. n. (Protozoa, Amoebina), with special reference to its mitosis and phylogenetic relations. J. Morphol. **113**, 139–147.

Hazard, E. I., Andreadis, T. G., Joslyn, D. J., Ellis, E. A., 1979: Meiosis and its implication in the life cycles of *Amblyospora* and *Parathelohania* (Microspora). J. Parasitol. **65**, 117–122.

Heath, I. B., 1975: The possible significance of variations in the mitotic systems of the aquatic fungi (Phycomycetes). BioSystems **7**, 351–359.

– 1980: Variant mitoses in lower eukaryotes: indicators of the evolution of mitosis? Int. Rev. Cytol. **64**, 1–80.

– 1981: Nucleus-associated organelles in fungi. Int. Rev. Cytol. **69**, 191–221.

Heckmann, K., 1966: *Tokophrya lemnarum* (Suctoria). Nahrungsaufnahme und Schwärmerbildung. Publ. Inst. Wiss. Film (Göttingen) **1A**, 475–482.

Hedin, H., 1975: On the ultrastructure of *Favella ehrenbergii* (Claparède et Lachmann) and *Parafavella gigantea* (Brandt) (Protozoa, Ciliata, Tintinnida). Zoon **3**, 11–18.

Hedley, R. H., Ogden, C. G., 1973: Biology and fine structure of *Euglypha rotunda* (Testacea: Protozoa). Bull. Brit. Mus. Nat. Hist., Zool. **25**, 119–137.

– – 1974: Observations on *Trinema lineare* Penard (Testacea: Protozoa). Bull. Brit. Mus. Nat. Hist., Zool. **26**, 187–200.

– Parry, D. M., Wakefield, J. St. J., 1967: Fine structure of *Shepheardella taeniformis* (Foraminifera: Protozoa). J. Roy. Microscop. Soc. **87**, 445–456.

– Wakefield, J. St. J., 1969: Fine structure of *Gromia oviformis* (Rhizopodea: Protozoa). Bull. Brit. Mus. Nat. Hist., Zool. **18**, 69–89.

Heimke, J. W., Starr, R. C., 1979: The sexual process in several heterogamous *Chlamydomonas* strains in the subgenus *Pleiochloris*. Arch. Protistenk. **122**, 20–42.

Heller, G., 1969: Elektronenmikroskopische Untersuchungen an *Aggregata eberthi* aus dem Spiraldarm von *Sepia officinalis* (Sporozoa, Coccidia). I. Die Feinstrukturen der Merozoiten, Makrogameten und Sporen. Z. Parasitenk. **33**, 44–64.

– 1970: Elektronenmikroskopische Untersuchungen an *Aggregata eberthi* aus dem Spiraldarm von *Sepia officinalis* (Sporozoa, Coccidia). II. Die Entwicklung der Mikrogameten. Z. Parasitenk. **33**, 183–193.

Heller, I., Kopac, M., 1955: Cytochemical reactions of normal and starving *Amoeba proteus*. I. Localization of thiols, DNA, RNA and other organic phosphates. Exper. Cell Res. **8**, 62–75.

Helson, L., Pecora, P., Hirshfield, H. I., 1959: Macronuclear changes in a strain of *Blepharisma undulans* during the divisional cycle. J. Protozool. **6**, 131–135.

Henneré, E., 1967: Étude cytologique des premiers stades du dévéloppement d'une Coccidie: *Myriosporides amphiglenae*. J. Protozool. **14**, 27–39.

Hepler, P. K., Huff, C. G., Sprinz, H., 1966: The fine structure of the exoerythrocytic stages of *Plasmodium fallax*. J. Cell Biol. **30**, 333–358.

Herlan, G., Eckert, W. A., Kaffenberger, W., Wunderlich, F., 1979: Isolation and characterization of an RNA-containing nuclear matrix from *Tetrahymena* macronuclei. Biochemistry **18**, 1782–1788.

– Wunderlich, F., 1976: Isolation of a nuclear protein matrix from *Tetrahymena* macronuclei. Cytobiologie **13**, 291–296.

Hermolin, J., Zimmerman, A. M., 1976: RNA synthesis in division synchronized *Tetrahymena*: macronuclear and cytoplasmic RNA. J. Protozool. **23**, 594–600.

Herrick, G., Wesley, R. D., 1978: Isolation and characterization of a highly repetitious inverted terminal repeat sequence from *Oxytricha* macronuclear DNA. Proc. Nat. Acad. Sci. U.S.A. **75**, 2626–2630.

Hertwig, R., 1879: Der Organismus der Radiolarien. Jena: G. Fischer.

– 1889: Über die Conjugation der Infusorien. Abhandl. Königl. Bayer. Akad. Wiss., Math.-Phys. Cl. **17**, 151–233.

Herzog, M., Soyer, M.-O., 1981: Distinctive features of dinoflagellate chromatin. Absence of nucleosomes in a primitive species *Prorocentrum micans* E. Europ. J. Cell Biol. **23**, 295–302.

Heywood, P., 1974: Mitosis and cytokinesis in the chloromonadophycean alga *Gonyostomum semen*. J. Phycol. **10**, 355–358.

– 1976: Ultrastructural characterization of the interphase nucleus of *Gonyostomum semen*. Cytobios **17**, 79–86.

– 1978: Ultrastructure of mitosis in the chloromonadophycean alga *Vacuolaria virescens*. J. Cell Sci. **31**, 37–51.

– Godward, M. B. E., 1972: Centromeric organization in the chloromonadophycean alga *Vacuolaria virescens*. Chromosoma **39**, 333–339.

– – 1974: Mitosis in the alga *Vacuolaria virescens*. Amer. J. Bot. **61**, 311–338.

– Weinman, D., 1978: Mitosis in the hemoflagellate *Trypanosoma cyclops*. J. Protozool. **25**, 287–292.

– – Lipman, M., 1974: Fine structure of *Trypanosoma cyclops* in noncellular cultures. J. Protozool. **21**, 232–238.

Hibberd, D. J., 1970: Observations on the cytology and ultrastructure of *Ochromonas tuberculatus* sp. nov. (Chrysophyceae) with special reference to the discobolocysts. Brit. Phycol. J. **5**, 119–143.

– 1971: Observations on the cytology and ultrastructure of *Chrysamoeba radians* Klebs (Chrysophyceae). Brit. Phycol. J. **6**, 207–223.

– 1975: Observations on the ultrastructure of the choanoflagellate *Codosiga botrytis* (Ehr.) Saville-Kent with special reference to the flagellar apparatus. J. Cell Sci. **17**, 191–219.

– 1976: Observations on the ultrastructure of three new species of *Cyathobodo* Petersen et Hansen (*C. salpinx*, *C. intricatus*, and *C. simplex*) and on the external morphology of *Pseudodendromonas vlkii* Bourrely. Protistologica **12**, 249–261.

Higashinakagawa, T., Sezaki, M., Kondo, Sh., 1979: Isolation of nucleoli from *Tetrahymena pyriformis*. Devel. Biol. **69**, 601–611.

Hildebrand, H. F., 1974: Observations ultrastructurales sur le stade plasmodial de *Metchnikovella wohlfarthi* Hildebrand et Vivier 1971, Microsporidie hyperparasite de la Grégarine *Lecudina tuzetae*. Protistologica **10**, 5–15.

– 1976: Elektronenmikroskopische Untersuchungen an den Entwicklungsstadien des Trophozoiten von *Didymophyes gigantea* (Sporozoa, Gregarinida). I. Die Feinstruktur des Proto- und Epimeriten und die Beziehung zwischen Wirt und Parasit. Z. Parasitenk. **49**, 193–215.

Hildebrand, H. F., 1978: Elektronenmikroskopische Untersuchungen an den Entwicklungsstadien des Trophozoiten von *Didymophyes gigantea* (Sporozoa, Gregarinida). II. Die Feinstruktur des Deutomeriten mit besonderer Berücksichtigung der Kernteilungen und des Golgi-Apparates. Z. Parasitenk. **55**, 9–27.

– Vivier, E., 1971: Observations ultrastructurales sur le sporoblaste de *Metchnikovella wohlfarthi* n. sp. (Microsporidies), parasite de la Grégarine *Lecudina tuzetae*. Protistologica **7**, 131–139.

Hill, F. G., Outka, D. E., 1974: The structure and origin of mastigonemes in *Ochromonas minuta* and *Monas* sp. J. Protozool. **21**, 299–312.

Himes, M., Beam, C. A., 1975: Genetic analysis in the dinoflagellate *Crypthecodinium* (*Gyrodinium*) *cohni*: evidence for unusual meiosis. Proc. Nat. Acad. Sci. U.S.A. **72**, 4546–4549.

Hinchey, M. C., 1937: Mitosis in the giant amoeba, *Chaos chaos* Linnaeus. Biol. Bull. **73**, 368–369.

Hodge, F. A., McClellan, J. F., 1966: Deoxyribonucleic acid synthesis in the giant amoeba (*Pelomyxa carolinensis*). J. Protozool. **13**, suppl., 26–27.

Hollande, A., 1942: Étude cytologique et biologique de quelques Flagellés libres. Volvocales, Cryptomonadines, Eugleniens, Protomastigines. Arch. Zool. Expér. Génér. **83**, 1–268.

– 1972: Le déroulement de la cryptomitose et les modalités de la ségregation des chromatides dans quelques groupes de Protozoaires. Année Biol. **11**, 427–466.

– 1974a: Étude comparée de la mitose syndinienne et de celle des Péridiniens libres et de Hypermastigines. Infrastructure et cycle évolutif des Syndinidés parasites de Radiolaires. Protistologica **10**, 413–451.

– 1974b: Données ultrastructurales sur les isospores des Radiolaires. Protistologica **10**, 567–572.

– 1979: Complexes parabasaux et symétrie bilatérale du rostre chez *Joenia annectens*. Protistologica **15**, 407–426.

– 1980: Identification du parasome (Nebenkern) de *Janickina pigmentifera* à un symbionte (*Perkinsiella amoebae* nov. gen., nov. sp.) apparenté aux Flagellés kinetoplastidiés. Protistologica **16**, 613–625.

– 1981: Le problème de la sporogenèse des Phaeodariés. Protistologica **17**, 161–176.

– Cachon, J., Cachon-Enjumet, M., 1963: Considérations sur la structure et l'ultrastructure du nucléole de quelques Péridiniens libres. C. r. Acad. Sci. Paris **256**, 3193–3195.

– – Cachon, M., 1969: La dinomitose atractophorienne à fuseau endonucléaire chez les Radiolaires Thalassophysidae. Son homologie avec la mitose des Foraminifères et avec celle des levures. C. r. Acad. Sci. Paris **D 269**, 179–182.

– – Valentin, J., 1967: Infrastructure des axopodes et organisation générale de *Sticholonche zanclea* (Radiolaire Sticholonchidea). Protistologica **3**, 155–166.

– Cachon-Enjumet, M., 1959a: La polyploidie du noyau végétatif des Radiolaires. C. r. Acad. Sci. Paris **248**, 2641–2643.

– – 1959b: Origine, structure et évolution des nucléoles chez les Radiolaires (Collodaires et Sphaerellaires). C. r. Acad. Sci. Paris **249**, 167–169.

– Carré, D., 1974: Les Xanthelles des Radiolaires Sphaerocollides, des Acanthaires et de *Velella velella*: infrastructure – cytochimie – taxonomie. Protistologica **10**, 573–601.

– Carruette-Valentin, J., 1970a: Appariement chromosomique et complexes synaptonématiques dans les noyaux en cours de dépolyploidisation chez *Pyrsonympha flagellata*: le cycle évolutif des Pyrsonymphines symbiontes de *Reticulitermes lucifugus*. C. r. Acad. Sci. Paris **D 270**, 2550–2553.

Hollande, A., Carruette-Valentin, J., 1970b: Interprétation générale des structures rostrales des Hypermastigines et modalités de la pleuromitose chez les Flagellés du genre *Trichonympha*. C. r. Acad. Sci. Paris **D 270**, 1476–1479.

– – 1971: Les atractophores, l'induction du fuseau et la division cellulaire chez les Hypermastigines. Étude infrastructurale et révision systématique des Trichonymphines et des Spirotrichonymphines. Protistologica **7**, 5–100.

– – 1972: Le problème du centrosome et la cryptopleuromitose atractophorienne chez *Lophomonas striata*. Protistologica **8**, 267–278.

– Desportes, I., Pyne, Ch., 1980: Visualisation de la transcription chez un Protozoaire, le Radiolaire *Thalassicolla nucleata*. Biol. Cellulaire **38**, No. 3, 28a.

– Enjumet, M., 1953: Contribution à l'étude biologique des Sphaerocollides (Radiolaires Collodaires et Radiolaires Polycyttaires) et de leurs parasites. I. Thalassicollidae, Physematidae, Thalassophysidae. Ann. Sci. Natur., Zool., sér. 11, **15**, 99–183.

– – 1960: Cytologie, évolution et systématique des Sphaeroidés (Radiolaires). Arch. Mus. Nat. Hist. Natur., sér. 7, **7**, 1–134.

– – Manciet, J., 1953: Les Péridiniens parasites des Phaéodariés et le problème de la sporogénèse chez les Radiolaires. C. r. Acad. Sci. Paris **236**, 1607–1609.

– Nicholas, G., Escaig, J., 1981: Vêture glycostylaire et ultrastructure d'une amibe marine libre (*Mayorella pussardi* nov. sp.: Paramoebidae) observée après congélation ultrarapide suivie de cryosubstitution. Protistologica **17**, 147–154.

– Valentin, J., 1967a: Interprétation des structures dites "centriolaires" chez les Hypermastigines symbiontes des Termites et du *Cryptocercus*. C. r. Acad. Sci. Paris **D 264**, 1868–1871.

– – 1967b: Relations entre cinétosomes, "atractophores" et complexe fibrillaire axostyloparabasal, chez les Hypermastigines du genre *Barbulanympha*. C. r. Acad. Sci. Paris **D 264**, 3020–3022.

– – 1967c: Morphologie et infrastructure du genre *Barbulanympha*, Hypermastigine symbiontique de *Cryptocercus punctulatus* Scudder. Protistologica **3**, 257–267.

– – 1968a: Données critiques sur la pleuromitose et affinités entre Trichomonadines et Joeniides. C. r. Acad. Sci. Paris **D 267**, 1383–1386.

– – 1968b: Infrastructure des centromères et déroulement de la pleuromitose chez les Hypermastigines. C. r. Acad. Sci. Paris **D 266**, 367–370.

– – 1968c: Infrastructure du complexe rostral et origine du fuseau chez *Staurojoenina caulleryi*. C. r. Acad. Sci. Paris **D 266**, 1283–1286.

– – 1969: Appareil de Golgi, pinocytose, lysosomes, mitochondries, Bacteries symbiontiques, atractophores et pleuromitose chez les Hypermastigines du genre *Joenia*. Affinités entre Joeniides et Trichomonadines. Protistologica **5**, 39–86.

Holz, G. G., 1958: Mercaptoethanol and *Tetrahymena*. Biol. Bull. **115**, 354.

Honigberg, B. M., Daniel, W. A., Mattern, C. F. T., 1972: Fine structure of *Trichomitus batrachorum* (Perty). J. Protozool. **19**, 446–453.

– Mattern, C. F. T., Daniel, W. A., 1968: Structure of *Pentatrichomonas hominis* (Davaine) as revealed by electron microscopy. J. Protozool. **15**, 419–430.

Hoppe, G., 1974: La formation des mérozoites chez la coccidie *Eimeria tenella* (Raillet et Lucet 1891). Étude au microscope électronique. Protistologica **10**, 185–205.

– 1976: The ultrastructure of early generation development and later schizonts of *Eimeria tenella* in the chicken. Protistologica **12**, 169–181.

Houssay, D., Prenant, M., 1970: *Thecamoeba sphaeronucleolus*. Données physiologiques et étude morphologique en microscopie optique et électronique. Arch. Protistenk. **112**, 228–251.

Hovasse, R., 1945: Contribution à l'étude des Chloromonadines: *Gonyostomum semen* Diesing. Arch. Zool. Expér. Génér. **84**, 239–269.

HOVASSE, R., 1965a: Contribution à la connaissance des Heliozoaires: Étude de microscopie électronique sur *Raphidiophrys elegans* H. et L. Arch. Zool. Expér. Génér. **105**, 179–184.

– 1965b: Ultrastructure comparée des axopodes chez les Héliozoaires des genres *Actinosphaerium*, *Actinophrys*, et *Raphidiophrys*. Protistologica **1**, No. 1, 81–88.

HOWELL, S. H., 1972: The differential synthesis and degradation of ribosomal DNA during the vegetative cell cycle in *Chlamydomonas reinhardi*. Nature New Biol. **240**, 264–267.

HOWELLS, R. E., DAVIES, E. E., 1971: Nuclear division in the oocyst of *Plasmodium berghei*. Ann. Trop. Med. Parasitol. **65**, 451–459.

HUTH, W., 1913: Zur Entwicklungsgeschichte der Thalassicollen. Arch. Protistenk. **30**, 1–124.

HUTNER, S. H., CORLISS, J. O., 1976: Search for clues to the evolutionary meaning of ciliate phylogeny. J. Protozool. **23**, 48–56.

INABA, F., KUDO, N., 1972: Electron microscopy of the nuclear events during binary fission in *Paramecium multimicronucleatum*. J. Protozool. **19**, 57–63.

– SOTOKAWA, Y., 1968: Electron-microscopic observation on nuclear events during binary fission in *Blepharisma wardsi* (Ciliata, Heterotrichida). Japan. J. Genetics **43**, 335–348.

– SUGANUMA, Y., 1966: Electron microscopy of the nuclear apparatus of *Urostyla grandis*, a hypotrichous ciliate. J. Protozool. **13**, 137–143.

INJEYAN, H. S., HUEBNER, E., 1979: The ultrastructure of *Entamoeba* sp. (Laredo isolate). Observations on thin sections and freeze-fracture preparations. Canad. J. Zool. **57**, 1723–1735.

– – MEEROVITCH, E., 1979: Studies on a morphologically distinct colchicine-resistant variant of *Entamoeba* sp. J. Protozool. **26**, 253–259.

INOKI, S., NAKANISHI, K., NAKABAYASHI, T., 1959: Observations on *Trichomonas vaginalis* by electron microscopy. Biken's Journ. **2**, 21–24.

INOUÉ, S., RITTER, H., 1978: Mitosis in *Barbulanympha*. II. Dynamics of a two-stage anaphase, nuclear morphogenesis, and cytokinesis. J. Cell Biol. **77**, 655–684.

– SATO, H., 1967: Cell motility by labile association of molecules. The nature of mitotic spindle fibers and their role in chromosome movement. J. Gen. Physiol. **50**, 259–288.

ISAACKS, R. E., SANTOS, B. G., 1973: Studies on nuclei of *Paramecium aurelia*. II. Amino acid composition and electrophoretic properties of the chromosomal basic proteins. J. Protozool. **20**, 482–489.

ITO, J., LEE, Y. C., SCHERBAUM, O. H., 1968: Intranuclear microtubules in *Tetrahymena pyriformis* GL. Exper. Cell Res. **53**, 85–93.

IVANOV, A. V., 1968: The origin of Metazoa. Leningrad: Nauka Press (in Russian).

– 1970: On the independent and covergent origin of sexual process. Monit. Zool. Ital. n. ser., **4**, 183–189.

IVASHCHENKO, I. M., 1980: Morphology and fine structure of the nuclear apparatus of the freshwater holotrichous ciliate *Spathidium* sp. Tsitologiya (Leningrad) **22**, 387–393 (in Russian; English summary).

IWAMURA, Y., SAKAI, M., MITA, T., MURAMATSU, M., 1979: Unequal gene amplification and transcription in the macronucleus of *Tetrahymena pyriformis*. Biochemistry **18**, 5289–5294.

JADIN, J.-M., 1971: Cytologie et cytophysiologie des Trypanosomidae. Acta Zool. Pathol. Antverpiensia No. 53, 5–168.

– CREEMERS, J., 1966: L'ultrastructure des formes en rosaces de *Leishmania tropica* WRIGHT 1903. Ann. Soc. Belge Méd. Trop. **46**, 349–354.

– – 1967: Ultrastructure et biologie des toxoplasmes. I. Formes libres et formes intracellulaires. Ann. Soc. Belge Méd. Trop. **47**, 385–390.

– – 1968: Étude de l'ultrastructurale et de la biologie de *Toxoplasma gondii* NICOLLE et MANCEAUX 1908. Louvain Méd. **87**, 361–373.

– – LE RAY, D., 1968: Ultrastructure et biologie de *Plasmodium vinckei* RODHAIN 1952. Ann. Soc. Belge Méd. Trop. **48**, 497–502.

JAECKEL-WILLIAMS, R., 1978: Nuclear divisions with reduced numbers of microtubules in *Tetrahymena*. J. Cell Sci. **34**, 303–319.

JAMES, E. A., 1967: Regeneration and division in *Stentor coeruleus:* the effect of microinjected and externally applied actinomycin D and puromycin. Devel. Biol. **16**, 577–592.

JAMESON, A., 1920: The chromosome cycle of gregarines, with special reference to *Diplocystis schneideri*. Quart. J. Microscop. Sci. **64**, 207–225.

JANKOWSKI, A. V., 1966: Conjugation processes in *Paramecium putrinum* CLAP. et LACHM. IX. On "necrochromatin" and the functional significance of macronuclear fragmentation. Tsitologiya (Leningrad) **8**, 725–735 (in Russian; English summary).

– 1972: Nuclear apparatus of the Ciliophora. I. Micronuclear polyploidy in ciliates as a species character. Genetika (Moscow) **8**, 78–84 (in Russian; English summary).

– 1973: Subclass Chonotricha. (Series "Fauna SSSR", Infusoria, new ser., No. 103.) 355 pp. Leningrad: Nauka Press (in Russian).

JANOVY, J., LEE, K. W., BRUMBAUGH, J. A., 1974: The differentiation of *Herpetomonas megaseliae:* ultrastructural observations. J. Protozool. **21**, 53–58.

JARDINE, N. J., LEAVER, J. L., 1977: The isolation of nuclei and histones from *Euglena gracilis*. Exper. Cell Res. **106**, 423–426.

JAREÑO, M. A., 1974: Études sur les chromosomes polytènes de *Stylonychia mytilus:* appariement somatique. Protistologica **10**, 527–532.

– ALONSO, P., PÉREZ-SILVA, J., 1972: Identification of some puffed regions in the polytene chromosomes of *Stylonychia mytilus*. Protistologica **8**, 237–243.

JEFFREY, S. W., VESK, M., 1976: Further evidence for a membrane-bound endosymbiont within the dinoflagellate *Peridinium foliaceum*. J. Phycol. **12**, 450–455.

JENKINS, R. A., 1967: Fine structure of division in ciliate Protozoa. I. Micronuclear mitosis in *Blepharisma*. J. Cell Biol. **34**, 463–481.

– 1970: The fine structure of a nuclear envelope associated endosymbiont of *Paramecium*. J. Gen. Microbiol. **61**, 355–359.

– 1977: The role of microtubules in macronuclear division of *Blepharisma*. J. Protozool. **24**, 264–275.

– 1978: The dynamics of micronuclear mitosis in *Spirostomum teres*. J. Protozool. **25**, No. 3 (part 1), 13A–14A.

JERKA-DZIADOSZ, M., FRANKEL, J., 1970: The control of DNA synthesis in macronuclei and micronuclei of a hypotrich ciliate: a comparison of normal and regenerating cells. J. Exper. Zool. **173**, 1–22.

JOHNSON, B., WESTERGAARD, O., 1976: Accumulation of replicative DNA intermediates in *Tetrahymena* after excision-reparable damage to DNA. Europ. J. Biochem. **62**, 345–352.

JOHNSON, U. G., PORTER, K. R., 1968: Fine structure of cell division in *Chlamydomonas reinhardi*. Basal bodies and microtubules. J. Cell Biol. **38**, 403–425.

JONES, E. E., 1951: Encystment, excystment and the nuclear cycle in the ciliate *Dileptus anser*. J. Elisha Mitchell Scient. Soc. **67**, 205–218.

JOSEPH, H., 1907: Beobachtungen über die Kernverhältnisse von *Loxodes rostrum* O.F.M. Arch. Protistenk. **8**, 344–368.

JOYON, L., 1963: Contribution à l'étude cytologique de quelques Protozoaires flagellés. Ann. Fac. Sci. Univ. Clermont (Biol. Anim.), No. 22, 3–96.

– 1965: Compléments a la connaissance ultrastructurale des genres *Haematococcus pluvialis* FLOTOW et *Stephanosphaera pluvialis* COHN. Ann. Fac. Sci. Univ. Clermont (Biol. Anim.), No. 26, 57–69.

– CHARRET, R., 1962: Sur l'ultrastructure du thécamoebien *Hyalosphenia papilio* (LEIDY). C. r. Acad. Sci. Paris **255**, 2661–2663.

JOYON, L., FOTT, B., 1964: Quelques particularites infrastructurales du plaste des *Carteria* (Volvocales). J. Microscopie **3**, 159–166.

– LOM, J., 1966: Sur l'ultrastructure de *Costia necatrix* LECLERCQ (Zooflagellé); place systématique de ce Protiste. C. r. Acad. Sci. Paris **D 262**, 660–663.

– – 1969: Étude cytologique, systématique et pathologique d'*Ichthyobodo necator* (HENNEGUY 1883) PINTO 1928 (Zooflagellé). J. Protozool. **16**, 703–719.

JURAND, A., 1976: Some ultrastructural features of micronuclei during conjugation and autogamy in *Paramecium aurelia*. J. Gen. Microbiol. **94**, 193–203.

– BEALE, G. H., YOUNG, M. R., 1962: Studies on the macronucleus of *Paramecium aurelia*. I. With a note on ultraviolet micrography. J. Protozool. **9**, 122–131.

– – – 1964: Studies on the macronucleus of *Paramecium aurelia*. II. Development of macronuclear anlagen. J. Protozool. **11**, 491–497.

– BOMFORD, R., 1965: The fine structure of the parasitic suctorian *Podophrya parameciorum*. J. Microscopie **4**, 509–522.

– JACOB, J., 1969: Studies on the macronucleus of *Paramecium aurelia*. III. Localization of tritiated thymidine by electron microscope autoradiography. Chromosoma **26**, 355–364.

– SELMAN, G., 1969: Anatomy of *Paramecium aurelia*. London: Macmillan.

– – 1970: Ultrastructure of the nuclei and intranuclear microtubules of *Paramecium aurelia*. J. Gen. Microbiol. **60**, 357–364.

JUST, E., 1973: Untersuchungen zur zeitlichen Ordnung der Syntheseleistungen in der Interphase von *Paramecium bursaria*. Arch. Protistenk. **115**, 22–68.

KACZANOWSKI, A., 1968: Mitosis and polyploidy of nuclei of *Opalina ranarum*. Experientia **24**, 846–847.

– 1971: *Opalina ranarum* PURKINJE et VALENTIN: meiosis and dimorphism of nuclear behaviour during meiosis. Acta Protozool. **9**, 105–106.

KAFFENBERGER, W., ECKERT, W. A., 1980: Regulation of rRNA metabolism in *Tetrahymena pyriformis*. II. Nutritional shift-up. Europ. J. Cell Biol. **21**, 200–207.

KALLINIKOVA, V. D., 1981: The kinetoplast as a cell organelle. Int. Rev. Cytol. **69**, 105–156.

KANEDA, M., 1960: The structure and reorganization of the macronucleus during the binary fission of *Chlamydodon pedarius*. Japan. J. Zool. **12**, 477–491.

– 1961 a: On the division of macronucleus in the living gymnostome ciliate, *Chlamydodon pedarius*, with special reference to the behaviours of chromonemata, nucleoli and endosome. Cytologia **26**, 89–104.

– 1961 b: Fine structure of the macronucleus of the gymnostome ciliate, *Chlamydodon pedarius*. Japan. J. Genetics **36**, 223–234.

– 1961 c: On the interrelations between the macronucleus and the cytoplasm in the gymnostome ciliate *Chlamydodon pedarius*. Cytologia **26**, 408–418.

KANESHIRO, E. S., HOLZ, G. G., 1976: Observations on the ultrastructure of *Uronema* sp., marine scuticociliates. J. Protozool. **23**, 503–517.

KARADZHAN, B. P., 1975: Ultrastructure of the macronuclear anlagen of *Didinium nasutum* (Holotricha, Gymnostomata). Tsitologiya (Leningrad) **17**, 881–886 (in Russian; English summary).

– 1976: Ultrastructure of the sessile ciliate *Cavichona elegans* (Chonotricha). I. Non-dividing animals. Acta Protozool. **15**, 315–330 (in Russian; English summary).

– 1977: Ultrastructure of the micronucleus of *Didinium nasutum* (Ciliophora, Gymnostomata) during meiosis. Tsitologiya (Leningrad) **19**, 1327–1332 (in Russian; English summary).

– RAIKOV, I. B., 1977: Fine structure of the nuclear apparatus of *Didinium nasutum* (Ciliophora, Gymnostomata) in interphase and during binary fission. Protistologica **13**, 15–29.

KARADZHAN, B. P., RAIKOV, I. B., 1979: Ultrastructural and autoradiographic investigation of the resorption of the old macronucleus and of the development of the new macronucleus following conjugation in *Didinium nasutum*. Protistologica **15**, 507–519.

KARAWAIEW, W., 1895: Beobachtungen über die Struktur und Vermehrung von *Aulacantha scolymantha* HAECK. Zool. Anz. **18**, 286–301.

KARRER, K. M., GALL, J. G., 1976: The macronuclear ribosomal DNA of *Tetrahymena pyriformis* is a palindrome. J. Molec. Biol. **104**, 421–453.

KASANZEFF, W., 1910: Zur Kenntnis von *Loxodes rostrum*. Arch. Protistenk. **20**, 79–96.

– 1928: Beitrag zur Kenntnis der Großkerne der Ciliaten. Trudy Osoboi Zool. Lab. AN SSSR, ser. 2, No. 11, 1–30.

KATASHIMA, R., 1968: Micronuclear differences between stocks of *Euplotes patella* (Ciliata). Chromosome Inform. Service No. 9, 11–13.

KATTAR, M. R., 1975: Données complémentaires sur l'ultrastructure de *Neobursaridium gigas* BALECH 1941 (Ciliatea, Hymenostomatida, Peniculina). Acta Protozool. **14**, 179–184.

KATZEN, A. L., CANN, G. M., BLACKBURN, E., 1981: Sequence-specific fragmentation of macronuclear DNA in a holotrichous ciliate. Cell **24**, 313–320.

KAWAKAMI, H., YAGIU, R., 1963: The electron microscopical study of the change of fine structures in the ciliate, *Colpoda cucullus*, during its life cycle. I. The trophic stage. Zool. Mag. **72**, 89–96.

KAY, M. W., 1946: Studies on *Oxytricha bifaria* STOKES. III. Conjugation. Trans. Amer. Microscop. Soc. **65**, 132–148.

KEIDING, J., ANDERSEN, H. A., 1978: Regulation of ribosomal RNA synthesis in *Tetrahymena pyriformis*. J. Cell Sci. **31**, 13–24.

KELLENBERGER, E., 1962: The study of natural and artificial DNA-plasms by thin sections. In: The interpretation of ultrastructure (HARRIS, R. J. C., ed.), pp. 233–247. New York-London: Academic Press.

KELLEY, G. L., HAMMOND, D. M., 1972: Fine structural aspects of early development of *Eimeria ninakohlyakimovae* in cultured cells. Z. Parasitenk. **38**, 271–284.

– – 1973: Fine structural aspects of nuclear division and merogony of *Eimeria ninakohlyakimovae* in cultured cells. J. Parasitol. **59**, 1071–1079.

KENNEDY, J. R., 1965: The morphology of *Blepharisma undulans* STEIN. J. Protozool. **12**, 542–561.

KEPKA, O., SCHOLTYSECK, E., 1970: Weitere Untersuchungen der Feinstruktur von *Frenkelia* spec. (= M-Organismus, Sporozoa). Protistologica **6**, 249–266.

KHAN, M. A., 1969: Fine structure of *Ancistrocoma pelseneeri* (CHATTON et LWOFF), a rhynchodine thigmotrichid ciliate. Acta Protozool. **7**, 29–47.

– 1970: Electron microscopic studies of *Thigmophrya macomae* CH. et LW., an arhynchodine thigmotrichid ciliate. Acta Protozool. **8**, 67–77.

KIDDER, G. W., 1933: On the genus *Ancistruma* STRAND (*Ancistrum* MAUPAS). I. The structure and division of *A. mytili* QUENN. and *A. isseli* KAHL. Biol. Bull. **64**, 1–20.

– CLAFF, C. L., 1938: Cytological investigations of *Colpoda cucullus*. Biol. Bull. **74**, 178–197.

KILLBY, V. A. A., SILVERMAN, P. H., 1969: Fine structural observations of the erythrocytic stages of *Plasmodium chabaudi* LANDAU 1965. J. Protozool. **16**, 354–370.

KIMBALL, R. F., 1953: The structure of the macronucleus of *Paramecium aurelia*. Proc. Nat. Acad. Sci. U.S.A. **39**, 345–347.

– BARKA, T., 1959: Quantitative cytochemical studies on *Paramecium aurelia*. II. Feulgen microspectrophotometry of the macronucleus during exponential growth. Exper. Cell Res. **17**, 173–182.

KIMBALL, R. F., PERDUE, S. W., 1962: Quantitative cytochemical studies on *Paramecium aurelia*. V. Autoradiographic studies of nucleic acid syntheses. Exper. Cell Res. **27**, 405–415.

– PRESCOTT, D. M., 1962: Deoxyribonucleic acid synthesis and distribution during growth and amitosis of the macronucleus of *Euplotes*. J. Protozool. **9**, 88–92.

– VOGT-KÖHNE, L., CASPERSSON, T. O., 1960: Quantitative cytochemical studies on *Paramecium aurelia*. III. Dry weight and ultraviolet absorption of isolated macronuclei during various stages of the interdivision interval. Exper. Cell Res. **20**, 368–377.

KIMMEL, A. R., GOROVSKY, M. A., 1976: Numbers of 5S and tRNA genes in macro- and micronuclei of *Tetrahymena pyriformis*. Chromosoma **54**, 327–337.

– – 1978: Organization of the 5S RNA genes in macro- and micronuclei of *Tetrahymena pyriformis*. Chromosoma **67**, 1–20.

KIRNOS, M. D., MERKULOVA, N. A., BORCHSENIUS, S. N., VANYUSHIN, B. F., 1980: The character of macronuclear DNA methylation in the protozoan, *Tetrahymena pyriformis*. Doklady Akad. Nauk SSSR **255**, 225–227 (in Russian).

KLASS, M. R., SMITH-SONNEBORN, J., 1976: Studies on DNA content, RNA synthesis, and DNA template activity in aging cells of *Paramecium aurelia*. Exper. Cell Res. **98**, 63–72.

KLEI, T. R., 1972: The fine structure of *Haemoproteus columbae* sporozoites. J. Protozool. **19**, 281–286.

KLITZKE, M., 1915: Ein Beitrag zur Kenntnis der Kernentwicklung bei den Ciliaten. Arch. Protistenk. **36**, 215–235.

KLOETZEL, J. A., 1970: Compartmentalization of the developing macronucleus following conjugation in *Stylonychia* and *Euplotes*. J. Cell Biol. **47**, 395–407.

– 1981 a: Nuclear roles in the postconjugant development of the ciliated protozoan *Euplotes aediculatus*. 1. Evidence for sequential roles of the differentiating macronucleus in exconjugant development. Devel. Biol. **83**, 20–32.

– 1981 b: Ibid. 2. Experimentally induced regeneration of old macronuclear fragments. J. Protozool. **28**, 108–116.

KLUSS, B. C., 1962: Electron microscopy of the macronucleus of *Euplotes eurystomus*. J. Cell Biol. **13**, 462–465.

KLUYEVA, T. S., CHENTSOV, YU. S., POLYAKOV, V. YU., 1978: Electron microscopic study of the nuclei of *Euglena gracilis*. Tsitologiya (Leningrad) **20**, 987–991 (in Russian; English summary).

KOFOID, CH., SWEZY, O., 1925: On the number of chromosomes and the type of mitosis in *Endamoeba dysenteriae*. Univ. Calif. Publs Zool. **26**, 331–352.

KOŚCIUSZKO, H., 1965: Karyologic and genetic investigations in syngen 1 of *Paramecium aurelia*. Folia Biol. (Kraków) **13**, 339–370.

KOVALEVA, V. G., 1972: Meiosis and some stages of conjugation in the holotrichous ciliate *Trachelonema sulcata*. Protistologica **8**, 83–90.

– RAIKOV, I. B., 1970: Ultrastructure of the micronuclei and of the diploid macronuclei of the lower ciliate *Trachelonema sulcata*. Tsitologiya (Leningrad) **12**, 1446–1455 (in Russian; English summary).

– – 1973: Ultrastructure de l'appareil nucléaire de *Trachelonema sulcata* KOVALEVA, Cilié holotriche gymnostome à macronoyaux diploïdes. Protistologica **9**, 471–480.

– – 1978: Diminution and re-synthesis of DNA during development and senescence of the "diploid" macronuclei of the ciliate *Trachelonema sulcata* (Gymnostomata, Karyorelictida). Chromosoma **67**, 177–192.

– – VINNIKOVA, N. V., 1977: Fine structure of the macronucleus of the marine psammophilic ciliate *Helicoprorodon gigas* (KAHL). Tsitologiya (Leningrad) **19**, 592–597 (in Russian; English summary).

KOVALEVA, V. G., VINNIKOVA, N. V., RAIKOV, I. B., 1979: Electron microscopical study of the nuclear apparatus of the lower ciliate *Tracheloraphis totevi.* Tsitologiya (Leningrad) **21**, 280–286 (in Russian; English summary).

KRASCHENINNIKOW, S., SCHOLTYSECK, E., 1975: Beobachtungen an der Feinstruktur des Kernapparates von *Balantidium coli* (Ciliata, Holotricha, Trichostomatida). Protistologica **11**, 297–303.

KRAVTSOV, V. A., 1977: Ultrastructure of the sporozoite of *Eimeria acervulina* (Coccidiida, Eimeriidae). Parasitologiya (Leningrad) **11**, 162–165 (in Russian; English summary).

KRAWCZYŃSKA, W., 1979: Ultrastructure of isolated *Paramecium aurelia* macronuclei. Acta Protozool. **18**, 231–236.

– 1980: Localization of ribonucleoproteins in the macronucleus of *Paramecium aurelia.* Acta Protozool. **19**, 253–260.

KUBAI, D. F., 1972: Chromosome segregation in *Trichonympha.* J. Cell Biol. **55**, 143A.

– 1973: Unorthodox mitosis in *Trichonympha agilis:* kinetochore differentiation and chromosome movement. J. Cell Sci. **13**, 511–552.

– 1975: The evolution of the mitotic spindle. Internat. Rev. Cytol. **43**, 167–227.

– RIS, H., 1969: Division in the dinoflagellate *Gyrodinium cohnii* (SCHILLER). A new type of nuclear reproduction. J. Cell Biol. **40**, 508–528.

KUDO, R. R., 1947: *Pelomyxa carolinensis* WILSON. II. Nuclear division and plasmotomy. J. Morphol. **80**, 93–144.

– DANIELS, E. W., 1963: An electron microscope study of the spore of a microsporidian, *Thelohania californica.* J. Protozool. **10**, 112–120.

KUDRJAVTSEV, B. N., 1966: Changes of the DNA content in macro- and micronucleus of *Paramecium putrinum* in the interdivision phase. Acta Protozool. **4**, 51–57.

KUMAR, A., 1970: Ribosome synthesis in *Tetrahymena pyriformis.* J. Cell Biol. **45**, 623–634.

KUSEL, J. P., MOORE, K. E., WEBER, M. M., 1967: The ultrastructure of *Crithidia fasciculata* and morphological changes induced by growth in acriflavin. J. Protozool. **14**, 283–296.

LADDA, R. L., 1969: New insights into the fine structure of rodent malaria parasites. Military Med. **134**, 825–864.

LA FOUNTAIN, J. R., DAVIDSON, L. A., 1979: An analysis of spindle ultrastructure during prometaphase and metaphase of micronuclear division in *Tetrahymena.* Chromosoma **75**, 293–308.

LAINSON, R., BAKER, J. R., BIRD, R. G., GARNHAM, P. C. C., HEALEY, P., 1961: Electron micrographs of *Lankesterella* (*Atoxoplasma*) from English sparrows. Trans. Roy. Soc. Trop. Med. Hyg. **55**, 9–18.

LANGRETH, S. G., JENSEN, J. B., REESE, R. T., TRAGER, W., 1978: Fine structure of human malaria *in vitro.* J. Protozool. **25**, 443–452.

– TRAGER, W., 1973: Fine structure of the malaria parasite *Plasmodium lophurae* developing extracellularly in vitro. J. Protozool. **20**, 606–613.

LARSSON, R., 1980a: Ultrastructural study of *Toxoglugea variabilis* n. sp. (Microsporida: Thelohaniidae), a microsporidian parasite of the biting midge *Bezzia* sp. (Diptera: Ceratopogonidae). Protistologica **16**, 17–32.

– 1980b: Insect pathological investigations on Swedish Thysanura. II. A new microsporidian parasite of *Petrobius brevistylis* (Microcoryphia, Machilidae); description of the species and creation of two new genera and a new family. Protistologica **16**, 85–101.

LASMAN, M., 1977: Light and electron microscopic observations on encystment of *Acanthamoeba palestinensis* REICH. J. Protozool. **24**, 244–248.

LAUTH, M. R., SPEAR, B. B., HEUMANN, J., PRESCOTT, D. M., 1976: DNA of ciliated Protozoa: DNA sequence diminution during macronuclear development of *Oxytricha*. Cell **7**, 67–74.

LAVAL, M., 1971: Ultrastructure et mode de nutrition du Choanoflagellé *Salpingoeca pelagica* sp. nov. Comparaison avec les choanocytes des Spongiaires. Protistologica **7**, 325–336.

– 1972: Ultrastructure de *Petalotricha ampulla* (FOL), comparaison avec d'autres Tintinnides et avec les autres ordres de Ciliés. Protistologica **8**, 369–386.

LAVETTE, A., 1970: Sur le genre *Projoenia* et les affinités des Joeniidae (Zooflagellés Metamonadina). C. r. Acad. Sci. Paris **D 270**, 1695–1698.

– 1971: *Stephanonympha chagasi* n. sp., Flagellé trichomonadine et ses Bactéries symbiotiques. C. r. Acad. Sci. Paris **D 272**, 1785–1788.

– 1973: Sur l'ultrastructure et les affinités systématiques de *Microrhopalodina inflata*, Flagellé symbiotique du Termite à cou jaune, *Calotermes flavicollis*. C. r. Acad. Sci. Paris **D 276**, 1309–1311.

– 1975: Nouvelles recherches sur l'ultrastructure et la biologie de quelques Flagellés symbiotiques de Termites. Ann. Sci. Natur., Zool., sér. 12, **17**, 179–214.

LAWN, R. M., 1977: Gene-sized DNA molecules of the *Oxytricha* macronucleus have the same terminal sequence. Proc. Nat. Acad. Sci. U.S.A. **74**, 4325–4328.

– HEUMANN, J. M., HERRICK, G., PRESCOTT, D. M., 1978: The gene-sized DNA molecules of *Oxytricha*. Cold Spring Harbor Symp. Quant. Biol. **42**, 483–492.

LEADBEATER, B. S. C., 1977: Observations on the life-history and ultrastructure of the marine choanoflagellate *Choanoeca perplexa* ELLIS. J. Mar. Biol. Ass. U.K. **57**, 285–301.

– DODGE, J. D., 1966: The fine structure of *Woloszynskia micra* sp. nov., a new marine dinoflagellate. Brit. Phycol. Bull. **3**, 1–17.

– – 1967: An electron microscope study of nuclear and cell division in a dinoflagellate. Arch. Mikrobiol. **57**, 239–254.

– MORTON, C., 1974: A microscopical study of a marine species of *Codosiga* JAMES-CLARK (Choanoflagellata) with special reference on the ingestion of bacteria. Biol. J. Linn. Soc. **6**, 337–347.

LEBEDEW, W., 1909: Über *Trachelocerca phoenicopterus* COHN. Arch. Protistenk. **13**, 71–114.

LE CALVEZ, J., 1938: Recherches sur les Foraminifères. I. Développement et reproduction. Arch. Zool. Expér. Génér. **80**, 163–333.

– 1950: Recherches sur les Foraminifères. II. Place de la meiose et sexualité. Arch. Zool. Expér. Génér. **87**, 211–243.

– 1953: Ordre des Foraminifères (Foraminifera D'ORBIGNY 1826). In: Traité de Zoologie (GRASSÉ, P. P., ed.), Vol. I, fasc. 2, pp. 149–265. Paris: Masson et Cie.

LÉCHER, P., 1972: Fusorial fibers and kinetochores during the division of *Aulacantha* (Radiolaria). J. Protozool. **19**, suppl., 69.

– 1978: The synaptonemal complex in the bipartition division of the radiolaria *Aulacantha scolymantha*. Canad. J. Genet. Cytol. **20**, 85–95.

LEDOIGT, G., CALVAYRAC, R., 1979: Phénomènes périodiques, métaboliques et structuraux chez un protiste, *Euglena gracilis*. J. Protozool. **26**, 632–643.

LEE, D. L., MILLARD, B. J., 1971: Fine structure of the schizonts of *Eimeria praecox*. Intern. J. Parasitol. **1**, 37–41.

LEE, J. J., FREUDENTHAL, H. D., KOSSOY, V., BÉ, A., 1965: Cytological observations on two planktonic foraminifera, *Globigerina bulloides* D'ORBIGNY 1826 and *Globigerinoides ruber* (D'ORBIGNY 1839) CUSHMAN 1927. J. Protozool. **12**, 531–542.

Lee, K. A., Kemp, C. L., 1975: Microspectrophotometric analysis of DNA replication in *Eudorina elegans* (Volvocales, Chlorophyta). Phycologia **14**, 247–252.

Lee, Y. C., Scherbaum, O. H., 1966: Nucleohistone composition in stationary and division synchronized *Tetrahymena* cultures. Biochemistry **5**, 2067–2075.

Leedale, G. F., 1958a: Nuclear structure and mitosis in the Euglenineae. Arch. Mikrobiol. **32**, 32–64.
– 1958b: Mitosis and chromosome numbers in the Euglenineae (Flagellata). Nature **181**, 502–503.
– 1959a: Amitosis in three species of *Euglena*. Cytologia **24**, 213–219.
– 1959b: The time-scale of mitosis in the Euglenineae. Arch. Mikrobiol. **32**, 352–360.
– 1966: The Euglenophyceae. In: The chromosomes of the algae (Godward, M. B. E., ed.), pp. 78–95. London: E. Arnold.
– 1967: Euglenoid flagellates. Englewood Cliffs, N. J.: Prentice-Hall.
– 1968: The nucleus in *Euglena*. In: The biology of *Euglena* (Buetow, D. E., ed.), Vol. 1, pp. 185–243. New York-London: Academic Press.
– 1969: Observations on endonuclear bacteria in euglenoid flagellates. Österr. Bot. Z. **116**, 279–294.
– 1970: Phylogenetic aspects of nuclear cytology in the algae. Ann. N.Y. Acad. Sci. **175**, 429–453.
– 1974: Special cytology: morphology and morphogenesis of algal cells. Fortschr. Bot. **36**, 30–44.
– 1975: Preliminary observations on nuclear cytology and ultrastructure in carbon-starved streptomycin-bleached *Euglena gracilis*. Colloq. Internat. CNRS No. 240, 285–290.
– Buetow, D. E., 1976: Observations on cytolysome formation and other cytological phenomena in carbon-starved *Euglena gracilis*. J. Microscop. Biol. Cell. **25**, 149–154.
– Leadbeater, B. S. C., Massalski, A., 1970: The intracellular origin of flagellar hairs in the Chrysophyceae and Xanthophyceae. J. Cell Sci. **6**, 701–719.
– Meeuse, B. J. D., Pringsheim, E. G., 1965: Structure and physiology of *Euglena spirogyra*. Arch. Mikrobiol. **50**, 68–102.

Leer, J. C., Gocke, E., Nielsen, O. F., Westergaard, O., 1978: Isolation of transcriptionally active nucleoli and r-chromatin from *Tetrahymena*. J. Protozool. **25**, No. 3 (part 1), 44A.

Leeson, T. S., Bhatnagar, R., 1975a: *Amoeba proteus:* the nuclear periphery. Cell Differentiation **4**, 79–86.
– – 1975b: Microfibrillar structures in the nucleus and cytoplasm of *Amoeba proteus*. J. Exper. Zool. **192**, 265–270.

Leick, V., 1969: Formation of subribosomal particles in the macronucleus of *Tetrahymena pyriformis*. Europ. J. Biochem. **8**, 221–228.
– Bro, B., Bruns, P., 1979: Macronuclear development: a model system to study specific gene expression in nuclear differentiation. In: Specific eukaryotic genes (Alfred Benzon Symposium No. 13) (Engberg, J., Klenow, H., Leick, V., eds.), pp. 342–350. Copenhagen: Munksgaard.
– Plesner, P., 1968: Formation of ribosomes in *Tetrahymena pyriformis*. Biochim. Biophys. Acta **169**, 398–408.
– Tønnesen, T., Engberg, J., Nilsson, J. R., 1976: Nucleolar reorganization in *Tetrahymena pyriformis* GL during a nutritional shift-up; timing of ribosomal RNA transcription and ribosomal DNA replication. In: Proc. 9th Congr. Nordic Soc. Cell Biol., pp. 205–211. Odense: Odense Univ. Press.

LENGSFELD, A. M., 1969: Zum Feinbau der Foraminifere *Allogromia laticollaris*. I. Zellen mit ausgestreckten und eingezogenen Rhizopodien. Helgoländer Wiss. Meeresuntersuch. **19**, 230–261.

LERMAN, L. S., 1974: Chromosomal analogues: long-range order in ψ-condensed DNA. Cold Spring Harbor Sympos. Quant. Biol. **38**, 91–96.

LEVINE, N. D., CORLISS, J. O., COX, F. E. G., DEROUX, G., GRAIN, J., HONIGBERG, B. M., LEEDALE, G. F., LOEBLICH, A. R., LOM, J., LYNN, D., MERINFELD, E. G., PAGE, F. C., POLJANSKY, G., SPRAGUE, V., VAVRA, J., WALLACE, F. G., 1980: A newly revised classification of the Protozoa. J. Protozool. **27**, 37–58.

LEVY, M. R., ELLIOTT, A. M., 1968: Biochemical and ultrastructural changes in *Tetrahymena pyriformis* during starvation. J. Protozool. **15**, 208–222.

– GOLLON, C. E., ELLIOTT, A. M., 1969: Effects of hyperthermia on *Tetrahymena*. I. Localization of acid hydrolases and changes in cell ultrastructure. Exper. Cell Res. **55**, 295–305.

LEWIS, L. M., 1974: The ultrastructure of the nuclear events of binary fission in *Paramecium bursaria* and the effects of colchicine on cell division. Proc. Electron Microscopy Soc. Amer. **32**, 276–277.

– 1975: The evolutionary significance of ultrastructural variations in the micronuclear spindle apparatus in the genus *Paramecium*. BioSystems **7**, 380–385.

– 1977: Microfilaments, microlamellae, and microtubules in the micronucleus of *Paramecium bursaria* during Prophase I of meiosis. Abstr. V Internat. Congr. Protozool., p. 50. New York.

– WITKUS, E. R., VERNON, G. M., 1976: The role of microtubules and microfilaments in the micronucleus of *Paramecium bursaria* during mitosis. Protoplasma **89**, 203–219.

LIEDER, H.-J., RUTHMANN, A., 1976: Cytologische und cytophotometrische Untersuchungen am Makronucleus eines hypotrichen Ciliaten der Gattung *Oxytricha*. Arch. Protistenk. **118**, 268–284.

LIESCHE, W., 1938: Die Kern- und Fortpflanzungsverhältnisse von *Amoeba proteus* (PALL.). Arch. Protistenk. **91**, 135–186.

LIPPS, H. J., 1974: Die Oberflächenantigene bei dem hypotrichen Ciliaten *Euplotes minuta* und eine antiserumabhängige Makronukleus-Degeneration. Arch. Protistenk. **116**, 106–131.

– 1980: *In vitro* aggregation of the gene-sized DNA molecules of the ciliate *Stylonychia mytilus*. Proc. Nat. Acad. Sci. U.S.A. **77**, 4104–4107.

– HANTKE, K. G., 1975: Studies on the histones of the ciliate *Stylonychia mytilus*. Chromosoma **49**, 309–320.

– MORRIS, N. R., 1977: Chromatin structure in the nuclei of the ciliate *Stylonychia mytilus*. Biochem. Biophys. Res. Comm. **74**, 230–234.

– NOCK, A., RIEWE, M., STEINBRÜCK, G., 1978: Chromatin structure of the macronucleus of *Stylonychia mytilus*. Nucleic Acid Res. **5**, 4699–4709.

– STEINBRÜCK, G., 1978: Free genes for rRNA in the macronuclear genome of the ciliate *Stylonychia mytilus*. Chromosoma **69**, 21–26.

LIPSCOMB, D. L., 1980: The curious case of *Stephanopogon*: implications in ciliate and flagellate taxonomy. J. Protozool. **27**, No. 3, 29A.

LIU, T. P., DAVIES, D. M., 1972: Fine structure of developing spores of *Thelohania bracteata* (STRICKLAND 1913) (Microsporidia, Nosematidae) emphasizing polar-filament formation. J. Protozool. **19**, 461–469.

LIVOLANT, F., BOULIGAND, Y., 1978: New observations on the twisted arrangement of dinoflagellate chromosomes. Chromosoma **68**, 21–44.

– – 1980: Double helical arrangement of spread dinoflagellate chromosomes. Chromosoma **80**, 97–118.

LOEBLICH, A. R., 1976: Dinoflagellate evolution: speculation and evidence. J. Protozool. **23**, 13–28.

LOM, J., 1969: Notes on the ultrastructure and sporoblast development in fish-parasitizing Myxosporidians of the genus *Sphaeromyxa*. Z. Zellforsch. **97**, 416–437.

– CORLISS, J. O., 1967: Ultrastructural observations on the development of the microsporidian protozoon *Plistophora hyphessobryconis* SCHÄPERCLAUS. J. Protozool. **14**, 141–152.

– KOZLOFF, E. N., 1967: The ultrastructure of *Phalacrocleptes verruciformis*, an unciliated ciliate parasitizing the polychaete *Schizobranchia insignis*. J. Cell Biol. **33**, 355–364.

– – 1969: Ultrastructure of the cortical regions of ancistrocomid ciliates. Protistologica **5**, 173–192.

– DE PUYTORAC, P., 1965: Studies on the myxosporidian ultrastructure and polar capsule development. Protistologica **1**, 53–65.

LOTT, J. N. A., WILSON, J. J., VOLLMER, C. M., 1977: Temperature-induced changes in the nuclear envelope of *Euglena* and *Scenedesmus*. J. Ultrastr. Res. **61**, 1–9.

LOUBÈS, C., 1979: Recherches sur la meïose chez les microsporidies: conséquences sur les cycles biologiques. J. Protozool. **26**, 200–208.

– AKBARIEH, M., 1977: Étude ultrastructurale de *Nosemoides simocephali* n. sp. (Microsporidie), parasite intestinal de la Daphnie *Simocephalus vetulus* (MÜLLER 1776). Z. Parasitenk. **54**, 125–137.

– – 1978: Étude ultrastructurale de la microsporidie *Baculea daphniae* n. g., n. sp., parasite de l'épithelium intestinal de *Daphnia pulex* LEYDIG 1860 (Crustacé, Cladocère). Protistologica **14**, 23–28.

– MAURAND, J., 1975: Étude ultrastructurale de *Gurleya chironomi* n. sp. Microsporidie parasite des larves *d'Orthocladius* (Diptera-Chironomidae). Protistologica **11**, 233–244.

– – 1976: Étude ultrastructurale de *Pleistophora debaisieuxi* JIROVEC 1943 (Microsporidia): son transfert dans le genre *Tuzetia*. Protistologica **12**, 577–591.

– – GASC, C., BOUIX, G., 1976a: Étude ultrastructurale de *Glugea habrodesmi* n. sp. (Microsporida Glugeidae), parasite des Myriapodes *Habrodesmus falx* COOK et *Oxydesmus granulosus* PALISOT DE BEAUVOIS (Myriapoda – Polydesmidae). Protistologica **12**, 435–450.

– – ORMIÈRES, R., 1979: Étude ultrastructurale de *Spraguea lophii* (DOFLEIN 1898), Microsporidie parasite de la Baudroie; essai d'interprétation du dimorphisme sporal. Protistologica **15**, 43–54.

– – ROUSSET-GALANGAU, V., 1976b: Présence de complexes synaptonématiques dans le cycle biologique de *Gurleya chironomi* LOUBÈS et MAURAND 1975: un argument en faveur d'une sexualité chez les Microsporidies? C. r. Acad. Sci. Paris **D 282**, 1025–1027.

LUDVIK, J., 1956: Vergleichende elektronenoptische Untersuchungen an *Toxoplasma gondii* und *Sarcocystis tenella*. Zbl. Bakteriol., Abt. I (Orig.) **166**, 60–65.

– 1958a: Morphology of *Toxoplasma gondii* in electron microscope. Věst. Českosl. Společ. Zool. **22**, 130–136.

– 1958b: Elektronenoptische Befunde zur Morphologie der Sarcosporidien (*Sarcocystis tenella* RAILLET 1886). Zbl. Bakteriol., Abt. I (Orig.) **172**, 330–350.

– 1960: The electron microscopy of *Sarcocystis miescheriana* KUHN 1865. J. Protozool. **7**, 128–135.

– 1963: Electron microscopic study of some parasitic Protozoa. In: Progress in protozoology (Proc. I Internat. Congr. Protozool.), pp. 387–392. Prague: Publ. House ČSAV.

LUPORINI, P., DINI, F., 1975: Relationships between cell cycle and conjugation in 3 hypotrichs. J. Protozool. **22**, 541–544.

Luporini, P., Magagnini, G., 1970: Recherches sur la structure fine et la biologie reproductive du Cilié hypotriche *Swedmarkia arenicola* Dragesco. Protistologica **6**, 113–125.

Luykx, P., 1970: Cellular mechanisms of chromosome distribution. Internat. Rev. Cytol., Suppl. 2, 1–173.

Lwoff, A., 1923: Sur un Infusoire cilié homocaryote à vie libre. Son importance taxonomique. C. r. Acad. Sci. Paris **177**, 910–913.

– 1936: Le cycle nucléaire de *Stephanopogon mesnili* Lw. (Cilié homocaryote). Arch. Zool. Expér. Génér. **78**, 117–132.

Lykkesfeldt, A. E., 1979: Extrusion of DNA from the macronucleus of *Tetrahymena pyriformis* GL. J. Protozool. **26**, 586–588.

Lynch, M. J., Leake, R. E., O'Connell, K. M., Buetow, D. E., 1975: Isolation, fractionation and template activity of the continuously-condensed chromatin of *Euglena gracilis*. Exper. Cell Res. **91**, 349–357.

Lynn, D. H., 1976a: Comparative ultrastructure and systematics of the Colpodida. An ultrastructural description of *Colpoda maupasi* Enriquez 1908. Canad. J. Zool. **54**, 405–420.

– 1976b: Comparative ultrastructure and systematics of the Colpodida. Structural conservatism hypothesis and a description of *Colpoda steinii* Maupas. J. Protozool. **23**, 302–314.

– 1976c: Comparative ultrastructure and systematics of the Colpodida. Fine structural specializations associated with large body size in *Tillina magna* Gruber 1880. Protistologica **12**, 629–648.

– 1976d: Comparative ultrastructure and systematics of the Colpodida (Ciliophora): structural differentiation in the cortex of *Colpoda simulans*. Trans. Amer. Microscop. Soc. **95**, 581–599.

MacDonald, C. M., Darbyshire, J. F., 1977: The morphology of a soil flagellate, *Spiromonas angusta* (Duj.) Alexeieff (Mastigophorea: Protozoa). Protistologica **13**, 441–450.

MacDougall, M. S., 1925: Cytological observations on gymnostomatous ciliates, with a description of the maturation phenomena in diploid and tetraploid forms of *Chilodon uncinatus*. Quart. J. Microscop. Sci. **69**, 361–384.

– 1936: Étude cytologique de trois espèces du genre *Chilodonella* Strand. Morphologie, conjugaison, réorganisation. Bull. Biol. France et Belgique **70**, 308–331.

Macgregor, H. C., Thomasson, P. A., 1965: The fine structure of two archigregarines, *Selenidium fallax* and *Ditrypanocystis cirratuli*. J. Protozool. **12**, 438–443.

Mackenzie, T. B., Stone, G. E., Prescott, D. M., 1966: The duration of G_1, S, and G_2 at different temperatures in *Tetrahymena pyriformis* HSM. J. Cell Biol. **31**, 633–635.

Magnaval, R., Bertaux, O., Valencia, R., 1979: Hetero- and euchromatin of synchronous *Euglena* cells. I. Physical fractionation of nuclei into differentially condensed chromatin. Exper. Cell Res. **121**, 251–265.

Makhlin, E. E., Kudryavtseva, M. V., Kudryavtsev, B. N., 1979a: Peculiarities of changes in DNA content of *Amoeba proteus* nuclei during interphase. Exper. Cell Res. **118**, 143–150.

– Kudryavtsev, B. N., Kudryavtseva, M. V., Ioffe, V. A., 1979b: Diet dependence of the nuclear DNA content in *Amoeba proteus*. Tsitologiya (Leningrad) **21**, 935–941 (in Russian; English summary).

Makino, F., Tsuzuki, J., 1971: Absence of histone in the blue-green alga *Anabaena cylindrica*. Nature **231**, 446–447.

Manton, I., 1964: Observations with the electron microscope on the division cycle in the flagellate *Prymnesium parvum* Carter. J. Roy. Microscop. Soc. **83**, 317–325.

MANTON, I., 1966: Observations on scale production in *Prymnesium parvum*. J. Cell Sci. **1**, 375–380.

– 1967a: Further observations on the fine structure of *Chrysochromulina chiton* with special reference to the haptonema, "peculiar" Golgi structure, and scale production. J. Cell Sci. **2**, 265–272.

– 1967b: Further observations on scale formation in *Chrysochromulina chiton*. J. Cell Sci. **2**, 411–418.

– 1975: Observations on the microanatomy of *Scourfieldia marina* THRONDSEN and *Scourfieldia caeca* (KORSCH.) BELCHER and SWALE. Arch. Protistenk. **117**, 358–368.

– HARRIS, K., 1966: Observations on the microanatomy of the brown flagellate *Sphaleromantis tetragona* SKUJA with special reference to the flagellar apparatus and scales. J. Linnean Soc. London (Bot.) **59**, 397–403.

– KOWALLIK, K., STOSCH, H. A. VON, 1969: Observations on the fine structure and development of the spindle at mitosis and meiosis in a marine centric diatom (*Lithodesmium undulatum*). I. Preliminary survey of mitosis in spermatogonia. J. Microscopy **89**, 295–320.

– LEEDALE, G. F., 1961: Observations on the fine structure of *Paraphysomonas vestita*, with special reference to the Golgi apparatus and the origin of scales. Phycologia **1**, 37–57.

– – 1969: Observations on the microanatomy of *Coccolithus pelagicus* and *Cricosphaera carterae*, with special reference to the origin and nature of coccoliths and scales. J. Marine Biol. Assoc. U.K. **49**, 1–16.

– OATES, K., PARKE, M., 1963: Observations on the fine structure of the *Pyramimonas* stage of *Halosphaera* and preliminary observations on three species of *Pyramimonas*. J. Marine Biol. Assoc. U.K. **43**, 225–238.

– PARKE, M., 1965: Observations on the fine structure of two species of *Platymonas* with special reference to flagellar scales and the mode of origin of the theca. J. Marine Biol. Assoc. U.K. **45**, 743–754.

– RAYNS, D. G., ETTL, H., 1965: Further observations on green flagellates with scaly flagella: the genus *Heteromastix* KORSHIKOV. J. Marine Biol. Assoc. U.K. **45**, 241–255.

MANUKYAN, I. A., 1967: Submicroscopic structure of the leptomonad form of *Leishmania tropica*. Biol. Zhurnal Armenii **19**, No. 10, 23–27 (in Russian).

MANWELL, R. D., 1928: Conjugation, division and encystment in *Pleurotricha lanceolata*. Biol. Bull. **54**, 417–454.

MARANO, F., 1976: Étude ultrastructurale de la division chez *Dunaliella*. J. Microscopie Biol. Cell **25**, 279–282.

MARCHAND, J., SPRAGUE, V., 1979: Ultrastructure de *Minchinia cadomensis* sp. n. (Haplosporida) parasite du Décapode *Rhithropanopeus harrisii tridentatus* Maitland dans le Canal de Caen à la Mer (Calvados, France). J. Protozool. **26**, 179–185.

MARCO, Y., ROCHAIX, J.-D., 1980: Organization of the nuclear ribosomal DNA of *Chlamydomonas reinhardii*. Molec. Gen. Genet. **177**, 715–723.

MARTINEZ-PALOMO, A., DA SILVA, P. P., CHAVEZ, B., 1976: Membrane structure of *Entamoeba histolytica*: fine structure of freeze-fractured membranes. J. Ultrastr. Res. **54**, 148–158.

MASHANSKY, V. F., 1963: Electron microscopical study of the macronuclei of some ciliates. In: Morfologiya i fiziologiya prostejshikh (Coll. of papers No. 3, Inst. of Cytology), pp. 3–8. Leningrad: Nauka Press (in Russian).

MATSUDO, H., 1966: A cytological study of a chonotrichous ciliate protozoan, *Lobochona prorates*, from the gribble. J. Morphol. **120**, 359–389.

MATTERN, C. F. T., DANIEL, W. A., HONIGBERG, B. M., 1969: Structure of *Hypotrichomonas acosta* (MOSKOWITZ) (Monocercomonadidae, Trichomonadida) as revealed by electron microscopy. J. Protozool. **16**, 668–685.

MATTERN, C. F. T., HONIGBERG, B. M., DANIEL, W. A., 1967: The mastigont system of *Trichomonas gallinae* (RIVOLTA) as revealed by electron microscopy. J. Protozool. **14**, 320–339.
– – – 1972: Structure of *Monocercomonas* sp. as revealed by electron microscopy. J. Protozool. **19**, 265–274.

MATTHYS-ROCHON, E., 1979: Évolution d'un Dinoflagellé libre au cours d'un cycle cellulaire. Biol. Cellul. **35**, 313–320.
– DURRAND-ROUSSY, A. M., 1974: Observations de contacts definis entre les chromosomes et la membrane nucléaire au niveau des invaginations de celle-ci dans le noyau en cours de division chez *Amphidinium carteri*. C. r. Acad. Sci. Paris **D 279**, 1163–1166.

MATTOX, K. R., STEWART, K. D., 1977: Mitosis in the scaly flagellate *Heteromastix* (Chlorophyta). Amer. J. Bot. **64**, 931.

MAUPAS, E., 1889: La rajeunissement karyogamique chez les Ciliés. Arch. Zool. Expér. et Génér. **7**, 149–517.

MAURAND, J., LOUBÈS, C., 1978: Les microsporidies des larves de Simuliés: données ultrastructurales. Z. Parasitenk. **56**, 131–146.
– VEY, A., 1973: Études histopathologique et ultrastructurale de *Thelohania contejeani* (Microsporida, Nosematidae), parasite de l'Ecrevisse *Austropotamobius pallipes* LEREBOULLET. Ann. Parasitol. Hum. Comp. **48**, 411–421.

MCCLELLAN, J. F., 1959: Nuclear division in *Pelomyxa illinoisensis* KUDO. J. Protozool. **6**, 322–331.
– 1964: Tritiated thymidine incorporation and utilization in amoeba. J. Protozool. **11**, suppl., 20–21.

MCCOY, J. W., 1978: New approaches to the problem of macronuclear assortment. J. Theor. Biol. **74**, 475–489.
– 1979: The functional units of macronuclear assortment in *Tetrahymena thermophila*. Genet. Res. **34**, 47–56.

MCDONALD, B. B., 1958: Quantitative aspects of deoxyribose nucleic acid (DNA) metabolism in an amicronucleate strain of *Tetrahymena*. Biol. Bull. **114**, 71–94.
– 1962: Synthesis of deoxyribonucleic acid by micro- and macronuclei of *Tetrahymena pyriformis*. J. Cell Biol. **13**, 193–203.
– 1973: Nucleic acids in *Tetrahymena* during vegetative growth and conjugation. In: Biology of *Tetrahymena* (ELLIOTT, A. M., ed.), pp. 287–306. Stroudsburg, Penn.: Dowden, Hutchinson, and Ross.

MCENERY, M. E., LEE, J. J., 1976: *Allogromia laticollaris*: a foraminiferan with an unusual apogamic metagenetic life cycle. J. Protozool. **23**, 94–108.

MCINTOSH, J. R., HEPLER, P. K., VAN WIE, D. G., 1969: Model for mitosis. Nature **224**, 659–663.

MCLAREN, D. J., 1969: Observations on the fine structural changes associated with schizogony and gametogony in *Eimeria tenella*. Parasitology **59**, 569–574.

MCLOUGHLIN, D. K., 1957: Macronuclear morphogenesis during division of *Blepharisma undulans*. J. Protozool. **4**, 150–153.

MCTAVISH, C., SOMMERVILLE, J., 1980: Macronuclear DNA organization and transcription in *Paramecium primaurelia*. Chromosoma **78**, 147–164.

MEHLHORN, H., 1972a: Elektronenmikroskopische Untersuchungen an Entwicklungsstadien von *Eimeria maxima* (Sporozoa, Coccidia). II. Die Feinstruktur der Mikrogameten. Z. Parasitenk. **40**, 151–163.
– 1972b: Elektronenmikroskopische Untersuchungen an Entwicklungsstadien von *Eimeria maxima* aus dem Haushuhn. III. Der Differenzierungsprozeß der Mikrogameten unter besonderer Berücksichtigung der Kernteilungen. Z. Parasitenk. **40**, 243–260.

MEHLHORN, H., HEYDORN, A. O., GESTRICH, R., 1975: Licht- und elektronenmikroskopische Untersuchungen an Cysten von *Sarcocystis ovicanis* HEYDORN *et al.* (1975) in der Muskulatur von Schafen. Z. Parasitenk. **48**, 83–93.
– MARKUS, M. B., 1976: Electron microscopy of stages of *Isospora felis* of the cat in the mesenteric lymph node of the mouse. Z. Parasitenk. **51**, 15–24.
– PETERS, W., HABERKORN, A., 1980: The formation of kinetes and oocyst in *Plasmodium gallinaceum* (Haemosporidia) and considerations on phylogenetic relationships between Haemosporidia, Piroplasmida and other Coccidia. Protistologica **16**, 135–154.
– SCHEIN, E., 1977: Electron microscopic studies of the development of kinetes in *Theileria annulata* DSCHUNKOWSKI and LUHS 1904 (Sporozoa, Piroplasmea). J. Protozool. **24**, 249–257.
– – WARNECKE, M., 1979: Electron-microscopic studies on *Theileria ovis* RODHAIN 1916: Development of kinetes in the gut of the vector tick, *Rhipicephalus evertsi evertsi* NEUMANN 1897, and their transformation within cells of the salivary gland. J. Protozool. **26**, 377–385.
– SCHOLTYSECK, E., 1973: Elektronenmikroskopische Untersuchungen an Cystenstadien von *Sarcocystis tenella* aus der Oesophagus-Muskulatur des Schafes. Z. Parasitenk. **41**, 291–310.
– SÉNAUD, J., SCHOLTYSECK, E., 1973: La schizogonie chez *Eimeria falciformis* (EIMER 1870), Coccidie Eimeriidae parasite de l'épithélium intestinal de la Souris (*Mus musculus*). Protistologica **9**, 269–291.
MEISTER, H., 1970: Über die Modifizierbarkeit der DNS-Verteilung im Makronucleus bei *Stentor coeruleus* EHRENBERG. Arch. Protistenk. **112**, 314–342.
MERCER, E. H., 1959: An electron microscopic study of *Amoeba proteus*. Proc. Roy. Soc. **B 150**, 216–232.
MERKULOVA, N. A., BORCHSENIUS, S. N., 1976: Replication and integration of fragments of new chains into the DNA of the ciliate *Tetrahymena pyriformis* GL. Molek. Biol. (Moscow) **10**, 1072–1077 (in Russian; English summary).
– IRLINA, I. S., 1974: Characterization of the RNA synthesized in *Tetrahymena pyriformis* cells in normal culture conditions and under starvation. In: Funkzion. Morfol., Genetika i Biochimiya Kletki (Coll. of papers No. 16, Inst. of Cytology), pp. 100–102. Leningrad: Institute of Cytology (in Russian).
MÉTÉNIER, G., 1979: À propos de la régulation de la teneur en ADN macronucléaire pendant le cycle cellulaire du Cilié *Tetrahymena paravorax* (forme microstome). C. r. Acad. Sci. Paris **D 289**, 497–500.
– 1981: Chronologie de la synthèse du DNA macronucléaire chez les formes prédatrices de *Tetrahymena paravorax*. Europ. J. Cell Biol. **24**, 252–258.
MEYER, G. F., LIPPS, H. J., 1980: Chromatin elimination in the hypotrichous ciliate *Stylonychia mytilus*. Chromosoma **77**, 285–297.
– – 1981: The formation of polytene chromosomes during macronuclear development of the hypotrichous ciliate *Stylonychia mytilus*. Chromosoma **82**, 309–314.
MIGNOT, J.-P., 1965: Étude ultrastructurale de *Cyathomonas truncata* FROM. (Flagellé cryptomonadine). J. Microscopie **4**, 239–252.
– 1966: Structure et ultrastructure de quelques Euglenomonadines. Protistologica **2**, No. 3, 51–117.
– 1967: Structure et ultrastructure de quelques Chloromonadines. Protistologica **3**, 5–23.
– 1970: Remarques sur le développement du réticulum endoplasmique et du système vacuolaire chez les Gymnodiniens. Protistologica **6**, 267–281.
– 1974a: Étude ultrastructurale d'un Protiste flagellé incolore: *Pseudodendromonas vlkii* BOURRELY. Protistologica **10**, 397–412.
– 1974b: Étude ultrastructurale des *Bicoeca*, Protistes flagellés. Protistologica **10**, 543–565.

MIGNOT, J.-P., 1975: A propos des particularités ultrastructurales du noyau de *Distigma proteus*, Euglenida. J. Protozool. **22**, No. 3, 67A.
— 1976: Compléments à l'étude des Chloromonadines. Ultrastructure de *Chattonella subsala* BIECHELER, Flagellé d'eau saumâtre. Protistologica **12**, 279–293.
— 1977: Étude ultrastructurale d'un Flagellé du genre *Spumella* CIENK. (= *Heterochromonas* PASCHER = *Monas* O. F. MÜLLER), Chrysomonadine leucoplastidiée. Protistologica **13**, 219–231.
— 1979: Étude ultrastructurale de la pédogamie chez *Actinophrys sol* (Héliozoaire). I. La division progamique. Protistologica **15**, 387–406.
— 1980a: Étude ultrastructurale de la pédogamie chez *Actinophrys sol* (Héliozoaire). II. Les divisions de maturation. Protistologica **16**, 205–225.
— 1980b: Étude ultrastructurale de la pédogamie chez *Actinophrys sol* (Héliozoaire). III. Gamétogénèse, fécondation, enkystement. Protistologica **16**, 533–547.
— BRUGEROLLE, G., 1975a: Étude ultrastructurale de *Cercomonas* DUJARDIN (*Cercobodo* KRASSILSTCHICK), Protiste flagellé. Protistologica **11**, 547–554.
— — 1975b: Étude ultrastructurale du Flagellé phagotrophe *Colponema loxodes* STEIN. Protistologica **11**, 429–444.
— SAVOIE, A., 1979: Observations ultrastructurales sur *Nuclearia simplex* CIENKOWSKI (Protozoa, Rhizopodea, Filosia). Protistologica **15**, 23–32.

MIKAMI, K., 1979: Internuclear control of DNA synthesis in exconjugant cells of *Paramecium caudatum*. Chromosoma **73**, 131–142.

MILDE, K., 1979: Light and electron microscopic studies on Isosporan parasites (Sporozoa) in sparrows (*Passer domesticus* L.). Protistologica **15**, 607–627.

MILDER, R., DEANE, M. P., 1967: Ultrastructure of *Trypanosoma conorhini* in the crithidial phase. J. Protozool. **14**, 65–72.

MILLECCHIA, L. L., RUDZINSKA, M. A., 1971: The ultrastructure of nuclear division in a suctorian, *Tokophrya infusionum*. Z. Zellforsch. **115**, 149–164.

MILLER, J. H., SWARTZWELDER, J. C., DEAS, J. E., 1961: An electron microscopic study of *Entamoeba histolytica*. J. Parasitol. **47**, 577–587.

MINASSIAN, I., BELL, L. G. E., 1976a: Studies on changes in the nuclear helices of *Amoeba proteus* during the cell cycle. J. Cell Sci. **20**, 273–288.
— — 1976b: Timing of nucleolar DNA replication in *Amoeba proteus*. J. Cell Sci. **22**, 521–530.

MITA, T., SHIOMI, H., IWAI, K., 1966: Isolation of nuclei from exponentially growing *Tetrahymena pyriformis*. Exper. Cell Res. **43**, 696–699.

MIYAKE, A., HECKMANN, K., GÖRTZ, H.-D., 1979: Meiosis in *Blepharisma japonicum*. Protistologica **15**, 473–486.

MOENS, P. B., PERKINS, F. O., 1969: Chromosome number of a small protist: accurate determination. Science **166**, 1289–1291.

MOESTRUP, Ø., THOMSEN, H. A., 1976: Fine structural studies on the flagellate genus *Bicoeca*. I. *Bicoeca maris* with particular emphasis on the flagellar apparatus. Protistologica **12**, 101–120.

MOLON-NOBLOT, S., 1976: Ultrastructure du centriole chez la grégarine *Grebnickiella gracilis* (Sporozoaire). Protistologica **12**, 431–434.
— 1977: Les gamètes de la Grégarine *Grebnickiella* (= *Nina*) *gracilis* (GREBNICKI) (Sporozoaire) parasite de la Scolopendre *Scolopendra cingulata* L.: premières données ultrastructurales. C. r. Acad. Sci. Paris **D 284**, 1179–1182.
— 1978: Étude ultrastructurale de la gamétogenèse chez la Grégarine *Grebnickiella gracilis* (GREBNICKI) (Sporozoaire), parasite de la Scolopendre *Scolopendra cingulata* L. Ann. Sci. Nat., Zool. sér. 12, **20**, 79–97.

Molon-Noblot, S., Desportes, I., 1977: Mise en évidence des complexes synaptonématiques dans le noyau méiotique d'un sporozoaire, la grégarine *Grebnickiella gracilis* (Grebnicki), parasite de la Scolopendre *Scolopendra cingulata*. C. r. Acad. Sci. Paris **D 285**, 217–219.

– – 1980: Étude ultrastructurale des mitoses gamogoniques de la grégarine *Grebnickiella gracilis* Gr., parasite de la scolopendre *Scolopendra cingulata* L. Considérations sur les mitoses schizogoniques des Sporozoaires (Apicomplexa). Protistologica **16**, 395–411.

Moltmann, U. G., 1981: Light and electron microscopic studies on *Klossia helicina* (Coccidia): development of gamonts is small kidney tissue cultures and sporogony in the natural host. Protistologica **17**, 185–197.

Molyneux, D. H., Bafort, J. M., 1970: The fine structure of *Babesia microti* in the natural host, *Microtus agrestis*. Z. Tropenmed. Parasitol. **21**, 393–401.

Moore, J., Cantor, M. H., Sheeler, P., Kahn, W., 1970: The ultrastructure of *Polytomella agilis*. J. Protozool. **17**, 671–676.

– Sinden, R. E., 1974: Fine structure of *Plasmodium mexicanum*. J. Parasitol. **60**, 825–833.

Morat, G., 1965: Étude cytophotométrique des teneurs en ADN chez quelques espèces de Ciliés astomes du genre *Metaradiophrya* Heid. Arch. Zool. Expér. Génér. **105**, 201–214.

– 1970: Études cytophotométrique et autoradiographique de l'appareil nucléaire de *Colpidium campylum* au cours du cycle cellulaire. Protistologica **6**, 83–95.

– 1973: La chromatine macronucléaire au cours du cycle chez la Cilié *Colpidium campylum*: études ultrastructurale et autoradiographique. In: Progress in Protozoology (Abstr. IV Internat. Congr. Protozool.), p. 283. Clermont-Ferrand: Université de Clermont.

– 1979: Observations sur les teneurs en ADN nucléaire d'un *Plagiotoma* parasite de deux Vers d'espèces différentes. J. Protozool. **26**, No. 3 (part 1), 65A.

– 1980: Régulation de la teneur en ADN macronucléaire et variation du rapport nucléo-cytoplasmique au cours de la croissance exponentielle, chez le Cilié *Colpidium campylum*. J. Protozool. **27**, No. 3, 85A–86A.

Morosoli, R., Lafontaine, J. G., 1976: Characterization of *Astasia longa* ribosomal DNA. Exper. Cell Res. **99**, 88–94.

Moses, M., J., 1950: Nucleic acids and proteins of the nuclei of *Paramecium*. J. Morphol. **87**, 493–536.

Mosevich, T. N., 1968: The fine structure of the macronucleus in *Ichthyophthirius multifiliis* at different stages of the life cycle. Protistologica **4**, 469–476.

– Cheissin, E. M., 1961: Some electron microscopic data on the merozoites of *Eimeria intestinalis* from the rabbit intestine. Tsitologiya (Leningrad) **3**, 34–39 (in Russian).

Moshkovski, S. D., 1957: On the nature of the Protozoa and the limits of protozoology. Trudy Leningrad. Obsch. Estestvoisp., Otdel. zool. **73**, 129–137 (in Russian).

– 1963: The levels of organization of living matter and the place of Protozoa in the system of organisms. In: Progress in Protozoology (Proc. I Internat. Congr. Protozool.), pp. 57–63. Prague: Publ. House ČSAV.

Moyne, G., Bertaux, O., Puvion, E., 1975: The nucleus of *Euglena*. I. An ultracytochemical study of the nucleic acids and nucleoproteins of synchronized *Euglena gracilis* Z. J. Ultrastr. Res. **52**, 362–376.

Mügge, E., 1957: Die Konjugation von *Vorticella campanula* (Ehrbg.). Arch. Protistenk. **102**, 165–208.

Mühlpfordt, H., Bayer, M., 1961: Elektronenmikroskopische Untersuchungen an Protozoen (*Trypanosoma gambiense*). Z. Tropenmed. Parasitol. **12**, 334–346.

Müller-Merz, E., Lee, J. J., 1976: Symbiosis in the larger foraminiferan *Sorites marginalis* (with notes on *Archaias* spp.). J. Protozool. **23**, 390–396.

Mulsow, K., 1911: Über Fortpflanzungserscheinungen bei *Monocystis rostrata* n. sp. Arch. Protistenk. **22**, 20–55.

Mulsow, W., 1913: Die Conjugation von *Stentor coeruleus* und *Stentor polymorphus*. Arch. Protistenk. **28**, 363–388.

Munitoli, F., Hirshfield, H. I., 1968: DNA synthesis cycle in *Blepharisma*. J. Protozool. **15**, 532–535.

Murti, K. G., 1972: "Nucleoli" in the micronucleus of *Tetrahymena*. Exper. Cell Res. **72**, 584–586.

– 1973: Electron-microscopic observations on the macronuclear development of *Stylonychia mytilus* and *Tetrahymena pyriformis* (Ciliophora – Protozoa). J. Cell Sci. **13**, 479–509.

– 1976a: Helices in the macronucleus of *Euplotes eurystomus*. Exper. Cell Res. **99**, 423–425.

– 1976b: Organization of genetic material in the macronucleus of hypotrichous ciliates. In: Handbook of genetics (King, R. C., ed.), pp. 113–137. New York-London: Plenum Press.

– Prescott, D. M., 1970: Micronuclear ribonucleic acid in *Tetrahymena pyriformis*. J. Cell Biol. **47**, 460–467.

– – 1978: Electron microscopic visualization of transcribed genes in the nucleus of *Amoeba proteus*. Exper. Cell Res. **112**, 233–240.

– – Pène, J. J., 1972: DNA in ciliated Protozoa. III. Binding of RNA polymerase and denaturation at the ends of the low molecular weight DNA in *Stylonychia*. J. Molec. Biol. **68**, 413–416.

Muse, K. E., Roberts, J. E., 1973: Microbodies in *Crithidia fasciculata*. Protoplasma **78**, 343–348.

Myers, E. H., 1935: The life history of *Patellina corrugata* Williamson, a Foraminifer. Bull. Scripps Inst. Oceanogr., Techn. ser. **3**, 355–392.

– 1936: The life-cycle of *Spirillina vivipara* Ehrenberg with notes on morphogenesis, systematics and distribution of the Foraminifera. J. Roy. Microscop. Soc. **56**, 120–146.

Nägel, W. C., Wunderlich, F., 1977: Effect of temperature on nuclear membranes and nucleo-cytoplasmic RNA transport in *Tetrahymena* grown at different temperatures. J. Membr. Biol. **32**, 151–164.

Nanney, D. L., 1964: Macronuclear differentiation and subnuclear assortment in ciliates. In: Role of chromosomes in development, pp. 253–273. New York: Academic Press.

– Preparata, R. M., 1979: Genetic evidence concerning the structure of the *Tetrahymena thermophila* macronucleus. J. Protozool. **26**, 2–9.

Naville, A., 1925: Recherches sur le cycle sporogonique des *Aggregata*. Rev. Suisse Zool. **32**, 125–179.

– 1927: Recherches sur le cycle évolutif et chromosomique de *Klossia helicis*. Arch. Protistenk. **57**, 427–474.

Netrawali, M. S., 1970: On the presence of histones in *Euglena gracilis* var. *bacillaris*. Exper. Cell Res. **63**, 422–425.

Ng, S. F., 1980: Elimination of the germ nuclei of *Paramecium tetraurelia* by laser microbeam irradiation. Exper. Cell Res. **127**, 475–478.

Nicklas, R. B., 1971: Mitosis. In: Advances in cell biology (Prescott, D. M., Goldstein, L., McConkey, E. H., eds.), Vol. 2, pp. 225–297. New York: Appleton.

Nielsen, M. H., Ludvik, J., Nielsen, R., 1966: On the ultrastructure of *Trichomonas vaginalis* Donné. J. Microscopie **5**, 229–250.

Niles, E. G., 1978: Isolation of a high specific activity 35 S ribosomal RNA precursor from *Tetrahymena pyriformis* and identification of its 5′-terminus ppAp. Biochemistry **17**, 4839–4844.

NILOVA, V. K., 1968: A study of DNA synthesis in the macronucleus of *Bursaria truncatella*. Tsitologiya (Leningrad) **10**, 1580–1587 (in Russian; English summary).
– SUKHANOVA, K. M., 1964: Synthesis of nucleic acids in *Opalina ranarum* EHRBG. Nature **204**, 459–460.
– – 1969: A study of nucleic acid synthesis in the life cycle of *Opalina ranarum*. Tsitologiya (Leningrad) **11**, 721–730 (in Russian; English summary).

NILSSON, J. R., 1969: The fine structure of *Neobursaridium gigas* (BALECH). C. r. Trav. Lab. Carlsberg **37**, 49–76.
– 1970: Suggestive structural evidence for macronuclear "subnuclei" in *Tetrahymena pyriformis* GL. J. Protozool. **17**, 539–548.
– 1973: An electron microscope autoradiographic study of the *Tetrahymena pyriformis* GL macronucleus after a short period of tritiated nucleoside labelling. C. r. Trav. Lab. Carlsberg **39**, 357–373.
– 1974: Effects of DMSO on vacuole formation, contractile vacuole function and nuclear division in *Tetrahymena pyriformis* GL. J. Cell Sci. **16**, 39–47.
– 1976a: Physiological and structural studies on *Tetrahymena pyriformis* GL. C. r. Trav. Lab. Carlsberg **40**, 215–355.
– 1976b: On structural nucleolar changes in *Tetrahymena pyriformis* GL. In: Proc. 9th Congr. Nordic Soc. Cell Biol., pp. 197–203. Odense: Odense Univ. Press.
– 1977: Effects of dimethyl sulfoxide on *Tetrahymena pyriformis* GL. Fine structural changes and their reversibility. J. Protozool. **24**, 275–283.
– LEICK, V., 1970: Nucleolar organization and ribosome formation in *Tetrahymena pyriformis* GL. Exper. Cell Res. **60**, 361–372.
– ZEUTHEN, E., 1974: Microscopical studies on the macronucleus of heat synchronized *Tetrahymena pyriformis* GL. C. r. Trav. Lab. Carlsberg **40**, 1–18.
– – 1980: On nucleolar DNA replication in *Tetrahymena pyriformis* GL-8: an electron microscope autoradiographic study of heat synchronized and exponentially multiplying cells. Carlsberg Res. Comm. **45**, 101–111.

NOBLE, E. R., 1938: The life cycle of *Zygosoma globosum* sp. nov., a gregarine parasite of *Urechis caupo*. Univ. Calif. Publs Zool. **43**, 41–66.

NOCK, A., 1981: RNA and macronuclear transcription in the ciliate *Stylonychia mytilus*. Chromosoma **83**, 209–220.

NOIROT-TIMOTHÉE, C., 1959: Recherches sur l'ultrastructure d'*Opalina ranarum*. Ann. Sci. Natur., Zool. sér. 12, **1**, 265–281.
– 1960: Étude d'une famille de Ciliés: les Ophryoscolecidae. Structures et ultrastructures. Ann. Sci. Natur., Zool. sér. 12, **2**, 527–718.
– 1967: Étude au microscope électronique des inclusions cytoplasmiques de *Cepedea dimidiata* (Protozoa, Opalinina) et plus spécialement des mitochondries ou "corps de Zeller". Protistologica **3**, 301–312.

NORRIS, R. E., PEARSON, B. R., 1975: Fine structure of *Pyramimonas parkeae* sp. nov. (Chlorophyta, Prasinophyceae). Arch. Protistenk. **117**, 192–213.

NOUZARÈDE, M., 1975: Sur un nouveau genre de Protozoaires ciliés géants mésopsammiques appartenant à la famille des Geleiidae KAHL. C. r. Acad. Sci. Paris **D 280**, 625–628.
– 1976: Cytologie fonctionelle et morphologie expérimentale de quelques Protozoaires ciliés mésopsammiques géants de la famille des Geleiidae (KAHL). Bull. Stat. Biol. d'Arcachon, n. sér. No. 28, suppl., 1–315.

NYBERG, D., 1976: Are macronuclear subunits in *Paramecium* functionally diploid? Genet. Res. **27**, 239–248.

NYHOLM, K.-G., NYHOLM, P.-G., 1975: Ultrastructure of monothalamous Foraminifera. Zoon **3**, 141–150.

Oakley, B. R., Bisalputra, T., 1977: Mitosis and cell division in *Cryptomonas* (Cryptophyceae). Canad. J. Bot. **55**, 2789–2800.
– Dodge, J. D., 1973: Mitosis in the Cryptophyceae. Nature **244**, 521–522.
– – 1974: Kinetochores associated with the nuclear envelope in the mitosis of a dinoflagellate. J. Cell Biol. **63**, 322–325.
– – 1976a: The ultrastructure of mitosis in *Chroomonas salina* (Cryptophyceae). Protoplasma **88**, 241–254.
– – 1976b: Mitosis and cytokinesis in the dinoflagellate *Amphidinium carterae*. Cytobios **17**, 35–46.
– – 1979: Evidence for a double-helically coiled toroidal chromonema in the dinoflagellate chromosome. Chromosoma **70**, 277–291.
– Heath, I. B., 1978: The arrangement of microtubules in serially sectioned spindles of the alga *Cryptomonas*. J. Cell Sci. **31**, 53–70.
Ockleford, C. D., Tucker, J. B., 1973: Growth, breakdown, repair, and rapid contraction of microtubular axopodia in the Heliozoan *Actinophrys sol*. J. Ultrastr. Res. **44**, 369–387.
O'Donnell, E. H. J., 1965: Nucleolus and chromosomes in *Euglena gracilis*. Cytologia **30**, 118–154.
Ogden, C. G., 1979: An ultrastructural study of division in *Euglypha* (Protozoa: Rhizopoda). Protistologica **15**, 541–556.
Old, K. M., Darbyshire, J. F., 1980: *Arachnula impatiens* Cienk., a mycophagous giant amoeba from soil. Protistologica **16**, 277–287.
Opas, M., Hrebenda, B., Tołłoczko, B., 1978: Contractility of glycerol-extracted nuclei of *Amoeba proteus*. Acta Protozool. **17**, 369–374.
Ord, M. J., 1968: The synthesis of DNA through the cell cycle of *Amoeba proteus*. J. Cell Sci. **3**, 483–492.
– 1969: Control of DNA synthesis in *Amoeba proteus*. Nature **221**, 964–966.
– 1971: The initiation, maintenance and termination of DNA synthesis: a study of nuclear DNA replication using *Amoeba proteus* as a cell model. J. Cell Sci. **9**, 1–21.
Orias, E., Flacks, M., 1975: Macronuclear genetics of *Tetrahymena*. I. Random distribution of macronuclear gene copies in *T. pyriformis*, syngen 1. Genetics **79**, 187–206.
– Newby, C. J., 1975: Macronuclear genetics of *Tetrahymena*. II. Macronuclear location of somatic mutations to cycloheximide resistance. Genetics **80**, 251–262.
Ormières, R., 1970: Formation des capsules polaires dans la spore de l'Actinomyxidie *Aurantiactinomyxon eiseniellae* Orm. Fréz. (Étude ultrastructurale). C. r. Acad. Sci. Paris **D 271**, 2326–2328.
– 1971: Une Grégarine paradoxale, *Gigaductus anchi* Tuz. Orm. 1966: ultrastructure des stades schizogoniques et position systématique des Gigaductidae Filipponi 1948. Protistologica **7**, 261–271.
– Baudoin, J., Brugerolle, G., Pralavorio, R., 1976: Ultrastructure de quelques stades de la Microsporidie *Octosporea muscaedomesticae* Flu, parasite de *Ceratitis capitata* (Wiedemann) (Diptère Trypetidae). J. Protozool. **23**, 320–328.
– Loubès, C., Maurand, J., 1977: Étude ultrastructurale de la Microsporidie *Perezia lankesteriae* Léger et Duboscq 1909, espèce-type: validation du genre *Perezia*. Protistologica **13**, 101–112.
– Porchet-Henneré, E., 1974: Données nouvelles sur la coccidie *Grasseella microcosmi* Tuz. Orm. 1960: démonstration, par l'étude ultrastructurale, de l'existence d'une schizogonie dans son cycle biologique. Bull. Soc. Zool. France **99**, 545–549.
– Puytorac, P. de, 1968: Ultrastructure des spores de l'Haplosporidie *Haplosporidium ascidiarum*, endoparasite du Tunicier *Sydnium elegans* Giard. C. r. Acad. Sci. Paris **D 266**, 1134–1136.

ORMIÈRES, R., SPRAGUE, V., 1973: A new family, new genus, and new species allied to the Microsporida. J. Invertebr. Pathol. **21**, 224–240.

OSADA, M., 1959: Electron-microscopic studies on Protozoa. I. Fine structure of *Entamoeba histolytica*. Keijo J. Med. **8**, 99–103.

– 1962: Electron-microscopic studies on Protozoa. II. Studies on *Trichomonas muris*. Keijo J. Med. **11**, 227–240.

OSSIPOV, D. V., 1981: Problems of nuclear heteromorphism in the unicellular organisms. 167pp. Leningrad: Nauka Press (in Russian).

– BORCHSENIUS, O. N., 1973: Peculiarities of micronuclear polyploidy in ciliates. Tsitologiya (Leningrad) **15**, 243–253 (in Russian; English summary).

– GROMOV, B. V., MAMKAEVA, K. A., RAUTIAN, M. S., SKOBLO, I. I., BORCHSENIUS, O. N., 1976: The use of symbiotic bacteria for analysis of the structure and function of nuclear apparatus of ciliates. In: Karyologiya i Genetika Prostejshikh (Series "Protozoology", No. 1), pp. 101–139. Leningrad: Nauka Press (in Russian; English summary).

– SKOBLO, I. I., 1968: The autogamy during conjugation in *Paramecium caudatum* EHRBG. II. The exautogamont stages of the nuclear reorganization. Acta Protozool. **6**, 33–52.

– – 1973: Morphogenetic activity of generative nucleus in the sexual cell reorganization of ciliates. Acta Protozool. **12**, 151–164.

– – BORCHSENIUS, O. N., RAUTIAN, M. S., PODLIPAEV, S. A., 1980: *Holospora acuminata*, a new species of symbiotic bacteria in the micronucleus of the ciliate *Paramecium bursaria*. Tsitologiya (Leningrad) **22**, 922–929 (in Russian; English summary).

OVCHINNIKOVA, L. P., 1970: Variability of DNA content in micronuclei of *Paramecium bursaria*. Acta Protozool. **7**, 211–220.

– SELIVANOVA, G. V., CHEISSIN, E. M., 1965: Photometric study of the DNA content in the nuclei of *Spirostomum ambiguum* (Ciliata, Heterotricha). Acta Protozool. **3**, 69–78.

– SUKHANOVA, K. M., 1976: Content of DNA and histones in the nuclei of *Euglena gracilis* in normal conditions and during adaptation to extreme environmental factors (a cytophotometric study). Tsitologiya (Leningrad) **18**, 482–489 (in Russian; English summary).

– – RIVEL, T. G., VLASSOVA, T. D., 1976: Cytophotometric study of nuclear DNA and histones in *Euglena gracilis* and *Astasia longa*. In: Karyologiya i genetika prostejshikh (Series "Protozoology", No. 1), pp. 151–158. Leningrad: Nauka Press (in Russian; English summary).

PACHECO, N. D., FAYER, R., 1977: Fine structure of *Sarcocystis cruzi* schizonts. J. Protozool. **24**, 382–388.

– VETTERLING, J. M., DORAN, D. J., 1975: Ultrastructure of cytoplasmic and nuclear changes in *Eimeria tenella* during first-generation schizogony in cell culture. J. Parasitol. **61**, 31–42.

PADNOS, M., 1962: Cytology of cold induced transformation of octogenic reproductive cysts to resting cysts in *Colpoda maupasi*. J. Protozool. **9**, 13–20.

PAGE, F. C., 1967a: Taxonomic criteria for limax amoebae, with descriptions of 3 new species of *Hartmannella* and 3 of *Vahlkampfia*. J. Protozool. **14**, 499–521.

– 1967b: Re-definition of the genus *Acanthamoeba* with description of three species. J. Protozool. **14**, 709–724.

– 1974: *Rosculus ithacus* HAWES 1963 (Amoebida, Flabellulidae) and the amphizoic tendency in amoebae. Acta Protozool. **13**, 143–154.

– 1978: An electron-microscopical study of *Thecamoeba proteoides* (Gymnamoebia), intermediate between Thecamoebidae and Amoebidae. Protistologica **14**, 77–85.

– 1979a: *Vexillifera armata* n. sp. (Gymnamoebia, Paramoebidae), an estuarine amoeba with distinctive surface structures and trichocyst-like bodies. Protistologica **15**, 111–122.

PAGE, F. C., 1979b: The fine structure of *Vexillifera bacillipedes* (Amoebida, Paramoebidae). Arch. Protistenk. **122**, 9–19.

– 1979c: Two genera of marine amoebae (Gymnamoebia) with distinctive surface structures: *Vannella* BOVEE 1965, and *Pseudoparamoeba* n. gen., with two new species of *Vannella*. Protistologica **15**, 245–257.

– 1980: A light- and electron-microscopical comparison of limax and flabellate marine amoebae belonging to four genera. Protistologica **16**, 57–78.

– BALDOCK, B. M., 1980: *Polychaos fasciculatum* (PENARD 1902) (Amoebida): a light- and electron-microscopic study. Arch. Protistenk. **123**, 211–235.

PAKES, S. P., SHADDUCK, J. A., CALI, A., 1975: Fine structure of *Encephalitozoon cuniculi* from rabbits, mice and hamsters. J. Protozool. **22**, 481–488.

PAN, W.-C., BLACKBURN, E. H., 1981: Single extrachromosomal ribosomal RNA gene copies are synthesized during amplification of the rDNA in *Tetrahymena*. Cell **23**, 459–466.

PAPPAS, G. D., 1956a: Helical structures in the nucleus of *Amoeba proteus*. J. Biophys. Biochem. Cytol. **2**, 221–222.

– 1956b: The fine structure of the nuclear envelope of *Amoeba proteus*. J. Biophys. Biochem. Cytol. **2**, suppl., 431–434.

– 1959: Electron microscope studies on amoebae. Ann. N.Y. Acad. Sci. **78**, 448–473.

PARKE, M., MANTON, I., 1965: Preliminary observations on the fine structure of *Prasinocladus marinus*. J. Marine Biol. Assoc. U.K. **45**, 525–536.

– – 1967: The specific identity of the algal symbiont in *Convoluta roscoffensis*. J. Marine Biol. Assoc. U.K. **47**, 445–464.

PARKER, J. W., GIESE, A. C., 1966: Nuclear activity during regeneration in *Blepharisma intermedium* BHANDARY. J. Protozool. **13**, 617–622.

PASTERNAK, J., 1967: Differential genic activity in *Paramecium aurelia*. J. Exper. Zool. **165**, 395–417.

PÄTAU, K., 1937a: Partiell heterochromatische Chromosomen bei Radiolarien. Arch. Protistenk. **89**, 393–404.

– 1937b: SAT-Chromosomen und Spiralbau der Chromosomen der extracapsulären Körper (*Merodinium* spec.) von *Collozoum inerme* MÜLLER. Cytologia **2**, 667–680.

PATTEN, R., 1935: The life history of *Merocystis kathae* in the whelk *Buccinum undatum*. Parasitology **27**, 399–430.

PATTERSON, D. J., 1979: On the organization and classification of the protozoon, *Actinophrys sol* EHRENBERG 1830. Microbios **26**, 165–208.

PAULIN, J. J., 1975: Intranuclear microtubules in dividing *Trypanosoma equiperdum*. BioSystems **7**, 308.

– BROOKS, A. S., 1975: Macronuclear differentiation during oral regeneration in *Stentor coeruleus*. J. Cell Sci. **19**, 531–541.

– KRASCHENINNIKOW, S., 1973: An electron microscopic study of *Balantidium caviae*. Acta Protozool. **12**, 97–103.

PEARLMAN, R. E., ANDERSSON, P., ENGBERG, J., NILSSON, J. R., 1979: Synthesis of ribosomal DNA in conjugating *Tetrahymena*. Exper. Cell Res. **123**, 147–155.

PEARSON, B. R., NORRIS, R. E., 1975: Fine structure of cell division in *Pyramimonas parkeae* NORRIS and PEARSON (Chlorophyta, Prasinophyceae). J. Phycol. **11**, 113–124.

PECK, R., PELVAT, B., BOLIVAR, I., HALLER, G. DE, 1975: Light and electron microscopic observations on the heterotrich ciliate *Climacostomum virens*. J. Protozool. **22**, 368–385.

PELSTER, B., 1973: Vergleichende elektronenmikroskopische Untersuchungen an den Makrogameten von *Isospora felis* und *I. rivolta*. Z. Parasitenk. **41**, 29–46.

PELSTER, B., PIEKARSKI, G., 1971: Elektronenmikroskopische Analyse der Mikrogametenentwicklung bei *Toxoplasma gondii*. Z. Parasitenk. **37**, 267–277.

PELVAT, B., DE HALLER, G., 1976: Macronuclear DNA in *Stentor coeruleus*: a first approach to its characterization. Genet. Res. **27**, 277–289.
– HYVERT, N., DE HALLER, G., 1973: Synthèse d'acides nucléiques dans le macronoyau de *Stentor coeruleus*: régénération et cycle de division. Protistologica **9**, 255–267.

PÉREZ-SILVA, J., ALONSO, P., 1966: Demonstration of polytene chromosomes in the macronuclear anlage of oxytrichous ciliates. Arch. Protistenk. **109**, 65–70.

PERKINS, F. O., 1968: Fine structure of the oyster pathogen *Minchinia nelsoni* (Haplosporida, Haplosporidiidae). J. Invertebr. Pathol. **10**, 287–307.
– 1969: Electron microscope studies of sporulation in the oyster pathogen, *Minchinia costalis* (Sporozoa: Haplosporida). J. Parasitol. **55**, 897–920.
– 1970: Formation of centriole and centriole-like structures during meiosis and mitosis in *Labyrinthula* sp. (Rhizopodea, Labyrinthulida). An electron microscope study. J. Cell Sci. **6**, 629–653.
– 1971: Sporulation in the trematode hyperparasite *Urosporidium crescens* DE TURK 1940 (Haplosporida: Haplosporidiidae)—an electron microscope study. J. Parasitol. **57**, 9–23.
– 1975: Fine structure of the haplosporidan Kernstab, a persistent, intranuclear mitotic apparatus. J. Cell Sci. **18**, 327–346.
– 1976: Ultrastructure of sporulation in the European flat oyster pathogen, *Marteilia refringens*—taxonomic implications. J. Protozool. **23**, 64–74.
– AMON, J. P., 1969: Zoosporulation in *Labyrinthula* sp.: an electron microscope study. J. Protozool. **16**, 253–257.
– CASTAGNA, M., 1971: Ultrastructure of the Nebenkörper or "secondary nucleus" of the parasitic amoeba *Paramoeba perniciosa* (Amoebida, Paramoebidae). J. Invertebr. Pathol. **17**, 186–193.

PERNIN, P., PUSSARD, M., 1979: Étude en microscopie photonique et électronique d'une Amibe voisine du genre *Acanthamoeba*: *Comandonia operculata* n. gen., n. sp. (Amoebida, Acanthamoebidae). Protistologica **15**, 87–102.

PESCHKOFF, M. A., 1929: Biology and morphology of *Bursaria truncatella* O. F. MÜLL. and its nuclear apparatus during division. Russkij Archiv Protistol. **8**, 1–16 (in Russian; German summary).
– 1966: Comparative cytology of blue-green algae, bacteria and actinomycetes. Moscow: Nauka Press (in Russian).
– AKINSHINA, G. T., 1963: Cytological studies of the rhizopod *Chlamydophrys major*. I. Fine structure of the *Chlamydophrys* trophont based on electron microscopy of ultrathin sections and *in vivo* observations. Tsitologiya (Leningrad) **5**, 554–564 (in Russian).
– TIKHOMIROVA, L. A., 1967: On the structure of the macronucleus of the parasitic ciliate *Ichthyophthirius multifiliis*. Doklady Akad. Nauk SSSR **175**, 211–212 (in Russian).

PESHKOVSKAYA, L. S., 1936: Changes of the nuclear apparatus of *Climacostomum virens* during conjugation. Biol. Zhurnal (Moscow) **5**, 207–220 (in Russian; English summary).
– 1941: Changes of the nuclear apparatus during conjugation in some holotrichous ciliates. Trudy Inst. Cytol., Histol. i Embryol. **1**, 19–27 (in Russian).
– 1948: Metamorphosis of the nuclear apparatus during the sexual process in two species of hypotrichous ciliates. Trudy Inst. Cytol., Histol. i Embryol. **3**, 201–208 (in Russian).

PICKETT-HEAPS, J. D., 1967: Ultrastructure and differentiation in *Chara* sp. II. Mitosis. Austral. J. Biol. Sci. **20**, 883–894.

PICKETT-HEAPS, J. D., 1969: The evolution of the mitotic apparatus: an attempt at comparative ultrastructural cytology in dividing plant cells. Cytobios **1**, 257–280.
– 1971: The autonomy of the centriole: fact or fallacy? Cytobios **3**, 205–214.
– 1972: Variation in mitosis and cytokinesis in plant cells: its significance in the phylogeny and evolution of ultrastructural systems. Cytobios **5**, 59–77.
– 1974: The evolution of mitosis and the eukaryotic condition. BioSystems **6**, 37–48.
– OTT, D. W., 1974: Ultrastructural morphology and cell division in *Pedinomonas*. Cytobios **11**, 41–58.
– TIPPIT, D. H., ANDREOZZI, J. A., 1978a: Cell division in the pennate diatom *Pinnularia*. I. Early stages of mitosis. Biol. Cellul. **33**, 71–78.
– – – 1978b: Cell division in the pennate diatom *Pinnularia*. II. Later stages in mitosis. Biol. Cellul. **33**, 79–84.
– – LESLIE, R., 1980a: Light and electron microscopic observations on cell division in two large pennate diatoms, *Hantzschia* and *Nitzschia*. I. Mitosis *in vivo*. Eur. J. Cell Biol. **21**, 1–11.
– – – 1980b: Light and electron microscopic observations on cell division in two large pennate diatoms, *Hantzschia* and *Nitzschia*. II. Ultrastructure. Eur. J. Cell Biol. **21**, 12–27.
PIEKARSKI, G., 1939: Cytologische Untersuchungen an einem normalen und einem micronucleuslosen Stamm von *Colpoda steinii* MAUPAS. Arch. Protistenk. **92**, 117–130.
– 1941: Endomitose beim Großkern der Ziliaten? Versuch einer Synthese. Biol. Zbl. **61**, 416–426.
PIERI, J., 1965: Interprétation cytophotométrique des phénomènes nucléaires au cours de la conjugaison chez *Stylonychia pustulata*. C. r. Acad. Sci. Paris **D 261**, 2742–2744.
– 1966: Étude cytophotométrique de la quantité relative de l'ADN micronucléaire au cours de la division binaire normale chez *Glaucoma scintillans*. C. r. Acad. Sci. Paris **D 262**, 775–777.
– VAUGIEN, C., 1967a: Analyse microspectrographique dans l'ultraviolet des zones de biosynthèse nucléaire chez *Euplotes*. C. r. Acad. Sci. Paris **D 264**, 1907–1910.
– – 1967b: Analyse microspectrographique dans l'ultraviolet d'une processus de biosynthèse nucléaire chez *Chilodonella*. C. r. Acad. Sci. Paris **D 265**, 256–259.
– – TROULLIER, M., 1968: Interprétation cytophotométrique des phénomènes micronucléaires au cours de la division binaire et des divisions prégamiques chez *Paramecium trichium*. J. Cell Biol. **36**, 664–668.
PITTAM, M. D., 1963: Studies on an amoebo-flagellate, *Naegleria gruberi*. Quart. J. Microscop. Sci. **104**, 513–529.
PITTILO, R. M., BALL, S. J., HUTCHISON, W. M., 1980: The ultrastructural development of the macrogamete of *Eimeria stiedai*. Protoplasma **104**, 33–41.
POLJANSKY, G., 1934: Geschlechtsprozesse bei *Bursaria truncatella* O. F. MÜLL. Arch. Protistenk. **81**, 420–546.
– RAIKOV, I. B., 1960: The role of polyploidy in protozoan evolution. Tsitologiya (Leningrad) **2**, 509–518 (in Russian).
– RAIKOV, I., 1961: Nature et origine du dualisme nucléaire chez les infusoires ciliés. Bull. Soc. Zool. France **86**, 402–411.
– RAIKOV, I. B., 1976: Polymerization and oligomerization phenomena in protozoan evolution. Trans. Amer. Microsc. Soc. **95**, 314–326.
– SERGEJEVA, G. I., 1977: Autoradiographic study of DNA synthesis during conjugation of *Bursaria truncatella* (Ciliophora, Heterotrichida). Abstr. V Internat. Congr. Protozool., p. 102. New York.
– – 1981: Autoradiographic investigation of the DNA synthesis during development of the new macronucleus of the ciliate *Bursaria truncatella*. Tsitologiya (Leningrad) **23**, 666–673 (in Russian; English summary).

Pomryaskinskaya, N., 1940: Observations on cysts of the hypotrichous ciliate *Oxytricha hymenostoma*. Uchenye Zapiski Leningr. Gos. Pedagog. Inst. im. Gerzena **30**, 93–132 (in Russian).

Porchet-Henneré, E., 1967a: Étude des premiers stades de développement de la Coccidie *Coelotropha durchoni*. Z. Zellforsch. **80**, 556–569.
– 1967b: Premières observations sur la structure fine du gamète mâle de la Coccidie *Coelotropha durchoni* (Sporozoaire Coccidiomorphe). C. r. Acad. Sci. Paris **D 264**, 2130–2133.
– 1970: La microgamétogenèse chez la Coccidie *Coelotropha durchoni* (Vivier–Henneré); étude au microscope électronique. Arch. Protistenk. **112**, 21–29.
– 1971: La fécondation et la sporogénèse chez la Coccidie *Coelotropha durchoni*. Étude en microscopie photonique et électronique. Z. Parasitenk. **37**, 94–125.
– 1972a: Observations en microscopie photonique et électronique sur la sporogénèse de *Dehornia sthenelais* (n. gen., sp. n.), Sporozoaire parasite de l'Annélide polychète *Sthenelais boa* (Aphroditides). Protistologica **8**, 245–255.
– 1972b: Considérations générales sur les processus de schizogonie chez les Sporozoaires à la lumière des données de la microscopie électronique. Année Biol. **11**, 413–426.
– 1977: Étude ultrastructurale de la schizogonie chez la Coccidie *Globidium gilruthi*. Protistologica **13**, 31–52.
– Fischer, A., 1973: *Diplauxis schreveli* n. sp., Grégarine parasite du coelome de l'Annélide polychète *Platynereis dumerilii*: cycle et étude ultrastructurale préliminaire. Protistologica **9**, 437–454.
– Ormières, R., 1973: Étude ultrastructurale des stades "hépatiques" (gamontes) de *Grasseella microcosmi*, Coccidie parasite du Tunicier *Pyura microcosmus*. Protistologica **9**, 187–211.
– Richard, A., 1969: Structure fine du sporoblaste immature uninucléé d'*Aggregata eberthi* Labbé (Sporozoaire Coccidiomorphe). C. r. Acad. Sci. Paris **D 269**, 1681–1683.
– – 1970a: Ultrastructure des stades végétatifs d'*Aggregata eberthi* Labbé: le trophozoite et le schizonte. Z. Zellforsch. **103**, 179–191.
– – 1970b: Structure fine des microgamètes d'*Aggregata eberthi* Labbé. Protistologica **6**, 71–81.
– – 1971a: La schizogonie chez *Aggregata eberthi*; étude en microscopie électronique. Protistologica **7**, 227–259.
– – 1971b: La sporogénèse chez la Coccidie *Aggregata eberthi*. Étude en microscopie électronique. J. Protozool. **18**, 614–628.
– Vivier, E., 1971: Ultrastructure comparée des germes infectieux (sporozoïtes, mérozoïtes, schizozoïtes, endozoïtes, etc.) chez les Sporozoaires. Année Biol. **10**, 77–113.

Porter, D., 1972: Cell division in the marine slime mold, *Labyrinthula sp.*, and the role of the bothrosome in extracellular membrane production. Protoplasma **74**, 427–448.

Prandtl, H., 1906: Die Konjugation von *Didinium nasutum* O.F.M. Arch. Protistenk. **7**, 229–258.

Preer, J. R., 1969: Genetics of the Protozoa. In: Research in protozoology (Chen, T. T., ed.), Vol. 3, pp. 129–278. Oxford-New York: Pergamon Press.
– 1976: Quantitative predictions of random segregation models of the ciliate macronucleus. Genet. Res. **27**, 227–238.
– Preer, L. B., 1979: The size of macronuclear DNA and its relationship to models for maintaining genic balance. J. Protozool. **26**, 14–18.

Preer, L. B., 1969: Alpha, an infectious macronuclear symbiont of *Paramecium aurelia*. J. Protozool. **16**, 570–573.

PRENSIER, G., 1970: Structure de la paroi du sporoblaste et origine du complexe membranaire interne du sporozoïte de *Diplauxis hatti* (Grégarine monocystidée), demontrées par la microscopie électronique. C. r. Acad. Sci. Paris **D 271**, 2329–2331.
– 1973: Étude de la mitose chez la Grégarine *Diplauxis hatti*. In: Progress in Protozoology (Abstr. IV Internat. Congr. Protozool.), p. 327. Clermont-Ferrand: Université de Clermont.
PRESCOTT, D. M., 1960: Relation between cell growth and cell division. IV. The synthesis of DNA, RNA, and protein from division to division in *Tetrahymena*. Exper. Cell Res. **19**, 228–238.
– 1962: Nucleic acid and protein metabolism in the macronuclei of two ciliated Protozoa. J. Histochem. Cytochem. **10**, 145–153.
– 1966: The synthesis of total macronuclear protein, histone, and DNA during the cell cycle in *Euplotes eurystomus*. J. Cell Biol. **31**, 1–9.
– BOSTOCK, C. J., MURTI, K. G., LAUTH, M. R., GAMOW, E., 1971: DNA of ciliated Protozoa. I. Electron microscopic and sedimentation analyses of macronuclear and micronuclear DNA of *Stylonychia mytilus*. Chromosoma **34**, 355–366.
– HEUMANN, J. M., SWANTON, M., BOSWELL, R. E., 1979: The genome of hypotrichous ciliates. In: Specific eukaryotic genes (ALFRED BENZON Symposium No. 13) (ENGBERG, J., KLENOW, H., LEICK, V., eds.), pp. 85–97. Copenhagen: Munksgaard.
– GOLDSTEIN, L., 1967: Nuclear-cytoplasmic interaction in DNA synthesis. Science **155**, 469–470.
– KIMBALL, R. F., 1961: Relation between RNA, DNA, and protein syntheses in the replicating nucleus of *Euplotes*. Proc. Nat. Acad. Sci. U.S.A. **47**, 686–693.
– – CARRIER, R. F., 1962: Comparison between the timing of micronuclear and macronuclear DNA synthesis in *Euplotes eurystomus*. J. Cell Biol. **13**, 175–176.
– MURTI, K. G., 1974: Chromosome structure in ciliated protozoans. Cold Spring Harbor Symp. Quant. Biol. **38**, 609–618.
– – BOSTOCK, C. J., 1973: Genetic apparatus of *Stylonychia* sp. Nature **242**, 576–600.
– STONE, G. E., 1966: Replication and function of protozoan nucleus. In: Research in protozoology (CHEN, T. T., ed.). Vol. 2, pp. 85–147. Oxford-New York: Pergamon Press.
PRESTON, T. M., 1969: The form and function of the cytostome-cytopharynx of the culture forms of the elasmobranch haemoflagellate *Trypanosoma raiae* LAVERAN et MESNIL. J. Protozool. **16**, 320–333.
PRZYBOŚ, E., 1980: The african strain of *Paramecium jenningsi*. Cytological and karyological investigations. Folia Biol. (Kraków) **28**, 391–397.
– KOŚCIUSZKO, H., KOMALA, Z., 1979: Karyological studies of strain 324 of *Paramecium triaurelia*. Folia Biol. (Kraków) **27**, 355–359.
PURRINI, K., 1980: *Malamoeba scolyti* sp. n. (Amoebidae, Rhizopoda, Protozoa) parasitizing the Bark beetles, *Dryocoetes autographus* RATZ., and *Hylurgops palliatus* GYLL. (Scolytidae, Coleoptera). Arch. Protistenk. **123**, 358–367.
PUSSARD, M., 1965: Cuticule et caryocinèse de *Gocevia fonbrunei* n. sp. (Cochliopodidae, Testacea, Lobosa). Arch. Zool. Expér. Génér. **105**, 101–117.
– 1966: Le genre *Acanthamoeba* VOLKONSKY 1931 (Hartmannellidae – Amoebida). Protistologica **2**, 71–93.
– 1973: Modalités de la division nucléaire et taxonomie chez les amibes (Amoebea, Protozoa). Révision des notions de promitose, mésomitose et métamitose. Protistologica **9**, 163–173.
– PONS, R., 1978: Étude de la "centrosphère" d'*Acanthamoeba echinulata* PUSSARD et PONS 1978. Protistologica **14**, 247–251.
– SÉNAUD, J., PONS, R., 1977: Observations ultrastructurales sur *Gocevia fonbrunei* PUSSARD 1965 (Protozoa, Rhizopodea). Protistologica **13**, 265–285.

Puytorac, P. de, 1959: Structures et ultrastructures nucléolaires et périnucléolaires du macronoyau intermitotique des Ciliés Haptophryidae. C. r. Acad. Sci. Paris **249**, 1709–1711.
– 1961: Observations sur l'ultrastructure d'*Anoplophrya commune* de Puyt., Cilié parasite du Ver *Eophila savignyi* (G. et H.). C. r. Soc. Biol. **155**, 783–786.
– 1963a: Contribution à l'étude des Ciliés astomes Haptophryidae Cépède 1903 (cytologie, ultrastructure, taxonomie). Ann. Sci. Natur., Zool. sér. 12, **5**, 173–210.
– 1963b: Observations sur l'ultrastructure du Cilié astome: *Mesnilella trispiculata* K. J. Microscopie **2**, 189–196.
– Blanc, J., 1967: Observations sur les modifications ultrastructurales des micronoyaux au cours de leurs transformations en macronoyaux chez *Paramecium caudatum*. C. r. Soc. Biol. **161**, 297–299.
– Detcheva, R., Grolière, C.-A., 1981: Le Cilié incertae sedis *Cardiostomatella vermiforme* Kahl 1928 est un Scuticociliatida Loxocephalidae Jank. 1964. Arch. Protistenk. **124**, 110–124.
– Kattar, M. R., 1969: Observations sur l'ultrastructure du Cilié *Helicoprorodon multinucleatum* Dragesco 1960. Protistologica **5**, 549–560.
– Njine, T., 1970: Sur l'ultrastructure des *Loxodes* (Ciliés holotriches). Protistologica **6**, 427–444.
– Savoie, A., 1968: Observations cytologiques et biologiques sur *Prorodon palustris* nov. sp. Protistologica **4**, 53–60.
Pyne, C. K., 1959: L'ultrastructure de *Cryptobia helicis* (Flagellé, fam. Bodonidae). C. r. Acad. Sci. Paris **248**, 1410–1413.
– 1967: Études préliminaires sur l'organisation de la chromatine pendant l'interphase chez *Euglena gracilis*. Protistologica **3**, 291–294.
– 1978: Electron microscopic studies on the macronuclear development in the ciliate *Chilodonella uncinata*. Cytobiologie **18**, 145–160.
– Gache, D., 1979: Electron microscopic autoradiographic studies on DNA synthesis during macronuclear development in the ciliate *Chilodonella uncinata*. Protistologica **15**, 67–77.
– Ruch, F., Leeman, U., Schneider, S., 1974: Development of the macronuclear anlage in the ciliate *Chilodonella uncinata*. I. Morphological and cytophotometric studies on the evolution of DNA. Chromosoma **48**, 225–238.
Raabe, H., 1947: L'appareil nucléaire d'*Urostyla grandis* Ehrbg. II. Appareil macronucléaire. Ann. Univ. M. Curie-Sklodowska **C 1**, 133–170.
Radzikowski, S., 1965: Changes in the heteromeric macronucleus in division of *Chilodonella cucullulus* (Müller). Acta Protozool. **3**, 233–238.
– 1967: Nuclear behaviour during conjugation and polytene chromosomes in the exconjugants of *Chilodonella cucullulus* (O. F. Müller). Bull. Acad. Polon. Sci., Cl. II, **15**, 749–751.
– 1969: Interdivisional nuclear changes. Microphotometric studies on DNA synthesis in *Chilodonella cucullulus* (O. F. Müller). Bull. Acad. Polon. Sci., Cl. II, **17**, 399–402.
– 1973a: Ultrastructure of the nuclear apparatus in *Chilodonella cucullulus* (O. F. M.) with observations on number and size of nucleoli during interphase. Acta Protozool. **12**, 23–28.
– 1973b: Die Entwicklung des Kernapparates und die Nukleinsäuresynthese während der Konjugation von *Chilodonella cucullulus* O. F. Müller. Arch. Protistenk. **115**, 419–428.
– 1976: DNA and RNA synthesis in the nuclear apparatus of *Chilodonella cucullulus* O. F. M. Acta Protozool. **15**, 47–56.
– 1979: Asynchronous replication of polytene chromosome segments of the new macronucleus anlage in *Chilodonella cucullulus* O. F. Müller. Protistologica **15**, 521–526.

RAE, P. M. M., 1970: The nature and processing of ribosomal ribonucleic acid in a dinoflagellate. J. Cell Biol. **46**, 106–113.
– 1976: Hydroxymethyluracil in eukaryote DNA: a natural feature in the Pyrrhophyta (Dinoflagellates). Science **194**, 1062–1064.
– SPEAR, B. B., 1978: Macronuclear DNA of the hypotrichous ciliate *Oxytricha fallax*. Proc. Nat. Acad. Sci. U.S.A. **75**, 4992–4996.

RAFALKO, J. S., 1947: Cytological observations on the amoebo-flagellate, *Naegleria gruberi*. J. Morphol. **81**, 1–44.
– 1951: Mitotic division in the amoebo-flagellate, *Tetramitus rostratus*. J. Morphol. **89**, 71–90.

RAIKHEL, N., PAULIN, J. J., SKARLATO, S. O., 1981: Mitosis of micronuclei during division and regeneration in the ciliate *Stentor coeruleus*. J. Protozool. **28**, 103–107.

RAIKOV, I. B., 1957a: Nuclear apparatus and its reorganization in the division cycle of the ciliates *Trachelocerca margaritata* (KAHL) and *T. dogieli*, sp. n. (Holotricha). Zool. Zhurnal (Moscow) **36**, 344–359 (in Russian, English summary).
– 1957b: Reorganization of the nuclear apparatus in ciliates and the problem of the origin of their binuclearity. Vestnik Leningrad. Univ. No. 15, 21–37 (in Russian, English summary).
– 1958: Der Formwechsel des Kernapparates einiger niederer Ciliaten. I. Die Gattung *Trachelocerca*. Arch. Protistenk. **103**, 129–192.
– 1959a: Der Formwechsel der Kernapparate einiger niederer Ciliaten. II. Die Gattung *Loxodes*. Arch. Protistenk. **104**, 1–42.
– 1959b: Cytological and cytochemical peculiarities of the nuclear apparatus and division in the holotrichous ciliate *Geleia nigriceps* KAHL. Tsitologiya (Leningrad) **1**, 566–579 (in Russian).
– 1960: Chromosomes in the macronucleus of the holotrichous ciliate *Nassula ornata* EHRBG. Tsitologiya (Leningrad) **2**, 598–601 (in Russian).
– 1962a: Der Kernapparat von *Nassula ornata* EHRBG. (Ciliata, Holotricha). Zur Frage über den Chromosomenaufbau des Makronucleus. Arch. Protistenk. **105**, 463–488.
– 1962b: Les Ciliés mésopsammiques du littoral de la Mer Blanche (U.R.S.S.), avec une description de quelques espèces nouvelles ou peu connues. Cahiers Biol. Marine **3**, 325–361.
– 1963a: Some stages of conjugation of the holotrichous ciliate *Trachelocerca coluber*. Tsitologiya (Leningrad) **5**, 685–689 (in Russian).
– 1963b: Ciliates of the mesopsammon of the Ussuri Gulf (Japan Sea). Zool. Zhurnal (Moscow) **42**, 1753–1767 (in Russian; English summary).
– 1963c: Nuclear apparatus and division of *Remanella granulosa* KAHL and *R. rugosa* KAHL (Holotricha, Gymnostomatida). In: Morfologiya i Fiziologiya Prostejshikh (Coll. of papers No. 3, Inst. of Cytology), pp. 20–34. Leningrad: Nauka Press (in Russian).
– 1963d: The nuclear apparatus of *Remanella multinucleata* KAHL (Ciliata, Holotricha). Acta Biol. Acad. Scient. Hung. **14**, 221–229.
– 1963e: The nuclear apparatus of holotrichous ciliates *Geleia orbis* FAURÉ-FREMIET and *G. murmanica* RAIKOV. Acta Protozool. **1**, 21–30 (in Russian; English summary).
– 1963f: On the origin of nuclear dualism in ciliates. In: Progress in Protozoology (Proc. I Internat. Congr. Protozool.), pp. 253–258. Prague: Publ. House ČSAV.
– 1964: DNA content of the nuclei and the nature of macronuclear chromatin strands of the ciliate *Nassulopsis elegans* (EHRBG.). Acta Protozool. **2**, 339–355.
– 1966: Elektronenmikroskopische Untersuchung des Kernapparates von *Nassula ornata* EHRBG. (Ciliata, Holotricha). Arch. Protistenk. **109**, 71–98.
– 1967a: Karyology of Protozoa. 260 pp. Leningrad: Nauka Press (in Russian).

RAIKOV, I. B.., 1967b: The nuclear apparatus and some cytoplasmic structures of *Helicoprorodon gigas* (Holotricha, Gymnostomatida). Acta Protozool. **5**, 49–58 (in Russian: English summary).

– 1969: The macronucleus of ciliates: In: Research in protozoology (CHEN, T. T., ed.), Vol. 3, pp. 1–128. Oxford-New York: Pergamon Press.

– 1972a: Ultrastructures macronucléaires et micronucléaires de *Tracheloraphis dogieli*, Cilié marin à macronoyaux diploïdes. C. r. Soc. Biol. **166**, 608–611.

– 1972b: The nuclear apparatus of the psammophilic ciliate *Kentrophoros fistulosum* FAURÉ-FREMIET: structure, divisional reorganization and ultrastructure. Acta Protozool. **10**, 227–248.

– 1972c: Ultrastructure des "capsules nucléaires" ("noyaux composés") du Cilié psammophile *Kentrophoros latum* RAIKOV 1962. Protistologica **8**, 299–313.

– 1972d: Nuclear phenomena during conjugation and autogamy in ciliates. In: Research in protozoology (CHEN, T. T., ed.), Vol. 4, pp. 147–289. Oxford-New York: Pergamon Press.

– 1973: Mitose intranucléaire acentrique du micronoyau de *Loxodes magnus*, Cilié holotriche. Étude ultrastructurale. C. r. Acad. Sci. Paris. **D 276**, 2385–2388.

– 1974: Fine structure of the nuclear apparatus of a lower psammobiotic ciliate, *Tracheloraphis dogieli* RAIKOV. Acta Protozool. **13**, 85–96.

– 1975a: Fine structure of the nuclear apparatus of the ciliate *Loxodes magnus*. I. Macronuclei, macronuclear anlagen and interphasic micronuclei. Tsitologiya (Leningrad) **17**, 1009–1017 (in Russian; English summary).

– 1975b: The problem of origin and evolution of meiosis. In: Tsitologiya i genetika meioza (KHVOSTOVA, V. V., BOGDANOV, YU. F., eds.), pp. 344–371. Moscow: Nauka Press (in Russian).

– 1976a: Fine structure of the nuclear apparatus of the ciliate *Loxodes magnus*. II. Micronuclear mitosis. Tsitologiya (Leningrad) **18**, 144–151 (in Russian; English summary).

– 1976b: Evolution of macronuclear organization. Annual Rev. Genet. **10**, 413–440.

– 1979: Fine structure of the nuclear apparatus of *Remanella multinucleata* KAHL (Ciliata, Gymnostomata). Arch. Protistenk. **121**, 1–19.

– 1981: Fine structure and cytochemical peculiarities of the nuclear apparatus of the ciliate *Geleia orbis* (Karyorelictida). In: Progress in Protozoology (Abstr. VI Internat. Congr. Protozool.), p. 302. Warsaw.

– AMMERMANN, D., 1974: The macronucleus of ciliates: recent advances. In: Actualités protozoologiques (Resumé Tables Rondes, 4ᵉ Congr. Internat. Protozool.), Vol. 1, pp. 143–159. Clermont-Ferrand: Université de Clermont.

– CHEISSIN, E. M., BUZE, E. G., 1963: A photometric study of DNA content of macro- and micronuclei in *Paramecium caudatum*, *Nassula ornata*, and *Loxodes magnus*. Acta Protozool. **1**, 285–300.

– DRAGESCO, J., 1969: Ultrastructure des noyaux et de quelques organites cytoplasmiques du Cilié *Tracheloraphis caudatus* DRAGESCO et RAIKOV (Holotricha, Gymnostomatida). Protistologica **5**, 193–208.

– KOVALEVA, V. G., 1978: Fine structure of the nuclear complex of the psammophilic gymnostome ciliate *Tracheloraphis phoenicopterus* (COHN). Protistologica **14**, 269–282.

– – 1980: Electron microscopic cytochemistry of the macro- and micronuclei of the lower ciliate *Tracheloraphis dogieli* (Karyorelictida). Tsitologiya (Leningrad) **22**, 1139–1145 (in Russian; English summary).

– – 1981: Fine structural morphology and cytochemistry of the nuclei of the lower marine ciliate *Trachelocerca geopetiti* (Karyorelictida). Protistologica **17**, 113–130.

– MORAT, G., 1977: Étude autoradiographique de la synthèse de l'ADN et de l'ARN dans les noyaux du Cilié *Loxodes magnus* STOKES. Protistologica **13**, 391–399.

RAO, M. V. N., 1964: Nuclear behavior of *Euplotes woodruffi* during conjugation. J. Protozool. **11**, 296–304.
– 1966: Conjugation in *Kahlia* sp. with special reference to meiosis and endomitosis. J. Protozool. **13**, 565–573.
– 1968a: Nuclear behavior of *Spirostomum ambiguum* during conjugation with special reference to macronuclear development. J. Protozool. **15**, 748–752.
– 1968b: Macronuclear development in *Euplotes woodruffi* following conjugation. Exper. Cell Res. **49**, 411–419.
– AMMERMANN, D., 1970: Polytene chromosomes and nucleic acid metabolism during macronuclear development in *Euplotes*. Chromosoma **29**, 246–254.
– PRESCOTT, D. M., 1967: Micronuclear RNA synthesis in *Paramecium caudatum*. J. Cell Biol. **33**, 281–286.
RATCLIFFE, H. L., 1927: Mitosis and cell division in *Euglena spirogyra* EHRENBERG. Biol. Bull. **53**, 109–122.
RAY, C., 1956: Meiosis and nuclear behavior in *Tetrahymena pyriformis*. J. Protozool. **3**, 88–96.
– 1958: Tetraploidy in *Tetrahymena pyriformis*. In: Proc. X Internat. Congr. Genet., Vol. 2, pp. 229–230. Montreal.
RAY, D., HAYES, R., 1954: *Hartmanella astronyxis*: a new species of free-living ameba. Cytology and life cycle. J. Morphol. **95**, 159–188.
REICHENOW, E., 1921: Die Hämococcidien der Eidechsen. I. Die Entwicklungsgeschichte von *Karyolysus*. Arch. Protistenk. **42**, 179–291.
– 1928: Ergebnisse mit der Nuclealfärbung bei Protozoen. Arch. Protistenk. **61**, 144–166.
RESHETNYAK, V. V., 1966: Deep-sea Phaeodaria of the North-West Pacific. (Series "Fauna SSSR". Radiolaria, new ser., No. 94). 208 pp. Moscow-Leningrad: Nauka Press (in Russian).
RIEDER, N., 1971: Elektronenoptische Untersuchungen an *Didinium nasutum* O. F. MÜLLER (Ciliata, Gymnostomata) in Interphase und Teilung. Forma et Functio **4**, 46–86.
RINGERTZ, N. R., BOLUND, L., BAULT, L. E. DE, 1967a: Isolation and chemical composition of macronuclei from *Tetrahymena*. Exper. Cell Res. **45**, 519–532.
– ERICSSON, J. L. E., NILSSON, O., 1967b: Macronuclear chromatin structure in *Euplotes*. Exper. Cell Res. **48**, 97–117.
– HOSKINS, G. C., 1965: Cytochemistry of macronuclear reorganization. Exper. Cell Res. **38**, 160–179.
RINGO, D. L., 1967: Flagellar motion and fine structure of the flagellar apparatus in *Chlamydomonas*. J. Cell Biol. **33**, 543–571.
RIS, H., 1961: Ultrastructure and molecular organization of genetic systems. Canad. J. Genet. Cytol. **3**, 95–120.
– 1962: Interpretation of ultrastructure of the cell nucleus. In: Interpretation of ultrastructure (HARRIS, R. J. C., ed.), Vol. 1, pp. 69–88. New York-London: Academic Press.
– 1969: Use of the high voltage electron microscope for the study of thick biological specimens. J. Microscopie **8**, 761–766.
– 1975: Primitive mitotic mechanisms. BioSystems **7**, 298–304.
– KORENBERG, J., 1979: Chromosome structure and levels of chromosome organization. In: Cell biology (PRESCOTT, D. M., GOLDSTEIN, L., eds.), Vol. 2, pp. 267–361. New York-London: Academic Press.
– KUBAI, D. F., 1974: An unusual mitotic mechanism in the parasitic protozoan *Syndinium* sp. J. Cell Biol. **60**, 702–720.
RITTER, H., INOUÉ, S., KUBAI, D., 1978: Mitosis in *Barbulanympha*. I. Spindle structure, formation, and kinetochore engagement. J. Cell Biol. **77**, 638–654.

Rizzo, P. J., 1979: RNA synthesis in isolated nuclei of the dinoflagellate *Crypthecodinium cohnii*. J. Protozool. **26**, 290–294.

– Burghardt, R. C., 1980: Chromatin structure in the unicellular algae *Olisthodiscus luteus*, *Crypthecodinium cohnii* and *Peridinium balticum*. Chromosoma **76**, 91–99.

– Noodén, L. D., 1972: Chromosomal proteins in the dinoflagellate alga, *Gyrodinium cohnii*. Science **176**, 796–797.

– – 1973: Isolation and chemical composition of dinoflagellate nuclei. J. Protozool. **20**, 666–672.

– – 1974a: Isolation and partial characterization of dinoflagellate chromatin. Biochim. Biophys. Acta **349**, 402–414.

– – 1974b: Partial characterization of dinoflagellate chromosomal proteins. Biochim. Biophys. Acta **349**, 415–427.

Roberts, T. M., Tuttle, R. C., Allen, J. R., Loeblich, A. R., Klotz, L. C., 1974: New genetic and physicochemical data on structure of dinoflagellate chromosomes. Nature **248**, 446–447.

Roberts, W. L., Hammond, D. M., 1970: Ultrastructural and cytologic studies of the sporozoites of four *Eimeria* species. J. Protozool. **17**, 76–86.

– – Anderson, L. C., Speer, C. A., 1970a: Ultrastructural study of schizogony in *Eimeria callospermophili*. J. Protozool. **17**, 584–592.

– – Speer, C. A., 1970b: Ultrastructural study of the intra- and extracellular sporozoites of *Eimeria callospermophili*. J. Parasitol. **56**, 907–917.

Rodrigues de Santa Rosa, M., 1976: Observations sur l'ultrastructure du Cilié *Coleps hirtus* Nitzsch 1817. Protistologica **12**, 205–216.

Rogers, T. D., Scholes, V. E., Schlichting, H. E., 1972: An ultrastructural comparison of *Euglena gracilis* Klebs, bleached *Euglena* and *Astasia longa* Pringsheim. J. Protozool. **19**, 133–139.

Ron, A., Prescott, D. M., 1969: The timing of DNA synthesis in *Amoeba proteus*. Exper. Cell Res. **56**, 430–434.

– Urieli, S., 1977: Qualitative and quantitative studies on DNA and RNA synthesis in *Loxodes striatus* nuclei. J. Protozool. **24**, 150–154.

Rondanelli, E. G., Carosi, G., Scaglia, M., Dei Cas, A., 1976: An electron microscope study of *Naegleria fowleri* (Carter 1970) in axenic "medium". The ameboid stage. Protistologica **12**, 25–30.

– Osculati, F., Gerna, G., Carosi, G., 1967a: Studio microelettronico di *Entamoeba moshkovskii* Tschalaia 1941. Boll. Ist. Sieroterap. Milan. **46**, 343–368.

– – – – 1967b: Indagini ultrastrutturali su *Entamoeba histolytica*, ceppo Laredo. Boll. Ist. Sieroterap. Milan. **46**, 455–465.

Roos, U.-P., 1975: Mitosis in the cellular slime mold *Polysphondilium violaceum*. J. Cell Biol. **64**, 480–491.

Roque, M., Puytorac, P. de, Savoie, A., 1965: *Ophryoglena bacterocaryon* sp. nov., Cilié holotriche péniculien (cytologie, ultrastructure, cycle). Arch. Zool. Expér. Génér. **105**, 309–344.

Rosenfeld, A., Buchanan, L., Chapman, G. B., 1969: Comparison of the fine structure of spores of three species of *Minchinia* (Haplosporida, Haplosporidiidae). J. Parasitol. **55**, 921–941.

Roskin, G. I., Ginsburg, A. S., 1944: Zymonucleic acid of the protozoan cell. Doklady Akad. Nauk SSSR **42**, 362–365 (in Russian).

– Schishliaiewa, S., 1928: Die Kernteilung bei Trypanosomen. Arch. Protistenk. **60**, 460–481.

– Struve, M. M., 1953: Cytochemical differences of the protoplasm of various protozoan species. Doklady Akad. Nauk SSSR **93**, 151–153 (in Russian).

ROSSOLIMO, L. L., 1916: Observations on *Loxodes rostrum* O. F. MÜLLER. Dnevnik Zool. Otdel. Imper. Obsch. Ljubit. Estestvozn., Antropol. i Etnogr., n. ser. **3**, No. 4, 1–18 (in Russian; French summary).
– JAKIMOWITSCH, K., 1929: Die Kernteilung bei *Conchophthirius steenstrupii* St. Zool. Anz. **84**, 323–333.
ROTH, L. E., 1957: An electron microscope study of the cytology of the protozoan *Euplotes patella*. J. Biophys. Biochem. Cytol. **3**, 985–1000.
– 1960: Observations on division stages in the protozoan hypotrich *Stylonychia*. In: Verhandl. IV Internat. Kongr. Elektronenmikr., Vol. 2, pp. 241–244. Berlin-Göttingen-Heidelberg: Springer.
– 1967: Electron microscopy of mitosis in amebae. III. Cold and urea treatments: a basis for tests of direct effects of mitotic inhibitors on microtubule formation. J. Cell Biol. **34**, 47–59.
– DANIELS, E. W., 1962: Electron microscopic studies of mitosis in amebae. II. The giant ameba *Pelomyxa carolinensis*. J. Cell Biol. **12**, 57–78.
– MINICK, O. T., 1961: Electron microscopy of nuclear and cytoplasmic events during division in *Tetrahymena pyriformis* strains W and HAM-3. J. Protozool. **8**, 12–21.
– OBETZ, S. W., DANIELS, E. W., 1960: Electron microscopic studies of mitosis in amebae. I. *Amoeba proteus*. J. Biophys. Biochem. Cytol. **8**, 207–220.
– SHIGENAKA, Y., 1964: The structure and formation of cilia and filaments in rumen Protozoa. J. Cell Biol. **20**, 249–270.
ROUILLER, C., FAURÉ-FREMIET, E., 1958: Structure fine d'un Flagellé Chrysomonadien: *Chromulina psammobia*. Exper. Cell Res. **14**, 47–67.
RUDZINSKA, M. A., 1956: Further observations on the fine structure of the macronucleus in *Tokophrya infusionum*. J. Biophys. Biochem. Cytol. **2**, suppl., 425–428.
– 1961: The use of a protozoan for studies on aging. II. The macronucleus in young and old organisms of *Tokophrya infusionum*: light and electron microscope observations. J. Gerontol. **16**, 326–334.
– 1969: The fine structure of malaria parasites. Internat. Rev. Cytol. **25**, 161–199.
– 1976: Ultrastructure of intraerythrocytic *Babesia microti* with emphasis on the feeding mechanism. J. Protozool. **23**, 224–233.
– D'ALESANDRO, P., TRAGER, W., 1964: The fine structure of *Leishmania donovani* and the role of the kinetoplast in the leishmania – leptomonad transformation. J. Protozool. **11**, 166–191.
– JACKSON, G. J., TUFFRAU, M., 1966: The fine structure of *Colpoda maupasi* with special emphasis on food vacuoles. J. Protozool. **13**, 440–459.
– PORTER, K. R., 1955: Observations on the fine structure of the macronucleus of *Tokophrya infusionum*. J. Biophys. Biochem. Cytol. **1**, 421–428.
– SPIELMAN, A., RIEK, R. F., LEWENGRUB, S. A., PIESMAN, J., 1979: Intraerythrocytic "gametocytes" of *Babesia microti* and their maturation in ticks. Canad. J. Zool. **57**, 424–434.
– TRAGER, W., 1959: Phagotrophy and two new structures in the malaria parasite *Plasmodium berghei*. J. Biophys. Biochem. Cytol. **6**, 103–112.
– – 1961: The role of the cytoplasm during reproduction in a malarial parasite (*Plasmodium lophurae*) as revealed by electron microscopy. J. Protozool. **8**, 307–322.
– – 1962: Intracellular phagotrophy in *Babesia rodhaini* as revealed by electron microscopy. J. Protozool. **9**, 279–288.
– – 1968: The fine structure of trophozoites and gametocytes in *Plasmodium coatneyi*. J. Protozool. **15**, 73–88.
RUFFOLO, J. J., 1975: Fine structure of *Euplotes* macronucleus. J. Protozool. **22**, No. 3, 44A.
– 1978a: Micronuclear envelope formation after telophase of mitosis in the ciliate *Euplotes eurystomus*. Trans. Amer. Microscop. Soc. **97**, 259–263.

RUFFOLO, J. J., 1978b: Fine structure of the macronucleus during the cell division cycle of *Euplotes eurystomus*, a ciliate protozoon. J. Morphol. **157**, 211–222.

– 1979: The intranuclear helices of *Euplotes:* their substructure, localization, and apparent migration to the cytoplasm. J. Morphol. **159**, 81–88.

RUTHMANN, A., 1963: Die Struktur des Chromatins und die Verteilung der Ribonucleinsäure im Makronucleus von *Loxophyllum meleagris*. Arch. Protistenk. **106**, 422–436.

– 1964: Autoradiographische und mikrophotometrische Untersuchungen zur DNS-Synthese im Makronucleus von *Bursaria truncatella*. Arch. Protistenk. **107**, 117–130.

– 1972: Division and formation of the macronuclei of *Keronopsis rubra*. J. Protozool. **19**, 661–666.

– 1973: Fine structure of the macronuclear anlage of the hypotrichous ciliate *Keronopsis rubra*. J. Protozool. **20**, 648–651.

– GRELL, K. G., 1964: Die Feinstruktur des intracapsularen Cytoplasmas bei dem Radiolar *Aulacantha scolymantha*. Z. Zellforsch. **63**, 97–119.

– HAUSER, M., 1974: Mitosis-like macronuclear division in a ciliate. Chromosoma **45**, 261–272.

– HECKMANN, K., 1961: Formwechsel und Struktur des Makronucleus von *Bursaria truncatella*. Arch. Protistenk. **105**, 313–340.

RYLEY, J. F., 1969: Ultrastructural studies on the sporozoite of *Eimeria tenella*. Parasitology **59**, 67–72.

RYLOV, W., 1923: Über den Einfluß des im Wasser gelösten Sauerstoffs und Schwefelwasserstoffs auf den Lebenszyklus und die vertikale Verteilung des Infusors *Loxodes rostrum* O. F. MÜLL. (Aspirotricha, Amphileptina). Internat. Rev. Ges. Hydrobiol. Hydrograph. **11**, 179–192.

RZEPCZYK, C., SCHOLTYSECK, E., 1976: Light and electron microscope studies on the *Sarcocystis* of *Rattus fuscipes*, an Australian rat. Z. Parasitenk. **50**, 137–150.

SAEDELEER, H. DE, 1930: La mitose de l'Amoebien *Cochliopodium* sp. en culture pure mixte (note préliminaire). Ann. Protistol. **3**, 12.

SAGER, R., PALADE, G. E., 1957: Structure and development of the chloroplast in *Chlamydomonas*. I. The normal green cell. J. Biophys. Biochem. Cytol. **3**, 463–488.

SAITO, M., 1961a: Studies in the mitosis of *Euglena*. I. On the chromosome cycle of *Euglena viridis* EHRBG. J. Protozool. **8**, 300–307.

– 1961b: A note on the duplication process of macronuclear chromosomes in a peritrichous ciliate, *Vorticella campanula*. Japan. J. Genet. **36**, 184–186.

– SATO, H., 1961a: Morphological studies on the macronuclear structure of *Paramecium caudatum*. II. Structural changes of the macronucleus during the division cycle. Zool. Mag. **70**, 73–80.

– – 1961b: Morphological studies on the macronuclear structure of *Paramecium caudatum*. III. On the development of the macronucleus in the exconjugant. Zool. Mag. **70**, 81–88.

SALAMONE, M. F., NACHTWEY, D. S., 1979: Formation and fate of extranuclear chromatin bodies in *Tetrahymena*. J. Protozool. **26**, 227–231.

SALVANO, P., 1974a: Étude cytophotométrique de la teneur en DNA des micronoyaux au cours de la division binaire et des divisions prégamiques chez un Cilié marin: *Euplotes crassus* (DUJ. 1842). Protistologica **10**, 253–259.

– 1974b: Analyse microspectrophotométrique dans l'ultraviolet du macronoyau de *Strombidium sulcatum* CLAP. et LACHM. 1859 (Cilié Oligotriche), pendant l'intercinèse. C. r. Acad. Sci. Paris **D 278**, 2445–2447.

– 1975: Comparaison du fonctionnement des bandes de réorganisation d'*Euplotes crassus* (DUJARDIN) et de *Strombidium sulcatum* CLAPARÈDE et LACHMANN après analyse microspectrographique en UV. J. Protozool. **22**, 230–232.

SAMPSON, J. R., HAMMOND, D. M., 1972: The fine structure of the early stages of *Eimeria alabamensis* in cell cultures. J. Parasitol. **58**, 311–322.

SAMUEL, C., MACKIE, J., SOMMERVILLE, J., 1981: Macronuclear chromatin organization in *Paramecium primaurelia*. Chromosoma **83**, 481–492.

SAMUELS, R., 1957: Studies of *Tritrichomonas batrachorum*. 2. Normal mitosis and morphogenesis. Trans. Amer. Microscop. Soc. **76**, 295–307.

– 1959: Studies of *Tritrichomonas batrachorum*. 3. Abnormal mitosis and morphogenesis. Trans. Amer. Microscop. Soc. **78**, 49–65.

SANABRIA, A., 1966: Ultrastructure of *Trypanosoma cruzi* in the rectum of *Rhodnius prolixus*. Exper. Parasitol. **19**, 276–299.

SANDERS, R. D., POINAR, G. O., 1973: Fine structure and life cycle of *Lankesteria clarki* sp. n. (Sporozoa, Eugregarinida) parasitic in the mosquito *Aedes sierrensis* (LUDLOW). J. Protozool. **20**, 594–602.

SAPRA, G. R., AMMERMANN, D., 1973: RNA synthesis and acquisition of actinomycin D insensitivity during conjugation in *Stylonychia mytilus*. Exper. Cell Res. **78**, 168–174.

– – 1974: An analysis of the developmental program in relation to RNA metabolism in the ciliate *Stylonychia mytilus*. Devel. Biol. **36**, 105–112.

– DASS, C. M. S., 1970: Organization and development of the macronuclear anlage in *Stylonychia notophora* STOKES. J. Cell Sci. **6**, 351–363.

– – 1978: Effect of 5′-Bromodeoxyuridine on the development and differentiation of exconjugants in *Stylonychia mytilus* EHRENBERG. Ind. J. Exper. Biol. **16**, 1–4.

– – AMMERMANN, D., 1976: Stable mRNA synthesis for micronuclear division during regeneration in *Stylonychia mytilus*. Ind. J. Exper. Biol. **14**, 219–222.

SARKISYAN, S. A., 1965: Fine structure of *Entamoeba moshkovskii* TSCHALAYA 1951. Izvest. Akad. Nauk Arm. SSR, Biol. Nauki **18**, No. 4, 63–68 (in Russian).

SARMA, Y. S. R. K., SHYAM, R., 1973: On certain aspects of mitotic division in *Eudorina elegans* EHRENBERG. Nucleus **16**, 93–100.

SASSUCHIN, D. N., 1929: Life conditions, structure and development of *Endamoeba blattae* (BÜTSCHLI) LEIDY (1879). Russkij Archiv Protistol. **8**, 163–244 (in Russian; German summary).

– 1930: Lebensbedingungen, Cytologie und Entwicklung von *Endamoeba blattae* (BÜT.) LEIDY 1879. Arch. Protistenk. **70**, 681–686.

– 1931: Lebensbedingungen der Mikrofauna in Sandanschwemmungen der Flüsse und im Triebsand der Wüsten. Arch. Hydrobiol. **22**, 369–388.

SATHANANTHAN, A. H., 1977: Cytochemical nature and fine structure of the gregarine *Zeylanocystis burti* DISSANAIKE, with special reference to microfilaments, microtubules and movement. J. Protozool. **24**, 233–243.

SATIR, B., DIRKSEN, E. R., 1971: Nucleolar aging in *Tetrahymena* during the cultural growth cycle. J. Cell Biol. **48**, 143–154.

SATO, H., SAITO, M., 1958: Morphological studies on the macronuclear structure of *Paramecium caudatum*. I. Macronuclear constituents in interphase. Zool. Mag. **67**, 249–258.

– – 1959: Morphological study on the macronuclear structure in interphase and prefission stage of *Tetrahymena geleii* W. Zool. Mag. **68**, 209–214.

SAWYER, T. K., GRIFFIN, J. L., 1971: *Acanthamoeba comandoni* and *A. astronyxis*: taxonomic characteristics of mitotic nuclei, "centrosomes" and cysts. J. Protozool. **18**, 382–388.

SAX, K., 1932: The cytological mechanism of crossing over. J. Arnold Arboretum **13**, 180–212.

SAXENA, D. M., JURAND, A., 1973: Studies on the macronucleus of *Paramecium aurelia*. IV. Acid phosphatase activity in macronuclear fragments: comparison between conjugation and autogamy. Arch. Protistenk. **115**, 403–407.

Saxena, R., Saxena, D. M., 1978a: Cytology and cell cycle of *Onychodromus grandis* Stein 1859 (Hypotrichida). Arch. Protistenk. **120**, 118–127.

– – 1978b: Cytology and DNA synthesis of a gymnostome ciliate, *Homalozoon vermiculare* (Stokes). Arch. Protistenk. **120**, 128–137.

– – Jurand, A., 1978: Ultrastructural studies on a gymnostome ciliate *Homalozoon vermiculare* (Stokes). II. The endoplasm and endoplasmic organelles. Arch. Protistenk. **120**, 378–392.

Scalzi, H. A., Bahr, G. F., 1968: An electron microscopic examination of erythrocytic stages of two rodent malarial parasites, *Plasmodium chabaudi* and *P. vinckei*. J. Ultrastr. Res. **24**, 116–133.

Schaechter, M., De Lamater, E. D., 1955: Mitosis of *Chlamydomonas*. Amer. J. Bot. **42**, 417–422.

Schein, E., Mehlhorn, H., Warnecke, M., 1978: Electron microscopic studies on the schizogony of four *Theileria* species of cattle (*T. parva*, *T. lawrencei*, *T. annulata*, and *T. mutans*). Protistologica **14**, 337–348.

Schensted, I. V., 1958: Model of subnuclear segregation in the macronucleus of ciliates. Amer. Naturalist **92**, 161–170.

Schewiakoff, W., 1888: Über die karyokinetische Kernteilung der *Euglypha alveolata*. Morphol. Jahrb. **13**, 193–258.

Schmitter, R. E., 1971: The fine structure of *Gonyaulax polyedra*, a bioluminescent marine dinoflagellate. J. Cell Sci. **9**, 147–173.

Scholtyseck, E., 1954: Beobachtungen über den Austritt von ganzen Karyosomen ins Zytoplasma bei *Eimeria maxima*. Naturwissenschaften **41**, 40.

– 1962: Electron microscope studies on *Eimeria perforans* (Sporozoa). J. Protozool. **9**, 407–414.

– 1963: Untersuchungen über die Kernverhältnisse und das Wachstum bei Coccidiomorphen unter besonderer Berücksichtigung von *Eimeria maxima*. Z. Parasitenk. **22**, 428–474.

– 1965a: Elektronenmikroskopische Untersuchungen über die Schizogonie bei Coccidien (*Eimeria perforans* und *Eimeria stiedae*). Z. Parasitenk. **26**, 50–62.

– 1965b: Die Mikrogametenentwicklung von *Eimeria perforans*. Z. Zellforsch. **66**, 625–642.

– 1979: Fine Structure of Parasitic Protozoa. (An atlas of micrographs, drawings, and diagrams.) 206 pp. Berlin-Heidelberg-New York: Springer.

– Gönnert, R., Haberkorn, A., 1969: Die Feinstruktur der Makrogameten des Hühnercoccids *Eimeria tenella*. Z. Parasitenk. **33**, 31–43.

– Hammond, D. M., Ernst, J. V., 1966: Fine structure of the macrogametes of *Eimeria perforans*, *E. stiedae*, *E. bovis*, and *E. auburnensis*. J. Parasitol. **52**, 975–987.

– Kepka, O., Piekarski, G., 1970: Die Feinstructur der Zoiten aus reifen Zysten des sog. M-Organismus (*Frenkelia* spec.). Z. Parasitenk. **33**, 252–261.

– Mehlhorn, H., Hammond, D. M., 1972: Electron microscope studies of microgametogenesis in coccidia and related groups. Z. Parasitenk. **38**, 95–131.

– – Müller, B. E. G., 1973: Identifikation von Merozoiten der vier cystenbildenden Coccidien (*Sarcocystis*, *Toxoplasma*, *Besnoitia*, *Frenkelia*) auf Grund feinstruktureller Kriterien. Z. Parasitenk. **42**, 185–206.

– Piekarski, G., 1965: Elektronenmikroskopische Untersuchungen an Merozoiten von Eimerien (*Eimeria perforans* und *E. stiedae*) und *Toxoplasma gondii*. Z. Parasitenk. **26**, 91–115.

– Spiecker, D., 1964: Vergleichende elektronenmikroskopische Untersuchungen an den Entwicklungsstadien von *Eimeria perforans* (Sporozoa). Z. Parasitenk. **24**, 546–560.

Schreil, W. H., 1964: Studies on the fixation of artificial and bacterial DNA-plasms for the electron microscopy of thin sections. J. Cell Biol. **22**, 1–20.

Schrével, J., 1971a: Contribution à l'étude des Selenidiidae, parasites d'Annélides polychètes. II. Ultrastructure de quelques trophozoites. Protistologica **7**, 101–130.

– 1971b: Observations biologiques et ultrastructurales sur les Selenidiidae et leurs conséquences sur la systématique des Grégarinomorphes. J. Protozool. **18**, 448–470.

– Asfaux-Foucher, G., Bafort, J. M., 1977: Étude ultrastructurale des mitoses multiples au cours de la sporogonie du *Plasmodium b. berghei*. J. Ultrastr. Res. **59**, 332–350.

– Dhainaut, A., 1968: Incorporation d'uridine-^{3}H et de leucine-^{3}H chez *Lecudina tuzetae* Schr., Grégarine parasite de l'Annélide polychète *Nereis diversicolor* O. F. Müller. C. r. Soc. Biol. **162**, 465–467.

Schubert, G., 1968: Elektronenmikroskopische Untersuchungen zur Sporenentwicklung von *Henneguya pinnae*. Z. Parasitenk. **30**, 57–77.

Schulman, S. S., 1966: Myxosporidia of the Fauna of USSR. 507 pp. Moscow-Leningrad: Nauka Press (in Russian).

Schuster, F. L., 1963: An electron microscope study of the amoebo-flagellate *Naegleria gruberi* (Schardinger). I. The amoeboid and flagellate stages. J. Protozool. **10**, 297–313.

– 1968: Ultrastructure of *Histomonas meleagridis* (Smith) Tyzzer, a parasitic amoebo-flagellate. J. Parasitol. **54**, 725–737.

– 1969: Intranuclear virus-like bodies in the amoebo-flagellate *Naegleria gruberi*. J. Protozool. **16**, 724–727.

– 1975a: Ultrastructure of mitosis in the amoebo-flagellate *Naegleria gruberi*. Tissue and Cell **7**, 1–12.

– 1975b: Ultrastructure of cysts of *Naegleria* spp.: a comparative study. J. Protozool. **22**, 352–359.

– 1976: Fine structure of the schizont stage of the testate marine ameba, *Trichosphaerium* sp. J. Protozool. **23**, 86–93.

– Dunnebacke, T. H., 1976: Development and release of virus-like particles in *Naegleria gruberi* EG. Cytobiologie **14**, 131–147.

– Goldstein, S., Hershenov, B., 1968: Ultrastructure of a flagellate, *Isonema nigricans* nov. gen. nov. sp., from a polluted marine habitat. Protistologica **4**, 141–149.

– Hershenov, B., 1974: *Khawkinea quartana*, a colorless euglenoid flagellate. I. Ultrastructure. J. Protozool. **21**, 33–39.

– Pollak, A., 1978: Ultrastructure of *Cercomonas* sp., a free-living ameboid flagellate. Arch. Protistenk. **120**, 206–212.

– Svihla, G., 1968: Ribonucleoprotein-containing vesicles in cysts of *Naegleria gruberi*. J. Protozool. **15**, 752–758.

Schwab, D., 1969: Elektronenmikroskopische Untersuchung an der Foraminifere *Myxotheca arenilega* Schaudinn. Z. Zellforsch. **96**, 295–324.

– 1970: Elektronenmikroskopische Untersuchungen an der Foraminifere *Allogromia laticollaris* Arnold. Der heranwachsende Agamont. Z. Zellforsch. **108**, 35–45.

– 1972: Electron microscopic studies on the foraminifer *Allogromia laticollaris* Arnold. Mitosis in agamonts. Protoplasma **75**, 79–89.

– 1973: Centrosomal bodies during meiosis in the foraminifer *Myxotheca arenilega* Schaudinn. Protoplasma **78**, 339–341.

– 1974: Elektronenmikroskopische Untersuchungen an der Foraminifere *Allogromia laticollaris* Arnold. Protoplasma **80**, 305–322.

– 1976: Gametogenesis in *Allogromia laticollaris*. J. Foram. Res. **6**, 251–257.

– 1977: Light and electron microscopic investigations on the monothalamous foraminifer *Boderia albicollaris* n. sp. J. Foram. Res. **7**, 189–195.

– Schwab-Stey, H., 1979: Plate-shaped nuclear pole bodies in the monothalamous foraminifer *Kibisidytes* sp. Protoplasma **98**, 355–361.

SCHWARTZ, V., 1935: Versuche über Regeneration und Kerndimorphismus bei *Stentor coeruleus* EHRBG. Arch. Protistenk. **85**, 100–139.
– 1956: Nukleolenformwechsel und Zyklen der Ribosenucleinsäure in der vegetativen Entwicklung von *Paramecium bursaria*. Biol. Zbl. **75**, 1–16.
– 1957: Über den Formwechsel achromatischer Substanz in der Teilung des Makronucleus von *Paramecium bursaria*. Biol. Zbl. **76**, 1–23.
– 1958: Chromosomen im Makronucleus von *Paramecium bursaria*. Biol. Zbl. **77**, 347–364.
– 1965: Die Teilungsspindel des Mikronucleus von *Paramecium bursaria*. Zool. Anz., suppl. **28**, 123–131.
– 1976: Der Zyklus der mitotischen Spindel von *Paramecium bursaria*. Arch. Protistenk. **118**, 243–258.
– 1978: Struktur und Entwicklung des Makronucleus von *Paramecium bursaria*. Arch. Protistenk. **120**, 255–277.
– MEISTER, H., 1975a: Die Extinktion der feulgengefärbten Makronucleusanlage von *Paramecium bursaria* in der DNS-armen Phase. Arch. Protistenk. **117**, 60–64.
– – 1975b: Einige quantitative Daten zum Problem des Alterns bei *Paramecium*. Arch. Protistenk. **117**, 85–109.

SCHWEIKHARDT, F., 1965: DNS-Verteilung im Makronucleus von *Stentor coeruleus* EHRBG. Naturwissenschaften **52**, 146–147.

SCOTT, N. S., 1973: Ribosomal RNA cistrons in *Euglena gracilis*. J. Molec. Biol. **81**, 327–336.

SELIVERSTOVA, V. G., 1969: Some data on the fine structure of the macrogametes of *Eimeria tenella* (Sporozoa, Coccidia). Acta Protozool. **7**, 49–56 (in Russian; English summary).

SELUKAITE, Z., SEMENOV, V. M., 1976: Some peculiarities of the fine structure of the trophonts of *Trichomonas muris*. Tsitologiya (Leningrad) **18**, 91–97 (in Russian; English summary).

SÉNAUD, J., 1967: Contribution à l'étude des Sarcosporidies et des Toxoplasmes. Protistologica **3**, 167–232.
– 1969a: Ultrastructure des formations kystiques des *Besnoitia jellisoni* (FRENKEL 1953), Protozoaire, Toxoplasmea, parasite de la souris (*Mus musculus*). Protistologica **5**, 413–430.
– 1969b: Sur l'ultrastructure des kystes de *Besnoitia jellisoni* FRENKEL 1953 (Sporozoa, Toxoplasmea) chez la souris (*Mus musculus*). C. r. Acad. Sci. Paris **D 268**, 816–819.
– AUGUSTIN, H., DOENS-JUTEAU, O., 1980: Observations ultrastructurales sur le développement sexué de la coccidie *Eimeria acervulina* (TYZZER 1929) dans l'épithelium interestinal du poulet: la microgamétogenèse et la macrogamétogenèse. Protistologica **16**, 241–257.
– ČERNÁ, Z., 1969: Étude ultrastructurale des mérozoïtes et de la schizogonie des Coccidies (Eimeriina): *Eimeria magna* (PÉRARD 1925) de l'intestin des lapins et *E. tenella* (RAILLIET et LUCET 1891) des coecums des poulets. J. Protozool. **16**, 155–165.
– – 1970: La microgamétogenèse chez *Eimeria pragensis* (ČERNÁ et SÉNAUD 1969), Sporozoa, Telosporea, Coccidia, Eimeriina, parasite de l'intestin de la souris: étude au microscope électronique. Protistologica **6**, 5–19.
– – 1978: Le cycle de développement asexué de *Sarcocystis dispersa* (ČERNÁ, KOLAROVA et SULC 1977) chez la souris: étude au microscope électronique. Protistologica **14**, 155–176.
– MEHLHORN, H., 1978: *Besnoitia jellisoni* FRENKEL 1953 (Sporozoa, Apicomplexa) en culture sur cellules de rein de chien (MDCK): étude au microscope électronique. Protistologica **14**, 5–14.

SERGEJEVA, G. I., 1964: Activity of acid phosphatase at various stages of the life cycle of *Nyctotherus cordiformis* (EHRBG.) STEIN (Ciliata). Acta Protozool. **2**, 163–171 (in Russian; English summary).

– 1976: Macronuclear activity during conjugation in *Bursaria truncatella* (Ciliophora, Heterotrichida). In: Karyologiya i genetika prostejshikh (Series "Protozoology", No. 1), pp. 159–168. Leningrad: Nauka Press (in Russian; English summary).

– 1977: On the structure of macronuclear chromatin in the ciliate *Bursaria truncatella*. Tsitologiya (Leningrad) **19**, 1146–1154 (in Russian; English summary).

– 1979: Ultrastructure of the macronucleus (Ma) in dividing *Bursaria truncatella*. J. Protozool. **26**, No. 3 (part 1), 35A.

– POLJANSKY, G. I., 1974: Acid phosphatase activity dynamics in the ciliate *Bursaria truncatella* in relation to feeding and conjugation. In: Funkzion. Morfol., genetika i biochimiya kletki (Coll. of papers No. 16, Inst. of Cytology), pp. 274–276. Leningrad: Institute of Cytology (in Russian).

SESHACHAR, B. R., 1950a: Desoxyribonucleic acid content of the ciliate micronucleus. Nature **165**, 848.

– 1950b: The nucleus and nucleic acids of *Chilodonella uncinatus* EHRBG. J. Exper. Zool. **114**, 517–544.

– 1953: Metachromasy of the ciliate macronucleus. J. Exper. Zool. **124**, 117–130.

– 1960: Effect of centrifugation on the macronucleus of *Spirostomum* and *Blepharisma*. Nature **186**, 333–334.

– 1964: Observations on the fine structure of the nuclear apparatus of *Blepharisma intermedium* BHANDARY (Ciliata: Spirotricha). J. Protozool. **11**, 402–409.

– 1965: The fine structure of the nuclear apparatus and the chromosomes of *Spirostomum ambiguum* EHRBG. Acta Protozool. **3**, 337–343.

– 1966: Further observations on the chromosomal filaments of the ciliate macronucleus. Cytologia **31**, 77–79.

– BAI, A. R. K., 1968: Conjugation in *Spirostomum ambiguum* EHRBG. Acta Protozool. **6**, 51–56.

– DASS, C. M. S., 1953: Macronuclear regeneration in *Epistylis articulata*. Quart. J. Microscop. Sci. **94**, 185–192.

– – 1954a: The macronucleus of *Epistylis articulata* FROM. during conjugation: a photometric study. Physiol. Zool. **27**, 280–286.

– – 1954b: Photometric study of desoxyribonucleic acid (DNA) synthesis in regenerating macronucleus of *Epistylis articulata* FROM. Proc. Nat. Inst. Sci. India **20**, 656–659.

– SAXENA, D. M., 1968: DNA-dependent RNA synthesis in isolated macronuclei of *Blepharisma intermedium* (Ciliata). J. Protozool. **15**, 697–700.

SEYFERT, H.-M., 1979: Evidence for chromosomal macronuclear substructures in *Tetrahymena*. J. Protozool. **26**, 66–74.

– PREPARATA, R. M., 1979: The regulation of amounts and proportions of genetic elements in the macronuclei of *Tetrahymena thermophila* strains of diverse karyotype. J. Cell Sci. **40**, 111–123.

SHEFFIELD, H. G., 1966: Electron microscope study of the proliferative form of *Besnoitia jellisoni*. J. Parasitol. **52**, 583–594.

– 1968: Observations on the fine structure of the "cyst stage" of *Besnoitia jellisoni*. J. Protozool. **15**, 685–693.

– 1970: Schizogony in *Toxoplasma gondii*: an electron microscope study. Proc. Helminth. Soc. Washington **37**, 237–242.

– GARNHAM, P. C. C., SHIROISHI, T., 1971: The fine structure of the sporozoite of *Lankesteria culicis*. J. Protozool. **18**, 98–105.

SHEFFIELD, H. G., HAMMOND, D. M., 1966: Fine structure of first-generation merozoites of *Eimeria bovis*, J. Parasitol. **52**, 595–606.

– MELTON, M., 1968: The fine structure and reproduction of *Toxoplasma gondii*. J. Parasitol. **54**, 209–226.

SHEPARD, D. C., 1965: Production and elimination of excess DNA in ultraviolet-irradiated *Tetrahymena*. Exper. Cell Res. **38**, 570–579.

SHIBALOVA, T. A., 1974: Electron microscope observations on the development of *Eimeria tenella* (Sporozoa, Coccidia) in tissue culture. I. The fine structure of the sporozoite. Acta Protozool. **12**, 313–318.

– MOROZOVA, T. I., 1979: Intranuclear development of the macrogametes of the coccidian *Tyzzeria parvula*. Tsitologiya (Leningrad) **21**, 969–972 (in Russian; English summary).

– – KRAVTSOV, V. A., 1974a: Electron microscope observations on the development of *Eimeria tenella* (Sporozoa, Coccidia) in tissue culture. II. The fine structure of the schizonts and merozoites. Acta Protozool. **13**, 123–129.

– – – 1974b: Ultrastructure of sporozoites and merozoites of *Eimeria tenella*. Parazitologiya (Leningrad) **8**, 266–271 (in Russian; English summary).

SHIVAJI, S., SAXENA, D. M., PILLAI, M. K. K., 1978a: Mode of action of alkylating agents using a ciliate protozoan as a model system. II. Effect of Metepa on RNA and protein synthesis in *Stylonychia notophora* (Ciliata: Hypotrichida). Arch. Protistenk. **120**, 354–365.

– – – 1978b: Mode of action of alkylating agents using a ciliate protozoan as a model system. Part III. Effects of Metepa on cell division and DNA synthesis in the ciliate *Blepharisma intermedium*. Ind. J. Exp. Biol. **16**, 450–454.

– – – 1978c: Mode of action of alkylating agents using a ciliate protozoan as a model system. Part IV. Effect of Metepa on the synthesis of RNA in the ciliate *Blepharisma intermedium*. Ind. J. Exp. Biol. **16**, 455–457.

SHYAM, R., SARMA, Y. S. R. K., 1975: Certain aspects of mitotic division in *Gonium pectorale* MÜLLER (Volvocales). Nucleus **18**, 129–137.

SIAU, Y., 1979: Observations en microscopie électronique de complexes synaptonématiques chez des Myxosporidies. C. r. Acad. Sci. Paris **D 288**, 403–404.

SIDDIQUI, W., RUDZINSKA, M. A., 1965: The fine structure of axenically grown trophozoites of *Entamoeba invadens* with special reference to the nucleus and helical ribonucleoprotein bodies. J. Protozool **12**, 448–459.

SIEBERT, A. E., WEST, J. A., 1974: The fine structure of the parasitic dinoflagellate *Haplozoon axiothellae*. Protoplasma **81**, 17–35.

SIMPSON, C. F., 1966: Electron microscopy of *Sarcocystis fusiformis*. J. Parasitol. **52**, 607–613.

– BILD, C. E., STOLIKER, H. E., 1963: Electron microscopy of canine and equine *Babesia*. Amer. J. Veter. Res. **24**, 408–414.

SIMPSON, L., 1972: The kinetoplast of the hemoflagellates. Int. Rev. Cytol. **32**, 139–207.

SINCLAIR, J. H., 1972: Buoyant density of ribosomal genes in *Chlamydomonas reinhardi*. Exper. Cell Res. **74**, 569–571.

SINGH, B. N., 1952: Nuclear division in nine species of small, free-living amoebae and its bearing on the classification of the order Amoebida. Philos. Trans. Roy. Soc. London **B 236**, 405–460.

SIU, C.-H., SWIFT, H., CHIANG, K.-S., 1976: Characterization of cytoplasmic and nuclear genomes in the colorless alga *Polytoma*. I. Ultrastructural analysis of organelles. J. Cell Biol. **69**, 352–370.

SKARLATO, S. O., 1977: Fine structure of the nuclear apparatus of the ciliate *Stentor coeruleus*. Tsitologiya (Leningrad) **19**, 478–483 (in Russian; English summary).

– 1978a: Fine structure of the macronuclear anlagen of the ciliate *Stentor coeruleus*. Doklady Akad. Nauk SSSR **242**, 1192–1193 (in Russian).

Skarlato, S. O., 1978b: Electron microscopic study of the changes of the macronucleus of the ciliate *Stentor coeruleus* during conjugation. Tsitologiya (Leningrad) **20**, 607–611 (in Russian; English summary).

– 1979a: Fine structure of the micronuclei of the ciliate *Stentor coeruleus* during meiosis. J. Protozool. **26**, No. 3 (part 1), 36A.

– 1979b: Ultracytochemical investigation of the macronucleus of the ciliate *Stentor coeruleus*. Tsitologiya (Leningrad) **21**, 292–294 (in Russian; English summary).

Skoczylas, O., 1958: Über die Mitose von *Ceratium cornutum* und einigen anderen Peridineen. Arch. Protistenk. **103**, 193–228.

Slankis, T., Gibbs, S. P., 1972: The fine structure of mitosis and cell division in the chrysophycean alga *Ochromonas danica*. J. Phycol. **8**, 243–256.

Sleigh, M., 1973: The biology of Protozoa. New York: American Elsevier.

Smetana, K., Busch, H., 1974: The nucleolus and nucleolar DNA. In: The cell nucleus (Busch, H., ed.), Vol. 1, pp. 73–147. New York-London: Academic Press.

Smith, B. F., Stewart, B. T., 1966: Fine structure of *Trichomonas vaginalis*. Exper. Parasitol. **19**, 52–63.

Smith-Sonneborn, J., Klass, M., 1974: Changes in the DNA synthesis pattern of *Paramecium* with increased clonal age and interfission time. J. Cell Biol. **61**, 591–598.

Snigirevskaya, E. S., 1969a: Electron microscopic study of schizogony in *Eimeria intestinalis*. Acta Protozool. **7**, 57–70 (in Russian; English summary).

– 1969b: Changes of some ultrastructures during microgametogenesis in rabbit coccidians. Tsitologiya (Leningrad) **11**, 382–385 (in Russian; English summary).

– 1969c: Electron microscopic study of the macrogametes of *Eimeria intestinalis* (Coccidia). Tsitologiya (Leningrad) **11**, 700–706 (in Russian; English summary).

Solari, A. J., 1980a: The 3-dimensional fine structure of the mitotic spindle in *Trypanosoma cruzi*. Chromosoma **78**, 239–255.

– 1980b: Function of the dense plaques during mitosis in *Trypanosoma cruzi*. Exper. Cell Res. **127**, 457–460.

Soldo, A. T., Brickson, S. A., Larin, F., 1981: The kinetic and analytical complexities of the DNA genomes of certain marine and fresh-water ciliates. J. Protozool. **28**, 377–383.

– Godoy, G. A., 1972: The kinetic complexity of *Paramecium* macronuclear deoxyribonucleic acid. J. Protozool. **19**, 673–678.

Solovjev, M. M., 1963: Cytochemical study of *Lamblia duodenalis*. Medizin. Parasitol. i Parasit. Bolezni (Moscow) No. 6, 675–678 (in Russian; English summary).

– Chentsov, Yu. S., 1966: Electron microscopic study of *Lamblia muris* in relation to the peculiarities of the parasite's ecology. Medizin. Parasitol. i Parasit. Bolezni (Moscow) No. 6, 667–672 (in Russian; English summary).

Sommer, J. R., Blum, J. J., 1965: Cell division in *Astasia longa*. Exper. Cell Res. **39**, 504–527.

Sonneborn, T., 1947: Recent advances in the genetics of *Paramecium* and *Euplotes*. Adv. Genet. **1**, 263–358.

Sousa e Silva, E., 1967: *Cochlodinium heterolobatum* n. sp.: structure and some cytophysiological aspects. J. Protozool. **14**, 745–754.

– 1969: Cytological aspects on multiplication of *Goniodoma* sp. Botanica Marina (Hamburg) **12**, 233–243.

– 1971: The "small form" in the life cycle of dinoflagellates and its cytological interpretation. In: Proc. II Plankt. Conf., pp. 1157–1167. Roma: Tecnoscienza.

– 1977: Some ultrastructural variations of the nucleus in dinoflagellates throughout the life cycle. Acta Protozool. **16**, 277–288.

– 1978: Endonuclear bacteria in two species of dinoflagellates. Protistologica **14**, 113–119.

SOYER, M.-O., 1969a: L'enveloppe nucléaire chez *Noctiluca miliaris* SURIRAY (Dinoflagellata). I. Quelques données sur son ultrastructure et son évolution au cours de la sporogenèse. J. Microscopie **8**, 569–580.
– 1969b: L'enveloppe nucléaire chez *Noctiluca miliaris* SURIRAY (Dinoflagellata). II. Rôle des ampoules nucléaires et de certains constituants cytoplasmiques dans la mécanisme mitotique. J. Microscopie **8**, 709–720.
– 1969c: Rapports existant entre chromosomes et membrane nucléaire chez un Dinoflagellé parasite du genre *Blastodinium* CHATTON. C. r. Acad. Sci. Paris **D 268**, 2082–2084.
– 1970a: Observations ultrastructurales sur la condensation sporogénétique des chromosomes chez *Noctiluca miliaris S.* (Dinoflagellé, Noctilucidae). C. r. Acad. Sci. Paris **D 271**, 1003–1006.
– 1970b: Étude ultrastructurale de l'endoplasme et des vacuoles chez deux types de Dinoflagellés appartenant aux genres *Noctiluca* (SURIRAY) et *Blastodinium* (CHATTON). Z. Zellforsch. **105**, 350–388.
– 1971: Structure du noyau des *Blastodinium* (Dinoflagellés parasites). Division et condensation chromatique. Chromosoma **33**, 70–114.
– 1972: Les ultrastructures nucléaires de la Noctiluque (Dinoflagellé libre) au cours de la sporogenèse. Chromosoma **39**, 419–441.
– 1973: Complement à l'étude ultrastructurale des Volvocales. Étude des colonies femelles de *Volvox aureus* E. Ann. Sci. Natur., Zool. sér. 12, **15**, 231–258.
– 1975: Chromosomic division and organization in some dinoflagellates. BioSystems **7**, 306–307.
– 1978: Particules de type viral et filaments trichocystoïdes chez les Dinoflagellés. Protistologica **14**, 53–58.
– ESCAIG, J., 1980: Les structures nucléaires et leurs modifications au cours de la division chez le Dinoflagellé libre *Prorocentrum micans* E. Étude en cryofracture. Protistologica **16**, 485–495.
– HAAPALA, O. K., 1973: Filaments extra-chromosomiques: variations et relations avec l'enveloppe nucléaire pendant la division chez les Dinoflagellés. J. Microscopie **18**, 267–270.
– – 1974a: Division and function of dinoflagellate chromosomes. J. Microscopie **19**, 137–146.
– – 1974b: Structural changes of dinoflagellate chromosomes by pronase and ribonuclease. Chromosoma **47**, 179–192.
– HERZOG, M., 1980: The behaviour of native whole-mounted dinoflagellate chromosomes subjected to digestion by RNase, DNases or proteinase. Eur. J. Cell Biol. **22**, 73.

SPEAR, B. B., 1980: Isolation and mapping of the rRNA genes in the macronucleus of *Oxytricha fallax*. Chromosoma **77**, 193–202.
– LAUTH, M. R., 1976: Polytene chromosomes of *Oxytricha:* biochemical and morphological changes during macronuclear development in a ciliated protozoan. Chromosoma **54**, 1–13.
– MCCARROL, R. M., SWANTON, M. T., 1980: The structure of free rRNA genes from macronuclei of four hypotrichous ciliated Protozoa. J. Cell Biol. **87**, No. 2 (part 2), 106A.
– PRESCOTT, D. M., 1980: Nuclear DNA in normal and refed *Amoeba proteus*. Exper. Cell Res. **130**, 387–392.

SPEER, C. A., DANFORTH, H. D., 1976: Fine structural aspects of microgametogenesis of *Eimeria magna* in rabbits and in kidney cell cultures. J. Protozool. **23**, 109–115.
– HAMMOND, D. M., YOUSSEF, N. N., DANFORTH, H. D., 1973: Fine structural aspects of macrogametogenesis in *Eimeria magna*. J. Protozool. **20**, 274–281.

SPETH, V., WUNDERLICH, F., 1970: The macronuclear envelope of *Tetrahymena pyriformis* GL in different physiological states. III. Appearance of freeze-etched nuclear pore complexes. J. Cell Biol. **47**, 772–777.

SPINDLER, M., ANDERSON, O. R., HEMLEBEN, C., BÉ, A. W. H., 1978: Light and electron microscopic observations of gametogenesis in *Hastigerina pelagica* (Foraminifera). J. Protozool. **25**, 427–433.

SPRAGUE, V., 1941: Studies on *Gregarina blattarum* with particular reference to the chromosome cycle. Illinois Biol. Monogr. **18**, 2–20.

– VERNICK, S. H., 1968: Light and electron microscope study of a new species of *Glugea* (Microsporida, Nosematidae) in the 4-spined stickleback *Apeltes quadracus*. J. Protozool. **15**, 547–571.

– – 1969: Light and electron microscope observations on *Nosema nelsoni* SPRAGUE 1950 (Microsporida, Nosematidae) with particular reference to its Golgi complex. J. Protozool. **16**, 264–271.

– – 1971: The ultrastructure of *Encephalitozoon cuniculi* (Microsporida, Nosematidae) and its taxonomic significance. J. Protozool. **18**, 560–569.

– – 1974: Fine structure of the cyst and some sporulation stages of *Ichthyosporidium* (Microsporida). J. Protozool. **21**, 667–677.

STACK, S. M., BROWN, W. V., 1969: Somatic pairing, reduction and recombination and evolutionary hypothesis of meiosis. Nature **222**, 1275–1276.

STEHBENS, W. E., 1966: The ultrastructure of *Lankesterella hylae*. J. Protozool. **13**, 63–73.

STEINBRÜCK, G., HAAS, I., HELLMER, K.-H., AMMERMANN, D., 1981: Characterization of macronuclear DNA in five species of ciliates. Chromosoma **83**, 199–208.

STEINERT, M., STEINERT, G., 1962: La synthèse de l'acide désoxyribonucléique au cours du cycle de division de *Trypanosoma mega*. J. Protozool. **9**, 203–211.

– VAN ASSEL, S., 1967: Réplications coordonnées des acides désoxyribonucléiques nucléaire et mitochondrial chez "*Crithidia luciliae*". Arch. Intern. Physiol. Biochim. **75**, 370–371.

STERLING, C. R., 1972: Ultrastructural study of gametocytes and gametogenesis of *Haemoproteus metchnikovi*. J. Protozool. **19**, 69–76.

– AIKAWA, M., 1973: A comparative study of gametocyte ultrastructure in avian Haemosporidia. J. Protozool. **20**, 81–92.

– DE GIUSTI, D. L., 1972: Ultrastructural aspects of schizogony, mature schizonts and merozoites of *Haemoproteus metchnikovi*. J. Parasitol. **58**, 641–652.

– – 1974: Fine structure of differentiating oocysts and mature sporozoites of *Haemoproteus metchnikovi* in its intermediate host *Chrysops callidus*. J. Protozool. **21**, 276–283.

STEVENS, A. R., 1963: Electron microscope autoradiography of DNA and RNA syntheses in *Euplotes eurystomus*. J. Cell Biol. **19**, 67A.

– 1967: Machinery for exchange across the nuclear envelope. In: The control of nuclear activity (GOLDSTEIN, L., ed.), pp. 189–271. Englewood Cliffs, N.J.: Prentice-Hall.

– PRESCOTT, D. M., 1965: Additional observations on helices in *Amoeba proteus*. Exper. Cell Res. **40**, 204–207.

– – 1966: RNA in the helices of the *Amoeba* nucleus. J. Cell Biol. **31**, 113A.

STEVENSON, I., 1967: A method for the isolation of macronuclei from *Paramecium aurelia*. J. Protozool. **14**, 412–414.

– 1972: Ultrastructure of nuclear division in *Paramecium aurelia*. III. Meiosis in the micronucleus during conjugation. Austral. J. Biol. Sci. **25**, 775–799.

– LLOYD, F. P., 1971a: Ultrastructure of nuclear division in *Paramecium aurelia*. I. Mitosis in the micronucleus. Austral. J. Biol. Sci. **24**, 963–975.

– – 1971b: Ultrastructure of nuclear division in *Paramecium aurelia*. II. Amitosis of the macronucleus. Austral. J. Biol. Sci. **24**, 977–987.

STEWART, K. D., MATTOX, K. R., CHANDLER, C. D., 1974: Mitosis and cytokinesis in *Platymonas subcordiformis*, a scaly green monad. J. Phycol. **10**, 65–79.

STEY, H., 1969: Elektronenmikroskopische Untersuchung an *Labyrinthula coenocystis* SCHMOLLER. Z. Zellforsch. **102**, 387–418.

STONE, G. E., 1969: Change in histone/DNA ratio during population growth in *Tetrahymena*. J. Cell Biol. **42**, 837–840.

STORMS, R., HASTINGS, P. J., 1977: A fine structural analysis of meiotic pairing in *Chlamydomonas reinhardi*. Exper. Cell Res. **104**, 39–48.

STROUT, R. G., SCHOLTYSECK, E., 1970: The ultrastructure of first-generation development of *Eimeria tenella* (RAILLIET and LUCET) FANTHAM 1909 in cell cultures. Z. Parasitenk. **35**, 85–96.

STUDITSKY, A. N., 1930: Materialien zur Morphologie von *Dileptus gigas* STEIN. Arch. Protistenk. **70**, 155–184.

SUGAI, T., HIWATASHI, K., 1974: Cytologic and autoradiographic study of the micronucleus at meiotic prophase in *Tetrahymena pyriformis*. J. Protozool. **21**, 542–548.

SUGANUMA, Y., 1969: Electron microscope studies on the mitosis of binary fission of micronucleus of the ciliate, *Urostyla*. J. Nara Med. Assoc. **20**, 323–335.

SUHR-JESSEN, P. B., STEWARD, J. M., RASMUSSEN, L., 1977: Timing and regulation of nuclear and cortical events in the cell cycle of *Tetrahymena pyriformis*. J. Protozool. **24**, 299–303.

SUKHANOVA, K. M., 1959: Cytophysiological characteristic of *Protoopalina caudata* ZELLER. Tsitologiya (Leningrad) **1**, 333–340 (in Russian).

– 1961: Morphological and cytochemical investigation of *Protoopalina intestinalis*. Tsitologiya (Leningrad) **3**, 577–585 (in Russian).

– OVCHINNIKOVA, L. P., RADCHENKO, A. I., VLASSOVA, T. D., 1979: Effect of external factors on DNA content in euglenae. Tsitologiya (Leningrad) **21**, 271–278 (in Russian; English summary).

SUMMERS, F. M., 1935: The division and reorganization of *Aspidisca lynceus*, *Diophrys appendiculata* and *Stylonychia pustulata*. Arch. Protistenk. **85**, 173–208.

SUNDARARAMAN, Y., DASS, C. M. S., 1973: Effect of β-mercaptoethanol on the binary fission in *Spirostomum ambiguum* MÜLL.-EHRBG. Ind. J. Exp. Biol. **11**, 71–75.

SWANSON, C. P., 1960: Cytology and cytogenetics. Englewood Cliffs, N.J.: Prentice-Hall.

SWANTON, M. T., GRESLIN, A. F., PRESCOTT, D. M., 1980a: Arrangement of coding and non-coding sequences in the DNA molecules coding for rRNAs in *Oxytricha* sp. Chromosoma **77**, 203–215.

– HEUMANN, J. M., PRESCOTT, D. M., 1980b: Gene-sized DNA molecules of the macronuclei in three species of hypotrichs: size distribution and absence of nicks. Chromosoma **77**, 217–227.

SWARCZEWSKY, B., 1928: Beobachtungen über *Spirochona elegans* n. sp. Arch. Protistenk. **61**, 185–222.

TAKIZAWA, H., VIVIER, E., PETITPREZ, A., 1973: Développement intranucléaire de la microsporidie *Nosema bombycis* dans les cellules de Vers à Soie après infestation expérimentale. C. r. Acad. Sci. Paris **D 277**, 1769–1772.

TAMURA, S., TSURUHARA, T., WATANABE, Y., 1969: Function of nuclear microtubules in macronuclear division of *Tetrahymena pyriformis*. Exper. Cell Res. **55**, 351–358.

TARTAR, V., 1953: Chimeras and nuclear transplantations in ciliates, *Stentor coeruleus* X *S. polymorphus*. J. Exper. Zool. **124**, 63–103.

– 1957: Equivalence of macronuclear nodes. J. Exper. Zool. **135**, 387–401.

– 1961: The biology of *Stentor*. Oxford-London-New York-Paris: Pergamon Press.

TAYLOR, C., GARNJOBST, L., 1941: Nuclear reorganization in resting cysts of *Colpoda duodenaria*. J. Morphol. **68**, 197–213.

Taylor, F. J. R., 1976: Flagellate phylogeny: a study in conflicts. J. Protozool. **23**, 28–40.

Taylor, J. H., 1962: Chromosome reproduction. Internat. Rev. Cytol. **13**, 39–70.

Terzakis, J. A., 1971: Transformation of the *Plasmodium cynomolgi* oocyst. J. Protozool. **18**, 62–73.

– Sprinz, H., Ward, R. A., 1967: The transformation of the *Plasmodium gallinaceum* oocyst in *Aedes aegypti* mosquitoes. J. Cell Biol. **34**, 311–326.

Tilney, L. G., Porter, K. R., 1965: Studies on microtubules in Heliozoa. I. The fine structure of *Actinosphaerium nucleofilum* (Barrett) with particular reference to the axial rod structure. Protoplasma **60**, 317–344.

Tippit, D. H., McDonald, K. L., Pickett-Heaps, J. D., 1975: Cell division in the centric diatom *Melosira varians*. Cytobiologie **12**, 52–73.

– Pickett-Heaps, J. D., 1976: Apparent amitosis in the binucleate dinoflagellate *Peridinium balticum*. J. Cell Sci. **21**, 273–290.

– – 1977: Mitosis in the pennate diatom *Surirella ovalis*. J. Cell Biol. **73**, 705–727.

– Pillus, L. A., Pickett-Heaps, J. D., 1980: Organization of spindle microtubules in *Ochromonas danica*. J. Cell Biol. **87**, 531–545.

Tomas, R. N., 1974: Cell division in *Gonyaulax catenella*, a marine catenate dinoflagellate. J. Protozool. **21**, 316–321.

– Cox, E. R., 1973: Observations on the symbiosis of *Peridinium balticum* and its intracellular alga. I. Ultrastructure. J. Phycol. **9**, 304–323.

– – Steidinger, K. A., 1973: *Peridinium balticum* (Levander) Lemmermann, an unusual dinoflagellate with a mesocaryotic and a eucaryotic nucleus. J. Phycol. **9**, 91–98.

Tønnesen, T., 1978: The genes of 5S ribosomal RNA and transfer RNA in *Tetrahymena pyriformis* (a review article). Cytobiologie **16**, 451–479.

– Andersen, H. A., 1977: Timing of tRNA and 5S rRNA gene replication in *Tetrahymena pyriformis*. Exper. Cell Res. **106**, 408–412.

Torch, R., 1961: The nuclear apparatus of a new species of *Tracheloraphis* (Protozoa, Ciliata). Biol. Bull. **121**, 410–411.

– 1964: Autoradiographic studies of nucleic acid synthesis in a gymnostome ciliate, *Tracheloraphis* sp. J. Cell Biol. **23**, 98A.

– 1965: The effects of actinomycin D on RNA synthesis in the brackish-water ciliate, *Tracheloraphis* sp. Biol. Bull. **129**, 394.

Torres, A., Morenza, C., Fedriani, C., Gutierrez-Navarro, A. M., 1979: Nuclear cycles and DNA contents in *Laurentia acuminata* (Hypotrichida, Oxytrichidae). Protistologica **15**, 133–138.

– Nieto, J. J., Perez-Silva, J., 1980: Cellular cycle in *Gastrostyla steini*. Acta Protozool. **19**, 225–232.

Triemer, R. E., Brown, R. M., 1974: Cell division in *Chlamydomonas moewusii*. J. Phycol. **10**, 419–433.

Tronchin, G., Schrével, J., 1977: Chronologie des modifications ultrastructurales au cours de la croissance de *Gregarina blaberae*. J. Protozool. **24**, 67–82.

Truett, M. A., Gall, J. G., 1977: The replication of ribosomal DNA in the macronucleus of *Tetrahymena*. Chromosoma **64**, 295–303.

Tschenzoff, B., 1916: Die Kernteilung der *Euglena viridis* Ehrenberg. Arch. Protistenk. **36**, 137–173.

Tsujita, M., Watanabe, K., Tsuda, S., 1954: Electron-microscopical studies on the inner structure of *Paramecium caudatum* by means of ultra-thin sections. Cytologia **19**, 306–316.

Tucker, J. B., 1967: Changes in nuclear structure during binary fission in the ciliate *Nassula*. J. Cell Sci. **2**, 481–498.

– Beisson, J., Roche, D. L. J., Cohen, J., 1980: Microtubules and control of macronuclear "amitosis" in *Paramecium*. J. Cell Sci. **44**, 135–151.

TUFFRAU, M., 1953: Les processus cytologiques de la conjugaison chez *Spirochona gemmipara* STEIN. Bull. Biol. France et Belgique **87**, 314–322.

– 1962: Les processus régulateurs de la "caryophthisis" du macronucleus de *Nassulopsis lagenula* FAURÉ-FREMIET 1959. I. Cycle évolutif. Arch. Protistenk. **106**, 201–210.

TURNER, J. P., 1930: Division and conjugation in *Euplotes patella* EHRBG. with special reference to the nuclear phenomena. Univ. Calif. Publs Zool. **33**, 193–258.

TUTTLE, R. C., LOEBLICH, A. R., 1974: Genetic recombination in the dinoflagellate *Crypthecodinium cohnii*. Science **185**, 1061–1062.

– – 1977: N-methyl-N'-nitro-N-nitrosoguanidine and UV induced mutants of the dinoflagellate *Crypthecodinium cohnii*. J. Protozool. **24**, 313–316.

UHLIG, G., 1964: Die Folliculiniden der Deutschen Bucht. I. *Magnifolliculina* n. g., eine interessante Formengruppe der Epifauna des Sandgrundes. Helgoländer Wiss. Meeresuntersuch. **11**, 92–109.

USPENSKAJA, A. V., 1969: Fine structure of some stages of *Myxidium gasterostei* NOBLE 1943. Acta Protozool. **7**, 71–79 (in Russian, English summary).

– 1976: The nuclear cycle of Myxosporidia according to cytophotometry. Acta Protozool. **15**, 301–313.

– 1979: Further investigation of the nuclear cycle of Myxosporidia with aid of cytophotometry. J. Protozool. **26**, No. 3 (part 1), 37A.

– 1981: Comparison of the results of cytophotometric and electron microscopic studies of the sporogenesis in Myxosporidia belonging to various systematic groups. Tsitologiya (Leningrad) **23**, 570–580 (in Russian; English summary).

– OVCHINNIKOVA, L. P., 1966: Changes of DNA and RNA content during the life cycle of *Ichthyophthirius multifiliis*. Acta Protozool. **4**, 127–141 (in Russian; English summary).

UTTANGI, J. C., DESAI, R. N., 1963: Chromosome cycle in the gregarine *Stylocephalus mesomorphi* DEVDHAR. Naturwissenschaften **50**, 361–362.

VANDERBERG, J., RHODIN, J., YOELI, M., 1967: Electron microscopic and histochemical studies of sporozoite formation in *Plasmodium berghei*. J. Protozool. **14**, 82–103.

VARGHESE, T., 1975: Fine structure of the endogenous stages of *Eimeria labbeana*. 1. The first generation merozoites. J. Protozool. **22**, 66–71.

– 1977: Fine structure of the endogenous stages of *Eimeria labbeana*. 5. Schizonts and merogony, with special reference to the rhoptry-microneme system. J. Protozool. **24**, 376–382.

VÁVRA, J., 1969: *Lankesteria barretti* n. sp. (Eugregarinida, Diplocystidae), a parasite of the mosquito *Aedes triseriatus* (SAY) and a review of the genus *Lankesteria* MINGAZZINI. J. Protozool. **16**, 546–570.

– JOYON, L., PUYTORAC, P. DE, 1966: Observation sur l'ultrastructure du filament polaire des Microsporidies. Protistologica **2**, No. 2, 109–112.

VETTERLING, J. M., PACHECO, N. D., FAYER, R., 1973: Fine structure of gametogony and oocyst formation in *Sarcocystis* sp. in cell culture. J. Protozool. **20**, 613–621.

– TAKEUCHI, A., MADDEN, P. A., 1971: Ultrastructure of *Cryptosporidium wrairi* from the Guinea pig. J. Protozool. **18**, 248–260.

VICKERMAN, K., 1962: Patterns of cellular organization in limax amoebae. An electron microscope study. Exper. Cell Res. **26**, 497–519.

– 1969: The fine structure of *Trypanosoma congolense* in its bloodstream phase. J. Protozool. **16**, 54–69.

– 1970: Ultrastructure of *Trypanosoma* and relation to function. In: The African trypanosomiases (MULLIGAN, H. W., ed.), pp. 60–66. London: G. Allen and Unwin.

– 1973: The mode of attachment of *Trypanosoma vivax* in the proboscis of the tsetse fly *Glossina fuscipes*: an ultrastructural study of the epimastigote stage of the trypanosome. J. Protozool. **20**, 394–404.

VICKERMAN, K., 1974: Nuclear division in lower zooflagellates. In: Actualités protozoologiques (DE PUYTORAC, P., GRAIN, J., eds.). Vol. 1, p. 368. Clermont-Ferrand: Université de Clermont.

– 1977: DNA throughout the single mitochondrion of a kinetoplastid flagellate: observations on the ultrastructure of *Cryptobia vaginalis* (HESSE 1910). J. Protozool. **24**, 221–233.

– COX, F. E. G., 1967: Merozoite formation in the erythrocytic stages of the malaria parasite *Plasmodium vinckei*. Trans. Roy. Soc. Trop. Med. Hyg. **61**, 303–312.

– PRESTON, T. M., 1970: Spindle microtubules in the dividing nuclei of trypanosomes. J. Cell Sci. **6**, 365–383.

VINCKIER, D., 1969: Organisation ultrastructurale corticale de quelques monocystidées parasites du Ver oligochète *Lumbricus terrestris* L. Protistologica **5**, 505–517.

– 1975: *Nosemoides* gen. n., N. *vivieri* (VINCKIER, DEVAUCHELLE et PRENSIER 1970) comb. nov. (Microsporidie): étude de le différenciation sporoblastique et genèse des différentes structures de la spore. J. Protozool. **22**, 170–184.

– DEVAUCHELLE, G., PRENSIER, G., 1970: *Nosema vivieri* n. sp. (Microsporidae, Nosematidae), hyperparasite d'une Grégarine vivant dans le coelome d'une Nemerthe. C. r. Acad. Sci. Paris **D 270**, 821–823.

– – – 1971: Étude ultrastructurale du développement de la Microsporidie *Nosema vivieri* (VINCKIER, DEVAUCHELLE et PRENSIER 1970). Protistologica **7**, 273–287.

VINNIKOVA, N. V., 1974a: Ultrastructure of the micronuclei of the ciliate *Dileptus anser* during interphase and mitosis. Tsitologiya (Leningrad) **16**, 504–508 (in Russian; English summary).

– 1974b: Conjugation of *Dileptus anser* (O. F. M.) (Gymnostomatida, Tracheliidae). Acta Protozool. **12**, 275–288 (in Russian, English summary).

– 1974c: Fine structure and division of the macronuclei of the ciliate *Dileptus anser*. Tsitologiya (Leningrad) **16**, 765–769 (in Russian; English summary).

– 1974d: Fine structural changes of the macronuclei of *Dileptus anser* (O. F. M.) during conjugation. Acta Protozool. **13**, 99–107 (in Russian; English summary).

– 1975: Fine structure of the micronuclei of *Dileptus anser* during meiosis. Tsitologiya (Leningrad) **17**, 504–508 (in Russian; English summary).

– 1976: Conjugation in the ciliate *Dileptus anser*. I. Ultrastructure of micronuclei during mitosis and meiosis. Protistologica **12**, 7–24.

– 1977: DNA content of the macro- and micronuclei of the ciliate *Dileptus anser*. Tsitologiya (Leningrad) **19**, 351–356 (in Russian; English summary).

– GOLIKOVA, M. N., 1978: Ultrastructure of developed polytene chromosomes in macronuclear anlagen of *Nyctotherus cordiformis* (Protozoa, Ciliata). Tsitologiya (Leningrad) **20**, 731–735 (in Russian; English summary).

– – 1979: Fine structural cytochemistry of the macronucleus of *Nyctotherus cordiformis* STEIN (Ciliophora, Heterotrichida). Arch. Protistenk. **122**, 185–200.

VISVESVARA, G. S., CALLAWAY, C. S., 1974: Light and electron microscopic observations on the pathogenesis of *Naegleria fowleri* in mouse brain and tissue culture. J. Protozool. **21**, 239–250.

VIVARÈS, C. P., 1974: Occurrence of paracrystalline masses of intranuclear particles in a parasitic gregarine (Protozoa, Sporozoa). In: III Internat. Congr. Parasitol., p. 354. München.

– 1975: Étude comparative faite en microscopies photonique et électronique de trois espèces de Microsporidies appartenant au genre *Thelohania* HENNEGUY 1892, parasites de Crustacés décapodes marins. Ann. Sci. Natur., Zool. sér. 12, **17**, 141–178.

– 1980: Étude ultrastructurale de *Thelohania maenadis* PÉREZ (Microspora, Microsporida) et données nouvelles sur le genre *Thelohania* HENNEGUY. Arch. Protistenk. **123**, 44–60.

VIVARÈS, C. P., BOUIX, G., MANIER, J.-F., 1977: *Ormieresia carcini* gen. n., sp. n., Microsporidie du crabe méditerranéen, *Carcinus mediterraneus* CZERNIAVSKY 1884: cycle évolutif et étude ultrastructurale. J. Protozool. **24**, 83–94.

VIVIER, E., 1960a: Cycle nucléolaire en rapport avec l'alimentation chez *Paramecium caudatum*. C. r. Acad. Sci. Paris **250**, 205–207.

– 1960b: Contribution à l'étude de la conjugaison chez *Paramecium caudatum*. Ann. Sci. Natur., Zool. sér. 12, **2**, 387–506.

– 1963a: Premières observations sur l'ultrastructure d'une Coccidie: *Coelotropha durchoni* VIVIER, parasite de *Nereis diversicolor* (Annélide polychète). Ann. Inst. Pasteur Lille **14**, 217–226.

– 1963b: Étude au microscope électronique des nucléoles dans le macronucléus de *Paramecium caudatum*. In: Progress in Protozoology (Proc. I Internat. Congr. Protozool.), p. 421. Prague: Publ. House ČSAV.

– 1965: Présence de microtubules intranucléaires chez *Metchnikovella hovassei* VIVIER. J. Microscopie **4**, 559–562.

– 1967: Observations ultrastructurales sur l'enveloppe nucléaire et ses "pores" chez les Sporozoaires. J. Microscopie **6**, 371–390.

– 1970: Observations nouvelles sur la reproduction asexuée de *Toxoplasma gondii* et considération sur la notion d'endogenèse. C. r. Acad. Sci. Paris **D 271**, 2123–2126.

– ANDRÉ, J., 1961: Existence d'inclusions d'ultrastructure fibrillaire dans le macronucléus de certaines souches de *Paramecium caudatum* EHRBG. C. r. Acad. Sci. Paris **252**, 1848–1850.

– HENNERÉ, E., 1964: Cytologie, cycle et affinités de la Coccidie *Coelotropha durchoni*, nomen novum (= *Eucoccidium durchoni* VIVIER), parasite de *Nereis diversicolor* O. F. MÜLLER (Annélide polychète). Bull. Biol. France et Belgique **98**, 153–206.

– – 1965: Ultrastructure des stades végétatifs de la Coccidie *Coelotropha durchoni*. Protistologica **1**, No. 1, 89–104.

– PETITPREZ, A., 1969: Observations ultrastructurales sur l'hématozoaire *Anthemosoma garnhami* et examen de critères morphologiques utilisables pour la taxonomie chez les Sporozoaires. Protistologica **5**, 363–379.

– – 1972a: Données ultrastructurales complémentaires, morphologiques et cytochimiques, sur *Toxoplasma gondii*. Protistologica **8**, 199–221.

– – 1972b: Étude du système vacuolaire de l'hématozoaire *Anthemosoma garnhami* à l'aide des coupes sériées et de reconstitutions tridimensionnelles. J. Ultrastr. Res. **41**, 219–237.

– – LANDAU, I., 1972: Observations ultrastructurales sur la sporoblastogenèse de l'Hémogrégarine, *Hepatozoon domerguei*, Coccidie Adeleidea. Protistologica **8**, 315–333.

– – PONCHEL, G., 1974: Multiplication nucléaire et schizogonie chez *Anthemosoma* (Piroplasme). Étude de coupes sériées à l'aide de la microscopie électronique. Protistologica **10**, 113–123.

– SCHRÉVEL, J., 1964: Étude au microscope électronique d'une Grégarine du genre *Selenidium*, parasite de *Sabellaria alveolata* L. J. Microscopie **3**, 651–670.

– – 1973: Étude en microscopie photonique et électronique de différents stades du cycle de *Metchnikovella hovassei* et observations sur la position systématique des Metchnikovellidae. Protistologica **9**, 95–118.

– VICKERMAN, K., 1974: Divisions nucléaires chez les Protozoaires. In: Actualités Protozoologiques (DE PUYTORAC, P., GRAIN, J., eds.), Vol. 1, pp. 161–177. Clermont-Ferrand: Université de Clermont.

VOLKMANN, B., 1967: Vergleichend elektronenmikroskopische und lichtmikroskopische Untersuchungen an verschiedenen Entwicklungsstadien von *Klossia helicina* (Coccidia, Adeleidea). Z. Parasitenk. **29**, 159–208.

Vorob'ev, V. I., Borchsenius, S. N., Belozerskaya, N. A., Merkulova, N. A., Irlina, I. S., 1975: DNA replication in macronuclei of *Tetrahymena pyriformis* GL. Exper. Cell Res. **93**, 253–260.

Voronova, M. N., 1971: Observations on the life cycle of the Far-Eastern foraminifer *Rosalina* sp. Vestnik Leningrad. Univ. No. 3, 19–27 (in Russian; English summary).

– 1976: On polyploidy of the somatic nuclei of the foraminifer *Cibicides lobatulus*. Tsitologiya (Leningrad) **18**, 509–512 (in Russian; English summary).

– 1978a: Meiosis in the foraminiferen *Cibicides lobatulus*. Tsitologiya (Leningrad) **20**, 1328–1331 (in Russian; English summary).

– 1978b: Nuclear dualism in the gamonts of the foraminiferan *Cibicides lobatulus*. Tsitologiya (Leningrad) **20**, 859–867 (in Russian; English summary).

– Selivanova, G. V., 1976: The generative nuclei of *Cibicides lobatulus* (Foraminifera) studied with the aid of cytophotometric method. In: Karyologiya i Genetika Prosteishikh (Series "Protozoology", No. 1), pp. 144–150. Leningrad: Nauka Press (in Russian; English summary).

Wada, R. K., Spear, B. B., 1980: Nucleosomal organization of macronuclear chromatin in *Oxytricha fallax*. Cell Differentiation **9**, 261–268.

Walker, G. K., 1976: The fine structure of micronuclear division in the hypotrich ciliate *Gastrostyla steinii*. Protistologica **12**, 271–278.

– Goode, D., 1976: The role of microtubules in macronuclear division in the hypotrich ciliates *Gastrostyla steinii* and *Stylonychia mytilus*. Cytobiologie **14**, 18–37.

– Maugel, T. K., Goode, D., 1975: Some ultrastructural observations on encystment in *Stylonychia mytilus* (Ciliophora: Hypotrichida). Trans. Amer. Microscop. Soc. **94**, 147–154.

– – – 1980: Encystment and excystment in hypotrich ciliates. I. *Gastrostyla steinii*. Protistologica **16**, 511–524.

Walker, M. H., Hinsch, G. W., 1972: Ultrastructural observations of a microsporidian protozoan parasite in *Libinia dubia* (Decapoda). I. Early spore development. Z. Parasitenk. **39**, 17–26.

Walker, P., Mitchison, J., 1957: DNA synthesis in two ciliates. Exper. Cell Res. **13**, 167–170.

Walsh, R. D., Jr., Callaway, C. S., 1969: The fine structure of the gregarine *Lankesteria culicis* parasitic in the Yellow fever mosquito *Aedes aegypti*. J. Protozool. **16**, 536–545.

Warner, F. D., 1968: The fine structure of *Rhynchocystis pilosa*. J. Protozool. **15**, 59–73.

Watson, L. P., Lee, C. M., 1975: Ultrastructure of *Trypanosoma duttoni*. Z. Parasitenk. **46**, 133–140.

Watson, M. L., 1959: Further observations on the nuclear envelope of the animal cell. J. Biophys. Biochem. Cytol. **6**, 147–155.

Weber, H., 1965: Über die Paarung der Gamonten und den Kerndualismus der Foraminifere *Metarotaliella parva* Grell. Arch. Protistenk. **108**, 217–270.

Wecke, J., Giesbrecht, P., 1970: Freeze-etching of the nuclear membrane of dinoflagellates. In: Proc. VII Internat. Congr. Electr. Microsc., pp. 233–234. Grenoble.

Weidner, E., 1970: Ultrastructural study of microsporidian development. I. *Nosema* sp. Sprague 1965 in *Callinectes sapidus* Rathbun. Z. Zellforsch. **105**, 33–54.

Weiser, J., Žižka, Z., 1975: Stages in sporogony of *Plistophora debaisieuxi* Jirovec (Microsporidia). Acta Protozool. **14**, 185–194.

Weisz, P. B., 1949a: A cytochemical and cytological study of differentiation in normal and reorganizational stages of *Stentor coeruleus*. J. Morphol. **84**, 335–364.

– 1949b: The role of specific macronuclear nodes in the differentiation and the maintenance of the oral area in *Stentor*. J. Exper. Zool. **111**, 141–156.

– 1950: A correlation between macronuclear thymonucleic acid concentration and the capacity of morphogenesis in *Stentor*. J. Morphol. **87**, 275–286.

WENRICH, D. H., 1944: Nuclear structure and nuclear division in *Dientamoeba fragilis* (Protozoa). J. Morphol. **74**, 467–491.
– 1954a: Sex in Protozoa. A comparative review. In: Sex in microorganisms (WENRICH, D. H., ed.), pp. 137–265. Washington: AAAS Publ.
– 1954b: Comments on the origin and evolution of "sex". In: Sex in microorganisms (WENRICH, D. H., ed.), pp. 335–346. Washington: AAAS Publ.
WERMEL, E. M., 1925: Beiträge zur Cytologie der *Amoeba hydroxena* ENTZ. Russkij Archiv Protistol. **4**, 95–120.
– 1928: Untersuchungen über *Chromidina elegans* (FOETT.) GOND. Arch. Protistenk. **64**, 419–445.
WERY, M., GROODT-LASSEEL, M. DE, 1966: Ultrastructure de *Trypanosoma cruzi* en culture sur milieu semi-synthétique. Ann. Soc. Belge Méd. Trop. **46**, 337–348.
WESCHENFELDER, R., 1938: Die Entwicklung von *Actinocephalus parvus* WELLMER. Arch. Protistenk. **91**, 1–60.
WESLEY, R. D., 1975: Inverted repetitious sequences in the macronuclear DNA of hypotrichous ciliates. Proc. Nat. Acad. Sci. U.S.A. **72**, 678–682.
WEYER, G., 1930: Untersuchungen über die Morphologie und Physiologie des Formwechsels der *Gastrostyla steinii* ENGELMANN. Arch. Protistenk. **71**, 139–228.
WICHTERMAN, R., 1937: Division and conjugation in *Nyctotherus cordiformis* (EHR.) STEIN (Protozoa, Ciliata) with special reference to the nuclear phenomena. J. Morphol. **60**, 563–611.
WILD, M. A., GALL, J. G., 1979: An intervening sequence in the gene coding for 25S ribosomal RNA of *Tetrahymena pigmentosa*. Cell **16**, 565–573.
WILFERT, M., 1972: Zytologische Untersuchungen an dem Ciliaten *Blepharisma americanum* SUZUKI 1954, Stamm Berlin (Heterotrichida, Spirostomatidae) sowie Bemerkungen zur Taxonomie und Systematik der Gattung *Blepharisma* PERTY 1849. Arch. Protistenk. **114**, 152–230.
WILLAERT, E., STEVENS, A. R., TYNDALL, R. L., 1978a: *Acanthamoeba royreba* sp. n. from a human tumor cell culture. J. Protozool. **25**, 1–14.
– – – 1978b: Identification of *Acanthamoeba culbertsoni* from cultured tumor cells. Protistologica **14**, 319–336.
WILLIAMS, J. B., FLECK, E. W., HELLIER, L. E., UHLENHOPP, E., 1978: Viscoelastic studies on *Tetrahymena* macronuclear DNA. Proc. Nat. Acad. Sci. U.S.A. **75**, 5062–5065.
WILLIAMS, N. E., WILLIAMS, R. J., 1976: Macronuclear division with and without microtubules in *Tetrahymena*. J. Cell Sci. **20**, 61–77.
WILLIE, J. J., 1972a: Selective amplification of repetitive DNA in the eukaryote *Tetrahymena*. Biochem. Biophys. Res. Comm. **46**, 692–698.
– 1972b: Physiological control of DNA nucleotide sequence redundancy in the eukaryote, *Tetrahymena pyriformis*. Biochem. Biophys. Res. Comm. **46**, 677–684.
WISCHNITZER, S., 1958: An electron microscope study of the nuclear envelope of amphibian oocytes. J. Ultrastr. Res. **1**, 201–222.
WISE, B. N., 1965: The morphogenetic cycle in *Euplotes eurystomus* and its bearing on problems of ciliate morphogenesis. J. Protozool. **12**, 626–648.
WISE, G. E., GOLDSTEIN, L., 1972: Localization and characterization of chromatin within the interphase nucleus of *Amoeba proteus* by means of *in situ* centrifugation and radioautography. Chromosoma **36**, 176–192.
– STEVENS, A. R., PRESCOTT, D. M., 1972: Evidence of RNA in the helices of *Amoeba proteus*. Exper. Cell Res. **75**, 347–352.
WITT, P. L., 1977: Unequal distribution of DNA in the macronuclear division of the ciliate *Euplotes eurystomus*. Chromosoma **60**, 59–67.
WOLCOTT, G. B., 1952: Mitosis in *Trypanosoma lewisi*. J. Morphol. **90**, 189–199.

WOLFE, J., 1967: Structural aspects of amitosis: a light and electron microscope study of the isolated macronuclei of *Paramecium aurelia* and *Tetrahymena pyriformis*. Chromosoma **23**, 59–79.
– 1973: Conjugation in *Tetrahymena:* the relationship between the division cycle and cell pairing. Devel. Biol. **35**, 221–231.
– HUNTER, B., ADAIR, W. S., 1976: A cytological study of micronuclear elongation during conjugation in *Tetrahymena*. Chromosoma **55**, 289–308.

WOLSTENHOLME, D. R., 1966: Electron microscope identification of the interphase chromosomes of *Amoeba proteus* and *Amoeba discoides*, with some observations on helices and intranuclear electron-opaque bodies. J. Cell Biol. **31**, 165A.

WONG, S. T. C., DESSER, S. S., 1976: Fine structure of oocyst transformation and the sporozoites of *Leucocytozoon dubreuili*. J. Protozool. **23**, 115–126.
– – 1978: Ultrastructural observations on renal schizogony of *Leucocytozoon dubreuili* in the American robin. J. Protozool. **25**, 302–314.
– – 1981: An ultrastructural study of the schizogonic development of *Leucocytozoon dubreuili* in the liver of the American robin, *Turdus migratorius*. Protistologica **17**, 11–27.

WOODARD, J., GELBER, B., SWIFT, H., 1961: Nucleoprotein changes during the mitotic cycle in *Paramecium aurelia*. Exper. Cell Res. **23**, 258–264.
– KANESHIRO, E., GOROVSKY, M., 1972: Cytochemical studies on the problem of macronuclear subnuclei in *Tetrahymena*. Genetics **70**, 251–260.
– WOODARD, M., GELBER, B., SWIFT, H., 1966: Cytochemical studies of conjugation in *Paramecium aurelia*. Exper. Cell Res. **41**, 55–63.

WOODRUFF, L. L., 1941: Endomixis. In: Protozoa in biological research (CALKINS, G. N., SUMMERS, F. M., eds.), pp. 646–665. New York: Columbia Univ. Press.

WUJEK, D. E., 1969: Ultrastructure of flagellated chrysophytes. I. *Dinobryon*. Cytologia **34**, 71–79.
– 1976: Ultrastructure of flagellated chrysophytes. II. *Uroglena* and *Uroglenopsis*. Cytologia **41**, 665–670.

WUNDERLICH, F., 1969a: The macronuclear envelope of *Tetrahymena pyriformis* GL in different physiological states. I. Quantitative structural data. Exper. Cell Res. **56**, 369–374.
– 1969b: The macronuclear envelope of *Tetrahymena pyriformis* GL in different physiological states. II. Frequency of central granules in the pores. Z. Zellforsch. **101**, 581–587.
– BATZ, W., SPETH, V., WALLACH, D. F. H., 1974: Reversible, thermotropic alteration of nuclear membrane structure and nucleocytoplasmic RNA transport in *Tetrahymena*. J. Cell Biol. **61**, 633–640.
– FRANKE, W., 1968: Structure of macronuclear envelopes of *Tetrahymena pyriformis* in the stationary phase of growth. J. Cell Biol. **38**, 458–462.
– HERLAN, G., 1977: A reversibly contractile nuclear matrix. Its isolation, structure, and composition. J. Cell Biol. **73**, 271–278.
– SPETH, V., 1970: Antimitotic agents and macronuclear division of ciliates. IV. Reassembly of microtubules in macronuclei of *Tetrahymena* adapting to colchicine. Protoplasma **70**, 139–152.

YAO, M.-C., BLACKBURN, E., GALL, J. G., 1979: Amplification of the rRNA genes in *Tetrahymena*. Cold Spring Harbor Symp. Quant. Biol. **43**, 1293–1296.
– – – 1981: Tandemly repeated C-C-C-C-A-A hexanucleotide of *Tetrahymena* rDNA is present elsewhere in the genome and may be related to the alteration of the somatic genome. J. Cell Biol. **90**, 515–520.

YAO, M.-C., GALL, J. G., 1977: A single integrated gene for ribosomal RNA in a eukaryote *Tetrahymena pyriformis*. Cell **12**, 121–132.

– – 1979: Alteration of the *Tetrahymena* genome during nuclear differentiation. J. Protozool. **26**, 10–13.

– GOROVSKY, M. A., 1974: Comparison of the sequences of macro- and micronuclear DNA of *Tetrahymena pyriformis*. Chromosoma **48**, 1–18.

– KIMMEL, A. R., GOROVSKY, M. A., 1974: A small number of cistrons for ribosomal RNA in the germinal nucleus of a eukaryote, *Tetrahymena pyriformis*. Proc. Nat. Acad. Sci. U.S.A. **71**, 3082–3086.

YARWOOD, E., 1937: The life cycle of *Adelina cryptocerci* sp. nov., a coccidian parasite of the roach *Cryptocercus punctulatus*. Parasitology **29**, 370–390.

YOKOYAMA, R., YAO, M.-C., 1980: Elimination of DNA sequences during macronuclear differentiation in *Tetrahymena thermophila*. J. Cell Biol. **87**, No. 2 (part 2), 104A.

YOUSSEF, N. N., HAMMOND, D. M., 1972: The ultrastructural cytology of the nucleolus of *Eimeria magna* gamonts. J. Parasitol. **58**, 669–679.

ZAZULINA, I. V., POLYAKOV, V. YU., 1976: Ultrastructure of the nuclear apparatus of the ciliate *Spirostomum ambiguum* in the cell cycle. Tsitologiya (Leningrad) **18**, 1171–1175 (in Russian; English summary).

ZECH, L., 1964: Zytochemische Messungen an den Zellkernen der Foraminiferen *Patellina corrugata* und *Rotaliella heterocaryotica*. Arch. Protistenk. **107**, 295–330.

– 1966a: Cytochemical studies on the distribution of DNA in the macronucleus of *Stentor coeruleus*. J. Protozool. **13**, 532–534.

– 1966b: Dry weight and DNA content in sisters of *Bursaria truncatella* during the interdivision interval. Exper. Cell Res. **44**, 599–605.

ZHIVAGO, P. I., 1948: On structure of the nucleolus and the functional changes of its secretory apparatus in *Amoeba polypodia*. Doklady Akad. Nauk SSSR **60**, 1249–1251 (in Russian).

ZINGMARK, R. G., 1970: Ultrastructural studies on two kinds of mesocaryotic dinoflagellate nuclei. Amer. J. Bot. **57**, 586–592.

ŽIŽKA, Z., 1969: The fine structure of the macrogametocytes of *Adelina tribolii* BHATIA 1937 (Eucoccidia, Telosporea) from the fat body of the beetle *Tribolium castaneum* HBST. J. Protozool. **16**, 111–120.

– 1978a: Fine structure of the neogregarine *Farinocystis tribolii* WEISER 1953. Developmental stages in merogony. J. Protozool. **25**, 50–56.

– 1978b: Fine structure of the neogregarine *Farinocystis tribolii* WEISER 1953. Free gametocytes. Acta Protozool. **17**, 255–259.

– 1978c: Fine structure of the neogregarine *Farinocystis tribolii* WEISER 1953. Syzygy and gametes formation. Protistologica **14**, 209–215.

ZUBAY, G., WATSON, M. R., 1959: The absence of histone in the bacterium *Escherichia coli*. J. Biophys. Biochem. Cytol. **5**, 51–54.

ZUELZER, M., 1909: Bau und Entwicklung von *Wagnerella borealis* Mereschk. Arch. Protistenk. **17**, 135–202.

ZYPEN, E. VAN DER, PIEKARSKI, G., 1967a: Die Endodyogenie bei *Toxoplasma gondii*. Eine morphologische Analyse. Z. Parasitenk. **29**, 15–35.

– – 1967b: Ultrastrukturelle Unterschiede zwischen der sog. Proliferationsform (RH-Stamm, BK-Stamm) und dem sog. Cysten-Stadium (DX-Stamm) von *Toxoplasma gondii*. Z. Bakteriol., Abt. I (Orig.) **203**, 495–517.

Subject Index*

* Italicized numbers refer to pages including figures.

Index of Protozoan Genera and Suprageneric Latin Names*

* Italicized numbers refer to pages including figures.

GPSR Compliance

The European Union's (EU) General Product Safety Regulation (GPSR) is a set of rules that requires consumer products to be safe and our obligations to ensure this.

If you have any concerns about our products, you can contact us on ProductSafety@springernature.com

In case Publisher is established outside the EU, the EU authorized representative is:

Springer Nature Customer Service Center GmbH
Europaplatz 3
69115 Heidelberg, Germany

Batch number: 10726298

Printed by Printforce, the Netherlands